T0190324

Lecture Notes in Artificial Intelligence   10937

Subseries of Lecture Notes in Computer Science

More information about this series at http://www.springer.com/series/1244

Dinh Phung · Vincent S. Tseng
Geoffrey I. Webb · Bao Ho
Mohadeseh Ganji · Lida Rashidi (Eds.)

Advances in Knowledge Discovery and Data Mining

22nd Pacific-Asia Conference, PAKDD 2018
Melbourne, VIC, Australia, June 3–6, 2018
Proceedings, Part I

 Springer

Editors
Dinh Phung
Deakin University
Geelong, VIC
Australia

Vincent S. Tseng
National Chiao Tung University
Hsinchu City
Taiwan

Geoffrey I. Webb (iD)
Monash University
Clayton, VIC
Australia

Bao Ho
Japan Advanced Institute
 of Science and Technology
Nomi, Ishikawa
Japan

Mohadeseh Ganji
University of Melbourne
Melbourne, VIC
Australia

Lida Rashidi
University of Melbourne
Melbourne, VIC
Australia

ISSN 0302-9743   ISSN 1611-3349 (electronic)
Lecture Notes in Artificial Intelligence
ISBN 978-3-319-93033-6   ISBN 978-3-319-93034-3   (eBook)
https://doi.org/10.1007/978-3-319-93034-3

Library of Congress Control Number: 2018944425

LNCS Sublibrary: SL7 – Artificial Intelligence

Printed on acid-free paper

This Springer imprint is published by the registered company Springer International Publishing AG
part of Springer Nature
The registered company address is: Gewerbestrasse 11, 6330 Cham, Switzerland

PC Chairs' Preface

With its 22nd edition in 2018, the Pacific-Asia Conference on Knowledge Discovery and Data Mining is the second oldest conference and a leading venue in the area of knowledge discovery and data mining (KDD). It provides a prestigious international forum for researchers and industry practitioners to share their new ideas, original and latest research results, and practical development experiences from all KDD-related areas, including data mining, data warehousing, machine learning, artificial intelligence, deep learning, databases, statistics, knowledge engineering, visualization, and decision-making systems.

This year, we received 592 valid submissions, which is the highest number of submissions in the past 10 years. The diversity and reputation of PAKDD were also evident from the various regions from which submissions came, with over 25 different countries, noticeably from North America and Europe. Our goal was to continue to ensure a rigorous reviewing process with each paper assigned to one Senior Program Committee (SPC) member and at least three Technical Program Committee (TPC) members, resulting in an ideal minimum number of reviews of four for each paper. Owing to the unusually large number of submissions this year, we had to increase almost doubling the number of committee members, resulting in 72 SPC members and 330 TPC members. Each valid submission was reviewed by three PC members and meta-reviewed by one SPC member who also led the discussion. This required a total of approximately 2,000 reviews. The program co-chairs then considered recommendations from the SPCs, the submission, and the reviews to make the final decision. Borderline papers were discussed intensively before final decisions were made. In some cases, additional reviews were also requested.

In the end, 164 out of 592 papers were accepted, resulting in an acceptance rate of 27.9%. Among them, 58 papers were selected for long presentation and 107 papers were selected for regular presentation. This year, we introduced a new track in Deep Learning for Knowledge Discovery and Data Mining. This track was particularly popular (70 submissions); however, in the end, the number of papers accepted as the primary category for this track was moderate (six accepted papers), standing at 8.8%. The conference program contained 32 sessions in total. Long presentations were allocated 25 minutes and regular presentations 15 mins. These two types of papers, however, are not distinguished in the proceedings.

We would like to sincerely thank all SPC members, TPC members, and external reviewers for their time, effort, dedication, and services to PAKDD 2018.

April 2018

Dinh Phung
Vincent S. Tseng

General Chairs' Preface

Welcome to the proceedings of the 22nd Pacific-Asia Conference on Knowledge Discovery and Data Mining (PAKDD). This conference has a reputable tradition in bringing researchers, academia, developers, practitioners, and industry together with a focus on the Pacific-Asian regions. This year, PAKDD was held in the wonderful city of Melbourne, Australia, during June 3–6, 2018.

The single most important element of PAKDD is the technical contributions and submissions in the area of KDD. We were very pleased with the number of submissions received this year, which was well close to 600, showing a significant boost in the number of submissions and the popularity of this conference. We sincerely thank the many authors from around the world who submitted their work to the PAKDD 2018 technical program as well as its data competition and satellite workshops. In addition, PAKDD 2018 featured three high-profile keynote speakers: Professor Kate Smith-Miles, Australian Laureate Fellow from Melbourne University; Dr. Rajeev Rastogi, Director of Machine Learning at Amazon; and Professor Bing Liu from the University of Illinois at Chicago. The conference featured three tutorials and five satellite workshops in addition to a data competition sponsored by the Fourth Paradigm Inc. and ChaLean.

We would like to express our gratitude to the contribution of the SPC, TPC, and external reviewers, led by the program co-chairs, Dinh Phung and Vincent Tseng. We would like to thank the workshop co-chairs, Benjamin Fung and Can Wang; the tutorial co-chairs, Wray Buntine and Jeffrey Xu Yu; the competition co-chairs, Wei-Wei Tu and Hugo Jair Escalante; the local arrangements co-chairs, Gang Li and Wei-Luo; the publication co-chairs, Mohadeseh Ganji and Lida Rashidi; the Web and content co-chairs, Trung Le, Uyen Pham, and Khanh Nguyen; the publicity co-chairs, De-Chuan Zhan, Kozo Ohara, Kyuseok Shim, and Jeremiah Deng; and the award co-chairs, James Bailey, Bart Goethals, and Jinyan Li.

We are grateful to our sponsors: Deakin University as the host institution and gold sponsor; Monash University as the gold sponsor, University of Melbourne, Trusting Social, and the Asian Office of Aerospace Research and Development/Air Force Office of Scientific Research as silver sponsors, Springer as the publication sponsor, and the Fourth Paradigm, CodaLab and ChaLearn as the data competition sponsors.

April 2017

Tu-Bao Ho
Geoffrey I. Webb

Organization

Organizing Committee

General Co-chairs

Geoffrey I. Webb	Monash University, Australia
Bao Ho	Japan Advanced Institute of Science and Technology, Japan

Program Committee Co-chairs

Dinh Phung	Deakin University, Australia
Vincent Tseng	National Chiao Tung University, Taiwan

Tutorial Co-chairs

Wray Buntine	Monash University, Australia
Jeffrey Xu Yu	Chinese University of Hong Kong, Hong Kong, SAR China

Workshop Co-chairs

Benjamin Fung	McGill University, Canada
Can Wang	Griffith University, Australia

Data Competition Co-chairs

Wei-Wei Tu	Fourth Paradigm Inc., China
Hugo Jair Escalante	INAOE Mexico, ChaLearn, USA

Publicity Co-chairs

De-Chuan Zhan	Nanjing University, China
Kozo Ohara	Aoyama Gakuin University, Japan
Kyuseok Shim	Seoul National University, South Korea
Jeremiah Deng	University of Otago, New Zealand

Publication Co-chairs

Mohadeseh Ganji	University of Melbourne, Australia
Lida Rashidi	University of Melbourne, Australia

Local Arrangements Co-chairs

Gang Li	Deakin University, Australia
Wei Luo	Deakin University, Australia

Web and Content Co-chairs

Trung Le	Deakin University, Australia
Uyen Pham	Vietnam National University, Vietnam

Award Co-chairs

James Bailey	University of Melbourne, Australia
Bart Goethals	University of Antwerp, Belgium
Jinyan Li	University of Technology Sydney, Australia

Steering Committee

Co-chairs

Ee-Peng Lim	Singapore Management University, Singapore
Takashi Washio	Institute of Scientific and Industrial Research, Osaka University, Japan

Treasurer

Longbing Cao	Advanced Analytics Institute, University of Technology, Sydney, Australia

Members

Ee-Peng Lim	Singapore Management University, Singapore (member since 2006, co-chair 2015–2017)
P. Krishna Reddy	International Institute of Information Technology, Hyderabad (IIIT-H), India (member since 2010)
Joshua Z. Huang	Shenzhen Institutes of Advanced Technology, Chinese Academy of Sciences, China (member since 2011)
Longbing Cao	Advanced Analytics Institute, University of Technology, Sydney (member since 2013)
Jian Pei	Simon Fraser University, Canada (member since 2013)
Myra Spiliopoulou	Otto von Guericke University Magdeburg, Germany (member since 2013)
Vincent S. Tseng	National Chiao Tung University, Taiwan (member since 2014)
Tru Hoang Cao	Ho Chi Minh City University of Technology, Vietnam (member since 2015)
Gill Dobbie	University of Auckland, New Zealand (member since 2016)
Kyuseok Shim	Seoul National University, South Korea

Life Members

Hiroshi Motoda	AFOSR/AOARD and Osaka University, Japan (member since 1997, co-chair 2001–2003, chair 2004–2006, life member since 2006)
Rao Kotagiri	University of Melbourne, Australia (member since 1997, co-chair 2006–2008, chair 2009–2011, life member since 2007, treasury Co-sign since 2006)
Huan Liu	Arizona State University, USA (member since 1998, treasurer 1998–2000, life member since 2012)
Ning Zhong	Maebashi Institute of Technology, Japan (member since 1999, life member since 2008)
Masaru Kitsuregawa	Tokyo University, Japan (member since 2000, life member since 2008)
David Cheung	University of Hong Kong, SAR China (member since 2001, treasurer 2005–2006, chair 2006–2008, life member since 2009)
Graham Williams	Australian National University, Australia (member since 2001, treasurer since 2006, co-chair 2009–2011, chair 2012–2014, life member since 2009)
Ming-Syan Chen	National Taiwan University, Taiwan, ROC (member since 2002, life member since 2010)
Kyu-Young Whang	Korea Advanced Institute of Science and Technology, South Korea (member since 2003, life member since 2011)
Chengqi Zhang	University of Technology Sydney, Australia (member since 2004, life member since 2012)
Tu Bao Ho	Japan Advanced Institute of Science and Technology, Japan (member since 2005, co-chair 2012–2014, chair 2015–2017, life member since 2013)
Zhi-Hua Zhou	Nanjing University, China (member since 2007, life member since 2015)
Jaideep Srivastava	University of Minnesota, USA (member since 2006, life member since 2015)
Takashi Washio	Institute of Scientific and Industrial Research, Osaka University (member since 2008, life member since 2016)
Thanaruk Theeramunkong	Thammasat University, Thailand (member since 2009)

Past Members

Hongjun Lu	Hong Kong University of Science and Technology, Hong Kong, SAR China (member 1997–2005)
Arbee L. P. Chen	National Chengchi University, Taiwan, ROC (member 2002–2009)
Takao Terano	Tokyo Insitute of Technology, Japan (member 2000–2009)

Senior Program Committee

Albert Bifet	Universite Paris-Saclay, France
Andrzej Skowron	University of Warsaw, Poland
Benjamin C. M. Fung	McGill University, Canada
Byung Suk Lee	University of Vermont, USA
Chandan Reddy	Virginia Tech, USA
Chuan Shi	Beijing University of Posts and Telecommunications, China
Dat Tran	University of Canberra, Australia
Dinh Phung	Deakin University, Australia
Eibe Frank	University of Waikato, New Zealand
Feida Zhu	Singapore Management University, Singapore
Gang Li	Deakin University, Australia
Geoff Holmes	University of Waikato, New Zealand
George Karypis	University of Minnesota, USA
Guozhu Dong	Wright State University, USA
Hanghang Tong	City University of New York, USA
Hu Xia	Texas A&M University, USA
Hui Xiong	Rutgers University, USA
Jae-Gil Lee	KAIST, South Korea
James Bailey	University of Melbourne, Australia
Jeffrey Xu Yu	Chinese University of Hong Kong, Hong Kong, SAR China
Jia Wu	Macquarie University, Australia
Jian Pei	Simon Fraser University, Canada
Jianyong Wang	Tsinghua University, China
Jiliang Tang	Michigan State University, USA
Jiuyong Li	University of South Australia, Australia
Joshua Huang	Shenzhen Institutes of Advanced Technology, Chinese Academy of Sciences, China
Kai Ming Ting	Federation University, Australia
Kamalakar Karlapalem	International Institute of Information Technology, Hyderabad, India
Krishna Reddy P.	International Institute of Information Technology, Hyderabad, India
Kyuseok Shim	Seoul National University, South Korea
Latifur Khan	University of Texas at Dallas, USA
Longbing Cao	University of Technology Sydney, Australia
Masashi Sugiyama	University of Tokyo, Japan
Michael Berthold	University of Konstanz, Germany
Ming Li	Nanjing University, China
Min-Ling Zhang	Southeast University, China
Nikos Mamoulis	University of Ioannina, Greece
Niloy Ganguly	IIT, Kharagpur, India
Nitin Agarwal	University of Arkansas at Little Rock, USA

Olivier DeVel	DST, Australia
Osmar Goethals	University of Antwerp, Belgium
Patrick Gallinari	LIP6, Université Pierre et Marie Curie, France
Paul Montague	DST, Australia
Peter Christen	Australian National University, Australia
R. K. Agarwal	Jawaharlal Nehru University, India
Rajeev Raman	University of Leicester, UK
Reza Haffari	Monash University, Australia
Sang-Wook Kim	Hanyang University, South Korea
Seungwon Hwang	Yonsei University, South Korea
Shengjun Huang	Nanjing University of Aeronautics and Astronautics, China
Takashi Washio	Institute of Scientific and Industrial Research, Osaka University, Japan
Trung Le	Deakin University, Australia
Truyen Tran	Deakin University, Australia
Tu Nguyen	Deakin University, Australia
U. Kang	Seoul National University, South Korea
Vincenzo Piuri	Università degli Studi di Milano, Italy
Wee Keong Ng	Nanyang Technological University, Singapore
Wei Wang	University of California, Los Angeles, USA
Weidong Cai	University of Sydney, Australia
Wen-Chih Peng	National Chiao Tung University, Taiwan
Xiangjun Dong	Qilu University of Technology, China
Xiaofang Zhou	University of Queensland, Australia
Xiaohua Hu	Drexel University, USA
Xindong Wu	University of Vermont, USA
Xing Xie	Microsoft Research Asia, China
Xintao Wu	University of Arkansas, USA
Xuan Vinh Nguyen	University of Melbourne, Australia
Xuan-Hong Dang	IBM T. J. Watson Research Center, USA
Yan Wang	Macquarie University, Australia
Yanchun Zhang	Victoria University, Australia
Yu Zheng	Microsoft Research Asia, China
Yue Xu	Queensland University of Technology, Australia
Zhao Zhang	Soochow University, China

Program Committee

Adriel Cheng	Defence Science and Technology Group, Australia
Aijun An	York University, Canada
Aixin Sun	Nanyang Technological University, Singapore
Akihiro Inokuchi	Kwansei Gakuin University, Japan
Angelo Genovese	Università degli Studi di Milano, Italy
Anne Denton	North Dakota State University, USA

Arnaud Giacometti	François Rabelais University, France
Arnaud Soulet	François Rabelais University, France
Arthur Zimek	University of Southern Denmark, Denmark
Athanasios Nikolakopoulos	University of Minnesota, USA
Bay Vo	Ho Chi Minh City University of Technology, Vietnam
Bettina Berendt	Katholieke Universiteit Leuven, Belgium
Bin Liu	IBM T. J. Watson Research Center, USA
Bing Xue	Victoria University of Wellington, New Zealand
Bo Jin	Dalian University of Technology, China
Bolin Ding	Microsoft Research, USA
Brendon Woodford	University of Otago, New Zealand
Bruno Cremilleux	Université de Caen Normandie, France
Bum-Soo Kim	Korea University, South Korea
Canh Hao Nguyen	Kyoto University, Japan
Carson Leung	University of Manitoba, Canada
Chao Lan	University of Wyoming, USA
Chao Qian	University of Science and Technology of China, China
Chedy Raissi	Inria, France
Chen Chen	Nankai University, China
Chengzhang Zhu	University of Technology Sydney, Australia
Chenping Hou	National University of Defence Technology, China
Chia Hui Chang	National Central University, Taiwan
Choochart Haruechaiyasak	National Electronics and Computer Technology Centre, NECTEC, Thailand
Chuan Shi	Beijing University of Posts and Telecommunications, China
Chulyun Kim	Sookmyung Women's University, South Korea
Chun-Hao Chen	Tamkang University, Taiwan
Dao-Qing Dai	Sun Yat-Sen University, China
Dat Tran	University of Canberra, Australia
David Anastasiu	San José State University, USA
David Taniar	Monash University, Australia
David Tse Jung Huang	University of Auckland, New Zealand
De-Chuan Zhan	Nanjing University, China
Defu Lian	University of Electronic Science and Technology of China, China
Dejing Dou	University of Oregon, USA
De-Nian Yang	Academia Sinica, Taiwan
Dhaval Patel	IBM T. J. Watson Research Center, USA
Dinh Quoc Tran	University of North Carolina at Chapel Hill, USA
Divyesh Jadav	IBM Research, USA
Dragan Gamberger	Rudjer Boskovic Institute, Croatia
Du Zhang	California State University, USA
Duc Dung Nguyen	Institute of Information Technology, Vietnam
Elham Naghizade	University of Melbourne, Australia
Enhong Chen	University of Science and Technology of China, China

Enrique Ballester	Università degli Studi di Milano, Italy
Fabio Stella	University of Milano-Bicocca, Italy
Fan Zhang	University of New South Wales, Australia
Florent Masseglia	Inria, France
Francois Petitjean	Monash University, Australia
Fuzhen Zhuang	Institute of Computing Technology, Chinese Academy of Sciences, China
Gang Li	Deakin University, Australia
Gillian Dobbie	University of Auckland, New Zealand
Giuseppe Manco	Università della Calabria, Italy
Goce Ristanoski	Data61, Australia
Guandong Xu	University of Technology Sydney, Australia
Guangzhong Sun	University of Science and Technology of China, China
Guansong Pang	University of Technology Sydney, Australia
Gustavo Batista	University of Sao Paulo, Brazil
Hady Lauw	Singapore Management University, Singapore
Heitor Gomes	Telecom ParisTech, France
Hiroaki Shiokawa	University of Tsukuba, Japan
Hiroyuki Toda	NTT Cyber Solutions Laboratories, NTT Corporation, Japan
Hong Shen	Adelaide University, Australia
Hongzhi Yin	University of Queensland, Australia
Hsuan-Tien Lin	National Taiwan University, Taiwan
Hui (Wendy) Wang	Stevens Institute of Technology, USA
Hung-Yu Kao	National Cheng Kung University, Taiwan
Huzefa Rangwala	George Mason University, USA
Ichigaku Takigawa	Hokkaido University, Japan
Ickjai Lee	James Cook University, Australia
Irena Koprinska	University of Sydney, Australia
Jaegul Choo	Korea University, South Korea
Jason Wang	New Jersey Institute of Technology, USA
Jean Paul Barddal	Pontifical Catholic University of São Paulo, PUC-PR, Brazil
Jeffrey Chan	RMIT University, Australia
Jeffrey Ullman	Stanford University, USA
Jen-Wei Huang	National Cheng Kung University, Taiwan
Jeremiah Deng	University of Otago, New Zealand
Jerry Chun-Wei Lin	Harbin Institute of Technology, China
Jiajun Liu	Renmin University of China, China
Jiamou Liu	University of Auckland, New Zealand
Jiangang Ma	James Cook University, Australia
Jianhua Yin	Shandong University, China
Jianmin Li	Tsinghua University, China
Jianxin Li	University of Western Australia, Australia
Jia-Yu Pan	Google, USA
Jie Liu	Nankai University, China

Jing Zhang	Nanjing University of Science and Technology, China
Jingrui He	IBM Research, USA
Jingwei Xu	Nanjing University, China
Jingyuan Yang	Rutgers University, USA
Joao Vinagre	LIAAD – INESC Tec, Porto, Portugal
Johannes Bloemer	University of Paderborn, Germany
Jörg Wicker	University of Auckland, New Zealand
Joyce Jiyoung Whang	Sungkyunkwan University, South Korea
Jun Gao	Peking University, China
Jun Luo	Lenovo, Hong Kong, SAR China
Junbin Gao	University of Sydney, Australia
Jundong Li	Arizona State University, USA
Jungeun Kim	KAIST, South Korea
Jun-Ki Min	Korea University of Technology and Education, South Korea
Junping Zhang	Fudan University, China
K. Selçuk Candan	Arizona State University, USA
Keith Chan	Hong Kong Polytechnic University, Hong Kong, SAR China
Kevin Bouchard	Université du Quebec a Chicoutimi, Canada
Khoat Than	Hanoi University of Science and Technology, Vietnam
Ki Yong Lee	Sookmyung Women's University, South Korea
Ki-Hoon Lee	Kwangwoon University, South Korea
Kitsana Waiyamai	Kasetsart University, Thailand
Kok-Keong Ong	La Trobe University, Australia
Kouzou Ohara	Aoyama Gakuin University, Japan
Krisztian Buza	University of Bonn, Germany
Kui Yu	University of South Australia, Australia
Kun-Ta Chuang	National Cheng Kung University, Taiwan
Kyoung-Sook Kim	Artificial Intelligence Research Centre, South Korea
Latifur Khan	University of Texas, USA
Le Wu	Hefei University of Technology, China
Lei Gu	Nanjing University of Post and Telecommunications, China
Leong Hou U	University of Macau, SAR China
Liang Hu	Jilin University, China
Liang Hu	University of Technology Sydney, Australia
Liang Wu	Arizona State University, USA
Lida Rashidi	University of Melbourne, Australia
Lijie Wen	Tsinghua University, China
Lin Liu	University of South Australia, Australia
Lin Wu	University of Queensland, Australia
Ling Chen	University of Technology Sydney, Australia
Lizhen Wang	Yunnan University, China
Long Yuan	University of New South Wales, Australia
Lu Zhang	University of Arkansas, USA

Luiza Antonie	University of Guelph, Canada
Maciej Grzenda	Warsaw University of Technology, Poland
Mahito Sugiyama	National Institute of Informatics, Japan
Mahsa Salehi	Monash University, Australia
Makoto Kato	Kyoto University, Japan
Marco Maggini	University of Siena, Italy
Marzena Kryszkiewicz	Warsaw University of Technology, Poland
Md Zahidul Islam	Charles Sturt University, Australia
Meng Chang Chen	Academia Sinica, Taiwan
Meng Jiang	University of Illinois, USA
Miao Xu	RIKEN, Japan
Michael E. Houle	National Institute of Informatics, Japan
Michael Hahsler	Southern Methodist University, USA
Ming Li	Nanjing University, China
Ming Tang	Chinese Academy of Sciences, China
Ming Yin	Microsoft Research and Purdue University, USA
Mingbo Zhao	Donghua University, China
Min-Ling Zhang	Southeast University, China
Miyuki Nakano	Advanced Institute of Industrial Technology, Japan
Mohadeseh Ganji	University of Melbourne, Australia
Mohit Sharma	Walmart Labs, USA
Mostafa Haghir Chehreghani	Telecom Paristech, France
Motoki Shiga	GIFU University, Japan
Muhammad Aamir Cheema	Monash University, Australia
Murat Kantarcioglu	University of Texas at Dallas, USA
Nam Huynh	Japan Advanced Institute of Science and Technology, Japan
Nayyar Zaidi	Monash University, Australia
Ngoc-Thanh Nguyen	Wroclaw University of Technology, Poland
Nguyen Le Minh	Japan Advanced Institute of Science and Technology, Japan
Noseong Park	University of North Carolina at Charlotte, USA
P Sastry	IISc, India
P. Krishna Reddy	International Institute of Information Technology Hyderabad, India
Pabitra Mitra	Indian Institute of Technology Kharagpur, India
Panagiotis Papapetrou	Stockholm University, Sweden
Patricia Riddle	University of Auckland, New Zealand
Peixiang Zhao	Florida State University, USA
Pengpeng Zhao	Soochow University, China
Philippe Fournier-Viger	Harbin Institute of Technology, China
Philippe Lenca	IMT Atlantique, France
Qi Liu	University of Science and Technology of China, China
Qiang Tang	Luxembourg institute of Science and Technology, Luxembourg

Qing Wang	Australian National University, Australia
Qingshan Liu	Nanjing University of Information Science and Technology, China
Ranga Vatsavai	North Carolina State University, USA
Raymond Chi-Wing Wong	Hong Kong University of Science and Technology, Hong Kong, SAR China
Reza Zafarani	Syracuse University, USA
Rong-Hua Li	Shenzhen University, China
Rui Camacho	University of Porto, Portugal
Rui Chen	Samsung Research America, USA
Sael Lee	SUNY, South Korea
Sangkeun Lee	Korea University, South Korea
Sanjay Jain	National University of Singapore, Singapore
Santu Rana	Deakin University, Australia
Sarah Erfani	University of Melbourne, Australia
Satoshi Hara	Osaka University, Japan
Satoshi Oyama	Hokkaido University, Japan
Shanika Karunasekera	University of Melbourne, Australia
Sheng Li	Adobe Research, USA
Shirui Pan	University of Technology Sydney, Australia
Shiyu Yang	University of New South Wales, Australia
Shoji Hirano	Shimane University, Japan
Shoujin Wang	University of Technology Sydney, Australia
Shu Wu	NLPR, China
Shu-Ching Chen	Florida International University, USA
Shuhan Yuan	University of Arkansas, USA
Shuigeng Zhou	Fudan University, China
Sibo Wang	University of Queensland, Australia
Silvia Chiusano	Polytechnic University of Turin, Italy
Simon James	Deakin University, Australia
Songcan Chen	Nanjing University of Aeronautics and Astronautics, China
Songlei Jian	University of Technology Sydney, Australia
Steven Ding	McGill University, Canada
Suhang Wang	Arizona State University, USA
Sunhwan Lee	IBM Research, USA
Sunil Gupta	Deakin University, Australia
Tadashi Nomoto	National Institute of Japanese Literature, Japan
Takehiro Yamamoto	Kyoto University, Japan
Takehisa Yairi	University of Tokyo, Japan
Tanmoy Chakraborty	University of Maryland, College Park, USA
Teng Zhang	Nanjing University, China
Tetsuya Yoshida	Nara Women's University, Japan
Thanh Nguyen	Deakin University, Australia
Thin Nguyen	Deakin University, Australia
Tho Quan	John Von Neumann Institute, Vietnam

Tong Xu   University of Science and Technology of China, China
Toshihiro Kamishima   National Institute of Advanced Industrial Science
                               and Technology, Japan
Trong Dinh Thac Do   University of Technology, Sydney, Australia
Tru Cao   Ho Chi Minh City University of Technology, Vietnam
Tuan-Anh Hoang   Leibniz University of Hanover, Germany
Tzung-Pei Hong   National University of Kaohsiung, Taiwan
Vien Ngo   Queen's University Belfast, UK
Viet Huynh   Deakin University, Australia
Vincenzo Piuri   University of Milan, Italy
Vineeth Mohan   Arizona State University, USA
Vladimir Estivill-Castro   Griffith University, Australia
Wai Lam   Chinese University of Hong Kong, Hong Kong,
                               SAR China
Wang-Chien Lee   Pennsylvania State University, USA
Wei Ding   University of Massachusetts Boston, USA
Wei Kang   University of South Australia, Australia
Wei Liu   UTS, Australia, Australia
Wei Luo   Deakin University, Australia
Wei Shen   Nankai University, China
Wei Wang   University of New South Wales, Australia
Wei Zhang   ECNU, China
Weiqing Wang   University of Queensland, Australia
Wenjie Zhang   University of New South Wales, Australia
Wilfred Ng   HKUST, China
Woong-Kee Loh   Gacheon University, South Korea
Xian Wu   Microsoft Research Asia, China
Xiangfu Meng   Liaoning Technical University, China
Xiangjun Dong   Qilu University of Technology, China
Xiangliang Zhang   King Abdullah University of Science and Technology,
                               Saudi Arabia
Xiangmin Zhou   RMIT University, Australia
Xiangnan He   National University of Singapore, Singapore
Xiangnan Kong   Worcestor Polytechnic Institute, USA
Xiaodong Yue   Shanghai University, China, China
Xiaofeng Meng   Renmin University of China, China
Xiaohui (Daniel) Tao   University of Southern Queensland, Australia
Xiaoying Gao   Victoria University of Wellington, New Zealand
Xin Huang   Hong Kong Baptist University, Hong Kong,
                               SAR China
Xin Wang   University of Calgary, Canada
Xingquan Zhu   Florida Atlantic University, USA
Xintao Wu   University of Arkansas, USA
Xiuzhen Zhang   RMIT University, Australia
Xuan Vinh Nguyen   University of Melbourne, Australia

Xuan-Hieu Phan	University of Engineering and Technology – VNUHN, Vietnam
Xuan-Hong Dang	UC Santa Barbara, USA
Xue Li	University of Queensland, Australia
Xuelong Li	Chinese Academy of Science, China
Xuhui Fan	University of Technology Sydney, Australia
Yaliang Li	University at Buffalo, USA
Yanchang Zhao	CSIRO, Australia
Yang Gao	Nanjing University, China
Yang Song	University of Sydney, Australia
Yang Wang	University of New South Wales, Australia
Yang Yu	Nanjing University, China
Yang-Sae Moon	Kangwon National University, South Korea
Yanjie Fu	Missouri University of Science and Technology, USA
Yao Zhou	Arizona State University, USA
Yasuhiko Morimoto	Hiroshima University, Japan
Yasuo Tabei	RIKEN Centre for Advanced Intelligent Project, Japan
Yating Zhang	RIKEN AIP Centre/NAIST, Japan
Yidong Li	Beijing Jiaotong University, China
Yi-Dong Shen	Chinese Academy of Sciences, China
Yifeng Zeng	Teesside University, UK
Yim-ming Cheung	Hong Kong Baptist University, Hong Kong, SAR China
Ying Zhang	University of New South Wales, Australia
Yi-Ping Phoebe Chen	La Trobe University, Australia
Yi-Shin Chen	National Tsing Hua University, Taiwan
Yong Guan	Iowa State University, USA
Yong Zheng	Illinois Institute of Technology, USA
Yongkai Wu	University of Arkansas, USA
Yuan Yao	Nanjing University, China
Yuanyuan Zhu	Wuhan University, China
Yücel Saygın	Sabancı University, Turkey
Yue-Shi Lee	Ming Chuan University, Taiwan
Yu-Feng Li	Nanjing University, China
Yun Sing Koh	University of Auckland, New Zealand
Yuni Xia	Indiana University – Purdue University Indianapolis (IUPUI), USA
Yuqing Sun	Shandong University, China
Zhangyang Wang	Texas A&M University, USA
Zhaohong Deng	Jiangnan University, China
Zheng Liu	Nanjing University of Posts and Telecommunications, China
Zhenhui (Jessie Li)	Pennsylvania State University, USA
Zhiyuan Chen	University of Maryland Baltimore County, USA
Zhongfei Zhang	Binghamton University, USA
Zhou Zhao	Zhejiang University, China

Zhu Xiaofeng	Guangxi Normal University, China
Zijun Yao	Rutgers University, USA
Zili Zhang	Deakin University, Australia
Josh Jia-Ching Ying	Feng Chia University, Taiwan
Ja-Hwung Su	Cheng Shiu University, Taiwan
Chun-Hao Chen	Tamkang University, Taiwan
Chih-Ya Shen	National Tsing Hua University, Taiwan
Chih-Hua Tai	National Taipei University, Taiwan
Chien-Liang Liu	National Chiao Tung University, Taiwan
Ming-Feng Tsai	National Chengchi University, Taiwan
Hon-Han Shuai	National Chiao Tung University, Taiwan
Hoang Trong Nghia	MIT, USA
Bo Dao	Deakin University, Australia
Dang Nguyen	Deakin University, Australia
Binh Nguyen	Deakin University, Australia

Sponsors

Contents – Part I

Classification and Supervised Machine Learning

Healthcare, BioInformatics and Related Topics (Application)

Human, Behaviour and Interactions (Application)

Opinion Mining and Sentiment Analysis

Classification and Supervised Machine Learning

Classifier Risk Estimation Under Limited Labeling Resources

Anurag Kumar$^{(\boxtimes)}$ and Bhiksha Raj

Carnegie Mellon University, Pittsburgh, PA 15213, USA
{alnu,bhiksha}@cs.cmu.edu

Abstract. Evaluating a trained system is an important component of machine learning. Labeling test data for large scale evaluation of a trained model can be extremely time consuming and expensive. In this paper we propose strategies for estimating performance of a classifier using as little labeling resource as possible. Specifically, we assume a labeling budget is given and the goal is to get a good estimate of the classifier performance using the provided labeling budget. We propose strategies to get a precise estimate of classifier accuracy under this restricted labeling budget scenario. We show that these strategies can reduce the variance in estimation of classifier accuracy by a significant amount compared to simple random sampling (over **65%** in several cases). In terms of labeling resource, the *reduction* in number of samples required (compared to random sampling) to estimate the classifier accuracy with only 1% error is high as **60%** in some cases.

1 Introduction

The process of applying machine learning for a problem is usually a two phase process; the *training* phase, which involves learning meaningful models using the training data and the *testing* phase where the learned models are evaluated on an unseen dataset to estimate how well they perform. For classification problems, this would involve training a classifier and then obtaining accuracy of the classifier on test data. Labeled data are required in both phases. Labeling data is a tedious and expensive procedure and it is desirable to reduce the amount of labeling effort as much as possible.

There have been concrete efforts to reduce the dependence on labeled data for training by developing unsupervised and semi-supervised machine learning algorithms [8]. However, irrespective of the method employed in the training phase, the *testing* phase always requires labeled data to compute classifier accuracy. In several cases, it is almost impossible to label the whole test data because of its enormous size. To reduce the labeling effort in this phase, we need methods to precisely estimate classifier accuracy by labeling only a small fraction of the test

Electronic supplementary material The online version of this chapter (https://doi.org/10.1007/978-3-319-93034-3_1) contains supplementary material, which is available to authorized users.

© Springer International Publishing AG, part of Springer Nature 2018
D. Phung et al. (Eds.): PAKDD 2018, LNAI 10937, pp. 3–15, 2018.
https://doi.org/10.1007/978-3-319-93034-3_1

data. This restricted labeling budget evaluation is important in this era of big data. It can be immediately applied to text classification or multimedia event classification systems deployed to work on the web.

The simplest solution to the above problem is to randomly select instances for labeling and then estimate the accuracy using the selected labeled set. To be able to do better than random sampling, a sampling or instance selection strategy is required. This might make it appear similar to active learning [17]. However, active learning is built around classifier training; the problem of classifier evaluation is very different from it. Sampling in active learning is done for classifier training such that classifier performance improves by using the new labeled sample. In our case, however, sampling needs to be done such that the *estimate* of classifier performance improves with new samples. The classifier is a black box and the training phase is immaterial in this problem. Moreover, this problem is completely different from cross validation or any such method employed to measure the goodness of the classifier during the *training* phase.

Considering the importance of this problem, very few efforts have been made to address the constraints posed by labeling costs during the classifier evaluation phase. Very few works have looked into it. Some attempts have been made towards unsupervised evaluation of multiple classifiers [6,10,14]. However, these methods are feasible only if multiple classifiers are present. In contrast, our focus is on the more general and practical case where the goal is to estimate the accuracy of a single classifier without the aid of any other classifier. Since, the labeling resources are limited, we need sampling strategies (for labeling) such that the accuracy estimated on the sampled set is a close approximation of true accuracy. *Simple random sampling* leads to high variance in estimates, which may result in large errors in estimation of classifier accuracy.

Few works have looked into sampling techniques for classifier evaluation [2,7, 11,15]. Our proposed solution to the above problem is *Stratified Sampling*, which is a well known concept in statistics [3]. In stratified sampling the idea is to divide the data into different strata and then sample a certain number of instances from each stratum. This process usually leads to lower variance of estimated variable. [2,7] also used stratification for estimating classifier accuracy. However, several important aspects are missing in these works, such as theoretical and empirical study of the variances of the estimators and thorough investigation into stratification and allocation strategies.

There are several *novel contributions* of this work. (1) We establish variance relationships for accuracy estimators for both random sampling and stratified sampling. The variance relations allow us to analyze stratified sampling for accuracy estimation in theory, and also helps in thorough analysis of results empirically; leading to a comprehensive understanding. (2) We propose 2 strategies for practically implementing *Optimal* allocation in stratified sampling. Our proposed novel iterative method for optimal allocation offers several advantages over the non-iterative implementation of optimal allocation policy. (3) On the stratification front, we employ panoply of stratification methods and analyze their effect on the variance of estimated accuracy. Besides stratification methods from statistics literature, we also propose to use clustering methods for stratification. We also delve into factors such as number of strata during stratification. (4) In our empirical analysis, we employ both probabilistic as well as non-probabilistic classifiers and show that our method can work on both. (5) Moreover, we also study how the true accuracy of classifier affects its estimation through sampling.

2 Problem Formulation

Let $\mathcal{D}$ be a dataset of N instances, $\boldsymbol{x}_i$ and l_i represents instances and its true label respectively. We want to estimate the accuracy of a trained classifier C on dataset $\mathcal{D}$. The score output of the classifier on $\boldsymbol{x}_i$ is $C(\boldsymbol{x}_i)$ and the label predicted by C for $\boldsymbol{x}_i$ is $\hat{l}_i$. Let a_i be instance specific correctness measure such that $a_i = 1$ if $l_i = \hat{l}_i$, otherwise $a_i = 0$. Then the true accuracy, A, of the classifier over $\mathcal{D}$ can be expressed as $A = \frac{\sum_{i=1}^{N} a_i}{N}$. This is nothing but the population mean of variable a_i where $\mathcal{D}$ represents the whole population. To compute A, we need to know l_i for all $i = 1\,to\,N$. Our problem is to estimate the true accuracy A of C under constrained labeling resources, meaning only a small number of instances, n, can be labeled. Hence, the samples for labeling should be chosen intelligently. Mathematically, we require an unbiased estimator of A with minimum possible variance for a given n.

3 Estimation Methods

3.1 Simple Random Sampling

The trivial solution for the problem described in Sect. 2, is to randomly select n instances or samples and ask for labels for these instances. This process is called simple random sampling which we will refer to as random sampling for convenience. The correctness measure a_i can be computed for the selected n instances, using which we can obtain an estimate of A.

The estimate of the accuracy is the mean of a_i over the sampled set, $\hat{A}^r = \frac{\sum_{i=1}^{n} a_i}{n}$. $\hat{A}^r$ is an unbiased estimator of A and the variance of $\hat{A}^r$ is given by $V(\hat{A}^r) = \frac{S^2}{n}$, $where\ S^2 = \frac{\sum_{i=1}^{N}(a_i - A)^2}{N-1}$. S^2 is the variance of a_i over $\mathcal{D}$. The variance formula above will include a factor $1 - \frac{n}{N}$ if sampling without replacement. For convenience we will assume sampling with replacement in our discussion and hence this term will not appear. The following lemma establishes the variance S^2 of a_i in terms of A.

Lemma 1. S^2 for a_i is given by $S^2 = \frac{N}{N-1} \cdot A(1-A)$. *Lemma 1 is fairly simple to establish and we escape the proof due to space constraints.*

Using Lemma 1 for $V(\hat{A}^r)$, the variance of random sampling based estimator of accuracy, $V(\hat{A}^r) = \frac{N\,A(1-A)}{(N-1)\,n}$. Since A is unknown, we need an unbiased estimate of $V(\hat{A}^r)$ for empirical evaluation of variance. An unbiased estimate of S^2 can be obtained from a sample of size n by $s^2 = \frac{\sum_{i=1}^{n}(a_i - \hat{A}^r)^2}{n-1}$. Following Lemma 1, we can obtain

$$s^2 = \frac{n}{n-1} \cdot \hat{A}^r(1 - \hat{A}^r) \tag{1}$$

Proposition 1. *The unbiased estimate of variance of accuracy estimator $\hat{A}^r$, is given by $v(\hat{A}^r) = \frac{\hat{A}^r(1-\hat{A}^r)}{n-1}$.*

Proposition 1 follows from Eq. 1. The estimated accuracy becomes more precise as n increases due to decrease in variance with n. The goal is to achieve more precise estimation or lower variance estimate for a given n.

3.2   Stratified Sampling

Let us assume that the instances in $\mathcal{D}$ have been stratified into K sets or strata. Let $\mathcal{D}_1, \ldots, \mathcal{D}_K$ be those strata. The stratification is such that $\mathcal{D}_1 \cup \mathcal{D}_2 \cup \ldots \cup \mathcal{D}_K = \mathcal{D}$ and $\mathcal{D}_j \cap \mathcal{D}_k = \emptyset, where, \ j \neq k, 1 \leq j \leq K, 1 \leq k \leq K$. All instances belong to only one stratum. The number of instances in strata $\mathcal{D}_k$ is N_k and $\sum_{k=1}^{K} N_k = N$. The simplest form of stratified sampling is *stratified random sampling* in which samples are chosen randomly and uniformly from each stratum. If the labeling resource is fixed at n then n_k instances are randomly chosen from each stratum such that $\sum_{k=1}^{K} n_k = n$.

In contrast to random sampling, the estimate of accuracy by *stratified* random sampling is given by $\hat{A}^s = \sum_{k=1}^{K} \frac{N_k}{N} \hat{A}_k^r = \sum_{k=1}^{K} W_k \hat{A}_k^r$. $\hat{A}_k^r = \frac{1}{n_k} \sum_{i=1}^{n_k} a_i$ and $W_k = N_k/N$ are the estimated accuracy in k^{th} stratum and weight of kth stratum respectively. The superscript r denotes that random sampling is used to select instances within each stratum.

It is straightforward to show that $\hat{A}^s$ is an *unbiased estimator* of A. Under the assumption of independent sampling for each stratum, the variance of $\hat{A}^s$ is $V(\hat{A}^s) = \sum_{k=1}^{K} W_k^2 V(\hat{A}_k^r)$. Since sampling within a stratum is random,

Proposition 2. *The variance of stratified random sampling estimator of accuracy, $\hat{A}^s$, is given by*

$$V(\hat{A}^s) = \sum_{k=1}^{K} W_k^2 \frac{S_k^2}{n_k} = \sum_{k=1}^{K} W_k^2 \frac{N_k A_k(1 - A_k)}{(N_k - 1) n_k} \tag{2}$$

$S_k^2 = \frac{N_k A_k(1-A_k)}{(N_k-1)}$ is the variance of the a_i's in k^{th} stratum. A_k is the true accuracy in the kth stratum and clearly, $\sum_{k=1}^{K} W_k A_k = A$. Similarly, Proposition 1 can be applied to obtain an unbiased estimator of $V(\hat{A}^s)$.

Proposition 3. *The unbiased estimate of variance of $\hat{A}^s$ is*

$$v(\hat{A}^s) = \sum_{k=1}^{K} W_k^2 \frac{s_k^2}{n_k} = \sum_{k=1}^{K} W_k^2 \frac{\hat{A}_k^r(1 - \hat{A}_k^r)}{(n_k - 1)} \tag{3}$$

The variance for stratified sampling is related to two important factors (1) Stratification of $\mathcal{D}$ (2) Allocation of labeling resource n to each strata (n_k). Since optimal stratification methods usually depend on the allocation method to be used, we discuss the allocation aspect first.

3.3   Allocation Methods for Stratified Sampling

Proportional (PRO) Allocation: In proportional allocation the total labeling resource n is allocated proportional to the weight of the stratum. This implies $n_k = W_k \times n$. Substituting this value in Eq. 2, the variance of $\hat{A}^s$ under proportional allocation, $V_{pro}(\hat{A}^s)$, is

$$V_{pro}(\hat{A}^s) = \frac{1}{n} \sum_{k=1}^{K} W_k S_k^2 = \frac{1}{n} \sum_{k=1}^{K} W_k \frac{N_k A_k (1 - A_k)}{(N_k - 1)} \tag{4}$$

The unbiased estimate of $V_{pro}(\hat{A}^s)$ can be similarly obtained. Stratified random sampling with proportional allocation is fairly easy to implement. We compute n_k and then sample and label n_k instances from k^{th} stratum to obtain an estimate of accuracy A_k.

Equal (EQU) Allocation: In Equal allocation the labeling resource is allocated equally among all strata. This implies $n_k = n/K$, again straightforward for practical purposes. Under equal allocation the variance of estimator $\hat{A}^s$ is

$$V_{equ}(\hat{A}^s) = \frac{K}{n} \sum_{k=1}^{K} W_k^2 S_k^2 = \frac{K}{n} \sum_{k=1}^{K} W_k^2 \frac{N_k A_k (1 - A_k)}{(N_k - 1)} \tag{5}$$

Optimal (OPT) Allocation: For a fixed labeling resource n, optimal allocation obtains the most precise estimate of accuracy using stratified sampling. The allocation of n minimizes the variance of the estimate [3]. It factors in both the stratum size and variance within stratum for labeling resource allocation [3]. The labeling resource allocated to each stratum is given by $n_k^{opt} = n \frac{W_k S_k}{\sum_{k=1}^{K} W_k S_k}$. Using this value in Eq. 2 the variance of $\hat{A}^s$ comes out as,

$$V_{opt}(\hat{A}^s) = \frac{\left(\sum_{k=1}^{K} W_k S_k \right)^2}{n} = \frac{\left[\sum_{k=1}^{K} W_k \left(\frac{N_k A_k (1 - A_k)}{(N_k - 1)} \right)^{\frac{1}{2}} \right]^2}{n} \tag{6}$$

A larger stratum or a stratum with higher variance of a_i or both is expected to receive more labeling resource compared to other strata. Thus, a homogeneous stratum (low variance) gets fewer samples for accuracy estimation compared to high variance stratum. However, practical implementation of optimal allocation is not as straightforward as the previous two methods. The true accuracies A_k's and hence S_k^2 are unknown, implying we cannot directly obtain values of n_k. We propose two methods for practical implementation of optimal allocation policy.

Algorithm 1. OPT-A2 Allocation Method

1: **procedure OPT-A2**$(\mathcal{D}_1, \ldots, \mathcal{D}_k, n_{ini}, n_{step})$
2:    Randomly Select and Label n_{ini} instances from each stratum
3:    Estimate A_k and S_k^2 for each strata
4:    $n_{rem} = n - (K * n_{ini})$
5:    **while** $n_{rem} > 0$ **do**
6:        $n_{curr} = min(n_{step}, n_{rem})$
7:        Allocate n_{curr} among strata using current estimate of S_k^2 for n_k^{opt}
8:        Select and label new instances from each stratum according to allocation
          of n_{curr} in previous step
9:        Update estimates of A_k and S_k^2 for all k
10:       $n_{rem} = n_{rem} - n_{curr}$
11:    **end while**
12: **end procedure**

OPT-A1 - In the first method, an initial estimate of all A_k's are obtained by spending some labeling resources on each stratum. n_{ini} instances are randomly chosen instances from each stratum for labeling. Then, an unbiased estimate of S_k^2 for k^{th} stratum is obtained by Eq. 1. These unbiased estimates are used to allocate rest of the labeling resource $(n - K * n_{ini})$ according to optimal allocation policy given by n_k^{opt}. Estimates of A_k are then updated by sampling the allocated labeling resource for each stratum.

In theory, optimal allocation is expected to give the minimum possible variance in accuracy estimation. However, allocation of n according to OPT-A1 heavily depends on initial estimates of S_k^2 in each stratum. If n_{ini} is small, a good estimate of S_k^2 might not be obtained, which might result in an allocation far from true optimal allocation policy. On the other hand, if n_{ini} is large, we essentially end up spending a large proportion of the labeling resource in a uniform fashion which is same as equal allocation. Hence, the gain in preciseness or reduction in variance might not be obtained. Practically, it leaves us wondering about value of parameter n_{ini}.

To address this problem we propose another method for optimal allocation called **OPT-A2**. OPT-A2 is an iterative form of OPT-A1. The steps for OPT-A2 are described in Algorithm 1. In OPT-A2 n_{ini} is a small reasonable value. Instead of allocating all of the remaining labeling resource in the next step, we adopt an iterative formalism. In this iterative formalism, we allocate a fixed n_{step} labeling resources among the strata in each step. This is followed by an update in estimate of A_k and S_k^2. The process is repeated till we exhaust our labeling budget. We later show that results for OPT-A2 are not only superior compared to OPT-A1 but also removes concerns regarding parameter tuning (n_{ini}).

3.4   Comparison of Variances

The variance results can be established under two cases; (1) $1/N_k \ll 1$ and (2) the first one does not hold. $1/N_k \ll 1$ is the more practical case which we are expected to encounter in classifier evaluation problems. For $1/N_k \ll 1$, it can be easily established that, $V(\hat{A}^r) \geq V_{pro}(\hat{A}^s) \geq V_{opt}(\hat{A}^s)$. However, no such theoretical guarantee can be established for equal allocation.

First, we consider the cases of $V(\hat{A}^r)$ and $V_{pro}(\hat{A}^s)$. For $1/N_k \ll 1$, $N_k/(N_k-1)$ and $N/(N-1) \approx 1$. Then the difference between $V(\hat{A}^r)$ and $V_{pro}(\hat{A}^s)$ is

$$V(\hat{A}^r) - V_{pro}(\hat{A}^s) = \tfrac{1}{n}[A(1-A) - \sum_{k=1}^{K} W_k A_k(1-A_k)] = \tfrac{1}{n}\sum_{k=1}^{K} W_k(A_k - A)^2 \quad (7)$$

The above relation uses the fact that $A = \sum W_k A_k$ and $\sum W_k = 1$. Stratification in which accuracy of the strata are very different from each other will have higher difference between $V(\hat{A}^r)$ and $V_{pro}(\hat{A}^s)$. Hence, stratification which results in higher variance of A_k will lead to higher reduction in the variance of accuracy estimator. The worst case is when $A_k = A \; \forall k$, and in this case stratified sampling with proportional allocation will not give any variance reduction over random sampling.

For stratified sampling, $V_{opt}(\hat{A}^s)$ by definition is the minimum possible variance of $\hat{A}^s$ for a fixed n. At best we can expect $V_{pro}(\hat{A}^s)$ and $V_{equ}(\hat{A}^s)$ to attain $V_{opt}(\hat{A}^s)$. Difference between $V_{pro}(\hat{A}^s)$ and $V_{opt}(\hat{A}^s)$ comes out as

$$V_{pro}(\hat{A}^s) - V_{opt}(\hat{A}^s) = \tfrac{1}{n}\left[\sum_{k=1}^{K} W_k S_k^2 - (\sum_{k=1}^{K} W_k S_k)^2\right] = \tfrac{1}{n}\sum_{k=1}^{K} W_k(S_k - S_M)^2 \quad (8)$$

In the second step (Eq. 8), $S_M = \sum_{k=1}^{K} W_k S_k$ is the weighted mean of the S_k's. The second equality in Eq. 8 uses the definition of S_M and $\sum_{k=1}^{K} W_k = 1$. $V_{pro}(\hat{A}^s) = V_{opt}(\hat{A}^s)$ if and only if $S_k = S_M$. Thus, a stratification of $\mathcal{D}$ such that S_k is constant for all k would result in proportional allocation being optimal in the sense of variance.

Now, assume $W_k S_k = S_{wc}$ for all k, where S_{wc} is a fixed constant value. If $W_k S_k$ is constant then from Eq. 5, $V_{equ}(\hat{A}^s) = \frac{K^2 S_{wc}^2}{n}$. Also from Eq. 6, $V_{opt}(\hat{A}^s) = \frac{K^2 S_{wc}^2}{n}$. Hence, a stratification such that $W_k S_k$ is a constant implies equal allocation is optimal. Hence, if it can be ensured that $W_k S_k = Constant$, then the simpler equal allocation can substitute optimal allocation.

Practical implementation of proportional and equal allocation methods are much simpler compared to optimal allocation where we need OPT-A1 or OPT-A2. Hence, conditions under which proportional and equal allocation can achieve optimal minimum variances are important. As far as equal allocation is concerned, practically it can work well in several situations (as shown in experiments), however, it does not come with a theoretical guarantee that worst case variance will be same as simple random sampling.

3.5   Stratification Methods

Let z be the stratification variable and let $f(z)$ be the density distribution of z. z_i, $i = 1 \, to \, N$ denotes the discrete values of stratification variable for instances in dataset $\mathcal{D}$. If the classifier outputs $C(\boldsymbol{x}_i)$ are probabilistic then we use $z_i =$

$p(\hat{l}_i/\boldsymbol{x}_i)$. If the classifier scores are non-probabilistic and the predicted label is given by $\hat{l}_i = sign(C(\boldsymbol{x}_i))$, we use $z_i = |C(\boldsymbol{x}_i)|$, that is the magnitude of the classifier output.

The optimum stratification (in the sense of minimum variance) usually depends on the allocation policy [4,16]. A large body of stratification literature consists of approximate methods for optimum stratification. We employ several such methods from stratification literature. We also introduce use of clustering and simpler rule based methods which are usually not found in stratification literature. To estimate the density distribution $f(z)$ of the stratification variable using z_i's, we use Kernel Density estimation methods [8] with Gaussian kernels.

cum $\sqrt{\mathbf{f}}$ (SQRT) and $\mathbf{f}^{\frac{1}{3}}$ (CBRT): This method proposed in [5] is perhaps the most popular and widely used method for stratification. The simple rule is to divide the cumulative of $\sqrt{f(z)}$ into equal intervals. The points of stratification, $z_1^s < z_2^s < .. < z_{K-1}^s$, correspond to the boundary points corresponding to these intervals. The kth stratum consists of the set of instances for which z lies between z_{k-1}^s and z_k^s. z_0^s and z_K^s can be set as max and min of z. CBRT is same as SQRT except that the cube root of $f(z)$ is used in place of square root [18].

Weighted Mean (WTMN): In this method the key idea is to make the weighted mean of the stratification variable constant [9].

All of the previous methods try to approximate optimum stratification. We propose to introduce other techniques as well, which while not tailor-made for stratified sampling, can nevertheless serve as a way for stratification.

Clustering Methods: Clustering is one of the simplest ways to group the data $\mathcal{D}$ into different strata. We use K-means (KM) and Gaussian Mixture Models (GMM) based clustering to construct strata using z.

Simple Score Based Partitioning: The stratification variable z is obtained from classifier scores and we propose two simple partitioning methods. The first one is called EQSZ (Equal Size) in which the instances in $\mathcal{D}$ are first sorted according to the stratification variable. Starting from the top, each stratum takes away an equal number N/K of instances. It is expected that variation of z within each strata will be small. We call the other method as EQWD (Equal Width). In this case the range of z for $\mathcal{D}$ ($r = max(z) - min(z)$) is divided into sub-ranges of equal width. The points of stratification are $z_k^s = min(z) + rk/K$, $k = 1\,to\,K$. $z_0^s = min(z)$ is used in this case.

4 Experiments and Results

The variance of stratified sampling depends on three important factors, Allocation Method, Stratification Method and number of strata. We perform a comprehensive analysis of all of these factors. Each allocation method is applied on all 7 stratification methods. We vary the number of strata from 2 to 10 to study the effect of K. The previous related work [11] uses the feature space of instances and are hence not compared with here.

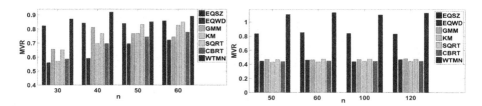

Fig. 1. MVR for different n on *rcv1* dataset (L: *Proportional*, R: *Equal*)

Datasets: We use three different dataset in our study. The first one, is the two-class form of the *rcv1* text categorization dataset [13]. The test set $\mathcal{D}$ consists of around 0.7 *million* instances. A logistic regression classifier is trained on the training set. The second one is the *epsilon* dataset from the Pascal Large Scale Challenge [1]. It contains 0.5 *million* instances of which we randomly selected 50,000 for training a linear SVM. The remaining 0.45 million instances are used as the test set $\mathcal{D}$. The last on is *News20 binary* dataset, the 2 class form of the text classification UCI News20 dataset [12]. It consist of a total of around 20000 instances from which 4000 randomly selected instances are used for training a logistic regression classifier and the rest are used as test set $\mathcal{D}$.

Evaluation Metrics: We will quantify our results in two ways. The first is the ratio of the variance of the stratified accuracy estimator to a random sampling estimator at a given n, VR $= V(\hat{A}^s)/V(\hat{A}^r)$. Unbiased estimates of $V(\hat{A}^r)$ and $V(\hat{A}^s)$ are used to measure VR. VR less than 1 is desirable; the lower it is the better it is. The second measure deals with absolute error (AE) percentage in estimating accuracy. Specifically, we look at the AE vs n plot and observe the amount of labeling resource required to achieve just 1% absolute error in accuracy estimates. We focus on % reduction, if any, in required n to achieve 1% error when using $\hat{A}^s$ in place of $\hat{A}^r$. All experiments are repeated for 3000 runs and Mean VR (MVR) and Mean AE (MAE) are used to report results. MAE vs. n plot are shown in *Supplementary* material which also contains detailed experimental discussion.

4.1 Proportional and Equal Allocation

Left panel in Fig. 1 shows the MVR values for each stratification method at different n on *rcv1* dataset. We can observe that EQWD is in general better compared to other methods leading to about **40–45%** reduction in variance for some cases. WTMN is the worst, showing only about 10% reduction in variances. The number of labeled instances required to achieve a 1% error in accuracy estimation goes down from 284 in random sampling to 218. This is about **23%** reduction in labeling resources.

The reduction in labeling resources for 1% error in accuracy estimation is about 12.5% for epsilon dataset and of 16% for news20 dataset. Figures not shown due to space constraints.

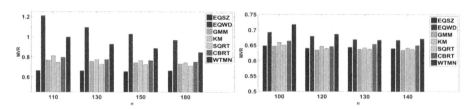

Fig. 2. MVR for different n. OPT-A1 Strat. Samp. (L: *epsilon*, R: *news20*)

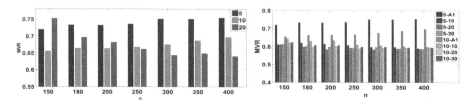

Fig. 3. Left: Effect of n_{ini} on OPT-A1. Selecting optimal n_{ini} can be a major problem. Right: OPT-A2 performs better and addresses problems of OPT-A1. Both hyper parameters n_{ini} and n_{step} are no longer critical. See Sect. 4.2 (OPT-A1 vs OPT-A2).

Strata size K is also an important factor in stratified sampling. Increasing K leads to better results. However, the general trend is that once K is large enough major variation in MVR values cannot be expected. The trend is same for all datasets.

The results on *rcv1* dataset for Equal Allocation are shown is right panel of Fig. 1. KM and SQRT stratification are in general better than other methods. Huge reduction in variance in the range of **55–60**% is observed, leading to very precise estimate of accuracy with very little labeling resource. Using KM based stratification, n required for 1% error margin is reduced by a substantial amount, close to **58.5**% (from 284 to 118). The labeling budget for 1% error in accuracy estimate reduced by about 16% for *epsilon* and 22% for *news20*.

4.2   Optimal Allocation

We observed that for *rcv1* dataset, Equal allocation resulted in a substantial reduction in the variance of estimate. The optimal allocation policy leads to further reduction in variance by only few percentage points (4–6% more) and hence we do not show the plots for *rcv1*. However, for *epsilon* and *news20* optimal allocation actually results in substantial reduction in variance. We use $n_{ini} = 10$ for OPT-A1 algorithm. For OPT-A1 mid range K such as $K = 6$ *or* 7 are better in general, especially at lower n. K affects the number of samples ($n_{ini} * K$) used up in initial estimation of S_k. Mid range K leads to good stratification and at the same time leave enough labeling resource which can be allocated optimally.

For epsilon dataset, left panel of Fig. 2 shows that EQSZ gives over **30–35**% reduction in variance compared to random sampling. n required for 1%

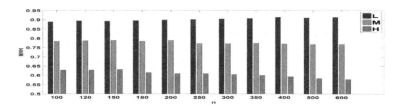

Fig. 4. Effect of True Accuracy on MVR. For epsilon dataset using OPT-A2 method.

error is reduced by **23%** using OPT-A1 which is about 10% and 7% higher over proportional and equal allocation respectively. For *news20* (right panel Fig. 2), the variance is reduced by more than 35% for several cases which is substantially higher than other two allocation methods. OPT-A1 leads to reduction in n by about **27%** for 1% error which is higher than that for proportional and equal by 11% and 5% respectively. SQRT stratification does well on both datasets.

OPT-A1 vs OPT-A2: We mentioned previously that setting n_{ini} in OPT-A1 might present practical difficulties. This is illustrated for *epsilon* dataset in left panel of Fig. 3 where we show MVR values for n_{ini} equal to 5,10 and 20. For sufficiently large n (>200), higher n_{ini} is better. This is expected as increasing n_{ini} results in better estimation of S_k and for large n, enough labeling resources are left to be allocated in an optimal sense to help achieve lower variance. However, problem occurs for lower n ($<=200$). MVR first reduces by increasing n_{ini} from 5 to 10 but then increases substantially when we increase it further to 20. Clearly, the optimal value of n_{ini} cannot be known a priori.

OPT-A2 gets around problems of OPT-A1. Right panel of Fig. 3 shows the efficiency and benefits of OPT-A2. Legend are in form $n_{ini} - n_{step}$. $n_{ini} - A1$ legends represent the corresponding MVR using OPT-A1. First, we observe that irrespective of the value of n_{ini} OPT-A2 results in lower MVR compared to OPT-A1. OPT-A2 leads to a further reduction in variance of estimated accuracy by upto **18%** in certain cases. Moreover, we note that setting n_{ini} is no more critical; $n_{ini} = 5$ works as good as $n_{ini} = 10$. Even more convenient is the fact that n_{step} does not affect MVR in any major way, which removes the role of any hyperparameter for OPT-A2. Hence, one can set n_{ini} to any small value such as 5 and any reasonable value of n_{step} such as 10 or 20 works fine.

4.3 Dependence on True Accuracy

It is expected that the value of true accuracy would have some effect on the MVR, which measures how well stratified sampling is doing compared to random sampling. Here, we try to empirically study the effect of actual value of true accuracy on the proposed accuracy estimation process. We use *epsilon* dataset for training 3 different classifiers (SVMs) with varying accuracies. The true accuracies of the classifiers are 88% (H), 77% (M) and 67% (L). The accuracies have been reduced by reducing the amount of training data used. Obviously, the test data $\mathcal{D}$ remains same for all 3 classifiers. Now we try to estimate these accuracies for the 3 classifiers by sampling from $\mathcal{D}$ and we observe the MVR values for different n. Figure 4 show the results for three cases using OPT-A2 with $n_{ini} = 5$ and $n_{step} = 10$. We observe that MVR follows an inverse trend with classifier accuracy. Thus, the better the classifier the more effective stratified sampling is in reducing the variance of accuracy estimate. Similar trend for OPT-A1 also exist.

5 Discussions and Conclusions

We presented a method for evaluating classifiers in a limited labeling budget scenario. We theoretically derived the variance of accuracy estimates for different cases and showed that *stratified sampling* can be used for reducing the variance of accuracy estimates. It can be applied for both probabilistic (logistic regression) as well non-probabilistic (support vector machines) classifiers. We observed that the labeling resource required to estimate accuracy with very low error can be reduced by as much as 60%. As far as stratification methods are concerned, it is worth noting that clustering methods in general perform as well as established stratification methods in sampling literature. We showed that implementation of optimal allocation is best done through the proposed OPT-A2 method. Performance of OPT-A2 is not only better compared to OPT-A1 but is almost independent of the parameters (n_{ini} and n_{step}) it takes as input, unlike, where OPT-A1 where n_{ini} plays a critical role.

References

1. Pascal large scale learning challenge (2008). largescale.ml.tu-berlin.de
2. Bennett, P.N., Carvalho, V.R.: Online stratified sampling: evaluating classifiers at web-scale. In: Proceedings of the 19th ACM International Conference on Information and Knowledge Management, pp. 1581–1584. ACM (2010)
3. Cochran, W.G.: Sampling Techniques. Wiley, New York (2007)
4. Dalenius, T., Gurney, M.: The problem of optimum stratification. II. Scand. Actuarial J. **1951**(1–2), 133–148 (1951)
5. Dalenius, T., Hodges Jr., J.L.: Minimum variance stratification. J. Am. Stat. Assoc. **54**(285), 88–101 (1959)
6. Donmez, P., Lebanon, G., Balasubramanian, K.: Unsupervised supervised learning i: estimating classification and regression errors without labels. J. Mach. Learn. Res. **11**, 1323–1351 (2010)
7. Druck, G., McCallum, A.: Toward interactive training and evaluation. In: Proceedings of the 20th ACM International Conference on Information and Knowledge Management, pp. 947–956. ACM (2011)
8. Hastie, T., Tibshirani, R., Friedman, J.: The Elements of Statistical Learning: Data Mining, Inference, and Prediction. SSS. Springer, New York (2009). https://doi.org/10.1007/978-0-387-84858-7
9. Hansen, H., Hurwitz, W., Madow, W.G.: Sample survey methods and theory (1953)

10. Jaffe, A., Nadler, B., Kluger, Y.: Estimating the accuracies of multiple classifiers without labeled data. arXiv preprint arXiv:1407.7644 (2014)
11. Katariya, N., Iyer, A., Sarawagi, S.: Active evaluation of classifiers on large datasets. In: 2012 IEEE 12th International Conference on Data Mining (ICDM), pp. 329–338. IEEE (2012)
12. Keerthi, S., DeCoste, D.: A modified finite Newton method for fast solution of large scale linear SVMs. J. Mach. Learn. Res. **6**, 341–361 (2005)
13. Lewis, D., Yang, Y., Rose, T., Li, F.: RCV1: a new benchmark collection for text categorization research. J. Mach. Learn. Res. **5**, 361–397 (2004)
14. Platanios, E., Blum, A., Mitchell, T.: Estimating accuracy from unlabeled data (2014)
15. Sawade, C., Landwehr, N., Bickel, S., Scheffer, T.: Active risk estimation. In: Proceedings of the 27th International Conference on Machine Learning (ICML 2010), pp. 951–958 (2010)
16. Sethi, V.: A note on optimum stratification of populations for estimating the population means. Aust. J. Stat. **5**(1), 20–33 (1963)
17. Settles, B.: Active learning. Synth. Lect. Artif. Intell. Mach. Learn. **6**(1), 1–114 (2012)
18. Singh, R.: Approximately optimum stratification on the auxiliary variable. J. Am. Stat. Assoc. **66**(336), 829–833 (1971)

Social Stream Classification
with Emerging New Labels

Xin Mu[1,2(✉)], Feida Zhu[2], Yue Liu[2], Ee-Peng Lim[2], and Zhi-Hua Zhou[1]

[1] National Key Laboratory for Novel Software Technology,
Nanjing University, Nanjing 210023, China
{mux,zhouzh}@lamda.nju.edu.cn
[2] School of Information Systems, Singapore Management University,
Singapore, Singapore
{fdzhu,yueliu,eplim}@smu.edu.sg

Abstract. As an important research topic with well-recognized practical values, classification of social streams has been identified with increasing popularity with social data, such as the tweet stream generated by Twitter users in chronological order. A salient, and perhaps also the most interesting, feature of such user-generated content is its never-failing novelty, which, unfortunately, would challenge most traditional pre-trained classification models as they are built based on fixed label set and would therefore fail to identify new labels as they emerge. In this paper, we study the problem of classification of social streams with emerging new labels, and propose a novel ensemble framework, integrating an instance-based learner and a label-based learner by completely-random trees. The proposed framework can not only classify known labels in the multi-label scenario, but also detect emerging new labels and update itself in the data stream. Extensive experiments on real-world stream data set from *Weibo*, a Chinese micro-blogging platform, demonstrate the superiority of our approach over the state-of-the-art methods.

Keywords: Stream classification · Emerging new labels
Model update

1 Introduction

Social stream classification has attracted an ever-increasing level of attention from both academia and industry due to the recent boom of social media platforms such as Twitter, in which the user-generated contents (i.e., tweets) naturally form a data stream by chronological order. As each item could assume one or multiple labels based on its content, classifying tweets into their corresponding labels serves as the foundation for profiling both users and information diffusion processes, in turn contributing to many real-life applications including targeted marketing, customer relationship management and credit risk evaluation.

© Springer International Publishing AG, part of Springer Nature 2018
D. Phung et al. (Eds.): PAKDD 2018, LNAI 10937, pp. 16–28, 2018.
https://doi.org/10.1007/978-3-319-93034-3_2

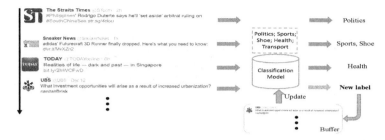

Fig. 1. An illustration of SSC-NL problem.

The basic process of social stream classification can be described as follows: with a set of social data which has been preprocessed and manually associated with concepts (labels), a classification method, such as SVMs or Random Forest, can be employed to train a model to predict labels for new incoming data. Despite the wealth of research efforts on social stream classification [1–3], most existing solutions, which are built with a fixed label set, face serious challenges when dealing with social data with emerging new labels due to its salient feature of topic novelty that is typical of social media content.

We therefore address in this paper a more challenging problem of Social Stream Classification with emerging New Labels (SSC-NL). Compared to previous social stream classification problems, the SSC-NL problem needs to accomplish three tasks simultaneously: (1) detecting emerging new labels; (2) classifying known labels in the multi-label scenario; and (3) updating the model with new labels identified. We illustrate this problem with a case for tweet stream classification in Fig. 1. We assume that the model is built initially with labels such as Politics, Sports, etc. This model is deployed in a tweet stream to classify each tweet with known labels, and correctly detect tweets with new labels as they emerge. These tweets with new labels are placed in a buffer, until the model update is triggered by some pre-defined criteria. Once the update is completed, the buffer is reset and the new model is ready for the next tweets in the stream.

To address the SSC-NL problem, we propose a novel ensemble framework named NL-Forest, which involves two cooperating models, an instance-based model and a label-based model, both composed of completely-random trees. NL-Forest can predict a ranking of known labels, and identify emerging new labels in the social streams. Furthermore, the models are to be updated once some criteria are met. We summarize some key contributions as follows: (1) The proposed method accomplishes three tasks simultaneously, including detection of new labels, classification for known labels, and model updating; (2) A straightforward approach for SSC-NL problem is to learn a known labels classifier and a new label detector like [4]. Compared to methods based on two distinct algorithm structures, completely-random trees are used as a single core to provide the solution to efficient prediction and updating as shown in Sect. 5.3. In addition, our model achieves better prediction performance and is more robust due to its ensemble strategy; (3) Experiments are conducted on both a real-world data stream and simulated streams where new labels appear under different scenarios. Our framework outperforms existing state-of-the-art methods.

The rest of this paper is organized as follows: Sect. 2 examines the related work. We introduce the proposed framework in Sect. 4 and the experimental evaluation is detailed in Sect. 5. We conclude the paper in Sect. 6.

2   Related Work

Social stream classification has been extensively studied in recent years. Zubiaga and Spina [3] analyzed social features, and then performed classification experiments with Support Vector Machines (SVMs). In [1], a text-based classification method and a network-based classification method were proposed for classifying social data topics. Other than these supervised classification models, unsupervised learning was also widely used. For instance, topic modeling is effective in grouping documents into a pre-defined number of coarse clusters based on inter-document similarity or the co-occurrence patterns of terms [5]. However, existing algorithms normally employ a classifier with a fixed label set, thus are unable to address the problem of emerging new labels. Though some online setting methods, such as [6], are able to tackle this problem, each item needs to be manually labeled before update, making them unsuitable for real data streams.

Class-Incremental Learning (C-IL), which is a branch of incremental learning [7], has attracted much attention recently. The SSC-NL problem is actually a C-IL problem in social data stream context. In recent years, a number of algorithms [8–11] have been developed for classification under emerging new classes. For instance, the ECSMiner [12] tackled the novel class detection and classification problem by introducing time constraints for delayed classification. Learning with Augmented Class (LAC) [13] was proposed for identifying emerging new classes, assuming the availability of an unlabeled dataset to help identifying these new classes. In [10], an isolation-based idea was used for new class detection. However, above-mentioned methods are tailored to the single label problem, and face serious challenges in identifying new labels if instances are with multiple labels. Although, a new effort, MuENL [4], includes one classifier based on regularized SVMs and one detector based on tree structure which can tackle the SSC-NL problem, two strategies for model updating are required, resulting in a high computation cost and being hard to implement in real-time problem.

Other relevant approaches include tree-based methods to address the multi-label classification problem [14] and the anomaly detection problem [15]. Indeed, creating models to cope with environment changes [16], is widely studied in the machine learning and data mining community. Solving the SSC-NL problem can be seen as a preliminary step in social stream context.

3   Preliminaries

In general, social data stream is in the form of continuous streams of text data [17]. Text representation is a fundamental component to represent text

Label set

Fig. 2. An example of ANL and PNL. The initial label set contains four labels, the second tweet shows the ANL, and the third tweet shows PNL.

into an amenable form. Here, we denote $x \in \mathbb{R}^d$ as a vector representation for each text data by using a representation model like [18]. The SSC-NL problem therefore is defined as follows: given a set of social data as the training set $D^T = \{(x_i, y_i)\}_{i=1}^m$, where $x_i \in \mathbb{R}^d$, $y_i \in \mathcal{Y} = \{-1, 1\}^c$ is the corresponding label vector, c is the number of labels. $y_{i,j} = 1$ iff the jth label is assigned to the example x_i and $y_{i,j} = -1$ otherwise. The streaming instance is from $D^S = \{(x_t, y_t')\}_{t=1}^\infty$, where $y' \in \mathcal{Y}' = \mathcal{Y} \bigcup \mathcal{Y}_{new}$, $\mathcal{Y}_{new} = \{-1, 1\}^a$, $a > 0$. The goal is to learn an initial model f with D^T, then f is used as a detector for emerging new labels ($\mathcal{Y}_{new}$) and a classifier for known labels ($\mathcal{Y}$) in the data stream. In addition, f can be updated when it maintains some criteria. The training set D^T is just used for building model at the beginning of the data stream and will then be discarded. Once update is completed, it is ready for the next instances in the data stream. Note that the model can detect instances of any number of emerging new labels, though they are grouped into one new meta-label.

Detecting emerging new labels in SSC-NL problem is a non-trivial task, because the instance with new labels is likely to contain known labels simultaneously. This is a distinct point of difference from the previous works [10,12]. To specify the form of new label emergence in the multi-label scenario, we define two types of instances with new label as follows, and Fig. 2 shows an illustration.

Definition 1. *[Absolutely New Label (ANL)]. Let $\mathcal{Y}$ be the known label space and $\mathcal{Y}_{new}$ be the new label space. An instance with absolutely new label is defined as (x, y), where $y \in \mathcal{Y}_{new}$ and $y \notin \mathcal{Y}$.*

Definition 2. *[Partially New Label (PNL)]. Let $\mathcal{Y}$ be the known label space and $\mathcal{Y}_{new}$ be the new label space. An instance with partially new label is defined as (x, y), where $y \in \mathcal{Y}_{new}$ and $y \in \mathcal{Y}$.*

4   The Proposed Framework

In this section, we propose a novel framework named NL-Forest, which is composed of two cooperating forests. The instance-based forest (I-F) is built on the whole training data set, and the label-based forest (L-F) consists of multiple sub-forests by considering label information. The details are provided as follows.

Algorithm 1. NL-Forest construction

Input: D - input data, Z - number of trees in I-F, z - number of trees in L-F, ψ, ϕ - sample size. **Output:** NL-Forest 1: **initialize:** I-F $\leftarrow$ {}, L-F $\leftarrow$ {}. 2: **for** $i = 1, \ldots, Z$ **do** 3: $D_1 \leftarrow sample(D, \psi)$ 4: I-F $\leftarrow$ I-F $\cup$ $Tree\,(D_1)$ 5: **end for** 6: **for** $j = 1, \ldots, c$ **do** 7: $D_2 \leftarrow \{(\boldsymbol{x}, \boldsymbol{y}) | \boldsymbol{x} \in D, \boldsymbol{y}_{(,j)} = 1\}$ 8: **for** $k = 1, \ldots, z$ **do** 9: $D_3 \leftarrow sample(D_2, \phi)$ 10: L-F(j) $\leftarrow$ L-F(j) $\cup$ $Tree(D_3)$ 11: **end for** 12: compute the threshold in L-F (j). 13: **end for** 14: **return** NL-Forest$\leftarrow$ L-F $\cup$ I-F	The function $\boldsymbol{Tree}(X)$ X - input data, $MinSize$ - minimum internal node size 1: **if** $|X| < MinSize$ **then** 2: return LeafNode$\{F[\cdot], center, v\}$. 3: **else** 4: let Q be a list of attributes in X. Ran- domly select an attribute $q \in Q$ and randomly select a split point p from max and min values of attribute q in X. 5: $X_L \leftarrow filter(X, q \le p)$ 6: $X_R \leftarrow filter(X, q > p)$ 7: return inNode$\{$Left $\leftarrow$ $Tree(X_L)$, 8: Right $\leftarrow$ $Tree(X_R)\}$ 9: **end if**

4.1   NL-Forest: Training Process

The training process is detailed in Algorithm 1. Steps 2–5 in the left side of Algorithm 1 show the process of building I-F. The function $sample(D, \psi)$ is defined as randomly sampling a subset with size ψ from the data set D. The function $Tree(\cdot)$ as shown in the right side is defined as building a completely-random tree, where a partition is produced by randomly selecting an attribute and its cut-point between the minimum and maximum values in the sample. The splitting is stopped when the number of instances is less than $MinSize$. Note that in each node, we just record the mean of instances as "centre", the label distribution $F[\cdot]$, and the average number of labels per instance as v. Steps 8–11 in the left side of Algorithm 1 show the L-F construction based on the label information, which is similar to building I-F.

In line 12, a threshold, which is used to measure new labels emerging in the data stream, is found in each sub-forest. The idea here is inspired by the model proposed in [15], wherein Liu et al. presented an isolation-based method. In the NL-Forest framework, each tree is actually built to isolate every instance from the rest of the instances in the input data set. Threshold determination is based on the fact that there exist "differences" between instances with new labels and original training instances, and thus instances with new labels are more susceptible to isolation than instances with only known labels. In other words, the instances with new labels will be isolated using fewer partitions[1] in a tree than instances with only known labels. To obtain threshold, because the sub-forest contains a subset of labels, we select the instances without this subset of labels in the training data to compute their average height in this sub-forest. The average height obtained will be finally used as the *threshold*.

---

[1] The fewer partitions means that instances with new labels are more likely to be of the shorter height in each tree.

4.2   NL-Forest: Deployment

Algorithm 2 describes the deployment of NL-Forest in a data stream. NL-Forest $(\boldsymbol{x})$ produces a label vector $(y_1, \ldots, y_c, y_{new})$ in line 2 and is defined as:

$$\text{NL-Forest}(\boldsymbol{x}) = \begin{cases} y_1, \ldots, y_c \leftarrow \text{I-F}(\boldsymbol{x}) & (1) \\ y_{new} \leftarrow \text{L-F}(\boldsymbol{x}, \boldsymbol{y}_{known}) & (2) \end{cases}$$

where c is the number of known labels and $\boldsymbol{y}_{new}$ is new label predicted. In Eq. (1), $\boldsymbol{x}$ falls into one node in each tree in I-F and the distribution $F[\cdot]$ is recorded. The output of I-F is the average of label vectors in $F[\cdot]$ as follow:

$$\text{I-F}(\boldsymbol{x}_{test}) = p(\boldsymbol{y}|\boldsymbol{x}_{test}) = \text{E}[\sum_{i=1}^{Z} p(\boldsymbol{y}|\boldsymbol{x}_{test}, F[i])] \qquad (3)$$

where $p(\boldsymbol{y}|\boldsymbol{x}_{test}, F[i])$ is the output of ith tree in I-F. I-F can also output an accurate number of labels by using the average results of v in each node. This is because previous works [19–21] have shown that, ensemble of completely-random trees can be successfully applied as a powerful classifier, and it is evident that the proposed method can be a classifier capable for classification task.

Equation (2) describes that L-F predicts a new label y_{new}. We first introduce a cooperating mechanism to detect instances with PNL. From the I-F outputs, we can obtain the probabilities of known labels in the form of a label vector, as indicated in Eq. (1). Thus, we generate a vector in descending order of known labels probabilities, denoted as $\boldsymbol{y}_{known}$. According to the order of labels in $\boldsymbol{y}_{known}$, we pass $\boldsymbol{x}_{test}$ to the corresponding sub-forests. The function L-F$(\cdot)$ is as follows:

$$\text{L-F}(\boldsymbol{x}_{test}, \boldsymbol{y}_{known}) = \text{E}[\sum_{i=1}^{u} p(y_{new}|\boldsymbol{x}_{test}, y_i)], \ y_i \in \boldsymbol{y}_{known} \qquad (4)$$

where $p(y_{new}|\boldsymbol{x}_{test}, y_i)$ is the output in one sub-forest in L-F and is defined as:

$$p(y_{new}|\boldsymbol{x}_{test}, y_i) = \begin{cases} 1, & if \ \Theta(\boldsymbol{x}_{test}) < threshold_i \\ 0, & otherwise \end{cases} \qquad (5)$$

where $\Theta(\cdot)$ is the average height of the instance in i^{th} sub-forest. Note that each sub-forest is able to partition instances with the specific label. If $\boldsymbol{x}_{test}$ contains new labels, it will be partitioned easier in this known label sub-forest, that is, $\boldsymbol{x}_{test}$ will have shorter height in this sub-forest. In Eq. (4), $\boldsymbol{x}_{test}$ is an instance with new label if the average height of instance $\boldsymbol{x}_{test}$ in i^{th} sub-forest is less than the $threshold_i$. We finally use the top u labels in $\boldsymbol{y}_{known}$ to predict whether new label is emerging in Eq. (3). We can use the predicted number of labels as a measure to guide the setup of u.

Because an instance with ANL is likely to differentiate from original instances in the training set, detecting ANL is equivalent to detecting new labels in single label setting. Hence, the instance with ANL can be directly isolated using fewer partitions in the I-F. Fortunately, some previous works employed random tree

Algorithm 2. NL-Forest deployment in the data stream

Input: NL-Forest, $\mathcal{B}$ - buffer, s - buffer size.
Output: prediction for each x in a data stream
1: **while** not end of data stream **do**
2:    $(y, y_{new}) \leftarrow$ NL-Forest(x)
3:    **if** $y_{new} > \frac{1}{2}$ **then**
4:        $\mathcal{B} \leftarrow \mathcal{B} \cup \{x\}$
5:        NewLable $\leftarrow$ 1
6:    **else**
7:        NewLabel $\leftarrow$ 0
8:    **end if**
9:    Output $\{y_1, \ldots, y_c, Newlabel\}$
10:    **if** $|\mathcal{B}| \geq s$ **then**
11:        Update (NL-Forest, $\mathcal{B}$)   # detailed in Sect. 4.3
12:        $\mathcal{B} \leftarrow$ NULL
13:    **end if**
14: **end while**

---

structures for new class detection and can naturally adapt to I-F. In this paper, we use the method in [10] to detect the ANL. In line 3–8 in Algorithm 2, NL-Forest outputs a positive decision if y_{new} is greater than a 0.5 threshold. This threshold corresponds intuitively to majority voting.

Model is updated when buffer $\mathcal{B}$ is full ($|\mathcal{B}| \geq s$) in line 10–13 in Algorithm 2. The size of buffer s is a user-defined parameter and can be set based on the memory space available[2]. Similar to [11], we only need to manually annotate instances with the true label in the buffer instead of labeling all instances in the data stream. In the following, we introduce two growing mechanisms.

4.3   NL-Forest: Model Update

Growing a subtree in I-F. Updating I-F is to update each leaf node in every tree using a random sample of size λ from $\mathcal{B}$. The update at each node involves either (1) a replacement with a simple update label distribution $F[\cdot]$ to include the new label y_{c+1} or (2) a newly grown subtree if the total number of instances falling into the same leaf node exceeds the limit. At each node, growing a subtree needs to generate pseudo instances in each node which have the same attribute-values as "centre". The number of pseudo instances is as recorded in $F[\cdot]$. The combined set of pseudo instances and the randomly selected instances which fall into this leaf node is used as input to build the subtree.

Growing a new sub-forest in L-F. A new sub-forest can be constructed using instances with the new label from $\mathcal{B}$. Once the new sub-forest is completed, a *threshold* is calculated as mentioned in Sect. 4.1 by using pseudo instances.

4.4   Model Complexity

In the training stage, the overall time complexity to construct random tree is $O(Z\psi \log \psi + cz\phi \log \phi)$. To predict an instance in the stream, it takes $O(Z \log \psi +$

---

[2] This is a trade-off parameter, the larger means method needs more memory. In practise, we use the value which is greater than ψ to guide the setup of this parameter.

$uz \log \phi$) time to traverse each of the Z trees in I-F and z trees in u L-F. During the update, growing a subtree using a s size buffer takes $O(Zs \log s)$ and growing a sub-forest takes $O(z\phi \log \phi)$. The total space required includes the buffer with size s and all centres and label distribution in leaf nodes in I-F. Thus, the space complexity is $O(s + Zd\psi)$.

5   Experiment

5.1   Experimental Setup

Data Sets. A summary of the data characteristics is provided in Table 1. The real streaming data is collected from *Sina Weibo*. This stream is about 220k items with 10 labels, and each item is preprocessed using *word2vec*[3] to produce a 300-dimension feature vector.

Competing Algorithms. Table 2 is a complete list of the methods used for new label detection and known label classification. It includes two multi-label supervised classifiers – binary relevance SVM (BR-SVM) [22], ML-KNN [23]; one supervised multi-label streaming classifier – SMART [6]; an existing solution for emerging new labels – MuENL [4]; an outlier detector as new label detector – iForest [15].

Experiment Settings and Evaluation Metrics. All methods are executed in the MATLAB environment with the following implementations: SVM is in the LIBSVM package [22]; MuENL, iForest and ML-KNN are the codes as released by the corresponding authors; SMART code is developed based on the original paper [6]. In NL-Forest, we set $Z = 200$, $z = 100$, ψ and ϕ are set by $0.6 * m$ and $0.6 * n_i$, where m and n_i are the sizes of D_1 and D_2 respectively. λ is set according to label balance in each tree. The trees stop growing when the total number of instances, which fall into a leaf node, exceeds the limit, e.g., $MinSize = 10$ in the simulated streams and $MinSize = 100$ in the real stream. BR-SVM trains a linear classifier for each label independently and parameters are set according to cross validation. In ML-KNN, K, the number of nearest neighbors is set as 10. In SMART, the tree height is $h = 30$, and the number of

Table 1. A summary of data sets.

	Emotions	Yeast	Enron	Weibo
#Attributes	72	103	1001	300
#Labels	6	14	53	10
Volume	593	2417	1702	220K

Table 2. Methods used in the experiments.

Method	Detector	Classifier
BR-SVM+iF	iForest	BR-SVM
ML-KNN+iF	iForest	ML-KNN
SMART+iF	iForest	SMART
MuENL	MuENLForest	MuENL$_{\text{MNL}}$
NL-Forest	NL-Forest	

---

[3] https://radimrehurek.com/gensim/index.html.

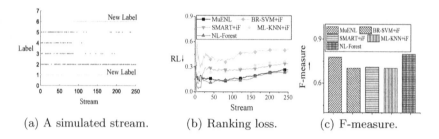

(a) A simulated stream.   (b) Ranking loss.   (c) F-measure.

Fig. 3. A simulated stream on *emotions* data set. (a) There are six labels, the blue points represent the known labels, and the red points represent the new labels. (b) and (c) the results of RL and F-measure. (Color figure online)

trees in the ensemble is $n_t = 30$. The number of trees in iForest is $n_t = 50$. In MuENL, the parameters in classification model are selected via cross validation, and the setup of detection model is same as iForest. We employed **Ranking Loss (RL $\downarrow^4$)** and **Average Precision (AP $\uparrow$)** for classification performances and **F-measure ($\uparrow$)** for detection results.

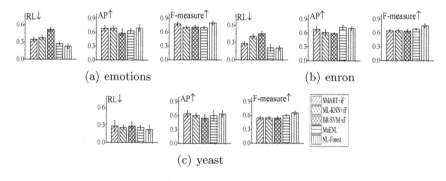

Fig. 4. Results of the simulated data streams.

5.2   Simulated Data Stream

To evaluate different scenarios under which new labels appear, we perform experiments in multiple simulated data streams which are generated from three benchmark multi-label data sets. In each data set, we first randomly select two labels as the new labels. The instances with any of these two labels are selected as set A, and the rest will be randomly divided into training set (80%) and testing set (20%). Then A is added to the testing set, and we simulate a steam by using the testing set. RL and AP are computed over the entire stream, F-measure is

---

[4] Here "$\downarrow$" means the smaller the value, the better the performance; and "$\uparrow$" means the larger the value, the better the performance.

computed when the buffer is full. An example is shown in Fig. 3. The simulation is repeated 30 times for each data set, the average results are reported in Fig. 4.

Detailed Analysis. The proposed NL-Forest has consistently produced better performance in all three data sets than any other methods for both emerging new labels detection and known labels classification. In terms of new label detection, NL-Forest produces higher F-measure than methods directly using an anomaly detector (i.e., iForest) which do not consider differences between new labels and anomaly when detecting the PNL. In terms of classification results (RL and AP), NL-Forest gives results comparable with state-of-the-art methods, e.g., MuENL$_{MNL}$ and ML-KNN. What has greatly contributed to the practical values of NL-Forest is the fact that it can be applied to a wide range of prediction problems and has fewer parameters to tune.

Compared to NL-Forest, MuENL consists of two independent models, i.e., a detector for new labels and a classifier for known labels. Despite its reasonable good classification and detection performances in simulations, it is not a good choice for the SSC-NL problem due to its high computational complexity in model update and practical difficulty in parameter determination. On the other hand, ML-KNN, BR-SVM and SMART are state-of-the-art multi-label classification methods for known labels, but they still require anther framework to detect new labels. In addition, BR-SVM often comes at high computational costs in an extensive parameter search.

5.3   Real Data Stream

In this section, we conduct experiments on a social stream and compare the proposed method with SMART+iForest and MuENL which focus on the streaming data problem. Figure 5(a) indicates label distribution in the stream. For convenience, we use "1st" to "10th" to represent the category "traffic safety" to "finance". We regard five labels (6th to 10th) as known labels and collect extra 15 K instances with them to initialize the model. The 1st to 5th labels are regarded as new labels which occur in the different periods. To be specific, the 2nd, 3rd and 5th label emerge at around point 0 to 50 K; at around point 50 K to 100 K, the 4th label appears; the 1st label emerges at last 100 K points. Note that when the buffer is full, model will be updated using buffer data with true label. Evaluation metrics are computed at different time points as shown in Figs. 5(b) and (c).

Figure 5(c) shows NL-Forest outperforms other methods in detecting new labels, and NL-Forest gives comparable results with other methods in classification in Fig. 5(b). We also show the time of processing 1000 data items and the average update time in the stream. In Fig. 5(d), the proposed method achieves the shorter running time than MuENL for the real data stream, and is comparable to the state-of-the-art method SMART, which also employs completely random trees. Figure 5(e) shows the proposed method can be more efficient in deploying in the real application with the faster update.

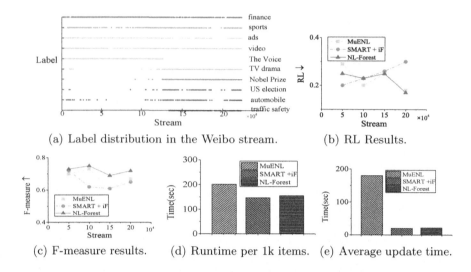

(a) Label distribution in the Weibo stream.   (b) RL Results.

(c) F-measure results.   (d) Runtime per 1k items.   (e) Average update time.

Fig. 5. Results of the real data stream.

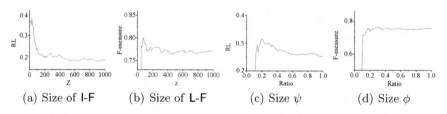

(a) Size of I-F   (b) Size of L-F   (c) Size ψ (d) Size ϕ

Fig. 6. Results of the sensitivity of parameters.

5.4   Sensitivity of Parameters

We study the influences of parameters in NL-Forest, i.e., z and Z (the number of trees), ψ and ϕ (the sampling sizes). We evaluate NL-Forest on the *emotions* data set with different settings of one parameter while the other parameters are fixed. Figures 6(a) and (b) show that the performance of NL-Forest is stable when we set the size of tree greater than 100. Therefore, in practice, model parameter setup can follow such guidelines. In Figs. 6(c) and (d), the X-axis represents a ratio between the sample size and original data size. Generally, the larger each random tree is, the better the performance is, but larger trees will consume memory. We observe that the RL or F-measure of NL-Forest converges at a small ψ or ϕ. Hence, the ratio set by half is safe and recommended in practise. Note that similar results are also observed on the other data sets.

6   Conclusion

This paper introduces a novel framework with an instance-based model and a label-based model to address the SSC-NL problem. The strength of NL-Forest is that the completely-random trees are used as a single core to effectively tackle emerging new labels detection and known labels classification, and provide the solution to efficient update. Evaluations on simulated streams and a real-world stream demonstrate the effectiveness of the proposed framework. In the future, the broader stream classification problem in real-world applications [24] including detection of concept drift, issues with outdated data, adaptation to the current state, and recurring contexts will be considered. It is also in our interest to explore the theoretical foundation for our model and extend the idea of this work to Multi-Instance Multi-Label learning (MIML) [25].

Acknowledgement. This research was supported by the National Research Foundation, Prime Minister's Office, Singapore under its International Research Centres in Singapore Funding Initiative; the NSFC (61333014) and Pinnacle lab for analytics at Singapore Management University.

References

1. Lee, K., Palsetia, D., Narayanan, R., Patwary, M.M.A., Agrawal, A., Choudhary, A.N.: Twitter trending topic classification. In: ICDM Workshops, pp. 251–258 (2011)
2. Tsai, M., Aggarwal, C.C., Huang, T.S.: Towards classification of social streams. In: SDM (2015) 649–657
3. Zubiaga, A., Spina, D., Martínez-Unanue, R., Fresno, V.: Real-time classification of twitter trends. JASIST **66**(3), 462–473 (2015)
4. Zhu, Y., Ting, K.M., Zhou, Z.H.: Multi-label learning with emerging new labels. In: ICDM, pp. 1371–1376 (2016)
5. Blei, D.M.: Probabilistic topic models. Commun. ACM **55**(4), 77–84 (2012)
6. Kong, X., Yu, P.S.: An ensemble-based approach to fast classification of multi-label data streams. In: 7th International Conference on Collaborative Computing: Networking, Applications and Worksharing, CollaborateCom, pp. 95–104 (2011)
7. Zhou, Z.H., Chen, Z.Q.: Hybrid decision tree. Knowl. Based Syst. **15**(8), 515–528 (2002)
8. Al-Khateeb, T., Masud, M.M., Khan, L., Aggarwal, C., Han, J., Thuraisingham, B.: Stream classification with recurring and novel class detection using class-based ensemble. In: ICDM, pp. 31–40 (2012)
9. Mu, X., Zhu, F., Du, J., Lim, E.P., Zhou, Z.H.: Streaming classification with emerging new class by class matrix sketching. In: AAAI, pp. 2373–2379 (2017)
10. Mu, X., Ting, K.M., Zhou, Z.H.: Classification under streaming emerging new classes: A solution using completely-random trees. IEEE TKDE **29**(8), 1605–1618 (2017)
11. Haque, A., Khan, L., Baron, M.: Sand: Semi-supervised adaptive novel class detection and classification over data stream. In: AAAI, pp. 1652–1658 (2016)
12. Masud, M., Gao, J., Khan, L., Han, J., Thuraisingham, B.M.: Classification and novel class detection in concept-drifting data streams under time constraints. IEEE TKDE **23**(6), 859–874 (2011)
13. Da, Q., Yu, Y., Zhou, Z.H.: Learning with augmented class by exploiting unlabeled data. In: AAAI, pp. 1760–1766 (2014)

14. Liu, F., Zhang, X., Ye, Y., Zhao, Y., Li, Y.: MLRF: multi-label classification through random forest with label-set partition. In: Huang, D.-S., Han, K. (eds.) ICIC 2015. LNCS (LNAI), vol. 9227, pp. 407–418. Springer, Cham (2015). https://doi.org/10.1007/978-3-319-22053-6_44
15. Liu, F.T., Ting, K.M., Zhou, Z.H.: Isolation forest. In: ICDM, pp. 413–422 (2008)
16. Zhou, Z.H.: Learnware: on the future of machine learning. Front. Comput. Sci. **10**(4), 355–384 (2016)
17. Aggarwal, C.C.: Mining text and social streams: a review. SIGKDD Explor. **15**(2), 9–19 (2013)
18. Mikolov, T., Chen, K., Corrado, G., Dean, J.: Efficient estimation of word representations in vector space. CoRR abs/1301.3781 (2013)
19. Fan, W., Wang, H., Yu, P.S., Ma, S.: Is random model better? on its accuracy and efficiency. In: ICDM, pp. 51–58 (2003)
20. Liu, F.T., Ting, K.M., Fan, W.: Maximizing tree diversity by building complete-random decision trees. In: Ho, T.B., Cheung, D., Liu, H. (eds.) PAKDD 2005. LNCS (LNAI), vol. 3518, pp. 605–610. Springer, Heidelberg (2005). https://doi.org/10.1007/11430919_70
21. Zhou, Z.H.: Ensemble Methods: Foundations and Algorithms. Chapman & Hall/CRC, Boca Raton, FL, USA (2012)
22. Chang, C.C., Lin, C.J.: LIBSVM: a library for support vector machines. ACM Trans. Intell. Syst. Technol. **2**, 27:1–27:27 (2011). http://www.csie.ntu.edu.tw/cjlin/libsvm
23. Zhang, M.L., Zhou, Z.H.: ML-KNN: a lazy learning approach to multi-label learning. Pattern Recogn. **40**(7), 2038–2048 (2007)
24. Gama, J.: Knowledge Discovery from Data Streams. Chapman and Hall / CRC Data Mining and Knowledge Discovery Series. CRC Press, Boca Raton (2010)
25. Zhou, Z.H., Zhang, M.L., Huang, S.J., Li, Y.F.: Multi-instance multi-label learning. Artif. Intell. **176**(1), 2291–2320 (2012)

Exploiting Anti-monotonicity of Multi-label Evaluation Measures for Inducing Multi-label Rules

Michael Rapp, Eneldo Loza Mencía$^{(\boxtimes)}$, and Johannes Fürnkranz

Technische Universität Darmstadt, Knowledge Engineering Group,
Hochschulstrasse 10, 64289 Darmstadt, Germany
michael.rapp90@gmail.com, {eneldo,fuernkranz}@ke.tu-darmstadt.de

Abstract. Exploiting dependencies between labels is considered to be crucial for multi-label classification. Rules are able to expose label dependencies such as implications, subsumptions or exclusions in a human-comprehensible and interpretable manner. However, the induction of rules with multiple labels in the head is particularly challenging, as the number of label combinations which must be taken into account for each rule grows exponentially with the number of available labels. To overcome this limitation, algorithms for exhaustive rule mining typically use properties such as anti-monotonicity or decomposability in order to prune the search space. In the present paper, we examine whether commonly used multi-label evaluation metrics satisfy these properties and therefore are suited to prune the search space for multi-label heads.

1 Introduction

Multi-label classification (MLC) is the task of learning a model for assigning a set of labels to unknown instances [16]. For example, newspaper articles can often be associated with multiple topics. This is in contrast to binary or multi-class classification, where single classes are predicted. As many studies show, MLC approaches that are able to take correlations between labels into account can be expected to achieve better predictive results (see [7, 11, 16]; and references therein).

In addition to statistical approaches that often rely on complex mathematical concepts, such as Bayesian or neural networks, rule learning algorithms have recently been proposed as an alternative, because rules are not only a natural and simple form to represent a learned model, but they are well suited for making discovered correlations between instance and label attributes explicit [11]. Especially for safety-critical application domains, such as medicine, power systems, autonomous driving or financial markets, where hidden malfunctions could lead to life-threatening actions or economic loss, the possibility of interpreting, inspecting and verifying a classification model is essential (cf. e.g., [9]). However, the algorithm of [11], which is based on the separate-and-conquer (SeCo) strategy, can only learn dependencies where the presence or absence of a single

© Springer International Publishing AG, part of Springer Nature 2018
D. Phung et al. (Eds.): PAKDD 2018, LNAI 10937, pp. 29–42, 2018.
https://doi.org/10.1007/978-3-319-93034-3_3

label depends on a subset of the instance's features. Especially co-occurrences of labels – a common pattern in multi-label data – are hence only representable by a combination of rules. Conversely, algorithms based on subgroup discovery were proposed which are able to find single rules that predict a subset of the possible labels [5]. However, this framework is limited in the sense that it relies on the adaptation of conventional rule learning heuristics for rating and selecting candidate rules and can thus not be easily adapted to a variety of different loss functions which are commonly used for evaluating multi-label predictions. Such an adaptation is not straight-forward, because it is not known whether these measures satisfy properties like anti-monotonicity that can ensure an efficient exploration of the search space of all possible rule heads – despite the fact that it grows exponentially with the number of available labels.

Thus, the main contribution of this work (presented in Sect. 3) is to formally define anti-monotonicity in the context of multi-label rules and to prove that selected multi-label metrics satisfy that property. Based on these findings, we present an algorithm that prunes searches for multi-label rules in Sect. 4. Said algorithm is not meant to set new standards in terms of predictive performance, but to serve as a starting point for developing more enhanced approaches. Nevertheless, we evaluate that it is able to compete with different baselines in terms of predictive and – more importantly – computational performance in Sect. 5.

2   Preliminaries

The task of MLC is to associate an instance with one or several labels λ_i out of a finite label space $\mathbb{L} = (\lambda_i, \ldots, \lambda_n)$ with $n = |\mathbb{L}|$ being the number of available labels. An instance X_j is typically represented in attribute-value form, i.e., it consists of a vector $X_j := \langle v_1, \ldots, v_l \rangle \in \mathbb{D} = A_1 \times \ldots \times A_l$ where A_i is a numeric or nominal attribute. Each instance is mapped to a binary *label vector* $Y_j \in \{0,1\}^n$ which specifies the labels that are associated with the example X_j. Consequently, the training data set of a MLC problem can be defined as a sequence of tuples $T := \langle (X_1, Y_1), \ldots, (X_m, Y_m) \rangle \subseteq \mathbb{D} \times \mathbb{L}$ with $m = |T|$. The model which is derived from a given multi-label data set can be viewed as a classifier function $g(.)$ mapping a single example X to a prediction $\hat{Y} = g(X)$.

2.1   Multi-label Rule Learning

We are concerned with learning multi-label rules $\mathbf{r} : \hat{Y} \leftarrow B$. The body B may consist of several conditions, the examples that are covered by the rule have to satisfy. In this work only conjunctive, propositional rules are considered, i.e., each condition compares an attribute's value to a constant by either using equality (nominal attributes) or inequalities (numerical attributes). It is also possible to include label conditions in the body [11,12]. This allows to expose and distinct between *unconditional* or *global* dependencies and *conditional* or *local* dependencies [7].

The head $\hat{Y}$ consists of one or several label attributes ($\hat{y}_i = 0$ or $\hat{y}_i = 1$) which specify the absence or presence of the corresponding label $\hat{y}_i$. Rules that contain a single label attribute in their head are referred to as *single-label head rules*, whereas *multi-label head rules* may contain several label attributes in their head.

A predicted label vector $\hat{Y}$ may have different semantics. We differentiate between *full predictions* and *partial predictions*.

- **Full predictions:** Each rule predicts a full label vector, i.e., if a label attribute $\hat{y}_i$ is not contained in the head, the absence of the corresponding label λ_i is predicted.
- **Partial predictions:** Each rule predicts the presence or absence of the label only for a subset of the possible labels. For the remaining labels the rule does not make a prediction (but other rules might).

We believe that partial predictions have several conceptual and practical advantages and therefore we focus on that particular strategy throughout the remainder of this work.

2.2 Bipartition Evaluation Functions

To evaluate the quality of multi-label predictions, we use bipartition evaluation measures (cf. [16]) which are based on evaluating differences between true (*ground truth*) and predicted label vectors. They can be considered as functions of two-dimensional *label confusion matrices* which represent the *true positive* (TP), *false positive* (FP), *true negative* (TN) and *false negative* (FN) label predictions. For a given example X_j and a label y_i the elements of an atomic confusion matrix C_i^j are computed as

$$C_i^j = \begin{pmatrix} TP_i^j & FP_i^j \\ FN_i^j & TN_i^j \end{pmatrix} = \begin{pmatrix} y_i^j \hat{y}_i^j & (1 - y_i^j)\hat{y}_i^j \\ (1 - y_i^j)(1 - \hat{y}_i^j) & y_i^j(1 - \hat{y}_i^j) \end{pmatrix} \tag{1}$$

where the variables y_i^j and $\hat{y}_i^j$ denote the absence (0) or presence (1) of label λ_i of example X_j according to the ground truth or the predicted label vector, respectively.

Note that for candidate rule selection we assess TP, FP, TN, and FN differently. To ensure that absent and present labels have the same impact on the performance of a rule, we always count correctly predicted labels as TP and incorrect predictions as FP, respectively. Labels for which no prediction is made are counted as TN if they are absent, or as FN if they are present.

Multi-label Evaluation Functions. In the following some of the most common bipartition metrics $\delta(C)$ used for MLC are presented (cf., e.g., [16]). They are surjections $\mathbb{N}^{2\times2} \to \mathbb{R}$ mapping a confusion matrix C to a heuristic value $h \in [0, 1]$. Predictions that reach a greater heuristic value outperform those with smaller values.

- **Precision:** Percentage of correct predictions among all predicted labels.

$$\delta_{prec}(C) := \frac{TP}{TP + FP} \qquad (2)$$

- **Hamming accuracy:** Percentage of correctly predicted present and absent labels among all labels.

$$\delta_{hamm}(C) := \frac{TP + TN}{TP + FP + TN + FN} \qquad (3)$$

- **F-measure:** Weighted harmonic mean of precision and recall. If $\beta < 1$, precision has a greater impact. If $\beta > 1$, the F-measure becomes more recall-oriented.

$$\delta_F(C) := \frac{\beta^2 + 1}{\frac{\beta^2}{\delta_{rec}(C)} + \frac{1}{\delta_{prec}(C)}} \ , \ \text{with } \delta_{rec}(C) = \frac{TP}{TP + FN} \text{ and } \beta \in [0, \infty] \qquad (4)$$

- **Subset accuracy:** Percentage of perfectly predicted label vectors among all examples. Per definition, it is always calculated using example-based averaging.

$$\delta_{acc}(C) := \frac{1}{m} \sum_j \left[Y_j = \hat{Y}_j \right] \ , \ \text{with } [x] - \begin{cases} 1, & \text{if } x \text{ is true} \\ 0, & \text{otherwise} \end{cases} \qquad (5)$$

Aggregation and Averaging. When evaluating multi-label predictions which have been made for m examples with n labels one has to deal with the question of how to aggregate the resulting $m \cdot n$ atomic confusion matrices. Essentially, there are four possible averaging strategies – either *(label- and example-based) micro-averaging, label-based (macro-)averaging, example-based (macro-) averaging* or *(label- and example-based) macro-averaging.* Due to the space limitations, we restrict our analysis to the most popular aggregation strategy employed in the literature, namely *micro-averaging.* This particular averaging strategy is formally defined as

$$\delta(C) = \delta \left(\sum_j \sum_i C_i^j \right) \equiv \delta \left(\sum_i \sum_j C_i^j \right) \qquad (6)$$

where the $\sum$ operator denotes the cell-wise addition of confusion matrices.

Relation to Conventional Association Rule Discovery. To illustrate the difference between measures used in association rule discovery and in multi-label rule learning, assume that the rule $\lambda_1, \lambda_2 \leftarrow B$ covers three examples $(X_1, \{\lambda_2\})$, $(X_2, \{\lambda_1, \lambda_2\})$ and $(X_3, \{\lambda_1\})$. In conventional association rule discovery the head is considered to be satisfied for one of the three covered examples (X_2), yielding a precision/confidence value of $\frac{1}{3}$. This essentially corresponds to subset accuracy. On the other hand, micro-averaged precision would correspond to the fraction of 4 correctly predicted labels among 6 predictions, yielding a value of $\frac{2}{3}$.

3   Properties of Multi-label Evaluation Measures

To induce multi-label head rules, we need to find the multi-label head $\hat{Y}$ which reaches the best possible performance

$$h_{max} = \max_{\hat{Y}} h(\mathbf{r}) = \max_{\hat{Y}} h(\hat{Y} \leftarrow B) \tag{7}$$

given an evaluation function $h(.)$ and a body B. In this section we consider rule evaluation functions that are based on micro-averaged atomic confusion matrices in a partial prediction setting, i.e., $h(\mathbf{r}) = \delta(C)$ where $\delta(C)$ is defined as in (6).

Due to the exponential complexity of an exhaustive search, it is crucial to prune the search for the best multi-label head by leaving out unpromising label combinations. The first property which can be exploited for pruning searches – while still being able to find the best solution – is *anti-monotonicity*.

Definition 1 (Anti-monotonicity). *Let $\hat{Y}_p \leftarrow B$ and $\hat{Y}_s \leftarrow B$ denote two multi-label head rules consisting of body B and heads $\hat{Y}_p$, respectively $\hat{Y}_s$. It is further assumed that $\hat{Y}_p \subset \hat{Y}_s$. A multi-label evaluation function h is* anti-monotonic *if the following condition is met, i.e., if no head Y_a that results from adding additional labels to Y_s may result in h_{max} being reached:*

$$\hat{Y}_p \subset \hat{Y}_s \wedge h(\hat{Y}_s \leftarrow B) < h(\hat{Y}_p \leftarrow B) \Longrightarrow h(\hat{Y}_a \leftarrow B) < h_{max}, \forall \hat{Y}_a : \hat{Y}_s \subset \hat{Y}_a$$

In addition to the adaptation of anti-monotonicity in Definition 1, we propose *decomposability* as a stronger criterion. It comes at linear costs, as the best possible head can be deduced from considering each available label separately. Due to its restrictiveness, if Definition 2 is met, Definition 1 is implied to be met as well.

Definition 2 (Decomposability). *A multi-label evaluation function h is* decomposable *if the following conditions are met:*

(i) *If the multi-label head rule $\hat{Y} \leftarrow B$ contains a label attribute $\hat{y}_i \in \hat{Y}$ for which the corresponding single-label head rule $\hat{y}_i \leftarrow B$ does not reach h_{max}, the multi-label head rule cannot reach that performance either (and vice versa).*

$$\exists i \left(\hat{y}_i \in \hat{Y} \wedge h(\hat{y}_i \leftarrow B) < h_{max} \right) \Longleftrightarrow h(\hat{Y} \leftarrow B) < h_{max}$$

(ii) *If all single label head rules $\hat{y}_i \leftarrow B$ which correspond to the label attributes of the multi-label head $\hat{Y}$ reach h_{max}, the multi-label head rule $\hat{Y} \leftarrow B$ reaches that performance as well (and vice versa).*

$$h(\hat{y}_i \leftarrow B) = h_{max} , \forall \hat{y}_i \left(\hat{y}_i \in \hat{Y} \right) \Longleftrightarrow h(\hat{Y} \leftarrow B) = h_{max}$$

In the following we examine selected multi-label metrics in terms of decomposability and anti-monotonicity to reveal whether they satisfy these properties when making partial predictions (cf. Section 2.1).

Theorem 1. *Micro-averaged precision is decomposable.*

Proof. We rewrite the performance calculation for a multi-label head rule $\mathbf{r}$: $\hat{Y} \leftarrow B$ with $h(\mathbf{r}) = h_{max}$ using the fact that the single label head rules $\mathbf{r}_i$: $\hat{y}_i \leftarrow B$ with $\hat{y}_i \in \hat{Y}$ share the same body B and therefore cover the same number of examples $|C|$.

$$
h(\mathbf{r}) = \frac{\displaystyle\sum_{\hat{y}_i \in \hat{Y}} \sum_{j} TP_i^j}{\displaystyle\sum_{\hat{y}_i \in \hat{Y}} \sum_{j} p_i^j} \ , \text{ with } p_i^j = TP_i^j + FP_i^j \text{ and } \sum_{j} p_i^j = |C| \ , \forall i
$$

(8)

$$
= \frac{\displaystyle\sum_{\hat{y}_i \in \hat{Y}} \sum_{j} TP_i^j}{|\hat{Y}| \cdot |C|} = \frac{1}{|\hat{Y}|} \sum_{\hat{y}_i \in \hat{Y}} \frac{\displaystyle\sum_{j} TP_i^j}{|C|} \equiv \frac{1}{|\hat{Y}|} \sum_{\hat{y}_i \in \hat{Y}} h(\mathbf{r}_i)
$$

Thus, the micro-averaged precision for $\mathbf{r}$ corresponds to the average of the micro-averaged precision of the single-label head rules $\mathbf{r}_i$. As we assume that $h(\mathbf{r})$ is maximal, it follows that $h(\mathbf{r}) = h(\mathbf{r}_i)$ for all single-label head rules $\mathbf{r}_i$.

Theorem 2. *Micro-averaged Hamming accuracy is decomposable.*

Proof. Similar to (8), we rewrite the micro-averaged Hamming accuracy of a multi-label head rule $\mathbf{r}$: $\hat{Y} \leftarrow B$ with $h(\mathbf{r}) = h_{max}$ in terms of averaging the performance of single-label head rules $\mathbf{r}_i$: $\hat{y}_i \leftarrow B$. This is possible as the performance for each label $\hat{y}_i$ calculates as the percentage of TP and TN among all m labels. For reasons of simplicity, we use the abbreviations $P_i^j = TP_i^j + FN_i^j$ and $N_i^j = FP_i^j + TN_i^j$.

$$
h(\mathbf{r}) = \frac{\displaystyle\sum_{\hat{y}_i \in \hat{Y}} \sum_{j} \left(TP_i^j + TN_i^j \right)}{\displaystyle\sum_{\hat{y}_i \in \hat{Y}} \sum_{j} \left(P_i^j + N_i^j \right)} \ , \text{ with } \sum_{j} \left(P_i^j + N_i^j \right) = m \ , \forall i
$$

$$
= \frac{\displaystyle\sum_{\hat{y}_i \in \hat{Y}} \sum_{j} \left(TP_i^j + TN_i^j \right)}{|\hat{Y}| \cdot m} = \frac{1}{|\hat{Y}|} \sum_{\hat{y}_i \in \hat{Y}} \frac{\displaystyle\sum_{j} \left(TP_i^j + TN_i^j \right)}{m} \equiv \frac{1}{|\hat{Y}|} \sum_{\hat{y}_i \in \hat{Y}} h(\mathbf{r}_i)
$$

(9)

Theorem 3. *Subset accuracy is anti-monotonic.*

Proof. In accordance with Definition 1, two multi-label head rules $\hat{Y}_p \leftarrow B$ and $\hat{Y}_s \leftarrow B$, for whose heads the subset relationship $\hat{Y}_p \subset \hat{Y}_s$ holds, take part in equation (10). The subscript notation $x|_{\hat{Y}}$ is used to denote that a left-hand expression x should be evaluated using the rule $\hat{Y} \leftarrow B$. The proof is based on writing subset accuracy in terms of TP and TN (cf. line 2).

$$\hat{Y}_p \subset \hat{Y}_s \wedge h(\hat{Y}_s \leftarrow B) < h(\hat{Y}_p \leftarrow B)$$

$$\Rightarrow \frac{1}{m} \sum_j \left[\sum_{\hat{y}_i \in \hat{Y}} \left(TP_i^j + TN_i^j \right) = |\hat{Y}| \right]\Bigg|_{\hat{Y}_s} < \frac{1}{m} \sum_j \left[\sum_{\hat{y}_i \in \hat{Y}} \left(TP_i^j + TN_i^j \right) = |\hat{Y}| \right]\Bigg|_{\hat{Y}_p} \leq h_{max}$$

$$\Rightarrow \exists j \left(0 = \left[\sum_{\hat{y}_i \in \hat{Y}} \left(TP_i^j + TN_i^j \right) = |\hat{Y}| \right]\Bigg|_{\hat{Y}_s} < \left[\sum_{\hat{y}_i \in \hat{Y}} \left(TP_i^j + TN_i^j \right) = |\hat{Y}| \right]\Bigg|_{\hat{Y}_p} = 1 \right)$$

$$\Rightarrow \exists \hat{y}_i \exists j \left(\hat{y}_i \in \hat{Y}_s \wedge \left(TP_i^j + TN_i^j \right) < |\hat{Y}|\Big|_{\hat{Y}_s} \right)$$

$$\Rightarrow \exists \hat{y}_i \exists j \left(\hat{y}_i \in \hat{Y}_a \wedge \left(TP_i^j + TN_i^j \right) < |\hat{Y}|\Big|_{\hat{Y}_a} \right), \forall \hat{Y}_a \left(\hat{Y}_s \subset \hat{Y}_a \right)$$

$$\Rightarrow \exists j \left(\left[\sum_{\hat{y}_i \in \hat{Y}} \left(TP_i^j + TN_i^j \right) = |\hat{Y}| \right]\Bigg|_{\hat{Y}_a} = 0 \right), \forall \hat{Y}_a \left(\hat{Y}_s \subset \hat{Y}_a \right)$$

$$\Rightarrow \frac{1}{m} \sum_j \left[\sum_{\hat{y}_i \in \hat{Y}} \left(TP_i^j + TN_i^j \right) = |\hat{Y}| \right]\Bigg|_{\hat{Y}_a} < h_{max}, \forall \hat{Y}_a \left(\hat{Y}_s \subset \hat{Y}_a \right)$$

$$\equiv h(\hat{Y}_a \leftarrow B) < h_{max}, \forall \hat{Y}_a \left(\hat{Y}_s \subset \hat{Y}_a \right)$$

(10)

In (10) it is concluded that when using the rule $\hat{Y}_s \leftarrow B$ the performance for at least one example Y_j is less than when using the rule $\hat{Y}_p \leftarrow B$. Due to the definition of subset accuracy, the performance for that example must be 0 in the first case and 1 in the latter (cf. line 3). As the performance only evaluates to 0 if at least one label is predicted incorrectly, the head $\hat{Y}_p$ must contain a label attribute $\hat{y}_i$ which predicts the corresponding label incorrectly (cf. line 4). When adding additional label attributes the prediction for that label will still be incorrect (cf. line 5). Therefore, for all multi-label head rules $\hat{Y}_a \leftarrow B$ which result from adding additional label attributes to the head $\hat{Y}_s$ the performance for the example Y_j evaluates to 0 (cf. line 6). Consequently, none of them can reach the overall performance of $\hat{Y}_p \leftarrow B$, nor h_{max} (cf. line 7 and 8).

Lemma 1. *Micro-averaged recall is decomposable.*

Proof. The *mediant* of fractions $\frac{a_1}{b_1}, \ldots, \frac{a_n}{b_n}$ is defined as $\frac{a_1 + \ldots + a_n}{b_1 + \ldots + b_n}$. The micro-averaged recall of a multi-label head rule $\mathbf{r} : \hat{Y} \leftarrow B$ is the mediant of the performances which are obtained for corresponding single-label head rules $\mathbf{r}_i : \hat{y}_i \leftarrow B$ with $\hat{y}_i \in \hat{Y}$ according to the recall metric.

$$h(\mathbf{r}) = \frac{\sum\limits_{\hat{y}_i \in \hat{Y}} \sum\limits_j TP_i^j}{\sum\limits_{\hat{y}_i \in \hat{Y}} \sum\limits_j \left(TP_i^j + FN_i^j \right)} \tag{11}$$

The *mediant inequality* states that the mediant strictly lies between the fractions it is calculated from, i.e., that $min\left(\frac{a_1}{b_1}, \ldots, \frac{a_n}{b_n}\right) \leq \frac{a_1 + \ldots + a_n}{b_1 + \ldots + b_n} \leq max\left(\frac{a_1}{b_1}, \ldots, \frac{a_n}{b_n}\right)$. This is in accordance with Definition 2.

Theorem 4. *Micro-averaged F-measure is decomposable.*

Proof. Micro-averaged F-measure calculates as the (weighted) harmonic mean $H(.)$ of micro-averaged precision and recall. This proof is based on the finding that both of these metrics fulfill the properties of decomposability (cf. Theorem 1 and Lemma 1). As multiple metrics take part in the proof, we use a superscript notation to distinguish between the best possible performances according to different metrics, e.g., h_{max}^F in case of the F-measure. Furthermore, we exploit the inequality $h_{max}^F \leq max\left(h_{max}^{rec}, h_{max}^{prec}\right)$.

$$
\begin{aligned}
&\exists i \left(\hat{y}_i \in \hat{Y} \wedge h_F(\hat{y}_i \leftarrow B) < h_{max}^F \leq h_{max}^{rec}\right) \\
\equiv &\exists i \left(\hat{y}_i \in \hat{Y} \wedge H\left(h_{rec}(\hat{y}_i \leftarrow B), h_{prec}(\hat{y}_i \leftarrow B)\right) < h_{max}^{rec}\right) \\
\Rightarrow &\exists i \left(\hat{y}_i \in \hat{Y} \wedge (h_{rec}(\hat{y}_i \leftarrow B) < h_{max}^{rec} \wedge h_{prec}(\hat{y}_i \leftarrow B) < h_{max}^{rec})\right. \\
&\left. \vee (h_{prec}(\hat{y}_i \leftarrow B) < h_{max}^{rec} \wedge h_{rec}(\hat{y}_i \leftarrow B) \leq h_{max}^{rec})\right) \\
\Rightarrow &\left(h_{rec}(\hat{Y} \leftarrow B) < h_{max}^{rec} \wedge h_{prec}(\hat{Y} \leftarrow B) < h_{max}^{rec}\right) \\
&\vee \left(h_{prec}(\hat{Y} \leftarrow B) < h_{max}^{rec} \wedge h_{rec}(\hat{Y} \leftarrow B) \leq h_{max}^{rec}\right) \\
\Rightarrow &H\left(h_{rec}(\hat{Y} \leftarrow B), h_{prec}(\hat{Y} \leftarrow B)\right) < h_{max}^F \leq h_{max}^{rec} \\
\equiv &h_F(\hat{Y} \leftarrow B) < h_{max}^F
\end{aligned}
\tag{12}
$$

In (12) the first property of Definition 2 is proved. As the premise of the proof, we assume w.l.o.g. that the best possible performance according to the recall metric is equal to or greater than the best performance according to precision, i.e., that the relation $h_{max}^{rec} \geq h_{max}^{prec}$ holds. We further assume that the F-measure of a single-label head rule $\mathbf{r}_i : \hat{y}_i \leftarrow B$ is less than the best possible performance h_{max} (cf. line 1 and 2). When rewriting the F-measure in terms of the harmonic mean of precision and recall, it follows that either recall or precision of $\mathbf{r}_i$ must be less than h_{max}^F, respectively h_{max}^{rec}. Due to the premise of the proof, h_{max}^{rec} can be considered as an upper limit for both recall and precision (cf. line 3). Furthermore, because precision and recall are decomposable, the multi-label head rule $\mathbf{r} : \hat{Y} \leftarrow B$ with $\hat{y}_i \in \hat{Y}$ cannot outperform h_{max}^F (cf. lines 5, 7 and 8). In order to prove the second property of decomposability to be met, the derivation in (13) uses a similar approach as in (12). However, it is not based on its premise.

$$h_F(\hat{y}_i \leftarrow B) = h_{max}^F \ , \ \forall \hat{y}_i \left(\hat{y}_i \in \hat{Y} \right)$$

$$\equiv H\left(h_{rec}(\hat{y}_i \leftarrow B), h_{prec}(\hat{y}_i \leftarrow B)\right) = h_{max}^F \ , \ \forall \hat{y}_i \left(\hat{y}_i \in \hat{Y} \right)$$

$$\Longrightarrow h_{rec}(\hat{y}_i \leftarrow B) = h_{prec}(\hat{y}_i \leftarrow B) = h_{max}^F \ , \ \forall \hat{y}_i \left(\hat{y}_i \in \hat{Y} \right) \tag{13}$$

$$\Longrightarrow h_{rec}(\hat{Y} \leftarrow B) = h_{prec}(\hat{Y} \leftarrow B) = h_{max}^F$$

$$\Longrightarrow H\left(h_{rec}(\hat{Y} \leftarrow B), h_{prec}(\hat{Y} \leftarrow B)\right) = h_{max}^F$$

$$\equiv h_F(\hat{Y} \leftarrow B) = h_{max}^F$$

4   Algorithm for Learning Multi-label Head Rules

To evaluate the utility of these properties, we implemented a multi-label rule learning algorithm based on the SeCo algorithm for learning single-label head rules by Loza Mencía an Janssen [11]. Both algorithms share a common structure where new rules are induced iteratively and the examples they cover are removed from the training data set if enough of their labels are predicted by already learned rules. The rule induction process continues until only few training examples are left. To classify test examples, the learned rules are applied in the order of their induction. If a rule fires, the labels in its head are applied unless they were already set by a previous rule.

For learning new multi-label rules, our algorithm performs a top-down greedy search, starting with the most general rule. By adding additional conditions to the rule's body it can successively be specialized, resulting in less examples being covered. Potential conditions result from the values of nominal attributes or from averaging two adjacent values of the sorted examples in case of numerical attributes. Whenever a new condition is added, a corresponding single- or multi-label head that predicts the labels of the covered examples as accurate as possible must be found.

Evaluating Possible Multi-label Heads. To find the best head for a given body different label combinations must be evaluated by calculating a score based on the used averaging and evaluation strategy. The algorithm performs a breadth-first search by recursively adding additional label attributes to the (initially empty) head and keeps track of the best rated head. Instead of performing an exhaustive search, the search space is pruned according to the findings in Sect. 3. When pruning according to anti-monotonicity unnecessary evaluations of label combinations are omitted in two ways: On the one hand, if adding a label attribute causes the performance to decrease, the recursion is not continued at deeper levels of the currently searched subtree. On the other hand, the algorithm keeps track of already evaluated or pruned heads and prevents these heads from being evaluated in later iterations. When a decomposable evaluation metric is used no deep searches through the label space must be performed. Instead, all possible single-label heads are evaluated in order to identify those that reach the highest score and merge them into one multi-label head rule.

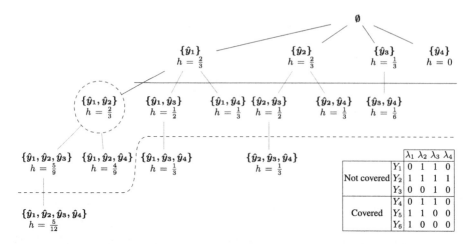

Fig. 1. Search through the label space $\mathbb{L} = (\lambda_1, \lambda_2, \lambda_3, \lambda_4)$ using micro-averaged precision of partial predictions. The examples corresponding to label sets Y_4, Y_5, Y_6 are assumed to be covered, whereas those of Y_1, Y_2, Y_3 are not. The dashed line (---) indicates label combinations that can be pruned with anti-monotonicity, the solid line (—) corresponds to decomposability.

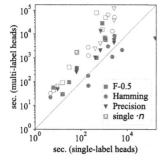

Fig. 2. Training times. (Color figure online)

$$\overline{red, green, blue, yellow, white} \leftarrow \text{colors} > 5, \text{stripes} \leq 3 \quad (65,0)$$
$$\overline{red}, green, \overline{blue}, yellow, white, \overline{black}, \overline{orange}$$
$$\leftarrow \text{animate, stripes} \leq 0, \text{crosses} \leq 0 \quad (11,0)$$

$yellow \leftarrow \text{colors} > 4$	(21,0)	$green \leftarrow \text{text}$	(11,0)
$\overline{red} \leftarrow yellow$	(21,0)	$\overline{orange} \leftarrow \text{saltires} < 1$	(1,0)
$blue \leftarrow \text{colors} > 5$	(14,0)	$\overline{black} \leftarrow \text{area} < 11$	(12,0)
$white \leftarrow blue$	(14,0)		

Fig. 3. Example of learned multi- and single-label head rule lists. TP and FP of respective rules are given in brackets.

Figure 1 illustrates how the algorithm prunes a search through the label space using anti-monotonicity and decomposability. The nodes of the given search tree correspond to the evaluations of label combinations, resulting in heuristic values h. The edges correspond to adding an additional label to the head which is represented by the preceding node. As equivalent heads must not be evaluated multiple times, the tree is unbalanced.

5   Evaluation

The purpose of the experimental evaluation was to demonstrate the applicability of the proposed SeCo algorithm despite the exponentially large search space. We did not expect any significant improvements in predictive performance since no enhancements in that respect were made to the original algorithm as proposed in [11].

Experimental Setup. We compared our multi-label head algorithm to its single-label head counterpart and also to the binary relevance method on 8 different data sets.[1] Following [11], we used Hamming accuracy, subset accuracy (only for multi-label heads), micro-averaged precision and F-measure (with $\beta = 0.5$) on partial predictions for candidate rule selection and also allowed negative assignments $\hat{y}_i = 0$ in the heads.

Predictive performance. Due to the space limitations, we limit ourselves to the results of the statistical tests (following [8]). The null hypothesis of the Friedman test ($\alpha = 0.05$, $N = 8$, $k = 10$) that all algorithms have the same predictive quality could not be rejected for many of the evaluation measures, such as subset accuracy and micro- and macro-averaged F1. In the other cases, the Nemenyi post-hoc test was not able to assess a statistical difference between the algorithms using the same heuristic.

Computational Costs. As expected, SeCo finds rules with a comparable predictive performance when searching for multi-label head rules. However, from the point of view of the proven properties of the evaluation measures, it was more interesting to demonstrate the usefulness of anti-monotonicity and decomposability regarding the computational efficiency. Figure 2 shows the relation between the time spent for finding single- vs. multi-label head rules using the same heuristic and data set. The empty forms denote the single-label times multiplied by the number of labels in the data set. Note that full exploration of the labels space was already intractable for the smaller data sets on our system. We can observe that the costs for learning multi-label head rules are in the same order of magnitude despite effectively exploring the full label space for each candidate body.

Rule Models. When analyzing the characteristics of the models which have been learned by the proposed algorithm, it becomes apparent that more multi-label head rules are learned when using the precision metric, rather than one of the other metrics. This is due to the fact that precision only takes TP and FP into account. Therefore, the performance of such a rule depends exclusively on the examples it covers. When using another metric, where the performance also depends on uncovered examples, it is very likely that the performance of a rule

---

[1] *scene* (6, 1.06), *emotions* (6, 1.87), *flags* (7, 3.39), *yeast* (14, 4.24), *birds* (19, 1.01), *genbase* (27, 1.25), *medical* (45, 1.24), *cal500* (174, 26.15), with respective number of labels and cardinality, from http://mulan.sf.net. Source code and results are available at https://github.com/keelm/SeCo-MLC.

slightly decreases when adding an additional label to its head. This causes single-label heads to be preferred. The inclusion of a factor which takes the head's size in account could resolve this bias and lead to heads with more labels.

Whether more labels in the head are more desirable or not highly depends on the data set at hand, the particular scenario and the preferences of the user, as generally do comprehensibility and interpretability of rules. These issues cannot be solved by the proposed method, nor are in the scope of this work. However, the proposed extension of SeCo to multi-label head rules can lay the foundation to further improvements, gaining better control over the characteristics of the induced model and hence better adaption to the requirements of a particular use case.

The extended expressiveness of using multi-label head rules can be visualized by the following example. Consider the rules in Fig. 3, learned on the data set *flags* which maps characteristics of a flag and corresponding country to the colors appearing on the flag. The shown rules all cover the flag of the US Virgin Islands. Whereas in this case the single-label heads allow an easier visualization of the pairwise dependencies between characteristics/labels and labels, the multi-label head rules allow to represent more complex relationships and provide a more direct explanation of why the respective colors are predicted for the flag.

6   Related Work

So far, only a few approaches to multi-label rule learning can be found in the literature. Most of them are based on association rule (AR) discovery. Alternatively, a few approaches use evolutionary algorithms or classifier systems for evolving multi-label classification rules [2–4]. Creating rules with several labels in the head is usually implemented as a post-processing step. For example, [15] and similarly [10] induce single-label ARs which are merged to create multi-label rules. By using a separate-and-conquer approach the step of inducing descriptive but often redundant models of the data is omitted and it is directly tried to produce predictive rules [11].

Most of the approaches mentioned so far have in common that they are restricted to expressing a certain type of relationship since labels are only allowed as the consequent of a rule. Approaches that allow labels as antecedents of an implication are often restricted to global label dependencies, such as the approaches by [6,13,14] that use the relationships discovered by AR mining on the label matrix for refining the predictions of multi-label classifiers.

The anti-monotonicity property is already well known from AR learning and subgroup discovery. For instance, it is used by the Apriori algorithm [1] to prune searches for frequent item sets. [5] already used anti-monotonicity for efficiently mining subgroups in multi-label problems. However, in contrast to our work, they have not considered evaluation measures that are commonly used in MLC, but instead adapted metrics that are commonly used in subgroup discovery. We believe that the anti-monotonicity property must be assessed differently in a multi-label context. This is because AR learning neglects partial matches and labels that are not present in the heads (cf. Sect. 2.2). In contrast, most MLC measures are much more sensitive in this respect. This is also demonstrated by the more restrictive property of decomposability which does not exist in common metrics for AR.

7 Conclusions

In this work, we formulated anti-monotonicity and decomposability criteria for multi-label rule learning and formally proved that several common multi-label evaluation measures meet these properties. Furthermore, we demonstrated how these results can be used to efficiently find rules with multi-label heads that are optimal with respect to commonly used multi-label evaluation functions. Our experiments showed that more work is needed to effectively combine such rules into a powerful rule-based theory.

Acknowledgments. We acknowledge support by the German Research Foundation (DFG) (grant number FU 580/11).

References

1. Agrawal, R., Mannila, H., Srikant, R., Toivonen, H., Verkamo, A.I.: Fast discovery of association rules. In: Advances in Knowledge Discovery and Data Mining, pp. 307–328 (1995)
2. Allamanis, M., Tzima, F.A., Mitkas, P.A.: Effective rule-based multi-label classification with learning classifier systems. In: Tomassini, M., Antonioni, A., Daolio, F., Buesser, P. (eds.) ICANNGA 2013. LNCS, vol. 7824, pp. 466–476. Springer, Heidelberg (2013). https://doi.org/10.1007/978-3-642-37213-1_48
3. Arunadevi, J., Rajamani, V.: An evolutionary multi label classification using associative rule mining for spatial preferences. Int. J. Comput. Appl. **3**(3), 28–37 (2011). Special Issue on Artificial Intelligence Techniques - Novel Approaches and Practical Applications. https://www.ijcaonline.org/specialissues/ait/number3, ISBN 978-93-80746-68-2
4. Ávila-Jiménez, J.L., Gibaja, E., Ventura, S.: Evolving multi-label classification rules with gene expression programming: a preliminary study. In: Corchado, E., Graña Romay, M., Manhaes Savio, A. (eds.) HAIS 2010. LNCS (LNAI), vol. 6077, pp. 9–16. Springer, Heidelberg (2010). https://doi.org/10.1007/978-3-642-13803-4_2
5. Bosc, G., Golebiowski, J., Bensafi, M., Robardet, C., Plantevit, M., Boulicaut, J.-F., Kaytoue, M.: Local subgroup discovery for eliciting and understanding new structure-odor relationships. In: Calders, T., Ceci, M., Malerba, D. (eds.) DS 2016. LNCS (LNAI), vol. 9956, pp. 19–34. Springer, Cham (2016). https://doi.org/10.1007/978-3-319-46307-0_2
6. Charte, F., Rivera, A.J., del Jesús, M.J., Herrera, F.: LI-MLC: a label inference methodology for addressing high dimensionality in the label space for multilabel classification. IEEE Trans. Neural Netw. Learn. Syst. **25**(10), 1842–1854 (2014)
7. Dembczyński, K., Waegeman, W., Cheng, W., Hüllermeier, E.: On label dependence and loss minimization in multi-label classification. Mach. Learn. **88**(1–2), 5–45 (2012)

8. Demšar, J.: Statistical comparisons of classifiers over multiple data sets. J. Mach. Learn. Res. **7**, 1–30 (2006)
9. Kayande, U., De Bruyn, A., Lilien, G.L., Rangaswamy, A., Van Bruggen, G.H.: How incorporating feedback mechanisms in a DSS affects DSS evaluations. Inf. Syst. Res. **20**(4), 527–546 (2009)
10. Li, B., Li, H., Wu, M., Li, P.: Multi-label classification based on association rules with application to scene classification. In: Proceedings of the 9th International Conference for Young Computer Scientists (ICYCS 2008), pp. 36–41. IEEE Computer Society (2008)
11. Loza Mencía, E., Janssen, F.: Learning rules for multi-label classification: a stacking and a separate-and-conquer approach. Mach. Learn. **105**(1), 77–126 (2016)
12. Malerba, D., Semeraro, G., Esposito, F.: A multistrategy approach to learning multiple dependent concepts. In: Nakhaeizadeh, G., Taylor, C.C. (eds.) Machine Learning and Statistics: The Interface, pp. 87–106. Wiley, London (1997)
13. Papagiannopoulou, C., Tsoumakas, G., Tsamardinos, I.: Discovering and exploiting deterministic label relationships in multi-label learning. In: Proceedings of the 21th ACM SIGKDD International Conference on Knowledge Discovery and Data Mining, pp. 915–924 (2015)
14. Park, S.H., Fürnkranz, J.: Multi-label classification with label constraints. In: Proceedings of the ECML PKDD 2008 Workshop on Preference Learning (PL 2008), pp. 157–171 (2008)
15. Thabtah, F., Cowling, P., Peng, Y.: Multiple labels associative classification. Knowl. Inf. Syst. **9**(1), 109–129 (2006)
16. Tsoumakas, G., Katakis, I., Vlahavas, I.: Mining multi-label data. In: Maimon, O., Rokach, L. (eds.) Data Mining and Knowledge Discovery Handbook. Springer, Boston (2009). https://doi.org/10.1007/978-0-387-09823-4_34

Modeling Label Interactions
in Multi-label Classification:
A Multi-structure SVM Perspective

Anusha Kasinikota[1], P. Balamurugan[2(✉)], and Shirish Shevade[3]

[1] DXC Technology, Bangalore, India
kasinikota.anusha@dxc.com
[2] IIT Bombay, Mumbai, India
balamurugan.palaniappan@iitb.ac.in
[3] Indian Institute of Science, Bangalore, India
shirish@iisc.ac.in

Abstract. Multi-label classification has attracted much interest due to its wide applicability. Modeling label interactions and investigating their impact on classifier quality are crucial aspects of multi-label classification. In this paper, we propose a multi-structure SVM (called MSSVM) which allows the user to hypothesize multiple label interaction structures and helps to identify their importance in improving generalization performance. We design an efficient optimization algorithm to solve the proposed MSSVM. Extensive empirical evaluation provides fresh and interesting insights into the following questions: (a) How do label interactions affect multiple performance metrics typically used in multi-label classification? (b) Do higher order label interactions significantly impact a given performance metric for a particular dataset? (c) Can we make useful suggestions on the label interaction structure? and (d) Is it always beneficial to model label interactions in multi-label classification?

1 Introduction

Given a set of training samples $\mathcal{T} = \{(x_i, \mathbf{y}_i)\}_{i=1}^n$, where $x_i \in \mathcal{X} \subseteq \mathbb{R}^m$, $\mathbf{y}_i \in \mathcal{Y} = \{+1, -1\}^L, L \geq 2$, multi-label classification aims to construct a rule $h : \mathcal{X} \to \mathcal{Y}$, such that for an arbitrary example $\hat{x} \in \mathcal{X}$, "all" the associated labels can be obtained using $\hat{\mathbf{y}} = h(\hat{x}) \in \mathcal{Y}$. An important and interesting problem is to identify the role of label interactions in learning a multi-label classifier. Existing approaches [3,4,9,12,15,32] assume that labels interact according to observed label co-occurrence information or pre-defined local and global probabilistic dependencies. Also label interactions are popularly assumed to *always* improve the classifier's performance. From a practical viewpoint, such fixed and hard-wired assumptions allow little or no freedom for the curious user to

A. Kasinikota—This work was done when the author was at IISc, Bangalore, India. A long version of this paper along with supplementary material is available at https://sites.google.com/site/pbalamuru/home/mssvm.

interact with the classifier with multiple label interaction structure assumptions (designed using domain-knowledge or an artificially intelligent hypothesis agent). When multiple label interaction structures are hypothesized, it becomes natural to study which structure or collection of structures really leads to improved performance. It is also of great interest to investigate the effect of label interactions on multiple performance metrics; *e.g.* Hamming loss, Jaccard accuracy, subset accuracy, F-measure etc. [21,38].

Contributions and Paper Organization: Catering to the aforementioned objectives (and motivated by [5]), we propose a multi-structure SVM (MSSVM) in Sect. 3, which allows the user to consider different prior assumptions on the nature of label interactions and provides an explicit interface to interact with the classifier and to investigate if such assumptions really hold for the given data. We illustrate an efficient optimization algorithm to solve MSSVM in Sect. 4. Extensive empirical evaluation of MSSVM is given in Sect. 5. Our experiments reveal that multi-label classification requires more flexibility in modeling label interaction structures; in fact, every dataset has its own flavor and different types of label interactions help in improving different performance metrics for different datasets. To the best of our knowledge, these important insights are a first of their kind. We conclude (Sect. 6) with some future perspectives.

2   A Quick Review of Existing Work

Perhaps the simplest label interaction structure is that where the label components of **y** are assumed to be independent of each other given the example x, leading to the well-known binary relevance algorithm [29]. However, exploiting the label interactions in multi-label classification resulted in improved performance in certain cases [9] and has since attracted significant attention. Generally, label interactions have been called using different names in the literature, the most popular being label dependencies [5,36] and label correlations [15]. Most of the existing works consider pair-wise interactions among labels [12,39] or co-occurrence counts of labels [11,14]. Strong assumptions on label interaction structures have been used before; *e.g.* chains and sequences of labels [33], trees [7], Ising models [11], graphs with restrictive clique structures [23,28], random ensembles of graphs [27], random spanning trees of Markov networks [22], etc. In these works, it is assumed that the structure assumptions strictly hold for all training and test samples. Certain algorithms introduce the influence of label interactions by several heuristic techniques; examples include stacking [10], chains of classifiers [26], classifiers built upon random subsets of labels [30] and upon ranked ensembles of label subsets [35], classifier trellises [25], label forests [32], etc. A suitable label covariance matrix is learned by some approaches [14,34]. Label interactions and missing labels have been jointly handled in [3]. Learning to predict multiple structured outputs has also been explored in fully supervised [19] and weakly supervised settings [13].

3 Multi-structure SVM

We use an equivalent set notation $\mathbf{y}_i = \{y_i^l\}_{l=1}^L \in \mathscr{Y}$, where $y_i^l \in \{+1, -1\}$ denotes the presence or absence of the l-th label component of $\mathbf{y}_i$. The training samples are assumed to be realizations of random variable pair (X, Y) from a fixed but unknown probability distribution P over $\mathscr{X} \times \mathscr{Y}$, and the notation $(x, \mathbf{y} = \{y^l\}_{l=1}^L)$ denotes a generic (example, multi-label) pair from the training set $\mathcal{T}$. A natural approach is to use the principle of empirical risk minimization [31] to minimize a suitable surrogate loss. The surrogate loss function we develop here will help us to capture the label interactions and identify if certain interesting structures lead to the best generalization performance; in addition, we will also develop an efficient algorithm for the resultant optimization problem.

If the output space $\mathscr{Y}$ is assumed to be endowed with a known label interaction structure s like a tree or sequence (where the components $\{y^l\}_{l=1}^L$ of each multi-label output $\mathbf{y} \in \mathscr{Y}$ interact according to structure s, denoted by $\mathbf{y} \rhd s$), we can use structural support vector machine (structural SVM) [7,33]. However, it is possible that for some sample $(x, \mathbf{y})$, label components of $\mathbf{y}$ do not obey structure s (denoted by $\mathbf{y} \not\rhd s$) or that multiple structures lead to the realization of $\mathbf{y}$. To handle such cases, we define an appropriate space of structures, $\mathcal{S} = \{s_1, s_2, \ldots, s_J\}$, where $J = |\mathcal{S}| < \infty$ is assumed for simplicity and computational efficiency. Unless advised by domain-specific knowledge, one can safely assume that $\mathbf{y} \rhd s, \forall s \in \mathcal{S}$ without losing generality. Regardless of the nature of structure specification (see Appendix A for a discussion), let us assume that every structure $s_j \in \mathcal{S}$ can be efficiently encoded using a corresponding joint-feature representation map $f_j : \mathscr{X} \times \mathscr{Y} \to \mathbb{R}^{d_j}, 0 < d_j < \infty$. By associating a corresponding parameter vector $\boldsymbol{w}_j \in \mathbb{R}^{d_j}$, we design the score function $F_j(\mathbf{y}; x, \boldsymbol{w}_j) = \boldsymbol{w}_j^\top f_j(x, \mathbf{y}), \forall j$. To construct a suitable surrogate loss, we first fix a structure $s_j \in \mathcal{S}$, and suppose that $\mathbf{y} \rhd s_j \forall \mathbf{y} \in \mathscr{Y}$. By fixing the structure to be s_j, we also assume the following: $\mathbf{y} \not\rhd s_k \forall s_k \in \mathcal{S}, s_k \neq s_j, \mathbf{y} \in \mathscr{Y}$. With an appropriate scaling factor $\delta(\mathbf{y}, s_k, \mathbf{y}_i, s_j; x_i) \geq 0$ to measure the dissimilarity between s_j and s_k, in addition to the dissimilarity between $\mathbf{y}$ and $\mathbf{y}_i$ for a given x_i, we get the following loss function corresponding to structure s_j:

$$\ell_{multi}^j(\mathbf{y}_i; x_i) = (\max\{\max_{\mathbf{y} \in \mathscr{Y}, \mathbf{y} \neq \mathbf{y}_i} \{F_j(\mathbf{y}; x_i, \boldsymbol{w}_j) + \delta(\mathbf{y}, s_j, \mathbf{y}_i, s_j; x_i)\},$$

$$\max_{k \neq j, \mathbf{y} \in \mathscr{Y}} \{F_k(\mathbf{y}; x_i, \boldsymbol{w}_k) + \delta(\mathbf{y}, s_k, \mathbf{y}_i, s_j; x_i)\}\} - F_j(\mathbf{y}_i; x_i, \boldsymbol{w}_j))_+ \quad (1)$$

where $(z)_+ = \max\{0, z\}$. Note that in (1), the score $F_j(\mathbf{y}_i; x_i, \boldsymbol{w}_j)$ for the actual output $\mathbf{y}_i$ with the assumed structure s_j, should be as large as the score for $\mathbf{y} \neq \mathbf{y}_i$, and $s_k \neq s_j$. Now, due to the uniform prior over the structures in $\mathcal{S}$, we can construct a distinct loss function $\ell_{multi}^j(\mathbf{y}_i; x_i), \forall j : s_j \in \mathcal{S}$. By minimizing the loss $\ell_{multi}(\mathbf{y}_i; x_i) = \sum_{j=1}^J \ell_{multi}^j(\mathbf{y}_i; x_i)$ with an appropriate regularization term, we arrive at $\min_{\{\boldsymbol{w}_j \in \mathbb{R}^{d_j}\}_{j=1}^J} \frac{1}{2} \sum_{j=1}^J \gamma_j \|\boldsymbol{w}_j\|_2^2 + \sum_{i=1}^n \sum_{j=1}^J \ell_{multi}^j(\mathbf{y}_i; x_i)$, with regularization constants $\gamma_j > 0, \forall j$. With minor assumptions on $\delta(\mathbf{y}, s_j, \mathbf{y}_i, s_j; x_i)$, this is equivalent to the following problem (see Appendix B for details), which we call the multi-structure SVM (MSSVM):

$$\min_{\{\boldsymbol{w}_j\}_{j=1}^J,\xi} \frac{1}{2}\sum_{j=1}^J \gamma_j\|\boldsymbol{w}_j\|_2^2 + \sum_{i=1}^n\sum_{j=1}^J \xi_{ij}$$

$$\text{s.t. } \boldsymbol{w}_j^\top f_j(x_i,\boldsymbol{y}_i) - \boldsymbol{w}_k^\top f_k(x_i,\boldsymbol{y}) \geq \delta(\boldsymbol{y}, s_k, \boldsymbol{y}_i, s_j; x_i) - \xi_{ij}$$

$$\forall i \in \{1,\ldots,n\}, \forall j, k : s_j, s_k \in \mathcal{S}, \forall \boldsymbol{y} \in \mathcal{Y}. \tag{2}$$

A natural rule to infer the output $\hat{\boldsymbol{y}} \in \mathcal{Y}$ for an arbitrary and possibly unseen example $\hat{x} \in \mathcal{X}$ is: $\hat{\boldsymbol{y}} = h(\hat{x}) = \underset{j:s_j\in\mathcal{S},\boldsymbol{y}\in\mathcal{Y}}{\arg\max} \; F_j(\boldsymbol{y}; \hat{x}, \boldsymbol{w}_j) = \underset{j:s_j\in\mathcal{S},\boldsymbol{y}\in\mathcal{Y}}{\arg\max} \; \boldsymbol{w}_j^\top f_j(\hat{x}, \boldsymbol{y})$.

4   Dual MSSVM and an Efficient Optimization Algorithm

The dual of MSSVM in (2) is as follows (see derivation in Appendix C):

$$\min_\theta D(\theta) = \frac{1}{2}\sum_{j=1}^J \frac{1}{\gamma_j}\|\boldsymbol{v}_j(\theta)\|^2 - \sum_{i,j,\boldsymbol{y},k} \theta_{ij\boldsymbol{y}k}\delta_{ij\boldsymbol{y}k},$$

$$\text{s.t. } \boldsymbol{v}_j(\theta) = \sum_{i,\boldsymbol{y}}\sum_{k=1}^J (\theta_{ij\boldsymbol{y}k}f_j(x_i,\boldsymbol{y}_i) - \theta_{ik\boldsymbol{y}j}f_j(x_i,\boldsymbol{y})), \; \forall j,$$

$$\sum_{\boldsymbol{y},k}\theta_{ij\boldsymbol{y}k} = 1, \forall\, (i,j), \; \theta_{ij\boldsymbol{y}k} \geq 0, \; \forall i,j,\boldsymbol{y},k. \tag{3}$$

Unless specified otherwise, we assume that the index $i \in \{1,\ldots,n\}$, $\boldsymbol{y} \in \mathcal{Y} = \{+1,-1\}^L$ and indices j, k represent the structures $s_j, s_k \in \mathcal{S}$. We also have $\delta_{ij\boldsymbol{y}k} = \delta(\boldsymbol{y}_i, s_j, \boldsymbol{y}, s_k; x_i)$ and let $\theta_{ij\boldsymbol{y}k}$ denote the dual variables. By exploiting the constraint set in (3), we now design an iterative sequential optimization method (in the spirit of [1]), which will depend on the following KKT-optimality conditions of problem (3): $\forall(i,j)$,

$$\eta_{ij} = \max_{(\boldsymbol{y},k):\theta_{ij\boldsymbol{y}k}>0} \nabla_{\theta_{ij\boldsymbol{y}k}}D(\theta) - \min_{(\boldsymbol{y},k)} \nabla_{\theta_{ij\boldsymbol{y}k}}D(\theta) \leq 0, \tag{4}$$

where $\nabla_{\theta_{ij\boldsymbol{y}k}}D(\theta)$ denotes the partial derivative of $D(\theta)$ with respect to $\theta_{ij\boldsymbol{y}k}$ variable and is given by $\frac{1}{\gamma_j}\boldsymbol{v}_j(\theta)^\top f_j(x_i,\boldsymbol{y}_i) - \frac{1}{\gamma_k}\boldsymbol{v}_k(\theta)^\top f_k(x_i,\boldsymbol{y}) - \delta_{ij\boldsymbol{y}k}$. However, solving (3) over the entire output space $\{+1,-1\}^L$ becomes quickly intractable. Hence, following [1,33], we consider a working set $A_i = \{\boldsymbol{y} : \theta_{ij\boldsymbol{y}k} > 0\}$ for each example i. Note that such a working set is naturally used in KKT conditions (4). Our proposed sequential optimization algorithm operates at two levels: at the first level, we fix a sample i, change $\theta_{ij\boldsymbol{y}k} = \theta_{ij\boldsymbol{y}k} + \lambda_{ij\boldsymbol{y}k} \; \forall j, \boldsymbol{y} \in A_i, k$, and optimize $\lambda_{ij\boldsymbol{y}k}$ by solving:

$$\min_\lambda G(\lambda) = -\sum_{j,\boldsymbol{y}\in A_i,k} \lambda_{ij\boldsymbol{y}k}\delta_{ij\boldsymbol{y}k} + \frac{1}{2}\sum_{j=1}^J \frac{1}{\gamma_j}\Big\|\boldsymbol{v}_j(\theta) - \sum_{\boldsymbol{y}\in A_i}\big(\sum_k \lambda_{ik\boldsymbol{y}j}\big)f_j(x_i,\boldsymbol{y})\Big\|^2$$

$$\text{s.t. } \sum_{\boldsymbol{y}\in A_i,k}\lambda_{ij\boldsymbol{y}k} = 0, \forall\, j, \; \lambda_{ij\boldsymbol{y}k} \geq -\theta_{ij\boldsymbol{y}k}, \; \forall j, \boldsymbol{y} \in A_i, k. \tag{5}$$

Algorithm 1. *Sequential Optimization Algorithm to solve MSSVM*

1: Input $\mathcal{T} = \{(x_i, \mathbf{y}_i)\}_{i=1}^n$, $\mathcal{S} = \{s_1, \ldots, s_J\}$, $\{\gamma_j\}_{j=1}^J$.
2: $t = 0$, $A_i = \{\mathbf{y}_i\}$, $\forall\ i = 1, 2, \cdots, n$.
3: $\{\mathbf{v}_j(\theta) = 0\}_{j=1}^J$, $\theta_{ij\mathbf{y}_i j} = 1$, $\forall i, j$, $\theta_{ij\mathbf{y}k} = 0$, $\forall i, j, \mathbf{y} \notin A_i, k \neq j$.
4: **while** (4) not satisfied **do**
5:    **for** $i = 1, 2, \ldots, n$ **do**
6:       **for** $j : s_j \in \mathcal{S}$ **do**
7:          Find $(\hat{\mathbf{y}}, \hat{k}) = \underset{\mathbf{y}, k: s_k \in \mathcal{S}}{\operatorname{argmin}} \nabla_{\theta_{ij\mathbf{y}k}} D(\theta)$.
8:          $\eta_{ij} = \underset{\mathbf{y} \in A_i, k: s_k \in \mathcal{S}}{\max} \nabla_{\theta_{ij\mathbf{y}k}} D(\theta) - \nabla_{\theta_{ij\hat{\mathbf{y}}\hat{k}}} D(\theta)$.
9:          **if** $\eta_{ij} > 0$ **then**
10:             $A_i = A_i \cup \{\hat{\mathbf{y}}\}$; $\theta_{ij\hat{\mathbf{y}}k} = 0$, $\forall k : s_k \in \mathcal{S}$.
11:             Solve (5) to get $\lambda_{ij\mathbf{y}k}, \forall\ \mathbf{y} \in A_i$, $\forall k : s_k \in \mathcal{S}$.
12:             $\theta_{ij\mathbf{y}k} = \theta_{ij\mathbf{y}k} + \lambda_{ij\mathbf{y}k}, \forall\ \mathbf{y} \in A_i, \forall k : s_k \in \mathcal{S}$.
13:          **end if**
14:       **end for**
15:       Update $\mathbf{v}_j(\theta) \leftarrow \mathbf{v}_j(\theta) - \sum_{\mathbf{y} \in A_i} (\sum_k \lambda_{ik\mathbf{y}j}) f_j(x_i, \mathbf{y}), \forall j : s_j \in \mathcal{S}$.
16:    **end for**
17:    $t = t + 1$.
18: **end while**
19: Output $\{\mathbf{w}_j = \frac{1}{\gamma_j} \mathbf{v}_j(\theta)\}_{j=1}^J$.

Since the subproblem (5) has a summation constraint, we use a variant of sequential minimal optimization (SMO) method [24] to solve (5) until the following KKT-optimality conditions are satisfied: $\forall j$,

$$\max_{(\mathbf{y} \in A_i, k): \lambda_{ij\mathbf{y}k} > -\theta_{ij\mathbf{y}k}} \nabla_{\lambda_{ij\mathbf{y}k}} G(\lambda) - \min_{(\mathbf{y} \in A_i, k)} \nabla_{\lambda_{ij\mathbf{y}k}} G(\lambda) \leq 0, \tag{6}$$

where $\nabla_{\lambda_{ij\mathbf{y}k}} G(\lambda)$ denoting the partial gradient of $G(\lambda)$ with respect to $\lambda_{ij\mathbf{y}k}$ is given by $-\delta_{ij\mathbf{y}k} - \frac{1}{\gamma_k} (\mathbf{v}_k(\theta) - \sum_{\hat{\mathbf{y}} \in A_i} (\sum_{\hat{k}} \lambda_{i\hat{k}\hat{\mathbf{y}}k}) f_k(x_i, \hat{\mathbf{y}}))^\top f_k(x_i, \mathbf{y})$.

At the next level of our algorithm: we fix index j corresponding to the structure s_j and solve (5) using SMO by repeatedly finding two variables $\lambda_{ij\mathbf{y}_1 k_1}$ and $\lambda_{ij\mathbf{y}_2 k_2}$ corresponding to the pairs $(\mathbf{y}_1, k_1) = \arg\max_{(\mathbf{y} \in A_i, k): \lambda_{ij\mathbf{y}k} > -\theta_{ij\mathbf{y}k}} \nabla_{\lambda_{ij\mathbf{y}k}} G(\lambda)$ and $(\mathbf{y}_2, k_2) = \operatorname{argmin}_{(\mathbf{y} \in A_i, k)} \nabla_{\lambda_{ij\mathbf{y}k}} G(\lambda)$ in KKT condition (6). If the condition (6) is not satisfied, we associate a change $\lambda_{ij\mathbf{y}_1 k_1} + \Delta$ and $\lambda_{ij\mathbf{y}_2 k_2} - \Delta$ to the variables and optimize Δ by solving:

$$\min_{\Delta} H(\Delta) = \frac{q_1}{2} \Delta^2 + q_2 \Delta \ \text{s.t.} - \lambda_{ij\mathbf{y}_1 k_1} - \theta_{ij\mathbf{y}_1 k_1} \leq \Delta \leq \lambda_{ij\mathbf{y}_2 k_2} + \theta_{ij\mathbf{y}_2 k_2}, \tag{7}$$

where $q_1 = \frac{1}{\gamma_{k_1}} \|f_{k_1}(x_i, \mathbf{y}_1)\|^2 + \frac{1}{\gamma_{k_2}} \|f_{k_2}(x_i, \mathbf{y}_2)\|^2$, $q_2 = \nabla_{\lambda_{ij\mathbf{y}_1 k_1}} G(\lambda) - \nabla_{\lambda_{ij\mathbf{y}_2 k_2}} G(\lambda)$. The solution Δ^* to (7) is easily obtained in closed form as: $\Delta^* = \max(-\lambda_{ij\mathbf{y}_1 k_1} - \theta_{ij\mathbf{y}_1 k_1}, \min(-\frac{q_2}{q_1}, \lambda_{ij\mathbf{y}_2 k_2} + \theta_{ij\mathbf{y}_2 k_2}))$. This procedure is repeated until the KKT conditions (6) are satisfied for every structure $s_j \in \mathcal{S}$. After solving the subproblem (5), the following updates are made $\mathbf{v}_j(\theta) \leftarrow \mathbf{v}_j(\theta) - \sum_{\mathbf{y} \in A_i} (\sum_k \lambda_{ik\mathbf{y}j}) f_j(x_i, \mathbf{y}), \forall j : s_j \in \mathcal{S}$. The overall algorithm is stopped when the KKT conditions (4) hold for every (example, label interaction structure) pair (i, j). The optimization procedure is illustrated in Algorithm 1. See also Appendix D for a detailed discussion on the 2-level approach to solve dual MSSVM The following result establishes the sub-linear convergence of Algorithm 1 (see Appendix E for proof details).

Theorem 1. *Let $\{\theta^t\}_{t\geq 1}$ be the sequence of iterates generated by Algorithm 1 and let θ^* denote the optimal solution. Then the number of iterations of Algorithm 1 to attain $D(\theta^t) - D(\theta^*) \leq \epsilon$ is of the order $t = O(1/\epsilon)$.*

Label interaction structures, feature vector construction and inference procedures: The construction of hypothesis space $\mathcal{S}$ of structures depends on the following crucial assumptions: (a) for a given training sample $(x, \mathbf{y})$ and a hypothesized structure $s_j \in \mathcal{S}$, computing the transformation $f_j(x, \mathbf{y}) \in \mathbb{R}^{d_j}$ takes polynomial time; (b) finding $\arg\max_{\mathbf{y}\in\mathcal{Y}} \boldsymbol{w}_j^\top f_j(x, \mathbf{y}), j : s_j \in \mathcal{S}$ takes polynomial time. Here, we consider the following simple structures for a sample $(x, \mathbf{y} = \{y^l\}_{l=1}^L)$. The structure s_1 represents the case where the occurrence of a label component y^l in $\mathbf{y}$ does not affect the occurrence of another label component $y^{\hat{l}}$, $\hat{l} \neq l$. When no label interactions are assumed ($j = 1$), the joint feature vector of dimension $d_1 = 2Lm$ is of the form $f_1(x, \mathbf{y}) = [\mathcal{I}(y^1 == 1)x^\top \ldots \mathcal{I}(y^L == -1)x^\top]^\top$, where $\mathcal{I}(z)$ is 1 if z is true and zero otherwise. The model vector w_1 consists of the corresponding components $[w_{1y^1=+1}^\top \ldots w_{1y^L=-1}^\top]^\top$ and the inference rule for an example $\hat{x}$ can be decomposed label-wise to get $\hat{y}_l = \arg\max_{y\in\{+1,-1\}} w_{1y^l=y}^\top \hat{x} \; \forall l = 1, \ldots, L$. A graphical model and a possible inference output for s_1 is given in Fig. 1(a) (the dependence on x is suppressed for brevity). Structure s_2 represents pairwise (or first-order) interactions among the label components along a linear chain structure with a fixed label permutation, which results in efficient inference using the Viterbi algorithm [8]. The graphical model for a linear chain with label permutation $y^1 - y^2 - y^3$ is given in Fig. 1(c). The feature vector $f_2(x, \mathbf{y})$ for Fig. 1(c) is constructed by appending the pair-wise features $\mathcal{I}(y^l == v^l \& y^{l+1} == v^{l+1}), v^l, v^{l+1} \in \{+1, -1\}, \forall l = 1, \ldots, L - 1$ to $f_1(x, \mathbf{y})$. Note that the dimension of $f_2(x, \mathbf{y})$ is $d_2 = 2Lm + 4(L - 1)$. However, when we consider all pair-wise label interactions, the computational complexity of inference increases due to cyclic dependencies [33] (see graphical model in Fig. 1(b)). Hence we make simple linear chain structure assumption in s_2. We further consider structure s_3 representing the extension to second-order label interactions along a chain structure and a fixed label permutation, as shown in Fig. 1(d). The feature vector $f_3(x, \mathbf{y})$ consists of $f_2(x, \mathbf{y})$ appended with second-order features $\mathcal{I}(y^l == v^l \& y^{l+1} == v^{l+1} \& y^{l+2} == v^{l+2}), v^l, v^{l+1}, v^{l+2} \in \{+1, -1\}, \forall l = 1, \ldots, L-2$. The dimension of $f_3(x, \mathbf{y})$ is thus $d_3 = d_2 + 8(L-2)$. Efficient extensions of the Viterbi algorithm [6] can be used for inference on a chain structure with second-order label interactions. Note that finding an approximately optimal label permutation (similar to that pursued in [20]) is not our goal in this work. We will show in the next section that simple linear chain structure and fixed label permutation assumptions give comparable results without much computational overhead. We will also highlight those specific cases where finding a suitable label permutation might lead to possible improvements. We further provide experiments in the next section where we consider MSSVM with a spanning forest structure assumption. The results show the importance of hypothesizing distinct structure assumptions in multi-label classifier design.

Iteration complexity: The complexity of each iteration of Algorithm 1 is dominated by $O(nJ(\tau_{inf} + \tau_{smo}))$, where τ_{inf} denotes the complexity of inference procedure and τ_{smo} denotes the complexity of SMO. Note that τ_{inf} is $O(Ld_1)$, $O((d_2 + 4)L)$, $O((d_3 + 8)L)$ respectively for $j = 1, 2, 3$. SMO has sub-linear convergence [2] and hence $\tau_{smo} = O(1/\epsilon)$ to attain an ϵ optimal solution.

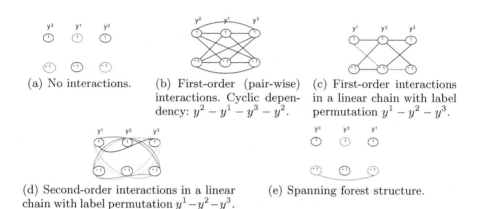

(a) No interactions.

(b) First-order (pair-wise) interactions. Cyclic dependency: $y^2 - y^1 - y^3 - y^2$.

(c) First-order interactions in a linear chain with label permutation $y^1 - y^2 - y^3$.

(d) Second-order interactions in a linear chain with label permutation $y^1 - y^2 - y^3$.

(e) Spanning forest structure.

Fig. 1. Different label interaction structures in multi-label classification. A possible inference structure is shown highlighted in blue. Colors in Fig. 1(d) indicate different second-order interaction combinations. (Best viewed in color.) (Color figure online)

5   Experiments

We now provide results from an extensive empirical evaluation of MSSVM. The experiments were run on a shared Linux box with 128 GB main memory. For MSSVM, δ_{ijyk} was defined to be a weighted Hamming loss. We compared our method with seven state-of-the-art methods for which code is publicly available: Binary Relevance (BR) [29], ML-kNN [37], RAndom k-labELsets (RAKEL) [30], MLLOC [15], Multilabel tree ensemble method (ML-Forest) [32], Probabilistic Classifier Chains (PCC) [5] and Clique Generation Machine (CGM) [28], and experimented with 11 benchmark datasets (more details are given in Appendix F). We used 6 performance measures: Hamming loss, Jaccard accuracy, subset accuracy, example based F-measure, label based micro F-measure and macro F-measure [21] to evaluate the predictive performance. For fair comparison of all methods we used only linear kernels.

Dataset preparation: We follow MLLOC [15] to add clustering-based additional features for the training examples. These additional features were constructed by k-means clustering algorithm (with $k = 15$), grouping the multilabel vectors into k clusters. The corresponding test set features were constructed using support vector regression. It is known that clustering-based additional features improve the generalization performance for classification tasks [17]. During training, MLLOC has an added advantage of tuning these additional clustering-based features [15]. However, since other methods cannot tune these features as such, we simply fixed them during their training.

Experimental results with no label interaction structure and pair-wise, triplet-wise label interactions along a linear chain:
For this set of experiments, we used the proposed MSSVM with three different label interaction structures: s_1 for zero-order label interactions (meaning no label interactions), s_2 and s_3 for pair-wise (first-order) and triplet-wise (second-order) label interactions respectively, along a linear chain structure with a fixed label permutation. The feature vector construction was performed according to the procedure given in Sect. 4. We tuned the $\{\gamma_j\}_{j=1}^3$ parameters using a multi-grid 3-fold cross-validation approach where we let $\gamma_j \in \{10^{-2}, 10^{-1}, 1, 10, 10^2, 10^3, 10^4, 10^5\}, j = 1, 2, 3$ and chose the best combination distinctly for each performance measure. Note that this can be done efficiently in parallel using multiple machines. We present in Table 1, the results on Hamming loss, Jaccard accuracy, example based F-measure and label based macro F-measure. Additional results on subset accuracy and label based micro F-measure are given in Appendix F. In our cross-validation experiments, we noticed that multiple combinations of $\{\gamma_j\}_{j=1}^3$ resulted in the best performance, indicating the usefulness of multiple label interaction structure combinations. Hence we highlight those structures which helped to achieve the best generalization performance distinctly for each performance metric using their indices, in the last column of Table 1. A novel aspect of MSSVM is its ability to explicitly identify the structures which help to achieve the best performance.

Note however that the results presented should be reflected in perspective of the simplistic label interaction structure assumptions along a fixed label permutation, made in Figs. 1(c) and (d). Even with such simple assumptions, we see that our results are very comparable to benchmark results. Our goal is not *just* to show superiority over existing methods; more importantly, we try to understand the role of label interaction structures from these results. We note the following interesting and insightful observations:

How do label interactions affect multiple performance metrics typically used in multi-label classification?
For many datasets, s_1 structure denoting zero-order label interactions achieves the best results for Jaccard accuracy and F-measure (both example based and label based). On almost all datasets, MSSVM requires a combination of zero, first and second order label interactions to achieve a comparable Hamming loss.

Table 1. *Comparison of multilabel classification algorithms.* Best results are indicated in bold. Second best results are indicated in italic fonts. Here s_1: no label interactions; s_2 and s_3 correspond to first-order and second-order label interactions along a linear chain with fixed label permutation. We use $1, 2, 3$ to denote s_1, s_2, s_3. *** indicates that we did not get reasonable results after 3 weeks of running the program. **Hamming loss:** the smaller the better, **other performance measures:** the larger the better.

Perf. Measure	Dataset	BR	ML-kNN	RAKEL	MLLOC	PCC	CGM	ML-Forest	MSSVM Best Perf.	Best Label Interaction Structures
Hamming Loss	Flags	0.3055	0.3187	0.2989	**0.2725**	0.4791	*0.2901*	0.3275	0.2923	1
	Emotions	*0.2541*	0.2772	0.2624	0.2665	0.4183	0.3243	0.2591	**0.2525**	{2, 3}
	Scene	0.0877	0.0832	*0.0803*	0.0842	0.0831	**0.0779**	0.0896	0.0842	2
	Yeast	0.2010	0.2062	0.1989	**0.1949**	0.2074	0.2017	0.2155	*0.1962*	2
	Genbase	0.0013	0.0013	**0.0007**	0.0011	**0.0007**	**0.0007**	0.003	*0.0009*	{1, 2},{1, 3}
	Medical	0.0127	0.0180	0.0123	0.0126	**0.0119**	*0.0120*	0.014	0.0149	{1, 2, 3}
	Enron	*0.0460*	0.0504	0.0538	0.0466	0.0497	**0.0458**	0.0579	0.0498	{2, 3}
	Bibtex	**0.0126**	0.0138	0.0146	0.0146	0.0141	*0.0129*	***	0.0130	1
	CAL500	0.1372	0.1366	**0.1359**	*0.1360*	0.1734	0.1363	0.151	0.1370	1,2,{1, 2}
	Bookmarks	**0.0088**	*0.0089*	0.0124	***	0.0129	0.0090	***	**0.0088**	1
	TMC2007	**0.0562**	0.0594	0.0584	***	*0.0575*	***	***	0.0596	2,{2, 3}
Jaccard Accuracy	Flags	0.5474	0.4995	0.5436	*0.5676*	0.4934	0.5612	0.4977	**0.6038**	1
	Emotions	0.3375	0.3205	0.3288	0.3535	0.3810	0.0165	*0.4851*	**0.5032**	{1, 2, 3}
	Scene	0.6533	0.7181	0.7198	0.6819	0.7044	0.7326	**0.7428**	*0.7338*	{1, 2}
	Yeast	0.5196	*0.5330*	0.5300	0.5117	0.5419	0.5134	0.518	**0.5436**	1
	Genbase	0.9791	0.9883	*0.9908*	0.9874	**0.9916**	**0.9916**	0.9732	*0.9899*	{1, 2},{1, 2, 3}
	Medical	0.6364	0.5371	0.6624	0.6354	0.6765	0.6736	**0.7189**	*0.7126*	{1, 2, 3}
	Enron	0.4387	0.3480	0.4006	0.4357	**0.4838**	0.4263	0.4251	*0.4776*	{1, 2}
	Bibtex	0.2750	0.1538	0.3070	*0.3088*	0.3000	0.2748	***	**0.3552**	1
	CAL500	0.2013	0.2083	0.2201	0.2076	**0.3322**	0.1968	0.2577	*0.2798*	1
	Bookmarks	0.2000	0.1778	0.2383	***	*0.2468*	0.2101	***	**0.2603**	1
	TMC2007	0.5547	0.5200	0.5423	***	*0.5584*	***	***	**0.5690**	{1, 2}
F-Measure (Example based)	Flags	0.6808	0.6403	0.6771	0.6821	0.6446	*0.6962*	0.6320	**0.7408**	1
	Emotions	0.4008	0.3967	0.3947	0.4201	0.5184	0.0188	*0.5778*	**0.6306**	{1, 2, 3}
	Scene	0.6690	0.7368	0.7356	0.7085	0.7252	0.7489	*0.7649*	**0.7765**	1
	Yeast	0.6230	0.6272	0.6341	0.6177	*0.6485*	0.6188	0.6218	**0.6523**	1
	Genbase	0.9824	0.9921	*0.9940*	0.9915	0.9816	**0.9941**	**0.9941**	0.9931	{1, 2}, {1, 2, 3}
	Medical	0.6620	0.5653	0.6869	0.6582	**0.7671**	0.7018	0.6971	*0.7638*	{1, 2, 3}
	Enron	*0.5479*	0.4477	0.5090	0.5423	0.5999	0.5335	0.5376	**0.5987**	{1, 2}
	Bibtex	0.3247	0.1867	0.3719	*0.3722*	0.3598	0.3281	***	**0.4266**	1
	CAL500	0.3310	0.3408	0.3550	0.3391	**0.4919**	0.3249	0.4028	*0.4308*	1
	Bookmarks	0.2081	0.1821	0.2653	***	*0.2697*	0.2253	***	**0.2910**	1
	TMC2007	0.6347	0.5938	0.6250	***	*0.6422*	***	***	**0.6618**	{1, 2}
Macro F-measure (Label based)	Flags	0.4631	0.4259	0.5308	*0.6475*	0.6046	0.5727	0.5313	**0.6777**	1
	Emotions	0.4532	0.3484	0.4103	0.4319	0.5242	0.0245	*0.5735*	**0.6332**	{1, 2, 3}
	Scene	0.7370	0.7545	0.7711	0.7528	0.7664	**0.7782**	0.7644	*0.7727*	1
	Yeast	0.4093	0.4403	0.4059	0.3796	0.4355	*0.4469*	0.4259	**0.4676**	1
	Genbase	0.7352	0.6840	**0.8519**	0.7359	**0.8519**	0.7778	0.6799	0.7763	1
	Medical	0.2493	0.1901	**0.4370**	0.2527	*0.4042*	0.2790	0.3405	0.3297	1
	Enron	0.1400	0.0832	0.2060	0.1558	*0.2084*	0.1991	0.1889	**0.2242**	1
	Bibtex	0.1968	0.0817	*0.2866*	0.2855	0.2489	0.1974	***	**0.3281**	1
	CAL500	0.0582	0.0581	0.1195	0.0626	**0.2102**	0.0396	0.1096	*0.1297*	2
	Bookmarks	0.0866	0.0641	*0.1700*	***	0.1580	0.1184	***	**0.1814**	1
	TMC2007	0.5715	0.4749	0.5695	***	*0.5727*	***	***	**0.6155**	1

Do higher order label interactions significantly impact a given performance metric for a particular dataset?

For all performance metrics, most of the best results are obtained from zero-order label interactions (s_1) or a combination of zero-order and first-order interactions (s_1 and s_2). Second-order label interactions (s_3) help in improving some

performance measures, when in combination with zero-order or first-order interactions (see especially Hamming loss, Jaccard accuracy results in Table 1).

**Experimental results with no label interaction structure, pair-wise
interactions along a linear chain and a spanning forest structure:**
For this set of experiments, we retained the structures s_1 and s_2 used in
the previous experiments. We let s_3 to denote a spanning forest structure (see
Fig. 1(e)). For s_3, we used a feature vector $f_3(x, \mathbf{y})$ constructed by appending to
$f_1(x, \mathbf{y})$, all pair-wise features $I(y^p==v^p \& y^q==v^q)$, $v^p, v^q \in \{+1, -1\}$, $\forall p, q =
1, \ldots, L, p \neq q$. Note that the dimension of $f_3(x, \mathbf{y})$ for a spanning forest structure
is $d_3 = 2L(m + L - 1)$. The inference was performed using a variant of Kruskal's
maximum spanning tree algorithm [16]. We followed a tuning procedure to choose
$\{\gamma_j\}_{j=1}^3$ values similar to that used in the previous section. The results on three
datasets are given in Table 2, where the spanning forest structure assumption
leads to superior results when compared to the existing best results of MSSVM.
Thus, we are able to answer the next important questions:

Can we make useful suggestions on the label interaction structure?
The notion of *best structure assumption* is relative to the hypothesis space $\mathcal{S}$
considered for the particular task. The results in Table 2 indicate that hypothesizing suitable label interaction structures play a pivotal role in improving certain
performance measures for some datasets. Note that for many practical applications, suitable structures can be designed using domain expertise. When the size
of $\mathcal{S}$ is large, powerful computational machines can be employed to speed up
training. We believe (and empirically checked) that label permutation plays an
important role in datasets like *Scene, Genbase* and *Medical,* when considering
Hamming loss and Micro F-measure. Adapting MSSVM to automatically choose
the best label permutation is a future work.

*Is it always beneficial to model label interactions in multi-label clas-
sification?*
Some best results for various performance measures are obtained when no
label interactions are assumed. Hence, contrary to popular belief, modeling label
interactions might not be always useful for all datasets. However, we should
be quick to add that it is always useful to hypothesize as many structures as
possible according to the assumptions given in Sect. 4 and check if they lead to
an improvement in performance. We believe that the proposed MSSVM serves
as a helpful tool in this regard.

Table 2. Results for MSSVM where spanning forest structure s_3 yields better performance. Superscripts a: $\{3\}$ b: $\{1, 3\}$ c: $\{2, 3\}$ d: $\{1, 2, 3\}$ denote the various combinations
of label interaction structures s_1, s_2 and s_3.

Dataset	Hamming loss	Subset accuracy	Micro F-measure
Scene	0.0818^a	0.6614^a	—
Medical	0.0140^b	0.5752^d	—
Enron	0.0495^d	0.1209^b	0.5903^b

Discussion: Note that BR considers L different binary classification problems (for L labels) and tunes parameters independently for each label. MSSVM rather uses a structured hinge loss ℓ_{multi}; moreover γ_1 for structure s_1 is common to all labels. Interestingly, we also observe that methods which claim to optimize a particular performance measure do not always achieve the best results for that performance measure on all datasets despite vigorous parameter tuning (see Hamming loss for BR and F-measure for PCC in Table 1).

Runtime comparison: Since different algorithms use different programming languages (RAKEL, ML-Forest, PCC use Java; BR, ML-kNN, MLLOC, CGM use Matlab; MSSVM uses C) for implementation, CPU times could not be directly compared. In terms of wall-clock times, MSSVM was comparable to BR, RAKEL, ML-kNN, PCC and was faster than MLLOC, CGM and ML-Forest.

6 Conclusion

We formulated a multi-structure SVM (MSSVM) for learning a multi-label classifier using multiple label interaction structures, and provided an efficient optimization algorithm to solve it. The proposed MSSVM allows the user to design a suitable hypothesis space of label interaction structures and explicitly identifies which structures are useful in achieving the best generalization performance. We believe that MSSVM will aid the practitioner to better understand the role of label interactions in multi-label classification. It will be an interesting future direction to incorporate other loss functions [5,18] in MSSVM framework.

Acknowledgments. The authors thank anonymous reviewers of the current and earlier versions of the paper for their useful comments. The second author thanks Prof. Francis Bach for the discussion.

References

1. Balamurugan, P., Shevade, S., Sundararajan, S., Keerthi, S.S.: A sequential dual method for structural svms. In: SDM (2011)
2. Beck, A.: The 2-coordinate descent method for solving double-sided simplex constrained minimization problems. JOTA **162**(3), 892–919 (2014)
3. Bi, W., Kwok, J.T.: Multilabel classification with label correlations and missing labels. In: AAAI (2014)
4. Dembczyński, K., Jachnik, A., Kotłowski, W., Waegeman, W., Hüllermeier, E.: Optimizing the f-measure in multi-label classification: plug-in rule approach versus structured loss minimization. In: ICML (2013)
5. Dembszynski, K., Waegeman, W., Cheng, W., Hüllermeier, E.: On label dependence and loss minimization in multi-label classification. Mach. Learn. **88**(1–2), 5–45 (2012)
6. Engelbrecht, H.A.: Efficient decoding of high-order hidden markov models. Ph.D. thesis, University of Stellenbosch (2007)
7. Finley, T., Joachims, T.: Training structural svms when exact inference is intractable. In: ICML (2008)

8. Forney, G.D.: The Viterbi algorithm. Proc. IEEE **61**, 268–278 (1973)
9. Ghamrawi, N., McCallum, A.: Collective multi-label classification. In: CIKM (2005)
10. Godbole, S., Sarawagi, S.: Discriminative methods for multi-labeled classification. In: PAKDD (2004)
11. Gonçalves, A., Zuben, F.J.V., Banerjee, A.: Multi-label structure learning with ising model selection. In: IJCAI (2015)
12. Guo, Y., Gu, S.: Multi-label classification using conditional dependency networks. In: IJCAI (2011)
13. Guzman-Rivera, A., Batra, D., Kohli, P.: Multiple choice learning: learning to produce multiple structured outputs. In: NIPS (2012)
14. Hariharan, B., Vishwanathan, S.V., Varma, M.: Efficient max-margin multi-label classification with applications to zero-shot learning. Mach. Learn. **88**(1–2), 127–155 (2012)
15. Huang, S.J., Zhou, Z.H.: Multi-label learning by exploiting label correlations locally. In: AAAI (2012)
16. Kruskal, J.B.: On the shortest spanning subtree of a graph and the traveling salesman problem. Proc. AMS **7**, 48–50 (1956)
17. Kyriakopoulou, A., Kalamboukis, T.: Using clustering to enhance text classification. In: SIGIR (2007)
18. Lafferty, J., McCallum, A., Pereira, F.C.N.: Conditional random fields: probabilistic models for segmenting and labeling sequence data. In: ICML (2001)
19. Lampert, C.H.: Maximum margin multi-label structured prediction. In: NIPS (2011)
20. Liu, W., Tsang, I.: On the optimality of classifier chain for multi-label classification. In: NIPS (2015)
21. Manning, C.D., Raghavan, P., Schütze, H.: Introduction to Information Retrieval. Cambridge University Press, New York (2008)
22. Marchand, M., Su, H., Morvant, E., Rousu, J., Shawe-Taylor, J.S.: Multilabel structured output learning with random spanning trees of max-margin markov networks. In: NIPS (2014)
23. Mirzazadeh, F., Ravanbakhsh, S., Ding, N., Schuurmans, D.: Embedding inference for structured multilabel prediction. In: NIPS (2015)
24. Platt, J.C.: Fast training of support vector machines using sequential minimal optimization. In: Advances in Kernel Methods, pp. 185–208 (1999)
25. Read, J., Martino, L., Olmos, P.M., Luengo, D.: Scalable multi-output label prediction: from classifier chains to classifier trellises. Pattern Recogn. **48**(6), 2096–2109 (2015)
26. Read, J., Pfahringer, B., Holmes, G., Frank, E.: Classifier chains for multi-label classification. In: Buntine, W., Grobelnik, M., Mladenić, D., Shawe-Taylor, J. (eds.) ECML PKDD 2009. LNCS (LNAI), vol. 5782, pp. 254–269. Springer, Heidelberg (2009). https://doi.org/10.1007/978-3-642-04174-7_17
27. Su, H., Rousu, J.: Multilabel classification through random graph ensembles. In: ACML (2013)
28. Tan, M., Shi, Q., van den Hengel, A., Shen, C., Gao, J., Hu, F., Zhang, Z.: Learning graph structure for multi-label image classification via clique generation. In: CVPR (2015)
29. Tsoumakas, G., Katakis, I.: Multi-label classification: an overview. Int. J. Data Warehous. Min. **3**, 1–13 (2007)

30. Tsoumakas, G., Vlahavas, I.: Random k-Labelsets: An Ensemble Method for Multilabel Classification. In: Kok, J.N., Koronacki, J., Mantaras, R.L., Matwin, S., Mladenič, D., Skowron, A. (eds.) ECML 2007. LNCS (LNAI), vol. 4701, pp. 406–417. Springer, Heidelberg (2007). https://doi.org/10.1007/978-3-540-74958-5_38

31. Vapnik, V.N.: Statistical Learning Theory. Wiley-Interscience, New York (1998)

32. Wu, Q., Tan, M., Song, H., Chen, J., Ng, M.K.: Ml-forest: a multi-label tree ensemble method for multi-label classification. IEEE TKDE **28**, 2665–2680 (2016)

33. Yu, C.N.J., Joachims, T.: Learning structural SVMs with latent variables. In: ICML (2009)

34. Yu, H.F., Jain, P., Kar, P., Dhillon, I.S.: Large-scale multi-label learning with missing labels. In: ICML (2014)

35. Zhai, S., Zhao, C., Xia, T., Wang, S.: A multi-label ensemble method based on minimum ranking margin maximization. In: ICDM (2015)

36. Zhang, M.L., Zhang, K.: Multi-label learning by exploiting label dependency. In: ACM SIGKDD (2010)

37. Zhang, M.L., Zhou, Z.H.: ML-KNN: a lazy learning approach to multi-label learning. Pattern Recogn. **40**, 2038–2048 (2007)

38. Zhang, M.L., Zhou, Z.H.: A review on multi-label learning algorithms. IEEE TKDE **26**(8), 1819–1837 (2014)

39. Zhang, X., Graepel, T., Herbrich, R.: Bayesian online learning for multi-label and multi-variate performance measures. In: AISTATS (2010)

Sentiment Classification Using Neural Networks with Sentiment Centroids

Maoquan Wang[(✉)], Shiyun Chen[(✉)], and Liang He[(✉)]

School of Computer Science and Software Engineering,
East China Normal University, Shanghai, China
{maoquanwang,sychen}@ica.stc.sh.cn,
lhe@cs.ecnu.edu.cn

Abstract. Neural networks (NN) have demonstrated powerful ability to extract text features automatically for sentiment classification in recent years. Although semantic and syntactic features are well studied, global category information has been mostly ignored within the NN based framework. Samples with the same sentiment category should have similar vectors in represent space. Motivated by this, we propose a novel global sentiment centroids based neural framework, which incorporates the sentiment category features. The centroids assist NN to extract discriminative category features from a global perspective. We apply our approach to several real large-scale sentiment-labeled datasets, and the extensive experiments show that our model not only obtains more powerful sentiment feature representations, but also achieves some state-of-the-art results with a simple neural network structure.

Keywords: Sentiment classification · Sentiment centroids
Deep neural network

1 Introduction

There is a large volume of sentiment rich text data in social websites with the forms of reviews, comments, tweets, and so on. Sentiment classification technologies play an important role in analyzing these kinds of texts. Recently, neural networks (NN) [10], such as recursive neural network (RENN) [21], convolutional neural network (CNN) [7,8], recurrent neural network (RNN) [22,23], have been attracting much attention in various natural language processing (NLP) tasks including sentiment classification. Many neural models are designed to learn deep features automatically. However, it still remains challenges of revealing sentiment categories for social review texts. And feature representation is a key point towards achieving the best possible accuracy.

NN is trained to represent given data by sentiment latent vector for sentiment classification task where semantics are relevant. For a more intuitive explanation of learned features, we provide a 2D view for mapped samples (deep features). As Fig. 1 shown, we illustrate two data distributions which got by HAN model [28]

© Springer International Publishing AG, part of Springer Nature 2018
D. Phung et al. (Eds.): PAKDD 2018, LNAI 10937, pp. 56–67, 2018.
https://doi.org/10.1007/978-3-319-93034-3_5

and our model separately. Five kinds of colors refer to five sentiment categories. Our observation from Fig. 1 shows same labeled points tend to cluster together around its own unique *sentiment centroid* in the form of real-valued vector. At the same time, there are also many overlapped points (color cycles overlapping parts) for different classes. For these overlapped points, it usually hard to separate them and it should to keep its distance from other categories, intuitively.

Points in the same class can be allocate together as close as possible and keep away from the other categories points as far as possible in the latent semantic space. To achieve this purpose, we propose a Sentiment Centroid (SC) based neural framework to help construct text embeddings and make feature vector more discernable. We use a global vector to represent the global sentiment features for each category. The main idea is: First, we assign a sentiment centroid to each sentiment category, which is originated from the work [18]. Rocchio uses average vectors as class centroid and assigns test vectors to the class with maximum centroid similarity. Second, sentiment centroids is used to help NN to learn text features, which are closer with the other same sentiment category embeddings. In order to achieve the second goal, we add a global centroids constraint to optimization target.

We constrain sentiment centroids with the same dimension of text representations to enhance the constructing process of deep feature learning. We conduct extensive experiments on five popular datasets. Our model achieves competitive results compared with the state-of-the-art approaches, and provides strong baselines for sentiment classification. Our main contributions are as follows:

- To the best of our knowledge, we are the first to introduce sentiment centroids to sentiment classification. Sentiment centroids are used to capture global sentiment category features of corpora.
- Also, We design a novel framework for sentiment classification. The entire model is trained end-to-end with batch gradient descent, where sentiment centroids are combined with a joint loss.
- The experimental results on several public datasets show that our approach outperform most state-of-the-art baselines. The sentiment centroids have been proved to be effective to prefer the sentiment classification performance.

The reminder of the paper is organized as follows: Related work is introduced in Sect. 2. Section 3 describes the proposed models and our joint objective function. In Sect. 4, five real-world datasets are discussed in detail, and experimental results are presented. Finally, the paper is concluded in Sect. 5.

2   Related Work

2.1   Sentiment Features Learning

Sentiment features are important to sentiment analysis. For word representation, some prior studies [25,26] have reported that words with similar embeddings may have opposite sentiment polarities. Tang [25] uses a supervised way to capture the sentiment information of word level. Ren [17] uses multi-prototype to improve word embeddings for twitter sentiment classification. To enhance the features of sentiment information, they typically apply an objective function to optimize word embeddings. And the enhanced word vectors have improved the performance of sentiment classification.

Unlike most previous studies that focus on word level sentiment information, we learn discriminative sentiment features for text representation. The sentiment features are formed as sentiment centroids.

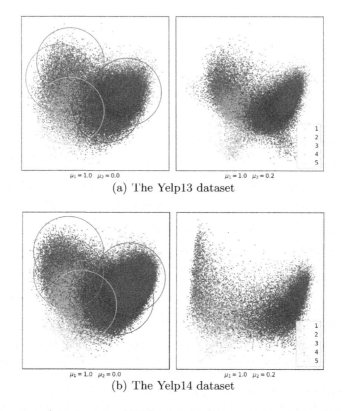

(a) The Yelp13 dataset

(b) The Yelp14 dataset

Fig. 1. 2D views (obtained by PCA) of data distributions of the Yelp13/Yelp14 datasets. Five kinds of colors refer to five sentiment categories (very negative, negative, neutral, positive, very positive). For each subfigure (a, b), distribution of the left part is extracted by HAN model, and the right part is extracted by our approach using global sentiment centroids. Color cycles are drawn by respective 2D centers and the overlapped areas contain many overlapped points. It is clear to see that our proposed centroids help encoder to get more separable text points. (Color figure online)

2.2   Neural Networks for Sentiment Classification

Sentiment classification via deep learning has achieved promising results in these years. The existing neural network methods can be divided into two groups: learning word embedding from text corpus and learning semantic representation of texts.

For learning word embeddings, Word2Vec [13] is an effective tool to build word vector which contains CBOW and Skip-Gram models. Global vectors [15] use statistics of word co-occurrence to train word representation which contains global information of corpus. SSWE [26] has been proposed to combine context and sentence-level sentiment evidence to learn sentiment-specific word embedding which proved effective for Twitter sentiment classification. For learning semantic composition, Recursive autoencoder neural networks (RANN) has achieved remarkable results [3,11,15,21] which based on syntactic analysis tree. Convolutional neural network (CNN) is another useful model for sentiment classification [6,7,20] which use filers to capture n-gram information. Long short-term memory model is a more powerful model for sequential signal [30]. There are also some other works that use hybrid structure in sequence generation which can be seen in [23,27,28].

Different from existing neural network approaches that only focus on embedding text sequence, we take consideration of global sentiment centroids for different categories.

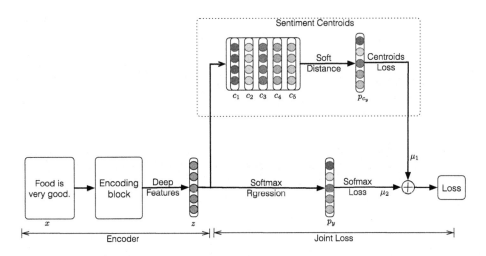

Fig. 2. Global centroids based framework for text sentiment classification.

3   Our Approach

To deal with challenges faced by previous models, we introduce sentiment centriods to represent the corpus global sentiment information. Leveraging global centriods to generate discriminative features, our approach has two main steps: firstly, encode a text sequence to a dense vector z and use the centroids as constrain of the text representation to enhance global sentiment features. Figure 2 shows the framework of our sentiment centroids based model for sentiment classification. The dashed part is our main work.

3.1   Text Sequence Encoder Models

In order to display the generalization of our approach and encode variable-length text to low dimensional vector z, we experimented with three models of increasing complexity.

Considering a word sequence, $x = \langle x_1, x_2, \cdots, x_n \rangle$, we define $e(*)$ as the word lookup function to map word x_i to $\boldsymbol{v}^{x_i} = e(x_i)$.

AVG-Based Sequence Encoder. The AVG method is a average of the word embeddings and project it to an another vector using a full connection layer:

$$z = f(W_p(\frac{1}{n}\sum_i^n \boldsymbol{v}^{x_i}) + \boldsymbol{b}_p) \tag{1}$$

where W_p is the projection matrix, $\boldsymbol{b}_p$ is a bias term, f is the nonlinear activation function. We refer to AVG-based encoder with sentiment centriods as SCAVG.

GRU-Based Sequence Encoder. To gain some order and dependency information between words within x, GRU [1] is used to model it:

$$\boldsymbol{h}_t = GRU(\boldsymbol{c}_{t-1}, \boldsymbol{h}_{t-1}, \boldsymbol{v}^{x_t}) \tag{2}$$

Where $\boldsymbol{c}_t \in \mathbb{R}^d$ is an additional memory cell, and the last hidden vector $\boldsymbol{h}_{-1}$ is used to represent the text sequence $z = \boldsymbol{h}_{-1}$. We refer to GRU-based encoder with sentiment centriods as SCGRU.

Attention-Based Sequence Encoder. The hierarchical attention network (HAN) [28] uses an attention mechanism over the hidden states of words and sentences to generate a representation of a document. In this work, the same mechanism is used to encode input text $\boldsymbol{x}$ to $\boldsymbol{z}$. The attention mechanism is defined as:

$$z = \sum_{i=1}^n \alpha_i \boldsymbol{h}_i \tag{3}$$

$$\alpha_i = \frac{exp(\boldsymbol{u}_i^T \boldsymbol{u}_w)}{\sum_{t=1}^n exp(\boldsymbol{u}_i^T \boldsymbol{u}_w)} \tag{4}$$

$$\boldsymbol{u}_i = tanh(W_w \boldsymbol{h}_i + b_w) \tag{5}$$

where α_i is the attention weight for hidden state $\boldsymbol{h}_i$, $\boldsymbol{u}_i$ is a hidden representation of $\boldsymbol{h}_i$, and $\boldsymbol{u}_w$ is a context vector. W_w and b_w are nonlinear projection parameters. We refer to Attention-based encoder with sentiment centriods as SCHAN.

3.2   Sentiment Centriods Constraint

In this section, we explain the proposed sentiment centroids constraint which is used to capture global category information and enhance sentiment features of text. We define the following loss function.

Softmax Loss. Most deep neural network based classifier use a full connection layer to convert text representation to category probability $\boldsymbol{p}_y$ with a softmax function, and use cross-entropy error between gold and predicted distribution as the objective function. The normal direct classification softmax loss is as below:

$$
\begin{aligned}
L_1 &= -\frac{1}{m} \sum_{i=1}^{m} log(\boldsymbol{p}_y) \\
&= -\frac{1}{m} \sum_{i=1}^{m} log(softmax(W^T \boldsymbol{z}^{(i)} + \boldsymbol{b})) \\
&= -\frac{1}{m} \sum_{i=1}^{m} log \frac{exp(W_{y_i}^T \boldsymbol{z}^{(i)} + b_{y_i})}{\sum_{j=1}^{K} exp(W_j^T \boldsymbol{z}^{(i)} + b_j)}
\end{aligned}
\tag{6}
$$

where m denotes the number of training samples, $\boldsymbol{z}^{(i)}$ refers to the text representation with sentiment category y_i. W_j is j-th column of multinomial logistic regression parameter $W \in \mathbb{R}^{d \times k}$ and d is dimensionality of text representation, k is number of categories.

Sentiment Centroids Loss. Give a corpus, we define its *sentiment centroids* as $\mathbb{C} = \{\boldsymbol{c}_i\}$, here $\boldsymbol{c}_i \in \mathbb{R}^d$ ($i = 1, \cdots, k$), and d is centroid vector size, k is number of sentiment categories.

In order to enhance the sentiment feature of review text representations, it is natural to think of making the text representation vector to closer with corresponding sentiment centroid which we believe different categories have different class centers in representation space. To this end, we can use a kernel to measure the distance between a review text representation and the corresponding sentiment centroid vector, and try to reduce the distance to reinforce sentiment information. There are many distance measurement methods to choose, such as Euclidean distance, Kullback-Leibler divergence, etc.

Triplet loss is proposed to improve face recognition and clustering [19]. About text classification, a weakly-supervised model [4] introduced to identify customer review sentiment categories, which use a rating lost to obtain a good enough sentence representation. The normal loss function of triplet loss is:

$$
L_t = -\frac{1}{m} \sum_{i}^{m} \left[dist(\boldsymbol{z}_i, \boldsymbol{c}_{y_i}) - dist(\boldsymbol{z}_i, \boldsymbol{c}_{y_i'}) + \alpha \right]_+
\tag{7}
$$

where $dist(\cdot)$ denotes distance between pairwise. In our work, we use Euclidean function to measure the distance between centroids and text presentations. α is a margin that is enforced the pairwise, c_{y_i} denotes sentiment center which has a same dimensionality with representation z_i. $c_{y_i'}$ is used to represent corroded centroids which is randomly chosen in the other categories respectively.

Student's t-distribution loss is proposed by [10]. In this work, we also use the Student's t-distribution as a distance kernel to measure the similarity between the text point z and corresponding sentiment centroid c_y:

$$p_{c_y} = \frac{(1 + \|z_i - c_y\|^2 / \alpha)^{-\frac{\alpha+1}{2}}}{\sum_{j=1}^{k}(1 + \|z_i - c_j\|^2 / \alpha))^{-\frac{\alpha+1}{2}}} \tag{8}$$

where α is the degrees of freedom of the Student's t-distribution, and we let $\alpha = 1$ for all experiments. The soft sentiment centroids lost can be computed as:

$$L_c = -\frac{1}{k} \sum_{i=1}^{k} log(p_{c_y}) \tag{9}$$

Joint Loss. To make the whole neural network become an end-to-end model, we use two hyper parameters μ_1 and μ_2 to joint softmax lost and sentiment centroid loss. In the training phase, we update the review text embedding part and the sentiment center representation synchronously. The joint loss L formulated as below:

$$L = \mu_1 L_1 + \mu_2 L_2 + \frac{\lambda_\theta}{2} \| \theta \|^2 \tag{10}$$

where L_2 can be L_t or L_c, and $\mu_1 \in (0, 1.0]$, $\mu_2 \in [0.0, 1.0]$, λ_θ is a regularization parameter, θ denote trainable parameters.

In the prediction phase, we use softmax function as a classifier for the following reason: Softmax functions have more power of discrimination ability [5], while centroid vector method is unable to handle the multi-model category problem.

4   Experiments

4.1   Datasets

Table 1 lists the characteristics of the datasets used in the experimental study. MR and SST are used for sentence-level classification. IMDB and Ylep13/14 are used for document-level classification. The description of each dataset in detail as below:

- **MR** are Movie Review [14] where each instance is a sentence. The objective is to identify sentiment polarity (positive or negative) of movie reviews.
- **SST1/2** are Stanford Sentiment Treebank [21]. We use SST1 for fine-grained sentiment classification while SST2 for binary.
- **IMDB** are obtained from [24]. The ratings range from 1 to 10.
- **Yelp13/14**. Yelp13 and Yelp14 are review datasets derived from Yelp Dataset Challenge[1] of year 2013 and 2014 respectively. The ratings range from 1 to 5. We obtained it from [24].

---

[1] http://www.yelp.com/dataset_challenge.

Table 1. Statistical information of the data sets. K is the number of target classes. N counts the number of instances. $|W|$ means the average number of words in each instances. Train/Dev/Test is a division of datasets (CV means 10-fold cross-validation).

Data		K	N	$|W|$	Train/Dev/Test
Sentence	**MR**	2	10,662	20	CV
	SST1	5	11,855	18	8,544/1,101/2,210
	SST2	2	9,613	19	6,920/872/1,821
Document	**IMDB**	10	84,919	395	67,426/8,381/9,112
	Yelp13	5	78,966	189	62,522/7,773/8,671
	Yelp14	5	231,163	197	183,019/22,745/25,399

4.2 Evaluation Metrics

We use Accuracy (Acc) and Root Mean Squared Error (RMSE) as evaluation metrics:

$$Acc = \frac{T}{M} \tag{11}$$

$$RMSE = \sqrt{\frac{\sum_{i=1}^{N}(p_i - g_i)^2}{M}} \tag{12}$$

where we use T to account the prediction correct number, and M to account the whole number of samples in the test. Acc is a standard metric to measure the overall classification result. RMSE is used to figure out the divergences between predicted sentiment classes p_i and ground truth classes g_i.

4.3 Training Settings

Update of Sentiment Centroids. Inspired by k-means, we employ mini-batch based algorithm to update sentiment centroids as:

$$c_j^{new} = c_j^{old} - \lambda_c \Delta c_j \tag{13}$$

$$\Delta c_j = \frac{\sum_{i=1}^{m} \delta(y_i = j)(c_{y_i} - z_j)}{1 + \sum_{i=1}^{m} \delta(y_i = j)} \tag{14}$$

$$\delta(cond) = \begin{cases} 1, & \text{if cond is True;} \\ 0, & \text{if cond is False.} \end{cases} \tag{15}$$

where λ_c is the updating rate of sentiment centroids, Δc_j is a mean-shift direction of mini-batch, m is the size of a mini-batch sample, *cond* refers to condition.

Other Settings. Both sentiment centroids c_k and text embedding z are limited to same dimensions over $\{50, 100, 200, 300\}$. For all the experiments, we trained the models for 10 epochs. We tuned on two optimization schemes: AdaGrad [29] and Adam [9] with learning rage over $\{0.001, 0.005, 0.01, 0.05\}$. Batch-size is tuned over $\{25, 50, 100\}$.

For sentence-level sentiment classification, we follow Kim [8] to preprocess the datasets (MR, SST1, SST2) and initialized the word embeddings with GloVe vectors [16].

For document-level sentiment classification, we use Stanford's CoreNLP [12] to split documents into sentences and tokenize words of sentences. We utilize word2vec [13] to obtain word embeddings and limit the dimensions of word embeddings to 300. We remove the words if their occurrence less than 5 times both in the training and validation datasets. For unknown words in the test datasets, we remove them.

4.4 Sentence-Level Classification

The results on MR as well as SST1 and SST2 are shown in Table 2. We report the performance of our three models: SCAVG, SCGRU, SCHAN, and compare them to three baselines: AVG, GRU, HAN.

Table 2. Accuracy of sentiment prediction in Sentence-level datasets. For all of our models, L_s loss is used. The best performances of our models are in bold.

Dataset	AVG	SCAVG	GRU	SCGRU	HAN	SCHAN
MR	0.787	0.801	0.792	**0.810**	0.780	0.804
SST1	0.421	0.429	0.437	**0.469**	0.440	0.454
SST2	0.848	0.861	0.858	**0.876**	0.852	0.865

The results are very clear that our global sentiment centroids significantly improve the accuracy of sentiment classification. The SCGRU model shows strong performance in three datasets, while the SCHAN model has next best performance. For SST2, our SCGRU gets almost 2.1% improvement of accuracy. And it also gets a competitive result with the previous state-of-art score 0.881 [8] while our approach has a simpler structure and easy to reproduce.

4.5 Document-Level Classification

The experiment results are shown in Table 3. Our observation shows that the proposed model works well in predicting review sentiment. Our model achieves best results in RMSE indicators, and competitive results in accuracy rates. The RMSE is a important performance for multi-classification task to measure the degree of predict category and real category. The lower the RMSE is, the closer the predicting sentiment category to the true sentiment category.

Table 3. Document-level sentiment classification results. The best performances are in bold. Results are grouped as follows: (a) baseline methods with our own implementation; (b) other competitive neural network based methods; (c) ours. In order to encode document-level text, hierarchical model HAN is used as a base encoder. SCHAN-t is L_t based model of SCHAN, SCHAN-c is L_c based model of SCHAN.

Model	IMDB		Yelp13		Yelp14	
	Acc	RMSE	Acc	RMSE	Acc	RMSE
AVG	0.300	1.996	0.523	0.900	0.530	0.896
GRU	0.410	1.605	0.581	0.808	0.591	0.804
HAN	0.470	1.440	0.629	0.713	0.636	0.686
RNTN + Recurrent [23]	0.400	1.764	0.574	0.804	0.582	0.821
NSC [2]	0.443	1.465	0.627	0.701	0.637	0.686
NSC+LA [2]	0.487	1.381	0.631	0.706	0.630	0.715
SCHAN-t	0.477	1.363	**0.641**	**0.677**	0.640	0.679
SCHAN-c	**0.491**	**1.327**	0.638	0.683	**0.641**	**0.678**

For IMDB, our SCHAN-c model achieves a great accuracy, and gets a 6.7% lower RMSE compared with NSC+LA model. It indicates that our approach is helpful with the task of sentiment classification. For Yelp13, our SCHA-t model achieves best results with 1.6% improvement of accuracy and 3.2% lower of RMSE. For Yelp14, our SCHAN-c model outperforms the NSC+LA model. It gets 1.74% accuracy improvement and 5.7% RMSE drop.

4.6 The Effect of Sentiment Centroids

The hyper parameter μ_1 means strength of softmax loss and μ_2 controls the constrain strength of sentiment centroids loss. To investigate the sensitiveness of parameters μ_1 and μ_2, we conduct a experiment on SST2.

We fix μ_1 to 1.0 and vary μ_2 from 0.0 to 0.5 to learn different models. The accuracy of these models is shown in Fig. 3(a). It is clear that a carefully choosing of μ_2 can improve the accuracy of sentiment classification. As the same, we fix μ_2 to 0.05 and vary μ_1 from 0.5 to 1.0 to learn different models. The accuracy of these models is shown in Fig. 3(b).

5 Conclusion and Future Work

We propose a general sentiment centeroids based model to predict sentiment of review text and a novel joint lost function to reinforce global category information. Experiments on the sentiment labeled dataset show that the global sentiment centroids help to understand the sentiment features, improve association of inter-class, and improve the accuracy of classification.

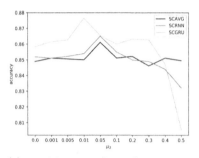 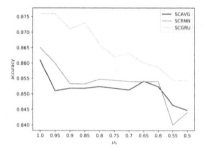

(a) sensitiveness of μ_2 where $\mu_1 = 1.0$ (b) sensitiveness of μ_1 where $\mu_2 = 0.05$

Fig. 3. Effect of sentiment centroids.

However, there is still much work to do: (a) apply the centroids mechanism to the general task of text classification; (b) explore other more powerful neural networks to encoding text.

References

1. Bahdanau, D., Cho, K., Bengio, Y.: Neural machine translation by jointly learning to align and translate (2014). arXiv preprint: arXiv:1409.0473
2. Chen, H., Sun, M., Tu, C., Lin, Y., Liu, Z.: Neural sentiment classification with user and product attention. In: Proceedings of EMNLP (2016)
3. Dong, L., Wei, F., Zhou, M., Xu, K.: Adaptive multi-compositionality for recursive neural models with applications to sentiment analysis. In: AAAI, pp. 1537–1543 (2014)
4. Guan, Z., Chen, L., Zhao, W., Zheng, Y., Tan, S., Cai, D.: Weakly-supervised deep learning for customer review sentiment classification
5. Joachims, T.: A probabilistic analysis of the rocchio algorithm with tfidf for text categorization. Technical report, DTIC Document (1996)
6. Johnson, R., Zhang, T.: Effective use of word order for text categorization with convolutional neural networks (2014). arXiv preprint: arXiv:1412.1058
7. Kalchbrenner, N., Grefenstette, E., Blunsom, P.: A convolutional neural network for modelling sentences (2014). arXiv preprint: arXiv:1404.2188
8. Kim, Y.: Convolutional neural networks for sentence classification (2014). arXiv preprint: arXiv:1408.5882
9. Kingma, D., Ba, J.: Adam: A method for stochastic optimization (2014). arXiv preprint: arXiv:1412.6980
10. LeCun, Y., Bengio, Y., Hinton, G.: Deep learning. Nature **521**(7553), 436–444 (2015)
11. Li, J.: Feature weight tuning for recursive neural networks (2014). arXiv preprint: arXiv:1412.3714
12. Manning, C.D., Surdeanu, M., Bauer, J., Finkel, J.R., Bethard, S., McClosky, D.: The stanford CoreNLP natural language processing toolkit. In: ACL (System Demonstrations), pp. 55–60 (2014)
13. Mikolov, T., Sutskever, I., Chen, K., Corrado, G.S., Dean, J.: Distributed representations of words and phrases and their compositionality. In: Advances in Neural Information Processing Systems, pp. 3111–3119 (2013)

14. Pang, B., Lee, L.: Seeing stars: exploiting class relationships for sentiment categorization with respect to rating scales. In: Proceedings of the 43rd Annual Meeting on Association for Computational Linguistics, pp. 115–124. Association for Computational Linguistics (2005)
15. Paulus, R., Socher, R., Manning, C.D.: Global belief recursive neural networks. In: Advances in Neural Information Processing Systems, pp. 2888–2896 (2014)
16. Pennington, J., Socher, R., Manning, C.D.: Glove: Global vectors for word representation. In: EMNLP, vol. 14, pp. 1532–1543 (2014)
17. Ren, Y., Zhang, Y., Zhang, M., Ji, D.: Improving twitter sentiment classification using topic-enriched multi-prototype word embeddings. In: AAAI, pp. 3038–3044 (2016)
18. Rocchio, J.J.: Relevance feedback in information retrieval (1971)
19. Schroff, F., Kalenichenko, D., Philbin, J.: Facenet: a unified embedding for face recognition and clustering. In: Proceedings of the IEEE Conference on Computer Vision and Pattern Recognition, pp. 815–823 (2015)
20. Severyn, A., Moschitti, A.: Twitter sentiment analysis with deep convolutional neural networks. In: Proceedings of the 38th International ACM SIGIR Conference on Research and Development in Information Retrieval, pp. 959–962. ACM (2015)
21. Socher, R., Perelygin, A., Wu, J.Y., Chuang, J., Manning, C.D., Ng, A.Y., Potts, C.: Recursive deep models for semantic compositionality over a sentiment treebank. In: Proceedings of the Conference on Empirical Methods in Natural Language Processing (EMNLP), vol. 1631, p. 1642. Citeseer (2013)
22. Tai, K.S., Socher, R., Manning, C.D.: Improved semantic representations from tree-structured long short-term memory networks (2015). arXiv preprint: arXiv:1503.00075
23. Tang, D., Qin, B., Liu, T.: Document modeling with gated recurrent neural network for sentiment classification. In: Proceedings of the 2015 Conference on Empirical Methods in Natural Language Processing, pp. 1422–1432 (2015)
24. Tang, D., Qin, B., Liu, T.: Learning semantic representations of users and products for document level sentiment classification. In: Proceedings of ACL (2015)
25. Tang, D., Wei, F., Qin, B., Yang, N., Liu, T., Zhou, M.: Sentiment embeddings with applications to sentiment analysis. IEEE Trans. Knowl. Data Eng. **28**(2), 496–509 (2016)
26. Tang, D., Wei, F., Yang, N., Zhou, M., Liu, T., Qin, B.: Learning sentiment-specific word embedding for twitter sentiment classification. In: ACL, vol. 1, pp. 1555–1565 (2014)
27. Wang, C., Jiang, F., Yang, H.: A hybrid framework for text modeling with convolutional RNN. In: Proceedings of the 23rd ACM SIGKDD International Conference on Knowledge Discovery and Data Mining, pp. 2061–2069. ACM (2017)
28. Yang, Z., Yang, D., Dyer, C., He, X., Smola, A., Hovy, E.: Hierarchical attention networks for document classification. In: Proceedings of the 2016 Conference of the North American Chapter of the Association for Computational Linguistics: Human Language Technologies (2016)
29. Zeiler, M.D.: Adadelta: an adaptive learning rate method (2012). arXiv preprint: arXiv:1212.5701
30. Zhu, X., Sobhani, P., Guo, H.: Long short-term memory over recursive structures. In: Proceedings of the 32nd International Conference on Machine Learning, pp. 1604–1612 (2015)

Random Pairwise Shapelets Forest

Mohan Shi, Zhihai Wang, Jidong Yuan[(✉)], and Haiyang Liu

School of Computer and Information Technology,
Beijing Jiaotong University, Beijing 100044, China
yuanjd@bjtu.edu.cn

Abstract. Shapelet is a discriminative subsequence of time series. An advanced time series classification method is to integrate shapelet with random forest. However, it shows several limitations. First, random shapelet forest requires a large training cost for split threshold searching. Second, a single shapelet provides limited information for only one branch of the decision tree, resulting in insufficient accuracy and interpretability. Third, randomized ensemble causes interpretability declining. For that, this paper presents Random Pairwise Shapelets Forest (RPSF). RPSF combines a pair of shapelets from different classes to construct random forest. It is more efficient due to omit of threshold search, and more effective due to including of additional information from different classes. Moreover, a discriminability metric, Decomposed Mean Decrease Impurity (DMDI), is proposed to identify influential region for every class. Extensive experiments show that RPSF improves the accuracy and training speed of shapelet forest. Case studies demonstrate the interpretability of our method.

Keywords: Time series classification · Shapelet · Random forest
Interpretability

1 Introduction

Time series is ubiquitous. It is produced everyday and everywhere in real world, such as ECG recordings, financial data, industrial observations, etc. Time series classification is an important subject in the field of data mining. Unlike general classification tasks, it takes attribute order into account. Recent studies have shown that the 1NN with dynamic time warping (DTW) remains among the most competitive classification approaches [4]. However, this method has drawbacks of high classification time complexity and lack of interpretability.

Shapelet is the most discriminant, phase independent subsequence in time series [15]. It is proposed to detect phase-independent localized similarity within the same class. In preliminary works, researchers use various methods to extract shapelet, and to embed it into decision tree directly. The shapelet-based approach has the following characteristics. First, shapelet shows local features, which is its main difference from 1NN. Second, since only the comparison with shapelet is needed, it is not only faster in the classification stage, but also needs less storage

© Springer International Publishing AG, part of Springer Nature 2018
D. Phung et al. (Eds.): PAKDD 2018, LNAI 10937, pp. 68–80, 2018.
https://doi.org/10.1007/978-3-319-93034-3_6

space. Third, shapelet figures out the key points of classification process and provides better interpretability. Nevertheless, this method suffers low accuracy and slow training process though several acceleration strategies are raised [6,11,17].

Random forest can achieve good performance through integrating a series of weak classifiers [1]. Shapelet-based random forest has attracted significant attention and research effort recently. Renard et al proposed a method of randomly extracting shapelet to build decision trees [14]. Karlsson et al introduced random shapelet forest. It selects both training instances and shapelet candidates randomly [10]. Experiment shows that effective result can be achieved when the amount of selected shapelet candidates is less than 1% of the full set, which greatly saves time. In response to the interpretability decline caused by randomization, a contribution metric Mean Decrease Impurity (MDI) is introduced [8]. Random shapelet forest has also been extended to multivariate time series forest, applied successfully to ECG classification [7] and early classification problem [9]. Deng et al proposed employing a combination of entropy gain and distance measure to evaluate the node split in forest [3]. Cetin et al proposed a shapelet discovery technique that allows efficient candidates evaluation in multivariate time series forest [2].

However, some shortcomings can be seen in shapelet forest. First, a single shapelet often cannot provide enough information to distinguish different classes. Second, a time-consuming split threshold searching is needed to evaluate candidate shapelet. Third, randomization and ensemble lead to interpretation declining easily.

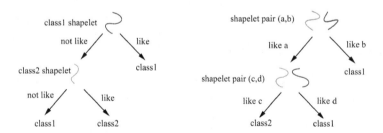

Fig. 1. (*left*) Classic shapelet tree structure; (*right*) Pairwise shapelets tree structure.

For these challenges above, this paper proposes an ensemble algorithm that combines a pair of shapelets in decision tree node, called **R**andom **P**airwise **S**hapelets **F**orest (RPSF). We also describe an effective metric for identifying influential data series regions for specific class. Our method is more accurate, faster and provides better interpretability in comparison to existing shapelet-based forest. Our main contributions are as follows.

– Better accuracy. The RPSF method provides more information by combining a pair of shapelets from different classes, each tree node is split according to subsequence distances between instances to the two shapelets. Figure 1 compares classic shapelet tree structure with the proposed one in a simple binary classification task, where like or not is measured by distance between instance and shapelet. It is easy to observe that our tree extracts discriminant features for both classes. This helps the classifier to sharpen the contrast between the two classes. Besides, since pairwise combination has more possible candidates than a single shapelet, it enhances the diversity of the ensemble model. Extensive experiments show that our approach improves the accuracy of the classifier.

- Faster. RPSF no longer needs to find the split threshold, which saves computing resources. This is especially true when introducing entropy early pruning to speed up [15] (constantly evaluating shapelet under limited information and abandoning apparent inadequate candidates in advance).
- Better interpretability. RPSF provides better interpretability at two levels rather than existing methods. First, each individual decision tree has better interpretability. We extract information for both sides of the binary decision tree. This is beneficial for researchers to understand the profound mechanism of model. Second, the Decomposed Mean Decrease Impurity (DMDI) proposed to illustrate the importance of each attribute provides better interpretability for RPSF. This method evaluates the contribution of each time series attribute for each class, and considers those with higher scores hold better discrimination. Existing MDI provides only a global score, which is determined by its tree structure [8]. On the contrary, the proposed tree structure allows us to decompose the contribution into different classes. Figure 2 shows results of the two methods on ECGFiveDays dataset, both methods point to similar subsequences (red), while the DMDI method also indicates additional discriminant sequences (blue) for the other class. This result is in line with medical conclusion (see [12]). In Sect. 5, we will further discuss the performance of DMDI on other datasets.

The remainder of the paper is organized by the following. In the next section, we present RPSF algorithm in detail. Section 3 explains the DMDI forest

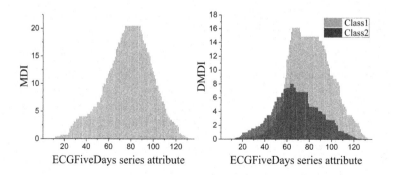

Fig. 2. (*left*) Existing MDI interpretation; (*right*) our DMDI interpretation. Both from ECGFiveDays dataset. The right one indicates discriminative features for both classes.

interpretation method. In Sect. 4, the experimental setup and results from the empirical investigation are described. Case studies are shown in Sect. 5. Finally, we summarize the main contributions in Sect. 6.

2   Random Pairwise Shapelets Forest

RPSF trains multiple pairwise shapelet-based trees to build an ensemble model, we will provide a detailed elaboration to the proposed algorithm in this section.

2.1   Providing More Information by Combination

Shapelet-based decision tree makes judgement according to subsequence distance between instance and shapelet. It can be summarized as follows. If the instance is similar with shapelet (the subsequence distance is less than the split threshold) from a specific class, it is assigned to that class. If not, assigned to another class. However, similarity with features (shapelet) from only one class often cannot accurately characterize the information in the data. Just as each branch of a traditional decision tree indicates the splitting attribute value under the branch, we also hope that each branch of shapelet tree clearly shows what the current branch means. Under this notion, we propose the idea of combining a pair of shapelets from different classes in a decision tree, so that "like and not like" becomes "like A or like B". Decision can be made according to distances to the two shapelets in the pair rather than distance to a single shapelet and a threshold. This improves the accuracy of the classifier due to the additional information. Besides, diversity of the ensemble model can be improved since pairwise combination has more possible candidates than single one. At the same time, this idea also allows researchers to more clearly realize the basis of node division.

Pairwise-based method also has advantage in terms of training time cost. Because of the combination, the split threshold calculation, which is a significant time over-head, is omitted. This is especially true with introduction of entropy early pruning, where the previously calculated distance information can be reused, but the split thresholds, which is avoided by RPSF, must be recalculated each time.

2.2   Proposed Algorithm

RPSF trains a set of trees by taking several parameters: the shapelet length interval l, u, the number of decision trees p, and the number of candidate shapelets pairs in tree node r. For each tree, we use a Bootstrap sample to generate a training set and construct the decision tree on the sampled dataset.

The Algorithm 1 shows the process of training each pairwise shapelets tree. At the beginning, the creation of leaf nodes is determined according to whether the data is pure enough (entropy < 0.1) (lines 1–3). Then subsequences are extracted randomly from two random classes r times to form a candidate set.

Algorithm 1. RandomPairwiseShapeletsTree $(\mathbf{D}, l, u, p, r)$

Input: The training set $\mathbf{D}$, the lower shapelet length l, the upper shapelet length u, the number of shapelet trees p, and the number of inspected shapelets pair r

Output: Pairwise shapelets tree ST, and shapelets pair (s_1, s_2)

1: **if** IsTerminal($\mathbf{D}$) **then**
2:    **return** MakeLeaf($\mathbf{D}$)
3: **end if**
4: $S \leftarrow \varnothing$
5: **for** $i = 0$ to r **do**
6:    $S \leftarrow S \cup$ SampleShapeletsPair($\mathbf{D}, l, u$)
7: **end for**
8: $(s_1, s_2) \leftarrow$ BestShapeletsPair($\mathbf{D}, S$)
9: $(\mathbf{D}_1, \mathbf{D}_2) \leftarrow$ Split($\mathbf{D}, s_1, s_2$)
10: $ST_L \leftarrow$ RandomPairwiseShapeletsTree $(\mathbf{D}_1, l, u, r)$
11: $ST_R \leftarrow$ RandomPairwiseShapeletsTree $(\mathbf{D}_2, l, u, r)$
12: **return** (s_1, s_2, ST_L, ST_R)

To be accurate, two time series are randomly selected from two randomly selected classes, then two lengths and starting points are randomly selected to form a pair of shapelets: s_i and s_j (lines 4–7). After that we assess candidates and find the best pair (line 8, details in next paragraph). The best pair (s_1, s_2) is used to split the dataset $\mathbf{D}$ to two subsets $\mathbf{D}_1$ and $\mathbf{D}_2$. Since there is no split threshold, we calculate the distances between a training instance $D[m]$ and (s_1, s_2) as d_1, d_2. If d_1 is less than or equal to d_2, it means that $D[m]$ is closer to s_1, and $D[m]$ is added to $\mathbf{D}_1$, otherwise $D[m]$ is added to $\mathbf{D}_2$ (line 9). Finally, the algorithm recursively calls itself on $\mathbf{D}_1$ and $\mathbf{D}_2$ to construct subtrees (lines 10–12).

Information gain and split interval are typically used when shapelet assessment is needed [10]. For a shapelets pair (s_1, s_2), we try to split training data by calculating the subsequence distance between training instance and s_1, s_2 separately, and assigning each instance to its closer side (similar with splitting process in previous paragraph). When this process completed, information gain and split interval can then be calculated for this partition as a measure of the quality of the pair. A pair with greater information gain and split interval is considered preferentially. Entropy early pruning is introduced in this process to abandon apparently inadequate candidate [15].

After training, each internal tree node consists of a shapelets pairs (s_1, s_2) and the left, right subtree. The leaf node records the class value. To classify a test instance T, we begin from the root node. If the distance between T and s_1 is less than the distance between T and s_2, the left subtree is recursively used. Otherwise, the right subtree is recursively traversed. The process runs repeatedly until it reaches a leaf node and gets a prediction. We have p trees, the final result is obtained by majority voting.

3   Decomposed Mean Decrease Impurity

The main advantage of shapelet-based decision tree is its interpretability. However, it is eroded by randomized ensemble since the interpretation of each tree is different or even contradictory. To solve this problem, we define an attribute contribution scoring strategy for pairwise shapelets forest. Compared with existing the method in [8], the improvement is that thanks to the tree structure of RPSF, attribute scores for each class could be explored so that discriminative patterns could be discovered.

We decompose all node's information gain according to shapelet's contribution (Eq. 1), then add the contribution to attributes that form the shapelet. For shapelets pair (s_1, s_2), if the dataset attracted by shapelet s_1 cause greater entropy reduction (Eq. 2), then it is assumed that s_1 contributes more. Contribution of different classes shapelet is accumulated respectively. Based on the above idea, we define DMDI. Given pairwise shapelet forest $\mathbf{R} = \{ST_1, ST_2, \ldots, ST_n\}$, where ST is a pairwise shapelet tree. Each tree has multiple nodes, each node corresponds to a shapelets pair (s_1, s_2). Given a training set $\mathbf{D}$ with series length m, for time series attribute k and class c, Decomposed Mean Decrease Impurity DMDI(k, c) is defined as follows.

$$DMDI(k,c) = \sum_n (\sum_{node} {}_{(k \in s_1 \wedge class(s_1)=c)} CV(node, s_1) + \sum_{node} {}_{(k \in s_2 \wedge class(s_2)=c)} CV(node, s_2))$$

where CV is the contribution value of one shapelet. It is obtained from the decomposition of total information gain of the node. Let the dataset that inputted to the node be D_0, the dataset obtained by dividing the D_0 be D_1, D_2. Assume $I_{(s1,s2)}(D_0)$ is the information gain of the tree node, then

$$CV(node, s_i) = \frac{ER(node, s_i)}{ER(node, s_1) + ER(node, s_2)} * I_{(s_1,s_2)}(D_0) \qquad (1)$$

$$ER(node, s_i) = E(D_0) - E(D_i) \qquad (2)$$

where $E(D)$ is the entropy of D, $ER(node, s_i)$ is the entropy reduction caused by a shapelet. We cannot guarantee that ER is positive. It is set to zero in negative case. If the two terms in denominator are both zero, the node is discarded.

For every class, DMDI searches all nodes that embed a shapelet from it, decomposes the information gain of these nodes, and adds the contribution of shapelet to attributes forming it. Eventually we recognize which attributes in the sequence contribute more for a particular class.

4   Experiment and Evaluation

In this part, we experimentally evaluate the performance of proposed RPSF algorithm in terms of accuracy and time consumption.

4.1   Experimental Setup

We select 43 datasets for experiments. All of them are UCR datasets and are widely used in studies. We discard datasets that cannot finish 10 times training in 48 h.

Several algorithms are used for comparison. The nearest neighbor method (1NN) is a widely-used benchmark whose performance can be improved with DTW (DTW1NN). We also include Euclidean distance based nearest neighbor (ED1NN). The ShapeletTree (ST) algorithm is a classic shapelet-based method proposed in [15]. FastShapelet (FS) refers to the decision tree algorithm proposed in [12] where getting the approximate shapelet quickly through SAX. Learning-Shapelets (LTS) is an algorithm proposed in [5] that search shapelet by using optimization approach. gRSF is the state-of-the-art shapelet-based random forest algorithm proposed in [10].

For parameter settings, shapelet length interval of ST, gRSF, and RPSF are set to 25% to 67% of the total length of corresponding time series, which covers a larger range and is a relatively safe value. The number of decision trees for the latter two algorithms is set to 50. The number of candidate pairs in each node is set to 1% of the possible candidate shapelets. For algorithms involving randomization, the results are the average of ten runs.

4.2   Predictive Performance

In this section, we demonstrate that RPSF is competitive in term of classification accuracy compared to state-of-the-art algorithms in literature.

We first compare RPSF to single tree based algorithms, ST, LTS, and FS. Since ST and LTS are time consuming, we finish the experiment on 22 relatively small datasets (with fewer instances and attributes). Figure 3 shows the results of this experiment and the average accuracy of each classifier. Ensemble based and nearest neighbour algorithms are faster, we compared the performance of RPSF, gRSF, and 1NN on 43 datasets. Figure 4 gives the accuracy comparison of those classifiers. The area bellows the diagonal line indicates that RPSF is better.

In Fig. 3, the two methods, LTS and RPSF, show outstanding performances. Although LTS performs better on a large proportion of the datasets (13 of 22),

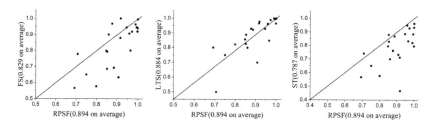

Fig. 3. Accuracy of RPSF compared with FS, LTS and ST

its ad-vantages are not obvious. In contrast, as we can see that three points at the lower triangular area of Fig. 3 (middle) are far away from the line, indicating that RPSF shows overwhelming advantage on these datasets. For example, it is 20.8% higher than the LTS on the OliveOil dataset, and 20.7% higher on the Wine dataset. Therefore, due to the advantage in average accuracy rate, RPSF overtakes LTS to be the best algorithm in this experiment. This experiment suggests that, when processing classification tasks, RPSF should be considered as a prior selection.

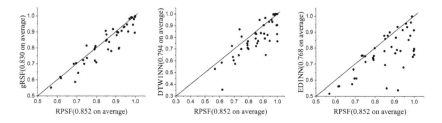

Fig. 4. Accuracy of RPSF compared with gRSF, DTW1NN and ED1NN

Figure 4 gives results of the two random shapelet forest methods as well as 1NN method. RPSF achieves enhancement in two-thirds of the datasets (29 of 43) compared to gRSF, and it is obviously superior to the nearest neighbour method on the vast majority of datasets.

Figure 5 gives the critical differences diagram for the accuracy of individual algorithms ($p = 0.05$). Although LTS outperforms RPSF in term of ranking, there is no significant difference between them, and RPSF holds better average accuracy. RPSF beats other algorithms and is significantly better than FS, ST (tested on 22 datasets), and 1NN (tested on 43 datasets).

Fig. 5. Average rank for methods. Groups of classifiers not significantly different (at $p = 0.05$) are connected. (*left*) 4 methods on 22 small datasets. (*right*) 4 methods on 43 datasets.

4.3 Computational Performance

Another advantage of RPSF is that it omits split threshold searching. This part shows the significant increase in time performance. Note RPSF approach is easy to parallelize, several times of further acceleration can be achieved using parallel computing.

Versus Other Algorithms. We compare RPSF with other time series classi-fication algorithms in terms of training time. Since 1NN is a lazy algorithm, it is not included. Similar with previous settings, ST, FS, and LTS are tested on the smaller datasets, while shapelet-based forests are compared on 43 datasets. Parameter settings are the same as the previous one. Figure 6 shows boxplots of the relative time consumption using RPSF as benchmark to make it more intuitive.

As you can see in Fig. 6, FS is faster than other methods, which is the major ad-vantage of this approach. RPSF method is significantly better than ST and LTS on the vast majority of datasets. It even appears tens of times faster on some datasets. It is also noticeable that the single shapelet-based forest (gRSF) is slower than RPSF on almost all datasets. This result verifies our idea of omitting the calculation of the split threshold for time saving. In the next part, we will discuss it further.

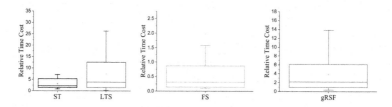

Fig. 6. Relative time consumption of LTS, ST, FS and gRSF compared with RPSF.

Stage Analysis. This part divides RPSF and gRSF into two main stages and analyzes the time consumption of them. We will show that omitting split interval indeed save computation resources.

The decision tree of RPSF algorithm combines a pair of shapelets while gRSF is based on a single shapelet. In terms of time, the main difference is that while assessing candidates, on the one hand RPSF needs to calculate subsequence distances between two shapelets and all training instance, which is twice as much as that of single shapelet based method; on the other hand, in the process of evaluating a single shapelet, a split threshold needs to be found. This is omitted by RPSF. Note that with entropy early pruning [15], threshold searching will be executed multiple times during candidate assessment, while subsequence distance will only be calculated once since distance information can be reused. This expands the advantage of RPSF.

The process of evaluating shapelets is divided into two main parts: calculat-ing subsequence distance and information gain (used in threshold searching and candidate assessment). Table 1 selects several datasets that behave differently in terms of time performance and compares their relative time cost in the two parts.

Table 1. Relative time cost of RPSF and gRSF in different stages

Dataset	RPSF			gRSF		
	Distance	Information gain	Total	Distance	Information gain	Total
Coffee	0.86	0.03	1.00	0.43	0.19	0.70
CBF	0.74	0.14	1.00	0.36	0.98	1.48
ECGFiveDays	0.67	0.10	1.00	0.62	1.18	2.15
FaceFour	0.91	0.03	1.00	0.44	0.14	0.62
ArrowHead	0.91	0.04	1.00	0.57	0.93	1.60
GunPoint	0.86	0.07	1.00	0.42	1.41	1.87
SyntheticControl	0.21	0.44	1.00	0.13	8.61	8.76

As the relative time cost shown in Table 1, the four bolded datasets particularly validate our hypothesis. RPSF's distance computation consumes approximately twice as much as that of the gRSF, while the time spent for calculating information gain is apparently smaller. It is specifically noticeable on SyntheticControl dataset. For dataset ECGFiveDays, the time spent to compute subsequence distances does not conform to the hypothesis, which may be caused by entropy early pruning. It also indicates that in some cases the difference between the two subsequence distance calculations is trivial. In addition, it is clear that the time consumption of information gain on RPSF is obviously less than gRSF, and there is no doubt that RPSF is generally more efficient than gRSF, especially on relatively larger datasets.

5 Case Studies

As discussed in Sect. 3, the pairwise shapelets and DMDI enhance interpretability by providing explanation for each possible class. We briefly show the profit of DMDI on ECGFiveDays dataset in previous. More detailed real-world examples will be included to demonstrate the usefulness of our DMDI and RPSF model in this section.

5.1 GunPoint

GunPoint is a dataset that has been studied extensively in literature. The 150-length dataset describes the action curve of an actor with or without a gun when doing an action (as shown in Fig. 7). The key discriminant pattern for this dataset is around 100–120 time stamps. Since inertia carries actors hand a little too far and she is forced to correct it in Point case [15], in this location Point instances bear a slight dip, while Gun instances mostly do not. Another discriminant pattern is near 40–60 time stamps. Some of Point instances are relatively flat near this range, while all Gun instances are in rising state. This

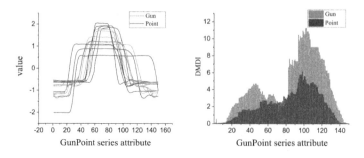

Fig. 7. DMDI interpretation for GunPoint dataset

is because the actor can be more agile when she does not hold a gun [13]. The accuracy of RPSF on this dataset is 99.9%.

The DMDI metric is applied to the dataset, and the results are shown in Fig. 7. It is clear that the discriminant scores of the two classes arrive at their highest value near 100–120 time stamps, showing that our method succeeds in identifying the distinguishing features. This is consistent with the conclusion of [8,15,16]. In addition, the score of Gun reaches a local peak among the 40–60 indexes, it also matches the range of another discriminant pattern.

5.2   ArrowHead

Arrows classification is an important topic in archaeology. ArrowHead is a multi-class dataset with 251 attributes. Arrows can be divided into three classes according to their place, age, and the race belonged: Avonlea, Clovis, and Mix. Through starting from arrow tip, moving round and recording distance to the centroid, the time series describes the outline of the arrows. Figure 8 briefly depicts the series on the left part. The differences among these three arrows are that, the Clovis arrow has an unnotched hafting area near the bottom connected by a deep concave bottom end (near 125 time stamp), while Avonlea and Mix differ in a

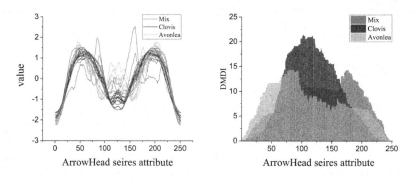

Fig. 8. DMDI interpretation for ArrowHead dataset

small notched hafting area connected by a shallow concave bottom end (near 90, 160 time stamp). These conclusions are confirmed by UCR anthropologists [15].

Our DMDI metric for this dataset is shown on the right side of Fig. 8. As described in the figure, Clovis peaks near 125 time stamp, and the other two peaks at 90 and 160 separately. It is obvious that shapelets pairwise forest and DMDI accurately capture the key points of real problem. In addition, for the multi-class classification problem, existing MDI can only output a global solution, it cannot provide a subtle explanation for every class. This example fully illustrates the superiority of DMDI.

6 Conclusion

In this paper, we present an effective and efficient random forest combining shapelets from different classes randomly. The model diversity, interpretability and classification accuracy are improved by including more information in a node. Due to the fact that pairwise shapelets do not have to search the split threshold, the time consumption is optimized. In addition, a novel forest interpretation method DMDI is proposed to evaluate the contribution of each attribute and to explain the proposed model. Extensive experiments and case studies show that our method outperforms state-of-the-art random shapelet forest.

Acknowledgment. This work is supported by National Natural Science Foundation of China (No. 61672086, 61702030, 61771058), Beijing Natural Science Foundation (No. 4182052), China Postdoctoral Science Foundation (2018M631328) and Fundamental Research Funds for the Central Universities (2018JBM014).

References

1. Breiman, L.: Random forests. Mach. Learn. **45**(1), 5–32 (2001)
2. Cetin, M.S., Mueen, A., Calhoun, V.D.: Shapelet ensemble for multi-dimensional time series. In: Proceedings of the 2015 SIAM International Conference on Data Mining, pp. 307–315. SIAM (2015)
3. Deng, H., Runger, G., Tuv, E., Vladimir, M.: A time series forest for classification and feature extraction. Inf. Sci. **239**, 142–153 (2013)
4. Ding, H., Trajcevski, G., Scheuermann, P., Wang, X., Keogh, E.: Querying and mining of time series data: experimental comparison of representations and distance measures. Proc. VLDB Endow. **1**(2), 1542–1552 (2008)
5. Grabocka, J., Schilling, N., Wistuba, M., Schmidt-Thieme, L.: Learning time-series shapelets. In: Proceedings of the 20th ACM SIGKDD International Conference on Knowledge Discovery and Data Mining, pp. 392–401. ACM (2014)
6. Hills, J., Lines, J., Baranauskas, E., Mapp, J., Bagnall, A.: Classification of time series by shapelet transformation. Data Min. Knowl. Disc. **28**(4), 851–881 (2014)
7. Karlsson, I., Papapetrou, P., Asker, L.: Multi-channel ECG classification using forests of randomized shapelet trees. In: Proceedings of the 8th ACM International Conference on PErvasive Technologies Related to Assistive Environments, pp. 43. ACM (2015)

8. Karlsson, I., Papapetrou, P., Boström, H.: Forests of randomized shapelet trees. In: Gammerman, A., Vovk, V., Papadopoulos, H. (eds.) SLDS 2015. LNCS (LNAI), vol. 9047, pp. 126–136. Springer, Cham (2015). https://doi.org/10.1007/978-3-319-17091-6_8

9. Karlsson, I., Papapetrou, P., Boström, H.: Early random shapelet forest. In: Calders, T., Ceci, M., Malerba, D. (eds.) DS 2016. LNCS (LNAI), vol. 9956, pp. 261–276. Springer, Cham (2016). https://doi.org/10.1007/978-3-319-46307-0_17

10. Karlsson, I., Papapetrou, P., Boström, H.: Generalized random shapelet forests. Data Min. Knowl. Disc. **30**(5), 1053–1085 (2016)

11. Mueen, A., Keogh, E., Young, N.: Logical-shapelets: an expressive primitive for time series classification. In: Proceedings of the 17th ACM SIGKDD International Conference on Knowledge Discovery and Data Mining, pp. 1154–1162. ACM (2011)

12. Rakthanmanon, T., Keogh, E.: Fast shapelets: a scalable algorithm for discovering time series shapelets. In: Proceedings of the 2013 SIAM International Conference on Data Mining, pp. 668–676. SIAM (2013)

13. Ratanamahatana, C.A., Keogh, E.: Making time-series classification more accurate using learned constraints. In: Proceedings of the 2004 SIAM International Conference on Data Mining, pp. 11–22. SIAM (2004)

14. Renard, X., Rifqi, M., Erray, W., Detyniecki, M.: Random-shapelet: an algorithm for fast shapelet discovery. In: IEEE International Conference on Data Science and Advanced Analytics (DSAA), 36678, pp. 1–10. IEEE (2015)

15. Ye, L., Keogh, E.: Time series shapelets: a new primitive for data mining. In: Proceedings of the 15th ACM SIGKDD International Conference on Knowledge Discovery and Data Mining, pp. 947–956. ACM (2009)

16. Yuan, J., Wang, Z., Han, M., Sun, Y.: A lazy associative classifier for time series. Intell. Data Anal. **19**(5), 983–1002 (2015)

17. Yuan, J.-D., Wang, Z.-H., Han, M.: A discriminative shapelets transformation for time series classification. Int. J. Patt. Recognit. Artif. Intell. **28**(06), 1450014 (2014)

A Locally Adaptive Multi-Label
k-Nearest Neighbor Algorithm

Dengbao Wang[1], Jingyuan Wang[1], Fei Hu[1], Li Li[1(✉)], and Xiuzhen Zhang[2]

[1] College of Computer and Information Science, Southwest University,
Chongqing, China
lily@swu.edu.cn
[2] School of Computer Science and Information Technology, RMIT University,
Melbourne, Australia

Abstract. In the field of multi-label learning, ML-kNN is the first lazy learning approach and one of the most influential approaches. The main idea of it is to adapt k-NN method to deal with multi-label data, where maximum a posteriori rule is utilized to adaptively adjust decision boundary for each unseen instance. In ML-kNN, all test instances which get the same number of votes among k nearest neighbors have the same probability to be assigned a label, which may cause improper decision since it ignores the local difference of samples. Actually, in real world data sets, the instances with (or without) label l from different locations may have different numbers of neighbors with the label l. In this paper, we propose a locally adaptive Multi-Label k-Nearest Neighbor method to address this problem, which takes the local difference of samples into account. We show how a simple modification to the posterior probability expression, previously used in ML-kNN algorithm, allows us to take the local difference into account. Experimental results on benchmark data sets demonstrate that our approach has superior classification performance with respect to other kNN-based algorithms.

1 Introduction

1.1 Background

Multi-Label classification has received considerable attention over the past several years. In multi-label classification, each instance in the dataset is associated with a set of labels, and the task of multi-label classification problem is to output a label set whose size is unknown for each test instances. Multi-label problems are ubiquitous in the real world, for example, in image categorization, each image can be associated with multiple labels, such as *sea, desert* and *mountain* [1]; in text categorization, each text may belong to a set of topics, such as *economics, poetry* and *health* [2]; in bioinformatics, a gene may be related to multiple functions, such as *metabolism* and *protein synthesis* [3].

Formally, let $\mathcal{X} = \mathcal{R}^d$ denote the d-dimensional feature space and $\mathcal{Y} = \{0,1\}^L$ be the label space with L possible labels, then the goal of multi-label classifier

© Springer International Publishing AG, part of Springer Nature 2018
D. Phung et al. (Eds.): PAKDD 2018, LNAI 10937, pp. 81–93, 2018.
https://doi.org/10.1007/978-3-319-93034-3_7

is to learn a function $f: \mathcal{X} \mapsto \mathcal{Y}$. Given a multi-label dataset $\mathcal{D}$, we can divide it into feature space $\mathcal{X}$ and label space $\mathcal{Y}$. An instance x_i is associated with a subset of labels $Y_i \subseteq \mathcal{Y}$ (finite set of labels), and a multi-label dataset is composed of m examples $(x_1, Y_1), (x_2, Y_2), \ldots, (x_n, Y_n)$ [4].

Given a multi-label learning task, it can be transformed into other well-established learning tasks. This category of approaches is formally defined as *Problem Transformation* method. In this way, we can decompose a multi-label problem into multiple single-label problems, and each single-label problems can be tackled by a binary classifier. Thus, the multi-label classification function can be represented in another form $f = \{f_1, f_2, \ldots, f_L\}$ in this way. Problem Transformation is widely used in multi-label learning problems for its greater flexibility [8,9,11]. Another way to tackle multi-label classification problems is so called *Algorithm Adaptation* method [5]. This category of approaches tackles multi-label learning problem by adapting existing popular learning approaches such as AdaBoost, Neural Networks or kNN to deal with the multi-label problems directly [2,12,13].

According to the idea of Algorithm Adaptation, Zhang and Zhou [6] proposed Multi-Label k-Nearest Neighbor (ML-kNN). It is the first lazy learning approach and one of the most influential multi-label classification approaches. The basic idea of this approach is to adapt the classic kNN algorithm to deal with multi-label classification problems, where maximum a posteriori (MAP) rule is utilized to adaptively adjust decision boundary for each new instance. In this method, the test instances which get the same number of votes among k nearest neighbors have the same probability to be assigned a label. It may cause improper decision since it ignores the local difference of samples. Actually, in real world data sets, the instances with (or without) label l from different locations may have different numbers of neighbors with the label l. Thus, in this paper, we propose a locally adaptive Multi-Label k-Nearest Neighbor method to address this problem.

1.2   Motivation

We begin by conducting a simple experiment to try to show the local difference of samples. The local difference here means the instances with (or without) the l-th label from different locations may have different numbers of neighbors with the l-th label.

For a dataset, we first find the k nearest neighbors of each instance x and denote as $\mathcal{N}(x)$. Then we can count the number of neighbors of x with label l. The counting vector can be defined as:

$$C_x(l) = \sum_{(x^*, Y^*) \in \mathcal{N}(x)} Y^*(l), l \in \mathcal{Y} \tag{1}$$

After calculating above statistics, we can figure out if the distribution of $C_x(l)$ is related to the location information. In our experiments, we separate the dataset into five clusters, and use the cluster index to represent location information. For each cluster S_j and each label l, we calculate: (1) the average

$C_x(l)$ of the instances with label l (defined as Eq. (2)); (2) the average $C_x(l)$ of the instances without label l (defined as Eq. (3)).

$$\overline{C}(S_j, l) = \frac{1}{|S_j^{l1}|} \sum_{(x,Y) \in S_j^{l1}} C_x(l) \tag{2}$$

$$\overline{C^*}(S_j, l) = \frac{1}{|S_j^{l0}|} \sum_{(x,Y) \in S_j^{l0}} C_x(l) \tag{3}$$

where $S_j^{l1} = \{(x,Y)|(x,Y) \in S_j, Y(l) = 1\}$ and $S_j^{l0} = \{(x,Y)|(x,Y) \in S_j, Y(l) = 0\}$.

We conduct the experiment on an image data set *scene*, which has 2407 instances and 6 labels. The results are shown in Table 1. The first part and the second part respectively show $\overline{C}(S_j, l)$ and $\overline{C^*}(S_j, l)$ of each cluster S_j and each label l. As is shown in Table 1, for a same label l, the $\overline{C}(S_j, l)$ and $\overline{C^*}(S_j, l)$ of different clusters may vary tremendously.

Table 1. Each cell of the table means the average $C_x(l)$ of the instances with (or without) label l in each cluster.

	Label	beach	sunset	fall	field	mountain	urban
$\overline{C}$	Cluster 1	1.432	0.926	1.470	2.000	1.785	5.532
	Cluster 2	1.600	1.500	2.047	6.188	2.256	1.000
	Cluster 3	1.250	5.265	6.055	1.571	0.500	1.090
	Cluster 4	1.044	1.607	1.214	1.936	4.761	2.000
	Cluster 5	4.863	1.333	1.000	2.333	1.444	1.956
$\overline{C^*}$	Cluster 1	0.401	0.029	0.083	0.176	1.073	4.503
	Cluster 2	0.221	0.005	0.219	4.462	1.116	0.167
	Cluster 3	0.098	0.248	1.696	0.201	0.134	0.198
	Cluster 4	0.346	0.067	0.123	0.556	3.759	1.369
	Cluster 5	3.543	0.006	0.195	0.430	0.891	1.084

The above results hint that the distribution of $C_x(l)$ is significantly related to the location information. In ML-kNN, however, the local difference of samples is ignored, which may cause the improper decision. To take the local difference into account, we propose a locally adaptive Multi-Label k-Nearest Neighbor method in this paper. In our approach, the test instances which get the same number of votes among k nearest neighbors may have different probabilities to be assigned a label if they come from different regions. Experimental results on benchmark data sets demonstrate that our approach has superior classification performance with respect to previous ML-kNN algorithm, especially on large scale data sets[1].

---

[1] The code available at https://github.com/DENGBAODAGE/LAMLKNN.

1.3   Paper Organization

The rest of this paper is organized as follows. The related work is discussed in Sect. 2. The details of our approach are proposed in Sect. 3. After that, the experiment results are reported in Sect. 4. Finally, the conclusion is summarized in Sect. 5.

2   Related Work

The k-nearest neighbors (kNN) rule [7] is one of the oldest and simplest methods for pattern classification. For traditional single-label classification problems, the kNN rule usually classifies each unlabeled instance by the majority label among its k nearest neighbors in the training data. The kNN-based methods often yield competitive results and have been widely used in practical applications mainly due to its implementation simplicity. However, for multi-label classification, the traditional kNN rule is inappropriate mainly due to the severe class-imbalance issue.

ML-kNN was proposed based on the traditional kNN algorithm to deal with multi-label classification problems. Rather than classifying new instance by the majority label among its k nearest neighbors, ML-kNN employs maximum a posteriori (MAP) principle to predict the set of labels of the new instance.

$$
\begin{aligned}
Y_t(l) &= \arg\max_{b \in \{0,1\}} P(H_b^l | E_{C_t(l)}^l) \\
&= \arg\max_{b \in \{0,1\}} P(H_b^l) P(E_{C_t(l)}^l | H_b^l)
\end{aligned}
\tag{4}
$$

where $Y_t(l)$ is the label vector for the new instance t. $C_t(l)$ is the same as described previously. H_1^l represents the event that t has label l, while H_0^l represents the event that t doesn't have label l. $E_{C_t(l)}^l$ denotes the event that, among the k nearest neighbors of t, there are exactly $C_t(l)$ instances which have label l. The prior probability $P(H_b^l)$ and the conditional probability $P(E_{C_t(l)}^l | H_b^l)$ in Eq. (4) can all be estimated from the training dataset in advance.

The reported experiment results show that ML-kNN performed well on several real world data sets. However, it ignores the local difference when using utilizing maximum a posteriori rule, and we think the location information of the new instance is helpful especially for large scale data sets.

There are also some other kNN based approaches to handle multi-label classification problems. Note that ML-kNN is a first-order approach which reasons the relevance of each label separately. Considering that this method is ignorant of exploiting label correlations, a dependent multi-label classification method derived from ML-kNN is proposed in [14], which takes into account the dependencies between labels. In order to exploit the non-parametric property of classical kNN method, Wang et al. [15] further developed classical KNN method, and proposed a Class Balanced K-Nearest Neighbor (BKNN) approach for multi-label classification. This method picks up the most representative training data points from every class with equal number, such that the label of a test data point is determined via the information from all the classes in a balanced manner. In [18], a kNN based ranking approach is proposed to solve the multi-label classification problem. This approach exploits a ranking model to learn which neighbor's labels are more trustable candidates for a weighted KNN-based strategy, and then assigns higher weights to those candidates when making weighted-voting decisions.

3   Methodology

As described in previous sections, we try to take the local difference into account by modifying the posterior probability expression used in ML-kNN algorithm. How to exploit the location information when using MAP principle to assign labels to a new instance? In this section, we introduce a Locally Adaptive Multi-Label k-Nearest Neighbor algorithm to address this problem.

Inspired by the results presented in Sect. 1.2, we firstly separate the training data into m groups $S_1, S_2, \ldots, S_m$ via clustering, where the average $\boldsymbol{C}_x(l)$ of instances in the different clusters may vary tremendously. For each test instance t, we can identify which group should it be assigned to by measuring the distance between the test instance and each cluster center.

$$w_t = \underset{1 \leq j \leq m}{\arg\min} \|x_t - c_j\|^2 \tag{5}$$

where w_t is the index of cluster to which should the test instance t assign. c_j stands for the center point of cluster S_j.

Therefore we can get two important information of the test instance t: $\boldsymbol{C}_t$ (records the numbers of x's neighbors with each label) and w_t (stands for the index of cluster to which should the test instance t assign). Then based on the membership counting vector $\boldsymbol{C}_t$ and the location information w_t, the category vector Y_t can be determined using the following maximum a posteriori principle:

$$Y_t(l) = \underset{b \in \{0,1\}}{\arg\max} P(H_b^l | E_{\boldsymbol{C}_t(l)}^l, W_{w_t}) \tag{6}$$

where H_b^l and $E_{\boldsymbol{C}_t(l)}^l$ is the same as described in Sect. 2. W_{w_t} denotes the event that the test instance t can be assigned to the cluster S_{w_t}. Based on Bayes theorem, we have:

$$Y_t(l) = \underset{b \in \{0,1\}}{\arg\max} P(H_b^l) P(E_{\boldsymbol{C}_t(l)}^l, W_{w_t} | H_b^l) \tag{7}$$

The prior probability $P(H_b^l)$ and the likelihood $P(E_{\boldsymbol{C}_t(l)}^l, W_{w_t} | H_b^l)$ can be estimated from the training data.

Equation (6) can also be rewritten by another way (based on Bayes theorem):

$$Y_t(l) = \underset{b \in \{0,1\}}{\arg\max} \, P(H_b^l, E_{\boldsymbol{C}_t(l)}^l, W_{w_t})$$

$$= \underset{b \in \{0,1\}}{\arg\max} \, P(H_b^l, W_{w_t}) P(E_{\boldsymbol{C}_t(l)}^l | H_b^l, W_{w_t}) \qquad (8)$$

$$= \underset{b \in \{0,1\}}{\arg\max} \, P(W_{w_t}) P(H_b^l | W_{w_t}) P(E_{\boldsymbol{C}_t(l)}^l | H_b^l, W_{w_t})$$

where $P(W_{w_t})$ represents the prior probability that W_{w_t} holds. $P(H_b^l|W_{w_t})$ represents the conditional probability that H_b^l holds when W_{w_t} holds. Furthermore, the conditional probability $P(E_{\boldsymbol{C}_t(l)}^l|H_b^l, W_{w_t})$ represents the likelihood that the instance x has $\boldsymbol{C}_t(l)$ neighbors with label l when H_b^l and W_{w_t} both hold.

By comparing Eqs. (8) and (4) (in Sect. 2), it is intuitive to understand how we exploit the location information by involving W_{w_t} in posterior probability expression. In our method, the category vector Y_t of the new instance t depends on the membership counting vector $\boldsymbol{C}_t$ as well as the location information w_t. Unlike ML-kNN, our approach can derive different probabilities of assigning a label to new instances which get the same number of votes among k nearest neighbors but come from different regions. Actually, ML-kNN can be regarded as a special case of our approach with $m = 1$. Note that Eqs. (7) and (8) we described above are actually equivalent. We choose the latter version in our implementation.

All the three terms in Eq. (8) can be estimated from the training data. Firstly, the prior probability $P(W_{w_t})$ is estimated by calculating the proportion of the cluster S_{w_t} in training data:

$$P(W_{w_t}) = \frac{|S_{w_t}|}{|S_{train}|} \qquad (9)$$

where $|S_{w_t}|$ and $|S_{train}|$ is the size of cluster w_t and training dataset.

Then the conditional probability $P(H_b^l|W_{w_t})$ are estimated by counting the number of training examples associated with each label in each cluster:

$$P(H_1^l|W_{w_t}) = \frac{s + \sum_{(x,Y) \in S_{w_t}} Y(l)}{2 \times s + |S_{w_t}|} \qquad (l \in \mathcal{Y})$$

$$P(H_0^l|W_{w_t}) = 1 - P(H_1^l|W_{w_t}) \qquad (l \in \mathcal{Y}) \qquad (10)$$

where s is the smoothing parameter controlling the effect of uniform prior on the estimation [6].

Finally, the estimation process for likelihoods $P(E_{\boldsymbol{C}_t(l)}^l|H_b^l, W_{w_t})$ is involved. For each label l, we calculate:

$$\mathcal{K}_l(r) = \sum_{(x,Y) \in \boldsymbol{S}_{w_t}} Y(l) \cdot [\![\boldsymbol{C}_x(l) = r]\!] \qquad (l \in \mathcal{Y}, \, 0 \le r \le k)$$

$$\mathcal{K}_l'(r) = \sum_{(x,Y) \in \boldsymbol{S}_{w_t}} (1 - Y(l)) \cdot [\![\boldsymbol{C}_x(l) = r]\!] \quad (l \in \mathcal{Y}, \, 0 \le r \le k) \qquad (11)$$

$\mathcal{K}_l(C)$ counts the number of training examples which have label l and have exactly C neighbors with label l, while $\mathcal{K}'_l(C)$ counts the number of training examples which don't have label l and have exactly C neighbors with label l. For any $\cdot$, $[\![\cdot]\!]$ equals 1 if $\cdot$ holds and 0 otherwise. After calculate $\mathcal{K}_l(C)$ and $\mathcal{K}'_l(C)$, we can estimate the likelihood in Eq. (8):

$$P(E^l_{C_t(l)}|H^l_1, W_{w_t}) = \frac{s + \mathcal{K}_l(C_t(l))}{s \times (k+1) + \sum_{r=0}^{k} \mathcal{K}_l(r)}$$
$$P(E^l_{C_t(l)}|H^l_0, W_{w_t}) = \frac{s + \mathcal{K}'_l(C_t(l))}{s \times (k+1) + \sum_{r=0}^{k} \mathcal{K}'_l(r)} \tag{12}$$

The following pseudo-code illustrates the complete description of our method. In training phase, we estimate the prior probability $P(W_j)$, the conditional probabilities $P(H^l_1|W_j)$, $P(H^0_1|W_j)$, the statistics $\mathcal{K}_l(r)$, and $\mathcal{K}'_l(r)$ (steps from 5 to 13). In classifying phase, the predicted label set of test instance t can be determined using the maximum a posteriori principle (by substituting Eqs. (9), (10) and (12) into (8)).

Train(S_{train}, k, m)

1 Divide training data into m clusters $\{S_1, S_2 \ldots, S_m\}$ with k-means

2 **for** $i = 1$ to $|S_{train}|$ **do:**

3   Identify k nearest neighbors $\mathcal{N}(x_i)$ for x_i

4 **end**

5 **for** $j = 1$ to m **do:**

6      $P(W_j) = \dfrac{|S_j|}{|S_{train}|}$

7   **for** $l = 1$ to L **do:**

8          $P(H^l_1|W_j) = \dfrac{s + \sum_{(x,Y) \in S_j} Y(l)}{2 \times s + |S_j|}$

9          $P(H^l_0|W_j) = 1 - P(H^l_1|W_j)$

10         $\mathcal{K}_l(r) = \displaystyle\sum_{(x,Y) \in S_j} Y(l) \cdot [\![C_x(l) = r]\!] \qquad (0 \le r \le k)$

11         $\mathcal{K}'_l(r) = \displaystyle\sum_{(x,Y) \in S_j} (1 - Y(l)) \cdot [\![C_x(l) = r]\!] \qquad (0 \le r \le k)$

12   **end**

13 **end**

Classify(t, k)

---

1   Identify S_{w_t} (the cluster should t be assigned to) using Equation (5)

2   Identify k nearest neighbors $\mathcal{N}(t)$ for t

3   **for** $l = 1$ to L **do**:

4   Calculate $C_t(l)$ according to Equation (1)

5   Estimate $P(E^l_{C_t(l)}|H^l_1, W_{w_t})$ and $P(E^l_{C_t(l)}|H^l_0, W_{w_t})$ according to (12)

6        $Y_t(l) = \arg\max_{b \in \{0,1\}} P(H^l_b|E^l_{C_t(l)}, W_{w_t})$

            $= \arg\max_{b \in \{0,1\}} P(W_{w_t})P(H^l_b|W_{w_t})P(E^l_{C_t(l)}|H^l_b, W_{w_t})$

7   **end**

---

4   Experiment

We compare the our proposed method with other multi-label lazy learning algorithms on several data sets. In the following sections, we first describe the experiment setup including the data sets, the evaluation metrics, and the compared algorithms; Then we discuss the experiment results.

4.1   Experiment Setup

Data Sets: We evaluated the algorithm presented in the previous section on twelve data sets[2] of varying size and difficulty. The statistics of the data sets are shown in Table 2. Six regular-scale data sets (first part) as well as six large-scale data sets (second part) are included (the data sets are roughly ordered by the number of instances). There are two additional properties [10] to measure the density of labels:

- The cardinality of a dataset S is the mean of the number of labels of the instances that belong to S, defined as:

$$cardinality(S) = \frac{1}{n}\sum_{i=1}^{n}|Y_i| \tag{13}$$

- The density of S is the mean of the number of labels of the instances that belong to S divided by L, defined as:

$$density(S) = \frac{1}{n}\sum_{i=1}^{n}\frac{|Y_i|}{L} \tag{14}$$

Metrics: In multi-label learning, the evaluation is more complicated than that in single-label learning. Various evaluation metrics have been proposed to measure

---

[2] Data sets were downloaded from http://mulan.sourceforge.net/datasets.html and http://meka.sourceforge.net/#datasets.

Table 2. Multi-label data sets used in experiments.

name	domain	instances	dimension	labels	cardinality	density
Emotions	Music	593	72	6	1.869	0.311
Birds	Audio	645	260	19	1.014	0.053
Enron	Text	1702	1001	53	3.378	0.064
Scene	Image	2407	294	6	1.074	0.179
Yeast	Biology	2417	103	14	4.237	0.3003
Slashdot	Text	3782	1079	22	1.181	0.054
bibtex	Text	7395	1836	159	2.402	0.015
corel5k	Image	5000	499	374	3.522	0.009
corel16k (1)	Image	13766	500	153	2.859	0.019
corel16k (2)	Image	13761	500	164	2.882	0.018
corel16k (3)	Image	13760	500	154	2.829	0.018
Ohsumed	Text	13929	1002	23	1.663	0.072

the performance of multi-label classifier. We use five commonly used metrics: *hamming loss*, *ranking loss*, *coverage*, *one error* and *average precision* [17]. These above five metrics evaluate the performance of a multi-label classifier from different horizon. Note that for average precision, the larger the values the better the performance, while for other four metrics, the smaller the values the better the performance.

Compared Algorithms: We compare the performance of our proposed method with that of three other kNN-based multi-label approaches: BRkNN, ML-kNN and DML-kNN. BRkNN [16] is an adaptation of the kNN algorithm that is conceptually equivalent to using BR method in conjunction with the traditional kNN algorithm. As we discussed in Sect. 2, DML-kNN is an extension approach based on ML-kNN, which takes into account the dependencies between labels.

4.2 Results

Following the experiment setup described above, we conduct the comparison experiments. The experimental results of each algorithm on each data set are respectively reported in Tables 3 and 4. For each algorithm, the k value is determined by cross-validation. We can see that our proposed method LAML-kNN outperform the compared methods in most cases. Furthermore, the advantages of our approach are more obvious on the large-scale data sets (in Table 4) than that on the regular-scale data sets (in Table 3).

The experimental results on benchmark data sets and diverse evaluation metrics validate the superior effectiveness of our approach to existing kNN-based multi-label approaches. Meanwhile, the experimental results demonstrate the number of clusters does not significantly affect the classifier's performance on

Table 3. Experimental results of each algorithm on regular-scale data sets.

Metrics	Algorithms	Emotions	Birds	Enron	Scene	Yeast	Slashdot
Hamming loss	LAMLkNN	0.197	0.045	**0.050**	0.097	**0.198**	**0.050**
	MLkNN	0.191	**0.044**	0.051	**0.096**	0.198	0.053
	BRkNN	0.193	0.045	0.058	0.105	0.203	0.090
	DMLkNN	**0.187**	0.045	0.051	0.097	0.198	0.051
Ranking loss	LAMLkNN	0.151	**0.093**	**0.088**	0.090	**0.170**	**0.157**
	MLkNN	**0.145**	0.102	0.093	0.096	0.171	0.168
	BRkNN	0.151	0.119	0.152	0.106	0.183	0.242
	DMLkNN	0.147	0.101	0.092	**0.083**	0.170	0.161
OneError	LAMLkNN	**0.243**	**0.709**	**0.252**	**0.230**	**0.236**	**0.610**
	MLkNN	0.253	0.728	0.280	0.233	0.242	0.645
	BRkNN	0.267	0.726	0.459	0.291	0.242	0.891
	DMLkNN	0.253	0.721	0.282	0.238	0.237	0.612
Coverage	LAMLkNN	0.307	**0.138**	**0.240**	0.093	**0.454**	**0.172**
	MLkNN	**0.298**	0.147	0.249	0.096	0.455	0.184
	BRkNN	0.303	0.172	0.382	0.105	0.472	0.253
	DMLkNN	0.300	0.145	0.246	**0.086**	0.455	0.176
Avg-Precision	LAMLkNN	**0.818**	**0.609**	**0.654**	0.856	**0.759**	**0.530**
	MLkNN	0.818	0.578	0.640	0.852	0.757	0.502
	BRkNN	0.810	0.570	0.564	0.824	0.754	0.334
	DMLkNN	0.816	0.580	0.643	**0.857**	0.758	0.526

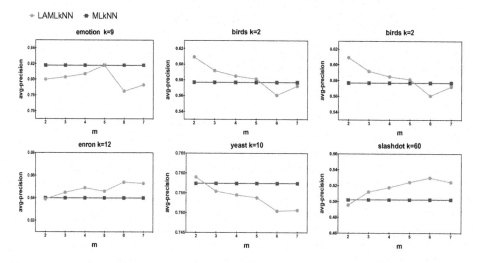

Fig. 1. Comparison results on six regular-scale data sets.

Table 4. Experimental results of each algorithm on large-scale data sets.

Metrics	Algorithms	bibtex	corel5k	corel16k(1)	corel16k(2)	corel16k(3)	Ohsumed
Hamming loss	LAMLkNN	**0.014**	**0.009**	**0.019**	**0.016**	**0.017**	**0.070**
	MLkNN	0.014	0.009	0.019	0.016	0.017	0.071
	BRkNN	0.015	0.010	0.019	0.016	0.017	0.072
	DMLkNN	0.014	0.009	0.019	0.016	0.017	0.071
Ranking loss	LAMLkNN	**0.145**	**0.118**	**0.160**	0.180	0.184	**0.214**
	MLkNN	0.217	0.127	0.175	0.181	0.183	0.231
	BRkNN	0.297	0.292	0.268	0.279	0.259	0.277
	DMLkNN	0.208	0.127	0.174	**0.179**	**0.179**	0.231
OneError	LAMLkNN	**0.542**	**0.670**	**0.698**	**0.731**	**0.732**	**0.613**
	MLkNN	0.578	0.706	0.736	0.782	0.769	0.639
	BRkNN	0.680	0.742	0.771	0.917	0.769	0.706
	DMLkNN	0.576	0.722	0.729	0.767	0.764	0.640
Coverage	LAMLkNN	**0.222**	**0.272**	**0.312**	**0.316**	**0.331**	**0.292**
	MLkNN	0.354	0.298	0.342	0.326	0.336	0.311
	BRkNN	0.431	0.591	0.493	0.475	0.476	0.361
	DMLkNN	0.332	0.299	0.339	0.327	0.332	0.311
Avg-Precision	LAMLkNN	**0.395**	**0.288**	**0.305**	**0.276**	**0.267**	**0.470**
	MLkNN	0.349	0.275	0.288	0.255	0.253	0.442
	BRkNN	0.268	0.210	0.200	0.170	0.222	0.394
	DMLkNN	0.350	0.265	0.291	0.266	0.258	0.441

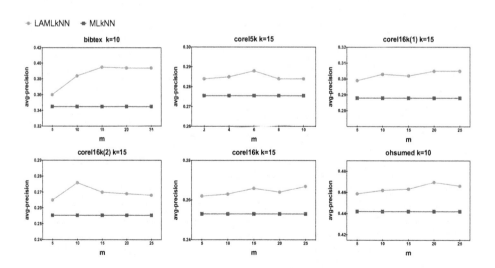

Fig. 2. Comparison results on six large-scale data sets.

large-scale data sets. We fix the k value as well as change m (the number of clusters) for our proposed approach, then compare the *average precision* of each case with that of ML-kNN. From Fig. 2 we can see, on these six large-scale data sets, across all the m value, our approach superior to ML-kNN. But the performance of our approach is sensitive to the cluster number m on small-scale data sets (see in Fig. 1). The proposed approach may inferior to ML-kNN if we select improper m for small-scale data sets (e.g. *emotions* and *yeast*). We think one possible reason may be due to lack of prior acknowledge when the size of each cluster is too small.

5   Conclusion

To achieve more effective multi-label classification using lazy learning method, in this paper, we introduced an original kNN-based multi-label classification algorithm. We show how to take into account the local difference of samples by modifying the posterior probability expression previously used in ML-kNN algorithm. The experimental results on benchmark data sets demonstrate effective classification of our approach, especially on large scale data sets.

Acknowledgement. It was supported by NSF Chongqing China (cstc2017zdcy-zdyf0366). Li Li is the corresponding author for the paper.

References

1. Wang, J., Yang, Y., Mao, J., Huang, Z., Huang, C., Xu, W.: CNN-RNN: a unified framework for multi-label image classification. In: Proceedings of CVPR, pp. 2285–2294 (2016)
2. Nam, J., Kim, J., Loza Mencía, E., Gurevych, I., Fürnkranz, J.: Large-scale multi-label text classification — revisiting neural networks. In: Calders, T., Esposito, F., Hüllermeier, E., Meo, R. (eds.) ECML PKDD 2014. LNCS (LNAI), vol. 8725, pp. 437–452. Springer, Heidelberg (2014). https://doi.org/10.1007/978-3-662-44851-9_28
3. Elisseeff, A., Weston, J.: A kernel method for multi-labelled classification. In: Proceedings of NIPS, vol. 14, pp. 681–687 (2001)
4. Zhang, M.L., Zhou, Z.H.: A review on multi-label learning algorithms. IEEE Trans. Knowl. Data Eng. **26**(8), 1819–1837 (2014)
5. Tsoumakas, G., Katakis, I., Taniar, D.: Multi-label classification: an overview. Int. J. Data Warehous. Min. **3**(3), 1–13 (2007)
6. Zhang, M.L., Zhou, Z.H.: ML-KNN: a lazy learning approach to multi-label learning. Pattern Recognit. **40**(7), 2038–2048 (2007)
7. Cover, T.M., Hart, P.E.: Nearest neighbor pattern classification. IEEE Trans. Inf. Theor. **13**(1), 21–27 (1967)
8. Read, J., Pfahringer, B., Holmes, G., Frank, E.: Classifier chains for multi-label classification. Mach. Learn. **85**(3), 333 (2011)
9. Brinker, K.: Multilabel classification via calibrated label ranking. Mach. Learn. **73**(2), 133–153 (2008)

10. Tsoumakas, G., Katakis, I., Vlahavas, I.: Mining multi-label data. In: Maimon, O., Rokach, L. (eds.) Data Mining and Knowledge Discovery Handbook, pp. 667–685. Springer, Boston (2009). https://doi.org/10.1007/978-0-387-09823-4_34
11. Zhang, M.L.: LIFT: multi-label learning with label-specific features. IJCAI **37**, 1609–1614 (2011)
12. Zhang, M.L., Zhou, Z.H.: Multilabel neural networks with applications to functional genomics and text categorization. IEEE Trans. Knowl. Data Eng. **18**(10), 1338–1351 (2006)
13. Zhang, M.L.: ML-RBF: RBF neural networks for multi-label learning. Neural Process. Lett. **29**(2), 61–74 (2009)
14. Younes, Z., Abdallah, F., Denoeux, T., Snoussi, H.: A dependent multilabel classification method derived from the k-nearest neighbor rule. EURASIP J. Adv. Signal Process. **2011**(1), 1–14 (2011)
15. Wang, H., Ding, C. H. Q., Huang, H.: Multi-label classification: inconsistency and class balanced k-nearest neighbor. In: Proceedings of AAAI (2010)
16. Spyromitros, E., Tsoumakas, G., Vlahavas, I.: An empirical study of lazy multilabel classification algorithms. In: Darzentas, J., Vouros, G.A., Vosinakis, S., Arnellos, A. (eds.) SETN 2008. LNCS (LNAI), vol. 5138, pp. 401–406. Springer, Heidelberg (2008). https://doi.org/10.1007/978-3-540-87881-0_40
17. Wu, X.Z., Zhou, Z.H.: A unified view of multi-label performance measures. arXiv preprint arXiv: 1609.00288 (2016)
18. Chiang, T.H., Lo, H.Y., Lin, S.D.: A ranking-based KNN approach for multi-label classification. In: Proceedings of ACML, pp. 81–96 (2012)

Classification with Reject Option Using Conformal Prediction

Henrik Linusson[1(✉)], Ulf Johansson[2], Henrik Boström[3], and Tuve Löfström[2]

[1] Department of Information Technology, University of Borås, Borås, Sweden
henrik.linusson@hb.se
[2] Department of Computer Science and Informatics, Jönköping University,
Jönköping, Sweden
{ulf.johansson,tuve.lofstrom}@ju.se
[3] School of Electrical Engineering and Computer Science,
Royal Institute of Technology, Kista, Sweden
bostromh@kth.se

Abstract. In this paper, we propose a practically useful means of interpreting the predictions produced by a conformal classifier. The proposed interpretation leads to a classifier with a reject option, that allows the user to limit the number of erroneous predictions made on the test set, without any need to reveal the true labels of the test objects. The method described in this paper works by estimating the cumulative error count on a set of predictions provided by a conformal classifier, ordered by their confidence. Given a test set and a user-specified parameter k, the proposed classification procedure outputs the largest possible amount of predictions containing on average at most k errors, while refusing to make predictions for test objects where it is too uncertain. We conduct an empirical evaluation using benchmark datasets, and show that we are able to provide accurate estimates for the error rate on the test set.

1 Introduction

Conformal predictors [13] are predictive models that associate each of their predictions with a measure of statistically valid confidence. Given a test object x_j, a conformal classifier outputs a *prediction set*—a class label set $\Gamma_j^\epsilon \subseteq Y$—where the probability of making an erroneous prediction (i.e., excluding the correct class label y_j) is at most $\epsilon \in (0, 1)$, where ϵ is a user-specified significance level. Importantly, conformal predictors are automatically well-calibrated, in that the error probability ϵ is guaranteed to correspond with the empirical error asymptotically [13].

Due to their ability to provide users with accurate confidence measures, conformal predictors are particularly useful in risk-sensitive applications, where poor predictions might incur large costs (monetary or otherwise), e.g., stroke risk assessment [4], diagnosis of acute abdominal pain [9] or drug development [3].

However, while conformal predictors are able to supply users with an appropriate estimate of error probability, the validity of a conformal classifier holds

© Springer International Publishing AG, part of Springer Nature 2018
D. Phung et al. (Eds.): PAKDD 2018, LNAI 10937, pp. 94–105, 2018.
https://doi.org/10.1007/978-3-319-93034-3_8

only *a priori*, i.e., before the prediction is made. After observing a particular prediction, it is no longer automatically correct to interpret ϵ as a well-calibrated error probability for any particular prediction, which leads to conformal classifiers instead requiring predictions to be interpreted in a manner that is potentially counter-intuitive to a user less familiar with p-value statistics [6]. Specifically, some prediction regions are always guaranteed to be correct (because they contain all possible labels) whereas others are always guaranteed to be incorrect (because they contain no class labels); since the overall error rate is asymptotically ϵ, this leads to the more interesting prediction regions (containing, e.g., only a single class label) potentially having an error rate that is not immediately related to ϵ.

In [6], a method was proposed for providing a more practical interpretation of the predictions provided by a conformal classifier, by producing adjusted confidence values specifically for predictions containing only a single class label (in a binary classification context). The method proposed in [6] relied on using posterior information regarding the frequencies of predictions containing one, two or zero class labels (estimated from the test set, without knowledge of the true labels). While that method showed promising results, i.e., such that the updated estimates appeared empirically well-calibrated, it does show obvious limitations; specifically, it retains a particular dependency on ϵ that is far from intuitive.

In this paper, we further refine the work presented in [6], and propose a more flexible method of producing an intuitive interpretation of the predictions produced by a conformal classifier. We remove the dependency on ϵ, and replace it with a new parameter, k, that denotes the maximum expected number of errors that we wish the classifier to make on the test set. The result is a classifier that can accurately estimate the error rate for ordered subsets of the test set (without any need to reveal the true test set class labels); by choosing a value for k, we are able to output predictions for a subset of the test objects (while refusing to make predictions when the underlying conformal predictor is too uncertain), where the predictions that are made contain on average at most k errors.

In the next section, we briefly describe the conformal classification framework. In Sect. 3, we outline the proposed approach for making predictions with a bound on the expected number of errors. In Sect. 4, we empirically evaluate the approach using 20 publicly available datasets. Finally, in Sect. 5, we summarize the main findings and discuss some directions for future research.

2 Conformal Classifiers

In order to produce confidence predictions, a conformal classifier depends on a *nonconformity function*—a function $f(z, \zeta) \rightarrow \mathbb{R}$ that scores a pattern $z = (x, y)$ based on how well it corresponds with a sequence of patterns $\zeta = z_1, \ldots, z_n$, such that nonconforming (i.e., strange or unlikely) patterns obtain larger nonconformity scores than more common patterns. A standard way of defining nonconformity functions is to base them on the predictions made by a traditional classification model, as

$$f(z_i, \zeta) = \Delta\left[h(x_i), y_i\right], \tag{1}$$

where h is a classifier (often called the *underlying model*) induced from ζ, and Δ is a function that measure the prediction errors of h. A common choice of Δ, for classification problems, is the margin error function,

$$\Delta[h(x_i), y_i] = \max_{y \neq y_i} \hat{P}_h(y \mid x_i) - \hat{P}_h(y_i \mid x_i), \tag{2}$$

where $\hat{P}_h(y \mid x)$ denotes the probability estimate provided by h for the class y.

Once a suitable underlying model and nonconformity function have been selected, a conformal classifier can be constructed in a few different manners. One of the more popular conformal classifier variants is the *inductive conformal predictor* [8,10,13], premiered in particular for its low computational overhead. In order to train an inductive conformal predictor for classification, the following training procedure is used:

1. Divide the training set Z into two disjoint subsets:
   - A *proper training set* Z_t.
   - A *calibration set* Z_c, where $|Z| = q$.
2. Train a classifier h using Z_t as the training data.
3. Let $\{\alpha_1, \ldots, \alpha_q\} = \{f(z_i, Z_t) : z_i \in Z_c\}$.

When a new test object x_j is observed, the standard way of obtaining a prediction from the conformal classifier is to produce a prediction region $\Gamma_j^\epsilon \subseteq Y$ as follows:

1. Fix a significance level $\epsilon \in (0, 1)$.
2. For each class $\tilde{y} \in Y$:
   (a) Tentatively label x_j as $(x_j, \tilde{y})$.
   (b) Let $\alpha_j^{\tilde{y}} = f\left[(x_j, \tilde{y}), Z_t\right]$.
   (c) Calculate $p_j^{\tilde{y}}$ as

$$p_j^{\tilde{y}} = \frac{\left|\left\{z_i \in Z_c : \alpha_i > \alpha_j^{\tilde{y}}\right\}\right|}{q+1} + \theta_j \frac{\left|\left\{z_i \in Z_c : \alpha_i = \alpha_j^{\tilde{y}}\right\}\right| + 1}{q+1}, \tag{3}$$

   where $\theta_j \sim U[0, 1]$.
   (d) Let $\Gamma_j^\epsilon = \left\{\tilde{y} \in Y : p_j^{\tilde{y}} > \epsilon\right\}$.

The resulting class label set Γ_j^ϵ contains the true label y_j with probability $1 - \epsilon$, i.e., an error (meaning that $y_j \notin \Gamma_j^\epsilon$) occurs with probability ϵ.

An alternative way of producing predictions with a conformal classifier is to output what we will refer to as *confidence-credibility predictions* [8]. Here, the output for a test object x_j takes the form $(\hat{y}_j, \gamma_j, \mu_j)$, where

- $\hat{y}_j$ is the most likely class label (i.e., the class label for which $p_j^{\tilde{y}}$ is greatest),
- γ_j is the *confidence*, which is one minus the second largest p-value, and
- μ_j is the *credibility*, which is the largest p-value.

Here, we are effectively forcing the conformal predictor to output the most confident prediction set containing only a single class label (if we were to increase the confidence of the prediction, at least one other class label would be included). Credibility corresponds with the significance level at which all class labels are rejected, and the prediction becomes empty—if credibility is very low, the conformal classifier considers all potential class labels as unsuitable for the test object.

Since conformal predictors are unconditionally valid by default, there is often a need to take the true class labels into consideration when evaluating their predictions [6,7,12,13]; specifically, it is possible that the error probability of a conformal predictor is greater (or smaller) than ϵ, depending on the test object's true label. In practice, this effectively means that, depending on properties of the dataset, it is possible that the most confident predictions are made only for test objects pertaining to a particular class (usually the majority class). This behaviour is easily rectified by employing a label-conditional (Mondrian) conformal classifier [12,13], where the p-values are additionally conditioned on the class labels using

$$p_j^{\tilde{y}} = \frac{\left|\left\{z_i \in Z^{\tilde{y}} : \alpha_i > \alpha_j^{\tilde{y}}\right\}\right|}{|Z^{\tilde{y}}| + 1} + \theta_j \frac{\left|\left\{z_i \in Z^{\tilde{y}} : \alpha_i = \alpha_j^{\tilde{y}}\right\}\right| + 1}{|Z^{\tilde{y}}| + 1}, \tag{4}$$

where $Z^{\tilde{y}} \subseteq Z_c$ are the calibration patterns that belong to the class $\tilde{y}$.

3 Error Probabilities Using Posterior Information

The confidence measures supplied by a conformal classifier are valid in the sense that the observed error rate, over a test sequence, is guaranteed to converge to ϵ (when the predictor is allowed to output prediction sets). This probability is by default unconditional, in the sense that we might not make any assertions regarding the distribution of errors with regard to the problem space; this can be contrasted to, e.g., label-conditional validity, where we might assert that error probability is independent of y_j or object-conditional validity, where the error probability is independent of x_j [12,13].

More importantly, the confidence measure of a conformal predictor is valid in an *a priori* sense, meaning the error probability before making a prediction, and the error probability after observing a prediction, are not necessarily the same [6]. In particular, when a conformal predictor is applied in a batch prediction setting (i.e., we are making predictions for a test set whose size is greater than one, and those predictions are obtained simultaneously in a batch), it is easy to see that *a priori* and *a posteriori* error probabilities are unequal: consider a binary classification problem, where we are predicting the output labels of a test set containing 100 objects, with a significance level 0.1 (i.e., we are expecting an overall error rate of 10%); if 90 of our predictions contain both class labels, while the remaining 10 predictions contain only a single class label, we are likely to fool ourselves if we were to trust the 10 "interesting" singleton predictions— we are expecting the conformal classifier to make approximately 10 errors over the entire test set, and none of the predictions containing both class labels can possibly be erroneous.

Of course, this seemingly counter-intuitive result stems from a forced misunderstanding of the conformal prediction procedure (we are effectively trying to interpret p-values as true probabilities); nonetheless, it is not clear how an end-user should interpret the predictions in an appropriate manner.

In [6], an attempt was made to utilize posterior information (empirical estimates of the rates of empty, singleton and double predictions, coupled with the knowledge that empty and double predictions cannot be correct or erroneous, respectively) in order to produce more reliable estimates of the error probability of singleton predictions (which, in a binary classification scenario, arguably make out the most interesting predictions that can be made). An unconditional (w.r.t. labels and objects) adjusted estimate was defined as

$$\hat{\epsilon}_s = \frac{\epsilon}{P(s) + P(e)} \tag{5}$$

where $P(s)$ and $P(e)$ are the rates of singleton predictions and empty predictions observed in the test set (without any need to consider the true output labels of the test patterns). A label conditional variant was also developed, but is omitted here.

3.1   Getting Rid of ϵ

The adjusted estimates proposed in [6] were intended to provide a better assessment of the quality of singleton predictions; unfortunately, these estimates (both unconditional and label-conditional variants) retain a dependency on the user-specified ϵ-parameter, which is rather unintuitive, since the final estimate $\hat{\epsilon}$ is only loosely related to ϵ. As an alternative, we propose an updated procedure, that is not dependent on ϵ, but instead operates on top of predictions made on the confidence-credibility form.

Table 1. Example of confidence-credibility predictions (credibility scores are omitted).

Idx	0	1	2	3	4	5	6	7	8	9
$\hat{y}$	0	0	1	1	0	1	0	1	0	1
Confidence	0.60	0.62	0.63	0.65	0.72	0.78	0.82	0.90	0.97	0.99

Suppose we are given a batch of confidence-credibility predictions, where we have sorted the predictions with respect to their confidence, e.g., as in Table 1. The appropriate manner of interpreting these confidence scores is: all predictions with confidence at least $c \in (0,1)$ contain on average $n(1-c)$ errors, where n is the total number of predictions made (in this case 10). Hence, from the predictions in Table 1, we should expect approximately four errors across the entire test set, and approximately one error among the predictions for indices 7–9.

Table 2. Datasets used in the experiments. #inst denotes the number of instances contained in the dataset; #min and #maj denote the number of examples belonging to the minority and majority classes, respectively. %min is the percentage of examples that belong to the minority class.

Dataset	#inst	#min	#maj	%min	Dataset	#inst	#min	#maj	%min
balance-scale	576	288	288	50.0	hepatitis	155	32	123	20.6
breast-cancer	286	85	201	29.7	ionosphere	351	126	225	35.9
breast-w	699	241	458	34.5	kr-vs-kp	3196	1527	1669	47.8
credit-a	690	307	383	44.5	labor	57	20	37	35.1
credit-g	1000	300	700	30.0	liver-disorders	345	145	200	42.0
diabetes	768	268	500	34.9	mushroom	8124	3916	4208	48.2
haberman	306	81	225	26.5	sick	3772	231	3541	6.1
heart-c	303	138	165	45.5	sonar	208	97	111	46.6
heart-h	294	106	188	36.1	spambase	4601	1813	2788	39.4
heart-s	270	120	150	44.4	tic-tac-toe	958	332	626	34.7

Based on this information, we propose the following: given a test set $x_1, \ldots, x_n$, obtain from a conformal classifier the predicted labels and their confidence, $(\hat{y}_1, \gamma_1), \ldots, (\hat{y}_n, \gamma_n)$. For each prediction, compute $\hat{k}_j = n(1 - \gamma_j)$, and construct the tentative prediction set $\hat{Y} = (\hat{y}_1, \gamma_1, \hat{k}_1), \ldots, (\hat{y}_n, \gamma_n, \hat{k}_n)$; here $\hat{k}_j$ is the expected error rate for all predictions with confidence γ_j or greater. Note that $\hat{k}_j$ has a anti-monotonic property with respect to γ_j, i.e., $\gamma_i \leq \gamma_j \rightarrow \hat{k}_i \geq \hat{k}_j$. Finally, output the predictions

$$\left\{ \gamma_j \in \hat{Y} : \hat{k}_j \leq k \right\}, \tag{6}$$

where k is user-specified, and denotes the maximum number of expected errors that we allow on the test set. Any prediction where $\hat{k} > k$ is rejected.

The main reason for constructing this new estimate $\hat{k}_j$ is that the confidence value γ_j, by itself, has no clear intuitive interpretation; the formal interpretation given above, i.e., among all test objects, $n(1 - c)$ errors are distributed among those predictions where $\gamma_j \geq c$, is inherently dependent on n. Here, we are simply coding this information into the new estimate $\hat{k}_j$, so that a much more intuitive interpretation can be obtained.

4 Experiments

In order to assess how well our proposed procedure is able to estimate the error rate on the test set, an experimental evaluation was performed using 20 datasets taken from the UCI repository [1], listed in Table 2.

The underlying conformal predictor used a random forest classifier [2], containing 100 decision trees, with a margin error nonconformity function (Eq. 2). The experiments were implemented in Python using the scikit-learn machine

Table 3. Average number of predictions and errors made per iteration, over the entire dataset (using 10 folds), using an unconditional conformal classifier. #pred denotes the total number of predictions made and %pred denotes the size of the prediction set as a percentage of the total test set. Finally, #err denotes the number of erroneous predictions. k is the user-specified expected error count.

k	1			5			10		
Dataset	#pred	%pred	#err	#pred	%pred	#err	#pred	%pred	#err
balance-scale	105.0	18.2	1.0	461.7	80.2	4.4	496.3	86.2	10.2
breast-cancer	7.1	2.5	1.3	30.3	10.6	4.8	51.6	18.0	8.8
breast-w	109.8	15.7	0.8	519.0	74.2	4.7	615.4	88.0	8.4
credit-a	23.1	3.3	1.4	114.9	16.7	5.1	210.0	30.4	9.9
credit-g	21.8	2.2	0.9	103.7	10.4	4.5	171.3	17.1	8.8
diabetes	23.6	3.1	0.7	112.9	14.7	4.3	170.7	22.2	9.6
haberman	8.4	2.7	1.2	49.4	16.1	5.0	77.3	25.3	8.7
heart-c	20.2	6.7	0.6	92.6	30.6	4.4	143.6	47.4	9.6
heart-h	20.8	7.1	1.1	87.8	29.9	4.9	143.7	48.9	10.4
heart-s	17.5	6.5	0.9	81.2	30.1	4.1	129.5	48.0	9.9
hepatitis	18.6	12.0	1.2	82.0	52.9	4.4	106.4	68.6	8.9
ionoshere	49.8	14.2	1.0	221.9	63.2	4.8	286.8	81.7	9.5
kr-vs-kp	549.0	17.2	1.2	2482.2	77.7	5.6	2954.9	92.5	10.4
labor	13.1	23.0	0.5	53.4	93.7	3.6	56.6	99.3	4.7
liver-disorders	5.5	1.6	0.6	27.0	7.8	3.6	52.2	15.1	8.3
mushroom	1820.6	22.4	1.1	8124.0	100.0	2.5	8124.0	100.0	2.5
sick	708.9	18.8	0.7	3184.5	84.4	4.9	3444.8	91.3	9.9
sonar	24.6	11.8	1.3	108.0	51.9	5.1	131.3	63.1	10.1
spambase	229.3	5.0	1.0	1112.9	24.2	5.7	1761.6	38.3	11.4
tic-tac-toe	182.6	19.1	1.0	799.8	83.5	4.2	868.6	90.7	8.1
Mean	197.96	10.66	0.98	892.46	47.64	4.53	999.83	58.61	8.91

learning library [11], as well as the nonconformist[1] library for conformal prediction. In the experiments, a 10x10-fold cross-validation was performed, and the results presented are averaged across the 10 iterations. In each fold, 25% of the training data was used as the calibration set for the inductive conformal classifier, as suggested in [5].

Table 3 lists the number of predictions made, as well as the number of errors among those predictions, for the 20 datasets. Here, an unconditional conformal classifier is used (Eq. 3). Results are averaged over 10 iterations. The maximum number of predictions possible (per dataset) is given by #inst in Table 2. For each dataset, the procedure was applied with $k = 1$, $k = 5$ as well as $k = 10$,

---

[1] https://github.com/donlnz/nonconformist.

Table 4. Average number of predictions made per iteration, using an unconditional conformal classifier. #pred is the total number of predictions made, and #min is the number of predictions made for test objects where the true label is the minority class. %min is #min expressed as a percentage of #pred.

k	1			5			10		
Dataset	#pred	#min	%min	#pred	#min	%min	#pred	#min	%min
balance-scale	105.0	52.2	49.7	461.7	230.3	49.9	496.3	247.7	49.9
breast-cancer	7.1	1.5	21.1	30.3	5.0	16.5	51.6	8.9	17.2
breast-w	109.8	29.3	26.7	519.0	133.1	25.6	615.4	184.9	30.0
credit-a	23.1	8.9	38.5	114.9	46.5	40.5	210.0	82.7	39.4
credit-g	21.8	0.9	4.1	103.7	4.5	4.3	171.3	9.0	5.3
diabetes	23.6	1.1	4.7	112.9	6.6	5.8	170.7	14.9	8.7
haberman	8.4	1.1	13.1	49.4	5.0	10.1	77.3	8.9	11.5
heart-c	20.2	7.4	36.6	92.6	40.1	43.3	143.6	62.9	43.8
heart-h	20.8	4.1	19.7	87.8	19.6	22.3	143.7	39.6	27.6
heart-s	17.5	6.5	37.1	81.2	31.9	39.3	129.5	53.0	40.9
hepatitis	18.6	1.3	7.0	82.0	5.5	6.7	106.4	11.5	10.8
ionosphere	49.8	13.6	27.3	221.9	62.0	27.9	286.8	93.0	32.4
kr-vs-kp	549.0	260.6	47.5	2482.2	1159.6	46.7	2954.9	1399.3	47.4
labor	13.1	3.9	29.8	53.4	17.8	33.3	56.6	19.7	34.8
liver-disorders	5.5	1.5	27.3	27.0	9.1	33.7	52.2	18.4	35.2
mushroom	1820.6	866.0	47.6	8124.0	3916.0	48.2	8124.0	3916.0	48.2
sick	708.9	7.5	1.1	3184.5	38.9	1.2	3444.8	70.4	2.0
sonar	24.6	12.2	49.6	108.0	46.1	42.7	131.3	55.7	42.4
spambase	229.3	85.3	37.2	1112.9	423.7	38.1	1761.6	667.5	37.9
tic-tac-toe	182.6	51.9	28.4	799.8	222.0	27.8	868.6	264.7	30.5
Mean	198.0	70.8	27.7	892.5	321.2	28.2	999.8	361.4	29.8

i.e., we are asking to make the maximum number of predictions containing on average 1, 5 or 10 errors. From the results in Table 3, it is evident that the proposed procedure is able to estimate the error count on the test set rather well, although the estimates appear to be somewhat conservative in general, in particular as k increases. In all cases, the procedure is able to output a non-trivial number of predictions (i.e., the prediction set is substantially greater than k), while still limiting the number of erroneous predictions.

Since we are using an unconditional conformal classifier, it becomes interesting to evaluate not only the number of predictions output by our proposed process, but also the number of predictions output for the minority and majority class test objects, respectively. The results shown in Table 4 indicate that, while there appears to be a bias towards premiering the majority class among the output predictions (see, e.g., sick, credit-g, diabetes and hepatitis), this bias is never so strong as to cause the classifier to make predictions only for test objects belonging to one of the two possible classes. This behaviour—displaying

Table 5. Average number of predictions and errors made per iteration, over the entire dataset (using 10 folds), using a label-conditional conformal predictor. #pred denotes the total number of predictions made and %min denotes the percentage of predictions made for test objects belonging to the minority class. Finally, #err denotes the number of erroneous predictions. k is the user-specified expected error count.

k	1			5			10		
Dataset	#pred	%min	#err	#pred	%min	#err	#pred	%min	#err
balance-scale	54.2	47.2	1.0	275.4	49.7	4.3	487.7	49.7	8.7
breast-cancer	3.7	35.1	1.1	17.4	39.1	4.4	34.7	35.4	8.9
breast-w	57.1	35.7	1.0	291.2	39.9	4.5	512.3	34.7	10.0
credit-a	20.1	42.8	1.3	101.8	41.3	5.6	195.9	45.0	10.6
credit-g	12.8	32.0	1.4	59.9	31.2	4.7	109.6	29.7	10.2
diabetes	15.1	27.2	1.5	63.3	27.5	5.8	122.0	25.2	9.9
haberman	3.2	21.9	0.6	19.0	17.4	4.6	37.9	16.9	9.2
heart-c	14.9	45.6	0.7	73.8	46.3	6.1	134.9	48.6	10.6
heart-h	13.6	48.5	1.3	63.4	45.9	5.3	114.3	40.3	10.1
heart-s	13.8	48.6	1.0	70.7	46.0	5.3	128.0	49.1	11.5
hepatitis	7.5	37.3	1.2	33.7	30.3	4.3	64.4	23.8	9.9
ionosphere	30.1	47.8	0.6	149.3	48.6	4.5	247.8	43.5	9.8
kr-vs-kp	306.0	47.7	1.2	1560.4	46.1	5.5	2762.9	48.5	9.4
labor	7.2	40.3	1.1	35.1	39.6	3.4	51.8	36.1	5.3
liver-disorders	6.4	28.1	1.0	28.5	33.7	4.0	54.5	36.9	9.1
mushroom	910.4	49.6	1.2	4579.6	50.0	4.1	8124.0	48.2	5.3
sick	81.2	39.8	1.3	388.8	39.8	5.5	624.5	26.9	10.2
sonar	14.0	48.6	0.8	70.3	49.1	4.1	129.5	47.4	9.7
spambase	147.3	41.0	1.9	751.8	41.1	5.6	1388.4	44.0	10.0
tic-tac-toe	94.6	49.6	1.4	464.6	49.8	5.4	775.6	40.8	9.5
Mean	90.7	40.7	1.1	454.9	40.6	4.9	805.0	38.5	9.4

a (sometimes substantial) bias towards the majority class—is common in unconditional conformal predictors when the dataset is heavily imbalanced. The issue is easily alleviated however, by employing a label-conditional Mondrian conformal classifier (Eq. 4) instead; results from such a classifier are shown in Table 5.

Table 5 shows results analogous to those in Tables 3 and 4, but instead using an underlying label-conditional conformal classifier (Eq. 4). The results correspond well with what is normally expected from a label-conditional conformal predictor: the overall error rate remains relatively untouched (we are still seeing a good correspondence between k and the empirical error rate), but the sensitivity of the classifier is reduced (the model is able to output far fewer predictions). The main benefit shown by the label-conditional variant, however, is that there is a clear reduction in bias with respect to the true class labels of the test objects. The classifier is able to output a more even distribution of positive and negative predictions, without any substantial negative effect on the error count among the predictions that are made.

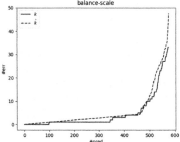

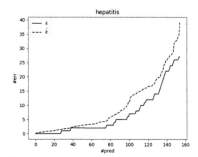

(a) Cumulative errors (real and predicted) on the balance-scale dataset; unconditional.

(b) Cumulative errors (real and predicted) on the hepatitis dataset; unconditional.

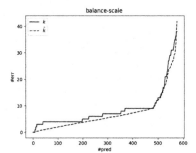

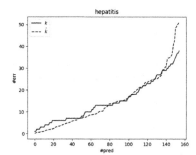

(c) Cumulative errors (real and predicted) on the balance-scale dataset; label-conditional.

(d) Cumulative errors (real and predicted) on the hepatitis dataset; label-conditional.

Fig. 1. Cumulative errors (real, k, solid lines; and predicted, $\hat{k}$, dashed lines) on the balance-scale and hepatitis datasets. Results are shown for a single iteration, with the x-axis showing the number of predictions made (with decreasing confidence) and the y-axis showing the cumulative error count among the output predictions.

Finally, in order to provide some insight into how well the proposed procedure functions over a larger selection of values for k, Fig. 1 shows the true and predicted cumulative error rates over the full test set for two of the datasets, taken from a single iteration. Figures 1a and b show results using an unconditional conformal classifier, and Figs. 1c and d show results using a label-conditional conformal classifier. The x-axis displays the number of predictions made (in order of decreasing confidence), and the y-axis displays the cumulative error count.

The two lines—dashed and solid—correspond to the predicted and actual cumulative error counts respectively. In each case, the predictive cumulative error count $(\hat{k})$ closely follows the true cumulative error count (k); with respect to their calibration, there is no clear difference between the unconditional and label-conditional variants.

5   Concluding Remarks

In this paper, we offer an interpretation of the conformal classification procedure, that is able to estimate the number of errors made by a classifier on the test set, without needing to reveal the true test set class labels. The procedure described results in a classifier with a reject option, that outputs predictions for a subset of the test set, where the expected error count is limited by a user-specified parameter k; given a test set and a choice of k, the proposed procedure outputs the largest possible number of predictions containing on average at most k errors.

We evaluate the procedure empirically using 20 benchmark datasets, and obtain very promising results, indicating that we are able to provide accurate estimates of the error rate on the test set.

It is not obvious how well the proposed procedure will perform on multi-class datasets or heavily imbalanced datasets; as such, evaluating the procedure on a more diverse selection of datasets would be of great interest. Naturally, it would also be of great interest to evaluate our proposed procedure to alternative methods for constructing classifiers with a reject option.

Additionally, it would be interesting to evaluate the proposed procedure with respect to specific applications—in particular heavily imbalanced problems where identifying the minority test patterns is the key objective. Extending the procedure so that errors are only allowed for one of two classes (normally the minority class) might be beneficial in several applications.

Acknowledgements. This work was supported by the Swedish Knowledge Foundation through the project Data Analytics for Research and Development (20150185).

References

1. Bache, K., Lichman, M.: UCI machine learning repository (2013). http://archive.ics.uci.edu/ml
2. Breiman, L.: Random forests. Mach. Learn. **45**(1), 5–32 (2001)
3. Eklund, M., Norinder, U., Boyer, S., Carlsson, L.: The application of conformal prediction to the drug discovery process. Ann. Math. Artif. Intell. **74**(1–2), 117–132 (2015)
4. Lambrou, A., Papadopoulos, H., Kyriacou, E., Pattichis, C.S., Pattichis, M.S., Gammerman, A., Nicolaides, A.: Assessment of stroke risk based on morphological ultrasound image analysis with conformal prediction. In: Papadopoulos, H., Andreou, A.S., Bramer, M. (eds.) AIAI 2010. IAICT, vol. 339, pp. 146–153. Springer, Heidelberg (2010). https://doi.org/10.1007/978-3-642-16239-8_21

5. Linusson, H., Johansson, U., Boström, H., Löfström, T.: Efficiency comparison of unstable transductive and inductive conformal classifiers. In: Artificial Intelligence Applications and Innovations, pp. 261–270. Springer (2014)
6. Linusson, H., Johansson, U., Boström, H., Löfström, T.: Reliable confidence predictions using conformal prediction. In: Bailey, J., Khan, L., Washio, T., Dobbie, G., Huang, J.Z., Wang, R. (eds.) PAKDD 2016. LNCS (LNAI), vol. 9651, pp. 77–88. Springer, Cham (2016). https://doi.org/10.1007/978-3-319-31753-3_7
7. Löfström, T., Boström, H., Linusson, H., Johansson, U.: Bias reduction through conditional conformal prediction. Intell. Data Anal. **9**(6) (2015)
8. Papadopoulos, H.: Inductive conformal prediction: theory and application to neural networks. Tools Artif. Intell. **18**(315–330), 2 (2008)
9. Papadopoulos, H., Gammerman, A., Vovk, V.: Reliable diagnosis of acute abdominal pain with conformal prediction. Eng. Intell. Syst. **17**(2), 127 (2009)
10. Papadopoulos, H., Proedrou, K., Vovk, V., Gammerman, A.: Inductive confidence machines for regression. In: Elomaa, T., Mannila, H., Toivonen, H. (eds.) ECML 2002. LNCS (LNAI), vol. 2430, pp. 345–356. Springer, Heidelberg (2002). https://doi.org/10.1007/3-540-36755-1_29
11. Pedregosa, F., Varoquaux, G., Gramfort, A., Michel, V., Thirion, B., Grisel, O., Blondel, M., Prettenhofer, P., Weiss, R., Dubourg, V., et al.: Scikit-learn: machine learning in python. J. Mach. Learn. Res. **12**, 2825–2830 (2011)
12. Vovk, V.: Conditional validity of inductive conformal predictors. Mach. Learn. **92**(2–3), 349–376 (2013)
13. Vovk, V., Gammerman, A., Shafer, G.: Algorithmic Learning in a Random World. Springer, DE (2006)

Target Learning: A Novel Framework to Mine Significant Dependencies for Unlabeled Data

Limin Wang[1(✉)], Shenglei Chen[2], and Musa Mammadov[3]

[1] Key Laboratory of Symbolic Computation and Knowledge Engineering of Ministry of Education, Jilin University, Changchun, China
wanglim@jlu.edu.cn
[2] School of Economics, Nanjing Audit University, Nanjing, China
tristan_chen@126.com
[3] Faculty of Science and Technology, Federation University, Ballarat, Australia
m.mammadov@federation.edu.au

Abstract. To mine significant dependencies among predictive attributes, much work has been carried out to learn Bayesian netwrok classifiers ($\text{BNC}_{\mathcal{T}}$s) from labeled training data set $\mathcal{T}$. However, if $\text{BNC}_{\mathcal{T}}$ does not capture the "right" dependencies that would be most relevant to unlabeled testing instance, that will result in performance degradation. To address this issue we propose a novel framework, called target learning, that takes each unlabeled testing instance as a target and builds an "unstable" Bayesian model $\text{BNC}_{\mathcal{P}}$ for it. To make $\text{BNC}_{\mathcal{P}}$ and $\text{BNC}_{\mathcal{T}}$ complementary to each other and work efficiently in combination, the same learning strategy is applied to build them. Experimental comparison on 32 large data sets from UCI machine learning repository shows that, for BNCs with different degrees of dependence target learning always helps improve the generalization performance with minimal additional computation.

Keywords: Bayesian network · Target learning · Unlabeled data

1 Introduction

The traditional formulation of the machine learning problem has been as a classification problem, the goal of which is producing a classifier that has good generalization performance on unlabeled testing instances in the problem domain. Bayesian networks (BNs) [1] have long been a popular medium for graphically representing the probabilistic dependencies in a probability distribution. BNs were considered as classifiers only after the discovery of naive Bayes [2], a very simple kind of BNs that assumes each attribute to be conditionally independent of every other attribute given the class variable. Although naive Bayes (NB) disregards conditional dependencies between predictive attributes and thus the structure complexity is 0-dependence, it is surprisingly effective and efficient

© Springer International Publishing AG, part of Springer Nature 2018
D. Phung et al. (Eds.): PAKDD 2018, LNAI 10937, pp. 106–117, 2018.
https://doi.org/10.1007/978-3-319-93034-3_9

for inference [3]. The success of NB has led to a recent furry of algorithms for learning Bayesian netwrok classifiers (BNCs) from data.

Numerous BNCs [4–9] learned from labeled training data, i.e., $\text{BNC}_\mathcal{T}$, have been proposed to alleviate NB's independence assumption and mine the significant conditional dependencies implicated in training data. Although some BNCs with sophisticated structures produce good results on some benchmark data sets, their advantages over those BNCs with simple structures are not obvious as expected. One plausible explanation is that, scientific data set can be massive, labeled training data may account for only a small portion. Correspondingly only a limited number of conditional dependencies, which are always the most significant, can be represented by $\text{BNC}_\mathcal{T}$. Moreover, the dependencies that exist in different unlabeled testing instances may differ greatly. It is impossible for $\text{BNC}_\mathcal{T}$ to capture all "right" dependencies that exist in each testing instance.

Semi-supervised learning methods, e.g., Self-training [10,11] or Co-training [12], generally use unlabeled data to either modify or re-prioritize hypotheses obtained from labeled data alone [13]. To achieve this goal, the unlabeled instance must be pre-assigned a class label first. Obviously, if the label is wrong, using such instance to re-train $\text{BNC}_\mathcal{T}$ will result in "noise propagation", and the negative effect may lead to the biased decision boundaries. To address this issue, in this paper we propose a novel semi-supervised learning framework, target learning, which builds a specific Bayesian model $\text{BNC}_\mathcal{P}$ for each unlabeled testing instance. That is, $\text{BNC}_\mathcal{P}$ and $\text{BNC}_\mathcal{T}$ will be built independently but work jointly. We reveal that $\text{BNC}_\mathcal{P}$ is indeed complementary to $\text{BNC}_\mathcal{T}$ and effective in further improving $\text{BNC}_\mathcal{T}$'s generalization performance with minimal additional computation. The following section introduces some state-of-the-art BNCs. Section 3 introduces the basic idea of target learning. Experimental study on 32 UCI machine learning data sets is presented in Sect. 4, including a comparison with 7 related algorithms. The final section draws conclusions and outlines some directions for further research.

2 Bayesian Network Classifiers

The structure of a Bayesian network on the random variables or attributes $\{X_1, \cdots, X_n\}$ is a directed acyclic graph (DAG), which represents each attribute in a given domain as a node in the graph and dependencies between these attributes as arcs connecting the respective nodes. Thus, independencies are represented by the lack of an arc connecting particular attributes. A node in the network for an attribute X_i represents the probability of X_i conditioned on the attributes that are immediate parents of X_i, denoted as $Pa(X_i)$. Nodes with no parents simply represent the prior probability for that attribute.

BNs are powerful tools for knowledge representation and inference under conditions of uncertainty. BNCs are special types of BNs designed for classification. A central problem is to estimate the underlying n-dimensional probability distributions from a finite number of instances. Let each instance $\mathbf{x}$ be characterized with n values $\{x_1, \cdots, x_n\}$ for attributes $\{X_1, \cdots, X_n\}$, and class label

$y \in \{y_1, \cdots, y_m\}$ is the value of class variable Y. Given the limitation in available instances and time complexity, the approximation of $P(\mathbf{x}, y)$ is transformed to a product of several of its component distributions of a lower order. The performance of these BNCs is, to a very great extent, determined by the approximations employed. That is to say, the estimation of conditional probability of lower order is reliable whereas the confidence level of the joint probabilities may differ greatly if the learning strategies of corresponding models differ greatly. As shown in Fig. 1(a), full Bayesian network classifier (FBC) [4] can fully model the joint probability distribution $P(\mathbf{x}, y)$ according to generative approach and thus will inevitably achieve optimal performance.

There exists $i - 1$ conditional dependencies between X_i and its parents. This general distribution can be captured in dependence tree structure at the most general extreme. The total number of conditional dependencies (TNCD) for FBC is $\text{TNCD}_{\text{FBC}} = 0 + 1 + \cdots + (n - 1) = \frac{n(n-1)}{2}$.

However, learning a FBC is very time consuming and quickly becomes an NP-hard problem as the number of predictive attributes grows. In contrast, NB represents the most restrictive extreme in attribute dependence spectrum based on the assumption that the predictive attributes are assumed to be conditionally independent given the class variable Y, i.e., $P_{\text{NB}}(\mathbf{x}|y) = \prod_{i=1}^{n} P(x_i|y)$. As shown in Fig. 1(b), NB strictly allows no dependencies between predictive attributes. Then the TNCD for NB is $\text{TNCD}_{\text{NB}} = 0$.

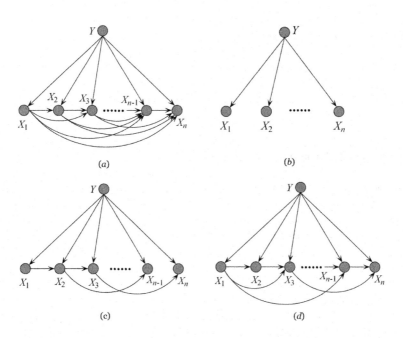

Fig. 1. Example of (a) full Bayesian network classifier, (b) Naive Bayes, (c) Tree augmented naive Bayes and (d) k-dependence Bayesian classifier.

Two state-of-the-art BNCs, Tree augmented naive Bayes (TAN) [5] and k-dependence Bayesian classifier (KDB) [6], allow to accommodate good trade-offs between bias and variance for different data quantities. TAN maintains the structure of NB and constructs maximum weighted spanning tree to represent 1-dependence relations between attributes. The conditional dependence between attributes X_i and X_j is measured by conditional mutual information $I(X_i; X_j|Y)$, which is defined as follows [14],

$$I(X_i; X_j|Y) = \sum_{x_i} \sum_{x_j} \sum_{y} P(x_i, x_j, y) log \frac{P(x_i, x_j|y)}{P(x_i|y)P(x_j|y)}. \tag{1}$$

The dependence tree structure of TAN is shown in Fig. 1(c). In contrast to Fig. 1(b), TAN allows each attribute node to have at most one parent. And the TNCD for TAN is $\text{TNCD}_{\text{TAN}} = 0 + 1 + \cdots + 1 = n - 1$.

Although TAN can represent the most significant 1-dependence relationships, Friedman provided no method to generalize to higher degree of dependence. In contrast, as shown in Fig. 1(d) KDB can represent arbitrary degree of dependence and also capture much of the computational efficiency of NB. KDB first sorts attributes by comparing mutual information $I(X_i; Y)$, which is defined as follows [14],

$$I(X_i; Y) = \sum_{x_i} \sum_{y} P(x_i, y) log \frac{P(x_i, y)}{P(x_i)P(y)} \tag{2}$$

Once X_i enters the model, its parents are selected from $min(i - 1, k)$ attributes already in the model with the highest values of $I(X_i; X_j|Y)$. And the TNCD for KDB is $\text{TNCD}_{\text{KDB}} = 0 + 1 + \cdots + (k-1) + k + k + \cdots + k = nk - \frac{k^2}{2} - \frac{k}{2}$. TNCD_{KDB} will grow as k grows and KDB can represent more dependencies than TAN.

Sahami [6] suggests that, if k is large enough to capture all "right" conditional dependencies that exist in a database, then a classifier would be expected to achieve optimal Bayesian accuracy. Taheri et al. [9] deal with the number k as a variable that is defined by solving some global optimization problem; accordingly, the resulting BN generates optimal dynamic structures by not specifying k a priori.

To perform accurate calculations for structure learning by using (1) and (2), the critical issue is to choose suitable distributions for the probabilities. We denote the training data by $\mathcal{T} = (\hat{x}_{i1}, \cdots, \hat{x}_{in}, \hat{y}_i)$, $i = 1, \cdots, N$, where N is the number of training instances, $\hat{y}_i \in \{y_1, \cdots, y_m\}$ is the class label of the i-th training instance (i.e., there are m classes).

The prior and joint probabilities in (1) and (2) will be estimated by the training data $\mathcal{T}$ as follows

$$\begin{cases} P(y) = \frac{1}{N} \sum_{k=1}^{N} \delta(y, \hat{y}_k) \\ P(x_j) = \frac{1}{N} \sum_{k=1}^{N} \delta(x_j, \hat{x}_{kj}) \\ P(x_j, y) = \frac{1}{N} \sum_{k=1}^{N} \delta(<x_j, y>, <\hat{x}_{kj}, \hat{y}_k>) \\ P(x_i, x_j, y) = \frac{1}{N} \sum_{k=1}^{N} \delta(<x_i, x_j, y>, <\hat{x}_{ki}, \hat{x}_{kj}, \hat{y}_k>) \end{cases} \tag{3}$$

where $< \cdot >$ denotes the set of attribute values and $\delta(\cdot)$ is a binary function, which is one if its two parameters are identical and zero otherwise. Then, $P(x_j|y)$ and $P(x_i, x_j|y)$ can be computed by (3):

$$\begin{cases} P(x_j|y) = \dfrac{P(x_j, y)}{P(y)} \\ P(x_i, x_j|y) = \dfrac{P(x_i, x_j, y)}{P(y)} \end{cases} \tag{4}$$

3   Target Learning

As Fig. 1 shows, these state-of-the-art algorithms described above, i.e., TAN and KDB, which learn from training data $\mathcal{T}$ and apply different strategies to build $\text{BNC}_\mathcal{T}$, may represent different conditional dependencies between attributes. It seems that $\text{BNC}_\mathcal{T}$s with higher degree of attribute dependence will more closely fit the training data and can achieve better generalization performance than those with lower degree of attribute dependence. However, higher degree of attribute dependence means more parameters, which increases the risk of overfitting. An overfitted model does not perform well on new (testing) samples.

Consider a particular testing instance $p = \{x_1, \cdots, x_n, Y =?\}$. To assign the class label to p, only a small number of "right" dependencies, which are set in $\text{BNC}_\mathcal{T}$, are necessary. The remaining "redundant" dependencies in $\text{BNC}_\mathcal{T}$ may counteract the effect of the necessary "right" dependencies. The proposed approach aims to give high priority the dependencies that related to the elements in p.

In what follows, we take p as a target and build a specific BNC for p that will be denoted by $\text{BNC}_\mathcal{P}$. In contrast to $\text{BNC}_\mathcal{T}$, $\text{BNC}_\mathcal{P}$ is defined by the conditional dependencies at the attribute values in p. Then, after training $\text{BNC}_\mathcal{T}$ and $\text{BNC}_\mathcal{P}$, an ensemble learning is generated that combines both predictions. Obviously, for different testing instances, $\text{BNC}_\mathcal{T}$ remains the same while $\text{BNC}_\mathcal{P}$ may differ greatly.

Given testing instance p, the corresponding $\text{BNC}_\mathcal{P}$ is constructed by using the conditional mutual information defined by the joint and conditional probabilities at the attribute values $X_i = x_i$, $X_j = x_j$; that is,

$$I_\mathcal{P}(X_i; X_j|Y) = \sum_y \hat{P}(x_i, x_j, y) log \frac{\hat{P}(x_i, x_j|y)}{\hat{P}(x_i|y)\hat{P}(x_j|y)} \tag{5}$$

$$I_\mathcal{P}(X_i; Y) = \sum_y \hat{P}(x_i, y) log \frac{\hat{P}(x_i, y)}{\hat{P}(x_i)\hat{P}(y)} \tag{6}$$

where

$$\begin{cases} \hat{P}(y) = \frac{1}{N+1}\left[\sum_{k=1}^N \delta(y, \hat{y}_k) + \frac{1}{m}\right] \\ \hat{P}(x_j) = \frac{1}{N+1}\left[\sum_{k=1}^N \delta(x_j, \hat{x}_{kj}) + 1\right] \\ \hat{P}(x_j, y) = \frac{1}{N+1}\left[\sum_{k=1}^N \delta(< x_j, y >, < \hat{x}_{kj}, \hat{y}_k >) + \frac{1}{m}\right] \\ \hat{P}(x_i, x_j, y) = \frac{1}{N+1}\left[\sum_{k=1}^N \delta(< x_i, x_j, y >, < \hat{x}_{ki}, \hat{x}_{kj}, \hat{y}_k >) + \frac{1}{m}\right] \end{cases} \tag{7}$$

Conditional probabilities can be estimated as follows

$$
\begin{cases}
\hat{P}(x_j|y) = \dfrac{\hat{P}(x_j, y)}{\hat{P}(y)} \\
\hat{P}(x_i, x_j|y) = \dfrac{\hat{P}(x_i, x_j, y)}{\hat{P}(y)}
\end{cases}
\tag{8}
$$

Similar to the Laplace correction [15], the main idea behind formula (7) is equivalent to creating a "pseudo" training set $\mathcal{P}$ by adding to the training set $\mathcal{T}$ a new instance $(x_1, \cdots, x_n)$ with multi-label by assuming that the probability that this new instance is in class y is $1/m$ for each $y \in \{y_1, \cdots, y_m\}$.

One advantage of BNC$_\mathcal{P}$ is its dependence on testing instance. For example, consider two unlabeled instances $p_1 = \{1, 1, \cdots, 0\}$ and $p_2 = \{1, 0, \cdots, 0\}$. A minor difference between p_1 and p_2 is the value of X_2. Then p_1 and p_2 are used to create pseudo training sets $\mathcal{P}_1$ and $\mathcal{P}_2$, respectively. The differences between $I_{\mathcal{P}_1}(X_i = x_i; X_2 = 0|Y)$ and $I_{\mathcal{P}_2}(X_i = x_i; X_2 = 1|Y)(i \neq 2)$, $I_{\mathcal{P}_1}(X_2 = 0; Y)$ and $I_{\mathcal{P}_2}(X_2 = 1; Y)$, may make the structures of BNC$_{\mathcal{P}_1}$ and BNC$_{\mathcal{P}_2}$ quite different. Breiman [16] revealed that ensemble learning brings improvement in accuracy only to those "unstable" learning algorithms, in the sense that small variations in the training set would lead them to produce very different models. BNC$_\mathcal{P}$ is obviously an example of such learners. Thus the negative effect caused by overfitting will be mitigated to a great extent.

Another advantage of BNC$_\mathcal{P}$ which makes it very suitable for data mining domains is its relatively small computational complexity. Computing the network structure BNC$_\mathcal{T}$ of TAN and KDB, requires $\mathcal{O}(n^2 N m v^2)$ time (dominated by the calculations of conditional mutual information [6]. Whereas computing the corresponding BNC$_\mathcal{P}$ takes only $\mathcal{O}(n^2 N m)$ time, where n is the number of attributes, N is the number of data instances, m is the number of class labels, and v is the maximum number of discrete values that an attribute may take.

To make BNC$_\mathcal{T}$ and BNC$_\mathcal{P}$ complementary to each other, the same learning strategy is applied to build them. In the following discussion, we clarify the basic idea of target learning by using TAN as the base classifier. The learning procedures of TAN$_\mathcal{T}$ and corresponding TAN$_\mathcal{P}$ can be described as follows:

Algorithm 1: The TAN$_\mathcal{T}$ algorithm

Input: training data $\mathcal{T}$.
Output: TAN$_\mathcal{T}$, network structure.

1. Calculate the prior and conditional probabilities by (3) and (4).
2. Calculate $I(X_i; X_j|Y)(i \neq j)$ between each pair of attributes by (1).
3. Build a complete undirected graph in which the vertices are the attributes $X_1, \cdots, X_n$. Annotate the weight of an edge connecting X_i to X_j by $I(X_i; X_j|Y)$ $(i \neq j)$.
4. Build the maximum weighting spanning tree.
5. Transform the resulting undirected tree to a directed one by choosing a root variable and setting the direction of all edges to be outward from it.

6. Construct a TAN$_T$ model by adding a vertex labeled by Y and adding an arc from Y to each X_i.

7. Return TAN$_T$.

Algorithm 2: The TAN$_P$ algorithm

Input: training data T and testing instance $p = \{x_1, \cdots, x_n\}$.
Output: TAN$_P$, network structure.

1. Calculate the prior and conditional probabilities by (7) and (8).

2. Calculate $I_P(X_i; X_j|Y)(i \neq j)$ between each pair of attribute values by (5).

3. Build a complete undirected graph in which the vertices are the attributes $X_1, \cdots, X_n$. Annotate the weight of an edge connecting X_i to X_j by $I_P(X_i; X_j|Y)$ $(i \neq j)$.

4. Build the maximum weighting spanning tree.

5. Transform the resulting undirected tree to a directed one by choosing a root variable and setting the direction of all edges to be outward from it.

6. Construct a TAN$_P$ model by adding a vertex labeled by Y and adding an arc from Y to each X_i.

7. Return TAN$_P$.

The decision of the ensemble should have better overall accuracy, on average, than any individual member. There exist numerous methods for model combination, e.g. linear combiner, the product combiner and the voting combiner. For subclassifier BNC, an estimate of the probability of class y given input $\mathbf{x}$ is $P(y|\mathbf{x}, \text{BNC})$. The linear combiner is used for models that output real-valued numbers, so is applicable for BNC. And the ensemble probability estimate for BNC$_T$ and BNC$_P$ is,

$$\hat{P}(y|\mathbf{x}) = w_T.P(y|\mathbf{x}, \text{BNC}_T) + w_P.P(y|\mathbf{x}, \text{BNC}_P).$$

If the weights $w_T = w_P = 1/2$, this is a simple uniform averaging of the probability estimates. The notation clearly allows for the possibility of a nonuniformly weighted average. If the classifiers have different accuracies on the data, a nonuniform combination could in theory give a lower error than a uniform combination. However, it is always difficult to determine the values of w_T and w_P. Thus in practice we use the uniformly rather than nonuniformly weighted average.

4   Experimental Study

We conduct experiments on 32 data sets (size> 1000) from the UCI machine learning repository [17]. Table 1 summarizes the characteristics of each data set, including the numbers of instances, attributes and classes. Missing values for qualitative attributes are replaced with modes and those for quantitative

Table 1. Data sets

Data set	# Instance	Att	Class	Data set	# Instance	Att	Class
Abalone	4177	8	3	Optical Recognition	5620	64	10
Adult	48842	14	2	Page Blocks	5473	10	5
Census-Income (KDD)	299285	41	2	Pen Recognition	10992	16	10
Connect-4	67557	42	3	Phoneme	5438	7	50
Dis	3772	29	2	Pioneer-1 Mobile	9150	36	57
Allhypo	3772	29	4	Poker Hand	1025010	10	10
Hypo	3163	25	2	Landsat	6435	36	6
IPUMS Census	88443	60	19	Statlog (Image)	2310	19	7
Chess	3196	36	2	Statlog (Shuttle)	58000	9	7
Letter Recognition	20000	16	26	Sick	3772	29	2
Localization	164860	5	11	Sign	12546	8	3
MAGIC Gamma	19020	10	2	Spambase	4601	57	2
Multiple Features	2000	6	10	Molecular Biology	3177	60	3
Mushroom	8124	22	2	Thyroid	9169	29	20
Musk (Version 2)	6598	166	2	Wall-Following Robot	5456	24	4
Nursery	12960	8	5	Waveform	5000	40	3

attributes are replaced with means from the training data. For each benchmark data set, numeric attributes are discretized using MDL discretization [18]. Sahami [6] proposed the notion of k-dependence BNC, which allows each attribute X_i to have a maximum of k attribute nodes as parents. The following techniques with different structure complexities are compared:

- NB (0-dependence).
- TAN (1-dependence).
- K_2DB (2-dependence), KDB with k=2.
- K_3DB (3-dependence), KDB with k=3.

These single-structure BNCs except NB learn from the labeled training data and are examples of $BNC_{\mathcal{T}}$. We can then build corresponding $BNC_{\mathcal{P}}$s by applying target learning. The final ensemble classifiers, TAN^e, K_2DB^e and K_3DB^e, will be compared with these $BNC_{\mathcal{T}}$s to verify the efficiency and effectiveness of target learning. 0-1 loss is the most common loss function to measure the classification performance. Kohavi and Wolpert [19] presented a bias-variance decomposition of 0-1 loss from sampling theory statistics for analyzing supervised learning scenarios. We perform a stratified 10-fold cross-validation to compare the performance of these seven learning algorithms.

We assess a difference as significant if the outcome of a one-tailed binomial sign test is less than 0.05. Cell$[i, j]$ in Table 2 contains the number of win/draw/loss (W/D/L) records for the classifier on row i against the classifier on column j. A win indicates that the algorithm has significantly higher classification accuracy than the comparator. A draw indicates that the differences in classification accuracy are not significant. Table 2 presents the win/draw/loss records of the above 7 algorithms with respect to 0-1 loss, bias and variance. K_3DB can represent the largest number of dependencies among all. However, as the dependence degree or structure complexity increases, K_3DB does not enjoy

Table 2. All pairwise comparisons of the seven BNCs for 32 domains.

	Classifier	NB	TAN	K_2DB	K_3DB	TAN^e	K_2DB^e
0-1 Loss	TAN	24-5-3	-	-	-	-	-
	K_2DB	26-3-3	9-20-3	-	-	-	-
	K_3DB	24-5-3	8-13-11	2-19-11	-	-	-
	TAN^e	26-2-4	**17-14-1**	14-10-8	15-10-7	-	-
	K_2DB^e	28-3-1	19-9-4	**18-12-2**	19-11-2	14-13-5	-
	K_3DB^e	27-3-2	14-11-7	11-12-9	**14-17-1**	9-13-10	4-16-12
Bias	TAN	27-0-5	-	-	-	-	-
	K_2DB	29-0-3	24-7-1	-	-	-	-
	K_3DB	29-0-3	22-9-1	8-17-7	-	-	-
	TAN^e	27-2-3	**8-15-9**	3-4-25	2-6-24	-	-
	K_2DB^e	29-1-2	22-9-1	**11-12-9**	10-14-8	25-5-2	-
	K_3DB^e	28-0-4	20-8-4	7-13-12	**8-14-10**	22-9-1	6-14-12
Variance	TAN	5-1-26	-	-	-	-	-
	K_2DB	8-0-24	6-7-19	-	-	-	-
	K_3DB	7-1-24	6-4-22	4-8-20	-	-	-
	TAN^e	9-4-19	**25-5-2**	25-4-3	27-0-5	-	-
	K_2DB^e	9-3-20	17-9-6	**25-4-3**	25-5-2	6-7-19	-
	K_3DB^e	9-1-22	18-7-7	22-5-5	**25-4-3**	7-7-18	10-9-13

significant advantage in classification over the other BNCs. K_2DB beats K_3DB in 11 domains and loses in only 2 and TAN even beats K_3DB in 11 domains and loses in 8. Thus overfitting may be the main reason why the classification performance degrades when $k = 3$.

When $BNC_{\mathcal{P}}$ is introduced for further discovery of more significant dependencies that exist in unlabeled instance, the application of target learning helps TAN^e, K_2DB^e and K_3DB^e possess significant advantage over corresponding base BNCs. For example, TAN^e beats TAN in 17 domains and loses in 1, K_2DB^e beats K_2DB in 18 domains and loses in 2, and K_3DB^e beats K_3DB in 14 domains and loses in 1. Although K_3DB performs the worst when compared to TAN and K_2DB, K_3DB^e performs better than TAN (14-11-7) and K_2DB (11-12-9). Then we compare these three ensemble classifiers, i.e., TAN^e and K_2DB^e and K_3DB^e, we can see that K_2DB^e performs much better than TAN^e (14-13-5) and TAN^e slightly better than K_3DB^e (10-13-9). Thus the advantage of ensemble classifier is greatly determined by corresponding base classifier.

We can also observe that, in terms of bias BNC^e may perform better or worse than corresponding BNC whereas the different is not significant. In terms of variance, NB performs the best among all because of its definite structure regardless of the change of training data. TAN^e beats TAN in 25 domains and loses in 2, K_2DB^e beats K_2DB in 25 domains and loses in 3, and K_3DB^e beats

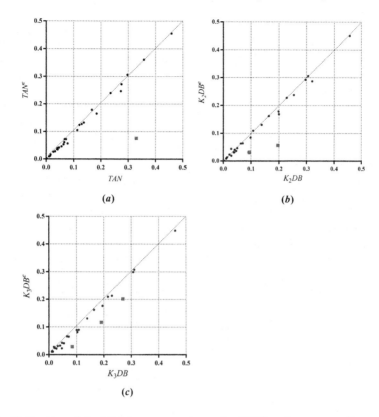

Fig. 2. Scatter plot of 0-1 loss comparisons for BNC and corresponding BNC^e.

K$_3$DB in 25 domains and loses in 3. Thus the advantage of BNC^e over corresponding BNC in terms of 0-1 loss can be attributed to the change in variance. The variance increases as the algorithm becomes more sensitive to the change in labeled training data. Obviously, target learning helps to alleviate the negative effect caused by overfitting.

To prove the complementary characteristic introduced by target learning, we compare the experimental results of BNC and BNC^e in terms of 0-1 loss in Figs. 2(a), 2(b) and 2(c) respectively, where squared symbols are used to indicate much indicate significant advantage of BNC^e over corresponding BNC. We can see that almost no points are far above the diagonal line, thus the negative effect caused by target learning is negligible. In contrast many more points are below the diagonal line. That means, target learning works effectively in most cases. Both BNC$_\mathcal{T}$ and BNC$_\mathcal{P}$ are strong rather than weak classifiers, they can independently achieve high classification accuracy. For 1-dependence BNC, e.g., TAN and TAN^e, only the most significant conditional dependencies can be represented, the structure similarity between them overwhelms the difference. However, as mentioned above the number of conditional dependencies to be represented will increase exponentially as k increases. Thus for relatively high-dependence BNCs, e.g., K_3DB and K_3DB^e, some non-significant dependencies can be represented. The structure difference between them overwhelms the similarity and that makes the complementary characteristic work. That can illustrate why there appear more points with squared symbol as dependence degree increases.

5   Conclusion

High degree of conditional dependencies may result in overfitting of the training data and lead to underfitting of unlabeled data. To address this issue, we propose target learning to build an unstable classifier $BNC_\mathcal{P}$ for each unlabeled instance p. To make $BNC_\mathcal{P}$ complementary to $BNC_\mathcal{T}$, we apply the same learning strategy to build them. Extensive experimental results show that target learning significantly improves the generalization performance of base classifiers. We explore reasons for the effectiveness of target learning on 32 data sets. Since $BNC_\mathcal{P}$ tries to directly represent the conditional dependencies that exist in unlabeled instance p, it is more appropriate to fit p than $BNC_\mathcal{T}$. Exploration of effective methods of optimizing the network structure of $BNC_\mathcal{P}$ to retain the most significant dependenices is a further area for future work.

6   Code

The code of the methods proposed in this work and detailed results of 0-1 loss, bias and variance can be obtained from the website, https://github.com/Bayes514/Target-Learning/tree/master/target-learning.

References

1. Pearl, J.: Probabilistic reasoning in intelligent systems: networks of plausible inference. Artif. Intell. **48**, 117–124 (1991). https://doi.org/10.1016/0004-3702(91)90084-w
2. Lewis, D.D.: Naive (Bayes) at forty: the independence assumption in information retrieval. In: Nédellec, C., Rouveirol, C. (eds.) ECML 1998. LNCS, vol. 1398, pp. 4–15. Springer, Heidelberg (1998). https://doi.org/10.1007/BFb0026666
3. Langley, P.: Induction of recursive Bayesian classifiers. In: Brazdil, P.B. (ed.) ECML 1993. LNCS, vol. 667, pp. 153–164. Springer, Heidelberg (1993). https://doi.org/10.1007/3-540-56602-3_134
4. Jiang. S., Harry, Z.: Full Bayesian network classifiers. In: 23rd International Conference on Machine Learning, Pittsburgh, Pennsylvania, pp. 897–904 (2006). https://doi.org/10.1145/1143844.1143957
5. Friedman, N., Geiger, D., Goldszmidt, M.: Bayesian network classifiers. Mach. Learn. **29**, 131–163 (1997). https://doi.org/10.1002/9780470400531.eorms0099
6. Sahami, M.: Learning limited dependence Bayesian classifiers. In: 2nd International Conference on Knowledge Discovery and Data Mining, Portland, United States, pp. 335–338 (1996). https://doi.org/10.1007/978-1-4471-0745-3_8

7. Zheng, Z.J., Webb, G.I.: Lazy learning of Bayesian rules. Mach. Learn. **41**, 53–84 (2000). https://doi.org/10.1007/978-1-4471-0745-3_8
8. Martínez, A.M., Webb, G.I., Chen, S.L., Zaidi, N.A.: Scalable learning of Bayesian network classifiers. J. Mach. Learn. Res. **17**, 1–30 (2016). https://doi.org/10.1145/1015330.1015339
9. Taheri, S., Mammadov, M.: Structure learning of Bayesian Networks using global optimization with applications in data classification. Optim. Lett. **9**, 931–948 (2015). https://doi.org/10.1007/s11590-014-0803-1
10. Triguero, I., García, S., Herrera, F.: Self-labeled techniques for semi-supervised learning: taxonomy, software and empirical study. Knowl. Inf. Syst. **42**, 245–284 (2015). https://doi.org/10.1007/s10115-013-0706-y
11. Nikos, F., Stamatis, K., Sotiris, K., Kyriakos, S.: Self-trained LMT for semisupervised learning. Comput. Intell. Neurosci. **2**, 1–13 (2016). https://doi.org/10.1155/2016/3057481
12. Didaci, L., Fumera, G., Roli, F.: Analysis of co-training algorithm with very small training sets. In: Gimel'farb, G., et al. (eds.) SSPR/SPR 2012. LNCS, vol. 7626. Springer, Heidelberg (2012). https://doi.org/10.1007/978-3-642-34166-3_79
13. Zhu, X.J.: Semi-supervised learning literature survey. Comput. Sci. **37**, 63–77 (2008). https://doi.org/10.7551/mitpress/9780262033589.003.0001
14. Shannon, C.E.: The Mathematical Theory of Communication. University of Illinois Press, Champaign (1949)
15. Park, S.H., Fürnkranz, J.: Efficient implementation of class-based decomposition schemes for Naive Bayes. Mach. Learn. **96**, 295–309 (2014). https://doi.org/10.21275/v4i11.nov151091
16. Breiman, L.: Bagging predictors. Mach. Learn. **24**, 123–140 (1996). https://doi.org/10.1007/bf00058655
17. UCI repository of machine learning databases (1995). http://www.ics.uci.edu/mlearn/MLRepository.html
18. Fayyad, U.M., Irani, K.B.: Multi-interval discretization of continuous valued attributes for classification learning. In: 5th International Joint Conference on Artificial Intelligence, France, Chambery, pp. 1022–1029 (1993). https://doi.org/10.1109/icmlc.2010.5581069
19. Kohavi, R., Wolpert, D.: Bias plus variance decomposition for zero-one loss functions. In: 13th International Conference on Machine Learning, Bari, Italy, pp. 275–283 (1996). https://doi.org/10.1007/978-0-387-09823-4_37

Automatic Chinese Reading Comprehension Grading by LSTM with Knowledge Adaptation

Yuwei Huang[1,2], Xi Yang[2(✉)], Fuzhen Zhuang[3,4(✉)], Lishan Zhang[2],
and Shengquan Yu[2]

[1] Beijing University of Chemical Technology, Beijing 100029, China
huangyw95@foxmail.com
[2] Beijing Advanced Innovation Center for Future Education,
Beijing Normal University, Beijing 100875, China
{xiyang85,lishan,yusq}@bnu.edu.cn
[3] Key Lab of Intelligent Information Processing of Chinese Academy of Sciences
(CAS), Institute of Computing Technology, CAS, Beijing 100190, China
zhuangfuzhen@ict.ac.cn
[4] University of Chinese Academy of Sciences, Beijing 100049, China

Abstract. Owing to the subjectivity of graders and the complexity of assessment standard, grading is a tough problem in the field of education. This paper presents an algorithm for automatic grading of open-ended Chinese reading comprehension questions. Due to the high complexity of feature engineering and the lack of consideration for word order in frequency based word embedding models, we utilize long-short term memory recurrent neural network to extract semantic feature in student answers automatically. In addition, we also try to impose the knowledge adaptation from web corpus to student answers, and represent the students' responses to vectors which are fed into the memory network. Along this line, the workload of teacher and the subjectivity in reading comprehension grading can both be reduced obviously. What's more, the automatic grading methods for Chinese reading comprehension will be more thorough. The experimental results on five Chinese and two English data sets demonstrate the superior performance over compared baselines.

Keywords: Automatic grading · Knowledge adaptation
Reading comprehension · LSTM · Text classification

1 Introduction

It is a tough problem in the field of education when grading student answers because of the subjectivity of graders and the complexity of assessment standards. Particularly, with the evolution of e-learning and online examination, the demand for assessment is increasing. It is apparent that hiring a great number of teachers is not a cost-efficient way. A growing number of researchers

© Springer International Publishing AG, part of Springer Nature 2018
D. Phung et al. (Eds.): PAKDD 2018, LNAI 10937, pp. 118–129, 2018.
https://doi.org/10.1007/978-3-319-93034-3_10

have engaged in automatic grading, especially automatic reading comprehension grading [1, 10, 16–18]. Nevertheless, there is little research on automatic open-ended Chinese reading comprehension grading. Therefore, it is fairly essential to develop automatic grading methods for open-ended Chinese reading comprehension questions.

Table 1. Sample of open-ended Chinese reading comprehension

Text Information
黄阿二的酒酿在古庙镇上老老少少都翘起大拇指 (…) 他那极有韵味的吆喝可以说已成了古庙镇的一种文化风景 (…) 黄阿二听后浑身一震, 他撑起身子说:"你们这一声吆喝, 对我来说, 比吃啥药都强, 这不, 毛病好了一半."
Text Information Translation
People in Gumiao town thumb up for the sweet ferment rice of Huang A'er(…) his lasting appeal of yo-heave-ho has became a kind of cultural landscape of Gumiao town, after hearing that(…) Huang A'er shocks his body and raises body to say: "your shouting is better than any other medicine for me, look, illness has been improved half."
Question
结合 "他那极有韵味的吆喝可以说已成了古庙镇的一种文化风景" 这句话, 结合你的生活体验, 请说一说你是怎么理解 "酒酿王的吆喝声" 的?
Question Translation
Combine with the sentence of "his lasting appeal of yo-heave-ho has became a kind of cultural landscape of Gumiao town" and your life experience, please express your understanding about the "yo-heave-ho of sweet ferment rice king".
Example of Full-Score Answer
我认为酒酿王的吆喝声是人的勤劳, 人与人间无形的热情. 生活中, 卖东西的小贩沿街叫卖, 无论天气季节, 都坚持和做交易时的样子, 都使人感到暖意.
Example of Full-Score Answer Translation
I think his yo-heave-ho shows diligent and intangible enthusiasm among people. Vendors' shouting along street and their insistence on trading no matter how is the weather and what is the season can make people feel warm-hearted in everyday life.

The reading comprehension question is the one that students should read a text and answer the questions about it. The questions can be designed with different openness. In this paper, we concentrate on the open-ended reading comprehension questions. As the example shown in Table 1, after reading a text, the students may be asked to express their understanding about the given text according to their actual experience. The students may answer this question differently since their different experiences and understandings. Moreover, the sentence semantic and words used by different students maybe divergent and diverse, which make the open-end reading comprehension actually no reference answers.

In the past few years, most of methods for automatic reading comprehension grading are based on two assumptions: (1) Reading comprehension is close-ended and can be graded by recognizing some specific words without considering word orders in student answers. (2) Reading comprehension questions would provide graders with reference answer. However, a large proportion of Chinese reading comprehension questions is open-ended and there is few specific words among student answers. Under this circumstance, automatic grading by specific words based on bag-of-words models without word orders would be invalid. What's worse, most of Chinese comprehension lack of reference answer, thus the automatic grading methods based on reference answers would not work.

Based on the analysis above, we have the following motivations to propose a new automatic reading comprehension grading model:

(1) These are enormous demands for automatic reading comprehension grading.
(2) It is significant to propose a framework for grading reading comprehension without reference answers.
(3) It is crucial to take word orders into consideration for automatic reading comprehension grading methods.

Along this line, we try to formalize the automatic open-ended Chinese reading comprehension grading problem as text classification. In this way, the algorithm can be conducted without reference answers. For word level, we represent each word as a vector trained by continuous bag-of-words model (CBOW) [20]. For sentence level, long-short term memory recurrent neural network (LSTM) [9] is used to model student answers by calculating the word embedding based on CBOW. LSTM is well approved to model sequence data, which can be used to extract word order features in student answers. We validate our model on seven data sets, including Chinese and English reading comprehension questions. The extensive experimental results demonstrate the superiority of our method over several state-of-the-art baselines in terms of QWKappa (Cohen's kappa with quadratic weight), accuracy, precision, recall, and F1-score.

The remainder of this paper is organized as follows. Section 2 introduces the framework and its solution details. The experimental results are reported in Sect. 3. Section 4 discusses the related work and finally Sect. 5 concludes.

2   Model for Automatic Open-Ended Chinese Reading Comprehension Grading

Figure 1 shows the framework and the training process of our proposed model, which is constructed based on continuous bag-of-words model (CBOW) [19] and long-short term memory recurrent neural network (LSTM) [9]. CBOW is a word embedding model, which represents each word as a vector. These word vectors are fed into LSTM in sequence. LSTM is a variant of recurrent neural network (RNN), and we take the advantage of LSTM to extract semantic information of each student answer. We use Adam in [14] to minimize the cross-entropy [4] loss function. In order to train the CBOW model more quickly and accurately, we utilize knowledge adaptation to transfer the external knowledge from web corpus to student answers (target corpus). Next, we will introduce the details of our model.

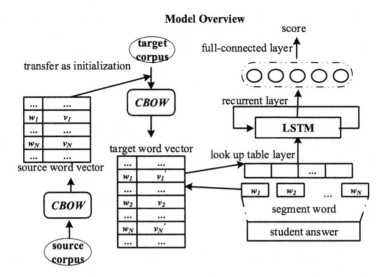

Fig. 1. Framework of our proposed model.

2.1 Negative Sampling Based Continuous Bag-of-Words Embedding

One hot encoding is a prevalent word representation method for neural network based natural language processing tasks. It encodes each word as a vector by marking at its index in vocabulary. However, encoding words in this way can not measure the distances among words. What's worse, it may be high-dimensional as the growth of vocabulary. Hence many word representation models are proposed for estimating continuous representations of words, including the prominent Latent Semantic Analysis (LSA) and Latent Dirichlet Allocation (LDA) [19]. In this paper, we use distributed representations of word learned by deep learning. Distributed representations proved to perform better than LSA for preserving linear regularities among words [19,21,28].

According to [19], the basic CBOW model is similar to the feed-forward neural network language model (NNLM) in [3], where the non-linear hidden layer is removed and the projection layer is shared for all words. In this paper, we also utilize negative sampling proposed in [20] for CBOW model that results in faster training and better vector representation for frequent words. Next, we would like to introduce the CBOW model more clearly.

Input Layer: For one of the target word w_k in a sentence, there are $2c$ words including c precedent words $w_{k-c}, ..., w_{k-2}, w_{k-1}$ and c posterior words $w_{k+1}, w_{k+2}, ..., w_{k+c}$ are used as context words of target word w_k, and we initialize the vector for each word in d-dimension randomly. These vectors will be updated by back propagation and used as input in look-up table layer.

Project Layer: Averaging all word vectors in each position to a vector v_k according to Eq. (1)

$$v_k = \frac{1}{2c} \sum_{i=k-c, i \neq k}^{k+c} v_i, \tag{1}$$

where v_i is the vector of word w_i.

Output Layer: For all words in the corpus, the CBOW model try to predict each target word w_k based on the context information v_k. The loss function is as Eq. (2):

$$\mathcal{L}_{CBOW} = log \prod_{k=1}^{|D|} \{\delta(v_k^T \theta^k) \prod_{j=1}^{|NEG_k|} [1 - \delta(v_k^T \theta^j)]\}, \tag{2}$$

where D is the whole word vocabulary for a corpus, NEG_k is the negative sampling set of target word w_k. Although the noise words in negative sampling set for each word is diverse, the total number of noise words $|NEG_k|$ are the same. $\delta(v_k^T \theta^k)$ is the likelihood of correct prediction, while $\delta(v_k^T \theta^j)$ is the likelihood of negative prediction and j is the word index in negative sampling set. The model maximizes the $\mathcal{L}_{CBOW}$, so that the likelihood of correct predication would be maximized and the likelihood of negative predication would be minimized at the same time. The back propagation would update context information v_k and the random initialized word vector during optimizing the CBOW model.

2.2   Knowledge Adaptation for Continuous Bag-of-Word Embedding

According to [25], the parameters in earlier layers of neural network which trained on large data sets are general to different tasks. Therefore, using existing parameters for initialization can benefit performance improvement and time-saving. Knowledge adaptation is a technique that aims to adapt pre-trained models to new natural language processing tasks.

As we have mentioned in the last subsection, the continuous bag-of-words (CBOW) model contains three layers, and weights of each layer are updated by back propagation. The word vectors initialized on input layer would be updated as well. Hence we first train CBOW model on large scale corpora (wikipedia) and transfer the existing word vectors to input layers, and add new initialized word vectors to target automatic grading task. In this way, the existing vectors transferred from large corpora would complete training quickly and the vectors for new words on target tasks would be trained precisely with accurate pre-trained context word vectors. Finally, the external knowledge learned from large scale corpora would be adapted for new automatic grading tasks.

2.3   Recurrent Layer

Frequency based word embedding models are well-known for statistic machine learning based automatic grading methods. Boolean vectorization is a model that reflects whether a word occurs in a document or not. And term frequency-inverse document frequency (TF-IDF) [13] is a numerical statistic that reflects how important a word is to a document in a collection or corpus. Since these models ignore context information in student answers, therefore we utilize the excellent sequence modeling technique LSTM to learn the semantic features of student answers.

LSTM contains special units called memory blocks in the recurrent hidden layer of RNN, which further contains memory cells with self-connections to store the temporal state of the network, and special multiplicative units called gates to control the information flow. Every block in the architecture contains an input gate and an output gate. The input gate controls the flow of input activations into memory cell, and the output gate controls the output flow of cell activations into the rest of the network [9]. The LSTM functions are as follows,

$$
\begin{aligned}
i_t &= \sigma(W_i x_t + U_i h_{t-1} + b_i), \quad f_t = \sigma(W_f x_t + U_f h_{t-1} + b_f) \\
c_t' &= \tanh(W_c x_t + U_c h_{t-1} + b_c), \quad c_t = i_t \circ c_t' + f_t \circ c_{t-1} \\
o_t &= \sigma(W_o x_t + U_o h_{t-1} + b_o), \quad h_t = o_t \circ \tanh(c_t)
\end{aligned}
\tag{3}
$$

where x_t and h_t are the input and output vectors at time t, W_i, W_f, W_c, W_o, U_i, U_f, U_c, and U_o are weight matrices and b_i, b_f, b_c, and b_o are bias vectors, $\circ$ is the element-wise multiplication, and σ represents the sigmoid function.

2.4   Fully-Connected Layer with Softmax Activation

After learning sequence features from student answers, we utilize a fully-connected layer and softmax activation [6] to calculate the output probability of each score. Assuming that there are K possible scores for each answer, and the output is a K-dimension vector as follows,

$$
h_\theta(x_i) = [P(y_i = 0|S_i; \theta), P(y_i = 1|S_i; \theta), \cdots, P(y_i = K - 1|S_i; \theta)]^\top \tag{4}
$$

where $P(y = k|S; \theta)(k = 0, 1, \cdots, K - 1)$ is the probability of each score for a student answer, and θ indicates the parameter in fully-connected layer. Finally, we use the Adam optimization algorithm [14] to minimize the cross-entropy [4] loss function on training data.

3   Experiments

To validate the effectiveness of our proposed model, we conduct experiments on seven data sets and compare it with several state-of-the-art baselines. Moreover, we also compare our proposed model with CBOW without knowledge adaptation, TCBOW train on student answers and SCBOW train on web corpus.

3.1   Data Sets and Preprocessing

The details of all data sets are showed in Table 2, where "Avg#word" denotes the average number of words for student answers, "#samples" denotes the total number of student answers for each data set, and "QWKappa" denotes the grading consistency between two graders. For Chinese data sets, we utilize Chinese word segmentation system jieba[1] to segment Chinese words. For English data sets, we use scikit-learn[2] to tokenize and extract answer features.

Table 2. Overview of all datasets

Problem	Avg#word	#samples	#Score level	Language	QWKappa
CRCC1	39	2579	0–4	Chinese	0.9847
CRCC2	33	2571	0–2	Chinese	0.9723
CRCC3	26	2382	0–3	Chinese	0.9427
CRCC4	27	2458	0–4	Chinese	0.9733
CRCC5	31	2538	0–3	Chinese	0.8319
ASAP-SAS3	47	2297	0–2	English	–
ASAP-SAS4	40	2033	0–2	English	–

CRCC Data Set: To evaluate our model in Chinese answers, we construct a Chinese Reading Comprehension Corpus (CRCC). In order to ensure the reliability of the grading label, two Chinese teachers were asked to grade the answer individually. The consistency of the two teachers' scoring is evaluated by QWKappa. The QWKappa scores of CRCC are shown in Table 2, which demonstrates the label of each data set is reliable (i.e., the value of QWKappa is required to larger than 0.8). Higher value of QWKappa indicates a higher consistency between the two teachers' scores. At last, two teachers will discuss to make agreements on those answers that with different scorings.

ASAP-SAS Data Set: There are ten data sets in Kaggle Automatic Student Assessment Prize: Short Answer Scoring (ASAP-SAS)[3] (denoted as ASAP-SASx, $x \in \{1, 2, \cdots, 10\}$). We combine the training and test data from the leaderboard solution to a complete data set. According to the data description on that competition web site, ASAP-SAS3, 4, 7, 8, 9 are normal reading comprehension questions, which belong to "English language arts" and "English" subjects. Students should read a text in these questions and extract information from it. However, only the openness of ASAP-SAS3 and 4 fit our openness definition. Therefore, we select them for experiments, which have about 2000 samples in each question and average number of words is from 40 to 50.

In total, we construct 7 automatic open-ended reading comprehension grading tasks for each automatic grading model, 5 for Chinese and 2 for English.

---

[1] https://github.com/fxsjy/jieba.
[2] http://scikit-learn.org/stable/index.html.
[3] https://www.kaggle.com/c/asap-sas/data.

3.2   Baselines

We compare our model, denoted as KAGrader, with several state-of-the-are baseline automatic scoring models, including MLR [17], ZNB [26], DJDT [7,12], HSVM [11].

Implementation Details: All these baselines of MLR, HSVM, DJDT and ZNB are implemented by scikit-learn. We utilize gensim [22][4] to train CBOW model and the size of window for posterior and precedent words is 5, and the words whose frequency below 5 are ignored. The size of negative sampling sets is set to 5, and the wikipedia corpus[5] is used for knowledge adaptation.

For TCBOW, SCBOW and KAGrader, LSTM are implemented by the deep learning library keras [5]. For CRCC and ASAP-SAS data sets, the numbers of nodes in recurrent layers are {128, 128} and batch size are {512, 1024}, respectively. Finally, we all use 5-fold cross validation to evaluate our approach and baselines.

3.3   Results and Analysis

Except for four baselines, we also investigate the performance of our model without knowledge adaptation. Specifically, "TCBOW" only uses corpus from student answers (target corpus), and "SCBOW" uses Chinese or English wikipedia corpus (source corpus). While our model "KAGrader" considers both wikipedia corpus and student answers for knowledge adaptation, and all the results are reported in Tables 3 and 4. From these results, we have the following observations,

(1) HSVM is still a strong baseline which outperforms the other baselines many times, indicating that bag-of-words models can be improved by carefully selecting competitive classifier.
(2) TCBOW has worse performance than SCBOW. We conjecture that the volume of corpus has great influence on CBOW training. Therefore, it is significant for us to utilize the large source corpus to help train the target corpus vectors.
(3) KAGrader outperforms all the baselines in terms of QWKappa and average accuracy, which indicates that our model can combine the advantages of TCBOW and SCBOW.

It is worth mentioning that for neural approaches, sometimes it is limited to use the source corpus word vectors since some keywords may not appear in large source corpus, which may lead to the loss of important information in target student answers and output pure performance. Also the vector training performance is influenced by the volume of data set.

To further compare the behavior between bag-of-words models and our proposed model, we choose several student answers for analyzing.

---

[4] https://radimrehurek.com/gensim/index.html.
[5] https://dumps.wikimedia.org/.

Table 3. QWKappa on all data sets

	MLR	ZNB	DJDT	HSVM	TCBOW	SCBOW	KAGrader
CRCC1	0.3697	0.1970	0.2959	0.4015	0.2213	0.4431	**0.4520**
CRCC2	0.3915	0.1729	0.2556	0.4254	0.3752	0.4825	**0.4983**
CRCC3	0.7913	0.6340	0.8108	0.8680	0.7276	0.8364	**0.8694**
CRCC4	0.5142	0.2954	0.4333	0.5789	0.5693	0.5612	**0.5911**
CRCC5	0.6270	0.4465	0.6288	0.6522	0.4214	0.6754	**0.7058**
ASAP-SAS3	0.5604	0.5046	0.4558	0.5905	0.5947	0.6126	**0.6430**
ASAP-SAS4	0.5482	0.5644	0.4433	0.5695	0.5655	0.5717	**0.6103**
Average	0.5432	0.4021	0.4748	0.5837	0.4964	0.5976	**0.6230**

answer 1. "酒酿王的吆喝声" 质朴、[勤劳]，在我的生活中，妈妈每天早上叫我起床，就是朴实的声音。

answer 2. "酒酿王的吆喝声" 代表了他的 [勤奋]，值得传承。

answer 3. [扁鹊] 进谏的方式太直白，[邹忌] 是以家事比作国事来委婉进谏，以喻设理，形象的比喻了，让人欣然接受。

answer 4. [邹忌] 进谏的方式太直白，[扁鹊] 是以家事比作国事来委婉进谏，以喻设理，形象的比喻了，让人欣然接受。

For these answers, HSVM and KAGrader grade the answer 1 successfully. However, HSVM failed in answer 2 because it lack of some specific words such as 质朴、勤劳、朴实 While KAGrader score it in a right way because 勤劳 and 勤奋 are closely in CBOW word vector space and KAGrader can recognize them to output a correct score. Furthermore, the figures of 扁鹊 and 邹忌 are exchanged in sentence 4, which leads to the failure of HSVM. In a word, HSVM can not recognize the wrong sequence order, while KAGrader can take advantage of LSTM to address this issue.

Table 4. Average accuracy, precision, recall and F1 on seven data sets

	MLR	ZNB	DJDT	HSVM	TCBOW	SCBOW	KAGrader
Accuracy	0.6724	0.6645	0.6367	0.6868	0.7000	0.7256	**0.7375**
Precision	0.6628	0.6436	0.6334	0.6871	0.6821	0.7144	**0.7281**
Recall	0.6724	0.6645	0.6367	0.6868	0.7000	0.7256	**0.7375**
F1	0.6634	0.6304	0.6334	0.6844	0.6745	0.7125	**0.7255**

3.4 Parameter Sensitivity

In this section, we discuss how the number of nodes in recurrent layer and the training batch size impact on the performance of our model. To tune the hyper-parameters, we randomly selected two Chinese and all English problems.

The number of nodes are sampled from {16, 32, 64, 128, 256}, and the training batch size is sampled from {64, 128, 256, 512, 1024}. From these results in Fig. 2, we finally set the numbers of nodes and training batch size to {128, 128, 128, 128} and {512, 512, 1024, 1024} for these four data sets. Furthermore, we tune the size of negative sampling sets NEG_k in CBOW models, and the size is sampled from {1, 5, 10, 15, 20}. From the results, we finally set the same size to 5 for both Chinese and English data sets.

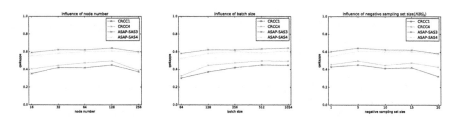

Fig. 2. Parameter sensitivity

4 Related Work

Short Answer and Reading Comprehension: NLP techniques are often used to extract various features from student answer to measure the similarity between the reference answer and student answers. Content Assessment Module (CAM) used features to measure the overlap of content on various linguistic levels [1]. The types of overlap include word unigrams, trigrams, text similarity thresholds etc. Madnani et al. in [17] used eight features based on the rubric (BLEU, coherence etc.) for summary assessment. After feature extracting process, these features are used to train various classification, regression or clustering models for grading new student answers automatically. Different machine learning models are utilized in ASAG task in [1,2,7,11,12,17,26]. Particularly, Zhang et al. [27] introduced Deep Belief Network (DBN) into short answer.

Automatic reading comprehension grading is regarded as an exception of ASAG, due to the fact that reading comprehension task need students to "understand" the reading text assuredly, not just "recall" the external knowledge. Meures et al. [18] considered that answers might show variation on different levels (lexical, morphological etc.). Horbach et al. [10] demonstrate that the use of text-based features can promote performance. Automatic reading comprehension grading was also investigated by Liu et al. [16] and Wang et al. [24].

Neural Network and Text Classification: A growing number of researchers applied neural network techniques in text classification which is a relevant topic for automatic grading. Graves et al. [8] applied LSTM into speech recognition. Tang et al. utilized GRU in sentiment classification [23]. Lai et al. used R-CNN in text classification [15]. Previous works reveal that neural network techniques perform well in natural language processing, which may have significant implications to automatic open-ended reading comprehension grading.

5 Conclusions

In this paper, we propose to combine continuous bag-of-words model (CBOW) and long-short term memory recurrent neural network (LSTM) for automatic open-ended Chinese reading comprehension grading. Our method does not rely on any reference answer due to the fact that reference answer is not always available for most open-ended reading comprehension questions. Based on CBOW and LSTM, our framework can extract semantic information automatically and effectively by considering the word orders in student response. Additionally, through knowledge adaptation, the external knowledge is transfered to present corpus. Experiments on seven data sets, including Chinese and English, demonstrate the effectiveness of the proposed method.

Acknowledgements. This work is supported by the National Natural Science Foundation of China (No. 61773361, 61473273), the Youth Innovation Promotion Association CAS 2017146, the China Postdoctoral Science Foundation (No. 2017M610054).

References

1. Bailey, S., Meurers, D.: Diagnosing meaning errors in short answers to reading comprehension questions. In: Proceedings of the Third Workshop on Innovative Use of NLP for Building Educational Applications, pp. 107–115. Association for Computational Linguistics (2008)
2. Basu, S., Jacobs, C., Vanderwende, L.: Powergrading: a clustering approach to amplify human effort for short answer grading. Trans. Assoc. Comput. Linguist. **1**, 391–402 (2013)
3. Bengio, Y., Ducharme, R., Vincent, P., Janvin, C.: A neural probabilistic language model. J. Mach. Learn. Res. **3**(6), 1137–1155 (2003)
4. Boer, P.T.D., Kroese, D.P., Mannor, S., Rubinstein, R.Y.: A tutorial on the cross-entropy method. Ann. Oper. Res. **134**(1), 19–67 (2005)
5. Chollet, F., et al.: Keras (2015). https://github.com/fchollet/keras
6. Dunne, R.A., Campbell, N.A.: On the pairing of the softmax activation and cross-entropy penalty functions and the derivation of the softmax activation function. In: Proceedings of the 8th Australian Conference on the Neural Networks, Melbourne, vol. 185, p. 181 (1997)
7. Dzikovska, M.O., Nielsen, R.D., Brew, C.: Towards effective tutorial feedback for explanation questions: a dataset and baselines. In: Proceedings of the Conference of the North American Chapter of the Association for Computational Linguistics: Human Language Technologies, pp. 200–210. Association for Computational Linguistics (2012)
8. Graves, A., Mohamed, A.r., Hinton, G.: Speech recognition with deep recurrent neural networks. In: IEEE International Conference on Acoustics, Speech and Signal Processing (ICASSP), pp. 6645–6649. IEEE (2013)
9. Hochreiter, S., Schmidhuber, J.: Long short-term memory. Neural Comput. **9**(8), 1735–1780 (1997)
10. Horbach, A., Palmer, A., Pinkal, M.: Using the text to evaluate short answers for reading comprehension exercises. In: * SEM@ NAACL-HLT, pp. 286–295 (2013)

11. Hou, W.-J., Tsao, J.-H., Li, S.-Y., Chen, L.: Automatic assessment of students' free-text answers with support vector machines. In: García-Pedrajas, N., Herrera, F., Fyfe, C., Benítez, J.M., Ali, M. (eds.) IEA/AIE 2010. LNCS (LNAI), vol. 6096, pp. 235–243. Springer, Heidelberg (2010). https://doi.org/10.1007/978-3-642-13022-9_24

12. Jimenez, S., Becerra, C.J., Gelbukh, A.F., Bátiz, A.J.D., Mendizábal, A.: Soft-cardinality: hierarchical text overlap for student response analysis. In: SemEval@ NAACL-HLT, pp. 280–284 (2013)

13. Jones, K.S.: A statistical interpretation of term specificity and its application in retrieval. J. Doc. **60**(5), 493–502 (2013)

14. Kingma, D., Ba, J.: Adam: A method for stochastic optimization. arXiv preprint arXiv:1412.6980 (2014)

15. Lai, S., Xu, L., Liu, K., Zhao, J.: Recurrent convolutional neural networks for text classification. AAAI **333**, 2267–2273 (2015)

16. Lui, A.K.-F., Lee, L.-K., Lau, H.-W.: Automated grading of short literal compre-hension questions. In: Lam, J., Ng, K.K., Cheung, S.K.S., Wong, T.L., Li, K.C., Wang, F.L. (eds.) ICTE 2015. CCIS, vol. 559, pp. 251–262. Springer, Heidelberg (2015). https://doi.org/10.1007/978-3-662-48978-9_23

17. Madnani, N., Burstein, J., Sabatini, J., O'Reilly, T.: Automated scoring of a summary-writing task designed to measure reading comprehension. In: BEA@ NAACL-HLT, pp. 163–168 (2013)

18. Meurers, D., Ziai, R., Ott, N., Bailey, S.M.: Integrating parallel analysis modules to evaluate the meaning of answers to reading comprehension questions. Int. J. Contin. Eng. Educ. Life Long Learn. **21**(4), 355–369 (2011)

19. Mikolov, T., Chen, K., Corrado, G., Dean, J.: Efficient estimation of word repre-sentations in vector space. arXiv preprint arXiv:1301.3781 (2013)

20. Mikolov, T., Sutskever, I., Chen, K., Corrado, G.S., Dean, J.: Distributed repre-sentations of words and phrases and their compositionality. In: Advances in Neural Information Processing Systems, pp. 3111–3119 (2013)

21. Mnih, A., Hinton, G.: Three new graphical models for statistical language mod-elling. In: Proceedings of the 24th International Conference on Machine Learning, pp. 641–648. ACM (2007)

22. Řehůřek, R., Sojka, P.: Software framework for topic modelling with large cor-pora. In: Proceedings of the LREC 2010 Workshop on New Challenges for NLP Frameworks, pp. 45–50. ELRA, Valletta, Malta, May 2010

23. Tang, D., Qin, B., Liu, T.: Document modeling with gated recurrent neural network for sentiment classification. In: EMNLP, pp. 1422–1432 (2015)

24. Wang, H.C., Chang, C.Y., Li, T.Y.: Assessing creative problem-solving with auto-mated text grading. Comput. Educ. **51**(4), 1450–1466 (2008)

25. Yosinski, J., Clune, J., Bengio, Y., Lipson, H.: How transferable are features in deep neural networks? In: Advances in Neural Information Processing Systems, pp. 3320–3328 (2014)

26. Zesch, T., Levy, O., Gurevych, I., Dagan, I.: UKP-BIU: similarity and entailment metrics for student response analysis, Atlanta, Georgia, USA, p. 285 (2013)

27. Zhang, Y., Shah, R., Chi, M.: Deep learning + student modeling + clustering: a recipe for effective automatic short answer grading. In: EDM, pp. 562–567 (2016)

28. Zhila, A., Yih, W.t., Meek, C., Zweig, G., Mikolov, T.: Combining heterogeneous models for measuring relational similarity. In: Proceedings of the 2013 Conference of the North American Chapter of the Association for Computational Linguistics: Human Language Technologies, pp. 1000–1009 (2013)

Data Mining with Algorithmic Transparency

Yan Zhou$^{(\boxtimes)}$, Yasmeen Alufaisan, and Murat Kantarcioglu

Erik Jonnson School of Engineering and Computer Science,
University of Texas at Dallas, Richardson, TX 75080, USA
{yan.zhou2,yxa130630,muratk}@utdallas.edu

Abstract. In this paper, we investigate whether decision trees can be used to interpret a black-box classifier without knowing the learning algorithm and the training data. Decision trees are known for their transparency and high expressivity. However, they are also notorious for their instability and tendency to grow excessively large. We present a classifier reverse engineering model that outputs a decision tree to interpret the black-box classifier. There are two major challenges. One is to build such a decision tree with controlled stability and size, and the other is that probing the black-box classifier is limited for security and economic reasons. Our model addresses the two issues by simultaneously minimizing sampling cost and classifier complexity. We present our empirical results on four real datasets, and demonstrate that our reverse engineering learning model can effectively approximate and simplify the black box classifier.

1 Introduction

The past decade has witnessed a rapid growth in the use of data mining techniques for better decision making. Statistical implications derived from data can help us understand and critically assess risks and uncertainties. However, the ubiquity of data and data mining techniques has also sparked new concerns on transparency, as has been emphasized in the recent report by PCAST (President's Council of Advisors on Science and Technology) [16]. Many proprietary intelligent software applications provide users with interfaces to the "smart algorithms" in their data analytics systems. The inner workings of these smart algorithms are often incomprehensible and opaque to ordinary users. Therefore, the information released to the end user is usually overly simplified, abstract, and untestable, which in return raises the problem of transparency and trustability. There are practical benefits of withholding the inner structure of knowledge the algorithms have learned, for example, protecting companies' information assets. However, data mining models can be discriminatory, making biased decisions on the basis of race, class, gender, etc. Recent work [22] has shown that some online ads are selected by intelligent advertising systems based on the racial background of the names used in search queries. This type of bias may be deeply and unconsciously hidden within data mining models. Increasing the transparency of these

© Springer International Publishing AG, part of Springer Nature 2018
D. Phung et al. (Eds.): PAKDD 2018, LNAI 10937, pp. 130–142, 2018.
https://doi.org/10.1007/978-3-319-93034-3_11

models can help users spot these caveats that would otherwise be hidden, and is important for ensuring trust and reducing potential abuses and biases.

Clearly, many issues need to be addressed to acquire truly transparent data mining models, for example, understanding the impact of the structure of data on a black-box classifier [11], and identifying subspaces where a black-box classifier does (not) work [6]. In this paper, we focus on learning a simple decision-tree equivalent of a given black box classifier with a small number of query samples. An immediate challenge is that decision trees are well known for their poor stabilities especially when the number of training samples is small [7]. We can build a decision tree or extract a set of decision rules from the kernel-based classifier with existing rule extraction algorithms [20]. However, rule extraction algorithms add additional complexity to existing kernel-based methods [5,9], and the output of rule extraction algorithms may still be incomprehensible [5].

In this paper, we present a black-box classifier reverse engineering approach as illustrated in Fig. 1. Our technique builds a kernel-based classifier and a decision tree classifier simultaneously. The kernel-based classifier is responsible for

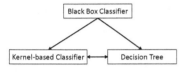

sampling under the maximum uncertainty constraint, and the decision tree classifier assists to curtail unnecessary growth in its own complexity. Our method provides: (a) a reverse engineering procedure with good stability and close similarity; (b) a cost and complexity-aware sampling technique; and (c) a human-comprehensible output.

Fig. 1. Our reverse engineering method.

We choose to use a kernel-based classifier for good stability and develop a new sampling technique to efficiently build a human-comprehensible decision tree counterpart of the black box classifier. Unlike existing kernel-based rule extraction algorithms, we do not operate on the kernel methods per se but instead focus on searching for data samples that naturally result in a simple decision tree equivalent of the black box classifier.

2   Related Work

Klivans et al. [13] study the learnability of convex bodies under the Gaussian distribution. They present a sub-exponential time algorithm for learning general convex bodies in the noise-free PAC-learning setting. Similarly, Rademacher and Goyal [17] consider learning a convex body in $\mathbb{R}^d$ given uniformly random samples from the convex body. The objective is to approximately learn the body with the fewest number of samples. They also show that it requires an exponential number of queries to learn the convex body. Also, Dyer et al. [8] present a random polynomial time algorithm for approximating the volume of a convex body in Euclidean space. Their algorithm requires a membership oracle and samples are selected nearly uniformly from within the convex body using a random walk.

Craven and Shavlik present an algorithm TREPAN to extract decision trees from artificial neural networks [3]. They modify the way a decision tree is built to limit the number of internal nodes. Henelius et al. [11] present a randomization approach to measuring the impact of groups of variables on a classification model. Duivesteijn and Thaele [6] present the SCaPE model class that highlights subspaces where the classifier performs particularly well or poorly. Both [6,11] require a sufficient number of queries to the black-box classifier for their models to work properly. Most recently, Ribeiro et al. [18] present a sparse linear model (LIME) for local exploration—providing interpretable representation locally faithful to the classifier. However, the global effectiveness of their model is questionable when the black-box classifier is highly non-linear. Datta et al. [4] present Quantitative Input Influence (QII) to measure the most influential inputs on the output of a classification model. QII also provides local transparency for a single instance or groups of instances. Unlike existing research discussed above, our technique reveals the global knowledge of the entire data domain by leveraging the inherent transparency and interpretability of decision trees. Three important aspects set our technique apart from the existing ones: (1) we assume querying the black-box classifier is limited; (2) our model can use any fast implementation of existing and future kernel methods and decision trees; and (3) our decision tree interpreter is global.

3   Problem Definition

We define the classifier reverse engineering learning problem as follows: given a black box classifier C and one random sample $x^c \in \mathbb{R}^d$ from each class $c = \{c_1, c_2, \dots c_K\}$, we would like to reverse engineer C with a finite set of samples S from a distribution D and transform C to a user understandable classification model C':

$$\underset{C',S}{\arg\min} \quad \ell(C', S)$$
$$s.t. \quad |S| < \delta_S$$
$$\underset{x \sim D}{\Pr}[C(x) \neq C'(x, S)] < \delta \tag{1}$$

where ℓ is a function that measures classifier complexity, and δ and δ_S are predefined constants. The problem has to be solved heuristically because of its exponential complexity [8,13,17]. At first glance, we can solve the problem using traditional active learning techniques [2,14,19,23]. However, existing active learning techniques cannot accomplish the task when it has to overcome both *cost* and *complexity* in the desired solution. This merits further research investigations on transparency inspired issues in terms of *cost* and *complexity*. We refer to *cost* as the total number of queries sent to the black box classifier and *complexity* as the size, that is, the number of leaf nodes in the decision tree.

4   The Reverse Engineering Approximate Learning (REAL) Model

For increasing human-understandability, we choose to use a decision tree to represent the approximation of the black box classifier. As mentioned earlier, decision trees are well known for their poor stabilities. Since the reverse engineering learning process begins with a small training set, the poor stability of the decision tree classifier may significantly impact the end result. Existing results show that when the underlying classifier is a decision tree, query-by-bagging [1] is more stable and more accurate compared to its competitors [7]. In our study, we implement a benchmark strategy referred to as *direct hypothesis formation* in which we adopt the `query-by-bagging` method in the query-by-committee strategy for data sampling.

To circumvent the instability problem of decision trees, we introduce an intermediate kernel-based learner that is more stable than a decision tree learner. The kernel-based learner is used in data sampling where query points are selected. After selecting data samples, we build a decision tree approximation of the black box classifier. We refer to this strategy as *indirect hypothesis formation*.

4.1   Minimum Cost and Complexity Sampling

To reverse engineer a black box classifier and transform it into a tree-structured classifier, we seek a set of training samples that is sufficient to construct a decision-tree counterpart of the black box classifier under the cost and complexity con-

Table 1. List of notations

$\mathcal{H}$	classifier built on existing training data
$\phi_\alpha(x)$	probability x is assigned to a leaf node α
$\phi_\alpha^*(x)$	maximum $\phi_\alpha(x)$ over all leaf nodes
μ_α	prototype of labeled samples in leaf node α
μ	set of prototypes of samples in all leaf nodes α
β	Lagrange multiplier in Gibbs distribution
ψ_u	function that measures uncertainty
c_i	the ith class label where $i = 1, \ldots, K$

straints. To minimize the sampling cost, we follow the principle of maximum uncertainty and select samples that maximally prune the version space. In the mean time, to limit the growth of the complexity of the decision tree classifier, we select samples that have a higher probability to be assigned to a leaf node with a large population of samples given the topology of the current decision tree. We provide a list of notations used throughout this section in Table 1.

Our sampling objective function is:

$$\arg\max_{x \sim D} \quad \phi_\alpha^*(x)$$
$$s.t. \quad \max_{x, i \in [1,K]} \Pr[\mathcal{H}(x) = c_i] < Pr(c_i) + \delta \qquad (2)$$

where $\mathcal{H}$ is the intermediate classifier built on existing labeled examples L, ϕ_α^* is the maximum probability that x falls in a leaf node of the decision tree built on L, and $\delta > 0$ is a small constant. The objective function selects a sample x for which its classification by $\mathcal{H}$ into any class $c_{i \in [1,K]}$ is no better than random guessing according to the prior when δ is very small, while in the mean time, the probability $\phi_\alpha^*(x)$ that x is assigned to a leaf node in the decision-tree counterpart of $\mathcal{H}$ is the greatest compared to other unlabeled samples. The selected sample achieves maximum homogeneity at leaf nodes, and therefore is unlikely to cause an internal node to split into new leaves. Our handling of classifier complexity during the sampling process draws a distinct line between our problem and the traditional active learning problem.

To estimate the probability that a sample x is assigned to a leaf node given the topology of a decision tree, we resort to the principle of maximum entropy. Let $X = \{x_i \in \mathbb{R}^d \mid i = 1, \ldots, N\}$ be a set of N unlabeled examples, and $\mu = \{\mu_\alpha \in \mathbb{R}^d \mid \alpha = 1, \ldots, J\}$ be a set of $J \ll N$ prototypes of labeled examples assigned to J leaf nodes in the current decision tree. Given no prior knowledge, the best way to relate an unlabeled data point in X and the representatives μ_α of the labeled data points is the maximum entropy distribution. Let $\phi_\alpha(x) = Pr(x \to \alpha)$ be the probability that data point x is assigned to leaf α, and we seek to optimize:

$$\max \sum_{\alpha=1}^{J} - Pr[x \to \alpha] \log Pr[x \to \alpha]$$

$$s.t. \quad \mathbb{E}(d_x) = \sum_{\alpha=1}^{J} Pr(x \to \alpha) d(x, \mu_\alpha)$$

where $\mathbb{E}(d_x)$ is the expected distance between x and the prototypes of all the leaf nodes, and $d(\cdot, \cdot)$ is a distance measure, for example, Euclidean distance. The solution is the Gibbs distribution:

$$Pr(x \to \alpha) = \frac{\exp(-\beta d(x, \mu_\alpha))}{\sum_j \exp(-\beta d(x, \mu_j))} \tag{3}$$

where β is the Lagrange multiplier that controls the degree of fuzziness of the probability distribution [10]. When $\beta = 0$, x is equally probable to be assigned to any leaf node. When β is large, the assignment of x conforms to the nearest neighbor philosophy. In a sequential sampling process, β can be incremented gradually in each iteration as more samples are used to estimate the leaf proto-types. $\phi_\alpha^*(x)$, the maximum probability of x over $\alpha = \{1, \ldots, J\}$, is:

$$\phi_\alpha^*(x) = \max_\alpha \frac{\exp(-\beta d(x, \mu_\alpha))}{\sum_j \exp(-\beta d(x, \mu_j))} \tag{4}$$

The optimization objectives with respect to decision tree complexity, specified in Eqs. (1) and (2), can be rewritten as: $\arg\max_{x \in X} \phi_\alpha^*(x)$, given the set of unlabeled examples X.

Putting everything together, we have a sampling technique uniquely designed for reverse engineering a black box classifier with minimum query cost and classification complexity. Without loss of generality, let $\mathcal{S}^t$ be the training set after the tth sample s_t has been added to the training set, where $t \geq 0$. When $t = 0$, $\mathcal{S}^t$ represents the initial training data we have at our disposal. Let the next query point be s^{t+1}. We estimate $\mu^t = \{\mu_\alpha^t \in \mathbb{R}^d \mid \alpha = 1, \ldots, J\}$, the set of J prototypes of $\mathcal{S}_\alpha^t \subset \mathcal{S}^t$ assigned to the J leaf nodes in the decision tree built on $\mathcal{S}^t$. We choose a query point s^{t+1} as follows:

$$s^{t+1} = \arg\max_{s \in \mathcal{S}_u^t} \phi_\alpha^*(s)$$

$$s.t. \quad \psi_u(s) > \psi_u(s' \in \mathcal{S}_u^t | s' \neq s)$$

where ψ_u is a function that measures uncertainty, ϕ_α^* is defined in Eq. (4), and $\mathcal{S}_u^t$ is the set of unlabeled points.

4.2   Direct Hypothesis Formation

As mentioned earlier, we can build a decision tree directly from the training set with the query points. Query-by-bagging [1] is more stable and more accurate in decision tree active learning. In query-by-bagging, a committee of decision tree classifiers is built on subsets of training data, and query points are selected if the committee has the largest variance on the predictions. We modify the sampling technique by incorporating the minimum classifier complexity objective. Let M be the number of component classifiers in the committee, and $h_{i|i=1\cdots M}$ be the ith component classifier. Let c be the total number of classes, at the tth step the total number of component classifiers that predict $s \in \mathcal{S}_u^t$ as $c_{k|k \in \{1,\cdots,K\}}$, denoted as T_k, is: $T_k(s) = |\{m \leq M | h_m(s) = c_k)\}|$ and $T(s) = [T_1(s), \cdots, T_K(s)]$ records the total number of component classifiers that classify s as $c_{k,\forall k=1,\cdots,K}$. We select $s^{t+1} \in \mathcal{S}_u^t$ by solving the following optimization problem:

$$s^{t+1} = \arg\max_{s \in \mathcal{S}_u^t} \phi_\alpha^*(s)$$

$$s.t. |\max(T(s)) - \min(T(s))| < |\max(T(s')) - \min(T(s'))| \quad \forall s' \in \mathcal{S}_u^t | s' \neq s.$$

where the largest variance constraint is equivalent to the maximum uncertainty constraint specified in Eq. (2).

4.3   Indirect Hypothesis Formation

In indirect hypothesis formation, we introduce an intermediate kernel-based classifier for selectively sampling query points. Let $\mathcal{S}^t$ be the training set after the tth round of sampling, we update the intermediate classifier and the decision tree classifier on $\mathcal{S}^t$. Let $\mathcal{S}_u^t$ be the set of unlabeled data from which query points are selected. In this study, we choose SVM as the intermediate learning algorithm. We estimate the uncertainty of a sample point using margin distance.

We now include the minimum complexity constraint in the sampling process. We relax the minimax margin constraint by including a small group of candidate points that are δ-*close* to the one that satisfies the minimax margin constraint:

$$\max \phi_D(s) < \min_{\forall s' \in \mathcal{S}_u^t} \max \phi_D(s') + \delta \tag{5}$$

where ϕ_D measures the margin distances of s to the positive and negative borders, and $\delta > 0$ is a small constant. We select a query point s^{t+1} from $\mathcal{S}_u^t$ such that

$$s^{t+1} = \arg\max_{s \in \mathcal{S}_u^t} \phi_\alpha^*(s)$$

$$s.t. \quad \max \phi_D(s) < \min_{\forall s' \in \mathcal{S}_u^t} \max \phi_D(s') + \delta.$$

where $\phi_\alpha^*(s)$ is the maximum probability s is consistent with a leaf node and the minimax margin constraint is equivalent to the maximum uncertainty constraint in Eq. (2). In general, given a dataset of m features and n instances, the time complexity of our algorithm is $O(mn^2)$ with a nonlinear kernel and the standard implementation of decision trees. It can be reduced to $O(mn)$ with a linear kernel [12] and a fast decision tree learning algorithm [21].

5   Experimental Results

We design a set of experiments to verify the applicability of our reverse engineering techniques for increasing transparency, with DT REAL referring to the direct hypothesis formation and SVM–DT REAL referring to the indirect hypothesis formation. The success is measured by examining the tree size and the **fidelity**—percentage of matching predictions by the reverse engineered classifier and the black-box classifier on independent *unseen* data sets. We run our experiments on four real data sets from the UCI repository [15]. δ in Eq. (5) is set to twice of the difference between the smallest and the second smallest margin. We clarify a few issues regarding our experimental setup in Table 2.

Table 2. Empirical study related issues

Issue	Clarification
(1) Should we use random sampling or active learning as the baseline?	In all our experiments, we choose to compare against the active learning baselines since they are significantly better than random sampling
(2) What learning algorithms should be used to train the black-box classifier?	Our reverse engineering learning model is classifier agnostic. It is not designed to gear towards any particular learning models
(3) Which design of decision tree should we use to train our classifier in the reverse engineering process?	We do not favor one type of decision tree over another in either our algorithmic design or our empirical study, because our algorithm is applicable to any decision tree design
(4) When should we terminate the reverse engineering process?	In many real applications, querying black-box for labels is not free (for example, getting credit score report). In addition, frequently querying actions may be considered as a suspicious abnormal behavior and would not be granted by companies' security standard. In practice, one can stop when either the budget or a desired fidelity measure has been reached. In our experiment, we allow the number of query points to be at most 10% of the size of the training data used to train the black-box classifier

5.1   Experiments on UCI Datasets

We test our techniques on four UCI Datasets: *Banknote Authentication, Cardiotocography, Phishing Websites* and *Human Activity Recognition with Smartphones* (referred to as *Smartphone* hereafter). The black-box classifiers are trained with support vector machine (SVM), logistic regression (LOGIT), decision tree (DT), naïve Bayes (NB), and neural network (NN). For SVM, we use Gaussian kernels with C = 10000; for NN, we set the number of nodes on the hidden layer to be 10. All the algorithms in our experiments are implemented in Matlab. All experiments are repeated 10 times and the average results are reported. *The accuracy of the black box classifier is shown as a dashed line in all figures as auxiliary information.* Detailed results are shown in Appendix A.

Banknote Authentication. The dataset has 1372 instances and two classes *genuine* or *forged*. We divide the data set equally into two parts: one for training the black box classifiers and the other is for active learning. The latter is further dived into two parts: one fifth is used for selecting query points to reverse engineer the black box classifier, and the rest is used for independent testing. The number of examples used at the beginning of reverse engineering is 1% of the size of the training data used to train the black box classifiers.

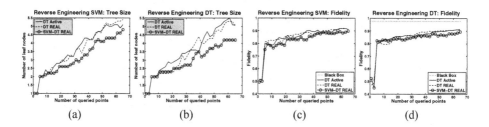

(a)   (b)   (c)   (d)

Fig. 2. Reverse engineering SVM and DT on the *Banknote Authentication* data set, using DT Active, DT REAL, and SVM-DT REAL decision tree learners.

Figure 2 shows the results of the two reverse engineering (RE) learners— DT REAL and SVM-DT REAL, and the baseline decision tree active learner—DT Active [1] with support vector machine (SVM) and decision tree (DT) as the black-box classifiers. Figures 2(a) and (b) show the growth of the complexity of the decision trees in terms of the number of leaf nodes as the number of queried samples increases. The solid line (—) is the baseline decision tree active learner (DT Active), the dashed line (- - -) is the decision tree active learner using our minimum cost and complexity sampling technique (DT REAL). The solid line with circular markers (–o–) is the SVM-DT RE classifier (SVM-DT REAL) also using the minimum cost and complexity sampling technique. It is clear that the complexity of the SVM-DT RE learner is consistently lower than that of the DT active learner. Although applying the same minimum cost and complexity sampling technique, the DT RE learner cannot effectively produce decision trees with lower complexity. Figures 2(c) and (d) show the fidelity of the three classifiers. Note that fidelity is the percentage of agreement between the predictions made by each classifier and the black box classifier. All three classifiers have comparable performance in terms of fidelity. The flat dashed lines show the accuracy of the black box classifiers. All three classifiers manage to predict similarly as the black box classifier more than 90% of the time, with a sample size less than 10% of the size of the training data used to train the black box classifiers. The results for the rest of the black-box classifiers are similar. Due to page limitations, we do not show the plots.

Cardiotocogram. The dataset consists of 2126 instances of fetal cardiotocograms. We select 21 features and classify a cardiotocogram to one of the three fetal states: {N, S, P} where N is `normal`, S is `suspect`, and P is `pathologic`. We again divide the data set equally into two subsets, one for training the black-box classifiers, and the other for query point sampling and testing (among which 20% is used as query data, and the rest is used as the independent test set).

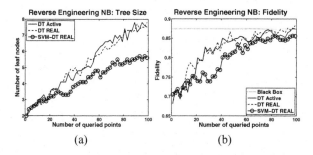

Fig. 3. Reverse engineering the *naïve Bayes* classifier on the *Cardiotocogram* data set, using `DT Active`, `DT REAL`, and `SVM-DT REAL` decision tree learners.

Figure 3 shows the results of `DT REAL`, `SVM-DT REAL`, and the baseline `DT Active` with *naïve Bayes* as the black-box classifier. Figure 3(a) shows the growth of the complexity of the decision trees in terms of the number of leaf nodes. Again, the complexity of the SVM-DT RE learner is consistently lower than that of the DT active learner and the DT RE learner. Figures 3(b) shows the fidelity of the three classifiers. The results for the rest of the black-box classifiers are similar. Due to page limitations, we do not show the plots.

Phishing Websites Dataset. The dataset has 30 attributes that characterize phishing websites. The learning task is a binary classification problem. There are 2456 instances in the data set. We again use 50% of the data for training the

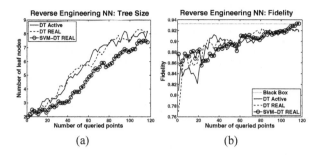

Fig. 4. Reverse engineering the *neural network* classifier on the *Phishing Website* data set, using `DT Active`, `DT REAL`, and `SVM-DT REAL` decision tree learners.

black box classifiers, and the other 50% for query point sampling and testing ($\frac{1}{5}$ as query data and the rest $\frac{4}{5}$ is used as independent test data).

Figure 4 shows the results of the three algorithms reverse engineering the *artificial neural network* black box classifier. Figure 4(a) shows the growth of the complexity of the decision trees in terms of the number of leaf nodes as the number of queried data points increases. Again, the complexity of the SVM-DT RE learner is consistently lower than that of the other two DT learners. Figure 4(b) shows the fidelity of the three classifiers. Except for the case where the black box classifier is trained with SVM, all three learners manage to exceed 90% fidelity with a sample size less than 10% of the training data used to train the black box classifiers. In the case of SVM as the black box classifier, the fidelity of the three learners is slightly less than 90% (above 88%). The results for the rest of the black-box classifiers are similar. Due to page limitations, we do not show the plots of the rest of the black box classifiers.

Smartphone. The dataset contains 10299 instances. Each instance has 562 attributes, and there are six class labels. This is the most complicated data set we used in our experiment. We randomly select 25% of the data for training the black box classifiers, 5% of the data for reverse engineering the black box classifiers, and then 25% random samples for independent testing.

Figure 5 shows the results of the three reverse engineering classifiers for the black box classifier: *logistic regression*. Figure 5(a) shows the growth of the complexity of the decision trees in terms of the number of leaf nodes. In this case, the complexity of the DT RE learner is in general lower than that of the other two learners. The SVM-DT RE learner is mostly comparable to the DT active learner in terms of complexity. Figure 5(b) shows the fidelity of the three classifiers. All three learners manage to achieve approximately 80% fidelity with a sample size less than 10% of the size of the training data for the black box classifier. The results for the rest of the black-box classifiers are similar. Due to page limitations, we do not show the plots of the other four black box classifiers.

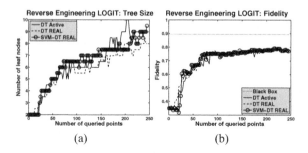

Fig. 5. Reverse engineering the *logistic regression* classifier on the *Smartphone* data set, using DT Active, DT REAL, and SVM-DT REAL decision tree learners.

6   Conclusions and Future Work

We investigate the feasibility of improving model transparency of data mining algorithms by reverse engineering a black-box classifier and transforming it to a decision tree. Our objective is to increase the transparency of the original black-box classifier with a small number of query points. We develop a reverse engineering learning technique that samples unlabeled data according to the principle of maximum uncertainty and minimum classifier complexity. Our experimental results demonstrate that our idea of reverse engineering classifiers is both feasible and practical. We also show that our reverse engineering model with indirect hypothesis formation is superior to traditional active learning with decision trees and SVMs. In the future, we would like to consider the problem in a game theoretic setting in which the black-box classifier employs a defense strategy against this type of reverse engineering, and the user counters the defense strategy with more sophisticated reverse engineering techniques.

Acknowledgement. The research reported herein was supported in part by NIH award 1R01HG006844, NSF awards CNS-1111529, CICI-1547324, and IIS-1633331 and ARO award W911NF-17-1-0356.

Appendix A Tree Size and Fidelity

Table 3 lists the results of all the test cases. Each row shows the tree size and the fidelity results of the last round right before the reverse engineering or active learning process terminates. As can be observed from the table, the SVM-DT REAL learner almost always produces smaller trees than the DT Active learner while producing comparable fidelity values. The accuracy results are very similar to the fidelity and therefore are eliminated from the paper because of page limitations.

Table 3. Reverse engineering results on all the datasets.

Banknote Authentication

BC	DT-A		DT-R		SVM-DT-R	
	Tree Size	Fidelity	Tree Size	Fidelity	Tree Size	Fidelity
SVM	5.0000 ± 0.8433	0.9009 ± 0.0162	5.5000 ± 1.0328	0.9024 ± 0.0304	4.7000 ± 1.1547	0.9005 ± 0.0308
Logit	5.3000 ± 0.9487	0.9035 ± 0.0348	5.4000 ± 0.8433	0.9089 ± 0.0274	4.0000 ± 1.2472	0.8949 ± 0.0256
DT	5.2000 ± 1.1972	0.9171 ± 0.0333	5.2000 ± 0.9944	0.9072 ± 0.0383	4.2000 ± 1.0801	0.9023 ± 0.0417
NB	4.2000 ± 1.3166	0.9322 ± 0.0143	3.9000 ± 0.8165	0.9335 ± 0.0189	3.7000 ± 0.8233	0.9379 ± 0.0139
ANN	5.8000 ± 1.1005	0.9172 ± 0.0298	5.6000 ± 0.9087	0.9029 ± 0.0364	4.1000 ± 0.9033	0.9076 ± 0.0141

Cardiotocogram

BC	DT-A		DT-R		SVM-DT-R	
	Tree Size	Fidelity	Tree Size	Fidelity	Tree Size	Fidelity
SVM	3.0000 ± 1.4907	0.9779 ± 0.0158	3.2000 ± 1.6193	0.9740 ± 0.0189	3.1000 ± 1.1972	0.9329 ± 0.0411
Logit	6.2000 ± 1.3984	0.9116 ± 0.0232	5.6000 ± 1.7764	0.9102 ± 0.0306	4.7000 ± 1.3375	0.9141 ± 0.0165
DT	5.5000 ± 0.9718	0.9629 ± 0.0222	5.5000 ± 1.0801	0.9574 ± 0.0222	4.6000 ± 1.0750	0.9267 ± 0.0312
NB	7.5000 ± 1.5811	0.8747 ± 0.0256	7.8000 ± 1.3984	0.8829 ± 0.0297	5.6000 ± 0.9661	0.8558 ± 0.0223
ANN	5.4000 ± 2.0656	0.9249 ± 0.0300	4.9000 ± 1.6633	0.9293 ± 0.0322	4.6000 ± 1.4298	0.9108 ± 0.0112

Phishing Websites

BC	DT-A		DT-R		SVM-DT-R	
	Tree Size	Fidelity	Tree Size	Fidelity	Tree Size	Fidelity
SVM	9.1000 ± 1.7288	0.8894 ± 0.0167	9.7000 ± 2.2632	0.8965 ± 0.0168	7.6000 ± 1.3499	0.8879 ± 0.0109
Logit	9.4000 ± 1.7764	0.9214 ± 0.0139	8.9000 ± 1.7920	0.9199 ± 0.0113	7.8000 ± 1.6193	0.9181 ± 0.0131
DT	8.9000 ± 1.3703	0.9189 ± 0.0192	9.3000 ± 1.3375	0.9198 ± 0.0260	7.6000 ± 0.5164	0.9187 ± 0.0119
NB	8.8000 ± 2.2010	0.9054 ± 0.0190	8.4000 ± 1.0750	0.9018 ± 0.0193	7.6000 ± 1.3499	0.9066 ± 0.0152
ANN	7.7000 ± 1.5670	0.9225 ± 0.0233	8.1000 ± 1.7288	0.9223 ± 0.0148	7.5000 ± 1.2693	0.9328 ± 0.0098

Smartphone

BC	DT-A		DT-R		SVM-DT-R	
	Tree Size	Fidelity	Tree Size	Fidelity	Tree Size	Fidelity
SVM	8.5000 ± 2.0790	0.7348 ± 0.0110	10.5000 ± 1.6364	0.7377 ± 0.0175	9.8000 ± 2.3118	0.7488 ± 0.0152
Logit	9.1000 ± 1.1785	0.7783 ± 0.0175	8.0000 ± 1.7029	0.7767 ± 0.0112	9.5000 ± 1.9120	0.7927 ± 0.0159
DT	10.9000 ± 1.3038	0.7693 ± 0.0087	10.5000 ± 2.1679	0.7689 ± 0.0152	11.0000 ± 1.0954	0.7634 ± 0.0086
NB	8.0000 ± 2.1145	0.8292 ± 0.0445	8.0000 ± 1.3012	0.8252 ± 0.0422	8.5000 ± 1.8166	0.8280 ± 0.0357
ANN	10.2000 ± 1.1145	0.7878 ± 0.0218	8.9000 ± 1.5166	0.7842 ± 0.0309	10.0000 ± 2.5100	0.7655 ± 0.0125

References

1. Abe, N., Mamitsuka, H.: Query learning strategies using boosting and bagging. In: ICML, pp. 1–9 (1998)
2. Beygelzimer, A., Dasgupta, S., Langford, J.: Importance weighted active learning. In: ICML, pp. 49–56 (2009)
3. Craven, M.W., Shavlik, J.W.: Extracting tree-structured representations of trained networks. In: Proceedings of the 8th International Conference on Neural Information Processing Systems, pp. 24–30 (1995)
4. Datta, A., Sen, S., Zick, Y.: Algorithmic transparency via quantitative input influence: theory and experiments with learning systems. In: 2016 IEEE Symposium on Security and Privacy (SP), pp. 598–617 (2016)
5. Diederich, J. (ed.): Rule Extraction from Support Vector Machines. Studies in Computational Intelligence, vol. 80. Springer, Heidelberg (2008). https://doi.org/10.1007/978-3-540-75390-2
6. Duivesteijn, W., Thaele, J.: Understanding where your classifier does (not) work - the scape model class for EMM. In: ICDM, pp. 809–814 (2014)
7. Dwyer, K., Holte, R.: Decision tree instability and active learning. In: Kok, J.N., Koronacki, J., Mantaras, R.L., Matwin, S., Mladenič, D., Skowron, A. (eds.) ECML 2007. LNCS (LNAI), vol. 4701, pp. 128–139. Springer, Heidelberg (2007). https://doi.org/10.1007/978-3-540-74958-5_15

8. Dyer, M., Frieze, A., Kannan, R.: A random polynomial-time algorithm for approximating the volume of convex bodies. J. ACM **38**(1), 1–17 (1991)
9. Fung, G., Sandilya, S., Rao, R.B.: Rule extraction from linear support vector machines. In: ACM SIGKDD, pp. 32–40 (2005)
10. Held, M., Buhmann, J.M.: Unsupervised on-line learning of decision trees for hierarchical data analysis. In: Advances in Neural Information Processing Systems, pp. 514–520 (1998)
11. Henelius, A., Puolamäki, K., Boström, H., Asker, L., Papapetrou, P.: A peek into the black box: exploring classifiers by randomization. Data Min. Knowl. Discov. **28**(5–6), 1503–1529 (2014)
12. Joachims, T.: Training linear SVMs in linear time. In: Proceedings of the 12th ACM SIGKDD International Conference on Knowledge Discovery and Data Mining, KDD 2006, pp. 217–226. ACM, New York (2006)
13. Klivans, A.R., O'Donnell, R., Servedio, R.A.: Learning geometric concepts via Gaussian surface area. In: Proceedings of the 2008 49th Annual IEEE Symposium on Foundations of Computer Science, pp. 541–550 (2008)
14. Lewis, D.D., Gale, W.A.: A sequential algorithm for training text classifiers. In: ACM SIGIR, pp. 3–12 (1994)
15. Lichman, M.: UCI machine learning repository (2013). http://archive.ics.uci.edu/ml
16. PCAST: Big data and privacy: a technological perspective (2014). https://www.whitehouse.gov/sites/default/files/microsites/ostp/PCAST/pcast_big_data_and_privacy_-_may_2014.pdf
17. Rademacher, L., Goyal, N.: Learning convex bodies is hard. In: COLT (2009)
18. Ribeiro, M.T., Singh, S., Guestrin, C.: "Why should i trust you?": explaining the predictions of any classifier. In: SIGKDD, pp. 1135–1144 (2016)
19. Roy, N., McCallum, A.: Toward optimal active learning through sampling estimation of error reduction. In: ICML, pp. 441–448 (2001)
20. Saad, E.W., Wunsch II, D.C.: Neural network explanation using inversion. Neural Netw. **20**(1), 78–93 (2007)
21. Su, J., Zhang, H.: A fast decision tree learning algorithm. In: Proceedings of the 21st National Conference on Artificial Intelligence, AAAI 2006, vol. 1, pp. 500–505. AAAI Press (2006)
22. Sweeney, L.: Discrimination in online ad delivery. Commun. ACM **56**(5), 44–54 (2013)
23. Tong, S., Koller, D.: Active learning for parameter estimation in Bayesian networks. In: NIPS, pp. 647–653 (2001)

Cost-Sensitive Reference Pair Encoding for Multi-Label Learning

Yao-Yuan Yang[1], Kuan-Hao Huang[1], Chih-Wei Chang[2], and Hsuan-Tien Lin[1(✉)]

[1] CSIE Department, National Taiwan University, Taipei, Taiwan
{b01902066,r03922062}@ntu.edu.tw,
htlin@csie.ntu.edu.tw
[2] Computer Science, Carnegie Mellon University, Pittsburgh, PA, USA
cwchang@cs.cmu.edu

Abstract. Label space expansion for multi-label classification (MLC) is a methodology that encodes the original label vectors to higher dimensional codes before training and decodes the predicted codes back to the label vectors during testing. The methodology has been demonstrated to improve the performance of MLC algorithms when coupled with off-the-shelf error-correcting codes for encoding and decoding. Nevertheless, such a coding scheme can be complicated to implement, and cannot easily satisfy a common application need of cost-sensitive MLC—adapting to different evaluation criteria of interest. In this work, we show that a simpler coding scheme based on the concept of a reference pair of label vectors achieves cost-sensitivity more naturally. In particular, our proposed cost-sensitive reference pair encoding (CSRPE) algorithm contains cluster-based encoding, weight-based training and voting-based decoding steps, all utilizing the cost information. Furthermore, we leverage the cost information embedded in the code space of CSRPE to propose a novel active learning algorithm for cost-sensitive MLC. Extensive experimental results verify that CSRPE performs better than state-of-the-art algorithms across different MLC criteria. The results also demonstrate that the CSRPE-backed active learning algorithm is superior to existing algorithms for active MLC, and further justify the usefulness of CSRPE.

Keywords: Multi-label Classification · Cost-sensitive
Active learning

1 Introduction

The *multi-label classification* (MLC) problem aims to map an instance to multiple relevant labels [6,11], which matches the needs of many real-world applications, such as object detection and news classification. Different applications generally require evaluating the performance of MLC algorithms with different criteria, such as the Hamming loss, 0/1 loss, Rank loss, and F1 score [14].

© Springer International Publishing AG, part of Springer Nature 2018
D. Phung et al. (Eds.): PAKDD 2018, LNAI 10937, pp. 143–155, 2018.
https://doi.org/10.1007/978-3-319-93034-3_12

Most existing MLC algorithms are designed to optimize one or few criteria. For instance, *binary relevance* (BR) [14] learns a binary classifier per label to predict its relevance, and naturally optimizes the Hamming loss. *Classifier chain* (CC) [12] extends BR by ordering the labels as a chain and using earlier labels of the chain to improve the per-label prediction, and optimizes the Hamming loss like BR. *Label powerset* (LP) [14] optimizes the 0/1 loss by solving a multi-class classification problem that treats each label combination as a hyper-class. These *cost-insensitive* algorithms cannot easily adapt to different criteria, and may suffer from bad performance when evaluated with other criteria.

Cost-sensitive MLC (CSMLC) algorithms are able to adapt to different criteria more easily. In particular, CSMLC algorithms take the criterion as an additional piece of input data and aim to optimize the criterion during the learning process. Two state-of-the-art CSMLC algorithms are *probabilistic classifier chain* (PCC) [3] and *condensed filter tree* (CFT) [7]. PCC estimates the conditional probability of the labels to infer the Bayes-optimal decision with respect to the given criterion. While PCC can tackle any criterion in principle, the Bayes-optimal inference step can be time-consuming unless an efficient inference rule of the criterion is derived in advance. CFT can be viewed as an extension of CC for CSMLC by re-weighting each example with respect to the criterion when training each binary classifier. Nevertheless, the re-weighting step depends on going back and forth within the chain, making CFT still somewhat time-consuming and hardly parallelizable.

The *multi-label error-correcting code* (ML-ECC) [4] framework is a more sophisticated algorithm that goes beyond the per-label classifiers to improve classification performance. ML-ECC uses error-correcting code (ECC) to transform the original MLC problem into a bigger MLC problem by adding error-correcting labels during encoding. Classifiers on those labels, much like ECC for communication, can be used to correct prediction errors made from the original per-label classifiers and improve MLC performance. While ML-ECC is successful in terms of the Hamming loss and 0/1 loss [4], it is not cost-sensitive and cannot easily adapt to other evaluation criteria. In fact, extending ML-ECC for CSMLC problem appears to be highly non-trivial and has not yet been deeply studied.

In this work, we study the potential of ECC for CSMLC by considering a special type of ECC, the one-versus-one (OVO) code, which is a popular code for multi-class classification [9]. We extend the OVO code to a cost-sensitive code, *cost-sensitive reference pair encoding* (CSRPE), which preserves the information of the criterion in each code-bit during encoding. We further propose a method to convert the criterion into instance weights during training, and a method to take the criterion into account during decoding. To make the whole CSRPE algorithm efficient enough to deal with exponentially many possible label vectors, we study the possibility of sampling the code-bits and zooming into a smaller subset of label vectors during prediction. The resulting algorithm is as efficient as a typical random forest (when coupled with decision trees) in training, and can be easily implemented in parallel. Extensive experimental results demonstrate that CSRPE outperforms existing ML-ECC algorithms and the state-of-the-art CSMLC algorithms across different criteria.

In addition, based on the proposed CSRPE, we design a novel algorithm for *multi-label active learning* (MLAL). Retrieving ground-truth labels is usually expensive in real-world applications [11]. The goal of MLAL is to actively query the labels for a small number of instances while maintaining good test MLC performance. Nevertheless, current MLAL algorithms [2,8,17] are not capable of taking the evaluation criterion into consideration when querying. In this paper, we formulate the *cost-sensitive multi-label active learning* (CSMLAL) setting, and propose a novel algorithm that leverages the code space computed by CSRPE to conduct cost-sensitive querying. Experimental results justify that the proposed algorithm is superior to other state-of-the-art MLAL algorithms.

This paper is organized as follows. First, we define CSMLC problem formally and introduce the ML-ECC framework in Sect. 2. Our proposed CSRPE algorithm is described in Sect. 3. In Sect. 4, we define the CSMLAL problem and solve it with a novel algorithm based on CSRPE. The empirical studies of both CSRPE and its active learning extension are presented in Sect. 5. Finally, we conclude the paper in Sect. 6.

2   Preliminary

The goal of a MLC problem is to map the feature vector $\mathbf{x} \in \mathcal{X} \subseteq \mathbb{R}^d$ to a label vector $\mathbf{y} \in \mathcal{Y} \subseteq \{0,1\}^K$, where $\mathbf{y}[k] = 1$ if and only if the k-th bit is relevant. During training, MLC algorithms use the training dataset $\mathcal{D} = \{(\mathbf{x}^{(n)}, \mathbf{y}^{(n)})\}_{n=1}^N$ to learn a classifier $f \colon \mathcal{X} \to \mathcal{Y}$. During testing, for any test example $(\mathbf{x}, \mathbf{y})$ drawn from the distribution that generated $(\mathbf{x}^{(n)}, \mathbf{y}^{(n)})$, the prediction $f(\mathbf{x})$ is evaluated with a cost function $C \colon \mathcal{Y} \times \mathcal{Y} \to \mathbb{R}$, where $C(\mathbf{y}, \hat{\mathbf{y}})$ represents the penalty of predicting $\mathbf{y}$ as $\hat{\mathbf{y}}$. The objective of MLC algorithms is to minimize the expected cost $\mathbb{E}_{(\mathbf{x},\mathbf{y})}[C(\mathbf{y}, f(\mathbf{x}))]$.

Traditional MLC algorithms are designed to optimize one or few cost functions. These algorithms may suffer from bad performance when other cost functions are used. On the contrary, *cost-sensitive multi-label classification* (CSMLC) algorithms take the cost function as an additional input and learn a classifier f from both $\mathcal{D}$ and C. Classifier f should adapt to different C easily.

The *multi-label error-correcting code* (ML-ECC) [4] framework is originally designed to optimize one cost function (the 0/1 loss). ML-ECC borrows the error-correcting code (ECC) from the communication domain. ML-ECC views the label vectors $\mathbf{y}^{(n)}$ as bit strings and encodes them to longer codes $\mathbf{b}^{(n)} = enc(\mathbf{y}^{(n)})$ with some ECC encoder $enc \colon \mathcal{Y} \to \{0,1\}^M$, where M is the code length. An MLC classifier h is trained on $\{(\mathbf{x}^{(n)}, \mathbf{b}^{(n)})\}$ to predict the codes instead of the label vectors. The code-bits store redundant information about the label vector to recover the intended label vector even when some bits are mispredicted by h. In prediction, the corresponding ECC decoder $dec \colon \{0,1\}^M \to \mathcal{Y}$, is used to convert the predicted vector from h back to the label vector $f(\mathbf{x}) = dec(h(\mathbf{x}))$. In other words, ML-ECC learns the classifier $f = dec \circ h$. Such an ECC decoder is often designed based on special nearest-neighbor search steps in the code space [4].

In the original work of ML-ECC [4], several encoder/decoder choices are discussed and experimentally evaluated. Nevertheless, none of them take the cost information into account. In fact, to the best of our knowledge, there is currently no work that deeply studies the potential of ECC for CSMLC. Next, we illustrate our ideas on making a special ECC cost-sensitive.

3 Proposed Approach

We start from a special cost-insensitive ECC, the *one-versus-one* (OVO) code. The OVO code is the core of the OVO meta-algorithm for *multi-class classification* (MCC). The meta-algorithm trains many binary classifiers, each representing the duel between *two* of the classes, and let the binary classifiers vote for the majority decision for MCC.

To study the OVO code for MLC, we can naïvely follow the label power-set algorithm [14] to reduce the MLC problem to MCC and then apply the OVO meta-algorithm to further reduce MCC to binary classification. As a consequence, each label vector $\mathbf{y} \in \mathcal{Y}$ is simply treated as a distinct hyper-class, and each binary classifier within the OVO meta-algorithm represents a duel between *two* label vectors. More specifically, the i-th classifier is associated with two label vectors $\mathbf{y}_\alpha^i$ and $\mathbf{y}_\beta^i$, called the reference label vectors. There are $\binom{2^K}{2}$ such classifiers, each can be trained with examples in D that match either $\mathbf{y}_\alpha^i$ and $\mathbf{y}_\beta^i$. During prediction, the $\binom{2^K}{2}$ binary classifiers can then vote for all the label vectors $\in \mathcal{Y}$ towards the majority decision.

The steps of applying OVO to MLC above can be alternatively described as a special ML-ECC algorithm, similar to how OVO is viewed as a special ECC for MCC [1]. OVO as ML-ECC encodes each label vector to a code of length $\binom{2^K}{2}$

with the following encoder $enc_{ovo}(\mathbf{y})[i] = \begin{cases} 1 & \text{if } \mathbf{y} = \mathbf{y}_\alpha^i \\ 0 & \text{if } \mathbf{y} = \mathbf{y}_\beta^i \\ 0.5 & \text{otherwise} \end{cases}$. The i-th bit in the

code represents whether the label vector matches either of the reference vectors. The special "bit" value of 0.5 represents other irrelevant label vectors. Then, decoding based on majority voting is equivalent to nearest-neighbor search in the code space over all possible encoded $\mathbf{y} \in \mathcal{Y}$ in terms of the Hamming distance (d_{ham}), as the Hamming distance is a linear function of the vote that each $\mathbf{y}$ gets. More precisely, denote the predicted code as $\hat{\mathbf{b}} = h(\mathbf{x})$, the decoder of OVO is simply $dec_{ovo}(\hat{\mathbf{b}}) = \operatorname{argmax}_{\mathbf{y} \in \mathcal{Y}}(d_{ham}(\hat{\mathbf{b}}, enc_{ovo}(\mathbf{y})))$.

The naïve OVO for ML-ECC above suffers from several issues. First, the code length $\binom{2^K}{2}$ is prohibitively long for large K, making it inefficient to compute. Second, many of the $\binom{2^K}{2}$ classifiers may not be associated with enough data during training. Last but not least, OVO is not cost-sensitive and cannot adapt to different cost functions easily. We resolve the issues in the designs below.

Cost-Sensitive Encoding. The OVO code is designed to optimize 0/1 loss $(C(\mathbf{y}, \hat{\mathbf{y}}) = [\![\mathbf{y} \neq \hat{\mathbf{y}}]\!]$, where $[\![\cdot]\!]$ is the indicator function) for MLC. In the OVO code, each bit of $enc_{ovo}(\mathbf{y})$ is learned from only the instances with $\mathbf{y}$ being exactly the same as $\mathbf{y}_{\alpha}^i$ or $\mathbf{y}_{\beta}^i$. For instances with $\mathbf{y}$ being neither $\mathbf{y}_{\alpha}^i$ nor $\mathbf{y}_{\beta}^i$, these instances will be dropped from training. This suits the design of optimizing $0/1$ loss. Now, we take a different perspective to view the OVO code.

When considering $0/1$ loss, what the OVO code does is to decide whether predict as $\mathbf{y}_{\alpha}^i$ or $\mathbf{y}_{\beta}^i$ suffers less $0/1$ loss. For the case that $\mathbf{y}$ is neither $\mathbf{y}_{\alpha}^i$ nor $\mathbf{y}_{\beta}^i$, the costs for predicting as $\mathbf{y}_{\alpha}^i$ and $\mathbf{y}_{\beta}^i$ are the same. That is why OVO code ignores these cases during training. However, for other cost functions, the costs for predicting $\mathbf{y}$ as $\mathbf{y}_{\alpha}^i$ and $\mathbf{y}_{\beta}^i$ can be different. Hence, even if the label vector $\mathbf{y}$ is neither $\mathbf{y}_{\alpha}^i$ nor $\mathbf{y}_{\beta}^i$, the vector can still provide information for training.

To generalize the encoding function towards cost-sensitivity, we hold the same idea that each bit should predict which reference label vector incurs less cost. The encoding function is designed as $enc_{cs}(\mathbf{y})[i] = \begin{cases} 1 & \text{if } C(\mathbf{y}, \mathbf{y}_{\alpha}^i) < C(\mathbf{y}, \mathbf{y}_{\beta}^i) \\ 0 & \text{if } C(\mathbf{y}, \mathbf{y}_{\alpha}^i) > C(\mathbf{y}, \mathbf{y}_{\beta}^i) \\ 0.5 & \text{otherwise} \end{cases}$

Training Classifiers for Cost-Sensitive Codes. With the encoding function defined, we learn a classifier h to predict the encoded vectors outputted from enc_{cs}. Although enc_{cs} gives the classifier a better ground truth, different label vectors are not equally important for the classifier. For example, if $C(\mathbf{y}, \mathbf{y}_{\alpha}^i)$ and $C(\mathbf{y}, \mathbf{y}_{\beta}^i)$ differ by a lot, there would be a high cost if the classifier gives the wrong prediction, thus making $\mathbf{y}$ very important. In contrast, if there exists a label vector $\mathbf{y}$ s.t. $C(\mathbf{y}, \mathbf{y}_{\alpha}^i) \approx C(\mathbf{y}, \mathbf{y}_{\beta}^i)$, then $\mathbf{y}$ is relatively unimportant because a misclassified $\mathbf{y}$ would not incur a high cost. Thus, we design a weight function to emphasize the importance for each label vector as $weight(\mathbf{y})[i] = |C(\mathbf{y}, \mathbf{y}_{\alpha}^i) - C(\mathbf{y}, \mathbf{y}_{\beta}^i)|$.

Dataset $\{(\mathbf{x}^{(n)}, enc_{cs}(\mathbf{y}^{(n)}), weight(\mathbf{y}^{(n)}))\}_{n=1}^N$ is used to train the classifier h to predict the encoded vector. Normally, h should be trained on the full-length encoded vectors. But the exponentially growing code length $\binom{2^K}{2}$ makes training on the full encoding infeasible. However, many classifiers would result in learning similar problems during training. This could allow us to use fewer bits and preserves the same amount of information. For example, let the i-th reference label vectors be $\mathbf{y}_{\alpha}^i = (1, 0, 1, 0)$ and $\mathbf{y}_{\beta}^i = (1, 0, 0, 1)$, and the j-th reference vectors be $\mathbf{y}_{\alpha}^j = (1, 1, 1, 0)$ and $\mathbf{y}_{\beta}^j = (1, 1, 0, 1)$. The i-th and j-th classifier are actually learning similar things: learning to predict whether the last two labels of the label vector should be $(1, 0)$ or $(0, 1)$. Observing the redundancy in the encoded vectors, it is clear that the length of the encoded vector can be decreased and thus learning becomes feasible. For simplicity, we uniformly sample some bits for from encoded vectors. In Sect. 5, we demonstrate that the number of needed bits are much smaller than $\binom{2^K}{2}$.

Cost-Sensitive Decoding. OVO code decodes by letting each bit votes on either of the reference label vectors. Following the idea for encoding, this is also a special case of decoding by considering the $0/1$ loss. To match with our proposed cost-sensitive encoding, the decoding approach is redesigned to utilize the information more effectively.

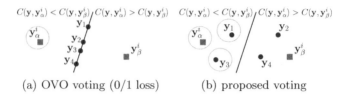

(a) OVO voting (0/1 loss)   (b) proposed voting

Fig. 1. An illustration of the decoding methods.

Figure 1 is an illustration of the relation between encoded vectors under OVO encoding and our cost-sensitive encoding. In 0/1 loss, all instances that are predicted incorrectly incur the same cost making all label vectors except $\mathbf{y}_\alpha^i$ and $\mathbf{y}_\beta^i$ on the decision boundary. Only $\mathbf{y}_\alpha^i$ and $\mathbf{y}_\beta^i$ are distinguishable under the current bit. Thus, original OVO voting only needs to be done on reference label vectors. When using our cost-sensitive encoding, all label vectors are generally separated into two groups by the boundary as Fig. 1(b): the group that is closer to $\mathbf{y}_\alpha^i$ (left) (in terms of cost) and the group that is closer to $\mathbf{y}_\beta^i$. A predicted encoded bit not only provides the information about the reference label vector, but also the information about all other label vectors in the same group. Following this thought, if the prediction is $\mathbf{y}_\alpha^i$, all label vectors $\mathbf{y}$ such that $C(\mathbf{y}, \mathbf{y}_\alpha^i) < C(\mathbf{y}, \mathbf{y}_\beta^i)$ should be voted as well. If predicted otherwise, all label vectors in the other group are voted. By this voting approach, we can use the information encoded within the vectors to decode more effectively.

In fact, this voting approach echoes the Hamming decoding for ECC [1]. More specifically, with the predicted encoded vector $\hat{\mathbf{b}} = h(\mathbf{x})$, the decoding function is written as $dec_{cs}(\hat{\mathbf{b}}) = \mathrm{argmax}_{\mathbf{y} \in \mathcal{Y}} \, d_{ham}(\hat{\mathbf{b}}, enc_{cs}(\mathbf{y}))$. With this formulation, dec_{cs} is formulated as the classic nearest neighbor search problem, where efficient algorithms exist to speed up the decoding process [10].

Despite the efficient decoding algorithm, the number of possible predictions $|\mathcal{Y}|$ equals 2^K, which makes it computationally infeasible. Inspired by [5], we propose to only work with a subset of label vectors that are more likely to be the prediction. We define a relevant set $\tilde{\mathcal{Y}} \subseteq \mathcal{Y}$, which contains a subset of the label vectors from the label space, on which we perform the nearest neighbor search. The decoding function is written as $dec_{cs}(\hat{\mathbf{b}}) = \mathrm{argmax}_{\mathbf{y} \in \tilde{\mathcal{Y}}} \, d_{ham}(\hat{\mathbf{b}}, enc_{cs}(\mathbf{y}))$.

The use of the $\tilde{\mathcal{Y}}$ introduces a trade-off between the number of possible predictions and the prediction efficiency. A reasonable choice of $\tilde{\mathcal{Y}}$ would be $\{\mathbf{y} | (\mathbf{x}, \mathbf{y}) \in \mathcal{D}\}$, which are the distinct label vectors in the training set. Given that the training and testing sets come from the same distribution, the label vectors that appear in the testing set are likely to have appeared in the training set. We justify this choice of $\tilde{\mathcal{Y}}$ in Sect. 5.

The algorithm that combines enc_{cs}, *weight* and dec_{cs} is called *cost-sensitive reference pair encoding* (CSRPE). Our design is inspired by a cost-sensitive extension of OVO for MCC problem called *cost-sensitive one-versus-one* [9], but is refined by our special ideas for encoding and decoding in the MLC problem.

4   Active Learning for CSMLC

CSRPE is able to preserve cost information in the encoded vectors. In this section, we design a novel active learning algorithm for MLC based on CSRPE.

MLC algorithms intend to learn a classifier from a fully labeled dataset, in which every feature vector is paired with a label vector. In many real-world applications, obtaining a label vector to the corresponding feature vector is very expensive [11]. This gives rise to a new problem, active learning, which investigates how to obtain good performance with as little data labeled as possible.

In this paper, we consider the pool-based *multi-label active learning* (MLAL) setting [13] and formulate the cost-sensitive extension of MLAL called *cost-sensitive multi-label active learning* (CSMLAL). In CSMLAL, the algorithm is presented with two sets of data, the labeled pool $\mathcal{D}_l = \{(\mathbf{x}^{(n)}, \mathbf{y}^{(n)})\}_{n=1}^{N_l}$ and the unlabeled pool $\mathcal{D}_u = \{\mathbf{x}^{(n)}\}_{n=1}^{N_u}$. During iterations $t = 1, \ldots, T$, the MLAL algorithm considers $\mathcal{D}_u$, $\mathcal{D}_l$, a MLC classifier f_t trained on $\mathcal{D}_l$ and cost function C to choose a instance $\mathbf{x}_t \in \mathcal{D}_u$ to query. After the queried label vector is retrieved as $\mathbf{y}_t$, $\mathbf{x}_t$ is removed from $\mathcal{D}_u$ and the pair $(\mathbf{x}_t, \mathbf{y}_t)$ is added to $\mathcal{D}_l$. With a small budget of T queries, the goal of the CSMLAL algorithm is to minimize the average prediction cost of f_t on the testing instances evaluated on C.

Many of the current MLAL algorithms are based on the idea of *uncertainty sampling*. They query the instance that current classifier f_t is most uncertain about. There are different uncertainty measures being developed. However, most of these measures consider only one specific C or even completely ignoring C. *Binary minimization* [2] was proposed to directly take the most uncertain bit in the label vector to represent the uncertainty of the whole instance. It queries based on one label at a time and arguably optimizes towards Hamming loss. Another work, in contrast, calculates an average over the uncertainty of all labels [17]. Yet another work uses the difference between the most uncertain relevant label and irrelevant label as an uncertainty measure [8]. This uncertainty is then combined with label cardinality inconsistency. However, this measure is designed heuristically and does not aim at any C.

We propose *cost-sensitive uncertainty* in the encoded vector space to evaluate the importance of instances. The cost-sensitive uncertainty can be separated into two parts, the *cost estimation uncertainty* and the *cost utility uncertainty*.

Cost Estimation Uncertainty. Cost estimation uncertainty measures how well CSRPE estimates the cost between label vectors. Let the predicted encoded vector $\hat{\mathbf{b}} = h(\mathbf{x})$ and $\tilde{\mathbf{b}} = enc_{cs}(dec_{cs}(\hat{\mathbf{b}}))$. Note that $\tilde{\mathbf{b}}$ is actually the nearest encoded vector of $\hat{\mathbf{b}}$. Ideally, if CSRPE estimates the cost information well, $\hat{\mathbf{b}}$ should be close to $\tilde{\mathbf{b}}$. If, unfortunately, the distance $d_{ham}(\hat{\mathbf{b}}, \tilde{\mathbf{b}})$ is large, this implies that CSRPE does not have a good cost estimation for this $\mathbf{x}$ and we hence need more information about it. In other words, we are uncertain about this $\mathbf{x}$. For this reason, we define $d_{ham}(\hat{\mathbf{b}}, \tilde{\mathbf{b}})$ as the *cost estimation uncertainty*.

Cost Utility Uncertainty. The cost utility uncertainty measures how uncertain the classifier f_t is under the current cost function. Let the prediction $\bar{\mathbf{y}} = f_t(\mathbf{x})$ and its encoding $\bar{\mathbf{b}} = enc_{cs}(\bar{\mathbf{y}})$. If the classifier f_t is certain about its prediction under current cost function, $\bar{\mathbf{b}}$ should be close to the cost estimation $\hat{\mathbf{b}} = h(\mathbf{x})$. If unfortunately, distance $d_{ham}(\hat{\mathbf{b}}, \bar{\mathbf{b}})$ is large, it implies that classifier f_t is uncertain under the current cost function. Therefore, we define $d_{ham}(\hat{\mathbf{b}}, \bar{\mathbf{b}})$ as the *cost utility uncertainty*.

The proposed cost-sensitive uncertainty is the combination of these two parts of uncertainty, namely $d_{ham}(\hat{\mathbf{b}}, \tilde{\mathbf{b}}) + d_{ham}(\hat{\mathbf{b}}, \bar{\mathbf{b}})$. The cost-sensitive uncertainty leads to a novel algorithm for CSMLAL. For each iteration, the algorithm selects the instance with the highest cost-sensitive uncertainty to query its label.

5   Experiments

We justify the proposed algorithm on ten public datasets [15] and three common evaluation criteria, including F1 score, Accuracy score and Rank loss. [14]. The experiment was run 20 times, each with a random 50–50 training-testing split. CSRPE has the flexibility to take any base learner. In CSMLC experiments, CSRPE is viewed as an ensemble MLC method, each bit with a binary classifier attached. Because ensemble of decision trees is arguably a popular ensemble method nowadays, we use decision trees as the base learner in these experiments. The parameters are searched with 3-fold cross-validation.

In CSMLAL experiments, the experiments are repeated for 10 runs. Since many competitors designed their algorithms based on linear base learners, the base learner is changed to logistic regression for fair comparison. The parameters are searched with 5-fold cross-validation using the initial dataset.

More detailed experimental setup can be found in the full version [18]. In the following experimental results, we use ↑ (↓) to indicate that a higher (lower) value for the criterion is better.

Effect of Code Length. To justify our claim in Sect. 3 that the code length can be reduced by sampling, we conduct experiments to analyzing the performance of CSRPE with respect to the code length.

Figure 2 shows the average performance and standard error versus code length. We select two of the datasets with larger label counts to showcase the effect of the code length on performance. The results of other datasets can be found in [18]. From the figure, CSRPE performs better as the number of bit increases. The performance of CSRPE generally converges when the code length reaches 3000 across all cost functions and datasets. The length is significantly smaller than the full encoding (2^K). This justifies our claim that full encoding is not needed to achieve top performance. In the following experiments, we set the code length as 3000.

Influence of the Relevant Set. In Sect. 3, we claim that a good choice for relevant set $\tilde{\mathcal{Y}}$ is all distinct label vectors in the training dataset. To justify our claim, we demonstrate that the possible downside of this choice, which is the inability to predict all possible label vectors, will not degrade the performance much. In particular, we compare CSRPE with CSRPE-ext, which is CSRPE-ext with a larger relevant set that includes label vectors that appeared in either the training set or the testing set.

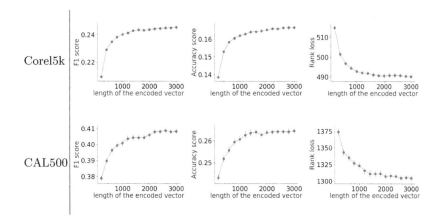

Fig. 2. Different criteria versus code length for CSRPE

Table 1. Experiment results (mean ± ste) of CSRPE and CSRPE-ext

Dataset	Rank loss ↓		F1 score ↑		Accuracy score ↑	
	CSRPE	CSRPE-ext	CSRPE	CSRPE-ext	CSRPE	CSRPE-ext
Corel5k	490.17 ± 1.20	**485.73 ± 0.88**	.2455 ± .0012	**.2492 ± .0011**	.1664 ± .0009	**.1674 ± .0009**
CAL500	1304.6 ± 4.57	**1303.4 ± 4.18**	.4083 ± .0017	**.4109 ± .0013**	.2645 ± .0013	**.2690 ± .0014**
bibtex	104.94 ± 0.38	**102.78 ± 0.32**	.4663 ± .0008	**.4695 ± .0009**	.3926 ± .0011	**.3946 ± .0010**
enron	34.32 ± 0.182	**33.47 ± 0.206**	.5911 ± .0014	**.5921 ± .0016**	.4772 ± .0016	**.4777 ± .0017**
medical	**5.330 ± 0.068**	5.415 ± 0.081	.8203 ± .0023	**.8204 ± .0023**	**.7939 ± .0024**	.7934 ± .0022
genbase	**0.353 ± 0.030**	0.360 ± 0.032	**.9878 ± .0009**	.9876 ± .0009	**.9836 ± .0010**	.9828 ± .0012
yeast	8.451 ± 0.030	**8.448 ± 0.026**	.6670 ± .0012	**.6679 ± .0012**	**.5653 ± .0012**	.5650 ± .0012
flags	**3.010 ± 0.047**	3.050 ± 0.050	**.7222 ± .0041**	.7192 ± .0043	**.6056 ± .0058**	.6028 ± .0052
scene	0.679 ± 0.008	**0.645 ± 0.006**	.7860 ± .0020	**.7913 ± .0014**	**.7620 ± .0020**	.7563 ± .0017
emotions	**0.591 ± 0.001**	0.592 ± 0.002	.6655 ± .0035	**.6673 ± .0030**	**.5775 ± .0037**	.5774 ± .0036

The results, which contain the mean and standard error (ste) of the criteria, are listed in Table 1. The results demonstrate that CSRPE-ext is slightly better performing, but the improvement is at best marginal and insignificant. Even in the CAL500 dataset, where all the label vectors in training and testing sets are different, there is only a small performance difference between CSRPE and CSRPE-ext. The result verifies that our choice of $\tilde{\mathcal{Y}}$ as all the distinct label vectors in the training set is sufficiently good.

Comparison with Other MLC Algorithms. In this experiment, we compare the performance of various MLC and CSMLC algorithms. For the MLC competitors, we include different codes applied within ML-ECC framework. The competing codes include the *Hamming on repetition code* (HAMR), *repetition code (REP)*, and *RAKEL repetition code (RREP)* [4]. REP and RREP are equivalent to BR [14] and RAKEL [16], respectively. In addition, CC [12] is added to serve as a baseline competitor together with REP and RREP. For CSMLC algorithms, we compete with PCC [3] and CFT [7].

The results are shown in Tables 2 and 3. The results show that CSMLC algorithms generally outperform traditional MLC algorithms. This justifies that it is important to take cost information into account. Among the CSMLC algorithms, CSRPE is superior over all other competitors with respect to F1 and Accuracy score. For Rank loss, PCC performs slightly better, but CSRPE still performs competitively with PCC and CFT. Such result justifies CSRPE as a top performing CSMLC algorithm.

Table 2. Experiment results (mean ± ste) on different criteria (best in bold)

Dataset	REP (BR)	RREP (RAKEL)	HAMR	CC	PCC	CFT	CSRPE
F1 score ↑							
Corel5k	.0683 ± .0011	.1028 ± .0010	.0608 ± .0008	.0661 ± .0009	.1759 ± .0008	.1708 ± .0017	**.2455 ± .0012**
CAL500	.3388 ± .0014	.3527 ± .0011	.3152 ± .0012	.3354 ± .0024	.3540 ± .0018	.3815 ± .0016	**.4083 ± .0017**
bibtex	.3636 ± .0009	.3761 ± .0010	.3658 ± .0008	.3569 ± .0009	.3736 ± .0011	.3957 ± .0015	**.4663 ± .0008**
enron	.5441 ± .0026	.5336 ± .0025	.5459 ± .0023	.5492 ± .0022	.5508 ± .0014	.5530 ± .0013	**.5911 ± .0014**
medical	.7883 ± .0028	.7757 ± .0034	.7877 ± .0031	.7924 ± .0035	.8131 ± .0023	.7970 ± .0031	**.8203 ± .0023**
genbase	.9897 ± .0012	.9893 ± .0014	.9896 ± .0012	.9896 ± .0012	**.9911 ± .0007**	.9845 ± .0009	.9878 ± .0008
yeast	.6119 ± .0014	.6130 ± .0011	.6171 ± .0015	.5968 ± .0018	.6013 ± .0013	.6111 ± .0024	**.6670 ± .0012**
flags	.6954 ± .0045	.6965 ± .0044	.7005 ± .0044	.6973 ± .0048	.7075 ± .0038	.6725 ± .0055	**.7222 ± .0041**
scene	.5895 ± .0026	.5926 ± .0019	.6365 ± .0021	.6547 ± .0019	.7306 ± .0016	.6592 ± .0027	**.7860 ± .0020**
emotions	.5968 ± .0038	.5773 ± .0047	.6100 ± .0021	.6205 ± .0035	.6384 ± .0033	.6015 ± .0043	**.6655 ± .0035**
Accuracy score ↑							
Corel5k	.0471 ± .0007	.0696 ± .0006	.0408 ± .0009	.0471 ± .0007	.1135 ± .0005	.0790 ± .0019	**.1664 ± .0009**
CAL500	.2097 ± .0010	.2179 ± .0008	.1925 ± .0007	.2085 ± .0018	.2209 ± .0012	.2425 ± .0015	**.2645 ± .0013**
bibtex	.3063 ± .0009	.3103 ± .0009	.3094 ± .0008	.3031 ± .0010	.2940 ± .0010	.3235 ± .0011	**.3926 ± .0011**
enron	.4303 ± .0023	.4215 ± .0022	.4344 ± .0024	.4437 ± .0021	.4259 ± .0013	.4363 ± .0018	**.4772 ± .0016**
medical	.7559 ± .0034	.7431 ± .0033	.7604 ± .0033	.7643 ± .0035	.7716 ± .0025	.7570 ± .0031	**.7939 ± .0024**
genbase	.9859 ± .0014	.9852 ± .0015	.9856 ± .0014	.9858 ± .0014	**.9873 ± .0009**	.9792 ± .0012	.9835 ± .0010
yeast	.5047 ± .0014	.5065 ± .0012	.5120 ± .0015	.4954 ± .0021	.4872 ± .0017	.5027 ± .0019	**.5653 ± .0012**
flags	.5849 ± .0047	.5860 ± .0046	.5913 ± .0051	.5908 ± .0057	.5974 ± .0041	.5616 ± .0059	**.6056 ± .0058**
scene	.5791 ± .0025	.5816 ± .0020	.6258 ± .0017	.6457 ± .0018	.6821 ± .0019	.6467 ± .0029	**.7620 ± .0020**
emotions	.5179 ± .0037	.4959 ± .0045	.5320 ± .0034	.5417 ± .0035	.5433 ± .0035	.5216 ± .0036	**.5775 ± .0037**
Rank loss ↓							
Corel5k	618.1 ± .6695	597.2 ± .6664	623.5 ± .6474	636.0 ± .5374	421.2 ± .6626	**300.7 ± .7848**	490.2 ± 1.1959
CAL500	1500. ± 5.023	1477. ± 4.835	1537. ± 4.488	1520. ± 6.155	1179. ± 4.498	**1122. ± 4.470**	1305. ± 4.574
bibtex	132.6 ± .2981	124.1 ± .2511	131.5 ± .2819	136.8 ± .2886	**69.10 ± .2454**	112.06 ± .2811	104.9 ± .3814
enron	43.39 ± .2919	44.06 ± .2810	43.40 ± .2540	43.56 ± .3000	27.94 ± .1681	**27.20 ± .1365**	34.32 ± .1815
medical	5.454 ± .1184	5.733 ± .1088	5.601 ± .1232	5.469 ± .0997	**3.058 ± .0603**	4.117 ± .0741	5.330 ± .0676
genbase	.2461 ± .0281	.2422 ± .0273	.2525 ± .0257	.2423 ± .0308	**.1976 ± .0178**	.4686 ± .0310	.3863 ± .0341
yeast	9.609 ± .0358	9.565 ± .0290	9.443 ± .0312	10.324 ± .0448	9.378 ± .0365	9.473 ± .0363	**8.451 ± .0298**
flags	3.123 ± .0434	3.139 ± .0383	3.078 ± .0352	3.120 ± .0450	3.012 ± .0490	3.363 ± .0504	**3.010 ± .0470**
scene	1.136 ± .0066	1.149 ± .0055	1.031 ± .0046	1.098 ± .0080	0.726 ± .0060	0.892 ± .0069	**0.679 ± .0083**
emotions	1.789 ± .0182	1.906 ± .0220	1.764 ± .0165	1.741 ± .0207	**1.563 ± .0176**	1.834 ± .0281	1.591 ± .0198

Table 3. CSRPE versus others based on *t*-test at 95% confident level

Criteria (win/tie/loss)	F1	Rank	Acc	Total
REP	9/1/0	7/2/1	9/0/1	27/7/6
RREP	9/1/0	9/0/1	9/1/0	31/5/4
HAMR	9/1/0	7/2/1	8/2/0	26/9/5
CC	9/1/0	7/2/1	8/2/0	30/6/4
CFT	9/1/0	6/1/3	9/1/0	30/4/6
PCC	9/0/1	2/2/6	8/1/1	22/7/11

Comparison with MLAL Algorithms. In this experiment, we evaluate the performance of CSRPE under the CSMLAL setting. We compare it with several state-of-the-art MLAL algorithms, which includes *adaptive active learning* (adaptive) [8], *maximal loss reduction with maximal confidence* (MMC) [17], and

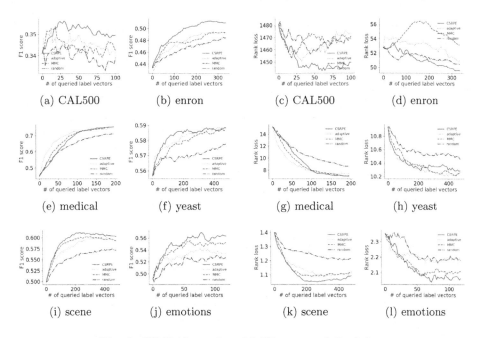

Fig. 3. CSMLAL results with F1 score and Rank loss

random sampling as a baseline algorithm. We do not include a comparison with *binary minimization* [2] since MMC and adaptive are reported to outperform it.

Figure 3 shows the performance with respect to the number of instances queried. For F1 score and Rank loss, CSRPE performs better than other strategies on four out of six datasets. These results indicate that CSRPE is able to consider the cost information, thus enabling it to outperform other competitors on most of the datasets across different evaluation criteria.

6   Conclusion

In this paper, we propose a novel approach for cost-sensitive multi-label classification (CSMLC), called *cost-sensitive reference pair encoding* (CSRPE). CSRPE is derived from the one-versus-one algorithm and can embed the cost information into the encoded vectors. Exploiting the redundancy of the encoded vectors, we use random sampling to resolve the training challenge of building so many classifiers. We also design a nearest-neighbor-based decoding procedure and use the relevant set to efficiently make cost-sensitive predictions. Extensive experimental results demonstrate that CSRPE achieves stable convergence respect to the code length and outperforms not only other encoding methods but also state-of-the-art CSMLC algorithms across different cost functions. In addition, we extend CSRPE to a novel multi-label active learning algorithm by designing a cost-sensitive uncertainty measure. Extensive empirical studies show that the proposed active learning algorithm performs better than existing active learning algorithms. The results suggest that CSRPE is a promising cost-sensitive encoding method for CSMLC for either supervised or active learning.

Acknowledgments. We thank the anonymous reviewers and the members of NTU CLLab for valuable suggestions. This material is based upon work supported by the Air Force Office of Scientific Research, Asian Office of Aerospace Research and Development (AOARD) under award number FA2386-15-1-4012, and by the Ministry of Science and Technology of Taiwan under MOST 103-2221-E-002-149-MY3 and 106-2119-M-007-027.

References

1. Allwein, E.L., Schapire, R.E., Singer, Y.: Reducing multiclass to binary: a unifying approach for margin classifiers. J. Mach. Learn. Res. **1**, 113–141 (2001)
2. Brinker, K.: On active learning in multi-label classification. In: Spiliopoulou, M., Kruse, R., Borgelt, C., Nürnberger, A., Gaul, W. (eds.) From Data and Information Analysis to Knowledge Engineering. Studies in Classification, Data Analysis, and Knowledge Organization, pp. 206–213. Springer, Heidelberg (2006). https://doi.org/10.1007/3-540-31314-1_24
3. Dembczynski, K., Cheng, W., Hüllermeier, E.: Bayes optimal multilabel classification via probabilistic classifier chains. In: ICML (2010)
4. Ferng, C.S., Lin, H.T.: Multilabel classification using error-correcting codes of hard or soft bits. IEEE Trans. Neural Netw. Learn. Syst. **24**(11), 1888–1900 (2013)
5. Huang, K.H., Lin, H.T.: Cost-sensitive label embedding for multi-label classification. Mach. Learn. **106**, 1725–1746 (2017)
6. Katakis, I., Tsoumakas, G., Vlahavas, I.: Multilabel text classification for automated tag suggestion. In: ECML PKDD Discovery Challenge, vol. 75 (2008)
7. Li, C.L., Lin, H.T.: Condensed filter tree for cost-sensitive multi-label classification. In: ICML (2014)
8. Li, X., Guo, Y.: Active learning with multi-label svm classification. In: IJCAI (2013)
9. Lin, H.T.: Reduction from cost-sensitive multiclass classification to one-versus-one binary classification. In: ACML (2014)
10. Liu, T., Moore, A.W., Gray, A.: New algorithms for efficient high-dimensional nonparametric classification. J. Mach. Learn. Res. **7**, 1135–1158 (2006)
11. Liu, Y.: Active learning with support vector machine applied to gene expression data for cancer classification. J. Chem. Inf. Comput. Sci. **44**, 1936–1941 (2004)
12. Read, J., Pfahringer, B., Holmes, G., Frank, E.: Classifier chains for multi-label classification. Mach. Learn. **85**(3), 333–359 (2011)
13. Settles, B.: Active learning literature survey. University of Wisconsin, Madison (2010)
14. Tsoumakas, G., Katakis, I., Vlahavas, I.P.: Mining multi-label data. In: Maimon, O., Rokach, L. (eds.) Data Mining and Knowledge Discovery Handbook, pp. 667–685. Springer, Boston (2010)
15. Tsoumakas, G., Spyromitros-Xioufis, E., Vilcek, J., Vlahavas, I.: Mulan: a java library for multi-label learning. J. Mach. Learn. Res. **12**, 2411–2414 (2011)
16. Tsoumakas, G., Vlahavas, I.P.: Random k -labelsets: An ensemble method for multilabel classification. In: ECML (2007)

17. Yang, B., Sun, J.T., Wang, T., Chen, Z.: Effective multi-label active learning for text classification. In: ICDM (2009)
18. Yang, Y.Y., Huang, K.H., Chang, C.W., Lin, H.T.: Cost-sensitive random pair encoding for multi-label classification. arXiv preprint arXiv:1611.09461 (2016)

Fuzzy Integral Optimization with Deep Q-Network for EEG-Based Intention Recognition

Dalin Zhang[1](✉), Lina Yao[1], Sen Wang[2], Kaixuan Chen[1], Zheng Yang[3], and Boualem Benatallah[1]

[1] School of Computer Science and Engineering, UNSW Sydney, Sydney, Australia
{dalin.zhang,kaixuan.chen}@student.unsw.edu.au,
{lina.yao,b.benatallah}@unsw.edu.au
[2] School of Information and Communication Technology,
Griffith University, Brisbane, Australia
sen.wang@griffith.edu.au
[3] School of Software, Tsinghua University, Beijing, China
hmilyyz@gmail.com

Abstract. Non-invasive brain-computer interface using electroen-cephalography (EEG) signals promises a convenient approach empow-ering humans to communicate with and even control the outside world only with intentions. Herein, we propose to analyze EEG signals using fuzzy integral with deep reinforcement learning optimization to aggre-gate two aspects of information contained within EEG signals, namely local spatio-temporal and global temporal information, and demonstrate its benefits in EEG-based human intention recognition tasks. The EEG signals are first transformed into a 3D format preserving both topologi-cal and temporal structures, followed by distinctive local spatio-temporal feature extraction by a 3D-CNN, as well as the global temporal feature extraction by an RNN. Next, a fuzzy integral with respect to the opti-mized fuzzy measures with deep reinforcement learning is utilized to integrate the two extracted information and makes a final decision. The proposed approach retains the topological and temporal structures of EEG signals and merges them in a more efficient way. Experiments on a public EEG-based movement intention dataset demonstrate the effec-tiveness and superior performance of our proposed method.

1 Introduction

Attempting to translate brain activities into commands for a computer or other devices always attracts great research interests due to its various potential appli-cations. Electroencephalography (EEG) signals reflecting the fluctuations of the voltages from the scalp are one of the most widely used tools for brain activity analysis. Despite many efforts have been devoted into EEG-based brain activity analysis, traditional methods not only separate feature extraction and classifica-tion stages, but also rely on handcrafted features which need domain knowledge

© Springer International Publishing AG, part of Springer Nature 2018
D. Phung et al. (Eds.): PAKDD 2018, LNAI 10937, pp. 156–168, 2018.
https://doi.org/10.1007/978-3-319-93034-3_13

and extensive experience, for example, determining which frequency bands are related to specific brain activities. Deep learning techniques have demonstrated advantages in automatic feature learning, and have dominated in many research fields [1]. Recently, many works have reported successful applications of deep learning in EEG analysis [2–4]. However, most neural network based approaches either still involve in complex preprocess stages or lack reasonable motivations in decoding EEG signals. There is still large room for in-depth research and improvement in terms of recognition accuracy and interpretability.

Since EEG signals are acquired on top of different cortical regions of human's head over a time period (see Fig. 1), effectively fusing these spatial and temporal information is crucial to identify uncertainties introduced by both inter- and intra-subject variability. The fuzzy integral has been proven an appropriate way of aggregating information from different sources according to their correlations in the human-computer interaction areas [5–8]. Compared with the simple weighted ensemble approach, the major advantage of fuzzy integral is the flexibility of fusing information with nonlinearity. Kim et al. have conducted extensive work using the fuzzy integral for human-robot interaction [6,7]. Cavrini [5] and Shoaie [8] also apply the fuzzy integral for a brain-computer interface application. However, the fuzzy measure for each information source often heavily depends on domain experts or massive experience. This is often unreliable, time-consuming and impractical to implement. Recent advances in deep reinforcement learning especially the deep Q-network (DQN) has shown promising capability of human-level control [9]. The reinforcement method simulates the process of a human brain interacting with an external environment. This enables the artificial intelligence to conduct tasks like human beings, and even beats human experts in certain areas [10].

In this paper, we present a novel ensemble method combining DQN and fuzzy logic to take advantages of both two techniques at the same time, wherein the DQN is utilized to tune the fuzzy measures in fuzzy integral for integrating the automatically extracted local spatio-temporal and global temporal information of EEG signals. The local spatio-temporal information represents the complex dependencies of adjacent sensory nodes, while the global temporal information is for the long term dependencies of the non-adjacent ones. The proposed model has good generalization in the cross-subject, multi-class scenario for brain activity analysis. The main contributions of this study can be summarized as follows:

- We propose to utilize the local spatio-temporal features and the global temporal features extracted by a 3D convolutional neural network (3D-CNN) and a recurrent neural network (RNN) respectively to enhance the EEG-based brain activity analysis.
- We develop an ensemble system with the fuzzy integral to combine both the 3D-CNN and the RNN classifiers. Rather than assigning the fuzzy measures heuristically, the deep reinforcement learning technique is employed to optimize the fuzzy measures of each integrated classifiers in the fuzzy integral.

– We evaluate the proposed approach on a public EEG-based movement intention dataset for the cross-subject, multi-class scenario analysis. The results
demonstrate that the proposed ensemble model is able to find the optimal
fuzzy measures of classifiers automatically and enhance the EEG-based human
intention recognition task.

2   Preliminaries

In this section we give a brief introduction of λ-fuzzy measure and Choquet integral, which we leverage in this study to fuse the local spatio-temporal information
and the global temporal information within EEG signals for human intention
recognition.

Let $X = \{x_1, \ x_2 \ ... \ x_n\}$ be a finite set represents n information sources and
the function $g_\lambda : 2^X \to [0,1]$ be the λ-fuzzy measure on X. The fuzzy measure
satisfies the following conditions:

1. $g_\lambda(\{X\}) = 1$, $g_\lambda(\{\varnothing\}) = 0$;
2. If $A, \ B \in 2^X$ and $A \cap B = \varnothing$, then $g_\lambda(\{A \cup B\}) = g_\lambda(\{A\}) + g_\lambda(\{B\}) + \lambda g_\lambda(\{A\})g_\lambda(\{B\})$

where $\lambda \in (-1, \infty)$ can be obtained through the following equation:

$$\lambda + 1 = \prod_{i=1}^{n}(\lambda g_\lambda(\{x_i\}) + 1). \tag{1}$$

So given the fuzzy measure density of one information set, the joint fuzzy measure
of any subsets can be achieved via the above axiom 2.

The Choquet integral with respect to fuzzy measure g is defined as

$$C_g(h) = \sum_{i=1}^{n}[h(x_i) - h(x_{i-1})]g(A_i), \tag{2}$$

where $A_i = \{x_i, \ x_{i+1} \ ... \ x_n\}$ is a subset of X, and $h(x_i)$ is the data/information
provided by the information source x_i. The $h(x_i)$ satisfies the monotonic prop-
erty, that is $h(x_1) \leqslant h(x_2) \leqslant h(x_3) \leqslant \ ... \ \leqslant h(x_n)$, and $h(x_0) = 0$. The joint
fuzzy measure of the subset A_i can be obtained by pre-defined fuzzy measure
rules. The fuzzy measure density, which is the fuzzy measure of each information source $g(\{x_i\})$, is usually heuristically assigned. In this study, the $g(\{x_i\})$
is proposed to be determined via the deep reinforcement learning technology.

3   Methodology

3.1   Local Spatio-Temporal Information Extraction

To represent the spatial topological structure of the EEG acquisition system, we
convert the traditional chain-like vectors to two-dimensional matrices according

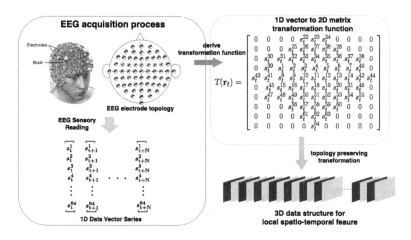

Fig. 1. EEG data acquisition and transformation process

to the EEG electrode topology. Let $\mathbf{r}_t = [s_t^1,\ s_t^2,\ s_t^i\ ...\ s_t^n]^T$ be the EEG sensory readings at time step $\mathbf{t}$, where s_t^i represents the ith electrode signal. The 1D vector to 2D matrix transformation function is defined as in Fig. 1. Through this equation each EEG reading vector $\mathbf{r}_t$ at time stamp $\mathbf{t}$ is converted to a EEG data matrix $\mathbf{m}_t$ of size 10×11. In the transformation function, the *null* electrode elements are set to zero as no effects on neural network. Some previous work has also proposed to apply topological transformation for spatial information representation [2]. However, our approach is fundamentally different from the previous conversion method by preserving both the raw data and spatial information, in contrast [2] employs complex data preprocess including frequency filter, data compression and interpolation to covert raw EEG data to images. On top of spatial representations, we use the sliding window technique to divide the converted streaming EEG matrices to individual clips for temporal information extraction. Each clip has fixed length of time slice of EEG matrices, and neighboring clips have 50% overlapping to keep the signal continuity. Overall, the raw streaming EEG readings are converted to clips with a 3D structure containing both temporal and spatial information:

$$[\mathbf{r}_t,\ \mathbf{r}_{t+1}\ ...\ ] \Rightarrow [\mathbf{C}_t,\ \mathbf{C}_{t+W}\ ...\ ],$$

where $\mathbf{C}_t = [\mathbf{m}_t,\ \mathbf{m}_{t+1}\ ...\ \mathbf{m}_{t+W-1}]$ is a 3D-structure clip with window size W starts at time step $\mathbf{t}$.

The 3D-CNNs have achieved great success in video processing applications, in which local spatio-temporal features are extracted for further analysis. Herein, we explore the 3D-CNNs for modeling the local spatio-temporal information of adjacent sensory nodes from the transformed 3D EEG data structures. The final local spatio-temporal information representation is the classification probabilities of each aimed brain intention task $\mathbf{A} = [a_1,\ a_2\ ...\ a_K]^T$ performing in the corresponding windowed period $[t,\ t+W-1]$:

$$\text{3D-CNN: } p_{tc} = C_{3D}([\mathbf{m}_t, \ \mathbf{m}_{t+1} \ ... \ \mathbf{m}_{t+W-1}]), \ p_{tc} \in \mathbb{R}^K.$$

In the 3D-CNN, we concatenate three convolutional layers directly without pooling operations. Although a convolutional operation is often followed by a pooling layer, this is not mandated. It is primarily introduced for balancing the information integrity and data complexity, while in this situation the data dimension is much smaller that we withdraw the pooling operation and keep all the information for better feature extraction. We utilize the *exponential linear unit (elu)* function as the activation function, which performs better than the commonly used rectified linear units (*ReLUs*) function in many cases. The kernel size of all 3D-CNN layers is set to 3×3 with a stride of 1, and the number of feature maps are 32, 64 and 128 respectively. Finally, a fully-connected layer with 1024 hidden units is applied on top of the 3D-CNN architecture, followed by a five-way softmax layer for final prediction.

3.2   Global Temporal Information Extraction

Since mental activities are temporal dynamic processes, modeling the evolution through time-series sequences with RNN, which has been demonstrated powerful in processing time series data in various fields, can provide important information about the ambiguity of brain activities. Long Short-Term Memory (LSTM) model is an improved RNN model with better capabilities discovering long-term dependencies. Two LSTM layers between two fully-connected layers are adopted for the global temporal information extraction. Only the output of the last LSTM unit after observing the whole time sequence is fed into the final fully-connected layer. To keep consistent with the local spatio-temporal information representations, we use the same window size for the global temporal feature extraction and 50% overlapping sliding window technique as well. We experiment with various sizes of hidden states in LSTM cell, and adopt the best result hidden state size of 1024. The RNN model takes the raw windowed EEG signal vectors $[\mathbf{r}_t, \ \mathbf{r}_{t+1} \ ... \ \mathbf{r}_{t+W-1}]$ as input, and makes the final prediction with a softmax layer:

$$\text{RNN: } p_{tr} = L([\mathbf{r}_t, \ \mathbf{r}_{t+1} \ ... \ \mathbf{r}_{t+W-1}]), \ p_{tr} \in \mathbb{R}^K.$$

Different from the 3D-CNN model, which just uses receptive fields to extract the local spatio-temporal features capturing the relations of the adjacent sensory nodes, the RNN considers the long-term temporal dependencies of the whole sensory values including non-adjacent ones. Thus the global temporal features extracted by the RNN model provides another informative description of raw EEG signals.

3.3   Choquet Integral with Deep Q-Network

The above local spatio-temporal and global temporal information extraction describes the EEG signals from different angles, thus aggregating the two aspects

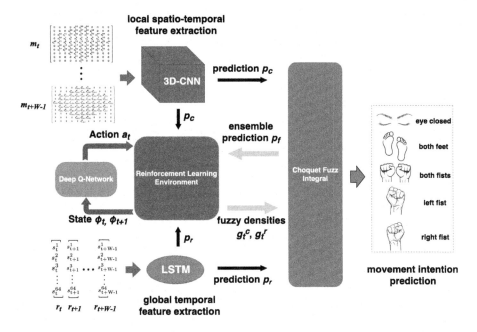

Fig. 2. Overall flowchart of the proposed approach

of data may enhance the EEG analysis tasks. The overall flowchart of the proposed approach is illustrated in Fig. 2. The two schools of information are integrated with respected to the λ-fuzzy measure by the Choquet integral to make final predictions. DQN is utilized to optimize the fuzzy measures of the 3D-CNN and the RNN instead of being selected heuristically in previous study.

Concretely, Algorithm 1 presents the pseudo-code of Choquet integral of 3D-CNN and RNN ensemble. The input is the probability predictions from the 3D-CNN and the RNN models, and the output is the aggregated results by Choquet integral. For the probability of a windowed instance belonging to one class, $p_c^{k,m}$ and $p_r^{k,m}$, the Choquet integral is applied to aggregate the two probabilities regarding their fuzzy measures, and calculate a final probability for this class. All the predictions from the 3D-CNN share the same fuzzy measure during the fuzzy fusion process and so do the predictions of the RNN model.

The overall procedure of fuzzy integral optimization with DQN is that an agent takes actions a_t tuning the fuzzy measures in a specific environment characterizing the fuzzy integral ensemble result and gets reward according to the tuning result. There are totally five candidate actions, namely, keeping unchanged, ascending or descending 3D-CNN fuzzy measure, and ascending or descending RNN fuzzy measure. The reward for one action is determined by the difference between the fusion accuracy before and after executing the action: $r_t = \delta \times (A_{t+1} - A_t)$, where $\delta = 10^5$ is the reward coefficient and A_t is the fusion accuracy at tuning step t. Thus the agent gets positive rewards when boosting the ensemble accuracy, negative rewards when descending the accuracy and zero

Algorithm 1. Ensemble 3D-CNN and RNN with Choquet fuzzy integral

Input: p_c and $p_r \in \mathbb{R}^{n \times K}$, are the classification probabilities of the 3D-CNN and the RNN, where n is the number of windowed instances in one dataset and K is the number of classes; g_c and g_r are their corresponding fuzzy measures.
Output: p_f is the classification probabilities after fuzzy fusion.

1: **function** ChoquetIntegral(p_c, p_r, g_c, g_r)
2:     Calculate all the joint fuzzy measures w.r.t fuzzy measure density
3:     **for** the prediction of each windowed time clip $p_c^k \in p_c$ and $p_r^k \in p_r$ **do**
4:         **for** the probability of each class $p_c^{k,m} \in p_c^k$ and $p_r^{k,m} \in p_r^k$ **do**
                // **perform Choquet integral**
5:             **if** $p_c^{k,m} \leqslant p_r^{k,m}$ **then**
6:                 $p_f^{k,m} = (p_c^{k,m} - 0) \times g_\lambda(\{cnn, \ rnn\}) + (p_r^{k,m} - p_c^{k,m}) \times g_\lambda(\{rnn\})$
7:             **else**
8:                 $p_f^{k,m} = (p_r^{k,m} - 0) \times g_\lambda(\{cnn, \ rnn\}) + (p_c^{k,m} - p_r^{k,m}) \times g_\lambda(\{cnn\})$
9:     **return** p_f

reward when no accuracy changes. We have also tried a fixed reward system with constant reward regardless of the accuracy fluctuation extent, and this method would easily lead the ensemble result to a local optimal value. We use how much the predictions deviate from the ground truth to characterize the state of the environment. Concretely, we define the state of a model ϕ as the class-wise summation of the absolute difference between the predicted probabilities and the ground truth:

$$\phi = |p[1] - T[1]| + |p[i] - T[i]| + \ \dots \ + |p[n] - T[n]|; \ p, \ T \in \mathbb{R}^{n \times K}, \phi \in \mathbb{R}^K$$

where $p[i]$ and $T[i]$ are the prediction and ground truth of the ith sample, respectively, and there are totally n samples over K classes. The states of 3D-CNN ϕ_t^c, RNN ϕ_t^r and the fuzzy integral ensemble model ϕ_t^f are stacked horizontally and normalized using the Z-score method to form a final representation of the environment: $\phi_t = [\phi_t^c, \ \phi_t^r, \ \phi_t^f]$, where $\phi_t \in \mathbb{R}^{3K}$.

The DQN based fuzzy measure optimization is presented in Algorithm 2. This procedure allows us to select optimal actions tuning the fuzzy measure for each information resource to achieve an optimized ensemble result. We first initialize the fuzzy measures g_1^c and g_1^r arbitrarily, and keep the same initialization for every episode. In one episode, there are total $T = 2000$ tuning steps updating the fuzzy measures 2000 times with an interval of 10^{-3}. The agent selects and executes actions according to an ϵ-greedy policy with ϵ annealed linearly from 1 to 0.01 over first 72 episodes, and fixed at 0.01 thereafter. It selects a random action with probability ϵ, otherwise select the action a_t with maximum Q value. Formally the action-value function:

$$Q^*(s, a) = \max_\pi \mathbb{E}[r_t + \gamma r_{t+1} + \gamma^2 r_{t+2} + \dots | s_t = s, a_t = a, \pi] \tag{3}$$

is approximated using a neural network called Q-network in DQN. $Q^*(s, a)$ is the maximum sum of rewards r_t with discounted factor γ at each tuning step.

Algorithm 2. Fuzzy measure optimized with deep Q-network

Input: predictions from the 3D-CNN p_c and RNN p_r; ground truth prediction T
Output: action-value network Q for action selection policy

1: Initialize replay memory D
2: Initialize action-value network Q with random weights θ
3: Initialize target action-value network $\hat{Q}$ with weights $\theta^- = \theta$
4: Initialize fuzzy measure density g_1^c and $g_1^r \in [0,1]$ arbitrarily
5: **for** episode = 1 to MaxEpisode **do**
6:      Observe the initial state of the environment $\phi_1(g_1^c, g_1^r)$
7:      **for** t = 1 to MaxStep **do**
8:          Select action $a_t = \begin{cases} \text{a random action} & \text{with probability } \epsilon \\ argmax_a Q(\phi_t, a; \theta) & \text{otherwise} \end{cases}$
9:          Execute action a_t
10:         Observe reward r_t, fuzzy measures g_{t+1}^c, g_{t+1}^r and next state ϕ_{t+1}
11:         Store transition $(\phi_t, a_t, r_t, \phi_{t+1})$ in replay memory D
            // **experience replay**
12:         Sample random minibatch of transitions $(\phi_j, a_j, r_j, \phi_{j+1})$ from D
13:         Set $y_i = \begin{cases} r_j & \text{if episode terminates at step } j+1 \\ r_j + \gamma max_{a'} \hat{Q}(\phi_{j+1}, a'; \theta^-) & \text{otherwise} \end{cases}$
14:         Perform gradient descent on $(y_j - Q(\phi_j, a_j; \theta))^2$ w.r.t Q-network parameter
15:         Every C steps reset $\hat{Q} = Q$. i.e. set $\theta^- = \theta$ // **target Q-network update**

The reward at each tuning step t is obtained by executing an action a selected according to state observation s and policy $\pi(a|s)$. This optimal action-value function obeys the following Bellman equation:

$$Q^*(s, a) = \mathbb{E}_{s'}[r + \gamma max_{a'} Q^*(s', a')|s, a], \qquad (4)$$

where $Q^*(s', a')$ is the optimal value of the state at the next tuning step. Then the optimal strategy is to select the action a' maximizing the optimal value at the next tuning step. A feedforward neural network with two hidden layers with 32 and 64 neurons respectively is used to approximate the optimal action-value function. We employ the experience replay and separate target Q-network techniques [9] to stabilize the Q-network training process. To perform experience replay, the transitions $(\phi_t, a_t, r_t, \phi_{t+1})$ are stored in a memory pool. During training process, stored transitions are randomly sampled using minibatches to feed the Q-network. The network is updated with the following loss function :

$$L_i(\theta_i) = [(r + \gamma max_{a'} Q^*(s', a'; \theta_i^-) - Q(s, a; \theta_i)]^2, \qquad (5)$$

as illustrated in Algorithm 2 line 14. A separate network with parameters θ_i^- is used to generate the target Q value and only updated with the Q-network parameters (θ_i) every C steps. Through the whole process, optimal fuzzy measures g^c and g^r are obtained for optimizing fuzzy integral based ensemble.

4 Experiments

4.1 Dataset and Model Implementation

To evaluate the proposed approach, we adopt the widely used EEG dataset *eeg-mmidb* from PhysioNet[1] for EEG-based movement intention recognition. The EEG data is collected using BCI2000 instrumentation system[2] [11] with 64 electrode channels and 160 Hz sampling rate. During the data acquisition process, one subject sits in front of a screen with prompts indicating the subject performing different movement intention tasks: imagine opening and closing left fist, right fist, both fists and both feet, and think nothing with eye closed. We select 20 subjects with 5 tasks to construct a cross-subject multi-task dataset.

During the preprocess stage, we carry out experiment with different sliding window size, namely 10, 20, 40 and 80 recordings per time window. The results show that 10 recordings in one time window gives the best performance. Thus we adopt the window size of 10 for all experimental setup. The 3D-CNN and RNN model are trained with Adam algorithm with a learning rate of 10^{-4} to minimize the cross-entropy loss function. Due to the large amount of parameters in the neural networks, we utilize dropout technique with 50% probabilities after the final fully connected layer in both models and L2 regularization in the 3D-CNN model to address the overfitting issue.

4.2 Compared Algorithms

All the methods are based on the same dataset with our model.

- **SR-FBCSP** [12]. The Shrinkage Regularized Filter Bank Common Spatial Patterns(SR-FBCSP) algorithm, which is based on the widely used FBCSP algorithm, outperforms FBCSP in classifying motor imagery tasks.
- **ICA+QDA** [13]. The independent component analysis (ICA) is for the feature extraction followed by quadratic discriminant analysis (QDA) for final classification.
- **Autoencoder+XGboost** [3]. The autoencoder is used for automatic EEG feature extraction. XGboost, which has been demonstrated competitive performance in many competitions, is used for final classification.
- **1D-CNN** [4]. The 1D-CNN is the traditional neural network based spatial filter for EEG signal analysis, and we apply it in our dataset.
- **3D-CNN**. The 3D-CNN is the model used in this study for classification based on the local spatio-temporal information of EEG signals.
- **RNN**. The RNN model is the model used in this study for classification based on the global temporal information of EEG signals.
- **Neural network ensemble**. We ensemble the 3D-CNN and RNN with a neural network based method. In this model, the 3D-CNN part and the RNN

---

[1] https://www.physionet.org/pn4/eegmmidb/.
[2] www.bci2000.org.

part have the same settings with the above described models. Their representations from the last fully-connected layer are concatenated together and fed into a softmax layer for final prediction. The training process and implementation tricks are also the same with the individual models.

- **Fuzzy Integral Ensemble**. We ensemble the 3D-CNN and RNN with Choquet integral with randomly selected fuzzy measure density, $g_c = 0.924$ and $g_r = 0.158$.
- **Fuzzy Integral Ensemble with DQN**. The proposed method in this study. The initial fuzzy measures are the same with the **Fuzzy Integral Ensemble** method, and the final measures are $g_c = 0.203$ and $g_r = 0.19$.

4.3   Experimental Result

Figure 3 gives the training information of the neural networks in this work. We stop model training at minimum validation loss. The performance of our proposed approach and the comparison models are summarized in Table 1. It is observed that the simple 1D-CNN spatial filter model outperforms the previous studies even without frequency band filter. This result is consistent with [14], in which it is shown the CNN can act as frequency band filters itself and achieves competitive performance. What's more, the 3D-CNN or RNN model performs better than the traditional 1D-CNN filter approach, demonstrating optimal data representation is capable of enhancing neural network performance and the local spatio-temporal and the global temporal information is favourable for successful EEG signal analysis. However, it is interesting to find that ensemble of the 3D-CNN and the RNN with neural network does not performs better performance. Although much evidence reveals that deeper network or combination of different kinds of neural network benefits the feature representation capabilities for final model performance, it is required careful parameter tuning or diverse implementation tricks. Complex models derived from suitable simple models may even suffer performance degradation problem [15]. It is not easy to optimize a complex neural network system.

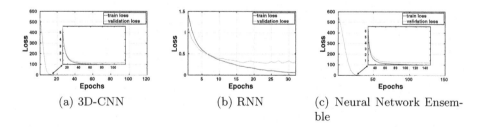

(a) 3D-CNN   (b) RNN   (c) Neural Network Ensemble

Fig. 3. Train and validation loss with training epochs

Thus in this work we adopt a traditional but widely used fusion approach, fuzzy integral, to aggregate the local spatio-temporal information and the global

Table 1. Comparison with previous studies and baseline models

Method	Multi-class	Validation	Accuracy
SR-FBCSP [12]	Binary	Intra-Sub	0.8206
Autoencoder+XGboost [3]	Multi(5)	Cross-Sub(20)	0.794
ICA+QDA [13]	Multi(3)	Cross-Sub(30)	0.8724
1D-CNN [4]	Multi(5)	Cross-Sub(20)	0.8909
3D-CNN	Multi(5)	Cross-Sub(20)	0.9006
RNN	Multi(5)	Cross-Sub(20)	0.9110
Neural network ensemble	Multi(5)	Cross-Sub(20)	0.9108
Fuzzy Integral Ensemble	Multi(5)	Cross-Sub(20)	0.9082
Fuzzy Integral Ensemble with DQN	**Multi(5)**	**Cross-Sub(20)**	**0.9302**

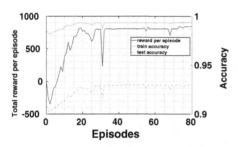

Fig. 4. Training curves tracking the total reward and accuracy

Fig. 5. Confusion matrix

temporal information from the 3D-CNN and the RNN models. Furthermore we propose to use the DQN to optimize the fuzzy measures instead of relying on domain knowledge or empirical selection. The results surpass both the single neural network methods and the neural network ensemble method. It is illustrated in the Table 1 that the fuzzy integral ensemble with randomly selected fuzzy measures does not provide an effective fusion scheme, thus can not enhance the overall performance. We also randomly select other fuzzy densities, and only a few sets aggregate information effectively. The Fig. 4 shows the process of the DQN optimization. It is observed that at the initial stage, the agent chooses actions according to the Q-network with little training, and large random probability ϵ, thus gets negative total rewards. As the training process proceeds, the total rewards increases and saturates. The test and train accuracy fluctuate in the same way as the total reward. In the final confusion matrix (Fig. 5), the model exhibits equally good performance for each class, with little imbalance.

5   Conclusion

In this study, we propose to employ the fuzzy integral, which is optimized by the deep reinforcement learning, to aggregate both the local spatio-temporal and the global temporal information within EEG signals for human intention recognition. To effectively select the fuzzy measures for each information sources, the DQN is utilized to search the optimal fuzzy measures. The developed model is further evaluated on movement intention recognition tasks in the cross-subject, multi-class scenario. The experimental results demonstrate the effectiveness of neural network ensemble using the fuzzy integral with respect to the optimized fuzzy measures with the deep Q-network technique.

References

1. LeCun, Y., Bengio, Y., Hinton, G.: Deep learning. Nature **521**(7553), 436–444 (2015)
2. Bashivan, P., Rish, I., Yeasin, M., Codella, N.: Learning representations from EEG with deep recurrent-convolutional neural networks. In: International Conference on Learning Representations (2016)
3. Zhang, X., Yao, L., Zhang, D., Wang, X., Sheng, Q., Gu, T.: Multi-person brain activity recognition via comprehensive EEG signal analysis. In: 14th International Conference on Mobile and Ubiquitous Systems: Computing, Networking and Services (2017)
4. Cecotti, H., Graser, A.: Convolutional neural networks for P300 detection with application to brain-computer interfaces. IEEE Trans. Pattern Anal. Mach. Intell. **33**(3), 433–445 (2011)
5. Cavrini, F., Bianchi, L., Quitadamo, L.R., Saggio, G.: A fuzzy integral ensemble method in visual P300 brain-computer interface. Comput. Intell. Neurosci. **2016**, 49 (2016)
6. Yoo, B.S., Kim, J.H.: Fuzzy integral-based gaze control of a robotic head for human robot interaction. IEEE Trans. Cybern. **45**(9), 1769–1783 (2015)
7. Yoo, J.K., Kim, J.H.: Fuzzy integral-based gaze control architecture incorporated with modified-univector field-based navigation for humanoid robots. IEEE Trans. Syst. Man Cybern. Part B (Cybern.) **42**(1), 125–139 (2012)
8. Shoaie, Z., Esmaeeli, M., Shouraki, S.B.: Combination of multiple classifiers with fuzzy integral method for classifying the EEG signals in brain-computer interface. In: International Conference on Biomedical and Pharmaceutical Engineering, ICBPE 2006, pp. 157–161. IEEE (2006)
9. Mnih, V., Kavukcuoglu, K., Silver, D., Rusu, A.A., Veness, J., Bellemare, M.G., Graves, A., Riedmiller, M., Fidjeland, A.K., Ostrovski, G., et al.: Human-level control through deep reinforcement learning. Nature **518**(7540), 529–533 (2015)
10. Silver, D., Huang, A., Maddison, C.J., Guez, A., Sifre, L., Van Den Driessche, G., Schrittwieser, J., Antonoglou, I., Panneershelvam, V., Lanctot, M., et al.: Mastering the game of go with deep neural networks and tree search. Nature **529**(7587), 484–489 (2016)
11. Schalk, G., McFarland, D.J., Hinterberger, T., Birbaumer, N., Wolpaw, J.R.: BCI 2000: a general-purpose brain-computer interface (BCI) system. IEEE Trans. Biomed. Eng. **51**(6), 1034–1043 (2004)

12. Shenoy, H.V., Vinod, A.P., Guan, C.: Shrinkage estimator based regularization for EEG motor imagery classification. In: 2015 10th International Conference on Information, Communications and Signal Processing (ICICS), pp. 1–5. IEEE (2015)
13. Sita, J., Nair, G.: Feature extraction and classification of EEG signals for mapping motor area of the brain. In: 2013 International Conference on Control Communication and Computing (ICCC), pp. 463–468. IEEE (2013)
14. Schirrmeister, R.T., Springenberg, J.T., Fiederer, L.D.J., Glasstetter, M., Eggensperger, K., Tangermann, M., Hutter, F., Burgard, W., Ball, T.: Deep learning with convolutional neural networks for EEG decoding and visualization. Hum. Brain Mapp. **38**(11), 5391–5420 (2017)
15. He, K., Zhang, X., Ren, S., Sun, J.: Deep residual learning for image recognition. In: Proceedings of the IEEE Conference on Computer Vision and Pattern Recognition, pp. 770–778 (2016)

Heterogeneous Domain Adaptation Based on Class Decomposition Schemes

Firat Ismailoglu[1], Evgueni Smirnov[2(✉)], Ralf Peeters[2], Shuang Zhou[3], and Pieter Collins[2]

[1] Department of Computer Engineering, Cumhuriyet Univertsity, Sivas, Turkey
`fismailoglu@cumhuriyet.edu.tr`
[2] Department of Data Science and Knowledge Engineering, Maastricht University, Maastricht, The Netherlands
`{smirnov,ralf.peeters,pieter.collins}@maastrichtuniversity.nl`
[3] College of Control Engineering, Chengdu University of Information Technology, Chengdu, People's Republic of China
`s.zhou@cuit.edu.cn`

Abstract. This paper introduces a novel classification algorithm for heterogeneous domain adaptation. The algorithm projects both the target and source data into a common feature space of the class decomposition scheme used. The distinctive features of the algorithm are: (1) it does not impose any assumptions on the data other than sharing the same class labels; (2) it allows adaptation of multiple source domains at once; and (3) it can help improving the topology of the projected data for class separability. The algorithm provides two built-in classification rules and allows applying any other classification model.

1 Introduction

Heterogeneous domain adaptation (HDA) for classification has recently received a significant attention [13]. It assumes a given target domain (the primary domain of interest) and at least one (auxiliary) source domain. These domains are represented by different input features and usually share the same class labels. The goal is to improve classification models in the target domain by utilizing data from the source domain(s).

The approaches to HDA can be either symmetric or asymmetric [13]. The symmetric approaches project the target and source data into a common feature space and train prediction models on the projected data. The asymmetric approaches project the source data into the target domain and train prediction models on the target data and projected source data. Below we describe the main approaches within each group.

Domain Adaptation Manifold Alignment (DAMA) [12] is one of the main symmetric approaches to HDA. DAMA is applicable to problems with a single target domain and $P - 1$ source domains that share the same class-label set. Assuming that each domain is a manifold, a manifold alignment algorithm learns

© Springer International Publishing AG, part of Springer Nature 2018
D. Phung et al. (Eds.): PAKDD 2018, LNAI 10937, pp. 169–182, 2018.
https://doi.org/10.1007/978-3-319-93034-3_14

P projection functions, one for each domain. It maintains a combinatorial graph Laplacian matrix to reflect inter-domain class similarity, a combinatorial graph Laplacian matrix to reflect inter-domain class dissimilarity, and a diagonal block-matrix of graph Laplacian matrices each reflecting instance similarities within a domain. The algorithm minimizes an objective function defined over the matrices so that the resulting P projection functions match same class instances and separate different class instances when projected. DAMA has its own prediction algorithm. It first trains regression models for the source data projected, and then adapts these models to the target domain using manifold regularization based on the projected target data.

Heterogeneous Feature Augmentation (HFA) [3] is another well-known symmetric approach to HDA. It is applied to problems with a target domain and a source domain that share the same class-label set. HFA augments the common latent feature space with the target-domain features and source-domain features. The projection functions are jointly represented by a matrix. To find the joint projection matrix, a problem to minimize the structural risk functional of SVMs is defined. The problem is tackled by an alternating optimization algorithm that solves the dual problem of SVMs and finds the corresponding optimal joint transformation matrix. The final classifier for HFA is the SVM classifier derived together with the transformation matrix.

Asymmetric Regularized Cross-Domain Transformation (ARC-t) [7] is one of the main asymmetric approaches to HDA. ARC-t is applied to problems with a target domain and a single source domain that share the same class-label set. The projection function is represented by a matrix that maps an instance from the source domain to an instance in the target domain. This matrix is learned in a non-linear Gaussian RBF kernel space. This is done by minimizing a matrix regularizer and a set of constraints imposed on any pair of a target instance and a projected source instance.

Sparse Heterogeneous Feature Representation (SHFR) [6] is another well-known asymmetric approach to HDA. It is applied to problems with a target domain and a single source domain that share the same class-label set. SHFR is similar ARC-t with a difference that the target data and source data are first represented in a code space based on a class decomposition scheme [2]. This allows a projection matrix to be learned by a nonnegative LASSO optimization.

We summarize the main HDA approaches in Table 1. We observe that an HDA algorithm is lacking that can adapt multiple source domains without imposing any domain assumptions. Such an algorithm has to preserve the data topology (at least locally) using class correspondence. The adapted datasets have to applicable for any classification model. Still, it is desirable that the algorithm can provide a specific classifier tailored to the projections employed.

In the rest of this paper we introduce a new HDA algorithm called class code alignment (CCA) algorithm that has all the characteristics listed above. In Sects. 2 and 3 we provide a problem formulation and background information. The new algorithm is given in Sect. 4. In Sects. 5 and 6 we present experiments and conclusions.

Table 1. HDA approaches

HDA method	# Source domains	Domain assumption	Class correspondence	Topology preservation	Classifier independence	Specific classifiers
DAMA	≥ 1	yes	yes	yes	yes	yes
HFA	1	no	yes	no	yes	yes
ARC-t	1	no	yes	yes	yes	no
SHFR	1	no	yes	yes	yes	no

2   Problem Formulation

We consider the problem of HDA in the context of classification. We define a classification domain as a triple that consists of an instance space X with d continuous features $X(j)$ ($j \in \{1, \ldots, d\}$), a finite set Y of $K = |Y|$ class labels, and an unknown probability distribution p over $X \times Y$. We assume the presence of a target domain and at least one source domain. The target domain is the domain of interest. It is given by a target instance space X^T with d_T features, a class-label set Y^T, and a target probability distribution p^T over $X^T \times Y^T$. The target training data D_T is a set of N_T instances $(x_i^T, y_i^T) \in X^T \times Y^T$ generated from p^T. Any source domain is an auxiliary domain. It is given by a source instance space X^S with d_S features, a class-label set Y^S, and a source probability distribution p^S over $X^S \times Y^S$. The source training data D_S is a set of N_S instances $(x_i^S, y_i^S) \in X^S \times Y^S$ generated from p^S.

In HDA, the instance spaces X^T and X^S are different. The classification problem in HDA is to provide a good estimate $\hat{y} \in Y^T$ of the true class of a target query instance $x_q^T \in X^T$ according to the target probability distribution p^T given the target and source training data D_T and D_S. In this paper we consider this problem under the assumptions that: (1) the target and source domains share the same label set Y, i.e.: $Y^T = Y^S = Y$, and (2) the number K of class labels in Y is larger than 2.

3   Class Decomposition Schemes and Coding Matrices

The HDA algorithm we propose is based on class decomposition schemes [2]. Such schemes consider any multi-class classification problem ($K > 2$) as a set of B binary classification problems [2] for some positive integer B. Any b-th binary classification problem is given by a binary class partition $P_b(Y)$ of the class-label set Y for $b \in \{1, \ldots, B\}$. The first (second) element $Y_b^+ \subseteq Y$ ($Y_b^- \subseteq Y$) of the partition $P_b(Y)$ stands for a positive (negative) binary super class. In this context *a class decomposition scheme* is a set $S(Y)$ of B binary class partitions $P_b(Y)$ such that for any two distinct classes $y_1, y_2 \in Y$ there exists a binary class partition $P_b(Y) \in S(Y)$ with super classes Y_b^- and Y_b^+ that separate y_1 and y_2; i.e. $\neg(y_1, y_2 \in Y_b^-) \wedge \neg(y_1, y_2 \in Y_b^+)$.

Any class decomposition scheme $S(Y)$ is represented by a binary coding matrix $C \in \{0, 1\}^{B \times K}$ [2]. For any index $b \in \{1, ..., B\}$ and class label $y \in Y$ the element $C(b, y)$ equals 1 if y belongs to the positive super class Y_b^+ of the class partition $P_b(Y)$. If y belongs to the negative super class Y_b^- of $P_b(Y)$, the element $C(b, y)$ equals 0.

Any binary class partition $P_b(Y)$ is represented by exactly one row $C(b, \cdot)$ in C that we call *the partition code word for* $P_b(Y)$. Any class label $y \in Y$ is represented by exactly one column $C(\cdot, y)$ in C that we call *the class code word for* y. Class code words $C(\cdot, y)$ are viewed as images of the class labels $y \in Y$ in a *code space* $\mathcal{C}$.

Definition 1 (Code Space). *Given a class decomposition scheme $S(Y)$ with B binary partitions $P_b(Y)$, the code space $\mathcal{C}$ is equal to $[0, 1]^B$.*

To solve a multi-class classification problem using a class decomposition scheme $S(Y)$ we take three steps. First, we train an encoding mapping h that can project any instance to the code space $\mathcal{C}$. The mapping h consists of binary classifiers h_b, one for each binary class partition $P_b(Y) \in S(Y)$. Second, we encode a query instance x using the mapping h; i.e., we create an instance code word consisting of bits $h_b(x)$ assigned by the binary classifiers h_b for that instance. Third, we decode the class of the instance x using the coding matrix C of $S(Y)$. We assign a class label $y \in Y$ of which the code word matches best with the instance code word of the instance x.

4   Class Code Alignment Algorithm

This section introduces our class code alignment (CCA) algorithm for the HDA classification problem from Sect. 2. The algorithm follows the symmetric approach to HDA. It projects the target and source data into a common feature space using any class decomposition scheme $S(Y)$. Given that the target and source domains share the same class-label set Y, the common feature space is the code space $\mathcal{C}$ associated with $S(Y)$. To project the data into $\mathcal{C}$ the algorithm builds the encoding mappings for the target data and source data as well as an additional encoding mapping for the class labels. This is done so that the target and source instances are projected close to the code words of their class labels in $\mathcal{C}$; i.e., they become class-code aligned.

4.1   Detailed Description

The CCA algorithm assumes that the class-label set Y consists of K standard unit vectors in $\mathbb{R}^K$ (i.e. $Y \subseteq \mathbb{R}^K$) so that the label of the k-th class ($k \in \{1, \ldots, K\}$) is given by a standard unit vector whose k-th bit equals 1. Given a class decomposition scheme $S(Y)$ with B class partitions, the algorithm builds three encoding mappings:

(1) $\sigma \circ T^T : \mathbb{R}^{d_T} \to [0, 1]^B$ from the target feature space X^T to the code space $\mathcal{C}$, and

(2) $\sigma \circ T^S : \mathbb{R}^{d_S} \to [0,1]^B$ from the source feature space X^S to the code space $\mathcal{C}$, and

(3) $\sigma \circ C : \mathbb{R}^K \to [0,1]^B$ from the set Y of K class labels to the code space $\mathcal{C}$,

where

- $\sigma : \mathbb{R}^B \to [0,1]^B$ is a multivariate logistic function defined for $b \in \{1, ..., B\}$ as $\sigma_b(w) = (1 + e^{-w(b)})^{-1}$, and
- $T^T : \mathbb{R}^{d_T} \to \mathbb{R}^B$ is a linear mapping given as a matrix in $\mathbb{R}^{B \times d_T}$, and
- $T^S : \mathbb{R}^{d_S} \to \mathbb{R}^B$ is a linear mapping given as a matrix in $\mathbb{R}^{B \times d_S}$, and
- $C : \mathbb{R}^K \to \mathbb{R}^B$ is a linear mapping given as a matrix in $\mathbb{R}^{B \times K}$.

The mappings $\sigma \circ T^T$ and $\sigma \circ T^S$ are encoding mappings of the class decomposition scheme used for the target domain and the source domain, respectively. More precisely, the rows of the target matrix T^T (the source matrix T^S) represent logistic regression models of the binary classifiers trained on the target data (the source data). The mapping $\sigma \circ C$ is a class-label encoding mapping: it determines the code word for each class label and is common for the target domain and source domain. We note that the matrix C is a real-value coding matrix of $S(Y)$ in contrast to the standard binary coding matrices. The CCA algorithm adjusts C to better fit the target and source domains.

Once the mappings T^T, T^S, and C are available, any target instance $x^T \in X^T$ is projected into $\sigma(T^T x^T) \in \mathcal{C}$, any source instance $x^S \in X^S$ is projected into $\sigma(T^S x^S) \in \mathcal{C}$, and any class label $y \in Y$ is projected into $\sigma(Cy) \in \mathcal{C}$ (see Fig. 1). Below we describe how to build the mappings so that the projected instances are grouped around the code words of their classes.

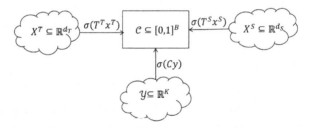

Fig. 1. Projection scheme

Any element p of the code space $\mathcal{C}$ is viewed as a parameter vector for a multivariate Bernoulli variable. If p is a true parameter vector and q is an approximate parameter vector in $\mathcal{C}$, to quantify the information gain from q to p we use KL divergence:

$$KL\,[p\|q] = \sum_{b=1}^{B} \left[p(b) \log \left(\frac{p(b)}{q(b)} \right) + (1 - p(b)) \log \left(\frac{1 - p(b)}{1 - q(b)} \right) \right]. \tag{1}$$

As we aim at mapping instances and their class labels into similar locations in the code space $\mathcal{C}$ we define a loss l for each instance label pair $(x, y) \in D_T \cup D_S$. Since our measure in $\mathcal{C}$ is the KL divergence, a loss l of the instance x for class label y for the mappings T and C can be defined as:

$$l(x, y \mid C, T) = KL\left[\sigma\left(Cy\right) \| \sigma\left(Tx\right)\right], \text{where } T = \begin{cases} T^T & \text{if } (x, y) \in D_T, \\ T^S & \text{if } (x, y) \in D_S. \end{cases}$$

Thus, the loss $\mathcal{L}$ for all N_T target instances $(x_i^T, y_i^T) \in D_T$ is defined as:

$$\mathcal{L}(D_T \mid C, T^T) = \sum_{i=1}^{N_T} l(x_i^T, y_i^T \mid C, T^T),$$

and the loss $\mathcal{L}$ for all N_S source instances $(x_i^S, y_i^S) \in D_S$ is defined as:

$$\mathcal{L}(D_S \mid C, T^S) = \sum_{i=1}^{N_S} l(x_i^S, y_i^S \mid C, T^S).$$

Minimizing $\mathcal{L}(D_T \mid C, T^T)$ for T^T and C forces the projected target instances $\sigma(T^T x_i^T)$ to be close to the code words $\sigma(Cy_i^T)$ of their class labels y_i^T in the code space $\mathcal{C}$. The same is the effect of minimizing $\mathcal{L}(D_S \mid C, T^S)$ for T^S and C for the projected source instances. Taken together, minimizing:

$$\mathcal{L}(D_T \mid C, T^T) + \mathcal{L}(D_S \mid C, T^S) \tag{2}$$

causes the same-class target and source instances to be close in the code space $\mathcal{C}$ independently on their domain. For this reason we analyze the mapping (matrix) C.

The matrix C has to have large separation properties [2]. The large column separation property means that for any class label $y_1 \in Y$ the code word Cy_1 has to be distant from the code Cy_2 of any other class label $y_2 \in Y$. If this property holds, then the code words $\sigma(Cy_1)$ and $\sigma(Cy_2)$ are distant in the code space $\mathcal{C}$; i.e. $KL\left[\sigma\left(Cy_1\right) \| \sigma\left(Cy_2\right)\right]$ and $KL\left[\sigma\left(Cy_2\right) \| \sigma\left(Cy_1\right)\right]$ are both large. Since the projected instances arrive close to the code words of their class labels, this causes the projected instances of different class labels to be distant in the code space $\mathcal{C}$.

The large row separation property means that for any $b_1 \in \{1, \ldots, B\}$ the partition code word $C(b_1, \cdot)$ has to be distant from the partition code word $C(b_2, \cdot)$ and its complement for any other $b_2 \in \{1, \ldots, B\} \setminus \{b_1\}$. If this property holds, the logistic regression models represented by rows $C(b_1, \cdot)$ and $C(b_2, \cdot)$ commit less errors simultaneously. This causes the loss $l(x, y \mid C, T)$ (defined in Eq. (1)) for any labeled instance (x, y) to decrease which in turn decreases the objective function (2) for all the projected target and source instances. The latter means that the projected instances are grouped more around the code words of their classes in the code space $\mathcal{C}$.

From the above it is clear that the large separation properties of the matrix C improves the topology of the data for class separability in the code space $\mathcal{C}$.

Thus, we initialize C using a reference coding matrix C_{ref} with known large separation properties. Since we learn C, we make sure that it is close to C_{ref}. Hence, we add the following *loss* term to the objective function (2) to penalize the *KL* divergence between the projections of the classes obtained by C and C_{ref} for the target and the source instances:

$$\mathcal{L}(D_T \mid C_{ref}, C) + \mathcal{L}(D_S \mid C_{ref}, C), \tag{3}$$

where

- $\mathcal{L}(D_T \mid C_{ref}, C) = \sum_{(x_i^T, y_i^T) \in D_T} KL\left[\sigma\left(Cy_i^T\right) \| \sigma\left(C_{ref}\, y_i^T\right)\right]$, and
- $\mathcal{L}(D_S \mid C_{ref}, C) = \sum_{(x_i^S, y_i^S) \in D_S} KL\left[\sigma\left(Cy_i^S\right) \| \sigma\left(C_{ref}\, y_i^S\right)\right]$.

The final objective function to minimize for T^T, T^S and C is defined as follows:

$$\mathcal{O}(D_T, D_S \mid T^T, T^S, C) = \mathcal{L}(D_T \mid C, T^T) + \mathcal{L}(D_S \mid C, T^S) \\ + \alpha\left(\mathcal{L}(D_T \mid C_{ref}, C) + \mathcal{L}(D_S \mid C_{ref}, C)\right), \tag{4}$$

where $\alpha > 0$ is a regularization parameter.

Once the objective function (4) is defined, we introduce the CCA algorithm. The algorithm learns the mappings T^T, T^S, and C from the target and source data by minimizing this function. It is given in Algorithm 1. The CCA algorithm is of *Alternating Minimization* type. It first improves the matrix T^T for the matrix C, then improves the matrix T^S for the matrix C, and, finally, improves the matrix C for the matrices T^T and T^S. The process is repeated to minimize the objective function (4).

Algorithm 1. Class Code Alignment Algorithm

Input: target data D_T, source data D_S, reference projection matrix C_{ref}, initial projection matrices T_0^T and T_0^S, step parameter $\beta \in (0,1)$, regularization parameter $\alpha > 0$, outer iteration number $M > 1$, and inner iteration number $N > 1$.

Output: projection matrices T^T, T^S, and C.

1: $C_0 := C_{ref}$;
2: **for** $t := 1$ **to** M **do**
3:    $T_t^T := \text{IMPROVE-T}(D_T, C_{t-1}, T_{t-1}^T, \beta, N)$;
4:    $T_t^S := \text{IMPROVE-T}(D_S, C_{t-1}, T_{t-1}^S, \beta, N)$;
5:    $C_t := \text{IMPROVE-C}(D_T, D_S, \alpha, C_{ref}, T_{t-1}^T, T_{t-1}^S)$;
6: **return** T^T, T^S, and C.

The IMPROVE-T function improves the mapping T^T (resp. T^S) for C (see Algorithm 2). It minimizes the loss term $\mathcal{L}(D_T | C, T^T)$ (resp. $\mathcal{L}(D_S | C, T^S)$) of the objective function (4). The IMPROVE-T function is a gradient descent algorithm with the backtracking line-search method. In each iteration it first computes the gradient matrix G and then executes the backtracking method using a parameter β to determine a step size η that reduces the objective function (4) (see lines 6–8 in IMPROVE-T). The function stops after N iterations and outputs the improved matrix T^T (resp. T^S).

Algorithm 2. IMPROVE-T Function

Input: data D, projection matrices C and T, step parameter $\beta \in (0,1)$, and
iteration number $N > 1$.
Output: projection matrix T.

1: let B and d be the sizes of T;
2: **for** $t := 1$ **to** N **do**
3:    $\eta := 1$;
4:    **for** $i := 1$ **to** B and $j := 1$ **to** d **do**
5:        $G(i,j) = \displaystyle\sum_{(x,y)\in D} x(j)\Big(\sigma\big(-C(i,\cdot)y\big)\sigma\big(T(i,\cdot)x\big) - \sigma\big(C(i,\cdot)y\big)\sigma\big(-T(i,\cdot)x\big)\Big)$;
6:        **while** $\mathcal{L}(D|C, T - \eta G) > \mathcal{L}(D|C,T) - \frac{\eta}{4}\|G\|^2$ **do**
7:            $\eta := \beta\eta$;
8:        $T := T - \eta \times G$;
9: **return** T.

Algorithm 3. IMPROVE-C Function

Input: target data D_T, source data D_S, regularization parameter $\alpha > 0$,
reference projection matrix C_{ref}, and projection matrices T^T and T^S.
Output: projection matrix C.

1: let B and K be the sizes of C_{ref};
2: **for** $i := 1$ **to** B and $j := 1$ **to** K **do**
3:    $D_{Tj} := \big\{(x^T, y) \in D_T \mid y = y_j \wedge y_j \in Y\big\}$;
4:    $D_{Sj} := \big\{(x^S, y) \in D_S \mid y = y_j \wedge y_j \in Y\big\}$;
5:    $C(i,j) := \frac{1}{1+\alpha}\left(\alpha C_{ref}(i,j) + \dfrac{\displaystyle\sum_{(x^T,y)\in D_{Tj}} T^T x^T + \sum_{(x^S,y)\in D_{Sj}} T^S x^S}{|D_{Tj}|+|D_{Sj}|}\right)$;
6: **return** C.

The IMPROVE-C function improves the mapping C for T^T and T^S by minimizing the term (3) in the objective function (4) (see Algorithm 3). Minimizing is analytical, since the function (4) has a unique stationary point for a fixed choice of T^T and T^S.

The objective function (4) can be extended for HDA classification problems with several source domains. In this case the IMPROVE-T function is called for each source domain and the IMPROVE-C function reflects the data from all the source domains.

4.2   CCA Classification of Target Instances

Once T^T, T^S, and C are available, any target instance $x^T \in X^T$ is projected into $\sigma(T^T x^T) \in C$, any source instance $x^S \in X^S$ is projected into $\sigma(T^S x^S) \in C$, and any label $y \in Y$ is projected into $\sigma(Cy) \in C$. This allows three CCA classification rules.

The first classification rule CCA.CDS is based on the class decomposition schemes [2]. It first projects a target query instance $x_q^T \in X^T$ into $\sigma(T^T x_q^T) \in C$. Then CCA.CDS assigns to x_q^T a class label $\hat{y}$ whose code word $\sigma(Cy)$ in C is closest to the projection $\sigma(T^T x_q^T)$; i.e. $\hat{y} = \underset{y \in Y}{\operatorname{argmin}} KL[\sigma(Cy) \parallel \sigma(T^T x_q^T)]$.

The second classification rule CCA.IDS is based on the instance decomposition schemes [5]. It first projects a target query instance $x_q^T \in X^T$ into $\sigma(T^T x_q^T) \in C$. Then CCA.IDS determines set NN of the nearest neighbors of $\sigma(T^T x_q^T)$ from the set of the projected target and source instances in C. It assigns to x_q^T a class label $\hat{y}$ that has a majority among the instances in NN; i.e. $\hat{y} = \underset{y \in Y}{\operatorname{argmax}} \#\{x_i \in NN | y_i = y\}$.

In addition any other classification rule is possible. Once the target and source data are in the code space C we can train any classifier on these data.

5   Experiments

This section provides the experiments of the CCA algorithm applied on three HDA datasets. The CCA generalization performance is compared with that of baseline classifiers trained on the target data only and two domain adaptation approaches.

5.1   Settings of the CCA Algorithm

The CCA algorithm was set up as follows. The reference coding matrix C_{ref} was set to be the One-vs-All coding matrix [10]. The regularization parameter α took values from the set $\{0.04, 0.02, 1, 5, 25\}$ and the results are reported for the α value that maximizes the accuracy. The iteration numbers M and N were set to 50 and 10, respectively. The parameter of the backtracking method β of the function IMPROVE-T was set to 0.5. The CCA.CDS and CCA.IDS classification rules were applied.

We note that the reference coding matrix C_{ref} was set equal to the One-vs-All coding matrix to make the setup of the CCA algorithm comparable with that of the baseline classifiers. The C_{ref} setup ensures a kind of worst-case generalization performance of the CCA.CDS and CCA.IDS rules due to small column and row separation of the One-vs-All matrices. This means that the comparison with the baseline classifiers is rather fair: the setup is unfavorable for the CCA algorithm.

5.2   Baseline Classifiers

We compare the CCA algorithm against two groups of baseline classifiers. The first group is the group of classifiers trained on the target data only[1]. It consists of:

- kNN.T, a first nearest neighbor classifier [4].
- SVM.T, a SVM classifier with a linear kernel and default setting from LIB-SVM. It employs the One-vs-All class decomposition scheme for multi-class classification.
- Bunching.T, the Bunching algorithm [1] with One-vs-All reference coding matrix C_{ref}, regularization parameter α equal to 1.0, and iteration number equal to 100.

The second group is the group of domain adaptation approaches trained on the target data and source data. It consists of:

- HFA, the Heterogeneous Feature Augmentation approach with default setting [3]: the regularization parameter was set to 1 and the parameter that controls the complexities of the transformation matrices was set to 100.
- SHFR, the Sparse Heterogeneous Feature Representation approach with default setting (One-vs-All coding matrix was used for asymmetric transformation) [7].

HFA and SHFR used one-vs-all SVM ensembles with linear kernel in common spaces.

5.3   Experiments on the Office Dataset

The Office dataset was introduced in [11]. It contains 4652 images from three domains: Amazon, dSLR, and webcam, that share the same 31 classes. The images in each domain were captured under different lighting conditions. We defined two classification problems. The first problem (resp. second problem) considers the dSLR domain as target domain and the amazon domain (resp. the webcam domain) as source domain. To estimate the accuracy of the classifiers we followed the hold-out protocol from [3,7]. 20 (resp. 8) training images were randomly selected for the source domain amazon (resp. webcam) and 3 training images were randomly selected from the target domain dSLR from each class. The remaining target dSLR images were served as target test instances. The hold-out evaluation was repeated 10 times. The results are given in Table 2.

5.4   Experiments on the Wikipedia Dataset

The Wikipedia dataset was introduced in [9]. It contains 5732 instances from a text domain and an image domain. Both domains have 2866 instances and share the same 10 classes. One classification problem was considered. It views the text

---

[1] This is indicated with postfix T.

Table 2. Classification accuracies (in percent) for the Office Data. Bold numbers indicate accuracies that are statistically greater than others based on a t-test on significance level of 0.05.

Target domain	Source domain	kNN.T	SVM.T	Bunching.T	HFA	SHFR	CCA.IDS	CCA.CDS
dSLR	amazon	46.12	53.13	49.17	54.7	52.66	54.66	**57.7**
	webcam				53.77	55.1	54.13	**58.65**

domain as target domain and the image domain as source domain. To estimate the accuracy of the classifiers we followed a hold-out protocol. For each class 10 training text instances were randomly selected as target instances and the remaining 2856 text instances were used as target test instances. All the image instances served as source instances. The hold-out evaluation was repeated 10 times. The results are given in Table 2.

Table 3. Classification accuracies (in percent) for the Wikipedia Data.

Target domain	Source domain	kNN.T	SVM.T	Bunching.T	HFA	SHFR	CCA.IDS	CCA.CDS
Text	Image	59.95	60.02	61.66	62.35	63.59	56.88	64.53

5.5 Experiments on the Multiple Feature Dataset

The Multiple Feature (Mfeat) dataset [8] contains 2000 hand-written images of ten numbers (classes) from 0 to 9. The images were preprocessed using different techniques which resulted in six domains: mfeat-fou (given with 76 Fourier coefficients), mfeat-fac (given with 216 profile correlation features), mfeat-kar (given with 64 Karhunen-Love coefficients), mfeat-pix (given with 240 pixel average features), mfeat-zer (given with 47 Zernike moment features), and mfeat-mor (given with 6 morphological features). The mfeat-mor domain is considered as target domain. In this context, we defined five classification problems so that each remaining domain is considered as source domain once. To estimate the accuracy of the classifiers we followed a hold-out protocol: for each number 10 training mfeat-mor instances were randomly selected as target instances and the remaining 1990 mfeat-mor instances were used as target test instances. All the instances of the corresponding source domain served as source instances. The hold-out evaluation was repeated 10 times. The results are given in Table 4.

5.6 Results and Discussions

Tables 2, 3 and 4 provide the accuracies of the CCA classification rules and the baseline classifiers. They show that the HDA methods outperform the classifiers trained on the target data. For example, the CCA.IDS rule improves the

Table 4. Classification accuracies (in percent) for the Mfeat Data. Bold numbers indicate accuracies that are statistically greater than others based on a t-test on significance level of 0.05.

Target domain	Source domain	kNN.T	SVM.T	Bunching.T	HFA	SHFR	CCA.IDS	CCA.CDS
mfeat-mor	mfeat-fou	44.71	43.7	50.14	63.03	64.22	**70.21**	62.92
	mfeat-fac				64.4	65.28	**70.91**	62.47
	mfeat-pix				63.07	64.45	**69.41**	58.12
	mfeat-kar				63.5	66.16	**71.1**	60.53
	mfeat-zer				63.51.	65.12	**70.01**	65.01

accuracy with 25% on the Mfeat dataset compared with the kNN.T classifier (a target-domain classifier with a similar classification rule) and the CCA.CDS rule improves the accuracy in the range of $[2.87\%, 14.87\%]$ compared with the Bunching.T classifier (a target-domain classifier similar to CCA). Thus, source data from different instances spaces can improve the classification accuracy on unseen target instances.

When comparing the CCA classification rules with the HDA methods, HFA and SHFR, we observe that they have a similar accuracy on the Wikipedia dataset while on the Office dataset and the Mfeat dataset one of the CCA classification rules is a winner (statistically significant improvement in accuracy is in the range of $1\% - 4\%$). Thus, the CCA classification rules are capable of outperforming the HFA and SHFR methods.

When comparing the CCA classification rules themselves, we observe that they are rather different. The CCA.CDS rule outperforms the CCA.IDS rule on the Office dataset and Wikipedia dataset. This is due to the fact that the projected instances are spread around the code words of their classes in the code space $\mathcal{C}$ for these datasets. For the Mfeat dataset, however, this does happen and in this case the CCA.IDS rule outperforms the CCA.CDS rule (the accuracy improvement is around 10%). Thus, the CCA classification rules correspond to very different states of HDA. The CCA.CDS rule is preferred when the minimized sub-loss (2) has relatively *low* values; i.e. the projected instances *are spread* around the code words of their classes. The CCA.IDS rule is preferred when the minimized sub-loss (2) has relatively *high* values; i.e. the projected instances *are not spread* around the code words of their classes.

6   Conclusion

This paper proposed the CCA algorithm for HDA classification that can use several source domains. The algorithm builds the encoding mappings for the target data, source data, and class labels by minimizing the total loss function (4). This aligns the projected target and source instances with the code words of their classes in the code space $\mathcal{C}$. To make different-class instances more separable in $\mathcal{C}$ the class encoding mapping is initialized using a coding matrix with well-presented separation properties [2]. This allows improving the topology of the projected data in the code space $\mathcal{C}$ for class separability. Once the encoding mappings have been learned, the CCA algorithm projects all the data and class labels in the common code space $\mathcal{C}$. In this context, we note that the CCA algorithm does not make any assumption on the underlying structure of the target and source domains. It only necessities common class labels for domain correspondence.

The CCA algorithm offers two built-in classification rules: the CCA.CDS rule and the CCA.IDS rule. CCA.CDS (resp. CCA.IDS) is preferable when the spread of the same class instances is relatively low (high) around the code words of their classes. In addition, any other classification rule can be trained on the projected data.

The CCA algorithm was experimentally tested. The experiments showed that the CCA algorithm is capable of outperforming standard HDA methods.

References

1. Dekel, O., Singer, Y.: Multiclass learning by probabilistic embeddings. Adv. Neural Inf. Process. Syst. **15**, 945–952 (2002)
2. Dietterich, T., Bakiri, G.: Solving multiclass learning problems via error-correcting output codes. J. Artif. Intell. Res. **2**, 263–286 (1995)
3. Duan, L., Xu, D., Tsang, I.: Learning with augmented features for HDA. In: Proceedings of the 29th Internatiobal Conference on Machine Learning (ICML 2012) (2012)
4. Duda, R., Hart, P., Stork, D.: Pattern Classification. Wiley, New York (2000)
5. Ismailoglu, F., Smirnov, E., Nikolaev, N., Peeters, R.: Instance-based decompositions of error correcting output codes. In: Schwenker, F., Roli, F., Kittler, J. (eds.) MCS 2015. LNCS, vol. 9132, pp. 51–63. Springer, Cham (2015). https://doi.org/10.1007/978-3-319-20248-8_5
6. Zhou, J., Tsang, I., Pan, S., Tan, M.: Heterogeneous domain adaptation for multiple classes. In: Proceedings of the Seventeenth International Conference on Artificial Intelligence and Statistics (AISTATS 2014), pp. 1095–1103 (2014)
7. Kulis, B., Saenko, K., Darrell, T.: What you saw is not what you get: domain adaptation using asymmetric kernel transforms. In: Proceedings of the 24th IEEE Conference on Computer Vision and Pattern Recognition (CVPR 2011), pp. 1785–1792. IEEE (2011)
8. Lichman, M.: UCI machine learning repository (2013)
9. Rasiwasia, N., Pereira, J., Coviello, E., Doyle, G., Lanckriet, G., Levy, R., Vasconcelos, N.: A new approach to cross-modal multimedia retrieval. In: Proceedings of the 18th International Conference on Multimedia (MM 2010), pp. 251–260. ACM (2010)
10. Rifkin, R., Klautau, A.: In defense of one-vs-all classification. J. Mach. Learn. Res. **5**, 101–141 (2004)
11. Saenko, K., Kulis, B., Fritz, M., Darrell, T.: Adapting visual category models to new domains. In: Daniilidis, K., Maragos, P., Paragios, N. (eds.) ECCV 2010. LNCS, vol. 6314, pp. 213–226. Springer, Heidelberg (2010). https://doi.org/10.1007/978-3-642-15561-1_16

12. Wang, C., Mahadevan, S.: Heterogeneous domain adaptation using manifold alignment. In: Proceedings of the 22th International Joint Conference on Artificial Intelligence (IJCAI 2011), pp. 1541–1546. AAAI Press (2011)
13. Weiss, K., Khoshgoftaar, T., Wang, D.D.: A survey of transfer learning. J. Big Data **3**(9), 45–85 (2016)

A Deep Neural Spoiler Detection Model Using a Genre-Aware Attention Mechanism

Buru Chang, Hyunjae Kim, Raehyun Kim, Deahan Kim,
and Jaewoo Kang[✉]

Department of Computer Science and Engineering,
Korea University, Seoul, South Korea
{buru_chang,hyunjae-kim,lhkim38,kdh5852,kangj}@korea.ac.kr

Abstract. The fast-growing volume of online activity and user-generated content increases the chances of users being exposed to spoilers. To address this problem, several spoiler detection models have been proposed. However, most of the previous models rely on hand-crafted domain-specific features, which limits the generalizability of the models. In this paper, we propose a new deep neural spoiler detection model that uses a genre-aware attention mechanism. Our model consists of a genre encoder and a sentence encoder. The genre encoder is used to extract a genre feature vector from given genres using a convolutional neural network. The sentence encoder is used to extract sentence feature vectors from a given sentence using a bi-directional gated recurrent unit. We also propose a genre-aware attention layer based on the attention mechanism that utilizes genre information for detecting spoilers which vary by genres. Using a sentence feature, our proposed model determines whether a given sentence is a spoiler. The experimental results on a spoiler dataset show that our proposed model which does not use hand-crafted features outperforms the state-of-the-art spoiler detection baseline models. We also conduct a qualitative analysis on the relations between spoilers and genres, and highlight the results through an attention weight visualization.

Keywords: Deep learning · Spoiler alert · Spoiler detection
Classification · Attention mechanism

1 Introduction

A spoiler is a description of any plot element of a fictional creative work such as a novel, movie, and TV program, and can reveal some important details on the climax or ending. Potential users of creative works generally do not want to be exposed to the spoiler materials beforehand because prior knowledge of the plot details can spoil the enjoyment. Spoilers are typically found in social media such as tweets and Facebook feeds, and review sites such as Rotten Tomato and IMDb. Due to the growing volume of online activities and user-generated contents, the chances of users being exposed to spoilers rapidly increase.

© Springer International Publishing AG, part of Springer Nature 2018
D. Phung et al. (Eds.): PAKDD 2018, LNAI 10937, pp. 183–195, 2018.
https://doi.org/10.1007/978-3-319-93034-3_15

To protect users from spoilers, some review sites alert users of the spoilers in reviews. When reviewers post their reviews that include spoilers, IMDb[1] and Metacritic[2] encourage the reviewers to specify the spoiler information. The review sites then use the provided information to warn the users about the existence of spoilers in the reviews. TV Tropes[3] is a wiki that collects descriptions of scenes in movies and TV programs. TV Tropes also encourages participants to provide spoiler information along with descriptions. Based on provided spoiler information, TV Tropes hides spoiler phrases in descriptions before presenting them to users. However, these manual spoiler alert approaches are limited because they rely on solely the active participation of reviewers. Some reviewers may be thoughtful enough to provide spoiler information but some others may not bother to provide information about spoilers.

To overcome the limitation of the manual spoiler alert systems, machine learning-based spoiler detection models have been proposed. Jeon et al. [6] proposed a spoiler detection model for tweets about a TV program. The authors consider four representative features of tweets such as named entity, frequency of verb, objectivity, and tense as significant indicators of spoilers. Also, Boyd-Graber et al. [2] proposed a spoiler detection model that utilizes meta-data on the genre of works and the country of production for TV Tropes. However, these models also have their own limitations. The models do not leverage the semantic relations of words in descriptions because the models employ discriminative representations such as binary vectors for representing input words. More importantly, these models require a considerable amount of feature engineering, which limits the generalizability of the models.

To address these problems, we propose a deep neural model for automatic spoiler detection which does not require labor-intensive feature engineering. Our proposed model mainly consists of two components: genre encoder and sentence encoder. We consider genre information as an important factor in detecting spoilers because different plot elements serve as spoilers in different genres. For example, a plot element involving "kill" is more likely to become a spoiler in the thriller genre than in the romance genre. Since a creative work can belong to multiple genres, the genre encoder captures the characteristics of the genre composition of the given work and encodes them into a dense continuous feature vector. A convolutional neural network (CNN) is used for the genre encoder. The sentence encoder extracts a sentence feature from an input sentence using a bi-directional gated recurrent unit (GRU). In this process, we employ a genre-aware attention layer which computes the attention weight of each word in the input sentence by calculating the similarity of the hidden state vector and the genre feature vector in each GRU time step. The computed attention weight enables the sentence feature generation process to focus on the words that are more likely to be spoilers in the given genres. Our proposed model uses the

---

[1] http://imdb.com/.

[2] http://www.metacritic.com/.

[3] http://tvtropes.org/.

extracted sentence feature vectors as inputs for the final fully-connected binary classifier to determine whether the given sentences are spoilers.

The contributions of this paper are summarized as follows.

- We introduce a novel deep neural network model that uses a genre-aware attention mechanism to utilize the characteristics of the spoilers, which vary by genre, for automatic spoiler detection.
- We analyze the relations between spoilers and genres in the TV Tropes spoiler dataset, and show the relationship between spoiler words and genres by visualizing the attention weights of the genre-aware attention layer.
- Without labor-intensive feature engineering, our proposed spoiler detection model outperforms the baseline models.

The remainder of the paper is organized as follows. In Sect. 2, we describe previously proposed spoiler detection models and several tasks utilizing attention mechanisms. In Sect. 3, we describe the TV Tropes spoiler dataset and investigate the characteristics of spoilers in the dataset. In Sect. 4, we describe our proposed spoiler detection model. In Sect. 5, we compare our proposed model to baseline models and analyze the experimental results. Last, in Sect. 6, we provide our concluding remarks of this paper.

2   Related Work

2.1   Spoiler Detection

To protect users from spoilers, several automatic spoiler detection models have been proposed. Jeon et al. [6] proposed a spoiler detection model that focuses on spoilers contained in Twitter posts. The authors crawled about 170,000 tweets about a reality TV show and manually labeled 5,618 tweets as spoilers. Based on Support Vector Machine (SVM), their model utilizes the following four features: named entities in a tweet, frequency of verbs, objectivity of the tweet, and tense of the tweet. Using a named entity recognizer [10], the model extracts named entities from a tweet. The information of the named entities may provide some important details about the reality TV show. The model also considers verbs frequently used in spoiler tweets to improve the performance. Since the authors observed that the spoiler tweets are more objective than the non-spoiler tweets, the model utilizes the objectivity of the tweets. Finally, the authors also observed that the spoilers of the reality TV show are written in the past tense. They extract the main tense of a tweet using a rule-based tense identifier [5], and utilize the tense to improve the performance.

However, their model has some limitations. The hand-crafted features cannot be applied to various creative works because the model is designed for the single TV reality program "Dancing with the Stars". Their manual labeling method is also unsuitable for handling a large amount of data from various creative works.

Boyd-Graber et al. [2] proposed a spoiler detection model that focuses on descriptions posted in TV Tropes. The authors labeled descriptions containing a spoiler phrase as spoilers. Their model employs a discriminative representation using a bi-gram to represent a description. Their model also utilizes meta-data on the running times, countries of production, and genres of works to improve the performance. Their model is more generalizable than the model proposed by Jeon et al. [6] as it does not limit the types of TV programs and uses primarily the content of descriptions. However, discriminative representations make it difficult for the model to exploit the semantics of words in descriptions.

To overcome the problems mentioned above, in this paper, we propose the deep neural spoiler detection model that utilizes the semantic information of words without using hand-crafted features.

2.2 Attention Mechanism

An attention mechanism is a tool that finds and emphasizes the most informative part in an input. As an example, an attention mechanism in Neural Machine Translation (NMT) [8] helps to determine which words in a source sequence to translate when generating the next word of a target sequence. An attention mechanism is as widely used for machine comprehension (MC) [11] as it is for NMT. In an MC task, it is important to understand the relationship between a document and a question. There are many structures that use an attention mechanism to find the most relevant part of a document for answering the given question. In addition to these examples above, an attention mechanism is used in many tasks such as document classification [13], speech recognition [4], and image caption generation [12].

Inspired by various studies using an attention mechanism, our model uses an attention mechanism so that it better focus on spoiler words for each genre.

3 Dataset Analysis

3.1 Dataset Description

In this paper, we use the available TV Tropes spoiler dataset used in Boyd-Graber et al. [2] to evaluate our proposed spoiler detection model and the baseline models.

Boyd-Graber et al. [2] constructed the dataset D^4 by collecting TV program descriptions on TV Tropes. TV Tropes, which is a wiki site about creative works, collects the descriptions of works from users. TV Tropes also collects spoiler information from the users. Figure 1 shows an example of a TV Tropes description. In the description, a sequence of spoiler words, which is called **span**, is hidden by the HTML span tag. The authors labeled 8,573 sentences containing the span as spoiler sentences. To create a balanced dataset, they randomly sampled 7,688 non-spoiler sentences that do not include the span. Finally, the authors divided the sentences into a training set (70%), two validation sets (10% each), and a test set (10%). Since the published dataset D does not contain genre information, we collected genre information from the IMDb page of each work.

---

[4] http://umiacs.umd.edu/jbg/downloads/spoilers.tar.gz.

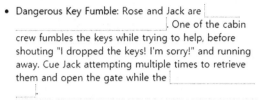

- Dangerous Key Fumble: Rose and Jack are ⌐
  ⌐   ¬. One of the cabin
  crew fumbles the keys while trying to help, before
  shouting "I dropped the keys! I'm sorry!" and running
  away. Cue Jack attempting multiple times to retrieve
  them and open the gate while the ⌐
  ⌐   ¬.

Fig. 1. An example of TV Tropes data. The spoiler words are hidden by an HTML span tag.

Table 1. KL divergence scores of genre pairs.

	Thriller	Action	Comedy	Romance
Thriller	-	0.0648	0.0793	0.1183
Action	0.0614	-	0.0276	0.0837
Comedy	0.0771	0.0259	-	0.0438
Romance	0.1116	0.0701	0.0378	-

3.2   Spoiler Characteristics Analysis

In the spoiler detection task, genre information is critical to improving model performance [2]. Here, we hypothesize that the contents of spoiler sentences in different genres differ significantly. To test our hypothesis on the dataset D, we use FrameNet [1]. Based on semantic frame theory, FrameNet is a tool that captures semantic components in sentences. For example, in the sentence "After Patrick walked into the room he killed him as well" we can obtain frames such as "Killing" and "Victim". It is important to note that FrameNet is not a word to frame pairwise matching model. FrameNet uses the structure of a sentence to find appropriate frames. In our example, "him" can be recognized as "victim" by the fact that he was killed by someone. Therefore, we assume that semantic frames represent the general tone of sentences. By analyzing occurrence patterns of semantic frames with respect to different genres, we can evaluate whether the contents of semantic frames have meaningful differences.

Using the SEMAFOR parser,[5] we first extracted semantic frames from spoiler sentences in the dataset D. We counted each frame in spoiler sentences from each genre. Then we divide each frame occurrence by the sum of all frame occurrence to obtain the normalized frame frequency (NFF) of each frame. We calculated KL Divergence scores of pairs of genres using their NFFs to compare the distributions of frame occurrences. Table 1 shows the resulting KL Divergence scores. As KL Divergence scores represent the difference between two distributions, a higher KL Divergence score indicates more differences in content exist. We can see that KL Divergence scores of inherently different genres are much higher.

For further analysis, we focused on the most contrasting genres, "Thriller" and "Romance". We calculated the difference of NFFs for each frame. Then we picked 10 frames with the most extreme values, which indicate that the frames appeared much more in one genre than others. Figure 2 shows the 10 most different frames from the genres. In thriller spoilers, frames associated with "kill" or "death" frequently appeared. However, for romance, we have many more frames which are mostly related to personal relationship.

Using frame based analysis, we showed that there exists a meaningful relationship between the genre and contents of a spoiler sentence. These relationships help our model more accurately detect spoilers with given genre information. In

---

[5] http://www.cs.cmu.edu/~ark/SEMAFOR/.

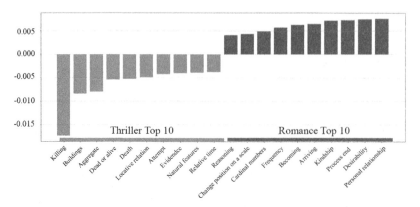

Fig. 2. Top10 differentially expressed frames between Thriller and Romance (NFF of Thriller frames - NFF of Romance frames).

our experimental section, we show how the attention mechanism of our model focused on different aspects of sentences in different genres.

4   Our Approach

In this section, we describe our proposed spoiler detection model. The architecture of our model is shown in Fig. 3.

Our proposed model consists of a genre encoder and sentence encoder. Since genres information can be used to understand the characteristics of spoilers, we design the genre encoder to extract genre features. Since a work can belong to multiple genres, it is necessary to capture the relationship between genres. For example, "Dawn of the Dead, 2004" is a horror and thriller movie, and "Shaun of the Dead, 2004" is a horror and comedy movie. Although the latter is a parody of the former, the two movies show different patterns of spoilers due to the differences between the horror-thriller genre pair and the horror-comedy genre pair. To consider these genre relationships, we employ CNN [7] as our genre encoder. The genre encoder uses various filter sizes for extracting genre feature vectors from genre combinations.

Since the textual context of the input sentence S provides important information that can be used to determine whether the sentence is a spoiler, we design a sentence feature vector to capture the textual context of the input sentence using a bi-directional GRU. GRU [3] is one of the gating mechanisms that prevents the vanishing gradient problem by information flow control. We also assume that there are different spoiler words in different genres. For example, the word "kill" is more likely to be a spoiler in the thriller genre rather than in the romance genre. To utilize these characteristics in the sentence feature extraction process, we propose a genre-aware attention layer. In the attention layer, our proposed model computes the attention weight of each input word by multiplying the

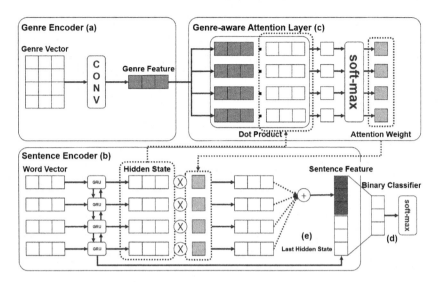

Fig. 3. Components of our proposed model that performs the spoiler detection task. (a), (b), (c), (d), and (e) indicate the genre encoder, sentence encoder, attention layer, binary classifier, and last hidden state respectively.

GRU hidden state and extracted genre feature. The sentence encoder generates sentence features by the attention weighted sum of the GRU hidden states. Finally, our proposed model classifies input sentences using a binary classifier on sentence features.

4.1   Task Formulation

The data d in spoiler detection dataset D consists of a sentence and meta-data on the genre, production country, and the number of episodes of a work. The sentence S is a sequence of words, and is denoted as $S = [w_1, w_2, w_3, ...]$. In this paper, we utilize and describe only the meta-data on genres. The input genre G is a sequence of genres, and is denoted as $G = [g_1, g_2, g_3, ...]$. When the sentence S and genre G are given, our spoiler detection model performs binary classification to determine whether the sentence S is a spoiler.

4.2   Genre Encoder

The genre encoder is illustrated in Fig. 3(a). We associate each genre in the dataset with a real-valued embedding vector $\mathbf{g} \in \mathbb{R}^r$ where r is the dimension of the genre embedding vector. We denote a set of genres of a work as x_g. All the input $x_g \in \mathbb{R}^{Max_g \times r}$ are constructed by juxtaposing a work's genre embedding vectors with zero padding. Max_g is the maximum number of genres of work. We generate a feature c_i by applying a convolutional operator f to rows $x_{g_{i:i+h-1}}$ with a filter $w \in \mathbb{R}^{h \times r}$ where h is a window size. We use Rectified Linear Unit

(ReLU) as our non-linear activation function in the convolution process. We create a feature map $c \in \mathbb{R}^{Max_g - h + 1}$ by sequentially applying the convolutional operation to the windows of an x_g. We create 50 filter maps each for window sizes 1, 2, and 3. To extract the genre feature $\mathbf{v}_g \in \mathbb{R}^{150}$, we apply max-pooling to create feature maps. The pooled feature maps are concatenated as a 150-dimensional genre feature vector $\mathbf{v}_g$.

4.3 Sentence Encoder

The sentence encoder is illustrated in Fig. 3(b). The sentence encoder uses a bi-directional GRU [3] which is one of the gating mechanisms in recurrent neural networks (RNNs). To apply the GRU, we first represent an input sentence S as $x_s \in \mathbb{R}^{Max_s \times d}$ where Max_s is the maximum length of a sentence in dataset D and $x_{s_i} \in \mathbb{R}^d$ is the word representation vector of the i-th word in input sentence S. We use word embedding vectors[6] pre-trained with GloVe [9]. Next, we compute the hidden state $h_t \in \mathbb{R}^e$ using the GRU at time t as follows $h_t = \overleftrightarrow{GRU}(x_{s_t})$.

We then use the genre-aware attention layer to capture the correlation between the input word and genre. The attention layer is illustrated in Fig. 3(c). We expect that focusing on spoiler words improves the spoiler detection performance. To compute the attention weight a_t of input word x_t, we first compute the attention feature q_t by multiplying the h_t and $\mathbf{v}_g$, as

$$q_t = h_t \cdot \mathbf{v}_g. \tag{1}$$

Using a soft-max function, we compute the attention weight a_t as

$$a_t = \frac{exp(q_t)}{\sum_{i=1}^{T} exp(q_i)}. \tag{2}$$

A high attention weight indicates that the hidden state is highly correlated with the genre feature. Finally, a sentence feature $\mathbf{v}_s \in \mathbb{R}^e$ is computed by the attention weighted sum of the hidden states.

$$\mathbf{v}_s = \sum_{i=1}^{T} a_t \cdot h_t, \tag{3}$$

The computed sentence feature captures how much each word in the input sentence S contributes to classifying the spoiler.

4.4 Binary Classifier

We design a binary classifier as shown in Fig. 3(e). We first concatenate the sentence feature vector $\mathbf{v}_s$ generated by the sentence encoder and the last hidden state vector h_T of the GRU (Fig. 3(d)). The last hidden state vector h_T provides

---

[6] https://nlp.stanford.edu/projects/glove/.

the textual context of an input sentence. By concatenating the two vectors, our proposed model can not only focus on the words containing crucial spoiler information but can also utilize the entire textual context. The concatenated vector x is denoted as $x = [\mathbf{v}_s, h_T]$. We then forward the concatenated vector x to the fully connected hidden layer with the ReLU activation function as follows

$$x_b = ReLU(xW_b + b_b), \tag{4}$$

where x_b is the hidden vector, and $W_b \in \mathbb{R}^{(2e) \times 2}$ and $b_b \in \mathbb{R}^2$ are weight and bias terms for the hidden layer respectively. Using the soft-max function, we calculate the probability that a given input sentence is a spoiler. We use the calculated probability distribution $\hat{y}$ as the output of our proposed model. For training our proposed model, we employ Binary Cross-Entropy (BCE) as our cost function which is represented by

$$BCE(\hat{y}) = -(y \cdot (log(\hat{y})) + (1 - y) \cdot (log(1 - \hat{y})), \tag{5}$$

where y is the true label of the input data (the one-hot vector with digit labels) and $\hat{y}$ is a spoiler probability produced by the model. The parameters of our proposed model are optimized in the training step.

We trained our proposed model using mini-batches of size 1024 and the Adam optimizer with a learning rate of 0.001. We set the word embedding and genre embedding as 300 and 25-dimensional vectors respectively. For the genre encoder, we use 50 filters each for window sizes 1, 2, and 3 respectively.

5 Experimental Evaluation

In this section, we conduct an experimental evaluation to compare the performance of our proposed model on the spoiler alert task with that of baselines. We evaluate our proposed model on the dataset D described in Sect. 3. We describe the metrics and baselines. Then, we show and analyze the experimental results.

5.1 Experimental Setup

Metric. To measure the performance of the spoiler alert models, we employ the following two evaluation metrics: **Accuracy** and **F1-Score**, which were used in Boyd-Graber et al. [2] and Jeon et al. [6] respectively. Accuracy is summarized by the following equation $Accuracy = \frac{1}{|D_{test}|} \sum_{d \in D_{test}} \mathbb{1}(\hat{y} = d.l)$ where $\mathbb{1}(.)$ is an indicator function, $\hat{y}$ is a candidate label produced by the model, and $d.l$ is the true spoiler label. F1-score is summarized by the following equation $F1 - score = \frac{2*Precision*Recall}{Precision+Recall}$.

Baselines. Since the spoiler alert model of Jeon et al. [6] designed for a single TV program (Sect. 2) cannot be applied to various works, it is unsuitable as a baseline. Therefore, we compare our proposed model with only the state-of-the-art spoiler alert model introduced by Boyd-Graber et al. [2]. To the best of

Table 2. Evaluation results of our model and baselines.

Approaches	Models	Accuracy	F1-score
SVM	Boyd-Graber et al.	0.6019	0.6947
	Boyd-Graber et al. + Genre	0.6777	0.6327
Deep learning based	CNN	0.7082	0.7351
	CNN + Genre Encoder	0.7400	0.7609
	Sentence Encoder of HAN	0.7231	0.7480
Our proposed models	Sentence Encoder (w/o attention) (Fig. 3(b))	0.7183	0.7584
	Sentence Encoder + Genre Average (Fig. 3(b)+(c))	0.7393	0.7671
	Sentence Encoder + Genre Encoder (Fig. 3(a)+(b)+(c))	0.7536	0.7682
	Sentence Encoder + Genre Encoder + Last Hidden State (Fig. 3(a)+(b)+(c)+(d))	0.7556	0.7847

our knowledge, our proposed model is the first deep learning-based spoiler alert model. For a fair comparison, we also use the deep-learning based sentence classification models [7,13] as deep learning-based baselines. The detailed descriptions of each baseline model are as follows.

SVM Approaches: The SVM model of Boyd-Graber et al. [2] uses a 130,534-dimensional bi-gram vector to represent an input sentence. The original model utilizes several meta-data but in our implementation, we use only genre data because the other meta-data are unavailable in our dataset D and it is shown that genre data has a more significant impact on the performance than other meta-data [2]. The genre data is represented as a 30-dimensional binary vector, and is concatenated with the sentence vector. We report both results produced with genre and without genre data.

Deep Learning Based Approaches: To evaluate the effectiveness of the genre-aware attention layer, we use deep learning based sentence classification models as baselines. The **CNN** model [7] was proposed to perform sentence sentiment classification. The model represents a given sentence as a sequence of word embedding vectors in the same way as our proposed model. We use 50 filters each for window sizes 3, 4, and 5 to create feature maps for each window. **HAN** [13] was proposed to classify documents using a hierarchical attention mechanism. Since the spoiler task is a sentence-level problem, we implement the sentence encoder of HAN as the deep learning-based baseline employing the attention mechanism.

With a grid search, we tune the hyper-parameters of each baseline model using validation set.

5.2   Results

Table 2 shows the experimental results. Our proposed model outperformed the baseline models in both evaluation metrics. Overall, deep learning based models achieved better performance than SVM models. Our proposed model achieved an accuracy and F1-Score at least 11.5% and 24.0% higher, respectively, than the SVM models. Also, the accuracy and F1-Score of our proposed model were at least 2.1% and 3.0% higher, respectively, than the deep learning-based models. When the sentence encoder utilizes the genre-aware attention layer, the performance is significantly increased. The genre features extracted by the genre encoder contributes more to improving performance than the average of genre embedding vectors. These results demonstrate that the genre encoder and the genre-aware attention layer are useful for capturing the characteristics of spoiler words categorized in various genres. We also observe that our proposed model performs better when we augment the hidden state of the last GRU cell. The result proves that the hidden state provides helpful information by capturing the context of input sentences.

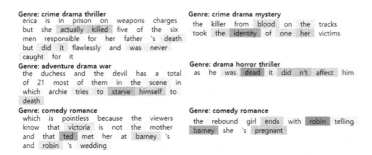

Fig. 4. Examples of visualized attention weights

Our proposed genre-aware attention layer plays an important role in spoiler detection. We visualize the attention weight of the attention layer to investigate how the layer works with respect to given genre. Figure 4 shows the examples of visualization. We observe that the genre-aware attention layer focuses on spoiler words. For example, in the case of a "crime" genre, the attention weight of the words related to the identity of the villain is high. In the case of a "thriller" genre, death related words such as "dead" and "kill" were highlighted to detect spoilers. On the other hand, in "romance" genre, the attention weights of the words related to "personal relationship" were high. These observations demonstrate that our proposed model is automatically trained to detect and classify spoiler words, without additional feature engineering.

6   Conclusion

In this paper, we proposed a novel deep neural network based spoiler alert model with a genre-aware attention layer. To capture the characteristics of a genre, we extracted the genre features using the CNN based genre encoder. We also used the GRU based sentence encoder to extract the sentence features. The sentence encoder utilizes the genre-aware attention layer which focuses on spoiler words providing crucial clues that help to determine whether a given sentence is a spoiler. Without feature engineering, our proposed spoiler alert model outperformed the spoiler alert baselines. The visualized attention weight demonstrated the effectiveness of the genre-aware attention layer. We believe that our proposed spoiler alert model can contribute to protecting users from spoilers on review sites and to promoting subsequent research in this area.

Acknowledgments. This work was supported by the National Research Foundation of Korea (NRF) grant funded by the Korea government (MSIP) (NRF-2017R1A2A1 A17069645, 2017M3C4A7065887).

References

1. Baker, C.F., Fillmore, C.J., Lowe, J.B.: The Berkeley framenet project. In: Proceedings of the 36th Annual Meeting of the Association for Computational Linguistics and 17th International Conference on Computational Linguistics, vol. 1, pp. 86–90. Association for Computational Linguistics (1998)
2. Boyd-Graber, J., Glasgow, K., Zajac, J.S.: Spoiler alert: machine learning approaches to detect social media posts with revelatory information. Proc. Assoc. Inf. Sci. Technol. **50**(1), 1–9 (2013)
3. Cho, K., Van Merriënboer, B., Bahdanau, D., Bengio, Y.: On the properties of neural machine translation: encoder-decoder approaches (2014). arXiv preprint: arXiv:1409.1259
4. Chorowski, J.K., Bahdanau, D., Serdyuk, D., Cho, K., Bengio, Y.: Attention-based models for speech recognition. In: Advances in Neural Information Processing Systems, pp. 577–585 (2015)
5. Gong, Z., Zhang, M., Tan, C., Zhou, G.: N-gram-based tense models for statistical machine translation. In: Proceedings of the 2012 Joint Conference on Empirical Methods in Natural Language Processing and Computational Natural Language Learning, pp. 276–285. Association for Computational Linguistics (2012)
6. Jeon, S., Kim, S., Yu, H.: Spoiler detection in TV program tweets. Inf. Sci. **329**, 220–235 (2016)
7. Kim, Y.: Convolutional neural networks for sentence classification (2014). arXiv preprint: arXiv:1408.5882
8. Luong, M.T., Pham, H., Manning, C.D.: Effective approaches to attention-based neural machine translation (2015). arXiv preprint: arXiv:1508.04025
9. Pennington, J., Socher, R., Manning, C.: Glove: global vectors for word representation. In: Proceedings of the 2014 Conference on Empirical Methods in Natural Language Processing (EMNLP), pp. 1532–1543 (2014)

10. Ritter, A., Clark, S., Etzioni, O., et al.: Named entity recognition in tweets: an experimental study. In: Proceedings of the Conference on Empirical Methods in Natural Language Processing, pp. 1524–1534. Association for Computational Linguistics (2011)
11. Xiong, C., Zhong, V., Socher, R.: Dynamic coattention networks for question answering (2016). arXiv preprint: arXiv:1611.01604
12. Xu, K., Ba, J., Kiros, R., Cho, K., Courville, A., Salakhudinov, R., Zemel, R., Bengio, Y.: Show, attend and tell: neural image caption generation with visual attention. In: International Conference on Machine Learning, pp. 2048–2057 (2015)
13. Yang, Z., Yang, D., Dyer, C., He, X., Smola, A.J., Hovy, E.H.: Hierarchical attention networks for document classification. In: HLT-NAACL, pp. 1480–1489 (2016)

Robust Semi-Supervised Learning on Multiple Networks with Noise

Junting Ye[1]([✉]) and Leman Akoglu[2]

[1] Department of Computer Science, Stony Brook University, Stony Brook, USA
`juyye@cs.stonybrook.edu`
[2] Heinz College of Information Systems and Public Policy,
Carnegie Mellon University, Pittsburgh, USA
`lakoglu@cs.cmu.edu`

Abstract. Graph-regularized semi-supervised learning has been effectively used for classification when (*i*) data instances are connected through a graph, and (*ii*) labeled data is scarce. Leveraging multiple relations (or graphs) between the instances can improve the prediction performance, however noisy and/or irrelevant relations may deteriorate the performance. As a result, an effective weighing scheme needs to be put in place for robustness.

In this paper, we propose iMUNE, a **robust** and effective approach for multi-relational graph-regularized semi-supervised classification, that is immune to noise. Under a **convex formulation**, we infer weights for the multiple graphs as well as a solution (i.e., labeling). We provide a careful analysis of the inferred weights, based on which we devise an algorithm that filters out irrelevant and noisy graphs and produces weights proportional to the informativeness of the remaining graphs. Moreover, iMUNE is **linearly** scalable w.r.t. the number of edges. Through extensive experiments on various real-world datasets, we show the **effectiveness** of our method, which yields superior results under different noise models, and under increasing number of noisy graphs and intensity of noise, as compared to a list of baselines and state-of-the-art approaches.

1 Introduction

Given (*i*) a network with *multiple* different relations between its nodes, and (*ii*) labels for a small set of nodes, how can we predict the labels of the unlabeled nodes in a *robust* fashion? Robustness is a key element especially when the data comes from sources with varying veracity, where some relations may be irrelevant or noisy for the prediction task.

This abstraction admits various real-world applications. For example, in fraud detection one may try to classify individuals as fraudulent or not based on the phone-call, SMS, etc. interactions. In biology, genes are classified as whether or not they perform a certain function through various similarity and interaction relations between them.

© Springer International Publishing AG, part of Springer Nature 2018
D. Phung et al. (Eds.): PAKDD 2018, LNAI 10937, pp. 196–208, 2018.
https://doi.org/10.1007/978-3-319-93034-3_16

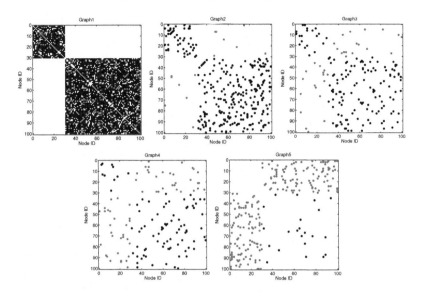

Fig. 1. A synthetic multi-relational graph ($n = 100$ nodes, $m = 5$ views), with 3 informative (top) and 2 noisy (bottom) graphs. Shown are adjacency matrices; red dots: cross-edges between nodes from two different classes, black dots: within-class edges. G_1–G_3 are in order of informativeness. G_4 depicts random noise. G_5 contains adversarial noise. Inferred weights (all graphs): $[25.17, 16.54, 12.79, 17.82, 27.68]$, Average Precision (AP) = 0.734. Weights after noisy graphs removed: $[0.5000, 0.3003, 0.1997, 0, 0]$, AP = 0.974. (Color figure online)

Accomplishing the above task requires addressing two main problems: (1) identifying and filtering out irrelevant and noisy relations, and (2) automatically weighing other relations by their informativeness for the task. Existing methods either are vastly affected in the presence of noise [1], produce locally optimal solutions due to their non-convex objective formulations [2–4], use only the labeled data [5–7], or are too expensive to compute [8–11].

In this work we introduce iMUNE, a robust, scalable, and effective graph-regularized semi-supervised classification approach for MUlti-relational NEtworks. In the example shown in Fig. 1, iMUNE recognizes and removes G_4 and G_5 as irrelevant/noisy, and estimates weights for relations G_1-G_3 so as to combine them effectively to achieve improved performance. Our contributions are as follows.

- **Model Formulation:** Under a convex formulation, we simultaneously estimate weights for the multiple relations (also graphs or views) as well as a solution (labeling) that utilizes a weighted combination of them. (Sect. 2)
- **Analysis of Weights:** We show that in the presence of noise, the inferred weights reflect the impact of different relations on the solution, where both dense informative and irrelevant/noisy graphs receive large weights. (Sect. 2.3)

- **Robust Algorithm:** Analysis of weights enable us to devise a robust algorithm that filters out irrelevant/noisy graphs, so as to produce weights proportional to the informativeness of graphs and yield improved performance. (Sect. 3)
- **Scalability:** Our proposed approach scales linearly w.r.t. the number of edges in the combined graph. (Sect. 3.1)
- **Effectiveness:** We show the efficacy of iMUNE on real-world multi-networks with (*i*) varying number of relevant/noisy graphs, (*ii*) under different noise models, and (*iii*) varying intensity of noise; where it outperforms six baseline approaches including the state-of-the-art. (Sect. 4)
- **Reproducibility:** We share the code of iMUNE and all datasets in experiments at http://www3.cs.stonybrook.edu/%7ejuyye/semi/semi.html

2   Problem Formulation

In this work we consider real-world problem settings in which (1) the problem is cast as a binary classification task, (2) data objects are related through multiple different relationships, and (3) ground-truth class labels are scarce. The data can be represented as a *multi-graph*, in which the nodes represent data objects and multiple sets of undirected edges capture associations implied by different relationships.

Using various relationships between data objects may provide more information for a given classification task, especially when input labels are scarce. Collectively, more accurate predictions can be made by combining these *multiple association networks*. However, it is not realistic to assume that all available relationships (i.e., graphs) would be equally, if at all, relevant for a given prediction task. Filtering irrelevant/intrusive relations is especially important when the data sources cannot be carefully controlled—for example, when data is collected from various repositories with varying veracity. In addition, the input graphs may have varying degree of relevance for a task, which necessitates a careful weighing scheme.

Overall, it is essential to build robust classification models that can effectively leverage multiple relationships by carefully weighing relevant graphs while filtering out the intrusive ones. Our work addresses this problem of Robust Semi-supervised Classification for MUlti-NEtworks (RSC-MUNE): Given a binary classification task, a multi-graph, and a small set of labeled objects, the goal is to build an effective classifier that is *robust to noisy and irrelevant data*. We give the formal problem definitions as follows.

Definition 1. MULTI-GRAPH: *A multi-relational graph (or a multi-graph) $\mathcal{G}(V, \mathcal{E})$ consists of a set of graphs (or relations) $\{G_1(V, E_1), G_2(V, E_2), \ldots, G_m(V, E_m)\}$, on the same node set V, $|V| = n$. Undirected (weighted) edges $\mathcal{E} = \{E_1, \ldots, E_m\}$ correspond to links implied by m different types of relations, where we denote $|\mathcal{G}| = m$.*

Definition 2. RSC-MUNE PROBLEM: *Given a multi-graph $\mathcal{G}(V, \mathcal{E})$, $|\mathcal{G}| = m$, and a set of labeled seed nodes $L \subset V$; devise a learning procedure to infer the labels of unlabeled nodes $V \backslash L$, which assigns a list of weights $\mathbf{w} = \{w_1, \ldots, w_m\}$ to individual graphs such that (i) intrusive graphs are filtered (i.e., $w_k = 0$), and (ii) relevant graphs receive weights relative to their informativeness.*

2.1 Graph-Based Semi-supervised Learning

There exist various objective formulations for graph-regularized semi-supervised classification provided a *single* graph [12–16]. Generalizing from those traditional semi-supervised learning objectives to multi-graphs, we can write

$$\arg\min_{\mathbf{f},\mathbf{w}} \|\mathbf{f} - \mathbf{y}\|_2^2 + \lambda \sum_k \mathbf{f}^\top w_k \mathbf{L_k} \mathbf{f}$$
$$s.t.\ w_k \geq 0,\ \sum_k w_k = 1 \tag{1}$$

where λ is a regularization parameter, $\mathbf{L}_k$ is the normalized Laplacian matrix of kth graph, w_k is the weight of $\mathbf{L_k}$, $\mathbf{y}$ is the input vector of known labels and $\mathbf{f}$ is the solution.

This objective function, however, is non-convex in both $\mathbf{f}$ and $\mathbf{w}$. To get around this, several previous approaches have proposed alternating optimization schemes for similar objectives [2,4]. However, these methods only produce locally optimal solutions.

2.2 Objective Formulation

In this work, inspired by the *TSS* approach [1], we introduce a scheme that infers $\mathbf{f}$ and $\mathbf{w}$ *together* under a *convex* setup. The graph weights we infer (i.e., w_k's) capture the *impact* that each graph has on the solution $\mathbf{f}$. Building on this interpretation, we devise a learning procedure that estimates $\mathbf{f}$ which is *robust* to intrusive graphs.

Our objective function is defined as in Eq. (2).

$$\min_{\mathbf{f},\xi} (\mathbf{f} - \mathbf{y})^\top (\mathbf{f} - \mathbf{y}) + c_0 \sum_{k=1}^{m} \xi_k$$
$$s.t.\ \mathbf{f}^\top \mathbf{L_k} \mathbf{f} \leq c + \xi_k,\ \xi_k \geq 0, \forall k = 1, \ldots, m \tag{2}$$

The dual form of Eq. (2) that estimates the graph weights as well as the final solution are respectively given in Eqs. (3) and (4) (derivations are omitted for brevity).

$$\min_{\mathbf{w}} \mathbf{y}^\top (\mathbf{I} + \sum_k w_k \mathbf{L_k})^{-1} \mathbf{y} + c\|\mathbf{w}\|_1$$
$$s.t.\ c_0 \geq w_k \geq 0, \forall k = 1, \ldots, m \tag{3}$$

$$\mathbf{f} = (\mathbf{I} + \sum_k w_k \mathbf{L_k})^{-1} \mathbf{y} \tag{4}$$

Handling Class Bias. In semi-supervised learning, only part of the nodes are labeled for training, and the rest are unlabeled (depicted with '0'). For each node type ('+1','0', '−1'), we assign a different penalty coefficient, c_+, c_u, c_- respectively. Let $\mathbf{C}$ be a $n \times n$ diagonal matrix, called the *class penalty matrix*, where $\mathbf{C}(i, i) = c_+$ if $y_i = 1$, c_- if $y_i = -1$, and c_u if $y_i = 0$. As such, the criterion in Eq. (2) can be reformulated:

$$\min_{\mathbf{f},\xi}(\mathbf{f} - \mathbf{y})^\top \mathbf{C}(\mathbf{f} - \mathbf{y}) + c_0 \sum_{k=1}^m \xi_k$$

The dual form and the solution are Eqs. (5) and (6).

$$\min_{\mathbf{w}} \mathbf{y}^\top \mathbf{C}(\mathbf{C} - \sum_k w_k \mathbf{L_k})^{-1} \mathbf{Cy} + c\|\mathbf{w}\|_1 \tag{5}$$

$$s.t. \ c_0 \geq w_k \geq 0, \forall k = 1, \ldots, m$$

$$\mathbf{f}^* = (\mathbf{C} + \sum_k w_k \mathbf{L_k})^{-1} \mathbf{Cy}. \tag{6}$$

The dual program in Eq. (5) is convex and can be solved (e.g., using the projected gradient descent method) to infer the graph weights $\mathbf{w}$. One can then plug in those weights directly into Eq. (6) to estimate $\mathbf{f}^*$. However, this procedure as we show in the experiments yields inferior results in the presence of irrelevant and noisy graphs.

2.3 Graph Weights Interpreted

Next we provide a detailed discussion on the interpretation of the inferred weights by Eq. (3) (instead of Eq. (5) for brevity). In a nutshell, we show that in the presence of intrusive graphs, the weights do *not* reflect the relative *informativeness* of individual graphs—but rather the relative *impact* of each graph on the solution.

Ideally, we want to infer a weight w_k for each graph G_k proportional to its informativeness for the task, where the weights for intrusive graphs are zero. For example, in Fig. 1 we illustrate a toy multi-graph with five views. The ideal weights would be $w_1 > w_2 > w_3 > w_4 = w_5 = 0$. As we show in the following, however, the estimated weights should be interpreted carefully when we have intrusive graphs.

G_k's with Larger $\mathbf{f'L_k f}$ Tend to Get Larger w_k. We have the dual problem $d(\mathbf{w})$ in (3) when learning the weights. We know from basic calculus that

$$\frac{\partial}{\partial x} Y^{-1} = -Y^{-1}(\frac{\partial}{\partial x} Y)Y^{-1}. \tag{7}$$

Thus we derive the derivative of $d(\mathbf{w})$ w.r.t w_k as

$$\frac{\partial d(\mathbf{w})}{\partial w_k} = -\mathbf{y}^\top (\mathbf{I} + \sum_{i=1}^m w_i \mathbf{L_i})^{-1} \mathbf{L_k} (\mathbf{I} + \sum_{i=1}^m w_i \mathbf{L_i})^{-1} \mathbf{y} + c \tag{8}$$

Since $\mathbf{f} = (\mathbf{I} + \sum_{i=1}^{m} w_i \mathbf{L_i})^{-1}\mathbf{y}$, we obtain

$$\frac{\partial d(\mathbf{w})}{\partial w_k} = -\mathbf{f}^\top \mathbf{L_k} \mathbf{f} + c \tag{9}$$

Based on (9), we make the following inference:

Both Dense Informative and Intrusive Graphs G_k has Large $\mathbf{f}^\top\mathbf{L_k}\mathbf{f}$—and Hence Large w_k. Consider a graph with no noisy edges (i.e., no edges between nodes from different classes) but with high edge density among nodes that belong to the same class. For such a graph, $\mathbf{f}^\top\mathbf{L_k}\mathbf{f} = \sum_{i,j \in V} \mathbf{W_k}(i,j)(\frac{f_i}{\sqrt{\mathbf{D_k}(i,i)}} - \frac{f_j}{\sqrt{\mathbf{D_k}(j,j)}})^2$ can be large due to the numerous non-zero (although likely small) quadratic terms in the sum. Importantly, it is not only the dense informative graphs that would have large $\mathbf{f}^\top\mathbf{L_k}\mathbf{f}$, but *also the intrusive graphs*. This is due to the many cross-edges that irrelevant and noisy graphs have between nodes from different classes, that would yield large quadratic terms. We demonstrate this through the inferred weights on our example multi-graph in Fig. 1. Notice that while the highly informative G_1 and G_2 receive large weight, the noisy graphs G_4 and G_5 also obtain comparably large weights.

3   iMUNE Algorithm

Our goal is to filter out the intrusive graphs. The main idea is to explore the search space through simulated annealing by carefully removing large-weighted graphs one at a time. Steps of our proposed algorithm is outline in Algorithm 1. We start with introducing a queue of graph-sets, which initially includes the set of all graphs (line 2). We process the graph-sets in the queue one by one until the queue becomes empty (line 3). For each graph-set $\mathcal{GS}$ that we dequeue (line 4), we compute its cross-validation performance cvP on the labeled data (line 5). In our experiments, we use average-precision (AP) as our performance metric. This metric is more meaningful than accuracy, especially in the face of class bias.

We record the best AP as $bestP$ during the course of our search (line 14). With probability $\exp(\frac{cvP - bestP}{t^{m-|\mathcal{GS}|+1}})$, we "process" the graph-set in hand (lines 7–13, which we will describe shortly), otherwise we discard it. In line 6, $t \leq 1$ is the temperature parameter of simulated annealing and $(m - |\mathcal{GS}|)$ denotes the number of removed graphs from the original set. If the graph-set $\mathcal{GS}$ in hand yields a cvP that is larger than $bestP$, we always process the set further, since when $(cvP - bestP) \geq 0$, $\exp(\frac{cvP - bestP}{t^{m-|\mathcal{GS}|+1}}) \geq 1$. On the other hand, if $\mathcal{GS}$ yields inferior performance, we still process it with some probability that is proportional to the size of the graph-set. That is, the probability of processing a set decreases as they have more graphs removed from the original set. The probability is also inversely proportional to the performance distance $(cvP - bestP)$. The larger the gap, the higher the chance that $\mathcal{GS}$ will be discarded.

Next we describe the steps to "process" a graph-set $\mathcal{GS}$. We first solve the optimization problem (5) using $\mathcal{GS}$ for the graph weights $\mathbf{w}_{\mathcal{GS}}$ and compute the solution using $\mathbf{w}_{\mathcal{GS}}$ in (6) (lines 7–8). Next we cluster the weights into two

---

Algorithm 1. iMUNE (proposed algorithm for robust semi-supervised classification for multi-graphs with noise)

Input: Multi-graph $\mathcal{G} = \{G_1, \ldots, G_m\}$, labeled nodes L, initial temperature t, class penalty matrix $\mathbf{C}$
Output: Label estimations $\mathbf{f}$
 1: Init $\mathbf{y}$ with L;
 2: $bestf \leftarrow \emptyset$, $bestP = 0$, $m = |\mathcal{G}|$, $Q \leftarrow \mathcal{G}$
 3: **while** Q is not empty **do**
 4:     $\mathcal{GS} \leftarrow dequeue(Q)$
 5:     cvP = Compute cross validation performance of $\mathcal{GS}$
 6:     **if** $rand(0,1) \leq \exp(\frac{cvP-bestP}{t^{m-|\mathcal{GS}|+1}})$ **then**
 7:         $\mathbf{w}_{\mathcal{GS}} \leftarrow$ Solve (5) using $\mathcal{GS}$ and input $\mathbf{C}$
 8:         $\mathbf{f}_{\mathcal{GS}} \leftarrow$ Compute solution using (6) and $\mathbf{w}_{\mathcal{GS}}$
 9:         Cluster the weights: $(W_s, W_l) \leftarrow 2\text{-}means(\mathbf{w}_{\mathcal{GS}})$
10:         **for each** $G_k \in \mathcal{GS}$ for which $w_k \in W_l$ **do**
11:             $v \leftarrow hash(\mathcal{GS}\backslash G_k)$
12:             **if** v is null **then** $Q \leftarrow Q \cup \mathcal{GS}\backslash G_k$
13:         **end for**
14:         **if** $cvP > bestP$ **then** $bestf \leftarrow \mathbf{f}_{\mathcal{GS}}$, $bestP = cvP$
15:     **end if**
16: **end while**
17: **return** $bestf$

---

groups, those with small weights W_s and those with large weights W_l (line 9). We know, through the analysis in Sect. 2.3, that intrusive graphs are *among* the large-weighted graphs. The issue is we do not know in advance which ones, as dense informative ones are likely to also belong to this group. As such, we create from $\mathcal{GS}$ candidate graph-sets that contain all but each large-weighted graph and add those to the queue. Note that we maintain a hash table of the candidate graph-sets (line 11), so that we avoid re-considering the same sets that might be generated through different removal paths. At the end, we return the solution $bestf$ with the $bestP$.

3.1   Complexity Analysis

At each node of our "search tree", we solve Eq. (5) using projected gradient descent, where the main computation involves computing the gradient (See Eq. (8) in Sect. 2.3). The gradient involves the term $(\mathbf{I} + \sum_{i=1}^{m} w_i \mathbf{L_i})^{-1}$, i.e., the inverse of a $(n \times n)$ matrix which is $O(n^3)$ if done naively. The same is true for the solution $\mathbf{f}$ which requires a similar inverse operation (See Eq. (4) or (6)). Importantly, however, we do not compute the inverse explicitly, because it always appears in vector form $\mathbf{x} = (\mathbf{I} + \sum_{i=1}^{m} w_i \mathbf{L_i})^{-1}\mathbf{y}$. We can obtain $\mathbf{x}$ as a solution of sparse linear systems [17], where the computational cost of the derivative is linear w.r.t. the number of non-zero entries of $\sum_{i=1}^{m} w_i \mathbf{L_i}$, i.e., proportional to the number of edges in the multi-graph.

Computing the dual objective then takes $O(s|\mathcal{E}|)$ for s number of gradient steps. All in all, total time complexity of an implementation that traverses each search path in parallel is $O(s|\mathcal{E}|m_u)$, which is linear on the total number of edges in the multi-graphs with small m_u (max. number of noisy graphs) and constant s.

4   Evaluation

4.1   Experiment Setup

Datasets. The multi-graphs used in our work are publicly available, and are listed in Table 1. *(i) RealityMining* [18] is a dataset collected through tracking activities on cell-phones. It contains 4 dif-

Table 1. Four real-world multi-graph datasets.

Dataset	#Graphs	#Nodes	#Pos.	#Neg.
RealityMining	4	78	27	51
Protein	5	3,588	306	3,282
Gene1	15	1,724	185	1,539
Gene2	15	3,146	214	2,932

ferent relations between two classes (MIT Sloan and CS students): phone call, SMS, friendship, and Bluetooth scans that capture proximity relations. *(ii) Protein* [1] consists of Yeast proteins, associated through 5 different relations. Those proteins with function *transport facilitation* constitute the positive class, and others are negative. *Gene1* and *Gene2* contain different sets of Yeast genes, each associated through 15 different genomic sources. The genes are labeled according to *Gene Ontology association* file from the *Saccharomyces Genome Database*. For *Gene1*, we choose the label with the maximum number of genes in *Cellular Component* (CC) domain as positive class. We construct *Gene2* in a similar way, where this time genes in *Molecular Function* (MF) domain are labeled as positive. See [5] for more details on datasets.

Baselines. We compare iMUNE against four state-of-the-art: *ClusDCA* [19], *TSS* [1], *RobustLP* [2], and *GeneMania* [5]. We also introduce two simple baselines, *EqlWght* that assigns equal weight to all graphs and *PerfWght* that assigns weights proportional to the cross-validation accuracy of individual graphs on labeled nodes. To make it a fair game, we use the same class-bias penalties described in Sect.,4.2 for the compared methods.

Noise-testing. To test the robustness of the methods, we injected intrusive graphs with varying level, model, and intensity of noise as described below.

- *Number of intrusive graphs*: We tested the effect of increasing noise level on classification performance by injecting 2, 4 and 6 intrusive graphs at a time.
- *Noisy graph models*: We adopted 3 strategies to generate intrusive graphs; (1) Erdos-Renyi random graphs (ER), (2) edge-rewired original graphs (RW), and what we call (3) adversarial graphs (AV) (where most edges are cross-edges between the different classes).

– *Noise intensity* (Low/High): Intensity reflects injected graph density (L: 5%, H: 50%) for ER, ratio of within-class edges randomly rewired to become cross-edges (L: 60%, H: 80%) for RW, and ratio of cross-edges (L: 60%, H: 80%) for AV model.

Overall, there are 3 different number of injected graphs, 3 noise models, and 2 noise intensities. Overall, the "noise-testing" involves 18 (3*3*2) different settings.

4.2   Parameters

Our algorithm expects two hyper parameters; the initial simulated annealing temperature t, and the class penalty matrix $\mathbf{C}$. We describe how we set these in the following. Note that our objective function in Eq. (5) has two further (hyper) parameters c and c_0, which are chosen by cross-validation.

Initial Temperature t. As we remove more and more graphs from the input multi-graph, the probability of further considering a set with inferior performance should decrease. That is when $d = (cvP - bestP) < 0$, $p = \exp(\frac{d}{t^{m-|\mathcal{GS}|+1}})$ should decrease as $r = (m - |\mathcal{GS}| + 1)$ increases. As such, we need $t \leq 1$. Assume that we have an expected range $[m_l, m_u]$ for the number of intrusive graphs in the data where m_l and m_u respectively denote the minimum and maximum number. We would then want the probability $p = \exp(\frac{d}{t^r})$ to approach zero as r gets closer to m_u even for a considerably small d. That is, as $r \to m_u$ and $0 > d \geq d_{min}$ for small d_{min}, we want $p_{max} > p > 0$ for small p_{max}. Since $t = (\frac{d}{\ln p})^{\frac{1}{r}}$, the range for t satisfying the above constraints can be given as $t \in [(\frac{d_{min}}{\ln p_{max}})^{\frac{1}{m_u}}, (\frac{d_{min}}{\ln p_{max}})^{\frac{1}{m_l}}]$. Empirically, we let $d_{min} = -0.1$ and $p_{max} = 0.01$. For example, if we expect $m_l = 5$ and $m_u = 10$, then the initial temperature is chosen randomly from $t \in [0.465, 0.682]$.

Class Penalty Matrix $\mathbf{C}$. As described in Sect. 2.2, we can normalize biased class distribution by assigning larger penalty to minority-class ('+1') misclassification. Recall that c_+, c_u, c_- denote penalty coefficients for classes '+1', '0' (unlabeled), and '−1', respectively. We set these parameters as $c_+ = 1 + sign(1-2p)*\gamma*\max(p, 1-p)$, $c_u = 1$, and $c_- = 1 + sign(2p-1)*\gamma*\max(p, 1-p)$, where γ is a constant drawn from $[0.5, 1]$, and p is the proportion of class '+1' instances in the labeled set. For e.g., for $\gamma = 0.7$ and $p = 0.1$, we would have $c_+ = 1.63, c_u = 1, c_- = 0.37$.

4.3   Evaluation Results

To perform semi-supervised classification, we label 5% of the nodes in *Protein*, *Gene1*, and *Gene2* and 30% in *RealityMining* which is a smaller dataset. We randomly sample the labeled set 10 times, and report the mean Average Precision (area under precision-recall curve) in Table 2 (notice in Table 1 that the datasets are class-imbalanced, hence accuracy is not a good measure to report). From the precision-recall plots in Fig. 2, we see that our method outperform baselines in

Table 2. *iMUNE consistently outperforms competing methods across various real-world and synthetic datasets.* Dataset *RM* is injected with 2, 4, and 6 intrusive graphs with various noise settings. Values depict mean Average Precision (10 runs).

Dataset	#Graph	Model	Intensity	iMUNE	PerfWght	EqlWght	TSS	RobustLP	Mania	ClusDCA
RM	4	——	——	**0.970**	0.944	0.939	**0.970**	0.947	0.951	0.933
	4+2	AV	Low	**0.970**	0.707	0.554	0.525	0.851	0.470	0.894
	4+2	AV	High	**0.970**	0.611	0.484	0.554	0.809	0.359	0.898
	4+2	RW	Low	**0.970**	0.695	0.669	0.718	0.873	0.505	0.912
	4+2	RW	High	**0.970**	0.563	0.537	0.65	0.824	0.290	0.927
	4+2	ER	Low	**0.970**	0.905	0.841	0.657	0.928	0.773	0.918
	4+2	ER	High	**0.970**	0.942	0.920	0.895	0.930	0.883	0.936
	4+4	AV	Low	**0.970**	0.531	0.372	0.390	0.427	0.260	0.866
	4+4	AV	High	**0.970**	0.383	0.297	0.576	0.339	0.215	0.846
	4+4	RW	Low	**0.930**	0.610	0.503	0.561	0.505	0.319	0.870
	4+4	RW	High	**0.907**	0.437	0.349	0.542	0.334	0.217	0.899
	4+4	ER	Low	**0.970**	0.867	0.770	0.482	0.869	0.698	0.895
	4+4	ER	High	**0.970**	0.942	0.917	0.659	0.930	0.834	0.933
	4+6	AV	Low	**0.970**	0.389	0.277	0.354	0.284	0.217	0.822
	4+6	AV	High	**0.970**	0.257	0.223	0.577	0.225	0.197	0.817
	4+6	RW	Low	**0.930**	0.468	0.396	0.597	0.371	0.235	0.845
	4+6	RW	High	**0.907**	0.292	0.267	0.571	0.264	0.202	0.903
	4+6	ER	Low	**0.970**	0.860	0.756	0.494	0.810	0.645	0.882
	4+6	ER	High	**0.970**	0.937	0.896	0.621	0.907	0.773	0.931
Protein	5	——	——	**0.457**	0.452	0.441	**0.457**	0.439	0.424	0.441
Gene1	15	——	——	**0.703**	0.658	0.632	0.648	0.628	0.509	0.651
Gene2	15	——	——	0.838	0.83	0.809	0.734	0.460	0.229	**0.907**

almost all cases, which is especially evident in the presence of noise, when the performance of other methods degrade considerably. Interestingly, the baselines appear to be more robust against random noise than the other noise models.

We further investigate the effect of noise using *RealityMining* as a running example, as in the absence of noise all methods perform similarly on this multi-graph. Figure 3 (left) shows how the performance of the methods change with increasing number of intrusive graphs (under rewiring and low-intensity). Figure 3 (right) shows the same with different noise intensity (under rewiring, 6 intrusive graphs). These show that iMUNE's performance remains near-stable, while the competing methods are relatively hindered by noise. In fact, as Fig. 5 shows iMUNE is robust under all settings: increasing level and intensity as well as different noise models.

We also analyze the inferred weights by each method (except *ClusDCA*, which adopts matrix factorization instead of learning graph weights). Figure 4 shows the normalized weights on *RealityMining* with 6 injected graphs, under AV with high intensity.

Notice that all competing methods give non-zero weights to all the injected graphs $G_5 - G_{10}$, which hinders their performance. In contrast, iMUNE puts non-zero weight only on the informative graphs $G_1 - G_4$, particularly large weights on the first two. These are in fact the well-structured and denser informative graphs.

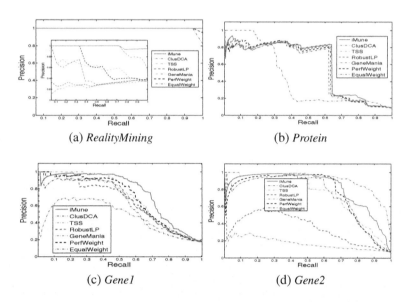

(a) *RealityMining*   (b) *Protein*

(c) *Gene1*   (d) *Gene2*

Fig. 2. *Noise hinders existing methods notably, whereas iMUNE remains near-stable.* Precision vs. Recall of competing methods in four real-world multi-graphs: (a) *RealityMining* (4 views), (b) *Protein* (5 views), (c) *Gene1* (15 views), and (d) *Gene2* (15 views). Inset plot in (a) shows performance when 6 rewired graphs with low intensity are injected to *RealityMining*

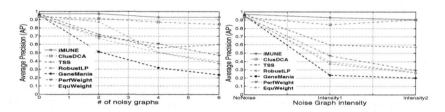

Fig. 3. *iMUNE performs better than all competitors* by (*left*) increasing number of intrusive graphs, and (*right*) increasing intensity of noise.

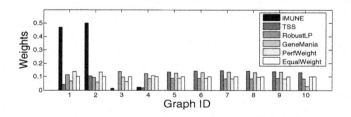

Fig. 4. *iMUNE filters out all noisy graphs, and gives large weights to most informative graphs. Competing methods are indifferent to noise and assign near-uniform weights.* Inferred graph weights on *RealityMining* (+6 injected noisy graphs G_5-G_{10} under AV and high-intensity).

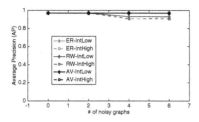
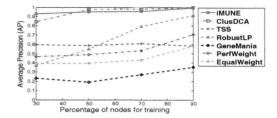

Fig. 5. iMUNE is robust under varying level (#graphs), intensity (low/high), models (ER,RW,AV) of noise.

Fig. 6. Performance vs. % labeled nodes. iMUNE maintains high performance with different ratio of training data.

Finally in Fig. 6 we show how the performance of the methods change when we increase the labeled set percentage in *RealityMining* from 30% up to 90% (6 injected graphs, under rewiring with low intensity; results are avg'ed over 10 runs).

As expected the performance improves for all methods with increasing labeled data. However, most competing methods cannot achieve improved robustness and reach the same performance level by iMUNE, even when they are provided 90% of the data labeled. While *ClusDCA* achieves comparable performance when 50% of data is labeled, it is not as robust to noise as iMUNE as shown in Table 2.

5 Conclusion

In this work we introduced iMUNE, for robust, scalable, and effective *semi-supervised transductive classification for multi-relational graphs*. The proposed method employs a **convex formulation** that estimates weights for individual graphs, along with a solution that utilizes a weighted combination of them. Based on the **analysis of weights**, we devise a new scheme that iteratively discards intrusive graphs to achieve **robust** performance. Moreover, iMUNE is **linearly** scalable w.r.t. the size of the combined graph. Extensive experiments on real-world multi-graphs show that iMUNE produces competitive results under varying level, intensity, and models of noise.

Acknowledgments. This research is sponsored by NSF CAREER 1452425 and IIS 1408287. Any conclusions expressed in this material are of the authors and do not necessarily reflect the views, expressed or implied, of the funding parties.

References

1. Tsuda, K., Shin, H., Schölkopf, B.: Fast protein classification with multiple networks. Bioinformatics **21**, 59–65 (2005)
2. Kato, T., Kashima, H., Sugiyama, M.: Robust label propagation on multiple networks. IEEE Trans. Neural Netw. **20**(1), 35–44 (2009)

3. Shin, H., Tsuda, K., Schölkopf, B.: Protein functional class prediction with a combined graph. Expert Syst. Appl. **36**(2), 3284–3292 (2009)
4. Wan, M., Ouyang, Y., Kaplan, L., Han, J.: Graph regularized meta-path based transductive regression in heterogeneous information network. In: SDM, SIAM (2015)
5. Mostafavi, S., Ray, D., Warde-Farley, D., Grouios, C., Morris, Q.: GeneMANIA: a real-time multiple association network integration algorithm for predicting gene function. Genome Biol. **9**(Suppl 1), S4 (2008)
6. Mostafavi, S., Morris, Q.: Fast integration of heterogeneous data sources for predicting gene function with limited annotation. Bioinformatics **26**(14), 1759–1765 (2010)
7. Luo, C., Guan, R., Wang, Z., Lin, C.: HetPathMine: a novel transductive classification algorithm on heterogeneous information networks. In: de Rijke, M., Kenter, T., de Vries, A.P., Zhai, C.X., de Jong, F., Radinsky, K., Hofmann, K. (eds.) ECIR 2014. LNCS, vol. 8416, pp. 210–221. Springer, Cham (2014). https://doi.org/10.1007/978-3-319-06028-6_18
8. Lanckriet, G.R.G., Bie, T.D., Cristianini, N., Jordan, M.I., Noble, W.S.: A statistical framework for genomic data fusion. Bioinformatics **20**(16), 2626–2635 (2004)
9. Argyriou, A., Herbster, M., Pontil, M.: Combining graph laplacians for semi-supervised learning. In: NIPS (2005)
10. Yu, G.X., Rangwala, H., Domeniconi, C., Zhang, G., Zhang, Z.: Protein function prediction by integrating multiple kernels. In: IJCAI (2013)
11. Wang, S., Jiang, S., Huang, Q., Tian, Q.: S3MKL: scalable semi-supervised multiple kernel learning for image data mining. In: ACM Multimedia, ACM, pp. 163–172 (2010)
12. Macskassy, S., Provost, F.: Classification in networked data: a toolkit and a univariate case study. J. Mach. Learn. Res. **8**, 935–983 (2007)
13. Blum, A., Chawla, S.: Learning from labeled and unlabeled data using graph mincuts. In: ICML, pp. 19–26 (2001)
14. Zhu, X., Ghahramani, Z., Lafferty, J., et al.: Semi-supervised learning using gaussian fields and harmonic functions. In: ICML (2003)
15. Zhou, D., Bousquet, O., Lal, T.N., Weston, J., Schölkopf, B.: Learning with local and global consistency. In: NIPS (2003)
16. Belkin, M., Matveeva, I., Niyogi, P.: Regularization and semi-supervised learning on large graphs. In: COLT (2004)
17. Spielman, D.A., Teng, S.H.: Nearly-linear time algorithms for graph partitioning, graph sparsification, and solving linear systems. In: STOC, ACM, pp. 81–90 (2004)
18. Eagle, N., Pentland, A.S., Lazer, D.: Inferring friendship network structure by using mobile phone data. PNAS **106**(36), 15274–15278 (2009)
19. Wang, S., Cho, H., Zhai, C., Berger, B., Peng, J.: Exploiting ontology graph for predicting sparsely annotated gene function. Bioinformatics **31**(12), i357–i364 (2015)

ε-Distance Weighted Support Vector Regression

Ge Ou[1], Yan Wang[1(⊠)], Lan Huang[1], Wei Pang[2(⊠)],
and George Macleod Coghill[2]

[1] Key Laboratory of Symbol Computation and Knowledge Engineering
of Ministry of Education, College of Computer Science and Technology,
Jilin University, Changchun 130012, China
ouge15@mails.jlu.edu.cn, {wy6868,huanglan}@jlu.edu.cn
[2] Department of Computing Science, University of Aberdeen,
Aberdeen AB24 3UE, UK
{pang.wei,g.coghill}@abdn.ac.uk

Abstract. We propose a novel support vector regression approach called
ε-Distance Weighted Support Vector Regression (ε-DWSVR). ε-DWSVR
specifically addresses a challenging issue in support vector regression:
how to deal with the situation when the distribution of the internal data
in the ε-tube is different from that of the boundary data containing support vectors. The proposed ε-DWSVR optimizes the minimum margin
and the mean of functional margin simultaneously to tackle this issue. To
solve the new optimization problem arising from ε-DWSVR, we adopt
dual coordinate descent (DCD) with kernel functions for medium-scale
problems and also employ averaged stochastic gradient descent (ASGD)
to make ε-DWSVR scalable to larger problems. We report promising
results obtained by ε-DWSVR in comparison with five popular regression methods on sixteen UCI benchmark datasets.

Keywords: Regression analysis · Support vector regression
Distance Weighted Support Vector Regression
Dual coordinate descent · Averaged stochastic gradient descent

1 Introduction

Support Vector Regression (SVR) has received a significant amount of attention
due to its competitive performance [2] compared with other regression methods,
including the method of least squares [13], Neural Networks (NN) [15], logistic
regression [9], and ridge regression [5]. However, the performance of SVR tends
to be sensitive to parameter values and easily affected by the boundary data. In
this research, the internal data indicates the data which are densely distributed
together in the ε-tube, and the boundary data indicates the data which are
distributed on the boundary of the ε-tube, which generally contain many support
vectors. In this paper, we present a novel SVR approach by considering recent
progress in support vector (SV) theory and addressing the above limitations.

© Springer International Publishing AG, part of Springer Nature 2018
D. Phung et al. (Eds.): PAKDD 2018, LNAI 10937, pp. 209–220, 2018.
https://doi.org/10.1007/978-3-319-93034-3_17

In general, SVR constructs decision functions in high-dimensional space for linear regression while the training data are mapped to a kernel Hilbert feature space. ε-SVR [4] was the first popular SVR strategy, which finds a function whose deviation from the actually observed values for all the training data is less than ε, thus forming the so-called ε-tube, to fit training data. To find the best fitting hyperplane, ε-SVR tries to maximize the minimum margin containing data in the ε-tube as much as possible, which is similar to Support Vector Machines (SVMs) [16]. However, ε-SVR is susceptible to the distribution of those boundary data. In fact, the optimization objective greatly depends on the margin between support vectors, and this makes the fitting function heavily reliant on the distribution of the boundary data: if the distribution of the internal data is very different from that of the boundary data, the final fitting function may not be reliable.

Recent progress in SV theory [11, 19] suggests that maximizing the minimum margin is not the only optimization goal to achieve better learning performance. Unlike traditional SVMs, Distance-weighted Discrimination (DWD) [11] maximize the mean of the functional margin (i.e. the harmonic mean of the distances of all data to the separating hyperplane), thus greatly improving the classification performance. Inspired by the idea of DWD, we can also improve the original optimization objective for our regression problems by introducing the concept of the mean of the functional margin in regression.

Considering the above limitations and recent progress in SV theory, we propose a novel SVR method called ε-Distance Weighted Support Vector Regression (ε-DWSVR), which optimizes the minimum margin and the mean of functional margin simultaneously. To solve the optimization problem, ε-DWSVR adopts the dual coordinate descent (DCD) [18] strategy with kernel functions on medium-scale problems, and it also employs the averaged stochastic gradient descent (ASGD) [17] strategy to improve its scalability. A comparison of ε-DWSVR with five popular regression methods (i.e. ε-SVR, linear regression, NN, logistic regression, and ridge regression) on sixteen UCI datasets indicates ε-DWSVR outperforms these methods: ε-DWSVR fits better the distribution of the internal data in most cases, especially for those datasets with strong interference noise.

2   Background

Let $S = (X, Y)$ be a training set of n instances. $X = [x_1, ..., x_n]$ are the input instances where $x_i \in R^m$, and $Y = [y_1, ..., y_n]$ are the output instances where $y_i \in R$. For classification problems, $Y = \{+1, -1\}$ is the label set. For regression problems, Y is the corresponding target values, where $y_i \in \{-\infty, +\infty\}$. The objective function is $f(x) = w \cdot \phi(x_i) + b$, where $x \in R^m$, $w \in R^m$, and $\phi(\cdot)$ is the mapping function induced by a kernel K, i.e., $K(x_i, x_j) = \phi(x_i) \cdot \phi(x_j)$.

2.1   Recent Progress in SV Theory

Recently, SV theory has made great progress. SVM aims to maximize the minimum margin, which denotes the smallest distances of all instances to the separating hyperplane [16]. The optimization problem is represented as follows:

$$\min_{w,\xi} \frac{1}{2}\|w\|^2 + C\sum_{i=1}^{n} \xi_i$$
$$s.t.\ y_i\left(w \cdot \phi(x_i) + b\right) \geq 1 - \xi_i,\quad \xi_i \geq 0,\quad i = 1, 2, ..., n,$$

where C is the regularization parameter and ξ measures the losses of instances.

DWD is proposed to solve data pilling problems [11], which uses a new criterion, that is, maximizing the mean of the functional margin, to replace the criterion of maximizing the minimum margin in SVM for solving the optimization problem [12]. DWD denotes the functional margin as $u_i = y_i(w \cdot \phi(x_i) + b)$ and let $r_i = u_i + \xi_i$ be the adjusted distance of the i-th data to the separating hyperplane, and the optimization problem is then given below:

$$\min_{w,b,\xi} \sum_{i=1}^{n} \left(\frac{1}{r_i} + C\xi_i\right)$$
$$s.t.\ r_i = y_i\left(w \cdot \phi(x_i) + b\right) + \xi_i,\quad r_i \geq 0,\quad \xi_i \geq 0,\quad \|w\|^2 \leq 1,\quad i = 1, 2, ..., n.$$

Since SVR [4] is the application of SV theory to regression problems, the fitting hyperplane is also affected by the distribution of the boundary data. When the distribution of the internal data is different from that of the boundary data, the fitting hyperplane produced by SVR may not be consistent with the actual data distribution, which is similar to the data piling problems. Therefore, we introduce recent progress in SV theory into the original optimization objective of SVR and hope that it will lead to better regression performance.

3   The Proposed ε-DWSVR

In this section, we propose the novel ε-DWSVR method, which applies the idea of the mean of the functional margin, and we adopt the DCD method to handle general conditions and employ the ASGD method to deal with larger problems.

3.1   The Formulation of ε-DWSVR

To simplify the complexity, we enlarge the dimension of the vectors w and $\phi(x_i)$ to handle the bias term b as in [6], i.e., $w \leftarrow [w, b]^T$, $\phi(x_i) \leftarrow [\phi(x_i), \mathbf{1}]$. Thus the regression function becomes $f(x) = w \cdot \phi(x)$. Then the margin in regression will be the distance of the data to the fitting hyperplane, i.e., $|w \cdot \phi(x_i) - y_i| / \|w\|$. Based on the concept of margin, we define the functional margin in regression.

Definition 1. *The functional margin in regression is defined as follows:* $\gamma = (w \cdot \phi(x_i) - y_i)^2,\quad i = 1, 2, ..., n.$

The functional margin in regression can describe the difference between the real values and the estimated ones. It also has a significant connection with the geometrical distance. If the value of w is determined, the ranking of all data to the fitting hyperplane with respect to the margin can be decided by the functional margin. Next, we define the mean of the functional margin in regression.

Definition 2. *The mean of the functional margin in regression is as follows:*

$$\bar{\gamma} = \frac{1}{n} \sum_{i=1}^{n} \left(w^T \phi(x_i) - y_i \right)^2 = \frac{1}{n} \left(w^T \phi(X) \phi(X)^T w - 2(\phi(X)Y)^T w + YY^T \right),$$

where $\phi(X) = [\phi(x_1), ..., \phi(x_n)]$ *and* $\phi(X)\phi(X)^T = \sum_{i=1}^{n} \phi(x_i)\phi(x_i)$.

Based on Definitions 1 and 2, we add the mean of the functional margin to ε-SVR objective problems. As in the soft-margin of ε-SVR [4] we also consider the soft-margin in our problem. So the final optimal function is as follows:

$$\begin{aligned}
&\min_{w,\xi,\xi^*} \frac{1}{2}\|w\|^2 + \lambda_1 \bar{\gamma} + C \sum_{i=1}^{n} (\xi_i + \xi_i^*) \\
&s.t.\ y_i - w \cdot \phi(x_i) \leq \varepsilon + \xi_i, \\
&\quad\quad w \cdot \phi(x_i) - y_i \leq \varepsilon + \xi_i^*, \quad \xi_i, \xi_i^* \geq 0, \quad i = 1, 2, ..., n,
\end{aligned} \tag{1}$$

where λ_1 is the parameter for achieving the trade-off between the mean of functional margin and the model complexity.

In our ε-DWSVR, we maximize the minimum margin and minimize the mean of the functional margin at the same time, to obtain a better tradeoff between the distribution of the internal data and that of the boundary data. ε-DWSVR considers the influence of all data to the fitting hyperplane, as this is closer to the actual distribution of the internal data, and it is more robust to noise.

To illustrate the robustness of ε-DWSVR to noise and the differences between ε-SVR and that of ε-DWSVR, we use an example for comparison among linear regression, ε-SVR, and ε-DWSVR on an artificial dataset. In Fig. 1, the green points represent the data in which the distribution of the internal data is different from that of the boundary data, and the purple points represent noise.

Obviously, the curve produced by linear regression largely deviates from the actual distribution of the dataset, which indicates the linear regression is more sensitive to noise. ε-SVR and ε-DWSVR are more robust with the presence of noise, so the grey dashed curve and the red solid curve are within the area of non-noisy data. However, ε-SVR is controlled by boundary data. Once the distribution of the internal data is different from that of the boundary data (which is the case in Fig. 1), ε-SVR may not achieve good performance. The grey dashed curve is different from the curve. Because ε-DWSVR considers the influence of all data to the fitting hyperplane, the red solid curve produced by ε-DWSVR is closer to the actual distribution of the internal data.

It is obvious that the optimization problem of (1) is more complicated than that of the original SVR. Thus, as mentioned before, to solve (1) and improve the scalability, we implement different methods for ε-DWSVR, that is, we adopt the DCD method with kernel functions for medium problems and the ASGD method for larger problems. These will be presented in the following sections.

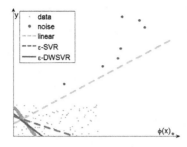

Fig. 1. The fitting curves produced by linear regression, ε-SVR, and ε-DWSVR. The data (green points) are composed of (1) 86.3% of all data which are evenly distributed across the line with a slope being -2 and $y \in [0, +\infty)$, $x \in [0, 10]$, and (2) 12.5% of all data which are evenly distributed on the line with a slope of 0 and $y \in [0, +\infty)$, $x \in [0, 40]$. This means the distribution of the internal data is different from that of the boundary data (those 12.5% of data). The rest 1.2% of data are noise (purple points). Due to noise, the cyan dashed curve produced by linear regression is very different from the rational one. The grey dashed curve produced by ε-SVR is adversely influenced by the distribution of the boundary data, while the red solid curve produced by ε-DWSVR better reflects the distribution of the internal data. (Color figure online)

3.2 The Regression of Medium Problems with Kernel Functions

Considering the mean of the functional margin $\bar{\gamma}$ in (1) and Definition 2, we can obtain the following form:

$$
\begin{aligned}
&\min_{w,\xi,\xi^*} \tfrac{1}{2}\|w\|^2 + \tfrac{\lambda_1}{n}\left(w^T\phi(X)\phi(X)^Tw - 2(\phi(X)Y)^Tw\right) + C\sum_{i=1}^{n}(\xi_i + \xi_i^*) \\
&\text{s.t. } y_i - w\cdot\phi(x_i) \le \varepsilon + \xi_i, \\
&\qquad w\cdot\phi(x_i) - y_i \le \varepsilon + \xi_i^*, \quad \xi_i,\xi_i^* \ge 0, \quad i = 1,2,...,n.
\end{aligned} \tag{2}
$$

Here we omit the term YY^T in $\bar{\gamma}$ (Definition 2) because it is regarded as a constant in an optimization problem. Obviously, the high dimensionality of $\phi(\cdot)$ and its complicated form makes (2) intractable. To simplify (2), we take the suggestion from [14] and the optimal solution w in [19]. We first give the following theorem which can be proved.

Theorem 1. *The optimal solution w for (2) can be represented as follows:*
$$w = \sum_{i=1}^{n}(\alpha_i - \alpha_i^*)\cdot\phi(x_i) = \phi(X)(\alpha - \alpha^*),\text{ where } \alpha = [\alpha_1,...,\alpha_n]^T \text{ and } \alpha^* = [\alpha_1^*,...,\alpha_n^*]^T \text{ are the parameters of } \varepsilon\text{-DWSVR.}$$

According to Theorem 1, (2) can be cast as

$$
\begin{aligned}
&\min_{\alpha,\alpha^*,\xi,\xi^*} \tfrac{1}{2}(\alpha - \alpha^*)^T Q(\alpha - \alpha^*) + p^T(\alpha - \alpha^*) + C\sum_{i=1}^{n}(\xi_i + \xi_i^*) \\
&\text{s.t. } y_i - (\alpha - \alpha^*)^T G_i \le \varepsilon + \xi_i, \\
&\qquad (\alpha - \alpha^*)^T G_i - y_i \le \varepsilon + \xi_i^*, \quad \xi_i,\xi_i^* \ge 0, \quad i = 1,2,...,n,
\end{aligned} \tag{3}
$$

where $G = \phi(X)^T \phi(X)$, G_i denotes the i-th column of G, $Q = 2\lambda_1 G^T G/n + G$, and $p = -2\lambda_1 GY/n$. Thus (3) can be transformed into a dual formulation with Lagrange multipliers, so the Lagrange function of (3) leads to

$$L = \tfrac{1}{2}(\alpha - \alpha^*)^T Q (\alpha - \alpha^*) + p^T (\alpha - \alpha^*) + C \sum_{i=1}^{n} (\xi_i + \xi_i^*) - \sum_{i=1}^{n} (\eta_i \xi_i + \eta_i^* \xi_i^*)$$
$$- \sum_{i=1}^{n} \beta_i \left(\varepsilon + \xi_i - y_i + (\alpha_i - \alpha_i^*)^T G \right) - \sum_{i=1}^{n} \beta_i \left(\varepsilon + \xi_i^* + y_i - (\alpha_i - \alpha_i^*)^T G \right), \tag{4}$$

where $\eta, \eta^*, \beta, \beta^*$ are Lagrange multipliers. To satisfy the KKT conditions [8], we set the partial derivatives of $(\alpha - \alpha^*)$ and $\xi^{(*)}$ to zero and thus obtain the following equations:

$$\frac{\partial L}{\partial (\alpha - \alpha^*)} = Q (\alpha - \alpha^*) + p - \sum_{i=1}^{n} (\beta_i - \beta_i^*) G_i = 0, \tag{5}$$

$$\frac{\partial L}{\partial \xi_i^{(*)}} = C - \beta_i^{(*)} - \eta_i^{(*)} = 0, \quad i = 1, 2, ..., n. \tag{6}$$

By substituting (5) and (6) into (4), and inspired by the work of [3], (4) can be written as follows to compute the values of $\begin{bmatrix} \beta \\ \beta^* \end{bmatrix}$ separately:

$$\min_{\beta, \beta^*} f(\beta, \beta^*) = \tfrac{1}{2} \begin{bmatrix} \beta^T, (\beta^*)^T \end{bmatrix} \begin{bmatrix} H & -H \\ -H & H \end{bmatrix} \begin{bmatrix} \beta \\ \beta^* \end{bmatrix}$$
$$+ \begin{bmatrix} \varepsilon e^T + \left(\tfrac{2\lambda_1}{n} HY - Y \right)^T, \varepsilon e^T - \left(\tfrac{2\lambda_1}{n} HY - Y \right)^T \end{bmatrix} \begin{bmatrix} \beta \\ \beta^* \end{bmatrix} \tag{7}$$
$$s.t. \ 0 \le \beta_i, \beta_i^* \le C, \quad i = 1, 2, ..., n,$$

where $H = GQ^{-1}G$, and e means the all-one vector.

We adopt the DCD method as in [18] to solve (7). This method continuously selects one variable for minimization and keeps others as constants at each iteration. In our situation, we minimize the variation of $f(\beta')$ by adjusting the value of β_k' with a step size of t while fixing other $\beta_{l \ne k}'$, where $\beta' = (\beta, \beta^*)^T$, and the following equation needs to be solved: $\min_t f(\beta' + t b_k)$ $s.t. \ 0 \le \beta_k' + t \le C$, $k = 1, 2, ..., 2n$, where b_k means the vector with 1 in the k-th element and 0's elsewhere. Then, we have the form of this sub-problem as follows:

$$f(\beta' + t b_k) = \frac{1}{2} h_{kk} t^2 + \nabla f(\beta')_k t + f(\beta'), \tag{8}$$

where h_{kk} is the diagonal entry of $\begin{bmatrix} H & -H \\ -H & H \end{bmatrix}$. It can be seen that $f(\beta')$ is independent of t, so we omit this term in (8).

Hence $f(\beta' + t b_k)$ is transformed into a simple quadratic function of t. Assume that $\beta_k'^{iter}$ is the value of β_k' at the $iter$-th iteration, then the value of β_k' at the

$(iter + 1)$-th iteration is $\beta_k^{'(iter+1)} = \beta_k^{'iter} + tb_k$. According to (8), the minimization of t which satisfies (8) is $t = -\frac{\nabla f(\beta^{'iter})_k}{h_{kk}}$. Considering the box constraint $0 \leq \beta_k' \leq C$, the minimization for $\beta_k^{'(iter+1)}$ has the following form:$\beta_k^{'(iter+1)} \leftarrow \min(\max(\beta_k^{'iter} - \frac{\nabla f(\beta^{'iter})_k}{h_{kk}}, 0), C)$. After β' converges, we can obtain $(\alpha - \alpha^*)$ according to (5) as follows: $(\alpha - \alpha^*) = Q^{-1}G\left(\frac{\lambda_1}{n}Y + (\beta - \beta^*)\right)$.

Therefore, the final fitting function becomes: $f(x) = \sum_{i=1}^{n} (\alpha_i - \alpha_i^*)K(x_i, x)$.

Algorithm 1 presents the steps of the DCD method for updating β'.

Algorithm 1. ε-DWSVR with Kernel Functions

Input: Dataset $X, Y, \lambda_1, C, \varepsilon, K$; **Output:** $\alpha - \alpha^*$; **Initialization:** $\beta' = 0, (\alpha - \alpha^*) = \frac{2\lambda_1}{n}Q^{-1}GY, A = Q^{-1}G, h_{kk} = b_k^T GQ^{-1}Gb_k$;

1: **while** β' not converges **do**
2:    **for** $k = 1, 2, ..., 2n$ **do**
3:       $\nabla f(\beta')_k \leftarrow \varepsilon + (G(\alpha - \alpha^*)b_k - y_k)$; if $k = 1, 2, ..., n$
4:       $\nabla f(\beta')_k \leftarrow \varepsilon - (G(\alpha - \alpha^*)b_k - y_{k-n})$; if $k = n+1, n+2, ..., 2n$
5:       $\beta_k^{'temp} \leftarrow \beta_k'; \beta_k' \leftarrow \min(\max(\beta_k' - \frac{\nabla f(\beta')_k}{h_{kk}}, 0), C)$;
6:       **for** $i = 1, 2, ..., n$ **do**
7:          $(\alpha_i - \alpha_i^*) \leftarrow (\alpha_i - \alpha_i^*) + \left(\beta_k' - \beta_k^{'temp}\right)Ab_k$; if $k = 1, 2, ..., n$
8:          $(\alpha_i - \alpha_i^*) \leftarrow (\alpha_i - \alpha_i^*) - \left(\beta_k' - \beta_k^{'temp}\right)Ab_k$; if $k = n+1, n+2, ..., 2n$
9:       **end for**
10:    **end for**
11: **end while**

3.3 The Regression of Larger Problems

In regression analysis, processing larger datasets may increase the time complexity. Although the DCD method can solve ε-DWSVR efficiently for medium problems, it is not the best strategy for larger problems. To improve the scalability of ε-DWSVR, we adjust the ASGD method to ε-DWSVR, which solves the optimization problem by computing a noisy unbiased estimate of the gradient, and it randomly samples a subset of the training instances rather than all data. Considering the constraints in (2), we reformulate (2) as follows:

$$\min_{w} g(w) = \frac{1}{2}\|w\|^2 + \frac{\lambda_1}{n}\left(w^T X^T X w - 2(XY)^T w\right)$$
$$+ C\sum_{i=1}^{n} \max\{0, y_i - w \cdot x_i - \varepsilon, w \cdot x_i - y_i - \varepsilon\} \tag{9}$$

Computing the gradient of w in (9) is time consuming because we need all the training instances for computation, especially when the dataset is large. Considering this issue, we use Stochastic Gradient Descent (SGD) [1] to reduce the computational time for larger problems. According to [1], the SGD method is expected to converge to the global optimal solution when the objective is convex.

Therefore, we give an unbiased estimation of the gradient $\nabla g(w)$ in our case. For representing the last term of (9) formally, we define a function $s(w)$ that has different values under different constraint conditions, as shown below:

$$s(w) = \begin{cases} -x_i, & i \in I_1 \\ x_i, & i \in I_2 \\ 0, & otherwise \end{cases}, \quad i = 1, 2, ..., n,$$

where $I_1 \equiv \{i \,|\, y_i - w \cdot x_i \leq \varepsilon\}$, and $I_2 \equiv \{i \,|\, w \cdot x_i - y_i \leq \varepsilon\}$. In order to obtain an unbiased estimation of the gradient $\nabla g(w)$, we first present the following theorem which can be proved for computing $\nabla g(w)$.

Theorem 2. *An unbiased estimate of the gradient $\nabla g(w)$ in (9) has the following form: $\nabla g(w, x_i) = 2\lambda_1 x_i x_i^T w + w - 2\lambda_1 y_i x_i + nC \cdot s(w)$, where (x_i, y_i) is an randomly sampled instance from the training set.*

Based on Theorem 2, the stochastic gradient can be updated as follows:

$$w_{t+1} = w_t - \varphi_t \nabla g_t (w_t, x_i), \tag{10}$$

where φ_t is the learning rate at the t-th iteration.

To make the solution to (9) more robust, we can adopt the ASGD method to solve the optimization problem in (9), which outperforms the SGD method [17]. In ASGD [17], a good choice for φ_t can be obtained by the form $\varphi_t = \varphi_0(1 + a\varphi_0 t)^{-c}$ to compute (10), where a, φ_0, and c are set by constant values as in [19]. In addition to updating the ordinary stochastic gradient in (10), we also compute $\bar{w}_t$ at each iteration as follows: $\bar{w}_t = \frac{1}{t-t_0} \sum_{i=t_0+1}^{t} w_i$, where t_0 is used to decide when we apply the averaging process. This average value can also be calculated in a recursive manner as follows: $\bar{w}_{t+1} = \delta_t w_{t+1} + (1 - \delta_t)\bar{w}_t$.

Finally, Algorithm 2 presents the detailed steps of the ASGD method for larger problems, where $T * n'$ determines the number of iterations. T is a coefficient for adjusting the number of iterations; n' is the sampling number from n instances. The settings of these two variable values follow those in [19].

Algorithm 2. ε-DWSVR for Larger Problems

Input: Dataset X, Y, λ_1, C, ε; **Output:** $\bar{w}$
Initialization: $w_0 = 0, \nabla g_0 = 0, t = 1$

1: **while** $t \leq T * n'$ **do**
2:    Randomly select one instance (x_i, y_i) from the training set;
3:    Compute $\nabla g_t (w_t, x_i); w_{t+1} \leftarrow w_t - \varphi_t \nabla g_t (w_t, x_i); \bar{w}_{t+1} \leftarrow \delta_t w_{t+1} + (1 - \delta_t)\bar{w}_t$;
4:    $t \leftarrow t + 1$;
5: **end while**

---

4 Experiments

In this section, we compare the fitting performance between ε-DWSVR and other regression methods on real datasets to assess the performance of our method.

4.1 Experimental Setup

We select sixteen datasets from UCI [10] to perform the evaluations on ε-DWSVR, ε-SVR, linear regression, NN, logistic regression, and ridge regression. This includes eight medium-scale datasets and eight larger datasets. The characteristics of all datasets are in Table 1. All the features of the datasets and target set are normalized into [0, 1] to balance the influence of each feature. After normalization, we use PCA with 95% for feature extraction to reduce the interference of irrelevant attributes. During the construction of the model, we divide the datasets into training sets and test sets by 5-fold cross validation. Parameters selections are processed on the test sets to obtain better results.

Table 1. The characteristics of benchmark datasets.

Scale	Datasets	Instances	Features	Datasets	Instances	Features
Medium	Slump	103	7	Housing	506	14
	Automobile	205	26	Stock	536	9
	Yacht	308	7	Concrete	1030	8
	Auto MPG	398	8	Music	1059	68
Larger	Crime	1994	128	Bike	17389	16
	SkillCraft	3338	18	ONP	39797	61
	CCPP	9568	4	CASP	45730	9
	Drift	13910	129	Buzz	140000	77

Finally, we use mean square error (MSE) [7] as the evaluation metric, and evaluations are also processed on the test sets. The experiments are repeated 30 times, and the average values of the evaluation metric are recorded. For medium-scale datasets, we evaluate both the linear and RBF kernels [16]. In addition, we record the computational time for larger datasets.

4.2 Results and Discussion

For medium-scale datasets, Table 2 shows the results of MSE on all methods, including linear and RBF kernel function for ε-DWSVR and ε-SVR. We can see that the performance of ε-DWSVR is much better than ε-SVR. Besides, the Housing dataset is ideal with less noise and a consistent distribution of overall data; thus linear regression works better on this dataset. The average MSE values on all datasets are shown in Table 2 and the best ones are indicated in bold.

For larger datasets, Fig. 2 summarizes the results of MSE on all methods. As one can see from Fig. 2, ε-DWSVR outperforms other methods on most datasets.

Table 2. The evaluation of average MSE on medium-scale datasets.

Datasets	ε-DWSVR (RBF)	ε-SVR (RBF)	ε-DWSVR (Linear)	ε-SVR (Linear)	LINEAR	NN	Logistic	Ridge
Slump	**0.0036**	0.0037	0.0047	0.0050	0.0063	0.0055	0.0054	0.0215
Automobile	**0.0057**	0.0063	0.0092	0.0102	0.0094	0.0129	0.0136	0.0232
Yacht	**0.0101**	0.0166	0.0154	0.0171	0.0180	0.0175	0.0171	0.0434
Auto MPG	**0.0133**	0.0137	0.0135	0.0136	0.0140	0.0148	0.0152	0.0380
Housing	0.0142	0.0170	0.0169	0.0176	**0.0117**	0.0199	0.0182	0.0178
Stock	**0.0080**	0.0083	0.0087	0.0088	0.0101	0.0111	0.0093	0.0148
Concrete	**0.0227**	0.0251	0.0256	0.0257	0.0262	0.0267	0.0261	0.0362
Music	**0.0306**	0.0348	0.0359	0.0360	0.0368	0.0388	0.0408	0.0594

In addition, the Drift dataset contains less noise, and there exists a consistent distribution of all data. So linear regression works better on this dataset. Besides, linear regression did not return the results on some datasets after 48 h.

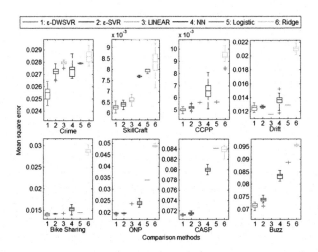

Fig. 2. The evaluation of MSE on larger datasets.

4.3   Parameter Effects

ε-DWSVR has three main parameters: λ_1, C, and ε. To further investigate the influence of these three parameters, we evaluate the MSE value by changing one of them on the medium-scale datasets and larger datasets, while fixing others.

Figures 3 and 4 show that the MSE on the medium-scale and larger datasets does not change significantly with the change of the parameters. This indicates that the performance of ε-DWSVR is not sensitive to parameter values, which demonstrates the robustness of ε-DWSVR.

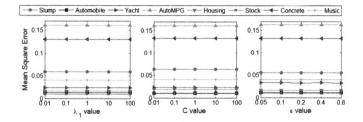

Fig. 3. Parameter influence on medium-scale datasets.

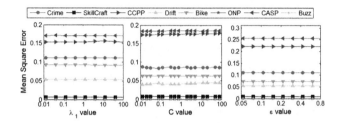

Fig. 4. Parameter influence on larger datasets.

4.4 Time Cost

We present a comparison of CPU time taken between ε-SVR and ε-DWSVR on each larger dataset in Fig. 5. For ε-SVR, C is set to 1; ε is set to 0.1. For ε-DWSVR, λ_1 is set to 1; C is set to 10; ε is set to 0.1. ε-SVR for larger problems was implemented by the LIBLINEAR [6] package and ε-DWSVR was implemented by ASGD. Figure 5 shows that ε-DWSVR cost less time than ε-SVR on most datasets, and it is only slightly slower than ε-SVR on two datasets.

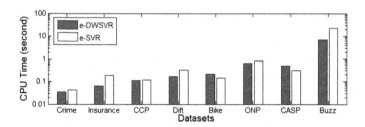

Fig. 5. The CPU time on larger datasets.

Acknowledgement. We gratefully thank Dr. Teng Zhang and Prof Zhi-Hua Zhou for providing the source code of "LDM", and their kind technical assistance. We also thank Prof Chih-Jen Lins team for providing the LIBSVM and LIBLINEAR packages and their support. This work is supported by the National Natural Science Foundation of China (Grant Nos.61472159, 61572227) and Development Project of Jilin Province of China (Grant Nos. 20160204022GX, 2017C033, 20180414012GH). This work is also partially supported by the 2015 Scottish Crucible Award funded by the Royal Society of Edinburgh and the 2016 PECE bursary provided by the Scottish Informatics & Computer Science Alliance (SICSA).

References

1. Bottou, L.: Large-Scale Machine Learning with Stochastic Gradient Descent. Physica-Verlag HD (2010)
2. Brown, J.D., Summers, M.F., Johnson, B.A.: Prediction of hydrogen and carbon chemical shifts from rna using database mining and support vector regression. J. Biomol. NMR **63**(1), 1–14 (2015)
3. Chang, C.C., Lin, C.J.: LIBSVM: a library for support vector machines. ACM Trans. Intell. Syst. Technol. **2**(3), 389–396 (2011)
4. Demir, B., Bruzzone, L.: A multiple criteria active learning method for support vector regression. Pattern Recogn. **47**(7), 2558–2567 (2014)
5. Dicker, L.H.: Ridge regression and asymptotic minimax estimation over spheres of growing dimension. Bernoulli **22**(1), 1–37 (2016)
6. Fan, R.E., Chang, K.W., Hsieh, C.J., Wang, X.R., Lin, C.J.: Liblinear: a library for large linear classification. J. Mach. Learn. Res. **9**(12), 1871–1874 (2010)
7. Guo, D.N., Shamai, S., Verdu, S.: Mutual information and minimum mean-square error in gaussian channels. IEEE Trans. Inf. Theory **51**(4), 1261–1282 (2005)
8. Izmailov, A.F., Solodov, M.V.: Karush-kuhn-tucker systems: regularity conditions, error bounds and a class of newton-type methods. Math. Program. **95**(3), 631–650 (2003)
9. Ke, Y., Fu, B., Zhang, W.: Semi-varying coefficient multinomial logistic regression for disease progression risk prediction. Stat. Med. **35**(26), 4764–4778 (2016)
10. Lichman, M.: UCI machine learning repository (2013). http://archive.ics.uci.edu/ml, http://archive.ics.uci.edu/ml. Accessed 01 Aug 2016
11. Marron, J.S.: Distance-weighted discrimination. J. Am. Stat. Assoc. **102**, 1267–1271 (2007)
12. Qiao, X.Y., Zhang, L.S.: Distance-weighted support vector machine. Stat. Interface **8**(3), 331–345 (2015)
13. Rajaraman, P.K., Manteuffel, T.A., Belohlavek, M., Mcmahon, E., Heys, J.J.: Echocardiographic particle imaging velocimetry data assimilation with least square finite element methods. Comput. Math. Appl. **68**(11), 1569–1580 (2016)
14. Scholkopf, B., Smola, A.: Learning with kernels: support vector machines, regularization, optimization, and beyond. J. Am. Stat. Assoc. **16**(3), 781–781 (2011)
15. Srivastava, N., Hinton, G., Krizhevsky, A., Sutskever, L., Salakhutdinov, R.: Dropout: a simple way to prevent neural networks from overfitting. J. Mach. Learn. Res. **15**(1), 1929–1958 (2014)
16. Vapnik, V.: The Nature of Statistical Learning Theory. Springer, New York (1995). https://doi.org/10.1007/978-1-4757-3264-1
17. Xu, W.: Towards optimal one pass large scale learning with averaged stochastic gradient descent. Comput. Sci. (2011). https://arxiv.org/abs/1107.2490
18. Yuan, G.X., Ho, C.H., Lin, C.J.: Recent advances of large-scale linear classification. Proc. IEEE **100**(9), 2584–2603 (2012)
19. Zhang, T., Zhou, Z.H.: Large margin distribution machine. In: Proceedings of the Twentieth ACM SIGKDD International Conference on Knowledge Discovery and Data Mining, pp. 313–322. ACM Press, Banff, Alberta, Canada (2014)

Healthcare, BioInformatics and Related Topics (Application)

Corrosion Prediction on Sewer Networks with Sparse Monitoring Sites: A Case Study

Jianjia Zhang[1(⊠)], Bin Li[2], Xuhui Fan[3], Yang Wang[1], and Fang Chen[1]

[1] Data61, CSIRO, Eveleigh, NSW 2015, Australia
{Jianjia.Zhang,Yang.Wang,Fang.Chen}@data61.csiro.au
[2] Fudan University, Shanghai, China
Libin@fudan.edu.cn
[3] The University of New South Wales, Sydney, Australia
Xuhui.Fan@unsw.edu.au

Abstract. Sewer corrosion is a widespread and costly issue for water utilities. Knowing the corrosion status of a sewer network could help the water utility to improve efficiency and save costs in sewer pipe maintenance and rehabilitation. However, inspecting the corrosion status of all sewer pipes is impractical. To prioritize sewer pipes in terms of corrosion risk, the water utility requires a corrosion prediction model built on influential factors that cause sewer corrosion, such as hydrogen sulphide (H_2S) and temperature. Unfortunately, monitoring sites of influential factors are very sparse on the sewer network such that a reliable prediction has often been hampered by insufficient observations – It is a challenge to predict H_2S distribution and sewer corrosion levels on the entire sewer network with a limited number of monitoring sites. This work leverages a Bayesian nonparametric method, Gaussian Process, to integrate the physical model developed by domain experts, the sparse H_2S and temperature monitored records, and the sewer geometry to predict corrosion risk levels on the entire sewer network. A case study has been conducted on a real data set of a water utility in Australia. The evaluation results well demonstrate the effectiveness of the model and admit promising applications for water utilities, including prioritizing high corrosion areas and recommending chemical dosing profiles.

1 Introduction

Sewer corrosion is a key issue in wastewater systems worldwide, particularly in warm climate countries such as Australia. Corrosion results in concrete loss, sewer pipe cracks and ultimately, structural collapse [8]. It gradually deteriorates sewer network, which is one of the most critical infrastructure assets for modern urban societies [7], and as a result of this, the value of public assets is being significantly diminished. The mitigation and renewal of corroded sewer pipes are highly costly. The cost of sewer corrosion in Australia is estimated to be hundreds of millions of AUD per year [20] (http://www.score.org.au) – This

© Springer International Publishing AG, part of Springer Nature 2018
D. Phung et al. (Eds.): PAKDD 2018, LNAI 10937, pp. 223–235, 2018.
https://doi.org/10.1007/978-3-319-93034-3_18

has not included those indirect costs, e.g., lost time and productivity caused by corrosion-related outages, delays, failures, and litigation [10]. Moreover, the cost is expected to increase in the future as the aging sewer pipes continue to corrode.

Considering the serious negative effect of sewer corrosion, measures should be taken to mitigate the corrosion process, e.g., dosing chemicals [3,6,7] or using protective coatings and liners [2,14,16,19]. A preliminary requirement of these preventive operations is to know the corrosion status of the sewer network. However, inspecting the corrosion status of all pipes is infeasible in practice. Firstly, there are a large number of sewer pipes (e.g., several thousands in Sydney) in a modern city. Vast human and material resources are required to inspect all these pipes, making it unaffordable for water utilities. Secondly, many pipes are not easily accessible because of their sizes, locations or hazardous conditions. Therefore, a water utility expects to inspect a small portion of the sewer pipes which are at high corrosion risk.

In this case, predicting sewer corrosion on the entire sewer network is a critical task for water utilities around the globe in order to improve efficiency and save costs in chemical dosing and sewer pipe rehabilitation. The water utility requires a corrosion prediction model built on influential factors that cause sewer corrosion, such as hydrogen sulphide (H_2S) and temperature. However reliable prediction of sewer corrosion has often been hampered by insufficient observations of influential factors (e.g., H_2S and temperature) and inspections of corrosion status as groundtruth for accurate modelling. As aforementioned, increasing the number of monitoring and inspection sites may be infeasible due to cost and accessibility. Therefore, modelling of sewer corrosion on the entire sewer network with a limited number of monitoring sites is nontrivial. Current study of corrosion rate prediction for concrete sewers is mainly conducted in very few testbeds deployed in the sewer system, with an array of coupons installed along with a variety of sensors for measuring different influential factors. However, the physical model [20] is calibrated in a certain testbed and it may not be versatile in any sewer system of any city due to very different environments and lack of measurements of the required factors.

This paper attempts to leverage a Bayesian nonparametric method to predict the sewer corrosion risk on the entire sewer network with a limited number of observations. Specifically, this is achieved in two steps: (1) Gaussian Process [13] is used to estimate the distributions of the two influential factors, H_2S and temperature, on the entire sewer network; (2) Based on the estimation results of influential factors, a second-level Gaussian Process is used to further predict the corrosion risk levels on the entire sewer network. Thanks to the Bayesian nonparametric method, the corrosion prediction model based on Gaussian Process is able to integrate the physical model developed by domain experts, the sparse H_2S and temperature monitored records, and the sewer geometry to predict corrosion risk levels on the entire sewer network. Because of incorporating physical model as prior knowledge, the hypothesis space of the model parameters can be regularized and the issue of insufficient observations can be mitigated.

The proposed method has the following desirable properties: (1) The proposed method is able to integrate expert domain knowledge (physical model) into the prediction model to alleviate the issue of insufficient data. The adopted data analytics technique is a Bayesian nonparametric method which provides a way to regularize the prediction with domain knowledge; (2) The proposed method is flexible. The prediction model in this work can readily incorporate more factors related to sewer corrosion. Therefore, the model can be easily improved by employing additional data collected in the future. In addition, the proposed model could well handle large-scale sewer networks, making it widely applicable; (3) The proposed model built on Gaussian Process not only predicts the sewer corrosion level quantitatively, but also estimates the uncertainty of the prediction. This uncertainty is an important measure in decision makings and cost-effective sewer operations. For example, it can be used to prioritize high corrosion areas, recommend chemical dosing locations, and suggest deployment of sensors.

A case study is conducted on real data set from a water utility in Australia. The empirical study demonstrates that the proposed method could achieve promising sewer corrosion prediction results. The results admit several promising further applications for water utilities, including prioritizing high corrosion areas and recommending chemical dosing profiles.

2   Case Study Background

This work is to collaborate with an Australian water utility to make use of data analytics techniques for sewer corrosion prediction. The water utility manages around 24,000 km of sewers, of which approximately 900 km is large concrete trunks up to 2 m in diameter. Sewer corrosion is a serious concern for the water utility, who spends about 40 million AUD per year on the rehabilitation of corroded sewer pipes. Therefore predicting sewer corrosion is a critical task for the utility to improve efficiency and save costs in sewer pipe rehabilitation and chemical dosing. This motivates a collaborative project between the utility and Data61, aiming to assess the feasibility of predicting corrosion in sewer network using data analytics. The data provided by the water utility include:

- **Sewer network geometry data:** Including the length and GPS coordinates of the sewer pipes in the sewer system;
- **H_2S observation data:** Including GPS coordinates of 17 observation sites the sewer system, as shown in Fig. 1 (left), and the H_2S records of these sites from Jan-2011 to Dec-2015 with a sampling frequency of 15 min;
- **Temperature observation data:** Including GPS coordinates of 13 observation sites, as shown in Fig. 1 (right), and the temperature records of these sites from Jan-2011 to Dec-2015 with a frequency of 15 min;
- **Traverse reports:** Including two batches of traverse reports conducted during 2007–2010 and 2011–2016, respectively. In each period, a set of sewer pipes are inspected and their corrosion risk levels (1–5) were recorded. The corrosion risk levels in the reports are shown in Fig. 2(a) and (b).

With the data provided above, sewer corrosion prediction aims to construct a mapping from two influential factors, H_2S and temperature, as input to the corrosion risk level as output. The challenge lies in the data sparsity, that is, the data sampling points on the sewer network is very sparse. Thus, developing a robust sewer corrosion prediction model requires techniques suitable for this particular problem. This study is an attempt to construct a prediction model for sewer corrosion on the entire sewer network.

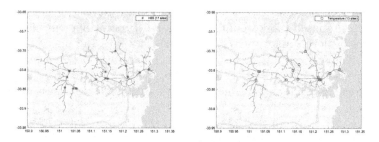

Fig. 1. H_2S (Left) and Temperature (Right) observation sites on the sewer network.

3   Preliminaries

3.1   Related Work on Sewer Corrosion

In the last decades, extensive research has been conducted on understanding and managing sewer corrosion [4,5,7,11,22]. It has been verified that the production and emission of hydrogen sulfide (H_2S) is a major cause of corrosion in sewer systems [1,15]. Sulfate-reducing bacteria residing in the sewer system could turn sulfate in the wastewater into sulfide when anaerobic conditions prevail in a sewer system. During this process, H_2S emits into the sewer atmosphere [8]. In a later stage, H_2S present in the sewer system will be consumed by bacteria and sulfuric acid will be generated in biological oxidation of H_2S [12,18]. The sulfuric acid generated in this stage causes internal cracking and pitting in the sewer pipe, which exposes more pipe surface for acid attack [8]. Step by step, mass corrosion of sewer pipe happens. Also, it has been found that the conversion rate of sulfuric acid in the sewer system from H_2S is proportional to the concentration of H_2S in the sewage [12].

Besides H_2S, temperature is also a marked factor affecting the rate of sewer corrosion since sulfuric acid generation is a biological phenomenon. Specifically, temperature plays an important role in the emission of H_2S from liquid to gas phase [21] and can affect various abiotic and biotic reaction rates important for corrosion [9]. It has been found in [9,12] that the generation rate of sulfide increases with the rise of temperature.

The findings above motivate this work to first conduct estimation of H_2S and temperature on the entire sewer network; and based on the estimation results, prediction of corrosion risk level on the entire sewer network is further carried out. This is because: (1) Both H_2S and temperature are well-verified influential factors to sewer corrosion in the literature; and (2) in comparison with inspecting corrosion status of the sewer network, H_2S concentration and temperature can be more conveniently monitored and collected by using electronic sensors and telecommunication techniques.

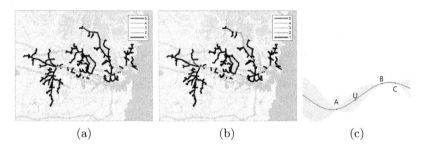

(a)   (b)   (c)

Fig. 2. (a) and (b): Two batches of traverse reports in (2007–2010) and (2011–2016). Five different colours denote five levels of corrosion risk 1–5 (1 for lowest risk and 5 for highest risk) while black lines denote those pipes without a traverse report. (c): Illustration of the prediction and the associated prediction uncertainty of a Gaussian Process on a segment of sewer. The red curve denotes the mean value of the prediction and the bandwidth denotes the uncertainty. The farther the prediction point away from the observation points (red dots), the more uncertain the prediction result is. (Color figure online)

3.2   Brief Introduction to Gaussian Process

Gaussian Process (GP) is a generic supervised learning method designed to solve regression and probabilistic classification problems. The general idea behind GP for regression is illustrated in Fig. 2(c). As seen, the unknown value of a certain type of measurement (in the following we take H_2S for example) at site U can be estimated as a weighted combination of values collected at the observation sites A, B and C.

$$V(U) = w_{A \to U} V(A) + w_{B \to U} V(B) + w_{C \to U} V(C) \tag{1}$$

where $V(U)$ denotes the predicted H_2S at any unknown point U (the green dot in Fig. 2(c)) on the sewer network while $V(A)$, $V(B)$, and $V(C)$ denote those points with observed H_2S (the three red dots in Fig. 2(c)). The weights $w_{A \to U}$, $w_{B \to U}$, $w_{C \to U}$ are learned automatically through the GP. By repeating this prediction for any unobserved point on the network, the estimation of H_2S on the entire network can be obtained, as shown by the red line in Fig. 2(c).

GP has several advantages: (1) GP enables integration of prior knowledge, such as the physical model developed by domain experts of sewer corrosion. This prior knowledge could regularise the hypothesis space of the prediction model; (2) The prediction of GP is a Gaussian distribution, such that one can compute empirical confidence interval using the variance of the Gaussian distribution and make decisions based on these confidence intervals. The confidence interval is illustrated in Fig. 2(c). As seen, the farther the prediction point away from the observation points (red dots), the more uncertain the prediction result is; (3) GP is flexible and versatile. Different regression objectives can be achieved by simply specifying different kernels (will be introduced in the following section). This enables both influential factor estimation and corrosion prediction on the entire sewer network in a similar framework.

A GP is a generalization of the Gaussian distribution in the infinite dimensional space. Similar to a Gaussian distribution, a GP is also fully specified by a mean function and a covariance function (also known as a kernel function). Therefore, the key to use GP is just to specify these two functions for our goals. The design of two functions for influential factor estimation and corrosion prediction will be introduced in the following section.

4 Methodology

The aim of this work is to develop a prediction model based on a Bayesian nonparametric method. A typical Bayesian model is in the form of "Prediction (posterior distribution) = Domain Knowledge (prior distribution) × Data Fitness (likelihood)", where "Domain Knowledge" provides a hypothesis space to the model such that the model is not only driven by the data (in terms of "Data Fitness") when data are sufficient, but also does not deviate too far from the domain expert's hypothesis when data are insufficient.

Through Bayesian modelling, we can thus (1) integrate domain experts' knowledge, for example, using the existing physical model as prior knowledge, and (2) conduct prediction as a posterior distribution, whose variance can be viewed as the uncertainty of the prediction. In the following we first introduce the Gaussian Process based prediction model and then elaborate how it is adapted to H_2S, temperature and corrosion prediction on the entire sewer network.

4.1 Gaussian Process Based Prediction Model

The prediction problem introduced above is essentially a regression problem on a network. To address this problem, we adopt a Bayesian nonparametric method - Gaussian Process (GP) [13] to achieve this goal due to its outstanding performance and desirable properties aforementioned. In order to make this paper self-contained, this section briefly introduces GP. GP assumes that all the training (observed) and test (unobserved) data can be represented as a joint multivariate Gaussian distribution:

$$\begin{bmatrix} \boldsymbol{y}_O \\ \boldsymbol{y}_U \end{bmatrix} \sim \mathcal{N} \left(\begin{bmatrix} \boldsymbol{\mu}_O \\ \boldsymbol{\mu}_U \end{bmatrix}, \begin{bmatrix} \boldsymbol{K}_{OO} & \boldsymbol{K}_{OU} \\ \boldsymbol{K}_{OU}^\top & \boldsymbol{K}_{UU} \end{bmatrix} \right) \tag{2}$$

where μ_O and μ_U denote the means of training and test points, respectively; K_{OO} denotes the covariance matrix of the training set, K_{OU} denotes the covariance matrix between the training set and the test set, and K_{UU} denotes the covariance matrix of the test set.

Given the values of the training set y_O, the conditional distribution of the test value y_U can be expressed as:

$$y_U | y_O \sim \mathcal{N}\left(\mu_U + K_{OU}^\top K_{OO}^{-1}\left(y_O - \mu_O\right), K_{UU} - K_{OU}^\top K_{OO}^{-1} K_{OU}\right) \qquad (3)$$

The optimal estimation of y_U is the mean of the above Gaussian distribution:

$$\hat{y}_U = \mu_U + K_{OU}^\top K_{OO}^{-1}\left(y_O - \mu_O\right) \qquad (4)$$

and the uncertainty of the estimation is reflected in its variance:

$$var(y_U) = K_{UU} - K_{OU}^\top K_{OO}^{-1} K_{OU} \qquad (5)$$

As discussed above, a GP is fully specified by its mean function to obtain μ_O and μ_U and covariance function $k(\cdot, \cdot)$ to calculate K_{OO}, K_{OU}, and K_{UU}. Please refer to [13] for more details about GP.

In our case study, μ_O and μ_U can be the observed and unobserved values of a factor (e.g., H$_2$S and temperature) or the corrosion risk level, respectively; and K_{OO}, K_{OU}, and K_{UU} are the covariance matrices between these observed and unoberserved sites. The specification of these functions will be introduced in the following.

4.2   Factor Estimation

This section introduces how mean and covariance functions of the GP-based prediction model are specified to estimate H$_2$S concentration and temperature on the entire sewer network.

Mean Function. The mean function for estimating H$_2$S is the output of the absorbing state random walks (ASRW)[1] [17]. ASRW is a widely used algorithm for interpolation and extrapolation on a network (e.g., electricity network). The input of an ASRW algorithm is the network structure represented as a directed graph and the values of some observed points on the network; the output of the ASRW is the interpolation and extrapolation results on the entire network. Since the interpolation and extrapolation are based on smoothing, the results can be naturally viewed as a coarse estimation of the mean values of H$_2$S (assigned with the results of ASRW), with the assumption that the real distribution of H$_2$S will not be far away from the mean function.

ASRW is adopted as the mean function of the prediction model for the following reasons: (1) ASRW has no specific assumption on the underlying graph

---

[1] The introduction to ASRW is out of the scope of this paper. Interested readers are referred to [17] for details.

structure and it can be easily applied to sewer networks, which usually have complicated graph structures; (2) The interpolation and extrapolation results of ASRW are smooth on the network and this coincides with the status of gas phase H_2S, which is smoothly distributed in sewer networks due to diffusion; and (3) ASRW is very efficient to compute which makes it applicable to large sewer networks.

Covariance Function. The commonly used exponential kernel function is employed as the covariance function of the GP-based prediction model. Due to the constraints of the network structure, H_2S can only diffuse along the sewer networks. Instead of the traditional Euclidean distance used in the exponential kernel, this work argues that geodesic distance should be used in the kernel function to incorporate the underlying network structure. Therefore, we need to first compute the shortest geodesic distance, denoted as d_{ij}, between any two points i and j on the network as the distance between these two points. Then, the exponential kernel can be defined in terms of the shortest geodesic distance as follows:

$$K_{i,j} = exp\left(\frac{-d_{ij}^2}{\sigma^2}\right) \qquad (6)$$

where $K_{i,j}$ denotes the i-th row and j-th column of the kernel matrix, and σ is the band-width of the exponential kernel. With this kernel function, the covariance matrices K_{OO}, K_{OU} and K_{UU} in Eq. (2) can be computed using training-training, training-test, and test-test data sets, respectively. Then the mean and variance of the H_2S concentration on the entire sewer network can be estimated by applying Eqs. (4) and (5), respectively. In this way, the spatial H_2S estimation is achieved. By repeating this process for the unknown points at any time point, the estimation of H_2S is finally obtained.

Similar to the H_2S estimation introduced above, the estimation of temperature on the entire sewer network can be obtained in the same manner with the observed temperature.

4.3 Corrosion Prediction

This section introduces how mean and covariance functions of the GP-based prediction model are specified to predict sewer corrosion rate on the entire sewer network.

Mean Function. The mean function, i.e., μ_O and μ_U in Eq. (2), is set as the physical model derived from [20]:

$$R_m = A \cdot H^{0.5} \cdot \frac{0.1602\eta - 0.1355}{1 - 0.977\eta} \cdot e^{\frac{-45,000}{RT}} \qquad (7)$$

where A is a constant calibrated empirically using the training data (coupons on the testbed), H denotes the H_2S concentration, η denotes the fractional relative humidity of the sewer atmosphere, which is set as the average humidity of several coupon sites in the sewer network, R denotes the universal gas constant, T denotes the absolute temperature; the result R_m denotes the corrosion rate.

As aforementioned, Eq. (7) was developed by the domain experts of sewer corrosion research [20] based on the coupons. Therefore, this mean function represents a hypothesis space of the corrosion rate based on domain knowledge such that the prediction of the proposed prediction model will not deviate domain experts' hypothesis too far. In other words, the mean function is used as prior knowledge to regularize the hypothesis space of the proposed prediction model.

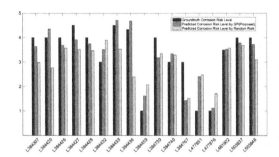

Fig. 3. Evaluation of sewer corrosion prediction. (Color figure online)

Covariance Function. Considering the fact that the sewer corrosion rate is closely related to the sewer network geometry, H_2S concentration and temperature [1,9,12,15], the kernel function of the GP-based prediction model for sewer corrosion is set as a linear combination of three kernels corresponding to these three factors, respectively:

$$\boldsymbol{K} = \alpha_1 \boldsymbol{K}^G + \alpha_2 \boldsymbol{K}^H + \alpha_3 \boldsymbol{K}^T \tag{8}$$

where $\boldsymbol{K}^G$ is the exponential kernel with pairwise geodesic distance defined in Eq. (6), $\boldsymbol{K}^H$ is an exponential kernel with pairwise difference of H_2S concentration between two points as the distance, $\boldsymbol{K}^T$ is another exponential kernel with pairwise difference of temperature between two points as the distance, and α_is are the linear combination coefficients for the three kernel matrices.

With the mean and kernel functions defined above, the corrosion rate prediction can be performed by applying Eqs. (4) and (5) on the entire sewer network at each time point, by assigning $\boldsymbol{\mu}_O$ with the observed corrosion rate calculated based on the two batch of traverse reports.

5   Case Study

In this section a case study is conducted to evaluate the proposed corrosion prediction model using the data provided by the water utility introduced in Sect. 2.

5.1   Evaluation

In order to predict the sewer corrosion, we first estimate H_2S and temperature distributions on the entire network over five years (2011–2015) using the method introduced in Sect. 4. The training data used in this estimation procedure are the H_2S and temperature records as described in Sect. 2. The hyper-parameters, including σ in Eq. (6), A in Eq. (7) and α_i in Eq. (8), are all tuned automatically by maximizing the log marginal likelihood [13].

The estimation of monthly average H_2S and temperature along with the sewer geometry data are then used as the input to predict corrosion rate as in Sect. 4.3. The ground-truth corrosion rates are derived from the two batches of traverse reports. Specifically, there are 17 sewer pipes having corrosion risk level records in both periods of traverse reports. This enables calculating the ground-truth average corrosion rates, denoted as R_g, of these 17 pipes using the following equation: $R_g = \frac{C(t_2)-C(t_1)}{t_2-t_1}$, where $C(t)$ denotes the corrosion risk level record at time t in the traverse reports.

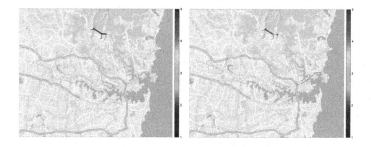

Fig. 4. Example sewer corrosion risk maps in Jan. 2011 (left) and Dec. 2015 (right). The corrosion risk levels on the maps are the prediction results of the proposed corrosion prediction model.

The case study adopts the commonly used leave-one-out (LOO) evaluation method. Specifically, the 17 sewer pipes with known corrosion rates are used in the evaluation. At each time, one of these 17 pipes is reserved for evaluation and the remaining 16 sewer pipes are used for training the GP-based prediction model. The trained model is then used to predict the corrosion rate of the reserved pipe. The evaluation is conducted in turn on each of the 17 sewer pipes and the final prediction accuracy is averaged over the 17 prediction results.

As the input of the GP-based corrosion prediction model is corrosion rate, the direct output of the model is also the corrosion rate, denoted as R_p. In order to obtain the corrosion risk level at a certain time, the following equation can be used: $\tilde{C}(t) = C(t_0) + R_p \cdot (t - t_0)$, where $C(t_0)$ denotes the known corrosion risk level at time t_0 while $\tilde{C}(t)$ denotes the predicted corrosion risk level at time t. Figure 3 plots the predicted corrosion risk levels (in green) for the 17 sewer pipes in comparison with the ground-truths (in blue). One can see that the majority of predictions have less than 10% difference comparing to the corresponding ground-truths. In average, the prediction error is less than 10% ($0.49/5 = 9.8\%$). The absorbing state random walks (ASRW) is also applied for a comparison with the proposed method. As seen in Fig. 3, ASRW (in yellow) has larger prediction errors than the proposed method for most pipes. Statistically, ASRW has an prediction error of 17.6% ($0.88/5 = 17.6\%$).

The proposed GP-based prediction model is able to perform corrosion prediction at any time on the entire sewer network as long as there are some observed H_2S and temperature records. The prediction results from Jan. 2011 to Dec. 2015 are illustrated in Fig. 4 for example. As seen, the corrosion risk levels are gradually increased from Jan. 2011 to Dec. 2015. It can also be found that the pipes in the same area often share similar corrosion risk levels while the pipes in different areas could vary much. This is probably because the corrosion rate of the pipes in the same area are similar since the two influential factors, H_2S and temperature, are likely to be similar; while these two factors could be significantly different in different areas. For example, the organic or chemical components in the wastewater released in industrial areas could accelerate the generation of H_2S and in turn leads to higher corrosion rates in comparison to residential areas.

5.2   Discussion

The above evaluation results have well demonstrated the high prediction accuracy of the proposed corrosion prediction model. Nevertheless, the model still has space to be improved in the following aspects: (1) the temporal patterns of H_2S or temperature; (2) Installing more H_2S and temperature sensors and collecting more H_2S and temperature records; and (3) collecting several other factors, e.g., humidity or pH.

Besides improving the model, more applications can be built on the corrosion prediction results. For example, smart chemical dosing is an on-going project tacking advantages of the H_2S and corrosion prediction results. A set of chemical dosing unites are installed to dose certain chemicals, e.g., Ferrous Chloride, to reduce H_2S concentration and sewer pipe corrosion. However, how much chemical should be dosed at each site to maximally reduce H2S is a challenging issue. The H_2S and corrosion prediction play an important role in optimizing the dosing strategy.

6   Conclusion

This paper proposed a corrosion prediction model based on a Bayesian nonparametric method, named Gaussian process, on the entire sewer network with confidence. The proposed corrosion prediction model was evaluated on a real data set of a water utility in Australia. The evaluation results have demonstrated the high prediction accuracy of the proposed model, with average corrosion risk level prediction error 9.8%, which has been well received by the domain expert in the water utility.

References

1. Boon, A., Lister, A.: Formation of sulphide in rising main sewers and its prevention by injection of oxygen. Prog. Wat. Tech. **7**(2), 289–300 (1975)
2. De Muynck, W., De Belie, N., Verstraete, W.: Effectiveness of admixtures, surface treatments and antimicrobial compounds against biogenic sulfuric acid corrosion of concrete. Cement Concr. Compos. **31**(3), 163–170 (2009)
3. Ganigue, R., Gutierrez, O., Rootsey, R., Yuan, Z.: Chemical dosing for sulfide control in australia: an industry survey. Water Res. **45**(19), 6564–6574 (2011)
4. Hernandez, M., Marchand, E.A., Roberts, D., Peccia, J.: In situ assessment of active thiobacillus species in corroding concrete sewers using fluorescent RNA probes. Int. Biodeterior. Biodeg. **49**(4), 271–276 (2002)
5. Ismail, N., Nonaka, T., Noda, S., Mori, T.: Effect of carbonation on microbial corrosion of concretes. Doboku Gakkai Ronbunshu **1993**(474), 133–138 (1993)
6. Jiang, G., Keating, A., Corrie, S., O'halloran, K., Nguyen, L., Yuan, Z.: Dosing free nitrous acid for sulfide control in sewers: results of field trials in Australia. Water Res. **47**(13), 4331–4339 (2013)
7. Jiang, G., Sun, J., Sharma, K.R., Yuan, Z.: Corrosion and odor management in sewer systems. Curr. Opin. Biotechnol. **33**, 192–197 (2015)
8. Jiang, G., Wightman, E., Donose, B.C., Yuan, Z., Bond, P.L., Keller, J.: The role of iron in sulfide induced corrosion of sewer concrete. Water Res. **49**, 166–174 (2014)
9. Joseph, A.P., Keller, J., Bustamante, H., Bond, P.L.: Surface neutralization and h 2 s oxidation at early stages of sewer corrosion: influence of temperature, relative humidity and h 2 s concentration. Water Res. **46**(13), 4235–4245 (2012)
10. Koch, G.H., Brongers, M.P., Thompson, N.G., Virmani, Y.P., Payer, J.H.: Corrosion cost and preventive strategies in the united states. Technical report (2002)
11. Okabe, S., Odagiri, M., Ito, T., Satoh, H.: Succession of sulfur-oxidizing bacteria in the microbial community on corroding concrete in sewer systems. Appl. Environ. Microbiol. **73**(3), 971–980 (2007)
12. Pomeroy, R., Bowlus, F.D.: Progress report on sulfide control research. Sew. Works J. **18**(4), 597–640 (1946)
13. Rasmussen, C.E.: Gaussian processes in machine learning. In: Bousquet, O., von Luxburg, U., Rätsch, G. (eds.) ML -2003. LNCS (LNAI), vol. 3176, pp. 63–71. Springer, Heidelberg (2004). https://doi.org/10.1007/978-3-540-28650-9_4
14. Redner, J.A., Hsi, R.P., Esfandi, E.J., Sydney, R., Jones, R., Won, D., Andraska, J.: Evaluation of protective coatings for concrete. County Sanitation Districes of Los Angeles County, Califiornia (1998)

15. Sharma, K., de Haas, D.W., Corrie, S., O'halloran, K., Keller, J., Yuan, Z.: Predicting hydrogen sulfide formation in sewers: a new model. Water **35**(2), 132–137 (2008)
16. Shook, W.E., Bell, L.W.: Corrosion control in concrete pipe and manholes. Technical Presentation, Water Environmental Federation, Florida (1998)
17. Snell, P., Doyle, P.: Random walks and electric networks. Free Software Foundation (2000)
18. Van Nguyen, L., Kodagoda, S., Ranasinghe, R., Dissanayake, G., Bustamante, H., Vitanage, D., Nguyen, T.: Spatial prediction of hydrogen sulfide in sewers with a modified Gaussian process combined mutual information. In: 2014 13th International Conference on Control Automation Robotics & Vision (ICARCV), pp. 1130–1135. IEEE (2014)
19. Vipulanandan, C., Liu, J.: Performance of polyurethane-coated concrete in sewer environment. Cem. Concr. Res. **35**(9), 1754–1763 (2005)
20. Wells, T., Melchers, R.: Concrete sewer pipe corrosion findings from an australian field study (2016)
21. Yongsiri, C., Vollertsen, J., Hvitved-Jacobsen, T.: Effect of temperature on air-water transfer of hydrogen sulfide. J. Environ. Eng. **130**(1), 104–109 (2004)
22. Zhang, L., De Schryver, P., De Gusseme, B., De Muynck, W., Boon, N., Verstraete, W.: Chemical and biological technologies for hydrogen sulfide emission control in sewer systems: a review. Water Res. **42**(1), 1–12 (2008)

CAPED: Context-Aware Powerlet-Based Energy Disaggregation

Jingyue Gao[1,2], Yasha Wang[1,3(✉)], Xu Chu[1,2], Yuanduo He[1,2],
and Ziqing Mao[1,2]

[1] Key Laboratory of High Confidence Software Technologies,
Ministry of Education, Beijing 100871, China
mzq5176@gmail.com
[2] School of Electronics Engineering and Computer Science, Peking University,
Beijing 100871, China
{gaojingyue1997,chu_xu,ydhe}@pku.edu.cn
[3] National Engineering Research Center of Software Engineering,
Peking University, Beijing 100871, China
wangyasha@pku.edu.cn

Abstract. Energy disaggregation is the task of decomposing a household's total electricity consumption into individual appliances, which becomes increasingly important in energy reservation research nowadays. In this paper, we propose a novel algorithm taking the context of disaggregation task into consideration. First, we design a new method to efficiently extract each appliance's typical consumption patterns, i.e. *powerlets*. When performing the disaggregation task, we model it as an optimization problem and incorporate context information into the cost function. Experiments on two public datasets have demonstrated the superiority of our algorithm over the state-of-the-art work. The mean improvements of disaggregation accuracy are about 13.7% and 4.8%.

1 Introduction

Energy efficiency has aroused more and more public concerns due to the massive use of fossil fuels in recent decades. To better monitor the electricity consumption, many urban households are equipped with smart meters [4]. Although smart meters can present detailed information about the total electricity consumption in real time, they fail to perform appliance-level monitoring of energy use. Hence there is now a significant interest in the research of energy disaggregation.

Energy disaggregation, also known as non-intrusive appliance load monitoring (NALM), tries to break down a household's aggregate electricity consumption into its component appliances [1], taking advantage of the fact that different appliances tend to show different consumption patterns. According to [2], if informed of appliance-level electricity consumption, residents will take steps to correct their consumption behavior, which may improve energy efficiency by 12%. Besides, disaggregated electricity consumption plays an important role in detecting malfunctioning of electrical appliances and forecasting demands [3].

© Springer International Publishing AG, part of Springer Nature 2018
D. Phung et al. (Eds.): PAKDD 2018, LNAI 10937, pp. 236–247, 2018.
https://doi.org/10.1007/978-3-319-93034-3_19

Research on energy disaggregation can be divided into unsupervised and supervised ones, depending on whether a training dataset of electricity consumption from individual appliances is used. Unsupervised work utilizes Factorial Hidden Markov Model (FHMM) to infer the state of each appliance. However, those approaches are highly task-dependent [7]. Besides, it is worthwhile spending some time collecting training data which usually guarantees more accurate disaggregation results [6]. Therefore, our algorithm works in a supervised setting.

Most supervised algorithms [5,6,8] first extract consumption patterns of each appliance and then use those patterns to decode the total electricity consumption. Usually they try to find a combination of devices' consumption patterns that sums up close to the total consumption. However, the context of disaggregation task is ignored by those works. Common context information includes the *hour of day, weekday or weekend, temperature, humidity* and so on. In fact, people's use of electrical appliances is heavily associated with those factors. For example, people who work during weekdays tend to watch TV at night or at weekends rather than at 11 am on a weekday. Similarly, air conditioner may be at one of its working patterns when it is hot while in contrast stove is more likely being used in a chilly weather. Disaggregation without considering contexts sometimes leads to unreasonable results. Therefore, it is promising to introduce context information to benefit energy disaggregation.

In this paper, we propose a context-aware powerlet-based algorithm (CAPED) for energy disaggregation. *Powerlets* are typical electricity consumption patterns of appliances during a short time interval [6]. For each appliance, we extract its powerlets from the training data containing a series of power values over some period. Then we estimate the occurrence probability of each powerlet under different contexts. When disaggregating the total electricity consumption, we try find a combination of powerlets with high occurrence probability under specific context. It is a challenging problem because of following two reasons:

- Frist, it is difficult to maximize context-aware occurrence probabilities of selected powerlets and meanwhile minimize the discrepancy between their sum and the real electricity consumption when doing disaggregation.
- Second, the state-of-the-art method learning powerlets is too time-consuming to be applied in practical training, which needs to be further improved.

To solve these challenges, we first classify candidate consumption patterns and apply a subset selection algorithm for each category to select representative ones as powerlets, which accelerates the training process. For disaggregation with learned powerlets, we model it as an optimization problem and take both the aforementioned discrepancy and context-aware occurrence probabilities of selected powerlets into consideration. Experimental results on two different datasets show the superiority of our proposed approach. On two datasets, the mean improvements of disaggregation accuracy over a previous state-of-the-art algorithm are 13.7% and 4.8% respectively.

Finally, the main contributions of this paper can be summarized as follows:

- We propose a context-aware powerlet-based algorithm, CAPED, which introduces context information into supervised energy disaggregation for the first
  time.
- We design a new powerlets learning method, which allows us to select representative consumption patterns of each appliance more efficiently.
- We conduct experiments on real-world datasets for evaluation and demonstrate that CAPED significantly outperforms baseline methods.

2   Related Work

The initial solution for energy disaggregation was first proposed by Hart [1]. He
assumes different electrical appliances generate distinct consumption signatures
and on-off events are sufficient to characterize the use of appliances. [1] models each appliance as a finite state machine and looks for sharp edges in power
signals to classify state transitions. However, appliances with similar power levels are almost indistinguishable. Subsequently, researchers incorporate transient
and harmonic information with high-frequency sampling to better distinguish
appliances [9]. But the requirement of installing expensive sensors makes such
methods costly and impractical.

Another category of energy disaggregation algorithms directly decomposes
electricity consumption into its appliances rather than detects on-off events.
Those approaches can be further divided into unsupervised and supervised
ones. Unsupervised work are mainly based on Factorial Hidden Markov Model
(FHMM) which estimates each appliance's hidden state using EM algorithm
[10]. [12] conducts energy disaggregation in an iterative way and [11] introduces
approximate inference into traditional FHMM. However, unsupervised work is
thought to be ill-defined and highly task-dependent [7]. Moreover, usage of EM
makes those algorithms dependent on initialization and get stuck in local optima
easily. Supervised work starts in [5]. It leverages nonnegative sparse coding to
learn each appliance's consumption patterns from training data and performs
disaggregation with those patterns. Similarly, [8] uses nonnegative tensor factorization to extract devices' patterns. However, patterns learned by [5,8] are weeklong, which need a large training dataset and fail to perform real-time energy
disaggregation. To solve this problem, Elhamifar proposes a powerlet-based approach where the duration of powerlets are usually tens of seconds [6], which is the
state-of-the-art work to the best of our knowledge. However, existing supervised
work neglects the significance of context in real-time energy disaggregation task
and this paper will focus on context-aware disaggregation.

3   Preliminary

In this section, we formulize the problem of energy disaggregation and describe
a commonly used form of required training data in earlier studies.

3.1   Problem Definition

There are K electrical appliances in one household. $x_i(t)$ is the power value of appliance i at time $t \in \{1, 2, \ldots, L\}$. $\hat{x}(t)$ denotes the aggregate power value recorded by a smart meter at time t. Thus, we can get

$$\hat{x}(t) = \sum_{i=1}^{K} x_i(t) \tag{1}$$

Energy Disaggregation: Given only total power consumption $\{\hat{x}(t)\}_{t=1}^{L}$, it aims to estimate $\{x_i(t)\}_{t=1}^{L}$ for $i \in \{1, 2, \ldots, K\}$, which decomposes the whole electricity consumption into all K appliances.

3.2   Description of a Typical Training Dataset

In a typical training dataset, we have already known $x_i(t)$ for $i \in \{1, 2, \ldots, K\}, t \in \{1, 2, \ldots, M\}$, which is consistent with [5,6]. M is the duration of the dataset. Each power value $x_i(t)$ has a time stamp t. Thus we can extract some self-contained types of contexts, including hour of day and weekday. Temperature, humidity and occupancy condition can also serve as context information if specially collected in the training dataset.

4   Proposed Approach: CAPED

The framework of CAPED consists of two parts: powerlets learning and context-aware signal decoding. We will describe each of the two parts in details.

4.1   Learning Powerlets

In the training dataset, from $\{x_i(t)\}_{t=1}^{M}, i = 1, 2, \ldots, K$, we learn a powerlets dictionary composed of each appliance's typical consumption patterns. In Fig. 1,

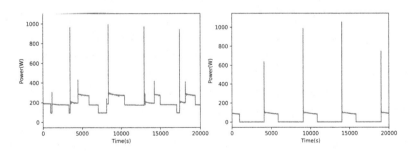

Fig. 1. Left: power signals of a freezor. Right: power signals of a fridge. Horizontal and vertical axes correspond to time and power consumption respectively.

one appliance has several distinct consumption patterns corresponding to different operation modes. We set the length of consumption window as ω (typically $\omega \ll M$) and thus get the fixed-length energy snippet of appliance i at time t, defined as:

$$y_i(t) = [x_i(t), x_i(t+1), \ldots, x_i(t+\omega-1)]^T \in \mathbf{R}^\omega \tag{2}$$

After traversing energy signals $\{x_i(t)\}_{t=1}^M$ with a sliding window, a collection of energy snippets $\{y_i(t)\}_{t=1}^{M-\omega+1}$ is generated. However, it is impractical to add all $M-\omega+1$ snippets into our powerlets dictionary. We need to select a representative subset of snippets, whose cardinality is denoted as N_i, standing for various operation modes of appliance i.

The state-of-the-art work [6] simply employs a dissimilarity-based sparse subset selection algorithm [13] to find typical snippets. To select representatives from φ items, [13] have to compute a $\varphi \times \varphi$ dissimilarity matrix and solve a convex optimization problem containing φ^2 parameters. To process a weeklong power series in the training dataset where $\varphi \gg 100000$, it takes hundreds of hours on a server with an Intel Xeon E5-2683 Octa-Core CPU and 64 GB memory, which is intractable for family deployment. To accelerate the above process, our approach will first partition energy snippets into several categories and then select representatives for each category independently.

Symbolizing Energy Signals. For appliance i, $x_i(t)$ records its power consumption at time t. According to [10,12], its power consumption is determined by its state and follows the Gaussian distribution when appliance i is at the same state. Thus $\{x_i(t)\}_{t=1}^M$ are subject to Gaussian Mixture Model (GMM), where each Gaussian distribution corresponds to one state of appliance i. A typical GMM model is formulated as:

$$p(x) = \sum_{h=1}^H \pi_h p(x|h) \tag{3}$$

H is the number of Gaussian distributions and π_h is the weight of h-th distribution in GMM model. The assumed GMM model is consistent with our obsearvation in Fig. 2. The parameter H is determined according to Davies-Bouldin index [15]. Given $\{x_i(t)\}_{t=1}^M$, we estimate each Guassian component in GMM model with EM algorithm [13]. We give each power value a symbol according to its associated Guassian component and thus symbolize the whole power series.

Selecting Powerlets for Classified Snippets. Since we have symbolized the whole power series, energy snippets $\{y_i(t)\}_{t=1}^{M-\omega+1}$ have their symbol representation respectively. Snippets with same symbol representation are assigned to the same category, indicating appliance i shows similar consumption behaviour when generating those snippets. For example, one category of snippets is generated when appliance i is at "on" state. Another category is generated when the appliance is transforming from "on" to "standby". After classifying snippets

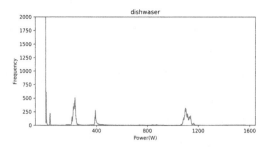

Fig. 2. The frequency of power values' occurrence in power series of a fridge. Horizontal and vertical axes correspond to value of power consumption and its frequency respectively.

by symbol representation, we employ [13] to select representative snippets for each category and construct the powerlets dictionary $\mathbf{B}_i^{\omega \times N_i}$ with all selected ones. N_i is the number of powerlets for appliance i. Combining powerlets of all appliances, we form the dictionary $\mathbf{B}$ as:

$$\mathbf{B} = [\mathbf{B}_1, \mathbf{B}_2, \ldots, \mathbf{B}_K] \in \mathbf{R}^{\omega \times N} \tag{4}$$

Analysis of Efficiency. As mentioned above, to select powerlets from φ energy snippets, tradional method has to compute an $\varphi \times \varphi$ dissimilarity matrix and solve an optimization problem with φ^2 parameters [6]. In our approach, we apply this procedure to each category respectively. If φ snippets have been equally divided into K categories, we need to compute K matrices whose size is $\frac{\varphi}{K} \times \frac{\varphi}{K}$ and solve K optimization problems with totally $\frac{\varphi^2}{K}$ parameters. Moreover, computation for K categories can be parallelized with multiple processors. Thus the efficiency of learning powerlets is greatly improved, making our method feasible in practical training.

4.2   Estimation of Context-Aware Occurrence Probability

The occurrence probabilities of powerlets under different contexts tend to be different. For example, the powerlets relating to "on" state of an oven are probable to occur near lunch hours but the probability is much lower in sleep hours. Thus we estimate the occurrence probability of powerlets under different contexts.

Taking *hour of day* as an example of context information, number of times n_{i*} that powerlets in $\mathbf{B}_i$ occur in each hour can be modeled as a multinomial distribution with N_i categories. The parameters of the underlying multinomial distribution, i.e. the occurrence probability can be estimated by the maximum likelihood estimator (MLE), which is the normalized frequency of powerlets in $\mathbf{B}_i$ occur in each hour.

$$\mathbf{p}_i(j) = \mathbf{normalize}([n_{i,1}, n_{i,2}, \ldots, n_{i,N_i}])^\mathrm{T} \tag{5}$$

With estimated probability distribution vector $\mathbf{p}_i(j) \in \mathbf{R}^{N_i}$ for each hour j, we build occurrence probability matrix $\mathbf{P}_i^{hour}$ for appliance i and subseqently all appliances in one household:

$$\mathbf{P}_i^{hour} = [\mathbf{p}_i(1), \mathbf{p}_i(2), \ldots, \mathbf{p}_i(24)]^T \in \mathbf{R}^{24 \times N_i}$$
$$\mathbf{P}^{hour} = [\mathbf{P}_1^{hour}, \mathbf{P}_2^{hour}, \ldots, \mathbf{P}_K^{hour}] \in \mathbf{R}^{24 \times N} \tag{6}$$

Similarly, we can estimate such matrices for other types of context information, like $\mathbf{P}^{weekday}$ for weekday or weekend, $\mathbf{P}^{temperature}$ for different temperature levels, $\mathbf{P}^{humidity}$ for different humidity levels and so on.

4.3 Context-Aware Signal Decoding

Having learned the powerlets dictionary $\mathbf{B}$ and related context-aware probability matrices such as $\mathbf{P}_{hour}$, we consider the problem of energy disaggregation. Given the aggregate signal $\{\hat{x}(t)\}_{t=1}^L$, we define household's total energy consumption snippet in the interval $[t, t + \omega - 1]$ as:

$$\hat{\mathbf{y}}(t) = [\hat{x}(t), \hat{x}(t+1), \ldots, \hat{x}(t + \omega - 1)] \in \mathbf{R}^\omega \tag{7}$$

To decompose electricity consumption into individual appliances, we search for a representation of $\hat{\mathbf{y}}(t)$ in the learned dictionary $\mathbf{B}$ which contains powerlets of all appliances. The representation can be formulated as:

$$\hat{\mathbf{y}}(t) \approx \mathbf{B}\mathbf{c}(t) \tag{8}$$

Here $\mathbf{c}(t) = [\mathbf{c}_1, \mathbf{c}_2, \ldots, \mathbf{c}_K]^T$ and $\mathbf{c}_i \in \mathbf{R}^{N_i}$ is the coeffecent vector of appliance i's powerlets dictionary $\mathbf{B}_i$. Each powerlet of in $\mathbf{B}_i$ corresponds to one typical operation mode of appliance i. In time t, appliance i must be in no more than one working mode. Thus, we put the following constraint on $\mathbf{c}_i(t)$

$$\mathbf{1}^T \mathbf{c}_i(t) \leq 1, \mathbf{c}_i(t) \in \{0,1\}^{N_i} \tag{9}$$

Let L denote the loss function which measures the difference between the $\hat{\mathbf{y}}(t)$'s representation and its true value. We need to find an optimal representation $\mathbf{B}\mathbf{c}(t)$ that approaches the aggregate signal $\hat{\mathbf{y}}(t)$. A simple method is to solve the optimization problem subject to constraint (9)

$$\min L(\hat{\mathbf{y}}(t) - \mathbf{B}\mathbf{c}(t)) \tag{10}$$

Incorporating Context Information. It is often the case that different combinations of powerlets from different appliances lead to similar aggregate signals. Under such circumstance, merely minimizing the representation error as (10) may result in unreasonable results. For example, (10) may decompose electricity consumption to stove just because of less representation error, even when it is at a hot summer noon. Therefore, CAPED leverages the context-aware occurrence probabilities of different powerlets to address the problem. We will use *hour of day* as an example of context information to explain our algorithm.

We have learned $\mathbf{P}_i^{hour} \in \mathbf{R}^{24 \times N_i}$ in the training process. j-th row in $\mathbf{P}_i^{hour}$ is the occurrence probability of appliance i's powerlets in j-th hour of day. Let $\mathbf{f}_{hour}$ be a function which maps time t to its one-hot representation. $\mathbf{f}_{hour}(t) = \mathbf{e}_j^T$ if time t is in j-th hour of day, where $\mathbf{e}_j \in \mathbf{R}^{24}$ denotes a vector whose j-th entry is one and the rest is zero. Considering context information(i.e. hour of day), the occurrence probability of appliance i's selected powerlet at time t is $\mathbf{f}_{hour}(t)\mathbf{P}_i^{hour}\mathbf{c}_i(t)$. For all K appliances, the probability of selected powerlets at time t can be formulated as

$$\Psi_{hour}(t) = \mathbf{f}_{hour}(t)\mathbf{P}^{hour}\mathbf{c}(t) \tag{11}$$

Similarly, we construct items of other available context information, including $\Psi_{weekday}, \Psi_{temperature}, \Psi_{humidity}$ and so on. In order to maximize the probability of selected powerlets as well as minimize the representation error, the optimization problem under constraint (9) can be extended to

$$\min L(\hat{\mathbf{y}}(t) - \mathbf{B}\mathbf{c}(t)) - \sum_{\tau \in S} \lambda_\tau \Psi_\tau(t) \tag{12}$$

where S is the set of available context information such as *hour of day*, *weekday* and *temperature*. It is worth noting that such context information can be obtained without much extra effort, which guaratees the feasibility of CAPED.

Moreover, there are some priors we can impose on the structure of $\mathbf{c}(t)$, following the research of [6]. Inter-Appliance priors fully use co-occurrence relationships between appliances. For example, kitchen appliances tend to work together while air-conditioner hardly works together with stoves. These two types of inter-appliance prior are denoted as ρ_a and ρ_b.

$$\rho_a(\mathbf{c}(t)) = \sum_{(i,j) \in \mathbb{A}} (\mathbf{c}_i(t)^T \mathbf{c}_i(t) - \mathbf{c}_j(t)^T \mathbf{c}_j(t))^2$$
$$\rho_b(\mathbf{c}(t)) = \sum_{(i,j) \in \mathbb{B}} (\mathbf{c}_i(t)^T \mathbf{c}_i(t))(\mathbf{c}_j(t)^T \mathbf{c}_j(t)) \tag{13}$$

where set $\mathbb{A}$ indicates the set of all pairs of appliances that work simultaneously and set $\mathbb{B}$ contains all pairs of appliances that usually do not work together. Our context-aware energy disaggregation can be formulated as:

$$\min L(\hat{\mathbf{y}}(t) - \mathbf{B}\mathbf{c}(t)) - \sum_{\tau \in S} \lambda_\tau \Psi_\tau(t) + \lambda_\rho(\rho_a(\mathbf{c}(t)) + \rho_b(\mathbf{c}(t)))$$
$$\text{s.t. } \mathbf{1}^T \mathbf{c}_i(t) \le 1, \ \mathbf{c}_i(t) \in \{0,1\}^{N_i} \tag{14}$$

where λ_τ and λ_ρ are parameters controlling the weight of each part in cost function. Once obtaining optimal $\mathbf{c}(t)$, we estimate the electricity consumption of appliance i in the interval $[t, t + \omega - 1]$ by $\tilde{\mathbf{y}}_i(t) = \mathbf{B}_i \mathbf{c}_i(t)$.

5   Experiments

In this section, we compare CAPED with baseline methods on two real world datsets for evaluation.

5.1   Experimental Setup

Datasets

- **ECO.** It consists of electricity consumption signals from 6 houses. For each house, the whole consumption data as well as appliance-level consumptions are recorded at the frequency of 1 Hz. Each record is associated with a UTC time stamp, from which we can extract context information like hour of day and weekday [16].
- **REDD.** It is another widely used energy disaggregation dataset, which is sampled at 15 kHz and contains consumption data from 6 houses. Each record is also associated with a time stamp [17].

Baseline Methods

- **Simple Mean.** It estimates the total consumption percentage of each appliance and predicts that the whole electricity signal breaks down according to this percentage at all times.
- **PED.** It is the state-of-the-art supervised algorithm for energy disaggregation to the best of our knowledge [6]. It first extracts powerlets for each appliance and searches for an optimal combination without considering context information.

Performance Metric. Similar to [6,16], We measure the disaggregation accuracy as:

$$Acc = 1 - \frac{\sum_{t \in \mathbb{W}} \sum_{i=1}^{M} \|\tilde{\mathbf{y}}_i(t) - \mathbf{y}_i(t)\|}{2 \sum_{t \in \mathbb{W}} \|\hat{\mathbf{y}}(t)\|_1} \tag{15}$$

where $\mathbb{W} \triangleq \{1, \omega + 1, 2\omega + 1, \ldots\}$. (15) has considered the problem of "double counting errors" by multiplying 2 in the denominator. $\tilde{\mathbf{y}}_i(t)$ and $\mathbf{y}_i(t)$ denote the estimated and actual energy consumption of appliance i at $[t, t + \omega - 1]$.

Parameter Settings. We set the window size of powerlets $\omega = 10$. Earlier research [6] demonstrates that $\omega \in [10, 50]$ is a suitable setting if the sampling rate is 1 Hz. As for the cost function in (14), we set loss function $L(\cdot) = \|\cdot\|_1$ to be robust to errors. After data filtering and completion on ECO and REDD, we use one week of recorded electricity consumption for training and the rest of available data for testing. Due to restrictions of datasets, we only incorporate two common contexts—*hour of day* and *weekday* when inplementing CAPED. Thus, $\lambda_{hour}, \lambda_{weekday}$ and λ_ρ are determined through a coarse-grained grid search. Parameters in baseline methods are determined similarly.

5.2   Performance Discussion

The disaggregation performance of different algorithms on ECO and REDD is shown in Table 1. The results of House 3 in ECO and House 5 in REDD are excluded due to low level of data quality [16,17]. Our algorithm significantly outperforms the naive Simple Mean and PED, achieving disaggregation accuracies of 75.1% on ECO and 75.6% on REDD. We attribute this improvement to the introduction of context information. People tend to show different energy consumption behaviors under different contexts, which is fully taken advantage of by context-aware energy disaggregation. Notice that we only use two common types of contexts for convenience of access. The framework of context-aware energy disaggregation has the potential of performance boosting if more context information like temperature and occupancy condition is available.

Table 1. Energy disaggregation accuracies on ECO and REDD dataset.

ECO	House 1	House 2	House 4	House 5	House 6	Total
Simple Mean	38.05%	37.02%	32.70%	63.16%	56.92%	45.57%
PED	61.62%	62.85%	59.64%	70.52%	52.26%	61.37%
CAPED	**67.39%**	**73.99%**	**77.64%**	**83.31%**	**73.14%**	**75.09%**
REDD	House 1	House 2	House 3	House 4	House 6	Total
Simple Mean	40.55%	51.08%	46.10%	54.36%	38.33%	46.08%
PED	76.27%	78.64%	61.41%	**59.46%**	78.14%	70.78%
CAPED	**79.92%**	**85.02%**	**68.97%**	57.76%	**86.31%**	**75.59%**

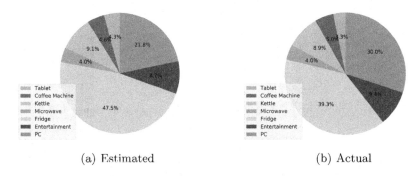

(a) Estimated   (b) Actual

Fig. 3. The electricity consumption proportion of each appliance in House 5 of ECO. The estimated result is shown in left and the actual one is in right.

According to the results of energy disaggregation, we further estimate the proportion of electricity consumption for each appliance. Fig. 3 shows the Pie charts corresponding to the actual and the estimated electricity consumption by CAPED for House 5 of ECO dataset over one week. The consumption estimated by CAPED is close to the actual one, achieving **90.56%** accuracy by Eq. (15). It shows that our algorithm can accurately give poeple an appliance-level feedback of electricity consumption.

5.3 Parameter Study

As we introduce two types of context information, there are two corresponding parameters λ_{hour} and $\lambda_{weekday}$. We first find a relatively reasonable parameter setting by grid search for all the parameters. To examine the impact of λ_{hour} and $\lambda_{weekday}$ on disaggregation accuracy, we fix the values of all other parameters and study how the accuracy changes with the value of the remaining one parameter. Fig. 4 shows the accuracy curves of varying parameters in House 4 of ECO. CAPED performs well for parameters $\lambda_{hour}, \lambda_{weekday} \in [10, 100]$, which demonstrates the robustness of our approach to parameter tuning. Notice that the disaggregation accuracy decreases for larger λ_{hour} and $\lambda_{weekday}$ because too much emphasis is put on context information while the representation error is neglected. Comparing two curves, we find λ_{hour} plays a more important role in disaggregation than $\lambda_{weekday}$. It is attributed to the fact that the regularity of people's consumption behaviors is more obvious daily than weekly. Moreover, the time span of testing data is too short for weekly regularity to be fully utilized.

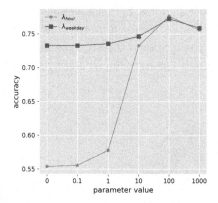

Fig. 4. Accuracy curves of varying λ_{hour} and $\lambda_{weekday}$ in House 4 of ECO.

6 Conclusion

In this paper, we propose a novel context-aware energy disaggregation algorithm, which incorporates context information of the disaggregation task into our framework. We also develop a method to efficiently learn typical consumption patterns of appliances. Experimental results on two datasets show the effectiveness of our approach. In the future, we will collect more types of context information to further improve the disaggregation accuracy.

Acknowledgments. This work was mainly funded by NSFC Grant (No. 61772045), Research Fund from China Electric Power Research Institute (No. JS71-16-005).

References

1. Hart, G.W.: Nonintrusive appliance load monitoring. Proc. IEEE **80**(12), 1870–1891 (1992)
2. Darby, S.: The effectiveness of feedback on energy consumption. A Review for DEFRA of the Literature on Metering, Billing and direct Displays **486**, 2006 (2006)
3. Froehlich, J., et al.: Disaggregated end-use energy sensing for the smart grid. IEEE Pervasive Comput. **10**(1), 28–39 (2011)
4. Navigant Research, Market Data: Smart Meters (2016)
5. Kolter, J.Z., Batra, S., Andrew, Y.Ng.: Energy disaggregation via discriminative sparse coding. In: NIPS (2010)
6. Elhamifar, E. Sastry, S.: Energy disaggregation via learning powerlets and sparse coding. In: AAAI (2015)
7. Wytock, M., Kolter, J.Z.: Contextually supervised source separation with application to energy disaggregation. In: AAAI (2014)
8. Figueiredo, M., Ribeiro, B., de Almeida, A.: Electrical signal source separation via nonnegative tensor factorization using on site measurements in a smart home. IEEE Trans. Instrum. Meas. **63**(2), 364–373 (2014)
9. Gupta, S., Reynolds, M.S., Patel, S.N.: ElectriSense: single-point sensing using EMI for electrical event detection and classification in the home. In: UbiComp (2010)
10. Kim, H., et al.: Unsupervised disaggregation of low frequency power measurements. In: Proceedings of SDM, SIAM (2011)
11. Kolter, J.Z., Jaakkola, T.: Approximate inference in additive factorial hmms with application to energy disaggregation. In: AISTATS (2012)
12. Parson, O., et al.: Non-intrusive load monitoring using prior models of general appliance types. In: AAAI (2012)
13. Elhamifar, E., Sapiro, G., Sastry, S.S.: Dissimilarity-based sparse subset selection. IEEE Trans. Pattern Anal. Mach. Intell. **38**(11), 2182–2197 (2016)
14. Dempster, A.P., Laird, N.M., Rubin, D.B.: Maximum likelihood from incomplete data via the EM algorithm. J. Royal Stat. Soc. Series B (Methodological), 1–38 (1977)
15. Davies, D.L., Bouldin, D.W.: A cluster separation measure. IEEE Trans. Pattern Anal. Mach. Intell. **1**(2), 224–227 (1979)
16. Beckel, C., et al.: The ECO data set and the performance of non-intrusive load monitoring algorithms. In: Proceedings of the 1st ACM Conference on Embedded Systems for Energy-Efficient Buildings. ACM (2014)
17. Kolter, J.Z., Johnson, M.J.: REDD: a public data set for energy disaggregation research. In: Sustkdd (2011)

Rolling Forecasting Forward by Boosting Heterogeneous Kernels

Di Zhang[1(✉)], Yunquan Zhang[1,2], Qiang Niu[3], and Xingbao Qiu[4]

[1] School of Computer Science, Communication University of China,
Beijing 100024, People's Republic of China
di.zhang@cuc.edu.cn
[2] State Key Lab of Computer Architecture, Institute of Computing Technology,
Chinese Academy of Sciences, Beijing 100190, People's Republic of China
zyq@ict.ac.cn
[3] Department of Mathematical Sciences, Xi'an Jiaotong-Liverpool University,
Suzhou 215123, People's Republic of China
qiang.niu@xjtlu.edu.cn
[4] China Mobile Communications Corporation,
Beijing 100032, People's Republic of China
qiuxingbao@sn.chinamobile.com

Abstract. The problem discussed in this paper stems from a project of cellular network traffic prediction, the primary step of network planning striving to serve the continuously soaring network traffic best with limited resource. The traffic prediction emphasizes two aspects: (1) how to exploit the potential value of physical and electronic properties for tens of thousands of wireless stations, which may partly determine the allocation of traffic load in some intricate way; (2) the lack of sufficient and high-quality historical records, for the appropriate training of long-term predictions, further aggravated by frequent reconfigurations in daily operation. To solve this problem, we define a general framework to accommodate several variants of multi-step forecasting, via decomposing the problem into a series of single-step vector-output regression tasks. They can further be augmented by miscellaneous attributive information, in the form of boosted multiple kernels. Experiments on multiple telecom datasets show that the solution outperforms conventional time series methods on accuracy, especially for long horizons. Those attributes

The work is partially supported by National Key R&D Program of China under Grant No. 2016YFB0200803, 2017YFB0202302, 2017YFB0202001, 2017YFB0202502, 2017YFB0202105; the National Natural Science Foundation of China under Grant No.61432018, No. 61521092, No. 61272136, No. 61402441, No. 61502450; the National High Technology Research and Development Program of China under Grant No. 2015AA011505; Key Technology Research and Development Programs of Guangdong Province under Grant No. 2015B010108006; the CAS Interdisciplinary Innovation Team of Efficient Space Weather Forecast Models; NSF of China under no. 11301420; NSF of Jiangsu province under no. BK20150373 and no. BK20171237; Suzhou science and technology program under no. SZS201613 and the XJTLU Key Programme Special Fund (KSF) under no. KSF-A-01.

D. Phung et al. (Eds.): PAKDD 2018, LNAI 10937, pp. 248–260, 2018.
https://doi.org/10.1007/978-3-319-93034-3_20

describing the macroscopic factors, such as the network type, topology, locations, are significantly helpful for longer horizons, whereas the immediate values in the near future are mainly determined by their recent records.

Keywords: Multi-dimensional time series · Multi-horizon prediction Multi-kernel learning · Network traffic prediction

1 Introduction

Nowadays, mobile communication has become a pivot ingredient for everyone's life and delivers the connectivity and infrastructure powering new digital economies and unleashing novel applications. The telecom operators must conduct network planning and optimization with limited budget at least one year ahead, to support the surging data traffic growing 18-fold for every five years in global [1]. The first step of network construction is to predict how traffic will evolve in a long-term view for tens of thousands of cell towers in the broad scope of a region. The underlying power of pushing the traffic blow up or fluctuate arises in two aspects: one is objective, including natural growth in consumption, the movement of population, and seasonal oscillation. The other is subjective, which means the reaction of user behavior towards the network change. Usually, given the higher bandwidth of a network, the users will intend to spend more time on enjoying the mobile apps, and vice versa.

The problem of traffic forecasting can be initially recognized as the geographical time series prediction, which has been extensively studied for decades, as summarized in [2]. These models include linear model, like ARIMA (Autoregressive Integrated Moving Average), VAR (Vector Autoregression), which have the advantage of simplicity and robust, and non-linear model, such as neural network and its deep learning variants [6], which sometimes can provide higher accuracy, but also need more training data and more computing resource.

We have adopted and tested these popular solutions to our problem, and pay attention to its specific difficulties: (1) long-term prediction of noisy data. The interested target in practice is the peak load of every day, not the summation or average, which results in the forecasting be much noisier. As the predicting horizon moves forward, the inherent noise of time series will accumulate with the increasing variation, and the possible bias will be amplified, which leads to the rapid deterioration of accuracy. (2) high dimensionality and scarceness of data. The number of cell towers serving as access points in a metropolis can reach up to more than ten thousand, which make the classic VAR vulnerable by the curse of dimensionality. Meanwhile, the traffic volumes are also determined by the layout and configuration of the network itself, and the model may benefit from these properties at inferring the underlying manifolds where series evolves, though some of them may be redundant and irrelevant. Though we can give an explicit explanation for every attribute as they are defined in a humanmade system, it is impossible to figure out, which part of them and how they influence the complex dynamics of user behavior and traffic characteristics. It seems that we must resort to some variable selection methods to find out a useful subset of attributes.

In the previous work of wireless network prediction, various popular technologies have been tested on this problem [2]. Two jobs worth to be noticed particularly. The work [9] arranges the historical measurements and values to be predicted (as zeros) into a matrix, which will be factorized based on a compressed sensing approach with spatial constraints. The work [5] makes use of the sensors' location based on CNN (Convolutional Neural Networks) in deep learning, via converting the traffic snapshot into images describing the spatial-temporal relation of traffic flows, and thus automatic feature extraction becomes viable. Although the advantage of these spatial-temporal models is proven, none of above works considers how to make use of miscellaneous properties other than locations, and inspect their potential benefit as forecasting horizons vary.

To explore this issue, we clarify several variations of long-term prediction and define a general framework to entangle one-step tasks in a cascading fashion. The properties of each entity are divided into small groups based on the business knowledge, and they are encoded by various kernels; all of these make the function space larger than an original linear regression could reach. The importance of attributes given by model can be used as an important reference for data collectors and system admins. The contributions of this paper include: (1) a customizable solution to transform, select and fuse properties containing context information, and apply them sequentially into a multi-step forecasting; (2) some practical skills are given, including a set of commonly used kernels and how they are combined; (3) effectiveness on multiple telecom sites, compared to commonly-used methods, are validated, and each kernel's contribution to different horizons are analyzed.

The left of paper is organized as follows: in Sects. 2 and 3, we briefly introduce the necessary background knowledge to understand the problem and summarize the works in related domains. Next, the solution is presented in detail, with its formalized model and kernel design. In Sect. 5, a set of experiments are executed and demonstrated with results. Finally, we conclude the whole article.

2   Background: Network Traffic and Device Configuration

The mobile network is a communication network where the last link is wireless. In a range of territory, the base stations are scattered with proper intervals to carry the network packets issued from a specific block of an area, shown in Fig. 1a. It can be further split into several sectors, each of which served by an individual cell transceiver installed on the same station tower, but with an independent antenna pointing towards a unique direction. In Fig. 1b, the high-level network architecture of such network is comprised of three main components: the user equipment, the radio access network composing of a bunch of base stations, and the core network offering routing and management services. Usually, the traffic meters are deployed at the interface between the base stations and the core network, and their readings are reported for every a predefined period. In practice, only the max of aggregated traffic for every clock hour, namely $\max_{t \in \{0,...,23\}} \sum_i \mathbb{I}(\mathrm{hour}(i) = t)\mathrm{len}(p_i)$ where $\mathrm{len}(p_i)$ is the length of the ith packet per day, will be studied for later engineering propose. This transformation makes the data much noisier and even harder to predict. For the cell transceivers,

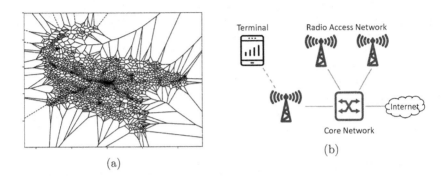

(a)

(b)

Fig. 1. (a) The base station layout of dataset D1 in Table 2. Each polygon roughly represents the land area a base station ought to cover, in the form of Voronoi diagram. (b) The concise architecture of the mobile network. The core network acts as the bridge of base stations and the Internet.

Table 1. The primary attributes in the Engineering Parameter table. Only the parameters relevant to our problem are kept, either based on domain knowledge or empirical tests.

Group	Attr.	Type	Desc.
Location	Longitude	Real	Acquired from GPS (Global Positioning System)
	Latitude	Real	
Topology	Cell id	Nominal	A hierarchical structure, where tens of nodes at a lower level are connected to one node at a higher level
	Station id	Nominal	
	District id	Nominal	
	Site id	Nominal	
Scene	Outdoor	Boolean	Indicate if the antenna is installed outdoor or indoor
Antenna	Azimuth	Real	The direction on horizontal plane
	Downtilt	Real	The direction on vertical plane
	Antenna type	Nominal	Manufacturers and versions
	Power	Real	The transmission power of electromagnetic signal
Extra	Converge	Real	A derived property, suggested by domain experts, approximately indicates the area size of a cell estimated by Voronoi method (Fig. 1(a))

there are many (suppose K) engineering parameters, to describe various configurations, whose names, data types, and business meaning are briefly given in Table 1.

3   Related Work

There are two purposes for traffic prediction: one is for planning&optimization, corresponding to month-level prediction; the other is for day-2-day maintenance or device controlling at minute-level. We focus on the former one. Many popular models have been tested on this problem, including ARIMA and RNN (Recurrent Neural Networks) [2]. If we take the purely multi-dimensional approach, there are already works based on VAR, sometimes considering with geographic information. Even deep learning methods are tested on this problem [6]. Models from other related domains, such as transportation traffic [7] and geostatistics [3], can also be immigrated onto this problem.

As for the long-term time series prediction, the paper [8] summarize the main strategies of forecasting, including direct, recursive, and hybrid. The most counter-intuitive result is that the recursive and direct are not necessarily equivalent especially under the nonlinear situation, derived from both theoretical and empirical results. The main reason is that the repetitive applying of the same non-linear generative function, even it is the ground truth for one-step prediction, will possibly result in asymptotical bias and cannot be eliminated during recursion [8]. Influenced by many factors, such as ground truth, data size, and optimization process, the question of which multi-step strategy best is an empirical one.

A guide for using multiple kernels can be found in [4]. Here we use the boosting methods [10], for its advantage in easy-to-implement and low demand on the computing resource.

4   Solution

4.1   Overall Process

The multi-horizon prediction can be formalized as:

$$\mathbf{Y}_{t+1:t+H} = \{\mathbf{y}_{t+1}, \ldots, \mathbf{y}_{t+H}\} = \mathbf{F}(\mathbf{Y}_{1:t}), \tag{1}$$

where the value of the series $\mathbf{Y}$, having N consecutive and D dimensional observations, at the future H steps, are determined by an unknown stochastic function $\mathbf{F}$, taking the $\mathbf{Y}_{1:t}$ as input. The auxiliary attributes $\{\mathbf{A}_{i,j} : i \in 1..D, j \in 1..K\}$ are not digested by $\mathbf{F}$ directly, but used for designing candidate kernels later in Sect. 4.2. The $\mathbf{F}$ can be designed to be the composition of a set of single horizon models $\mathbf{f}$, according to the four schemes in Fig. 2:

(1) Recursive, shown in Fig. 2a, only an one-step-forward model is trained and applied repeatedly to future steps:

$$\mathbf{y}_{t+h} = \mathbf{f}_1^{(h)}(\mathbf{X_t}) + \varepsilon; \tag{2}$$

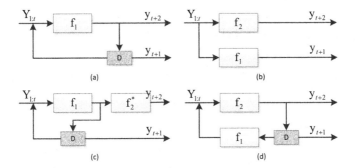

Fig. 2. Four approaches for composing a multiple-horizon model from single ones. Here we give the organization when $H = 2$. The D block means delay one step, i.e., passing through the input at next period.

(2) Direct, shown in Fig. 2b, one model for each step:

$$\mathbf{y}_{t+h} = \mathbf{f}_h(\mathbf{X_t}) + \varepsilon; \tag{3}$$

(3) Adjustable, shown in Fig. 2c, adding a rectification for each step before output based on (1), specially:

$$\mathbf{y}_{t+h} = \mathbf{f}_1^{(h)}(\mathbf{X_t}) + \mathbf{f}_h^*(\mathbf{X_{t+h}}) + \varepsilon, \text{ specially } \mathbf{f}_1^* \equiv \mathbf{0}; \tag{4}$$

(4) Multi-recursive, shown in Fig. 2d, overlay existing models at every step with one new recursive model:

$$\mathbf{y}_{t+h} = \sum_{i=1}^{h} \mathbf{f}_i^{(h)}(\mathbf{X_{t+i}}) + \varepsilon, \tag{5}$$

where $\mathbf{f}^{(h)}$ means applying $\mathbf{f}$ repeated along the time axis for h times, $\mathbf{X_{t+i}}$ denotes the concatenation of true measurements $\mathbf{Y}_{1:t}$ and estimated futures $\hat{\mathbf{Y}}_{t+1:t+i}(\mathbf{X_{t+i}} \triangleq [\mathbf{Y}_{1:t}, \hat{\mathbf{Y}}_{t+1:t+i}])$, and $\varepsilon \backsim \mathcal{N}(0, \sigma^2)$. After the entire model $\mathbf{F}$ has converted to a sequence of single-step tasks, at each step we are seeking a function $\mathbf{f}$ to solve the following optimization problem:

$$\hat{\mathbf{f}} = \min_{\mathbf{f}} L\left[\mathbf{y}_{t+h}, \mathbf{f}(\mathbf{X})\right], \tag{6}$$

where L is the squared error loss on $N - h$ samples, $\mathbf{X}$ is a unified denotation of all variable length of inputs from (1) to (4). The vector-output $\mathbf{f}$ can be solved under a gradient descent approach:

$$\mathbf{f}_m = \mathbf{f}_{m-1} - \rho \mathbf{g}_m, \tag{7}$$

where ρ is the step length, and $\mathbf{g}_m$ is the gradient residual at each data point. We use a weak learner $\mathbf{h}$, here is the multi-output ridge regression equipped with a matrix-valued kernel, to approximate the negative gradient signal, then the problem is transformed to seek a best-effort kernel representing the underlying correlation between every two observations:

$$\hat{\mathbf{K}}_m = \text{argmin}_{\mathbf{K}_m} \sum_{i=1}^{N-h} \left[-\mathbf{g}_{im} - \mathbf{h}(\mathbf{X}^{(i)}; \mathbf{K}_m) \right]^2, \tag{8}$$

$$\mathbf{h}(\mathbf{X}; \mathbf{K}_m) = \sum_{i=1}^{N-h} \alpha_{im} \mathbf{K}_m(\mathbf{X}, \mathbf{X}^{(i)}), \tag{9}$$

$$\alpha_{im} \triangleq (\mathbf{K}_m + \lambda \mathbf{I}_N)^{-1} \mathbf{g}_{im}, \tag{10}$$

where λ in the ridge regression $\mathbf{h}$ is set to avoid overfitting. Mentioned in Sect. 3, matrix-valued kernel $\mathbf{K}$ are usually learned from a linear combination of basic kernels. To make it more applicable, $\mathbf{K}$ is further assumed to be separated into the product of one composite kernel on the input-space, and the other composite kernel representing the correlations among the outputs. Moreover, we add the Dirac Delta kernel, or identity matrix in discrete form, to the pool of candidates for the outputs, which allows the evolvement of devices can be independent if possible. In all, we get the following form of combinations:

$$(\mathbf{K}(\mathbf{X}, \mathbf{X}'))_{i,j} = \left[\sum_{i=1}^{Q_1} \beta_i^{(I)} \kappa_i^{(I)}(\mathbf{X}_{:,i}, \mathbf{X}'_{:,j}) \right] \left[\beta_0^{(O)} \delta_{i,j} + \sum_{i=1}^{Q_2} \beta_i^{(O)} \kappa_i^{(O)}(\mathbf{A}_{:,i}, \mathbf{A}'_{:,j}) \right],$$
$$\tag{11}$$

where $\kappa^{(I)}$ and $\kappa^{(O)}$ are scalar-value kernels defined in the input and output spaces, and δ is the Dirac Delta kernel; Q_1 and Q_2 are the number of their candidates, and β is the coefficients for each kernel. Here we have $\sum_1^{Q_1} \beta_i^{(I)} = 1$ and $\sum_0^{Q_2} \beta_i^{(O)} = 1$. To simplify the computation and make the model at each iteration more sparse, the algorithm selects only one basic kernel, either for inputs or outputs, namely, the $\boldsymbol{\beta}_m \triangleq [\boldsymbol{\beta}_{1:Q_1}^{(I)}, \boldsymbol{\beta}_{1:Q_2}^{(O)}] = \hat{\beta}_m \mathbf{e}_j$ has only one non-zero entry at position p. At the start point of gradient descent, the algorithm firstly seeks the best position p_m at the range of $[1 : Q_1]$, and sets the Dirac Delta kernel as the initial kernel for outputs $[Q_1 + 1 : Q_1 + Q_2 + 1]$, which will be probably enriched by more non-zero values during later iterations.

4.2 Kernel Design

The kernels come from two groups of sources: historical records and device property. The similarity on time series can be further classified into two categories: inter-series and trans-series; the former includes AR, MA, and seasonality, inspired from the seasonal ARIMA model, and the latter is the DTW (Dynamic Time Warping) for capturing the nonlinear correlation, which is not easy to accomplish by VAR in the high dimensional setting. A practical method of converting the distance to a kernel is to encapsulate it in the form of RBF (Radial Basis Function):

$$\kappa(\boldsymbol{x}, \boldsymbol{x}'; d) = e^{\epsilon \cdot d(\boldsymbol{x}, \boldsymbol{x}')}. \tag{12}$$

One common heuristic for choosing ϵ is $1/D$, and the d can replace by the various definitions, such as the 1st–4th rows of Table 2. The usage of DTW is straightforward by substitution in Eq. 12, whereas the distance of two sub-series inside one series is determined by the L2 distance of their next step prediction values, based on AR, MA, or seasonal AR(1).

For devices, there are three kinds of data types: nominal, numerical, and nodes on a linked tree. The numerical and nominal can be simply handled by the RBF and Dirac Delta, while nodes on network topology need to compare their number of common ancestors:

$$\kappa_{Tree}(\mathbf{s}, \mathbf{s}') = \sum_{i=0}^{L} 2^{L-i} \delta_{\mathbf{s}(i), \mathbf{s}'(i)}, \tag{13}$$

where $\mathbf{s}$ is the nominal vector of a node's id along with the top-down path of a L-height topology tree. In all, we have five kinds of attributive kernels, listed in 5th–9th rows of Table 2.

Table 2. The kernels used in our solution. The names in the source column are kept consistent with the group column Table 1

ID	Space	Source	Kernel	Desc.
1	Input	Traffic Record	RBF of AR(2)	Short term correlation
2			RBF of MA(6)	Long term correlation
3			RBF of Seasonal AR(12, 1)	Year-on-year comparison
4	Output		RBF of DTW, window length = int(10% * N)	Nonlinear correlation
5		Location	Matérn	Distance between two points
6		Topology	Lowest common ancestor	#ancestors two nodes share
7		Scene	Dirac	Indoor or outdoor
8		Coverage	RBF	Land area a site serve
9		Antenna	RBF for numeric, Dirac for nominal	Parameters of hardware

5 Experiments

5.1 Data

We have collected three representative datasets from different types of sites, shown in Table 3. They are located in a western province of China and serves more than 250 million users with nearly 80k base stations. The dataset covers different standards of networks, ranging from 2G, 3G, and latest LTE (Long Term Evolution) networks, with different size of nodes. The earlier a network established, the longer records we have. The D2 contains a high rate of missing values, because of the owner's continuously large-scale reconstruction, when nearly 1/3 of stations were newly constructed or removed.

Table 3. Description of network traffic datasets.

Name	Unit	Type	Area	#unit	#month	#record	Missing rate	Mean	STD
D1	Base station	2G	Rural	5191	46	19915	8.34%	1.60E+02	1.94E+02
D2	Cell	3G	Metropolis	17097	35	151394	25.30%	1.01E+03	2.66E+03
D3	Base station	LTE	City	9946	27	36280	13.51%	3.87E+03	3.45E+04

There are some details on data preprocessing. The month level aggregation are taken from the average of top 3 days of every month. Before training, all data points are transformed by $\text{diff}_i(\log(\mathbf{Y}_{i,j}+1))$, to make it linearly predictable and stationary, and an inverse transform is needed before the accuracy evaluation. Miss values are filled by linear interpolation for each series.

5.2   Setup

The experiment composes of two parts: (1) comparative study of different models, including popular time series models as baselines and different strategies of multi-term prediction; (2) the contribution of properties or kernels at different horizons.

The ability of model needs to be evaluated with a proper train/test sets construction on genuine data. With a sliding window of length, $H = 12$ moving from right to left along the time axis, the section lying inside the window is used as test part, and the sub-series before the window is left as train part, also required to not shorter than $H = 12$. We're interested in the MAPE (Mean Absolute Percentage Error), which is calculated firstly by taking the mean of all trails' scores, and take their median for all entities at each horizon, namely $\text{median}_j(\text{mean}_i(\text{MAPE}_{i,j})), i = H + 1..N - H, j = 1..D$. The reason for using median is that some sites may be activated or deactivated over a given period, which makes the mean unreliable.

The model is implemented by writing a boosting framework and modifying the KernelRidge in the open source scikit-learn library (to support weights on samples). The ARIMA is implemented by auto.arima in R's forecast package, and the VAR is from the MTS package. The RNN is 3-layered and is built in the style of seq-2-seq, whose number of nodes in the middle layer is determined by grid search, and implemented by Keras with L1 regularization with other default parameters unchanged. Other default settings in experiments include, number of boosting step $= 100$, shrinkage rate $= 0.1$, ridge regularization strength $= 1.0$.

5.3   Comparison Results

Table 4 gives the experimental results of 8 models, which can be compared by 3 aspects: model, horizon, and dataset. For the model, we compare two kinds of models: the 1st–4th are popular models for benchmarks, including a naive way (as the 1st) using the current observation directly for output; the 5th–8th are the realizations of the 4 strategies illustrated in Sect. 4.1. The classical models relying on the recursive strategy, such as ARIMA, can achieve comparable results at short-term prediction, but deteriorate rapidly, especially for VAR, as steps go forward. The seq-2-seq RNN can do much better than others, but usually worse than our solution without the help of context information. In the 4 strategies, the direct and mixing can keep the errors growing much slower. The mixing strategies, including S3 and S4, are usually better than the direct strategy S1, while S4 is slightly better than S3 in 2 datasets.

Table 4. The comparison of MAPE for 8 models on 3 datasets. The 4 horizons are designed to observe the value change of short, mid, and long-term prediction.

Dataset	Model	Horizon			
		1 m	3 m	6 m	12 m
D1	Naive	0.285	0.511	0.869	1.367
	ARIMA	0.159	0.247	0.378	0.539
	VAR	0.131	0.185	0.261	0.341
	RNN	**0.108**	0.144	0.197	0.258
	S1-dir	0.111	0.126	0.157	0.187
	S2-rec	0.111	0.154	0.219	0.301
	S3-adj	0.111	0.125	0.152	0.170
	S4-mrec	0.111	**0.124**	**0.149**	**0.166**
D2	Naive	0.319	0.589	1.020	1.631
	ARIMA	**0.080**	0.127	0.204	0.315
	VAR	0.124	0.188	0.283	0.399
	RNN	0.082	0.125	0.192	0.281
	S1-dir	0.081	0.114	0.165	0.231
	S2-rec	0.081	0.128	0.204	0.313
	S3-adj	0.081	0.111	0.159	0.220
	S4-mrec	0.081	**0.107**	**0.148**	**0.196**
D3	Naive	0.425	0.781	1.349	2.154
	ARIMA	0.145	0.201	0.28	0.356
	VAR	0.278	0.382	0.538	0.732
	RNN	**0.137**	0.184	0.246	0.294
	S1-dir	0.150	0.176	0.225	0.267
	S2-rec	0.150	0.199	0.267	0.330
	S3-adj	0.150	**0.161**	**0.197**	**0.221**
	S4-mrec	0.150	0.165	0.201	0.225

In all situations, the long-term tasks are always harder than short ones. The single variable models (ARIMA, RNN), usually can achieve well enough results for the short-term prediction, even without the assistance of properties, which means it mainly depends on each series' own recent history. They are surpassed by the more sophisticated methods when the horizon goes into mid-term above. The recursive strategy is obviously not suitable for long terms.

The inherent intensity of noise and the missing rate of datasets result in the fundamental difference, even in applying the naive model. The length of data and the consistency of distribution limit the potential for accuracy improvement. The more aged network, the more stable their trends intend to be, and thus easier to predict. Anyway, the choice of strategy is still an empirical problem in practice.

5.4   Contribution of Kernels

We're interested in the effects of attributes when horizons and strategies change. At each horizon, we take the normalized weights of each basic kernel as their contributions and display them in the heatmap as Fig. 3. It can be discovered that the prediction for next month is mainly related to kernels from AR, MA, and seasonality, and meanwhile Dirac Delta kernel dominates the correlation of outputs, which implies these series can hardly correlate each other given a short period. When we switch to the mid-term, the weights start to shift to the DTW and topology, which means the model starts to refer the factors in a wider

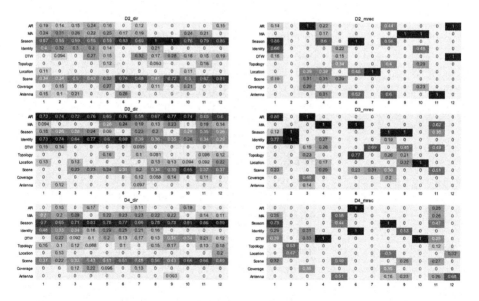

Fig. 3. The contribution of 9 kernels in boosting for the stages 1–12. Every row of sub-figures denotes a dataset, from D1 to D3; the left column denotes the S1 strategy, and the right column is the S4 strategy. In every sub-figure, the above 3 rows mean the kernel for inputs, while the others below are for outputs.

range. For long-term, the indoor/outdoor factor nearly dominates the decision, which is the most macroscopic factor we can find. The seasonality takes heavier proportion on D1, and D3 collected at higher station level. The weights of S4 are much sparser than S1, since the kernels found at previous stages will be applied repeatedly for later ones, and have much less possibility to appear again.

6 Conclusion

In this paper, we incorporated the context information into the multi-horizon network traffic prediction, and verify its effectiveness on long-term requests for the first time. The introduction of multi-kernel and multi-recursive is helpful to this problem, though with more development cost. This discovery builds a connection between the scales of temporal requests and the ranges of geographical terrain or the grains of device properties. This solution can further suggest the business operators find more macroscopic factors to support even longer forecasting, such as more traffic records from nearby provinces, or macroeconomic statistics of the whole country. The model may also be simplified or promoted by a solid theoretical analysis on the relationship between temporal scales and contextual information in future work.

References

1. Cisco Visual Networking Index: Global mobile data traffic forecast update, 2016–2021 white paper. https://www.cisco.com/c/en/us/solutions/collateral/service-provider/visual-networking-index-vni/mobile-white-paper-c11-520862.html. Accessed Mar 2017
2. Bui, N., Cesana, M., Hosseini, S.A., Liao, Q., Malanchini, I., Widmer, J.: A survey of anticipatory mobile networking: context-based classification, prediction methodologies, and optimization techniques. IEEE Commun. Surv. Tutor. (2017)
3. Das, A.K., Pathak, P.H., Chuah, C.-N., Mohapatra, P.: Contextual localization through network traffic analysis. In: Proceedings of INFOCOM 2014, pp. 925–933. IEEE (2014)
4. Gönen, M., Alpaydın, E.: Multiple kernel learning algorithms. J. Mach. Learn. Res. **12**, 2211–2268 (2011)
5. Ma, X., Dai, Z., He, Z., Ma, J., Wang, Y., Wang, Y.: Learning traffic as images: a deep convolutional neural network for large-scale transportation network speed prediction. Sensors **17**(4), 818 (2017)
6. Oliveira, T.P., Barbar, J.S., Soares, A.S.: Computer network traffic prediction: a comparison between traditional and deep learning neural networks. Int. J. Big Data Intell. **3**(1), 28–37 (2016)
7. Park, J., Raza, S.M., Thorat, P., Kim, D.S., Choo, H.: Network traffic prediction model based on training data. In: Gervasi, O., Murgante, B., Misra, S., Gavrilova, M.L., Rocha, A.M.A.C., Torre, C., Taniar, D., Apduhan, B.O. (eds.) ICCSA 2015. LNCS, vol. 9158, pp. 117–127. Springer, Cham (2015). https://doi.org/10.1007/978-3-319-21410-8_9

8. Taieb, S.B.: Machine learning strategies for multi-step-ahead time series forecasting. Ph.D. Thesis (2014)
9. Wen, Q., Zhao, Z., Li, R., Zhang, H.: Spatial-temporal compressed sensing based traffic prediction in cellular networks. In: 1st IEEE International Conference on Communications in China Workshops (ICCC), pp. 119–124. IEEE (2012)
10. Xia, H., Hoi, S.C.H.: MKBoost: a framework of multiple kernel boosting. IEEE Trans. Knowl. Data Eng. 25(7), 1574–1586 (2013)

IDLP: A Novel Label Propagation Framework for Disease Gene Prioritization

Yaogong Zhang[1], Yuan Wang[2(✉)], Jiahui Liu[1], Xiaohu Liu[1], Yuxiang Hong[1], Xin Fan[1], and Yalou Huang[1]

[1] College of Software, NanKai University, TianJin 300350, China
`huangyl@nankai.edu.cn`,
`{ygzhang,jiahui,liuxiaohu,hongyuxiang,nkufanxin}@mail.nankai.edu.cn`
[2] School of Computer Science and Information Engineering,
Tianjin University of Science and Technology, Tianjin 300222, China
`wangyuan23@tust.edu.cn`

Abstract. Prioritizing disease genes is trying to identify potential disease causing genes for a given phenotype, which can be applied to reveal the inherited basis of human diseases and facilitate drug development. Our motivation is inspired by label propagation algorithm and the false positive protein-protein interactions that exist in the dataset. To the best of our knowledge, the false positive protein-protein interactions have not been considered before in disease gene prioritization. Label propagation has been successfully applied to prioritize disease causing genes in previous network-based methods. These network-based methods use basic label propagation, i.e. random walk, on networks to prioritize disease genes in different ways. However, all these methods can not deal with the situation in which plenty false positive protein-protein interactions exist in the dataset, because the PPI network is used as a fixed input in previous methods. This important characteristic of data source may cause a large deviation in results. We conduct extensive experiments over OMIM datasets, and our proposed method IDLP has demonstrated its effectiveness compared with eight state-of-the-art approaches.

Keywords: Gene prioritization · Label propagation
Heterogeneous network · Bioinformatics

1 Introduction

Disease gene prioritization aims to identify potential implications of genes in disease susceptibility. The accurate identification of corresponding disease genetic information is the first step toward a systematic understanding of the molecular mechanisms of a complex disease. Also, it is essential to know disease-related genes for diagnosis and drug development [2]. However, identifying disease-related genes is not an easy work, which is still one of the major challenges in the field of bioinformatics.

© Springer International Publishing AG, part of Springer Nature 2018
D. Phung et al. (Eds.): PAKDD 2018, LNAI 10937, pp. 261–272, 2018.
https://doi.org/10.1007/978-3-319-93034-3_21

With the accumulation of studies on systems biology, researches have shown genes that are physically or functionally close to each other tend to be involved in the same biological pathways and have similar effects on phenotypes [4,12]. Based on such assumption, many network-based prioritization approaches have been developed to prioritize candidate genes [7–9,15,17,18]. Early algorithms prioritize candidate genes based on their similarity to known disease genes [8,15]. Though such type of methods perform well, they still have two limitations. The first limitation is caused by the fact that these methods only consider label propagation on homogeneous network (i.e. the PPI network). Thus, these methods easily fail when few disease-related genes are known. Later, methods that integrate heterogeneous networks have been proposed. Propagating label on both PPI network and phenotype similarity network [7,9,18], the prediction results have been boosted. Nevertheless, there is another limitation. Due to the alternating iterative learning approach they use, traditional methods still suffer from noise in data source and are limited to less satisfying performance. As we know, high-throughput technologies have produced vast amounts of protein-protein interaction data. However, imprecise measuring technology brings a large number of false-positives in the current available protein-protein interaction data [10,16]. The false positive interactions between proteins in the PPI network would introduce a bias while prioritizing disease genes by methods mentioned in previous research works [7–9,15,18].

To tackle these challenges, we propose an Improved Dual Label Propagation (IDLP) method. Firstly, we present a dual label propagation (DLP) framework on the heterogeneous network to prioritize disease genes. We construct a heterogeneous network by connecting the gene network and the phenotype similarity network with gene-phenotype associations. Then, we extend the basic label propagation (LP) [19] framework to the heterogeneous network. Target disease phenotypes and target disease genes are selected as seed nodes alternatively to propagate labels on the heterogeneous network. Secondly, an improved dual label propagation (IDLP) framework is proposed to reduce the bias introduced by false positive protein-protein interactions. To be specific, the PPI network adjacent matrix is considered as a variable to learn under IDLP framework, its values are amended from noise by optimizing the loss function of IDLP. In case of overfitting to the training data, an additional fitting term [19] is introduced to constrain the values in the PPI network matrix to be consistent with its initial values. The same fitting term is introduced to the phenotype similarity network as well. The target matrices are optimized by minimizing the loss function. Furthermore, we propose an effective closed-form solution to improve calculation efficiency.

Our contribution is constituted by two parts. (1) It's the first time that the basic label propagation (Zhou et al. Nips 2005) is extended from homogeneous networks to heterogeneous networks by directly modeling the label loss function between labeled data and unlabeled data, through which it's possible for us to take additional constraints into the loss function. On the contrary, alternating iteration strategy adopted by almost all previous works cannot deal with any constraints. (2) It's the first time that data bias regularization term has been taken into consideration, which greatly helps us to reduce the disturbance of data and improve the prediction accuracy in gene-phenotype prediction task as the bias is naturally and inevitably introduced by many false positive protein interactions (Tuan 2006, Christian 2002) in PPI network.

2   Materials and Methods

2.1   Materials

We downloaded two versions (Aug-2015 version and Dec-2016 version) of human gene-phenotype associations from OMIM database [5]. The Aug-2015 version consists of 5,117 associations between 4,392 phenotypes and 3,400 genes, and the Dec-2016 version contains 5,465 associations between 4,741 disease phenotypes and 3,638 genes. The human protein-protein interaction (PPI) network was obtained from BioGRID [1] in Aug. 2015. The PPI network contains 356,720 binary interactions between 19,511 genes. The disease phenotype network is an undirected graph with 8,004 vertices representing OMIM disease phenotypes, the disease phenotype similarity between two phenotypes is calculated by text mining [14]. After filtering interacting gene and disease phenotypes, we obtained 4,678/4,801 associations (Aug-2015/Dec-2016) between 4,120 disease phenotypes and 3,292 genes, corresponding PPI network and disease phenotype similarity network are extracted as well.

2.2   Notations

Let n be the number of genes, m be the number of phenotypes, and $W_1 \in \mathbb{R}^{n \times n}$ be the binary PPI network, and $W_2 \in \mathbb{R}^{m \times m}$ be the phenotype similarity network. The known gene-phenotype associations are represented by a binary matrix $\hat{Y}_{(n \times m)}$ with 1 for entries of known associations and 0 otherwise. W_1 and W_2 are used to construct a normalized network $\bar{S}_1 = D^{-\frac{1}{2}} W_1 D^{-\frac{1}{2}}$ and $\bar{S}_2 = D^{-\frac{1}{2}} W_2 D^{-\frac{1}{2}}$, and D is a diagonal matrix with the row-sum of corresponding W_i (i = 1, 2) on the diagonal entries. S_1, S_2 and Y are target matrices needed to learn.

2.3   Dual Label Propagation on Heterogeneous Network

Firstly, we introduce the conventional label propagation algorithm [19]. Given the PPI network W_1 and a target phenotype p, the objective of label propagation is to learn an assignment score for each gene with the query phenotype p as illustrated in Fig. 1(A). The score shows how close each gene is to the phenotype p. The target labeling $y = Y_{\bullet p}$ is the p-th column of the target association matrix Y, the initial labeling $\hat{y} = \hat{Y}_{\bullet p}$ is the p-th column of the known association matrix $\hat{Y}$. Label propagation assumes that genes should be assigned the similar labels

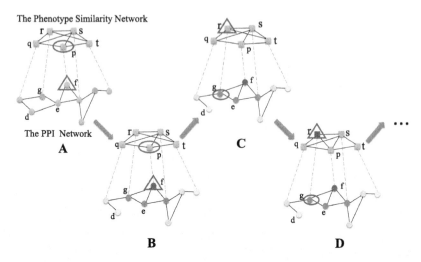

Fig. 1. Illustration of the IDLP framework. Square nodes represent phenotypes, all pairwise phenotype similarity relationships make up the phenotype similarity network. Circular nodes represent genes, all pairwise gene interactions make up the PPI network. Nodes surrounded by oval are query phenotypes (or genes), Nodes surrounded by triangle are seed genes (or phenotypes). (A) For a query phenotype p, the corresponding related genes are selected as seed nodes. (B) By modeling the noises in the PPI network, the interactions between gene nodes have been changed. In order to better explain the situation, we consider two extreme cases here, i.e., edge deletion and edge addition. During the optimization of IDLP, the interaction between gene g and f has been added, the interaction between gene d and e has been removed. The changes of the PPI network result in a high score on gene g, because gene g directly receive score from seed gene f. What's more, gene d no longer receives scores from gene e, which indirectly results in gene d receives more support from gene e. (C) For a query gene g, the corresponding related phenotypes are selected as seed nodes. (D) By modeling the noises in the phenotype network, the similarity scores between phenotypes have been changed. The edge addition between phenotype r and p and edge deletion between phenotype r and t result in a high score on phenotype p.

if they are connected in the PPI network, which leads to the following objective function,

$$\Psi(\boldsymbol{y}) = \sum_{i,j}(\boldsymbol{W}_1)_{ij}\left(\frac{y_i}{\sqrt{\boldsymbol{D}_{ii}}} - \frac{y_j}{\sqrt{\boldsymbol{D}_{jj}}}\right)^2 + \mu\sum_i(y_i - \hat{y}_i)$$
$$= \boldsymbol{y}^T(\boldsymbol{I} - \bar{\boldsymbol{S}}_1)\boldsymbol{y} + \mu||\boldsymbol{y} - \hat{\boldsymbol{y}}||^2,$$

(1)

Equation (1) can be extended to predict associations with all the phenotypes as Eq. (2),

$$\Psi_1(\boldsymbol{Y}) = tr(\boldsymbol{Y}^T(\boldsymbol{I} - \bar{\boldsymbol{S}}_1)\boldsymbol{Y}) + \mu||\boldsymbol{Y} - \hat{\boldsymbol{Y}}||_F^2.$$

(2)

In the same way, phenotypes should be assigned the similar labels if they have a high score in the phenotype similarity network for a given gene, as illustrated in Fig. 1(C). Let $z = Y_{q\bullet}$, i.e. the q-th row of the target association matrix Y, $\hat{z} = \hat{Y}_{q\bullet}$, i.e. the q-th row of the known association matrix $\hat{Y}$. Label propagation on phenotype similarity network for a given gene can be expressed as follows,

$$
\begin{aligned}
\Psi(z) &= \sum_{i,j}(W_2)_{ij}(\frac{z_i}{\sqrt{D_{ii}}} - \frac{z_j}{\sqrt{D_{jj}}})^2 + \zeta\sum_i(z_i - \hat{z}_i) \\
&= z(I - \bar{S}_2)z^T + \zeta||z - \hat{z}||^2,
\end{aligned}
\tag{3}
$$

Equation (3) can be extended to predict associations with all the genes as following Eq. (4),

$$
\Psi_2(Y) = tr(Y(I - \bar{S}_2)Y^T) + \zeta||Y - \hat{Y}||_F^2.
\tag{4}
$$

2.4   Improved Dual Label Propagation on Heterogeneous Network

To deal with false positive protein interactions, we consider S_1 as a variable needed to learn, and introduce fitting term $\sum_{i,j}((S_1)_{ij} - (\bar{S}_1)_{ij})^2$ to capture the noise in PPI network, where S_1 is the target normalized PPI network we need to learn, $\bar{S}_1$ is the normalized known PPI network. For a given phenotype, it leads to the following loss function,

$$
\begin{aligned}
\Psi'(y) &= y^T(I - S_1)y + \mu||y - \hat{y}||^2 + \nu\sum_{i,j}((S_1)_{ij} - (\bar{S}_1)_{ij})^2 \\
&= y^T(I - S_1)y + \mu||y - \hat{y}||^2 + \nu||S_1 - \bar{S}_1||_F^2,
\end{aligned}
\tag{5}
$$

Equation (5) can be extended to predict associations with all the phenotypes as follows,

$$
\Psi'_1(Y) = tr(Y^T(I - S_1)Y) + \mu||Y - \hat{Y}||_F^2 + \nu||S_1 - \bar{S}_1||_F^2.
\tag{6}
$$

To minimize the loss function in Eq. (6), an alternative iterative schema is adopted. It solves the problem with respect to one variable while fixing other variables. The loss function in Eq. (6) is not convex on Y and S_1 jointly, but it is convex on one variable with the other fixed.

The closed form solutions Y and S_1 can be expressed as,

$$
Y^* = \beta(I - \alpha S_1)^{-1}\hat{Y}
$$
$$
\alpha = \frac{1}{1+\mu}, \quad \beta = \frac{\mu}{1+\mu}
\tag{7}
$$
$$
S_1^* = \bar{S}_1 + \gamma YY^T, \quad \gamma = \frac{1}{2\nu}
$$

After the label propagation on the PPI network, the result is shown in Fig. 1(B). Besides the value of target gene-phenotype association matrix, the weight of each edge in the PPI network has been updated as well.

We introduce term $\sum_{i,j}((S_2)_{ij} - (\bar{S}_2)_{ij})^2$ to capture the noise in the phenotype similarity network. For a given gene, as shown in Fig. 1(C), it leads to the following loss function,

$$\Psi'(z) = z(I - S_2)z^T + \zeta||z - \hat{z}||^2 + \eta \sum_{i,j}((S_2)_{ij} - (\bar{S}_2)_{ij})^2 \qquad (8)$$
$$= z(I - S_2)z^T + \zeta||z - \hat{z}||^2 + \eta||S_2 - \bar{S}_2||_F^2,$$

For all the genes, Eq. (8) can be extended to predict associations as follows,

$$\Psi_2'(Y) = tr(Y(I - S_2)Y^T) + \zeta||Y - \hat{Y}||_F^2 + \eta||S_2 - \bar{S}_2||_F^2. \qquad (9)$$

The closed form solutions of Y and S_2 can be expressed as,

$$Y^* = \beta'\hat{Y}(I - \alpha'S_2)^{-1}$$
$$\alpha' = \frac{1}{1+\zeta}, \quad \beta' = \frac{\zeta}{1+\zeta} \qquad (10)$$
$$S_2^* = \bar{S}_2 + \gamma'Y^TY, \quad \gamma' = \frac{1}{2\eta}$$

Figure 1(D) shows the result after the label propagation on phenotype network. Besides the target gene-phenotype association matrix, the phenotype similarity network has also been updated.

The algorithm details of IDLP are shown in Algorithm 1. The illustration of the IDLP is shown in Fig. 1.

Algorithm 1. IDLP

Input:
 $\hat{S}_1$: normalized PPI network
 $\hat{S}_2$: normalized phenotype similarity network
 $\hat{Y}$: known binary gene-phenotype associations for training
 Y: initialized with random values
 $\alpha, \beta, \gamma, \alpha', \beta', \gamma'$: hyper-parameters
Output: model parameters Y, S_1, S_2
 1: **repeat**
 2:    $S_1 \leftarrow \bar{S}_1 + \gamma YY^T$
 3:    $Y \leftarrow \beta(I - \alpha S_1)^{-1}\hat{Y}$
 4:    $S_2 \leftarrow \bar{S}_2 + \gamma'Y^TY$
 5:    $Y \leftarrow \beta'\hat{Y}(I - \alpha'S_2)^{-1}$
 6: **until** convergence

---

3   Results

3.1   Baselines

We compare our methods to both classic and the state-of-the-art network-based algorithms. We give a brief introduction to the baselines used in our experiments.

CIPHER employs the regression model to quantify the concordance between the candidate gene and the target phenotype, then candidate genes are ranked by the concordance score [17]. RWR and DK (Diffusion Kernel) prioritize candidate genes by use of random walk from known genes for a given disease [8]. RWRH extends RWR algorithm to the heterogeneous network, it makes better use of the phenotypic data by using the target phenotypes and corresponding genes as seed nodes simultaneously [9]. PRINCE uses the known disease relationships to decide an initial set of genes that are associated with a query disease phenotype, then it performs label propagation on the PPI network to prioritize disease genes [15]. MINProp is based on a principled way to integrate three networks in an optimization framework and performs iterative label propagation on each individual subnetwork [7]. BiRW performs random walk on PPI network and phenotype similarity network alternatively to enrich genome-phenome association matrix, then prioritizes disease genes based on the enriched association matrix [18].

3.2 Experimental Settings

IDLP has four parameters, i.e. α, γ, $\alpha^{'}$, $\gamma^{'}$. Since the constraint $\alpha + \beta = 1$ and $\alpha^{'} + \beta^{'} = 1$, the value of β and $\beta^{'}$ are fixed when α and $\alpha^{'}$ are chosen. For the data of training in cross-validation, we select parameter values by using a usual manner of (5-fold) cross-validation: only a part (four folds) of the training dataset is used for getting model results of IDLP meanwhile the rest (one fold) for validation, this is done five times with each fold as validation set in turns. The average result of five folds is used for choosing best parameters. In this parameter value selection, we consider all combinations of the following values: $\{0.0001, 0.001, 0.01, 0.1, 1\}$ for α and $\alpha^{'}$, $\{1, 10, 100, 1,000, 10,000\}$ for γ and $\gamma^{'}$.

We implement all the baselines according to the descriptions in their papers. CIPHER doesn't have any parameters to tune, so it is applied to the test set directly. For RWR, DK, and PRINCE, they are network-based methods only walk on gene interaction network, the parameter α is chosen from $\{0.1, 0.3, 0.5, 0.7, 0.9\}$ by 5-fold cross-validation. For RWRH, MINProp and BiRW, they perform a random walk on a heterogeneous network of gene interactions and human diseases (i.e. OMIM phenotypes similarity network). We use the average version of BiRW which is shown to be the best among the three versions of BiRW proposed by Xie [18], and the left and right walk step is set to 4 as suggested by Xie. There is one parameter in BiRW, which is chosen from $\{0.1, 0.3, 0.5, 0.7, 0.9\}$ by cross-validation. There are two parameters in MINProp, which are chosen from $\{0.1, 0.3, 0.5, 0.7, 0.9\}$ by grid through cross-validation. There are three parameters in RWRH, which are all chosen from $\{0.1, 0.3, 0.5, 0.7, 0.9\}$ by grid search.

3.3 Evaluation

We evaluated the ranks of the tested genes with two metrics: (i) we calculated the area under the curve (AUC) [3,6] for each method and (ii) we calculated

Table 1. Average AUCs scores of gene prioritization on test set and validation set. We compared AUCs when the number of false positive genes are up to 20, 50, 100.

	Performance on test set			Performance on validation set		
	AUC20	AUC50	AUC100	AUC20	AUC50	AUC100
CIPHER_SP	0.0029	0.0046	0.0066	-	-	-
CIPHER_DN	0.0015	0.0027	0.0042	-	-	-
RWR	0.0075	0.0178	0.0283	0.0233	0.0358	0.0475
DK	0.0192	0.0255	0.0294	0.0211	0.0306	0.0399
RWRH	0.0916	0.1250	0.1664	**0.2009**	**0.2724**	**0.3288**
MINProp	0.0771	0.1266	0.1799	0.1963	0.2625	0.3104
BiRW	0.0421	0.0780	0.1142	0.1544	0.2180	0.26672
PRINCE	0.1117	0.1468	**0.2088**	0.1433	0.2137	0.2715
IDLP	**0.1123**	**0.1492**	0.1909	0.2004	0.2572	0.2990

the average precision and recall on test set at top-k positions (k = 20, 50, 100). The two metrics are complimentary: the AUC evaluates the entire rank of genes, while the top-k precision and recall emphasize the top-ranked genes.

Since the accuracy of top-ranked genes is more important than the lower ranked genes, we highlight a set of false positive cutoffs for the ROC curves and compare the corresponding average AUCs between methods. The higher AUC score, the better the performance. In this paper, AUC20 is chosen as a criteria for comparison between different methods.

Conventional cross-validation evaluation strategy, such as leave-one-out cross-validation strategy, does not necessarily reflect the property of novel gene-phenotype associations prediction. To address such cases, we adopt the strategy that has been utilized by [11,13,18], i.e. two versions of data are used in the experiments, the Aug-2015 version data are used as validation set to train the model, the newly added data accumulated between Aug-2015 and Dec-2016 are used as test set to measure the performance of the model. In the experiment, we split the known gene-disease associations of Aug-2015 version data into five folds. After doing 5 folds cross-validation, the average results of the five folds are used for selecting parameters for each method. Then, the methods are applied to predict the associations in an independent set of associations added into OMIM between Aug-2015 and Dec-2016.

3.4 Accuracy Evaluation

To quantitatively evaluate IDLP and other baseline methods, i.e. CIPHER, RWR, DK, RWRH, MINProp, BiRW, and PRINCE, these algorithms are applied to predict the disease genes for each phenotype.

The performance of IDLP and baseline methods on test set and cross-validation set are shown in Table 1. The performance results on cross-validation

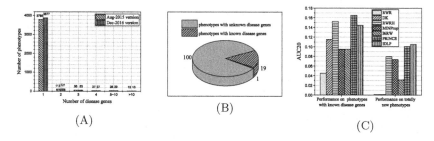

Fig. 2. Data analysis. (A) The phenotype distribution based on the genes it associates with. (B) The distribution of newly added phenotypes based on whether they have known disease causing gene(s). (C) The AUC20 scores of different methods in two situations: 1. Phenotypes with known disease genes are used as queries (left); 2. Phenotypes with unknown disease genes are used as queries (right).

are used for choosing parameters for each method. RWRH gets the best results on cross-validation set. However, the performance of RWRH on test set dramatically falls compared with that of IDLP. RWRH heavily depends on the completeness and correctness of PPI network and phenotype similarity network, which brings the serious overfitting. It can be seen that IDLP achieves the best performance under AUC20 and AUC50 on test set, which means the proposed IDLP can predict newly discovered gene-phenotype associations well. By introducing the dual label propagation framework and modeling the bias on the PPI network and phenotype similarity network into the framework, it successfully utilizes the information in the heterogeneous network and overcomes the interference of the noise in data source. This demonstrates the advantage of IDLP over other baselines.

In order to understand IDLP further, we give an analysis of the constitution of the data. Figure 2(A) shows the phenotype distribution of the two versions according to the disease genes they associate with. More specifically, there are 3785 phenotypes associated with one disease gene in Aug-2015 version data, the number of phenotypes increases to 3877 in Dec-2016 version data; the numbers of phenotypes which have been found with more than one disease genes change slightly. There are 123 newly added gene-phenotype associations. More specifically, as shown in Fig. 2(B), 100 phenotypes are newly added to Dec-2016 version data, which means there are 100 phenotypes with unknown disease genes in Aug-2015 version data. The remaining 23 associations can be divided into 2 categories, 19 phenotypes with known disease genes being added with one more disease gene and 1 phenotype with known disease genes being added with 4 new disease genes. From Fig. 2(A) and (B), we know the phenotypes involved in newly added gene-phenotype associations between Aug-2015 version and Dec-2016 version are mostly phenotypes with unknown disease genes in Aug-2015 version. Here we define these phenotypes without any known disease genes as *singleton* phenotypes. Since the number of singleton phenotypes accounts for a large percentage, it is important and necessary to explore the performance on singleton phenotypes.

Figure 2(C) shows the results when different associations are used as test set. The left histogram in Fig. 2(C) shows the performance when 23 associations with none singleton phenotypes are used as test set. The right histogram in Fig. 2(C) shows the performance when 100 associations with only singleton phenotypes are used as test set. Because the results of CIPHER_SP and CIPHER_DN are too small in the histogram, we ignore them in this discussion. Comparing these two histograms in Fig. 2(C), we can observe that predictions on phenotype queries that have known disease genes are more precise than phenotype queries that have non disease genes for each method. It is consistent with the intuition that enriched phenotypes (i.e. phenotypes with at least one known disease gene) are easier to find disease genes. RWRH, PRINCE, and IDLP have relatively high AUC20 scores on enriched phenotype queries. On the contrary, it's hard to identify disease genes for singleton phenotypes, because no known disease genes are discovered for these singleton phenotypes. That's why RWR and DK decrease to zero. Meanwhile, IDLP achieves best at this situation, which demonstrates IDLP's effectiveness on singleton phenotypes.

4   Robustness Evaluation of IDLP

We check the AUC20 performance result for each method under four disturbed PPI networks: (1) randomly delete 10% PPI data; (2) randomly delete 10% PPI data and add 10% PPI data; (3) randomly delete 20% PPI data; (4) randomly delete 20% PPI data and randomly add 20% PPI data. The best and the worst performance of these four situations are drawn as error bars on the histogram. Figure 3(A) shows the result when choosing all disease phenotypes as test set, and we can see that IDLP has a greatly stable performance under all kinds of disturbance. Figure 3(B) shows the result when total new disease phenotypes are chosen as test set. The advantage has become more obvious when we only consider the total new phenotypes (i.e. singleton phenotypes defined above) as test set. From the results in Fig. 3, we can conclude that IDLP has a good robustness.

The robustness comes from the design of the loss function of IDLP. More specifically, the update mechanism determines the robustness of IDLP. Let us go over the first two steps of Algorithm 1. At first, the target normalized PPI network matrix is updated by $S_1 \leftarrow \bar{S}_1 + \gamma Y Y^T$, then the target gene-phenotype associations matrix Y is updated by $Y \leftarrow \beta(I - \alpha S_1)^{-1}\hat{Y}$. After sufficient iterative update, $\gamma Y Y^T$ has much influence on S_1 and the influence is even stranger when γ becomes a large value.

5   Conclusion

A new framework IDLP is proposed based on network methods to prioritize candidate disease genes. IDLP effectively propagates the labels through out the

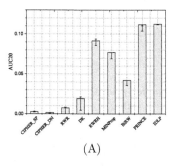
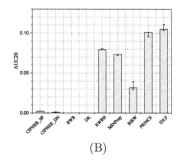

(A)   (B)

Fig. 3. Robustness of IDLP. Four disturbed PPI networks are applied into each algorithm: (1) randomly delete 10% PPI data; (2) randomly delete 10% PPI data and add 10% PPI data; (3) randomly delete 20% PPI data; (4) randomly delete 20% PPI data and add 20% PPI data. The best and the worse performance of these four situations are drawn as error bar on the histogram. (A) It shows the results when all diseases are chosen as test set. (B) It shows the results when totally new diseases are chosen as test set.

PPI network and phenotype similarity network, which avoids the method falling when few disease genes are known. Meanwhile, IDLP models the bias caused by false positive protein interactions and other potential factors by treating PPI network matrix and phenotype similarity matrix as target matrices to learn. By amending the noise in training matrices, it improves the performance results significantly. We also give a closed-form solution, which makes the algorithm more efficient. In our experiments, we find that IDLP has an outstanding performance for ranking top genes and a good robustness to deal with the noise in PPI network, which makes IDLP a better gene prioritization tool for biologists.

Acknowledgements. This work is supported by the National Natural Science Foundation of China (No. 61702367). The Research Project of Tianjin Municipal Commission of Education (No. 2017KJ033).

References

1. Chatr-Aryamontri, A., Breitkreutz, B.-J.: The BioGRID interaction database: 2015 update. Nucleic Acids Res. **43**, D470–8 (2015)
2. Chen, Y., Li, L.: Phenome-driven disease genetics prediction toward drug discovery. Bioinformatics **31**(12), i276–i283 (2015)
3. Fawcett, T.: An introduction to ROC analysis. Pattern Recogn. Lett. **27**(8), 861–874 (2006)
4. Gandhi, T.K.B., Zhong, J.: Analysis of the human protein interactome and comparison with yeast, worm and fly interaction datasets. Nat. Genet. **38**(3), 285–293 (2006)
5. Hamosh, A., Scott, A.F.: Online Mendelian Inheritance in Man (OMIM), a knowledge base of human genes and genetic disorders. Nucleic Acids Res. **33**(Database issue), D514–D517 (2004)

6. Hoehndorf, R., Schofield, P.N.: Analysis of the human diseasome using phenotype similarity between common, genetic, and infectious diseases. Sci. Rep. **5**, 10888 (2015)

7. Hwang, T., Kuang, R.: A heterogeneous label propagation algorithm for disease gene discovery. In: SIAM, p. 12 (2010)

8. Köhler, S., Bauer, S.: Walking the interactome for prioritization of candidate disease genes. Am. J. Hum. Genet. **82**(4), 949–958 (2008)

9. Li, Y., Patra, J.C.: Genome-wide inferring gene-phenotype relationship by walking on the heterogeneous network. Bioinformatics **26**(9), 1219–1224 (2010)

10. Genetic, T.N., Goodrich, J.A.: Protein-protein interaction assays: eliminating false positive interactions. Nat. Methods **3**(2), 135–139 (2006)

11. Ni, J., Koyuturk, M.: Disease gene prioritization by integrating tissue-specific molecular networks using a robust multi-network model. BMC Bioinform. **17**(1), 453 (2016)

12. Oti, M., Brunner, H.G.: The modular nature of genetic diseases. Clin. Genet. **71**(1), 1–11 (2006)

13. Petegrosso, R., Park, S.: Transfer learning across ontologies for phenome-genome association prediction. Bioinformatics **25** (2016). https://doi.org/10.1093/bioinformatics/btw649

14. van Driel, M.A., Bruggeman, J.: A text-mining analysis of the human phenome. Eur. J. Hum. Genet. **14**(5), 535–542 (2006)

15. Vanunu, O., Magger, O.: Associating genes and protein complexes with disease via network propagation. PLoS Comput. Biol. **6**(1), e1000641 (2010)

16. von Mering, C., Krause, R.: Comparative assessment of large-scale data sets of protein-protein interactions. Nature **417**(6887), 399–403 (2002)

17. Xuebing, W., Jiang, R.: Network-based global inference of human disease genes. Mol. Syst. Biol. **4**, 189 (2008)

18. Xie, M., Hwang, T., Kuang, R.: Prioritizing disease genes by bi-random walk. In: Advances in Knowledge Discovery and Data Mining, pp. 292–303 (2012)

19. Zhou, D., Bousquet, O.: Learning with local and global consistency. In: NIPS, vol. 1, pp. 595–602 (2004)

Deep Learning for Forecasting Stock Returns in the Cross-Section

Masaya Abe[1(✉)] and Hideki Nakayama[2]

[1] Nomura Asset Management Co., Ltd., Tokyo, Japan
m-abe@nomura-am.co.jp
[2] The University of Tokyo, Tokyo, Japan
nakayama@nlab.ci.i.u-tokyo.ac.jp

Abstract. Many studies have been undertaken by using machine learning techniques, including neural networks, to predict stock returns. Recently, a method known as deep learning, which achieves high performance mainly in image recognition and speech recognition, has attracted attention in the machine learning field. This paper implements deep learning to predict one-month-ahead stock returns in the cross-section in the Japanese stock market and investigates the performance of the method. Our results show that deep neural networks generally outperform shallow neural networks, and the best networks also outperform representative machine learning models. These results indicate that deep learning shows promise as a skillful machine learning method to predict stock returns in the cross-section.

Keywords: Deep learning · Stock returns · Cross-section · Forecasting
Neural networks · Industrial application

1 Introduction

Stock return predictability is one of the most important concerns for investors. In particular, many authors attempt to explain the cross-section of stock returns by using various factors, such as earnings–price ratio, company size and stock price momentum, and the efficacy of using such factors [1–3]. Conversely, the investors themselves must decide how to process and predict return, including selection and weighting of such factors.

One way to make investment decisions is to rely upon the use of machine learning models. This is a supervised learning approach that uses multiple factors to explain stock returns as input values and future stock returns as output values. Deep learning has attracted attention in recent years in the machine learning field because of its high performance in areas such as image recognition and speech recognition [4, 5]. Deep learning is a representation-learning method with multiple levels of representation. This method passes data through many simple but nonlinear modules. The data passes through many more layers than it does in conventional three-layer neural networks. This enables a computer to build complex concepts out of simpler concepts [4, 5].

By inputting data of multiple factors and passing them through many layers, deep learning could extract useful features, increase representational power, enhance

© Springer International Publishing AG, part of Springer Nature 2018
D. Phung et al. (Eds.): PAKDD 2018, LNAI 10937, pp. 273–284, 2018.
https://doi.org/10.1007/978-3-319-93034-3_22

performance, and improve the prediction accuracy for future stock returns. Currently, there have been few applications of deep learning to report on stock return predictability. Positive results of such applications could certainly be said to expand the versatility of the deep learning technique across multiple fields.

In this paper, we use deep learning to predict one-month-ahead stock returns in the cross-section in the Japanese stock market. We calculate predictive stock returns (scores) from the information of the past five points of time for 25 factors (features) for MSCI Japan Index constituents. As a measure of the performance, we use rank correlation between the actual out-of-sample returns and their predicted scores, directional accuracy, and performance of a simple long–short portfolio strategy. We compare with conventional three-layer neural networks and support vector regression and random forests as representative machine learning techniques.

2 Related Works

Many studies on stock return predictability have been reported on neural networks [6, 7]. Most of those are forecasts of stock market returns; however, forecasts of individual stock returns using the neural networks dealt with in this paper have also been conducted. For example, Olson and Mossman [8] attempted to predict one-year-ahead stock returns for 2,352 Canadian companies using 61 accounting ratios as input values and reported that neural networks outperform traditional regression techniques. As an application to emerging market, Cao et al. [9] predicted stock returns in the Chinese stock market. They showed that neural networks outperform the linear model. Besides those, Kryzanowski et al. [10] found that neural networks correctly classify 72% of the positive/negative returns to predict one-year-ahead stock returns by using financial ratios and macroeconomic variables.

Studies on deep learning have been recently undertaken due to the heightened attention toward this technique. Krauss et al. [11] used three different machine learning models, deep neural networks, gradient-boosted trees and random forests to predict one-day-ahead stock returns for the S&P500 constituents. As a result, they showed that combining the predictions of those three as an equal-weighted ensemble outperforms each individual model. Among each model, random forests outperform deep neural networks and gradient-boosted trees. Conversely, they stated that careful hyperparameter optimization may still yield advantageous results for the tuning-intensive deep neural networks. Outside the stock market, Dixon et al. [12] attempted to predict the direction of instrument movement for 5-min mid-prices for 43 CME listed commodity and FX futures. They showed 68% accuracy for the high ones. Moreover, in an application to a simple trading strategy, the best instrument has an annualized Sharpe Ratio of 3.29, indicating its high prediction ability. These studies were implemented for short investment horizons and do not use financial variables as input values. The present paper predicts one-month-ahead stock returns using multiple factors from both market and financial data as input values.

3   Data and Methodology

3.1   Dataset for MSCI Japan Universe

We prepare dataset for MSCI Japan Index constituents. The MSCI Japan Index comprises the large and mid-cap segments of the Japanese market. As of January 2017, the index is composed of 319 constituents and covers approximately 85% of the free float-adjusted market capitalization in Japan [13]. The index is also often used as a benchmark for overseas institutional investors investing in Japanese stocks. We use the 25 factors listed in Table 1. These are used relatively often in practice. In calculating these factors, we acquire necessary data from WorldScope, Thomson Reuters, I/B/E/S, EXSHARE, and MSCI. The actual financial data is acquired from WorldScope and Reuters Fundamentals (WorldScope priority). Taking into account the time when investors are actually available, we have a lag of four months. Forecast data is obtained from Thomson Reuters Estimates and I/B/E/S Estimates (Thomson Reuters priority). The data is used to calculate the factors from No. 2 to No. 8 and Nos. 16 and 17. Factors are calculated on a monthly basis (at the end of month) from December 1990 to November 2016. Note that factor calculation is not performed for Nos. 18 and 24. We directly use "Historical Beta" for No. 18 and "Predicted Specific Risk" for No. 24 from the MSCI Barra JPE4 model. Stock returns with dividends are acquired on a monthly basis (at the end of month).

Table 1. List of factors.

No.	Factor	No.	Factor
1	Book-to-market ratio	14	Investment growth
2	Earnings-to-price ratio	15	Investment-to-assets ratio
3	Dividend yield	16	EPS Revision(1 month)
4	Sales-to-price ratio	17	EPS Revision(3 months)
5	Cash flow-to-price ratio	18	Market beta
6	Return on equity	19	Market value
7	Return on asset	20	Past stock return(1 month)
8	Return on invested capital	21	Past stock return(12 months)
9	Accruals	22	Volatility
10	Sales-to-total assets ratio	23	Skewness
11	Current ratio	24	Idiosyncratic volatility
12	Equity ratio	25	Trading turnover
13	Total asset growth		

3.2 Problem Definition

To define the problem as a regression problem. For example, for stock i in MSCI Japan Index constituents at month T (end of month), 25 factors listed in Table 1 are defined by $x_{i,T} \in R^{25}$ and input values are defined by $v_{i,T} = \{x_{i,T}, x_{i,T-3}, x_{i,T-6}, x_{i,T-9}, x_{i,T-12}\} \in R^{125}$ using the past five points of time in three month intervals for 25 factors. The output value is defined by the next month's stock return, $r_{i,T+1} \in R$. As a more specific example, Fig. 1 shows the relationship between the input values and the output value for stock i from one set of training data at December 2001 as $T + 1$. The set consists of all stocks in MSCI Japan Index constituents at November 2001 (T). The input values are as follows: November 2001 (T), August 2001 ($T-3$), May 2001 ($T-6$), February 2001 ($T-9$), and November 2000 ($T-12$), as factors of past five points of time. The output value is the actual stock return at December 2001 ($T+1$). For data preprocessing, rescaling is performed so that each input value is maximally 1 (minimum ≈ 0) by ranking each input value in an ascending order by stock universe at each time point and then dividing by the maximum rank value. Similar rescaling is done for output values $r_{i,T+1}$, to convert to the cross-sectional stock returns (scores). Note that $v_{i,T}$ and $r_{i,T+1}$ are assumed to be the values after data preprocessing.

This procedure is extended to using the latest N months rather than the most recent set of training data (one training set). We use the mean squared error (MSE) as the loss function and define MSE_{T+1} when training the model at $T + 1$ as follows:

$$MSE_{T+1} = \frac{1}{K}\left\{ \sum_{t=T-N+1}^{T} \sum_{i \in U_t} \left(r_{i,t+1} - f\left(v_{i,t}; \theta_{T+1}\right) \right)^2 \right\} \quad (1)$$

In (1), K is the number of all training examples. U_t is the MSCI Japan Index universe at t. θ_{T+1} is the parameter calculated by solving (1) and makes the form of a function $f(.)$.

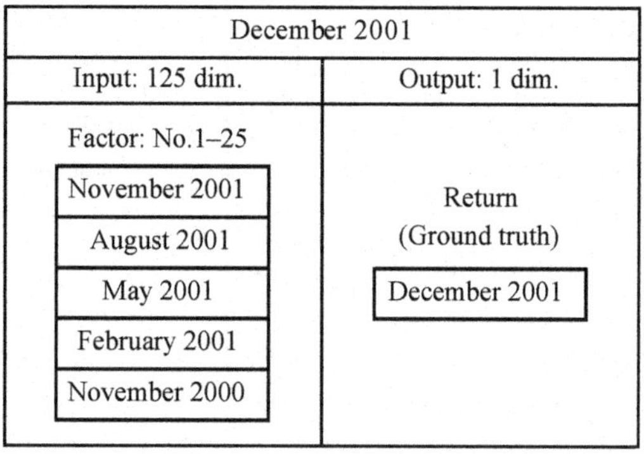

December 2001	
Input: 125 dim.	Output: 1 dim.
Factor: No.1–25	Return (Ground truth)
November 2001	
August 2001	
May 2001	December 2001
February 2001	
November 2000	

Fig. 1. Stock i from one set of training data at December 2001.

3.3   Training and Prediction

We train the model by using the latest 120 sets of training from the past 10 years. To calculate the prediction, we substitute the latest input values into the model after training has occurred. The cross-sectional predictive stock return (score) of stock i at time $T+2$ is calculated from time $T+1$ by (2) substituting $\mathbf{v}_{i,T+1}$ into the function $f(.)$ in (2) with the parameter $\boldsymbol{\theta}^*_{T+1}$, where $\boldsymbol{\theta}^*_{T+1}$ is calculated from (1) with $N = 120$:

$$Score_{i,T+2} = f\left(\mathbf{v}_{i,T+1}; \boldsymbol{\theta}^*_{T+1}\right) \tag{2}$$

For example, in order to calculate the prediction score at January 2002 ($T+2$) from December 2001 ($T+1$), the input values are as follows: December 2001 ($T+1$), September 2001 ($T-2$), January 2001 ($T-5$), March 2001 ($T-8$), December 2000 ($T-11$), as factors of the past five time points. The MSCI Japan Index constituents are from December 2001 ($T+1$). However, the prediction scores are not calculated for stocks with 63 or more missing input values, which is about half of the total number (125) of input values. For stocks with 62 or less missing input values, each missing value is replaced by the median value for the stocks that are not missing. For this series of processes, the model is updated by sliding one-month-ahead and carrying out a monthly forecast. The prediction period is 15 years: from January 2002 to December 2016 (180 months). An illustration of the flow of the processing is shown in Fig. 2, which shows the relationship between prediction and training data at each time point. For example, December 2001 in the "Training: 120 sets" is associated with Fig. 1 and January 2002 in the "Prediction: 1 set" represents the prediction for January 2002 from December 2001. The arrows indicate that the model is updated every month with the data sliding one-month-ahead.

3.4   Performance Measures

Rather than using the value of the loss function directly as a performance measure, we use the rank correlation coefficient (CORR) and directional accuracy (Direction) because these are more relevant measures of performance than the loss function. In addition, the performance of a simple long–short portfolio strategy is evaluated in

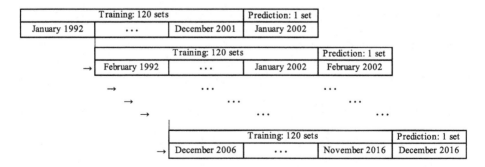

Fig. 2. Training-prediction set.

comparison with support vector regression and random forests. In practice, these are used as methods to evaluate the performance of the cross-sectional stock returns. CORR is Spearman's rank correlation coefficient between the actual out-of-sample returns (next month's returns) and the prediction scores, which is used to measure the prediction accuracy of the entire predicted stock excluding the influence of outliers of individual stock returns.

In an actual investment, there are many cases where the number of stocks is limited to those with higher prediction scores and those with lower prediction scores. We construct a portfolio comprising stock groups with top and bottom prediction scores. Direction is calculated by dividing the total number of the top stocks with high prediction scores that are above the cross-sectional median for next month's return on the stock universe and the bottom stocks with low prediction scores that are below the median by the total number of the top and bottom stocks.

The long–short portfolio strategy is a net zero investment strategy that buys the top stocks with equal weighting and sells the bottom stocks with equal weighting. To form into the top and bottom stock groups, we make two types of portfolios: tertile and quintile portfolios. These performance measures are calculated monthly during the prediction period. For example, at the evaluation starting point January 2002 (Prediction: 1 set in Fig. 2), these measures are calculated from the prediction scores for January 2002 from December 2001 and the actual out-of-sample returns at January 2002. Considering the stability of these evaluation results, it is necessary to consider a stock universe with at least a few dozen members in each category. Table 2 shows the monthly average numbers for the stock universe for the evaluation period from December 2001 to November 2016 with top and bottom stocks for both tertiles and quintiles. The total number of stocks exceeds 300; moreover, for quintiles, the total number of top and bottom stocks exceeds 100. Therefore, we consider that the size of the stock universe is adequate.

3.5   Compared Models

Neural Networks. All of the neural networks examined this paper are fully-connected feedforward neural networks. Table 3 shows all 16 types of the neural networks. The number in the "Hidden layers" column represents the number of units. For multiple layers, the layer numbers are connected with hyphens. We examine a total of 8 patterns of deep neural networks (DNN) with 8 layers (DNN8) and with 5 layers (DNN5). The dropout rate is set uniformly to 50%. The number of units in each layer is designed to decrease as the layer becomes closer to the output layer. The patterns of DNN5 are designed to exclude duplicated layers of DNN8. For the conventional three-layer architectures, there are 8 patterns in total, 4 patterns with dropout rate set to 50%

Table 2. Number of stocks (monthly average).

All	Tertile		Quintile	
	Top	Bottom	Top	Bottom
336.5	112.5	111.8	67.7	66.9

(NN3_DO) and 4 patterns with dropout rate set to 0% (NN3). For NN3_DO, the number of units of the hidden layer is adjusted so as to be approximately equal to the number of parameters (all weights including bias) of each pattern of DNN8. For example, the total number of parameters for NN3_DO_1, with 244 units in the hidden layer, is 30,989. This is approximately equal to 30,931 parameters for DNN8_1. For the number of units in the hidden layer of NN3, we select 4 large units in order from all the hidden layers of DNN8. As an intersection, we use the hyperbolic tangent as the activation function, Adam [14] for the optimization algorithm. The mini batch size is the size of the stock universe at each time point (approximately 300) with 100 epochs. We use TensorFlow for implementation. We initialize the biases to be 0 and generate the initial weight from TensorFlow's function "tf.truncated_normal" set to mean "0" and standard deviation "$1/\sqrt{M}$" (M is the size of the previous layer).

Support Vector Regression and Random Forests. Support vector regression (SVR) and random forests (RF) are implemented with scikit-learn. The problem definition for SVR is ε-SVR [15] which is implemented with the class "sklearn.svm. SVR". For hyper-parameters C, gamma, epsilon, we implement 24 patterns of combinations of C = {0.1, 1.0, 10.0}, gamma = {0.0001, 0.001, 0.01, 0.1}, epsilon = {0.01, 0.1}. As an intersection, we use Radial Basis Function (RBF) as the type of kernal. We also define RF as a regression problem [16], and implement with the class "sklearn.ensemble.RandomForestRegressor" in scikit-learn. For hyper-parameters max_features, max_depth, we implement 37 patterns added by 16 patterns of combinations of max_features = {5, 10, 15, 20}, max_depth = {3, 5, 7, 9} and 21 patterns of combinations of max_features = {25, 30, 35}, max_depth = {3, 5, 7, 9, 11, 15, 20}. As an intersection, we set n_estimators (number of trees) to 1,000.

Table 3. Architectures of neural networks.

Neural networks	Architectures		
	Number of layers	Hidden layers	Dropout
DNN8_1	8	100-100-50-50-10-10	50%
DNN8_2		100-100-70-70-50-50	
DNN8_3		120-120-70-70-20-20	
DNN8_4		120-120-80-80-40-40	
DNN5_1	5	100-50-10	50%
DNN5_2		100-70-50	
DNN5_3		120-70-20	
DNN5_4		120-80-40	
NN3_DO_1	3	244	50%
NN3_DO_2		322	
NN3_DO_3		354	
NN3_DO_4		399	
NN3_1	3	70	0%
NN3_2		80	
NN3_3		100	
NN3_4		120	

4 Experimental Results

4.1 Shallow Versus Deep Neural Networks

Table 4 shows results from the neural networks patterns listed in Table 3. All values are monthly averaged. We have conducted a one-sided test of H_0: $p = 50\%$ against H_1: $p > 50\%$ for Direction. The best value for each set of 4 patterns is shown in bold, and the best value in each column is also underlined.

First, we look at CORR. DNN8_3 has the highest value of 0.0591, NN3_3 is the lowest at 0.0437, and values tend to increase as the number of layers increases. It can be confirmed that the DNN group outperforms even NN3_DO, which has had the number of units in its hidden layer adjusted to approximately match the number of parameters in DNN8.

The results for Direction are generally consistent with those for CORR and tend to be better as the number of layers increased in both the tertile and the quintile groups. Direction values reject the null hypothesis at the 0.1% significance level for all the patterns. Top and bottom quintiles had Direction values that are 0.4 to 1.0% higher than those for top and bottom tertiles.

The values of MSE for the loss function are also shown for reference. Although there are no differences between DNN8 and DNN5, it can be seen that the values of MSE are larger when there are fewer layers, as can be seen when comparing against NN3_DO and NN3.

Table 5 shows the average of each category in order to see the tendency of the result by pattern more simply. We can easily confirm that the higher the number of layers, the higher the CORR and Direction.

4.2 Comparison with Support Vector Regression and Random Forests

Table 6 picks out each pattern with the highest CORR from the combination of hyper-parameters in SVR and RF, respectively, described in Sect. 3.5 and also picks out those pattern of neural networks from Table 4 that outperform the highest CORR of the SVR and RF patterns. The best values for each column are labeled in bold. The highest CORR from the combination of hyper-parameters in SVR is {C, gamma, epsilon} = {0.1, 0.01, 0.1}, and RF is {max_features, max_depth} = {25, 7}. For SVR and RF, we find that the tertile and quintile Direction reject the null hypothesis at the 0.1% significance level, and RF outperforms SVR including CORR. Four neural networks have been picked out, all of which are DNN, and three of which are the DNN8 patterns with the largest number of layers. In the rank relationship between CORR and Direction, DNN8_3 with the highest CORR is not completely correlated so that DNN8_3 is not the highest Direction. It is necessary to observe carefully for this in cases where CORR does not differ much. In comparison with SVR and RF, DNN patterns outperform SVR for almost all categories, but show little superiority to RF. These results can not completely indicate the superiority of DNN, including DNN patterns which are not picked up on Table 4. However, in terms of DNN8 patterns, three patterns out of four patterns outperform in CORR and the pattern of DNN8_4

Table 4. Rank correlation, directional accuracy, and mean squared error of neural networks.

Neural networks	CORR	Direction%		MSE
		Tertile	Quintile	
DNN8_1	0.0580	52.56***	53.36***	**0.0834**
DNN8_2	0.0568	52.49***	53.24***	0.0838
DNN8_3	**0.0591**	52.64***	53.37***	**0.0834**
DNN8_4	0.0587	**52.66*****	**53.48*****	0.0837
DNN5_1	**0.0582**	**52.43*****	**53.34*****	**0.0833**
DNN5_2	0.0555	52.25***	53.24***	0.0835
DNN5_3	0.0560	52.36***	53.22***	0.0835
DNN5_4	0.0557	**52.43*****	53.26***	0.0836
NN3_DO_1	**0.0537**	**52.35*****	**52.99*****	**0.0839**
NN3_DO_2	0.0520	52.15***	52.75***	0.0840
NN3_DO_3	0.0509	52.16***	52.94***	0.0841
NN3_DO_4	0.0527	52.24***	52.87***	0.0841
NN3_1	0.0450	52.09***	52.69***	**0.0856**
NN3_2	**0.0472**	52.10***	53.02***	**0.0856**
NN3_3	0.0437	51.79***	52.60***	0.0858
NN3_4	0.0445	**52.23*****	52.61***	0.0859

***$p < 0.001$, **$p < 0.01$, *$p < 0.05$.

Table 5. Rank correlation, directional accuracy, and mean squared error of neural networks (average).

Neural networks	CORR	Direction%		MSE
		Tertile	Quintile	
DNN8_Avg	**0.0582**	**52.59**	**53.36**	0.0836
DNN5_Avg	0.0563	52.37	53.27	**0.0835**
NN3_DO_Avg	0.0523	52.23	52.89	0.0840
NN3_Avg	0.0451	52.05	52.73	0.0857

outperforms RF in all items, hence deep learning promises to be one of the leading machine learning methods.

4.3 Ensemble

Next, we apply ensemble methodology to combine different machine learning models and to examine whether the results improve beyond each individual pattern of Table 6. The monthly prediction scores of SVR, RF and the DNN8_3 with the highest CORR are weighted equally to create the ensemble. Table 7 shows the CORR and Direction for the tertile and quintile portfolios. We find that CORR is the highest at 0.0604, which is higher than each of the three machine learning models before combination. This demonstrates the effectiveness of the ensemble approach. For Direction, on the

Table 6. Rank correlation and directional accuracy of SVR, RF, and DNN.

Machine learning	CORR	Direction%	
		Tertile	Quintile
SVR (best)	0.0569	52.53***	53.30***
RF (best)	0.0576	52.64***	53.44***
DNN8_1	0.0580	52.56***	53.36***
DNN8_3	**0.0591**	52.64***	53.37***
DNN8_4	0.0587	**52.66*****	**53.48*****
DNN5_1	0.0582	52.43***	53.34***

*** $p < 0.001$, ** $p < 0.01$, * $p < 0.05$.

other hand, only quintile portfolio is the highest for the ensemble, so the improvement gained through the ensemble technique is limited.

Table 7. Rank correlation and directional accuracy of Ensemble.

Machine learning	CORR	Direction%	
		Tertile	Quintile
Ensemble	0.0604	52.56***	53.50***

*** $p < 0.001$, ** $p < 0.01$, * $p < 0.05$.

4.4   Long–Short Portfolio Strategy

We have used CORR and Direction as performance measures so far, but in practice when investing based on this information, we need to analyze performance related to return more directly. We construct a portfolio strategy and use risk-adjusted return as a performance measure defined by Return/Risk (R/R) as return divided by risk. As described in Sect. 3.4, we construct a long–short portfolio strategy for a net-zero investment to buy top stocks and to sell bottom stocks with equal weighting in tertile and quintile portfolios. The transaction cost is not taken into account, and we examine the patterns described in Tables 6 and 7.

The results are shown in Table 8. Return is annualized from the monthly average, and Risk is also annualized. The highest R/R is shown in bold for each tertile and quintile portfolio. We find that the highest R/R is DNN in both portfolios, DNN8_3 is 1.24 in tertile and DNN5_1 is 1.29 in quintile. Let us focus on the quintile profiles to analyze the rank relationship between Direction in Tables 6 and 7 and R/R. RF, which is higher for Direction, is the lowest for R/R, and conversely, DNN5_1 which is lower for Direction is the highest for R/R. Thus we cannot make clear conclusions.

In Table 8, some DNN patterns do not outperform SVR, RF and the ensemble so that we cannot show the complete superiority of DNN, but we can note that the pattern which has the highest R/R for each tertile and quintile comes from the DNN patterns.

Table 8. Long–short portfolio strategy performance of SVR, RF, DNN, and Ensemble.

Machine learning	Tertile			Quintile		
	Return%	Risk%	R/R	Return%	Risk%	R/R
SVR (best)	8.38	7.46	1.12	11.36	9.56	1.19
RF (best)	9.04	7.37	1.23	10.72	9.59	1.12
DNN8_1	8.88	7.75	1.15	11.51	9.69	1.19
DNN8_3	9.52	7.70	**1.24**	11.59	9.52	1.22
DNN8_4	9.33	7.81	1.19	12.02	9.93	1.21
DNN5_1	8.41	7.56	1.11	12.32	9.51	**1.29**
Ensemble	8.96	7.45	1.20	11.59	9.85	1.18

5   Conclusions

In this paper, we implement deep learning techniques to predict one-month-ahead stock returns in the cross-section in the Japanese stock market. Our conclusions are as follows:

- In the comparison of different NN architectures, with more layers, the rank correlation coefficient (CORR) and the directional accuracy (Direction) are high. We find that DNN with greater numbers of layers could increase representational power by repeating nonlinear transformations and improve the prediction accuracy of the cross-sectional stock returns.
- In comparison with SVR and RF, there are 4 patterns of DNN that outperform the CORR of both, while the highest Directions in each tertile and quintile are DNN patterns. Ensemble gives a limited improvement. We also examine the performance of a simple long–short portfolio strategy and find that the best R/R in each tertile and quintile portfolio is selected from DNN patterns. These results cannot completely indicate the superiority of DNN, but deep learning promises to be one of the best machine learning methods.
- We examined only 8 DNN patterns consisting of 8 layers and 5 layers compared with 24 patterns of SVR and 37 patterns of RF, and applied simple fully-connected feedforward networks. Application of recurrent neural networks, which are designed to handle time series data, is a candidate for future research. We expect that an investigation of various deep learning models could further enhance the prediction accuracy of stock returns in the cross-section.

References

1. Subrahmanyam, A.: The cross-section of expected stock returns: what have we learnt from the past twenty-five years of research? Eur. Financ. Manag. **16**(1), 27–42 (2010)
2. Harvey, C.R., Liu, Y., Zhu, H.: … and the cross-section of expected returns. Review. Finan. Stud. **29**(1), 5–68 (2016)

3. McLean, R.D., Pontiff, J.: Does academic research destroy stock return predictability? J. Finan. **71**(1), 5–32 (2016)
4. LeCun, Y., Bengio, Y., Hinton, G.: Deep learning. Nature **521**(7553), 436–444 (2015)
5. Goodfellow, I., Bengio, Y., Courville, A.: Deep Learning. MIT Press, Cambridge (2016)
6. Atsalakis, G.S., Valavanis, K.P.: Surveying stock market forecasting techniques–Part II: Soft computing methods. Exper. Syst. Appl. **36**(3), 5932–5941 (2009)
7. Soni, S.: Applications of ANNs in stock market prediction: a survey. Int. J. Comput. Sci. Engineering. Technol. **2**(3), 71–83 (2011)
8. Olson, D., Mossman, C.: Neural network forecasts of Canadian stock returns using accounting ratios. Int. J. Forecast. **19**(3), 453–465 (2003)
9. Cao, Q., Leggio, K.B., Schniederjans, M.J.: A comparison between Fama and French's model and artificial neural networks in predicting the Chinese stock market. Comput. Oper. Res. **32**(10), 2499–2512 (2005)
10. Kryzanowski, L., Galler, M., Wright, D.: Using artificial neural networks to pick stocks. Finan. Anal. J. **49**(4), 21–27 (1993)
11. Krauss, C., Do, X.A., Huck, N.: Deep neural networks, gradient-boosted trees, random forests: Statistical arbitrage on the S&P 500. Eur. J. Oper. Res. **259**(2), 689–702 (2017)
12. Dixon, M., Klabjan, D., Bang, J. H.: Classification-based financial markets prediction using deep neural networks. CoRR(abs/1603.08604)
13. MSCI Inc. Tokyo branch.: Handbook of MSCI Index, MSCI Inc, February 2017. In Japanease
14. Kingma, D.P., Ba, J.: Adam: a method for stochastic optimization. CoRR(abs/1412.6980)
15. Smola, A.J., Schölkopf, B.: A tutorial on support vector regression. Stat. Comput. **14**(3), 199–222 (2004)
16. Breiman, L.: Random forests. Mach. Learn. **45**(1), 5–32 (2001)

Vine Copula-Based Asymmetry and Tail Dependence Modeling

Jia Xu[(✉)] and Longbing Cao[(✉)]

Advanced Analytics Institute, University of Technology Sydney, Ultimo, Australia
Jia.Xu-3@student.uts.edu.au, Longbing.Cao@uts.edu.au

Abstract. Financial variables such as asset returns in the massive market contain various hierarchical and horizontal relationships that form complicated dependence structures. Modeling these structures is challenging due to the stylized facts of market data. Many research works in recent decades showed that copula is an effective method to describe relations among variables. Vine structures were introduced to represent the decomposition of multivariate copula functions. However, the model construction of vine structures is still a tough problem owing to the geometrical data, conditional independent assumptions and the stylized facts. In this paper, we introduce a new bottom-to-up method to construct regular vine structures and applies the model to 12 currencies over 16 years as a case study to analyze the asymmetric and fat tail features. The out-of-sample performance of our model is evaluated by Value at Risk, a widely used industrial benchmark. The experimental results show that our model and its intrinsic design significantly outperform industry baselines, and provide financially interpretable knowledge and profound insights into the dependence structures of multi-variables with complex dependencies and characteristics.

1 Introduction

Modeling complex dependence structures of financial variables is a fundamental research problem in the financial domain, useful for a wide range of applications including economics prediction and risk management. Its extreme importance has been partially demonstrated in the 2008 global financial crisis (GFC). Existing studies are usually concerned with the degree of dependence rather than the other important respects of dependence – the dependence structure, especially the asymmetric and tail dependence characteristics. However, as demonstrated in GFC, it is useless when all stocks tend to fall as the market falls.

Asymmetric dependence between different markets can be easily seen from Fig. 1. Figure 1(a) shows the correlation of daily returns between the United States comprehensive index S&P500 and the index of Eurozone stocks STOXX50E, which indicates the strong negative dependence and normal positive dependence between them. The dependence between the United Kingdom comprehensive index FTSE100 and the foreign exchange rate GBP against the USD is shown in Fig. 1(b), which indicates strong dependence on both sides.

© Springer International Publishing AG, part of Springer Nature 2018
D. Phung et al. (Eds.): PAKDD 2018, LNAI 10937, pp. 285–297, 2018.
https://doi.org/10.1007/978-3-319-93034-3_23

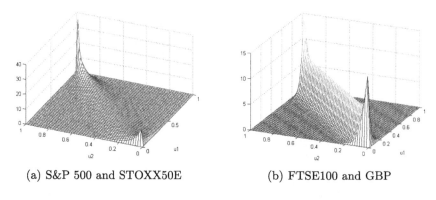

(a) S&P 500 and STOXX50E   (b) FTSE100 and GBP

Fig. 1. Asymmetric dependence across markets

These examples show that the financial market are not only dependent but also asymmetric.

The challenge of modeling dependence in financial market lies in the three major aspects concerning us in this paper. (1) As with any complex behavioral and social system, the cross-market dependence structure is often embedded with strong couplings [5] on high dimensionality. Flexible dependence structure without imposing any assumptions or restrictions are desired. (2) Financial variables, such as daily return, have been shown to follow non-normal distributions, which means dependence models should cover a wide range of dependencies in order to capture both positive and negative dependencies. (3) As discussed above, various lower and upper tail dependencies also need to be considered.

The dependence across markets has been studied by different communities, including statistics and machine learning. The typical approaches in the statistical community are joint distributions with Gaussian assumption and conditional correlation. The first method has been demonstrated that Gaussian assumption is inappropriate when studying either stock markets or exchange rate markets. The second one uses conditional correlation to calculate the covariance, which is generally used in empirical studies. As the current correlation depends on the previous one, the dependence structure is not flexible. The dependence studies in the machine learning community consist of hidden Markov models and graphical probability models. The hidden Markov models, however, could have a large number of hidden states when applied to a high dimensional case, which invariably leads to computational intractability in the algorithms when inferring the hidden states from observations. The graphical probability models, such as Bayesian logic program [12], impose unrealistic assumptions in constructing dependence structures.

In recent decades, a number of research works based on regular vine model capture the asymmetric dependence in currency markets and show decent effects [6,7,11,15]. A popular methodology is to combine the time series models (e.g. ARMA, GARCH) and copula to observe the joint distribution on multivariates. With this framework, we can simplify the problems about observing joint distribution into two parts: the marginal distribution of each variable and copula between variables. The multivariate Archimedean copula family has been studied in [2,13]. They show that Gaussian Copula models do not have lower and upper tail dependence, while the multivariate t copula does not have flexible tail dependence as the symmetric structure of t copula. Vine structures including canonical vine and D vine copula models [1,3] can implement a wide range of dependencies by decomposing the multivariate copula into different bivariate copulas. However, due to the structure assumptions, they do not have flexible dependence structures. Due to the assumptions imposed on dependence structures, their dependence structures may not reflect the actual dependence in real world.

In order to model the asymmetric dependencies in multivariate data with various dependence structures, we propose a new weighted partial regular vine copula model (WPRV) with asymmetric dependencies. WPRV is more powerful, because: (1) A new partial correlation-based algorithm constructs the regular vine structure. Our WPRV can uniquely determine the correlation matrix and is algebraically independent without any strong restriction on the dependence structure; The dependence structure is more flexible, since the current tree structure is independent of the established tree structure and bivariate copulas selection. (2) The bivariate copula with different types of tail dependencies (e.g., BB1, survival BB1, BB7 and survival BB7) are implemented to capture various tail dependencies between financial variables. (3) The moving trends of lower and upper tail dependence with the multivariate data structures and also the trends of lower and upper tail dependence during the dynamic period are analyzed.

The rest of paper is structured as follows. Section 2 introduces the related definitions of copula and different tail dependencies. Section 3 discusses how to construct our weighted partial regular vine copula model, copula family selection, and the parameter estimation in partial regular vine copula and marginal distribution. The case study results are shown in Sect. 4. Finally, Sect. 5 concludes the paper.

2 Preliminaries

2.1 Vine Copula

Vine theory was introduced in [4], which is one kind of graphical models. Let V, T, E and N represent vine structure, trees, edges, nodes respectively. The regular vine and its related definitions are given below.

Definition 1 (Regular Vine). *V is a regular vine on n variables if*

(1) T_1 is a tree with nodes $N_1 = 1, ..., n$ and a set of edges denoted by E_1;
(2) For $j = 2, ..., n - 1$, T_j is a tree with nodes $N_j = E_{j-1}$ and edge set E_j;
(3) (proximity condition) For $j = 2, ..., n-1$ and $a, b \in E_j$, $\#(a \triangle b) = 2$, where $\triangle$ denotes the symmetric difference operator and $\#$ denotes the cardinality.

Definition 2 (Complete Union, Conditioning and Conditioned Sets of an Edge). *The complete union of an edge* $e_i \in E_i$ *is the set* $U_{e_i} = \{n_1 \in N_1 \mid \exists e_j \in E_j, j = 1, 2, \ldots, i-1 \text{ with } n_1 \in e_1 \in e_2 \in \ldots \in e_{i-1} \in e_i\} \subset N_1$. *For* $e_i = \{a, b\} \in E_i, a, b \in N_i, i = 1, 2, \ldots, n-1$, *the conditioning set of an edge* e_i *is* $D_{e_i} = U_a \cap U_b$, *and the conditioned sets of an edge* e_i *are* $C_{e_i,a} = U_a \backslash D_{e_i}, C_{e_i,b} = U_b \backslash D_{e_i}$ *and* $C_{e_i} = C_{e_i,a} \cup C_{e_i,b} = U_a \triangle U_b$, *where* $A \triangle B := (A \backslash B) \cup (B \backslash A)$ *denotes the symmetric difference of two sets.*

Hence, U_{ei} is a set of all nodes in N_i that are connected by the edges e_i. By definition, $U_{ei}(1) = e_i$. The constraint set is defined below.

Definition 3 (Constraint Set). *The constraint set for V is a set:*
$$CV = \{(\{C_{e_a}, C_{e_b}\}, D_e) \mid e \in E_i, e = \{a, b\}, i = 1, \ldots, n-1\}$$

The edge e can be written as $\{Ce \mid De\}$, or $\{C_{e(a)}, C_{e(b)} \mid D_e, e = \{a, b\}\}$, where the conditioning set D_e is shown to the right of "|", and the conditioned set C_e to the left. $\{U_a \backslash D_e\}$ is the set which includes all variables in the set U_a, but excludes the variables in the set D_e.

2.2   Tail Dependencies

One important copula-based dependence measurement is *tail dependence coefficient*, which indicates the dependencies between extreme events. The extremal dependence of a multivariate distribution F can be described by various tail dependence parameters of its copula C. Suppose that random vector $(U_1, \ldots, U_n) := (F_1(x_1), \ldots, F_n(X_n))$ with standard uniform marginal distribution. The lower and upper tail dependence coefficients are defined as follows.

$$\lambda_L = \lim_{u \to 0} Pr\{U_1 \leq u, \ldots, U_n \leq u \mid U_n \leq u\}$$
$$= \lim_{u \to 0} \frac{C(u, \ldots, u)}{u}$$
$$\lambda_U = \lim_{u \to 0} Pr\{U_1 > 1 - u, \ldots, U_n > 1 - u \mid U_n > 1 - u\} \tag{1}$$
$$= \lim_{u \to 0} \frac{\overline{C}(1 - u, \ldots, 1 - u)}{u}$$

where $\overline{C}$ is the survival function of C. If λ_U exists and $\lambda_U \in (0, 1]$, then copula C has an upper tail dependence coefficient, but there is no upper tail dependence coefficient when $\lambda_U = 0$. Similarly, if λ_L exists and $\lambda_L \in (0, 1]$, then copula C has an upper tail dependence coefficient, but no upper tail dependence coefficient when $\lambda_L = 0$.

Frahm et al. [8] proposed a non-parametric method to obtain the nonparametric estimator of lower and upper tail dependence by using *Pickand's dependence function* [14]. One simple nonparametric estimator of tail dependence is the log estimator. which is denoted by :

$$\hat{\lambda}_L = 2 - \lim_{u^* \to 0} \frac{log\left(1 - 2(1 - u^*) + T_{-1}\sum_{t=1}^{T} \mathbf{1}\{U_1 \leq 1 - u^*, ..., U_n \leq 1 - u^*\}\right)}{log(1 - u^*)}$$

$$\hat{\lambda}_U = 2 - \lim_{u^* \to 0} \frac{log\left(T_{-1}\sum_{t=1}^{T} \mathbf{1}\{U_1 \leq 1 - u^*, ..., U_n \leq 1 - u^*\}\right)}{log(1 - u^*)}$$

(2)

In this work, the above nonparametric method is implemented for roughly analyzing the tail dependence coefficient before the regular vine model construction.

3   Our Weighted Partial Regular Vine Model

As highlighted in the introduction, our WPRV is centered on the bottom-to-top regular vine structure. According to the method proposed by Bedford and Cooke [4] for building vine structure by using partial correlation, for elliptical distributions, partial correlation is equal to the corresponding conditional correlation. We can thus use partial correlation instead of conditional correlation to measure the correlations on each node and every tree does not depend on the structure of the previous tree, which is more flexible.

3.1   Partial Regular Vine Construction

The regular vine on n variables shares several important properties (see details in [9]):

(1) There are $(j-1)$ and $(j+1)$ variables in the conditioning sets and constraint sets of an edge of the jth tree respectively;
(2) If two or more nodes have the same constraint sets, they are the same node;
(3) If variable i is a member of the conditioned set of an edge e in a regular vine, then i is a member of the conditioned set of exactly one of the m-child of e, and the conditioning set of an m-child is a subset of De.

According to the above properties, we derive two lemmas, which are important for constructing the partial regular vine tree structure. The two lemmas are given as follows.

Lemma 1. *Let* $I \in \{1, ..., n\}$, $x_1, x_2, y_1, y_2 \in I$ *and* $x_1 \neq x_2$, *the nodes of* T_j *be* $N_1 = \{x_1, y_1 ; I\backslash\{x_1, x_2, y_1\}\}$ *and* $N_2 = \{x_2, y_2 ; I\backslash\{x_1, x_2, y_2\}\}$. *For a regular vine on* n *variables, nodes* N_1 *and* N_2 *have a common m-child. If* $y_1 \neq y_2$, *the common m-child is* $\{y_1, y_2 ; I\backslash\{x_1, x_2, y_1, y_2\}\}$.

Proof. According to Definition 1, each node has two m-children. For N_1, the constraint set CV_{x_1} of its m-children are $\{x_1, I\backslash\{x_1, x_2, y_1\}\}$ and $\{y_1, I\backslash\{x_1, x_2, y_1\}\}$. For N_2, the constraint set CV_{x_2} of its m-children are $\{x_2, I\backslash\{x_1, x_2, y_2\}\}$ and $\{y_2, I\backslash\{x_1, x_2, y_2\}\}$. We can see that $\{y_1, I\backslash\{x_1, x_2, y_1\}\}$ and $\{y_2, I\backslash\{x_1, x_2, y_2\}\}$ are equal, but indexed by different variables in a conditioned set. According to Property (4), N_1 and N_2 have a common m-child. If $y_1 \neq y_2$, y_1 and y_2 should be in the conditioned set of the m-child.

Algorithm 1. Tree Structure Construction via A Bottom-to-Top Strategy

Require: Observations of n input variables
1: Calculate all values of partial correlation, and then allocate the smallest absolute value of partial correlation to the node in T_{n-1} (T_{n-1} is the bottom tree).
2: **for** $k = 1, \ldots, n-2$ **do**
3:    **for** $i = n-1, \ldots, \lceil \frac{n}{2} \rceil$ **do**
4:       **if** $T_i > T_k$ **then**
5:          Find variable combinations for nodes on both sides in tree T_i which can minimize the function $|\rho_{c:d}|$, where T_i indicates the ith tree and T_k is tree inverse level tree;
6:       **else**
7:          Find variable combinations for nodes on both sides in tree T_i which can minimize the function of $\sum ln(1 - \rho_{c:d}^2)$
8:       **end if**
9:    **end for**
10: **end for**
11: There will be $n-2$ regular vines as $k = 1, \ldots, n-2$.
12: **return** A group of partial regular vine dependence structure candidatures.

Lemma 2. *For a regular vine on n variables, $j = 2, \ldots, n-1$, the edge e in T_j has only two constraint sets of m-children in T_{j-1}, which are indexed by different variables in a conditioned set.*

Proof. Suppose there are three identical constraint sets indexed by different variables in a conditioned set, according to Property (4), nodes with the same constraint sets should be the same node. Based on Property (5), the variables in the conditioned set will still be in the conditioned set of its m-children. This means that the node will have three variables in its conditioned set, which violates Property (3) and the proximity condition in the regular vine definition. Therefore, one edge has only two constraint sets which are indexed by different variables in a conditioned set.

According to the above properties of regular vine and the two lemmas, we construct the partial regular vine by using Algorithm 1.

3.2   Vine Structure Selection

After building the $\lceil \frac{n}{2} \rceil$ candidate regular vines, the next step is to find the 'Best' regular vine among these candidates. In order to remove the bias by only selecting the strongest correlation on the top, giving a weight to each tree can enhance the influence of the trees on the top or at the bottom, and a balanced structure can be selected. We assume each level is a unit height and the tree inverse level k is the zero potential energy level. Hence, the weight of each level will increase from level k to level $n-1$ and level 1. Since the parameters m, g, K and T are constants in a given environment, Eq. (3) can be simplified as:

$$W = e^{-m_0 h} \tag{3}$$

where m_0 is the parameter and h is the distance from the level of each tree to the tree inverse level. To restrict the value of weight for each level in interval $[0, 1]$, we standardize the weight:

$$
W(h) = \begin{cases} 0.5 \times \dfrac{e^{-m_0(k-h)}}{\sum_{i=1}^{k} e^{-m_0(k-i)}}, & h \in [1, k]; \\[4mm] 0.5 \times \dfrac{e^{-m_0(h-k)}}{\sum_{i=k+1}^{N-1} e^{-m_0(i-k)}}, & h \in (k, N-1]. \end{cases} \tag{4}
$$

where N is the number of variables, k is the tree inverse level, h is the level of a tree and m_0 is a parameter which falls in interval $[0, 1]$.

The 'Best' regular vine structure maximizes the value of function $-ln(D)$, where D is the weighted determinant which is calculated by using:

$$
D = \prod_{i,j}(1 - W_i \rho^2_{i,j;d(i,j)}) \tag{5}
$$

where W_i is the corresponding weight and $d(i, j)$ is the conditioning set excluding variables i and j. The corresponding conditioned set is i and j.

3.3   Bivariate Copula Selection

Once the partial regular vine tree structure is identified, the next step is to select bivariate copulas for each edge in all trees. As discussed above, the partial correlation is equal to its corresponding conditional correlation for the elliptical family. This means our partial regular vine tree structure is built based on an elliptical copula family (i.e., Gaussian or t copulas). However, according to the following theorem, the limitation of partial correlation can be removed by mapping the partial regular vine tree structure to typical regular vine via conditional correlation.

Theorem 1. *For any regular vine on n variables, there is one-to-one correspondence between the set of $n \times n$ positive definite correlation matrices and the set of partial correlation specification of the vine.*

The proof of Theorem 1 can be referred to [4], which is omitted here. It shows that there is a one-to-one relationship between the partial regular vine specification and the correlation matrix, which ensures that we can map our partial regular vine tree structure to the typical conditional correlation based-regular vine tree structure. We can then choose bivariate copulas from a large number of copula family candidates, rather than the elliptical copula family. Hence, the limitation can be removed while selecting the bivariate copulas.

There are a huge of copula families, which have various tail dependencies. The detail of tail dependence of copula families are listed in Table 1. To capture the asymmetric characteristics, the BB1, S.BB1, BB7 and S.BB7 copulas are the best choice since they have various lower and upper tail dependencies, which can vary independently from 0 to 1.

Table 1. The tail dependence of copula family

	Lower tail dependence	Upper tail dependence
Gaussian	-	-
t	$2t_{\nu+1}(\mu(\nu,\phi))$	$2t_{\nu+1}(\mu(\nu,\phi))$
Gumbel	-	$2^{-1/\phi}$
Frank	-	-
Clayton	$2^{-1/\phi}$	-
Joe	-	$2 - 2^{1/\phi}$
BB1	$2^{-1/(\phi\delta)}$	$2 - 2^{1/\delta}$
S.BB1	$2 - 2^{1/\delta}$	$2^{-1/(\phi\delta)}$
BB6	-	$2 - 2^{1/(\delta\phi)}$
BB7	$2 - 2^{1/\delta}$	$2 - 2^{1/\phi}$
S.BB7	$2 - 2^{1/\phi}$	$2 - 2^{1/\delta}$
BB8	-	$2^{-1/\phi}$ when $\delta = 1$

S.BB1 and S.BB7 are survival BB1 and BB7 copula respectively. ϕ and δ are parameters of the corresponding copula family. For t copula, $\mu(\nu,\phi) = \left(-\sqrt{\nu+1}\sqrt{\frac{1-\phi}{1+\phi}}\right)$

3.4 Marginal Distribution Specification and Parameter Estimation

For the financial applications of the partial regular vine copula model, we use volatility models (i.e. $ARMA$-$GARCH$ models) as the margins. Typically, let $X_t(t = 0, 1, ..., \mathfrak{T})$ be a time series of the prices of a financial asset, such as the stock market index. The return of financial asset can be defined as $r_t = log(X_t/X_{t-1})$. If there are n assets with returns $r_{t,1}, ..., r_{t,n}$, we first select the appropriate marginal distribution of individual variables (i.e., returns of financial variables), which is a univariate distribution. Due to the characteristics of financial assets, such as volatility cluster, a common choice is $ARMA(1,1)$-$GARCH(1,1)$ with skewed student t innovations, which is defined as follows.

$$r_{t,j} = c_j + \Phi_j r_{j,t-1} + \Theta_j \varepsilon_{j,t-1} + \varepsilon_{j,t},$$
$$\varepsilon_{j,t} = \sigma_{j,t} \cdot Z_{j,t} \tag{6}$$
$$\sigma_{j,t}^2 = \omega_j + \alpha_j \varepsilon_{j,t-1}^2 + \beta_j \sigma_{j,t-1}^2$$

where $j = 1, ..., n, t = 1, ..., \mathfrak{T}$ and $Z_{j,t}$ is the innovations which follow skewed student t distribution.

Let $\theta_j^m = (c_j, \Phi_j, \Theta_j, \omega_j, \alpha_j, \beta_j)$ be the parameter set of marginal distribution, θ^c be the parameters of multivariate copula functions, the multivariate joint log-likelihood is given by:

$$L(\theta_1^m, ..., \theta_n^m, \theta^c) = \sum_{t=1}^{\mathfrak{T}} log f(r_{1,t}, ..., r_{n,t}; \theta_1^m, ..., \theta_n^m, \theta^c)$$

$$= \sum_{t=1}^{\mathfrak{T}} \log c(F_1(r_{1,t}), ..., F_n(r_{n,t}); \theta^c) + \sum_{t=1}^{\mathfrak{T}} \sum_{j=1}^{n} \log f_j(r_{j,t}; \theta_j^m) \tag{7}$$

where the multivariate $c(\cdot\,; \theta^c)$ is denoted as the regular vine model.

4   Case Study

4.1   Data and Marginal Distribution Specification

To evaluate the performance of our model, we use real-world data, involving 12 currency exchange rates against USD. These trading currencies are EUR, GBP, CHF, SEK, CAD, BRL, AUD, NZD, JPY, HKD, SGD, and INR, which are sequentially numbered from $v1$ to $v12$. They represent major currencies in the global market and can be arranged into portfolios. The training data set uses observations from 04/01/1999 to 27/08/2004, a total of 1298 daily returns. Observations from 6/09/2004 to 21/06/2013, a total 1912 daily returns are used for out-of-sample testing. All the data was downloaded from Yahoo Finance (http://finance.yahoo.com/).

As discussed in the last section, the standardized residuals are transferred to uniform data by using the empirical probability integral transformation, which is actually the input of partial regular vine. The raw returns are fitted with univariate $ARMA(1,1) - GARCH(1,1)$ models with the skewed student-t error distribution. The Ljung-Box (LB) test [10] is introduced to remove the autocorrelation among these financial returns. In this experiment, the corresponding p values of the LB test are all greater than the significant value 0.05.

4.2   Regular Vine Copula Structure Specification and Tail Dependence Analysis

The next step is to build our weighted partial regular vine copula model. Figure 2 shows the tree structure built by Algorithm 1. Due to space limitations, we only show the last three trees of our vine structure.

Typically, the selection of m_0 of the WPRV model is determined by the characteristics of data and domain knowledge. According to the discussion in Sect. 3.2, m_0 is restricted to interval $[0, 1]$. Table 2 shows the performance of the Log-likelihood of WPRV with parameter m_0 from 0.1 to 1.0. The high value of the Log-likelihood indicates good performance. According to Table 2, the WPRV model with parameter $m_0 = 0.7$ achieves the best performance.

Table 2. Log-likelihood Performance of WPRV with Parameter m_0

m_0	0.1	0.2	0.3	0.4	0.5	0.6	0.7	0.8	0.9	1.0
LL	2767.23	2767.23	2929.91	2929.91	3031.56	3031.56	3031.56	2929.91	2929.91	2767.23

[a] LL is short for Log-likelihood

Once the structure is identified, the next step is to choose the copula for each edge. As discussed above, the bivariate copula which can provide flexible lower and upper tail dependence is most appropriate to build the partial vine copula model with asymmetric dependence. Based on Sect. 3.3, BB1, S.BB1, BB7 and S.BB7 copulas can provide both lower and upper tail dependencies. Therefore, the BB1, BB7, S.BB1 and S.BB7 copulas are used to build vine copula model with asymmetric dependence to capture the asymmetric characteristics.

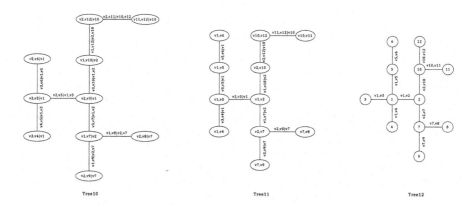

Fig. 2. Weighted partial R vine trees

The tail dependence in Tree 12 of Fig. 2 is shown in Table 3. The nonparametric and t copula results are listed as reference. From the table we can conclude that lower tail dependence of pairs in Tree 1 are less than their corresponding upper tail dependence. Although various bivariate copulas provide different results, similar conclusion can be drawn that lower tail dependencies are less than upper ones.

In order to investigate the tail dependence and its movement trend, two different fixed periods (24 months and 36 months) are used as the investigation

Table 3. Tail dependence analysis by using various copulas

	Non-para[a]		t		BB1		S.BB1[b]		BB7		S.BB7[b]	
	λ_L	λ_U	λ_L	λ_U	λ_L	λ_U	λ_L	λ_U	λ_L	λ_U	λ_L	λ_U
$\{v1, v2\}$	0.18	0.33	0.25	0.25	0.28	0.41	0.39	0.62	0.41	0.48	0.46	0.44
$\{v1, v3\}$	0.45	0.23	0.37	0.37	0.45	0.61	0.50	0.69	0.61	0.68	0.63	0.67
$\{v1, v4\}$	0.50	0.60	0.54	0.54	0.63	0.76	0.71	0.78	0.76	0.82	0.77	0.82
$\{v1, v5\}$	0.73	0.63	0.71	0.71	0.74	0.81	0.78	0.82	0.83	0.85	0.82	0.87
$\{v1, v6\}$	0.28	0.49	0.42	0.42	0.59	0.70	0.65	0.76	0.72	0.74	0.73	0.76
$\{v2, v7\}$	0.29	0.42	0.19	0.19	0.41	0.56	0.49	0.68	0.53	0.63	0.58	0.60
$\{v2, v10\}$	0.53	0.60	0.63	0.63	0.72	0.80	0.78	0.82	0.79	0.83	0.80	0.83
$\{v7, v8\}$	0.28	0.43	0.35	0.35	0.37	0.51	0.49	0.72	0.51	0.58	0.56	0.53
$\{v7, v9\}$	0.20	0.32	0.23	0.23	0.30	0.45	0.39	0.56	0.43	0.51	0.46	0.51
$\{v10, v11\}$	0.55	0.36	0.51	0.51	0.55	0.69	0.65	0.80	0.72	0.80	0.73	0.77
$\{v10, v12\}$	0.38	0.57	0.47	0.47	0.50	0.59	0.53	0.65	0.61	0.65	0.62	0.65

[a] Non-para means that the tail dependence coefficient is calculated via the non-parametric method;
[b] S.BB1 and S.BB7 are the survival BB1 and BB7 copula respectively

period of tail dependence to show the relationship between the length of period and the movement trend. Then, a moving window of 620 daily observations is introduced, from 07/02/2011 to 21/06/2013. The result of pair $\{v1, v2\}$ in tree 1 is shown in Fig. 3.

The gap in a short investigation period with 24 months is larger than those in a long investigation period with 36 months. It indicates that the difference between lower and upper tail dependence is more significant in a short investigated period than in a long one. However, the difference decreases when the length of investigation period increases.

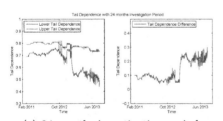

(a) 24 months investigation period

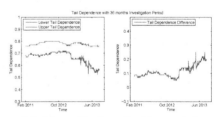

(b) 36 months investigation period

Fig. 3. Lower and upper tail dependence

4.3   Out-of-Sample Performance Analysis

The out-of-sample performance is evaluated by the Value at Risk (VaR), which a widely used industrial benchmark. Typically, backtesting methods based on

Log-likelihood ratios and a null hypothesis consist of unconditional and conditional coverage tests. A large p-value indicates that the VaR forecastings are accurate and reliable. However, the p-value should at least be greater than 0.05. In this experiment, we do not use any machine learning models since they do not directly support forecasting VaR. Table 4 presents the backtesting results of our WPRV, canonical vine and D vine with various copula. The bivariate copula selection is indicated in the second row. The results indicate the BB1 and S.BB1 copulas have the best performance, followed by the BB7 and S.BB7 copulas. The model with t copula is better than Clayton, Gumbel and BB6 copulas which have only one tail dependence.

In conclusion, the results of VaR forecasting indicate that the partial regular vine copula with asymmetric dependence is better than those with symmetric lower and upper tail dependence, and the models with two tail dependencies are better than those with only one tail dependence.

Table 4. The backtesting results of value at risk forecasting

	$1-\alpha$	Weighted partial regular vine							Canonical vine		D vine	
		BB1	S.BB1	BB7	S.BB7	t	Clayton	Gumbel	BB1	BB7	BB1	BB7
LR_{uc}	99%	0.002	0.002	0.100	0.100	0.048	0.354	3.74	0.048	0.248	0.100	0.100
		(0.964)	(0.964)	(0.751)	(0.751)	(0.626)	(0.552)	(0.503)	(0.826)	(0.618)	(0.751)	(0.751)
	95%	0.020	0.068	0.068	0.145	0.051	0.224	1.466	0.145	0.385	0.220	0.145
		(0.899)	(0.794)	(0.794)	(0.703)	(0.621)	(0.636)	(0.264)	(0.703)	(0.309)	(0.488)	(0.225)
	90%	0.473	0.919	0.023	0.175	1.101	1.101	0.258	0.357	0.385	0.423	0.175
		(0.492)	(0.338)	(0.880)	(0.676)	(0.294)	(0.294)	(0.212)	(0.550)	(0.535)	(0.338)	(0.676)
LR_{cc}	99%	0.282	0.282	0.341	0.341	0.369	0.859	0.934	0.369	0.569	0.641	0.541
		(0.869)	(0.869)	(0.843)	(0.843)	(0.831)	(0.651)	(0.609)	(0.831)	(0.831)	(0.467)	(0.869)
	95%	1.436	2.599	2.599	2.479	1.946	1.729	2.662	1.685	2.351	1.436	1.436
		(0.488)	(0.273)	(0.273)	(0.290)	(0.378)	(0.421)	(0.264)	(0.431)	(0.309)	(0.488)	(0.488)
	90%	1.467	2.316	0.862	0.837	1.613	1.997	1.023	1.633	1.374	2.316	1.387
		(0.480)	(0.314)	(0.650)	(0.658)	(0.446)	(0.369)	(0.312)	(0.442)	(0.503)	(0.394)	(0.363)

[a] LR_{uc} and LR_{cc} are short for the likelihood ratio of unconditional and conditional coverage respectively. The first row shows the value, while the corresponding p value is given the parenthesis in the following row. The critical value of LR_{uc} and LR_{cc} are 3.841 and 5.991

5   Conclusion and Future Work

Modeling the dependence between multivariate variables in asymmetric and tail-dependent data is a very challenging task in demanding applications related to big data and financial business, and existing methods cannot handle it well. This paper presents a weighted partial regular model to resolve the issue without imposing restrictions and on the dependence structures. The model is demonstrated through analyzing the complicated structures of portfolios in currency markets. The out-of-sample performance evaluation results highly outperform other methods from statistic and risk evaluation perspectives. Our future work will explore the performance of the regular vine with other high-dimensional time series data.

References

1. Aas, K., Czado, C., Frigessi, A., Bakken, H.: Pair-copula constructions of multiple dependence. Insur. Math. Econ. **44**(2), 182–198 (2009)
2. Abdous, B., Genest, C., Rémillard, B.: Dependence properties of meta-elliptical distributions. In: Duchesne, P., RËMillard, B. (eds.) Statistical Modeling and Analysis for Complex Data Problems, pp. 1–15. Springer, Boston (2005). https://doi.org/10.1007/0-387-24555-3_1
3. Almeida, C., Czado, C., Manner, H.: Modeling high dimensional time-varying dependence using D-vine scar models. arXiv preprint arXiv:1202.2008 (2012)
4. Bedford, T., Cooke, R.M.: Vines: a new graphical model for dependent random variables. Ann. Stat. **30**, 1031–1068 (2002)
5. Cao, L.: Non-IIDness learning in behavioral and social data. Comput. J. **57**, 1358–1370 (2013)
6. Czado, C., Brechmann, E.C., Gruber, L.: Selection of vine copulas. In: Jaworski, P., Durante, F., Härdle, W. (eds.) Copulae in Mathematical and Quantitative Finance. LNS, pp. 17–37. Springer, Heidelberg (2013). https://doi.org/10.1007/978-3-642-35407-6_2
7. Dissmann, J., Brechmann, E.C., Czado, C., Kurowicka, D.: Selecting and estimating regular vine copulae and application to financial returns. Comput. Stat. Data Anal. **59**, 52–69 (2013)
8. Frahm, G., Junker, M., Schmidt, R.: Estimating the tail-dependence coefficient: properties and pitfalls. Insur. Math. Econ. **37**(1), 80–100 (2005)
9. Kurowicka, D., Cooke, R.M.: Uncertainty Analysis with High Dimensional Dependence Modelling. Wiley, New York (2006)
10. Ljung, G.M., Box, G.E.: On a measure of lack of fit in time series models. Biometrika **65**(2), 297–303 (1978)
11. de Melo Mendes, B.V., Semeraro, M.M., Leal, R.P.C.: Pair-copulas modeling in finance. Financ. Mark. Portf. Manag. **24**(2), 193–213 (2010)
12. Neville, J., Jensen, D.: Relational dependency networks. J. Mach. Learn. Res. **8**, 653–692 (2007)
13. Nikoloulopoulos, A.K., Joe, H., Li, H.: Extreme value properties of multivariate t copulas. Extremes **12**(2), 129–148 (2009)
14. Pickands, J.: Multivariate extreme value distributions. In: Proceedings 43rd Session International Statistical Institute, vol. 2, pp. 859–878 (1981)
15. Riccetti, L.: A copula-garch model for macro asset allocation of a portfolio with commodities. Empir. Econ. **44**(3), 1315–1336 (2013)

Detecting Forged Alcohol Non-invasively Through Vibrational Spectroscopy and Machine Learning

James Large[1], E. Kate Kemsley[2], Nikolaus Wellner[2], Ian Goodall[3],
and Anthony Bagnall[1(✉)]

[1] School of Computing Sciences, University of East Anglia, Norwich, UK
{james.large,anthony.bagnall}@uea.ac.uk
[2] Quadram Institute, Norwich Research Park, Norwich, UK
{kate.kemsley,nikolaus.wellner}@quadram.ac.uk
[3] Scotch Whisky Research Institute, Research Avenue North, Edinburgh, UK
ian.goodall@swri.co.uk

Abstract. Alcoholic spirits are a common target for counterfeiting and adulteration, with potential costs to public health, the taxpayer and brand integrity. Current methods to authenticate spirits include examinations of superficial appearance and consistency, or require the tester to open the bottle and remove a sample. The former is inexact, while the latter is not suitable for widespread screening or for high-value spirits, which lose value once opened. We study whether non-invasive near infrared spectroscopy, in combination with traditional and time series classification methods, can correctly classify the alcohol content (a key factor in determining authenticity) of synthesised spirits sealed in real bottles. Such an experimental setup could allow for a portable, cheap to operate, and fast authentication device. We find that ethanol content can be classified with high accuracy, however methanol content proved difficult with the algorithms evaluated.

Keywords: Classification · Spectroscopy · Non-invasive
Authentication

1 Introduction

Up to 25% of licensed premises in some parts of the UK have been found to have counterfeit alcohol for sale[1]. Brown-Forman, the company that makes Jack Daniels, estimates that around 30% of all alcohol in China is fake[2].

Counterfeit alcohol poses a health risk to the consumer, as illegally produced spirits may contain harmful contaminants such as methanol, an economic risk

---

[1] http://www.bbc.co.uk/news/uk-12456360.
[2] https://www.theguardian.com/sustainable-business/2015/sep/16/china-fake-alcohol-industry-counterfeit-bathtub-booze-whisky.

© Springer International Publishing AG, part of Springer Nature 2018
D. Phung et al. (Eds.): PAKDD 2018, LNAI 10937, pp. 298–309, 2018.
https://doi.org/10.1007/978-3-319-93034-3_24

due to the avoidance of taxes, and a risk to brand integrity in cases where the fakes are being sold as named brands.

Forgeries can sometimes be detected through external appearance such as inconsistent labelling or bottling, but currently there is no way to conclusively tell whether spirits are forged without opening the bottle to analyse a sample directly. Breaking the seal and taking samples from a bottle is effectively a destructive process, because even if authenticity is confirmed the bottle cannot later be sold on store shelves or at auction, and collectors' whisky will be greatly devalued. Also, testing of samples can be an expensive and time consuming process that is not suitable for mass screening. No matter what process is used it will require one or more of: transport of the sample to a centralised lab; expert knowledge and handling; consumable materials used in the analysis; and time for methods such as chromatography. It is therefore desirable to develop a system that can non-invasively determine authenticity of a suspect bottle in a cheap, portable, simple and fast manner.

Vibrational spectroscopy in combination with modern machine learning techniques provides a promising potential solution to these problems. Ever improving computing power, spectroscopy equipment and algorithms mean that on-site classification using cost effective equipment is becoming evermore feasible.

The alcohol concentration of genuine spirits in the UK is tightly controlled. For example, Scotch whisky must contain the level stated on the bottle to within 0.3% (v/v). Forgeries typically do not have this level of quality control, with the alcohol content often being lower than reported. Alternatively, methanol and many higher alcohols and heavy metals have regulations prohibiting their presence in spirits to within certain maximal concentrations to ensure safe consumption, and are also tightly controlled. Both ethanol level and methanol level can in principle be characterised by vibrational spectroscopy, and ultimately determined with chemometric and machine learning techniques.

We wish to evaluate to what extent non-invasive determination of alcohol concentrations in arbitrary sealed bottles using vibrational spectroscopy is possible and worth pursuing. We describe experiments carried out on synthesised alcohol-water solutions, analysed through-bottle using near infrared spectroscopy (NIRS), and classified using a set of benchmark machine learning algorithms into 'genuine' and 'forged' categories based on their ethanol and methanol concentrations.

First, related work is reviewed in Sect. 2, and an overview of the data collection process and a high-level analysis of it is given in Sect. 3. The experimental and evaluation methods used are outlined in Sect. 4 and results presented in 5, before conclusions are drawn in Sect. 6.

2   Background

2.1   Spectroscopy

Vibrational spectroscopy (VS) is the term used to describe two complementary analytical techniques, infrared spectroscopy (IRS) and Raman spectroscopy (RS). These are non-destructive, non-invasive tools that provide information about the molecular composition of a sample by measuring the intensities of different vibrational interactions between a light source and sample. The spectra produced by VS acts like a fingerprint for the contents, and can be used qualitatively and quantitatively for identification, characterisation, and quality control.

VS methods, along with many competing techniques, are much researched within the food and drink sector [4,9] due to VS's non-invasive and relatively low operating-cost nature as an analytical technique. However, VS suffers from lower discriminatory power when compared with more time consuming and destructive techniques such as gas or liquid chromatography, one of which is often the technique used to determine the ground truth of studied samples.

As early as 2005, [10] carried out a comparison of NIR and Raman spectrometries for their suitability in combination with regression techniques for the determination of alcohol content in whisky and vodka contained within clear and coloured glass bottles. The study was conducted to evaluate the techniques for possible use in non-invasive, in-situ quality assurance in bottling plants.

Univariate regression models for each type of drink were calibrated for the Raman data in the first derivative spectrum, while a multivariate Partial Least Squares (PLS) model was calibrated for the NIR data. The latter calibration procedure involved some optimisations on the test data, and therefore the results specifically should be treated with caution. However, the higher level conclusions in terms of the relative difficulty of different aspects of the experiments are still insightful: that differences between bottles accounted for the greatest variation and difficulty in the analysis, relative to differences in bottle positioning and time of measurement. The authors concluded that both NIR and Raman were not suited to the analysis of samples within coloured glass in particular, due to the effect of large amounts of fluorescence on the spectra. They also found that for the doubly-transmitted NIR method, a signal could not be collected from the widest part (70 mm path length) of the largest bottles, whereas comparable signals to that of the smallest bottles could be found by measuring through the neck of the bottle (40 mm path length).

More recent work in this area is described in [7]. The ability for Raman spectroscopy to analyse and discriminate between certain Scotch whisky production factors from within their original containers is tested. 44 whisky samples, three of which had samples transferred to glass vials due to their original bottles being made of green glass, were measured directly through the glass walls using an Avantes Raman instrument. Although not detailed in full, the authors note that the location of measurement (from the neck, base or centre) had no influence on the quality of the readings. Furthermore, the stability of the sampling suggested excellent reproducibility, with normalised spectra being 'virtually identical'.

In an initial Principal Component Analysis (PCA) visualisation, separation could already be found between the type of cask each whisky was matured in. However, factors such as the source distillery and use of artificial caramel colourings could not be defined by the first three principal components (PC).

PLS Regression (PLSR) and Principal Component Regression (PCR) were subsequently evaluated. However PLSR reportedly delivered far better results and was therefore the only method discussed. Leave-one-out cross validation was used as an evaluation procedure. A quantitative analysis of important factors related to authentication was described: age; ethanol concentration; and a second attempt at the presence of artificial colourings. Age could be estimated within 0.42 years (root mean squared error (RMSE)), from the samples in the range 3–22 years. On average ethanol concentration could be estimated to within 0.44% (RMSE), which is only just outside the regulatory limits of Scotch Whisky (0.3%). These are very strong results, suggesting the feasibility of quantitatively determining key factors to whisky authentication.

Continuing on from these works, our own investigation into this problem focuses on portability, simplicity, and speed in all aspects of the analysis of a sample. The final aim is to allow a non-expert to determine the authenticity of an arbitrary spirit on-site and within seconds.

2.2   Classification

Classically, machine learning and chemometric methods handling spectral data have been linear regressive models built on top of (automatically or manually) selected attributes or PCA-transformed spaces. The physical interactions giving rise to the spectra are understood to be linear in nature, and the resonances of molecules being looked for are known to occur at certain wavelengths, even if the particular wavelengths are not known a priori. Given this, more complex systems may not have much room to increase predictive accuracy, be prone to overfitting, and in some cases may lose interpretability of results. Linear systems on reduced attribute spaces work satisfactorily for clean spectra collected under professional and standardised conditions. However, they may be unable to handle structural changes in the data.

The nature of the problem suggests a regression model. However, through consultation with industry, the ultimate use case designed to aid field use is a traffic light classification scheme; green (genuine), yellow (suspect), and red (forged). The confidence thresholds for each class can be set by the user in response to factors such as the costs of verification and screening.

The classification of spectra can be phrased as a time series classification (TSC) problem [1]. A time series is a set of (typically numerous) ordered and numeric attributes. While different sets of TSC data will have different underlying properties, the typical higher-order structures informing classification are the shapes and patterns of series and/or subseries.

Recent large scale evaluations on entire dataset archives in both traditional classification [6] and TSC [1] give indications as to the classification methods that could be suitable for this particular problem space. [1] found that for spectral datasets (of which there were 7 of 85 datasets in total) classifiers that considered the full series similarity were consistently better than those considering subseries similarity, frequency or distribution. Throughout both evaluations, the effectiveness of ensembling was clearly evident. In the benchmark experiments presented in this work we use a range of classifiers classically used in chemometrics, in addition to state of the art and ensemble classification methods.

3   Data

There are many experimental factors that could confound a non-invasive alcohol classification system in the field: ambient light and environmental conditions; variation in spectral hardware; and the measurement habits of different users may all cause variation in the resulting spectra. However, we believe one of the largest sources of variation which needs to be accounted for arises from the properties of the bottle a sample is contained in. Bottle shape and size, glass thickness and colour, and interfering labeling and embossing can all work to frustrate the collection of consistent, reliable spectra. Therefore, with these experiments, we primarily wish to determine the difficulty of measuring and classifying the alcohol content of samples in arbitrary bottles.

We have conducted experiments using 44 different examples of real, non-standardised bottles. While most of the bottles are transparent and cylindrical, some are coloured, rectangular or skewed. Using a single StellarNet BLACK-Comet-SR spectrometer, transmission near-infrared spectra over a one second integration time of ethanol, methanol and water solutions within each bottle were collected to form two datasets. For the ethanol concentration experiments, 40% ethanol (with the remainder being water) is taken to be the 'genuine' case, while concentrations of 35% and 38% ethanol are taken to be 'forgeries'. The second dataset is detecting the presence of methanol. With 40% total alcohol concentration being maintained, solutions with 1%, 2% and 5% methanol (v/v) form the forged class, while 0% methanol (i.e 40% ethanol) constitutes not forged. The two classification problems are therefore to determine from a spectra whether or not a solution within an arbitrary sealed bottle (1) has less than 40% alcohol or (2) contains dangerous levels of methanol. Information on the bottles and the raw data, including labels for bottle and concentration for each reading, can be downloaded at[3].

Three batches of each alcohol concentration were produced, and for each solution in each bottle the sample is placed, a spectra taken, and placed again for a total of three readings. A total of over 2000 readings were taken. Bottles were positioned such that the light travels through the widest part of the bottle while avoiding labelling, embossing and seals as much as possible. However, to mimic future conditions a precise recreation of the exact path on each placement was intentionally not attempted. For simplicity, and to mimic a possible portable sampling station, the geometry of the light source and receiver was fixed at 15 cm; enough to accommodate the widest bottles tested. Dark readings were subtracted from each spectra. Data collection took place over the course of multiple weeks by a single tester. Batches of each concentration were spread out over that time-frame, to reduce the chance of any patterns based on time of measurement forming. Spectra are presented in the wavelength range 876.5 nm–1101 nm, sampled every 0.5 nm, and each spectrum is standardised.

To help give an intuition of the classification problem, Fig. 1 shows the average series of each class to demonstrate their differences. The progressively shaded

---

[3] http://research.cmp.uea.ac.uk/DetectingForgedAlcohol/Data/.

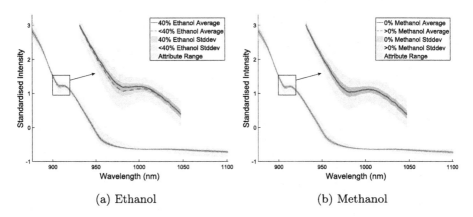

(a) Ethanol (b) Methanol

Fig. 1. Graphs showing the average series of each class, overall standard deviation and range for the ethanol and methanol concentration datasets. For each image, the main discriminatory region is zoomed.

regions show the overall standard deviation and range of intensities at each wavelength. The fact that these are difficult to distinguish by eye is itself quite telling. The overall variance in the dataset is very low, and the inter-class variance a fraction of that.

The zoomed regions show the wavelength ranges where alcohols are known to have a strong resonance. A clear separation between classes can be seen within the ethanol problem. However, for methanol the classes appear to be indistinguishable. Ethanol and methanol have overlapping resonances, and therefore the fact that the overall concentration of alcohol remains at 40% means any difference between the class values in the resulting spectra is drastically reduced.

Relative to the apparent differences in the average class spectra, individual series are greatly affected by noise introduced by a variety of means through the nature of the experiment, further increasing classification difficulty. For example, an individual series may be skewed by the lensing effects of a uniquely shaped bottle. This is evidenced by Fig. 2. It shows the first three PCs of the transformed ethanol dataset, which explain 95% of the total variance. In (a), the instances are categorised by their ethanol concentrations. While some separation is found between the two classes, this is observed mostly in the second and third PCs, which account for only 17% of the total variation. The first PC, as (b) shows, for the most part explains variance due to the bottles. This is in line with our expectations that bottle variation would be one of the larger obstacles to overcome for the final use case of an authentication system. While many bottles are clustered close together, there are some that form clear and separate clusters of their own. As might be expected, these are bottles that have some particularly non-standard bottle property, such as irregular shape or colour.

Promisingly, the PCA transform does suggest a good separation between ethanol concentrations within a particular type of bottle, as illustrated by the outlying bottle clusters when compared between figures. The equivalent figures

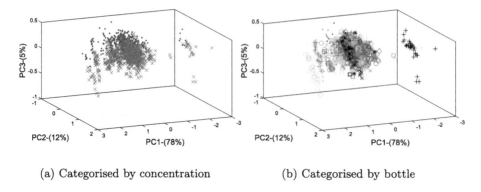

(a) Categorised by concentration   (b) Categorised by bottle

Fig. 2. Graphs of the top three PCs of the PCA-transformed ethanol forgery dataset, with samples categorised by (a) 'genuine' (blue dot) and 'forgery' (red cross), and (b) by bottle. (Color figure online)

for the methanol dataset are not included in this paper for the sake of readability and space, however they (and the source ethanol images including keys) are available on-line[4]. What they show is analogous to Fig. 1(b); that the PCA is almost entirely unable to distinguish between the alcohol concentrations, however trends by bottle type are largely the same.

4   Experimental Setup

For this application, our long-term hypothesis is that TSC methods that consider overall shape may be able to correct for structural defects in the spectra brought about by the many sources of noise involved with non-invasive spectra collection. For example, a linear method built on a small number of selected attributes may not be able to account for high-level structural changes in a new test case caused by an abnormally shaped bottle. Using the datasets that have been formed, we perform benchmark and exploratory evaluations with a wide variety of classification schemes.

The classifiers evaluated are: Logistic Regression (LR); Partial Least Squares Regression (PLSR); Multilayer Perceptron (MLP); 1-Nearest-Neighbour with Euclidean Distance (1NN); C4.5 Decision Tree (C45); linear SVM (SVML); quadratic SVM (SVMQ); radial basis function SVM (SVMRBF); Rotation Forest [11] (RotF); Random Forest [3] (RandF); Heterogeneous Ensemble of Standard Classification Algorithms [8] (HESCA), and for the TSC-specific classifiers: Random Interval Spectral Ensemble [2] (RISE); Bag of SFA Symbols [12] (BOSS); and Time Series Forest [5] (TSF).

We evaluated each classifier on the datasets using a leave-one-bottle-out (LOBO) cross validation. In this scheme, all samples contained within a single bottle are reserved for the test set, with the remainder forming the training

---

[4] http://research.cmp.uea.ac.uk/DetectingForgedAlcohol/FiguresAndTables/.

set. By evaluating in this manner, classifiers should not be able to leverage any discriminatory features caused by the bottle itself, focusing on alcohol level as the only commonly varying factor.

To avoid ambiguity, we stress that in all cases the training of a classifier, including any hyper-parameter tuning and model selection required, is performed independently on the train set of a given fold, and the trained classifier is evaluated exactly once on the corresponding test set. Our code[5] reproduces the splits used in this evaluation exactly, and results[6] are able to be recreated.

Our primary concern is generally accuracy (ACC) because of its ease of motivation and interpretability. However, in applications such as this the costs of measurement, verification, and misclassification externally influence the ways in which decisions need to be made. For example, if the costs of confirming the legitimacy of a suspect bottle are high, relative to the resources available to the tester, then the decision boundary may be skewed to favour the 'genuine' label. As a result, only samples that the device is more confident are fake will be seized or sent for further analysis. Accuracy cannot entirely capture these factors. Therefore balanced accuracy (BALACC) and measures that assess the quality of the classifiers' probabilistic outputs are also reported; the Log-Likelihood (LL) and the Area Under the Receiver Operating Characteristic (AUROC).

5 Results

Table 1 details the average accuracies achieved by each classifier on all datasets formed for the sake of space, however each subset of experiments is separately discussed in turn with the superscripts in column and section headers denoting the particular results being discussed.

5.1 Leave-one-bottle-out Cross Validation[a]

When considering the LOBO experiments on the original (time series form) data, two trends are immediately apparent: ethanol concentration, with the correct models, can be classified with high accuracy; determining methanol concentration in a constant overall alcohol level is much more difficult. Only some of the classifiers tested achieving higher than the minimum expected accuracy of 0.75, the proportion of the majority class.

To discover what classifiers are best for each evaluation statistics, we can perform statistical tests of difference over fold scores because all classifiers are evaluated on identical splits, which are reproducible with the published code. Figure 3 is a critical difference diagram over all folds of the LOBO-sampled ethanol and methanol datasets combined. Classifiers are ordered by average rank over fold scores, and those connected by a bar are pairwise not significantly different between each other, $p = 0.05$.

---

[5] http://research.cmp.uea.ac.uk/DetectingForgedAlcohol/Code/.

[6] http://research.cmp.uea.ac.uk/DetectingForgedAlcohol/Results/.

Table 1. Average accuracies (and standard deviations) over all folds of the alcohol datasets. The best classification accuracies on each dataset are bold.

Classifier	Ethanol LOBO[a]	Methanol LOBO[a]	PCA Ethanol LOBO[c]	PCA Methanol LOBO[c]	Bottle[b]
1NN	0.866(0.093)	0.672(0.103)	0.779(0.104)	0.627(0.101)	0.541(0.017)
BOSS	0.913(0.086)	0.786(0.050)	-	-	0.622(0.021)
C45	0.824(0.132)	0.658(0.098)	0.796(0.106)	**0.750**(0.001)	0.412(0.017)
HESCA	**0.965**(0.069)	0.843(0.079)	**0.818**(0.104)	**0.750**(0.001)	0.639(0.020)
LR	0.964(0.045)	0.809(0.100)	0.807(0.092)	0.744(0.030)	0.430(0.018)
MLP	0.960(0.068)	0.834(0.083)	0.813(0.108)	**0.750**(0.001)	0.617(0.027)
PLSR	**0.965**(0.053)	0.860(0.073)	0.801(0.089)	0.745(0.026)	0.061(0.010)
RandF	0.888(0.105)	0.758(0.047)	0.817(0.093)	0.714(0.060)	0.587(0.015)
RISE	0.776(0.115)	0.780(0.031)	-	-	0.622(0.016)
RotF	0.938(0.078)	0.839(0.049)	0.815(0.104)	**0.750**(0.001)	0.653(0.014)
SVML	0.945(0.075)	0.838(0.077)	0.801(0.094)	**0.750**(0.001)	0.517(0.017)
SVMQ	0.959(0.103)	**0.864**(0.102)	0.803(0.092)	**0.750**(0.001)	**0.656**(0.019)
SVMRBF	0.881(0.098)	0.841(0.092)	0.806(0.091)	**0.750**(0.001)	0.349(0.015)
TSF	0.868(0.112)	0.769(0.029)	-	-	0.635(0.018)

Because each test fold represents a single bottle, the accuracy on a fold gives an indication of the difficulty that a particular bottle adds to the classification problem. We took the top four classifiers on the ethanol problem (PLSR, LR, MLP, and HESCA) which all achieved similarly strong performances, and looked at which bottles were preventing perfect classification. On 34 of the 44 folds, at least one of these top four classifiers achieved an accuracy of 1. Where only a subset of the four classifiers met this criteria, the rest only ever misclassified only one or two test cases.

The worst fold accuracy represents the Bernheim Original Kentucky Straight wheat whiskey bottle, where the average accuracy across the four classifiers is 0.76. Only five bottles had an average accuracy of less than 0.9, and all of them are irregular in some way. This does lend credence to the idea that the determination of alcohol concentration cannot be done *entirely* irrespective of bottle. However, the fact that there is clearly some transferability (evidenced by better-than-guessing accuracy in this LOBO format) is promising.

In [7,10], coloured glass posed challenges for the collection of Raman spectra, which particularly struggles to handle fluorescence, but also for NIRS in [10]. Our experiments included three green-glass bottles, however on these no significant drop in predictive accuracy was observed in the same analysis of the top four classifiers. These three bottles also showed no clear separation from the largest central cluster in the PCA transform presented in Fig. 2b.

5.2   Classifying the Bottle[b]

The PCA transform of the ethanol dataset, shown in Fig. 2b, indicated that the majority of the variance corresponded with differences in the containing

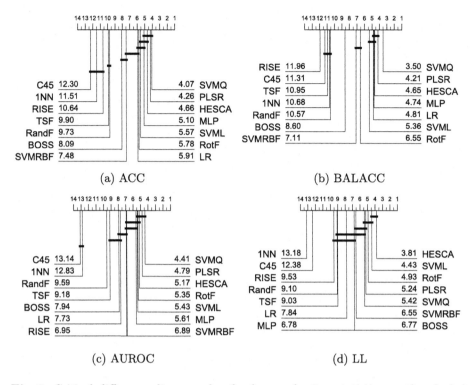

(a) ACC

(b) BALACC

(c) AUROC

(d) LL

Fig. 3. Critical difference diagrams for the four evaluation statistics on the alcohol datasets.

bottle's properties. Further, most of the first PC was caused by a small number of irregularly shaped bottles. The majority of bottles otherwise formed a dense cluster. To further investigate the extent to which features of the bottle are detectable in the spectra, we ran experiments with the same set of classifiers but with the containing bottle as the class label, instead of alcohol concentration. The full dataset was split 30 times using random stratified sampling with a 70/30 train/test split. We would expect the outlying bottles on the PCA transform to be the easiest to classify, while most standard bottles can only be guessed at. The best accuracies achieved were up to 0.656 (SVMQ), which on a 44 class problem is quite high. It is worth noting also that the non-linear methods and especially TSC methods make a relative improvement on this problem, signifying the different nature of the discriminatory features.

In the interest of finding where the classifiers were making their errors, we grouped bottles by whether they could be described as being standard (clear glass and cylindrical, 28 bottles) or irregular (coloured glass and/or non-cylindrical, 16 bottles). Considering the SVMQ's predictions, we counted the incorrect classifications for cases with a standard bottle label classified as a standard bottle, standard classified as irregular, irregular classified as standard, and irregular classified as irregular. The first of these four cases accounts for 69%

of the total error, while the remaining three account for a little over 10% each. When correcting for the number of possible ways to misclassify in each scenario, SVMQ was still twice as likely to make the first kind of error as the last.

These results have positive implications for the original goal of non-invasive alcohol level determination. The fact that it is relatively much easier to mistake one standard bottle for another suggests that a classifier could be reliably trained under the assumption that the test sample bottle has certain properties matching those in the train set. In terms of the practical use and production costs of a device, the worst case is that each individual type of bottle requires its own adequately populated training data for a model to learn on. While this or at least a two-stage classification procedure may still be needed for irregular bottles, a device that can effectively classify the contents of bottles within some particular range of properties is still a worthwhile improvement over the worst case.

5.3   PCA Transforms[c]

Lastly, we repeated the LOBO classification experiments again with PCA-transformed versions of the datasets (calculated and applied to each resample individually), maintaining components that explain 95% of the variance. Analysis of spectral data in the literature often involves a dimensionality-reducing transformation such as PCA, both to highlight discriminatory variance and reduce the computation time of analysis. However, in this case it appears to reduce accuracy relative to classification performed on the time series, in agreement with [7].

The methanol PCA transform seemingly cannot discriminate between concentrations at all, with all classifiers simply picking the majority class. For ethanol, all classifiers except 1NN achieve very similar accuracies. Referring to Fig. 2a, it would seem that most of the classifiers are forming almost identical decision boundaries, the same that a human naively would by eye.

6   Conclusions

We have demonstrated the feasibility of determining alcohol concentration non-invasively in arbitrary bottles using near infrared spectroscopy in combination with machine learning. While ethanol level could be classified with high accuracy, methanol concentration within a consistent overall alcohol level was much more difficult to detect. However, some classifiers demonstrated results significantly better than random guessing, suggesting that the discriminatory features are not entirely lost at the physical hardware level. There may still be room for improvement with different optical geometries and better-tailored data processing and model selection. Bottles with particularly unique properties introduced extra difficulty, but the contents of more standard bottles could be learned and determined with very good transferability.

Traditional methods within chemometrics such as Logistic and Partial Least Squares regression were strong. However, Principal Component Analysis led to significantly decreased performance. A quadratic support vector machine and simple neural network architecture also performed well. A larger computational investment for more thorough tuning would likely lead to improved results for these. A combination of tuning and ensembling these along with the regressive methods is a promising route to follow, especially due to their relatively fast prediction times once trained. Algorithms bespoke to time series classification did not provide an immediate increase in predictive power.

Acknowledgements. This work is supported by the Biotechnology and Biological Sciences Research Council [grant number BB/M011216/1], and the UK Engineering and Physical Sciences Research Council (EPSRC) [grant number EP/M015087/1]. Work was performed in collaboration with the Quadram Institute, Norwich and the Scotch Whisky Research Institute, Edinburgh. The classification experiments were carried out on the High Performance Computing Cluster supported by the Research and Specialist Computing Support service at the University of East Anglia.

References

1. Bagnall, A., Lines, J., Bostrom, A., Large, J., Keogh, E.: The great time series classification bake off: a review and experimental evaluation of recent algorithmic advance. Data Min. Knowl. Discov. **31**, 1–55 (2016)
2. Bagnall, A., Lines, J., Hills, J., Bostrom, A.: Time-series classification with COTE: the collective of transformation-based ensembles. IEEE Trans. Knowl. Data Eng. **27**, 2522–2535 (2015)
3. Breiman, L.: Random forests. Mach. Learn. **45**(1), 5–32 (2001)
4. Danezis, G.P., Tsagkaris, A.S., Camin, F., Brusic, V., Georgiou, C.A.: Food authentication: Techniques, trends & emerging approaches. TrAC - Trends in Analytical Chemistry (2016)
5. Deng, H., Runger, G., Tuv, E., Vladimir, M.: A time series forest for classification and feature extraction. Inf. Sci. **239**, 142–153 (2013)
6. Fernández-Delgado, M., Cernadas, E., Barro, S., Amorim, D.: Do we need hundreds of classifiers to solve real world classification problems? J. Mach. Learn. Res. **15**, 3133–3181 (2014)
7. Kiefer, J., Lynda Cromwell, A.: Analysis of single malt Scotch whisky using Raman spectroscopy. Anal. Methods **91**, 790–794 (2017)
8. Large, J., Lines, J., Bagnall, A.: The Heterogeneous Ensembles of Standard Classification Algorithms (HESCA): the Whole is Greater than the Sum of its Parts (2017)
9. Lohumi, S., Lee, S., Lee, H., Cho, B.K.: A review of vibrational spectroscopic techniques for the detection of food authenticity and adulteration. Trends Food Sci. Technol. **46**(1), 85–98 (2015)
10. Nordon, A., Mills, A., Burn, R., Cusick, F., Littlejohn, D.: Comparison of non-invasive NIR and Raman spectrometries for determination of alcohol content of spirits. Anal. Chim. Acta **548**(1–2), 148–158 (2005)
11. Rodriguez, J., Kuncheva, L., Alonso, C.: The random subspace method for constructing decision forests. IEEE Trans. Pattern Anal. Mach. Intell. **28**(10), 1619–1630 (2006)
12. Schäfer, P.: The BOSS is concerned with time series classification in the presence of noise. Data Min. Knowl. Discov. **29**(6), 1505–1530 (2015)

Research and Application of Mapping Relationship Based on Learning Attention Mechanism

Wanwan Jiang[✉], Lingyu Xu, Jie Yu, and Gaowei Zhang

Shanghai University, No. 99 Shangda Road, Baoshan District, Shanghai, China
jiangwanwan0327@163.com

Abstract. The study on the interactions between different or the same variables of financial markets is an interesting topic. Many efforts have been devoted to investigate this issue. However, there has been little work studying the relationship of the various attributes within the stock, while this relationship is essential for us to have a deeper understanding of stock's internal mechanisms. So in this paper, we explored using sequence-to-sequence model for extracting the relationship of arbitrarily two properties of the stock. We not only give a qualitative description of the relationship between stock's attributes, but also quantify the relationship through the model. The experimental results show that there are certain correlations between the internal attributes of the stock, among which the correlation between $Close\&\%Tuv$ and $\%Chg\&\%Tuv$ are more prominent. In addition, we also conducted the anomaly detection on network public opinion information, and found out the starting points of abnormal events combined with the network news information. By comparing the starting points of the events and the changes in the relationship between stock attributes, we concluded that there is a certain regularity between them.

Keywords: Stock transaction data · Mapping relationship
Time series · Neural network

1 Introduction

In recent years, more and more researchers have been getting to focus on stock market. With time goes by, numerous research methods on the relevance of the stock market have been put forward [1–10]. However, most of these methods have studied the relevance of the stock. There are few scholars, to the best of our knowledge, study the relevance of the various attributes within the stock. As important indicators of the stock market, the properties of the stock are interrelated and interact on each other [11]. So, it is necessary to study the correlation between them. For investors, they can determine whether there are abnormal phenomena based on changes in the relationship between the stock properties to efficiently grasp the trading opportunities to reduce investment

© Springer International Publishing AG, part of Springer Nature 2018
D. Phung et al. (Eds.): PAKDD 2018, LNAI 10937, pp. 310–321, 2018.
https://doi.org/10.1007/978-3-319-93034-3_25

risk. For government regulators, they can guard against stock market risk. Also, for the researchers, they can optimize the existing relational analysis method based on internal relations, and to propose new ideas and practical methods for the future research.

China's stock market is a complex system, which affected by political, economic and other factors. The traditional time series analysis methods, such as Auto-regressive model (AR), Moving-average model (MA) and Auto-regressive and Moving-average model (ARMA), have been slightly inadequate. At the same time, more and more methods of neural network are proposed, such as Deep Neural Network (DNN), Convolutional Neural Network (CNN) and Recurrent Neural Network (RNN) et al. But they can only be applied to problems whose inputs and targets can be sensibly encoded with vectors of fixed dimensionality.

In this paper, we use the sequence-to-sequence model for extracting the relationship among the internal properties of the stock. The advantage of this model is that it is good at dealing with the mapping problem of variable-length sequences, which is characterized by follows: (1) both input and output are sequences; (2) the length of the sequence is not fixed; (3) there is no correspondence between the input and output sequence length.

2 Problem Definition

Stock transaction data is a time series data, because it is sampled according to the order of time continuously. In this paper, we use the sequence-to-sequence learning methods to study the correlation between the internal properties of the stock.

2.1 Definition

Definition 2-1 Multivariate Time Series (MTS)

Multivariate time series refers to a sequence of multiple data at the same time. In general, it can be seen as a combination of multiple univariate time series. We define the MTS as:

$$X =< D_1, D_2, ..., D_k, ..., D_n > \tag{1}$$

where n expresses the number of attributes in X, and D_k is expressed as:

$$D_k =< (t_1, d_{k1}), (t_2, d_{k2}), ..., (t_j, d_{kj}), ..., (t_m, d_{km}) > , 1 \leq k \leq n. \tag{2}$$

In (2), D_k represents the k-th attribute time series in X, t_j represents the sampling time. $d_{kj} \in R$ represents the recorded value of D_k at time t_j, and m represents the length of the time series D_k. Usually the sampling interval $\Delta t = t_{i+1} - t_i$ is the same.

Definition 2-2 Segmented Time Series

A segmented time series is a sequence obtained by the division of the original time series according to the length of the segment. It is denoted as:

$$D_k^{(l)} =< H_1, H_2, ..., H_i, ..., H_{m-l+1} > \tag{3}$$

where l is the length of the segment, and H_i is expressed as follows:

$$H_i = <(t_i, d_{ki}), (t_{i+1}, d_{k(i+1)}), ..., (t_{i+l-1}, d_{k(i+l-1)}) > \qquad (4)$$

Therefore, the segmented time series D'_k corresponding to the original time series D_k can be defined as follows:

$$D'_k = < D_k^{(a)}, D_k^{(a+1)}, ..., D_k^{(l)}, ..., D_k^{(b)} > , a \le l \le b \qquad (5)$$

2.2 Evaluation Method

In this paper, we employed the Pearson correlation coefficient (ρ) to evaluate the correlation between two time series, and calculate mean square error (MSE) value to compare different forecasting series.

$$\rho_{X,Y} = \frac{COV(X,Y)}{\sigma X \sigma Y} = \frac{E(XY) - E(X)E(Y)}{\sqrt{E(X^2) - E^2(X)}\sqrt{E(Y^2) - E^2(Y)}} \qquad (6)$$

where X and Y represent two time series respectively. The value of $\rho_{X,Y}$ is between the interval $[-1,1]$, and $\rho_{X,Y} = 1$ means that X and Y are completely positive correlation; $\rho_{X,Y} = -1$ indicates a complete negative correlation between X and Y; $\rho_{X,Y} = 0$ means that X and Y are completely irrelevant. The larger the $|\rho_{X,Y}|$, the higher the degree of correlation between X and Y.

$$MSE = \frac{\sum_{t=1}^{n} |actual(t) - forecast(t)|^2}{n} \qquad (7)$$

where $actual(t)$ denotes the original series, $forecast(t)$ denotes the forecasting series, n is the length of time series. We define B_{12} as follow:

$$B_{12} = \frac{|MSE_1 - MSE_2|}{MSE_2} \qquad (8)$$

where MSE_1 is the MSE of time series 1, MSE_2 is the MSE of time series 2. When $MSE_1 < MSE_2$, B_{12} means how much MSE_1 is better than MSE_2.

2.3 Problem Definition

The purpose of this paper is to study the correlation between the internal properties of the stock. Hence, according to the above definition, we can describe the stock trading sequence as:

$$X = < Close, High, Low, Open, Pcl, \%Chg, \%Tuv, Vol > \qquad (9)$$

where X is a multivariate time series, and each attribute in X also represents a univariate time series respectively. The specific meaning of each variable in X is described in Table 1.

Table 1. Definition of each property of stock

Stock properties	Definition
Close	The closing price of stock
High	The highest price of stock
Low	The lowest price of stock
Open	The opening price of stock
Pcl	The closing price of the previous day of stock
%Chg	The change rate of stock
%Tuv	The turnover rate of stock
Vol	The volume of stock

Therefore, the research problem of this paper can be formally defined as follows: Given two stock attribute time series D_A and D_B, which are segmented according to the size of l to obtain their corresponding segmented time series sets D'_A and D'_B. Among them, A and B represent arbitrarily two properties of stock, such as A indicates *Close* and B indicates *High*. Take D'_A as the input sequence of the model, through training, predict the corresponding output sequence, which is expressed as D''_B. The $\rho_{D''_B, D'_B}$ for D''_B and D'_B is then calculated. According to the value of $\rho_{D''_B, D'_B}$, we determine whether there is a certain relationship between A and B, and the strength of the relationship.

3 Definition of Model

Our model takes the encoder-decoder-based learning framework accompanied by attention mechanism [12,13] to study the correlation between the internal properties of the stock.

As illustrated in Fig. 1, the model typically consists of three components: an encoder network, a decoder network, and an attention mechanism.

3.1 Encoder

Given a variable-length input sequence $X = (x_1, ..., x_{T_x})$, the encoder reads each symbol of X sequentially, until a special end-of-sequence symbol "$< eos >$" is read. When reading each symbol, the hidden state h_t of the encoder is changed by iterating the following equation:

$$h_t = f(h_{t-1}, x_t) \tag{10}$$

where f is a non-linear activation function (here is gated recurrent unit) that computes the current hidden state h_t given the previous hidden state h_{t-1} and the current input x_t.

In this paper, we take each subsequence of D'_A as model's input, where the end of each subsequence is marked by an end-of-sequence symbol.

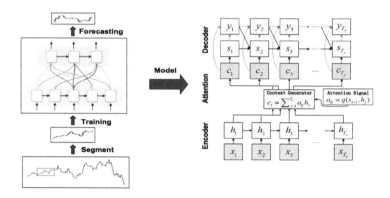

Fig. 1. The process of the experiment and the illustration of Encoder-Decoder framework.

3.2   Decoder

The decoder is trained to generate the output sequence $Y = (y_1, ..., y_{T_y})$ by predicting the next symbol y_t given the hidden state s_t. In this paper, Y refers to each subsequence of D''_B. Unlike the RNN, both y_t and s_t depend on previous symbol y_{t-1} and context vector c_t of the input sequence. Therefore, the hidden state at time t of the decoder is computed by the following equation:

$$s_t = f(s_{t-1}, y_{t-1}, c_t) \tag{11}$$

where f is a non-linear activation function (still GRU unit), and s_{t-1} denotes the hidden state at time $t-1$. What's more, the conditional distribution of the next symbol is

$$p(y_t | y_1, ..., y_{t-1}, \mathbf{x}) = g(y_{t-1}, s_t, c_t) \tag{12}$$

where g is a softmax activation function.

3.3   Attention Mechanism

For each output symbol y_i, the attention mechanism calculates the weight over each source hidden state h_i, and then decides how much attention should be paid to that hidden state. It is this feature that allows model to learn alignment between the input sequence and the partial output sequence automatically. In our model, the context vector c_t conditioned on a sequence of $annotations(h_1, ..., h_{T_x})$, which is obtained by the encoder.

The context vector c_t is computed as a weighted sum of hidden states:

$$c_t = \sum_{j=1}^{T_x} \alpha_{tj} h_j \tag{13}$$

where weighting factors α_{tj} is computed by

$$\alpha_{tj} = \frac{\exp(e_{tj})}{\sum_{k=1}^{T_x} \exp(e_{tk})} \tag{14}$$

$$e_{tj} = score(s_{t-1}, h_j) \tag{15}$$

is considered as a matching model. The higher the score, the greater the degree of alignment between s_{t-1} and h_j.

4 Experiments and Results

The purpose of our experiments is to extract the relationship among the internal properties of the stock. We prove this by testing different attributes pairs of stock by using our model. The detailed description of our experimental setup and results are given below.

4.1 Datasets

The stock data used in our experiment is downloaded from NetEase Finances website. And the experiment was carried out on twenty stocks. We first match any two stock attributes in pairs, and then delete the data with the value of None and normalize the entire data between 0–1 for unify comparison. Then we segment the data according to the segmentation rules described in Sect. 2 so as to obtain segmentation sets. At last, 70% of the data is treated as the training set, 10% of the data is treated as the validation set, and the rest data is treated as the testing set.

4.2 Results and Analysis

The encoder and decoder of our model have 3 hidden layers of GRU units respectively, and each layer has 32 hidden units. We initialize all the weights of our model with a distribution between -0.1 and 0.1 uniformly. For training, we use SGD (stochastic gradient descent) without momentum, with a start learning rate of 0.25. A decay of 0.99 is employed and learning rate will be decreased by this much if no improvement was seen over last three times.

4.2.1 The Discovery of the Mapping Relationship

Figure 2 depicts the average correlation coefficient matrix of the attribute pairs of six stocks when the length of the segmented time series is 7. Each small matrix represents the degree of correlation between an attribute pair, and the color is closer to yellow, indicating that the association is stronger. From the figure we can see that, in addition to the matrix on the main diagonal, there are also four other matrices close to yellow, while the other matrices are close to blue. And these four matrices represent $Vol\&Close$, $Pcl\&Open$, $Close\&\%Tuv$, and $\%Chg\&\%Tuv$ respectively. Therefore, we think that there are certain correlation between these four attribute pairs. The results of this experiment provide a basis for the expansion of our subsequent experiments.

Figure 3 shows the average correlation coefficient between two different attributes. P1 represents $Vol\&Close$, P2 represents $Close'\&Close$, P3 represents

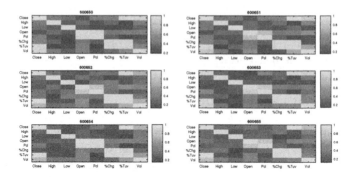

Fig. 2. The average correlation coefficient matrix of each attribute pair. (Color figure online)

$Pcl\&Open$, P4 represents $Open'\&Open$, P5 represents $Close\&\%Tuv$, P6 represents $\%Tuv'\&\%Tuv$, P7 represents $\%Chg\&\%Tuv$, P8 represents $\%Tuv'\&\%Tuv$, where $Close'$, $Open'$ and $\%Tuv'$ are all predicted by our model. From the figure, we can see that the average coef of P2, P4, P6 and P8 are always large, whenever the average coef between the two original time series is large (P3), medium (P1,P5) or small (P7). On the one hand, this shows that only using the correlation coefficient is not able to determine whether there is a correlation between the two stock attributes correctly; on the other hand, it proves that our model has the ability to discover potential relationships that can not be obtained by correlation coefficient.

Furthermore, with the comparison of P2, P4, P6 and P8, we can see that when $l = 3, 4$, the average coef is small. But when $l > 4$, it is greater than 0.6 (except for the point of $l = 8$). We believe that the above phenomenon may be due to the time dependence of stock data and the memory of the model. The model lacking ability to learn the knowledge when the length of the input sequence is small. However, with the increase of sequence length, the amount of information obtained by the model is more and more, so it can better to predict based on the information and reflect the correlation between attributes.

From the above experimental results, we can know that the proposed model has a certain ability to find out whether there is a potential relationship between attribute time series. In order to further validate this conclusion, we compare this model with LSTM and RNN. Figure 4 displays the predicted time series of one attribute generated by the time series of another attribute when using different models. Forecast 1, Forecast 2 and Forecast 3 are predicted by the proposed model, LSTM and RNN respectively. The MSE of each predicted time series is shown in Table 2, where the MSE_P is the MSE of forecast1, MSE_L is the MSE of forecast2, MSE_R is the MSE of forecast3, and the B_{PL}, B_{PR} are the B of MSE_P and MSE_L, MSE_P and MSE_R respectively. From the Fig. 4 and Table 2 we can know that the proposed model has a better performance in the exploration and prediction of attribute relations.

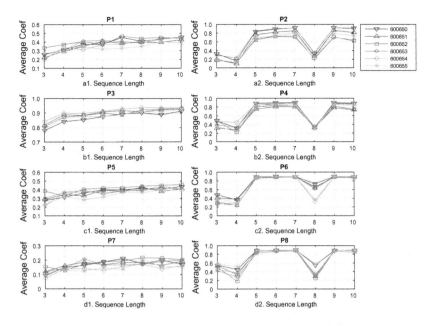

Fig. 3. The average correlation coefficient between two different attributes when l takes different values. The four charts (P1, P3, P5 and P7) on the left represent the average coef between the two original time series, while the four charts (P2, P4, P6 and P8) on the right represent the average coef between the time series predicted by our model and the original time series.

4.2.2   The Application of Mapping Relationship

When a hot event occurs, people's views and comments on this event will spread rapidly on the Internet, and form online public opinions with a certain scale and influence. At the same time, investors' sentiment is also easily affected by these online public opinions and other investor sentiment. Therefore, in this section, we choose the amount of post and click as indicators to analyze the event, and find the event by detecting the abnormal points in the time series of post and click. Then, we compare and analyze the trend of changes in the association of stock attributes with hot events to find out the relationship between them.

Table 2. The MSE of different models and B value. *stockcode*: 600652

Stock	MSE_P	MSE_L	MSE_R	B_{PL}	B_{PR}
$Vol\&Close$	0.000354	0.001744	0.002401	79.70%	85.26%
$Pcl\&Open$	0.000395	0.001179	0.001832	66.50%	78.44%
$Close\&\%Tuv$	0.000296	0.002170	0.002309	86.36%	87.18%
$\%Tuv\&\%Chg$	0.000401	0.001500	0.002160	73.27%	81.44%

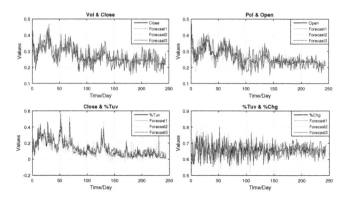

Fig. 4. The original series and the forecasting series with different models. (stock code: 600652)

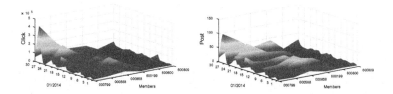

Fig. 5. The left subgraph shows the total number of click every day and the right subgraph shows the total number of post every day.

We crawled comments data of 20 stocks from 01/01/2014 to 12/31/2016 from East Wealth Network stock BBS. Figure 5 shows part of the click and post data of 6 stocks. From the figure we can observe that the trend of two maps is highly consistent, and the abnormal time periods of the events are all concentrated in 20–30. However, we can not find an accurate anomaly period for an event only by the observation of Fig. 5. Therefore, we use the EGADS (Extendible Generic Anomaly Detection System) proposed by Yahoo to detect the abnormal time period more precisely. Figure 6 shows the click time series, post time series, and abnormal point detection maps of stock 000799, where the point with the value 1 indicates the point is an abnormal point (the red point in the figure) and the point with the value 0 indicates that point is not an abnormal point. By comparing the outliers of click time series with the outliers of post time series, we find that the distribution of outliers in two time series is basically consistent. Then we match the anomalous starting points of these two sequences, as shown in Fig. 7, the red dashed line marks the start time of them. From the figure, we can see that there are nine abnormal starting points(A-I) marked on the map and eight of them are shared by Click and Post. According to these starting point, we conducted a event advanced search through Baidu News. The search results are shown in Table 3, where News indicates the start time of the event, Events indicates the specific event, and Amount indicates the total amount of related news within 5 days from the occurrence of the event. From the table, we can see that the four abnormal points labeled A, B, C and F correspond to a larger amount of news. We believe that this phenomenon may be caused by man-made invalid posts, personal emotions and other reasons. And Baidu News, which can truly reflect the hot news all the time, does not contain any artificial editing ingredients, so we define events that occurred at A, B, C and F as hot events. The third subgraph of Fig. 7 shows the changes in the relationship between $Close\&\%Tuv$ and $\%Chg\&\%Tuv$. From the graph, we can see that there are several obvious declines, and we mark them out in red circles. And the time of inflection points is highly matched with the time of the hot event that we defined above. For the other 19 stocks, we also can find this rule, and the matching rate is about 78.3%.

Based on the above experimental results, we draw the following conclusions: the occurrence of hot events will affect the relationship between the attributes of the stock. By observing the changes in the relationship between stock attributes,

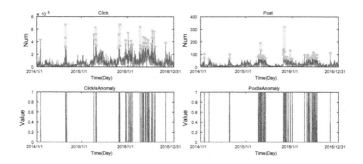

Fig. 6. The click time series, post time series, and their own abnormal point detection maps. (stock code: 000799) (Color figure online)

Table 3. A list of events that occurred during 2014–2016. *stockcode*: 000799

Points	News	Events	Amount
A	2014/01/27	Nearly 100 million yuan of Jiugui Liquor Company was stolen	414
B	2014/03/06	Jiugui Liquor Company trapped in plasticizers storm again	321
C	2014/08/19	COFCO may enter and host the Jiugui Liquor	251
D	2015/04/08	The first-season performance of Jiugui Liquor soared	20
E	2015/04/24	Jiugui Liquor Company loss of nearly 100 million, and helpless to morph "* ST Jiugui"	12
F	2015/10/23	Jiugui Liquor Company announced the termination of the reorganization of assets	122
G	2015/12/14	The chairman of Jiugui Liquor Company resign, and COFCO will take over	8
H	2016/04/07	*Jiugui apply to withdraw delisting risk warning	17
I	2016/10/26	The large amount flow of Jiugui Liquor ranked No. 6	21

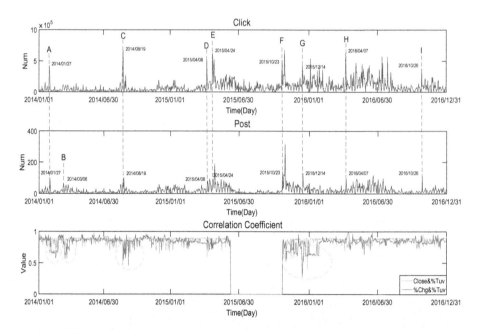

Fig. 7. Click: abnormal starting points of the click time series. Post: abnormal starting points of the post time series. Correlation Coefficient: the changes in the relationship between $Close\&\%Tuv$ and $\%Chg\&\%Tuv$, and the time period with a value of 0 indicates that the stock is in suspension. (Color figure online)

we can find the starting point of anomalous events effectively. This will help investors to identify valid information, form a comprehensive judgment on the stock market, and ultimately make effective investment decisions.

5   Conclusions and Future Work

In this paper, we have used sequence-to-sequence model to investigate the relationship among the internal properties of the stock. The experimental results show that there is a certain correlation between them, and the correlation between $Close\&\%Tuv$ and $\%Chg\&\%Tuv$ are the most obvious. Besides, we also applied the changes of correlation to the discovery of the starting points of the anomalous event. However, this paper only considers the impact of one stock attribute on another stock attribute and does not take into account the common effects of multiple stock attributes on another stock attribute. Therefore, in future work, we will be on this issue for further study.

Acknowledgments. This work is supported by National Key R&D Program of China (2016YFC1401902).

References

1. Cao, G., Zhang, M.: Extreme values in the Chinese and American stock markets based on detrended fluctuation analysis. Phys. A Stat. Mech. Appl. **436**, 25–35 (2015)
2. Wang, G.J., Xie, C., Chen, S., Yang, J.J., Yang, M.Y.: Random matrix theory analysis of cross-correlations in the us stock market: evidence from pearsons correlation coefficient and detrended cross-correlation coefficient. Phys. A Stat. Mech. Appl. **392**(17), 3715–3730 (2013)
3. Fu, T., Lee, K., Sze, D., Chung, F., Ng, C.: Discovering the Correlation between stock time series and financial news. In: IEEE/WIC/ACM International Conference on Web Intelligence and Intelligent Agent Technology, pp. 880–883. IEEE Computer Society (2008)
4. Yang, M., Jiang, Z.Q.: The dynamic correlation between policy uncertainty and stock market returns in China. Phys. A Stat. Mech. Appl. **461**, 92–100 (2016)
5. Zhang, J., Chen, Y., Zhai, D.: Network analysis of Shanghai sector in Chinese stock market based on partial correlation. In: IEEE International Conference on Information Management and Engineering, pp. 321–324. IEEE (2010)
6. Cao, G., Cao, J., Xu, L., He, L.Y.: Detrended cross-correlation analysis approach for assessing asymmetric multifractal detrended cross-correlations and their application to the Chinese financial market. Phys. A Stat. Mech. Appl. **393**(1), 460–469 (2014)
7. Hendahewa, C., Pavlovic, V.: Analysis of causality in stock market data. In: International Conference on Machine Learning and Applications, pp. 288–293. IEEE Computer Society (2012)
8. Kwon, Y.K., Choi, S.S., Moon, B.R.: Stock prediction based on financial correlation. In: Proceedings of GECCO, pp. 2061–2066 (2005)
9. Chen, Y., Zhang, L., Shi, Y.: Modeling return rate correlation between Shanghai and Shenzhen stock markets using copula function. In: IEEE/WIC/ACM International Conferences on Web Intelligence and Intelligent Agent Technology, pp. 20–24. IEEE (2013)
10. Yang, Z., Jia, H.: Network structure of the correlation between stock returns, pp. 5732–5736 (2011)
11. Tay, A.S., Ting, C.: Intraday stock prices, volume, and duration: a nonparametric conditional density analysis. Empir. Econ. **30**, 827–842 (2008)
12. Bahdanau, D., Cho, K., Bengio, Y.: Neural machine translation by jointly learning to align and translate. Comput. Sci. (2014)
13. Sutskever, I., Vinyals, O., Le, Q.V.: Sequence to sequence learning with neural networks. In: Advances in Neural Information Processing Systems, vol. 4, pp. 3104–3112 (2014)

Human Identification via Unsupervised Feature Learning from UWB Radar Data

Jie Yin[1(✉)], Son N. Tran[2], and Qing Zhang[2]

[1] Discipline of Business Analytics, The University of Sydney, Sydney, Australia
jie.yin@sydney.edu.au
[2] The Australian e-Health Research Centre, CSIRO, Brisbane, Australia
{son.tran,qing.zhang}@csiro.au

Abstract. This paper presents an automated approach to automatically distinguish the identity of multiple residents in smart homes. Without using any intrusive video surveillance devices or wearable tags, we achieve the goal of human identification through properly processing and analyzing the received signals from the ultra-wideband (UWB) radar installed in indoor environments. Because the UWB signals are very noisy and unstable, we employ unsupervised feature learning techniques to automatically learn local, discriminative features that can incorporate intraclass variations of the same identity, and yet reflect differences in distinguishing different human identities. The learned features are then used to train an SVM classifier and recognize the identity of residents. We validate our proposed solution via extensive experiments using real data collected in real-life situations. Our findings show that feature learning based on K-means clustering, coupled with whitening and pooling, achieves the highest accuracy, when only limited training data is available. This shows that the proposed feature learning and classification framework combined with the UWB radar technology provides an effective solution to human identification in multi-residential smart homes.

Keywords: Human identification · Unsupervised feature learning
UWB · Smart home

1 Introduction

The ability to recognize daily activities of residents is a core premise of smart homes for assisting with remote health monitoring. For a smart home with only one resident, by deploying various types of sensors in the living space and assuming all sensor data is generated by this only person, a resident's daily activities can be accurately recognized through various types of machine learning models. In reality, however, homes are often occupied by more than one residents. As a result, activity recognition models designed for single living environments fail to yield satisfactory results from environmentally deployed sensors, because of the absence of evidence with regards to which sensors are triggered by whom.

© Springer International Publishing AG, part of Springer Nature 2018
D. Phung et al. (Eds.): PAKDD 2018, LNAI 10937, pp. 322–334, 2018.
https://doi.org/10.1007/978-3-319-93034-3_26

Therefore, human identification is one of the most crucial problems faced by multi-residential smart homes to fully realize their functionality and potential.

Computer vision systems have been widely used to recognize human identity [12]. However, they have limited performance in poor visibility conditions (e.g., at night), and inevitably raise privacy concerns. This restricts them to be deployed in real-life smart homes that require to respect the privacy and comfort of residents under monitoring [10]. Another popular solution to human identification is the use of wearable devices that need to be carried by residents. In such wearable systems, each resident carries a tag which continuously advertises its unique ID through various types of wireless communication, such as Wi-Fi, Bluetooth, or RFID. These tags can be used as unique identifiers to help distinguish individuals. Each of these tags, however, is always assumed to be carried by a particular resident, and any of its presence is simply associated with that resident. These wearable systems require the residents to always remember and carry wearable devices all day, which leads to unsatisfactory acceptance by senior communities, let alone by people with neurodegenerative diseases.

In this work, we achieve the goal of detecting the identity of residents in smart homes through properly processing and analyzing the received signals from UWB radar. UWB radar systems can be installed in indoor environments in a non-intrusive way, offering many advantages such as high-resolution ranging, low power cost, and strong resistance to narrowband interference [2,5,19]. They have abilities to detect the changes within the vicinity of a UWB radar when people pass by. However, due to the multi-path effect, the received UWB signals are very noisy; they are very sensitive to changes in the environment, as well as differing walking modes of people. Thus, the major challenge is to find robust features that are invariant enough to incorporate naturally occurring intra-class variations, for example, resulting from differing walking modes of people or changes in surrounding environments, but discriminative enough to distinguish between different classes. Another significant challenge is to collect large amounts of labeled data for learning, which is very difficult, labor-intensive, and sometimes even impossible when considering practical operation of smart homes. Thus, it is desirable to build models that are both accurate and lightweight to enable effective and efficient human identification in real-life smart homes.

To cope with these challenges, we propose an unsupervised feature learning and classification framework to recognize the identity of multiple residents in smart homes. This framework utilizes unsupervised feature learning based on K-means clustering to automatically learn a sparse representation from the UWB signals. Specifically, we consider UWB signals in a two-dimensional space and discover discriminative local features that capture useful patterns while filtering out redundant noisy information. This way, the original UWB signals are locally transformed into sparse representations that have desirable properties for distinguishing different human identity. The derived new features are then used to train an SVM classifier to recognize the identity of individuals. We demonstrate the effectiveness of our proposed solution through extensive experiments using real UWB data collected from eight participants in real-life scenarios.

We compare our approach against several deep learning models, including sparse auto-encoders, sparse Restricted Boltzmann machines (RBMs), and convolutional neural networks (CNNs), and investigate the effect of several important factors (i.e., whitening, pooling, etc.) that affect their performance. We show that, despite its simplicity, K-means based feature learning achieves the highest accuracy, when only limited training data is available. This testifies its ability to discover discriminative patterns that are effective for human identification.

2   Related Work

The ultra-wideband technology has been widely used to identify targets in both military and civilian applications owing to its strong penetrability, high resolution, and anti-interference ability. A method for detecting human presence using UWB impulse-based radar in urban environments is proposed in [2]. Several other methods use UWB radar to detect and classify targets in foliage environments [5,19]. These methods rely on manually extracted features, such as energy, maximum amplitude, or excess delay of received signal, to perform classification. In contrast, we consider UWB signals in a two-dimensional space and automatically learn discriminative features for distinguishing human identity.

Much recent work in machine learning has focused on learning good feature representations from unlabeled input data to facilitate subsequent analytical tasks such as pattern recognition and classification. Deep architectures trained in an unsupervised manner have been proposed as an automatic method for extracting useful features. The focus has been on building different variants of deep neural networks with many hidden layers to learn multi-level representations, such as sparse auto-encoders [4], sparse RBM [6,15], and CNNs [16]. Although these recently introduced algorithms often yield better classification results on benchmark datasets like MNIST [14] and CIFAR [11] for handwriting/object recognition, they require careful selection of multiple hyper-parameters, such as learning rates, number of hidden nodes, etc., in search of better performance. Unfortunately, how to tune these hyper-parameters is a non-trivial task; their values are often carefully chosen via cross-validation, thus dramatically increasing the amount of training data and running times. Moreover, they require large amounts of labeled data to achieve the current state-of-the-art results; their performance may dramatically degrade with insufficient training data.

In the computer vision area, extensive research has been dedicated to extracting higher-level image features to achieve better performance on object recognition or scene categorization [13]. It has been shown that the classification of natural images can be significantly improved using a multi-stage architecture of feature learning [9]. Among others, K-means clustering, as the unsupervised learning module in these feature learning pipelines, can lead to excellent results, often beating state-of-the-art systems [3]. In this work, we analyze the UWB signals which are essentially different from natural images that are widely studied in computer vision; the UWB data is noisy, uncertain, and does not contain clear visual objects or edges, making it difficult to identify useful visual features. Thus, instead of working on the full-sized UWB data, we extract random patches and discover local patterns that are discriminative for recognizing human identity.

3   Characteristics of the UWB Data

Subject to reflection, refraction, diffraction, and even absorption by human body and the surrounding structures, UWB signal propagation suffers from sever multi-path effect; the impinging UWB electromagnetic wave scatters from different human body parts at different times with various amplitudes, depending on the distance to the body part, and the size and material of the reflecting part.

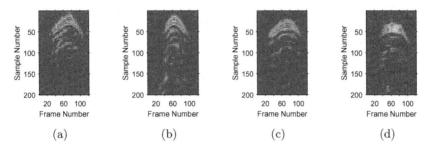

Fig. 1. Scattered UWB signals of two different subjects walking within the vicinity of a UWB radar. (a) and (b) indicate Subject A's straight and diagonal walking. (c) and (d) indicate Subject B's straight and random walking. It is evident that the local patterns of the UWB signals are quite different for different subjects, or the same subject but with differing walking modes.

Figure 1 shows scattered UWB signals of two different subjects walking within the vicinity of a UWB radar, where brighter colors indicate closer distance between the UWB radar and the target. Because the two subjects have different height and body shape, when they pass by the UWB radar, the UWB signals reflected from their bodies have different strength; the signals scattered from a taller subject would be stronger as the signals travel shorter distance to and back from his reflecting body (e.g., head, shoulder). As can be seen, scattered UWB signals of Subject A (Fig. 1(a)) are noticeably different from those of Subject B (Fig. 1(c)) when they both pass by the UWB radar along a straight line.

Moreover, due to the multi-path effect, the reflected UWB signals might vary a lot even when the same subject passes by using different walking modes; for example, a person might walk by randomly or diagonally, instead of strictly following a straight line. As an example of UWB radar's sensitivity to different walking modes, consider Fig. 1(c) and (d) that exhibit different local patterns of UWB signals when Subject B takes a straight/random walk.

4   Unsupervised Feature Learning and Classification for Human Identification

To cope with the noisy characteristics of the UWB signals, the major challenge is to discover class models that are invariant enough to handle intra-class variations of the same subject, such as differing walking modes or gaits, and yet discriminative enough to distinguish between different subjects. Because there is a lack of prior knowledge about what features are potentially useful, we resort to automatically learning useful features from the unlabeled data and use them for classification. We extract random patches from the UWB signals and discover local patterns that are discriminative for distinguishing human identity.

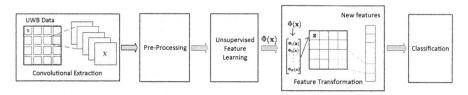

Fig. 2. The workflow of unsupervised feature learning and classification

The workflow of unsupervised feature learning and classification for human identification is shown in Fig. 2. It consists of five major components: (1) convolutional extraction module that generates random patches from the original UWB data, (2) pre-processing module that performs normalization and whitening on the input patches, (3) unsupervised feature learning module that learns a mapping function from an input patch to a new feature vector, (4) feature transformation module that converts an original UWB data into a new representation using the learned mapping function, and (5) classification module that takes the new representation as input and trains a classifier for identity recognition.

4.1   Convolutional Extraction

Given a set of input UWB data $U = \{u^{(1)}, \ldots, u^{(n)}\}$, we start by extracting random patches from unlabeled data, since we expect that the most useful, discriminative features are localized to a small region. Suppose each patch has dimension p-by-p, where p is referred to as the patch size. Each p-by-p patch can be represented as a vector $x^{(i)} \in \mathbb{R}^N$ of pixel intensity values, with $N = p \times p$. This way we construct a dataset of m patches $X = \{x^{(1)}, \ldots, x^{(m)}\}$ that are randomly sampled from the input UWB data U.

4.2   Pre-Processing

Before applying any feature learning algorithm, it is useful to perform several pre-processing steps on the input patches $x^{(i)}$. The first step is normalization which normalizes the brightness and contrast of the patches. For each patch $x^{(i)}$, we subtract out the mean of its elements and divide by the standard deviation, given by $x^{(i)} = (\tilde{x}^{(i)} - \mu_{\tilde{x}^{(i)}})/\sigma_{\tilde{x}^{(i)}}$, where $\mu_{\tilde{x}^{(i)}}$ and $\sigma_{\tilde{x}^{(i)}}$ are the mean and standard deviation of any unnormalized patch $\tilde{x}^{(i)}$. After normalization, we apply ZCA whitening [8] on the entire dataset X. This process is commonly used in deep learning to remove correlations between nearby pixels. We will later empirically assess the usefulness of whitening on the UWB data for identity recognition.

4.3 Unsupervised Feature Learning

After the pre-processing steps, we now apply an unsupervised learning algorithm on dataset X to extract discriminative features. Specifically, we aim to learn a function $\Phi : \mathbb{R}^N \to \mathbb{R}^K$, which maps an input vector $x^{(i)} \in \mathbb{R}^N$ to a new K-dimensional feature vector $z^{(i)} \in \mathbb{R}^K$. In this work, K-means clustering is used as the unsupervised feature learning module. The classic K-means algorithm finds cluster centroids that minimize the distance between data points and the nearest centroids. In our context, the data points are randomly extracted patches and the centroids are the filters that will be used to newly encode the data. From this perspective, K-means algorithm learns to construct a dictionary $\mathcal{D} \in \mathbb{R}^{N \times K}$ from the input vector $x^{(i)} \in \mathbb{R}^N$ for $i = 1, 2, \ldots, m$, such that the reconstruction error can be minimized. We use a modified version of K-means, similar to spherical K-means [7], that aims to find the dictionary $\mathcal{D}$ according to:

$$\min_{\mathcal{D},z} \sum_i \|\mathcal{D}z^{(i)} - x^{(i)}\|_2^2, \tag{1}$$

$$s.t. \ \|z^{(i)}\|_0 \leq 1, \forall i \ \ \|\mathcal{D}^{(j)}\|_2 = 1, \forall j$$

where $z^{(i)} \in \mathbb{R}^K$ is a code vector associated with the input $x^{(i)}$, and $\mathcal{D}^{(j)}$ is the j'th column of the dictionary $\mathcal{D}$. Here, the objective is to find a dictionary $\mathcal{D}$ and a new representation, $z^{(i)}$, of each input $x^{(i)}$, that can minimize the difference between $x^{(i)}$ and its reconstruction $\mathcal{D}z^{(i)}$. The objective function is optimized under two constraints. The first constraint, $\|z^{(i)}\|_0 \leq 1$, indicates that each $z^{(i)}$ is restricted to having at most one non-zero entry. The second constraint, $\|\mathcal{D}^{(j)}\|_2 = 1$, requires that each dictionary column has unit length, preventing them from becoming arbitrarily large or small. This objective function is similar in spirit to learning coding schemes, such as sparse coding [18], which requires to solve a convex optimization problem and thus is difficult to scale up. K-means, by contrast, has proved very useful for learning features due to its efficiency [3].

This optimization problem (1) can be easily solved by alternatively optimizing $\mathcal{D}$ and $z^{(i)}$ as follows:

$$z_j^{(i)} = \begin{cases} \mathcal{D}^{(j)\top} x^{(i)} & \text{if } j = \arg\max_l |\mathcal{D}^{(l)\top} x^{(i)}|, \\ 0 & \text{otherwise,} \end{cases} \tag{2}$$

$$\mathcal{D} = XZ^\top + \mathcal{D}, \tag{3}$$

$$\mathcal{D}^j = \frac{\mathcal{D}^{(j)}}{\|\mathcal{D}^{(j)}\|_2}, \tag{4}$$

where matrices $X \in \mathbb{R}^{N \times m}$ and $Z^{K \times m}$ have columns $x^{(i)}$ and $z^{(i)}$, respectively.

Finally, the columns of the dictionary $\mathcal{D}$, $\mathcal{D}^{(j)}$'s, are returned as K cluster centriods. Each patch $x^{(i)}$ is mapped to a new K-dimensional vector $z^{(i)} \in \mathbb{R}^K$, with each element being the "distance" to the corresponding cluster centroid.

4.4   Feature Transformation

After obtaining a new representation $z^{(i)} \in \mathbb{R}^K$ for each patch $x^{(i)}$, we can build a new representation of any given UWB data. Given a w-by-v UWB data, we define a $(w-p+1)$-by-$(v-p+1)$-by-K array of features by computing the representation z for every p-by-p patch of the input data. To alleviate invariance to small distortion, we further reduce the dimensionality of the new representation via average pooling that combines local regions of $z^{(ij)}$'s using an average operation. Concretely, we split the $z^{(ij)}$'s into s spatial regions, and compute the average $z^{(ij)}$'s in each region. This results in a reduced K-dimensional representation for each pooling region, and a total of $s \times K$ new features for each UWB data.

4.5   Classification

Given these pooled feature vectors for a set of training UWB data and their class labels, we adopt a linear support vector machine (SVM) algorithm to train a classifier for human identification. When an unseen UWB data is observed, we first apply the same feature transformation method described above to obtain a new representation, and then apply the learned classifier to predict its class label, that is, the corresponding human identity.

5   Experiments

In this section, we validate the performance of our proposed solution using the UWB data collected from a real-world scenario. We first describe the set-up for data collection and pre-processing, and then discuss experimental results.

5.1   Data Collection and Pre-Processing

We used the system as shown in Fig. 3 to collect the UWB data in an indoor environment [17]. The system mounts a UWB transmitter and a UWB receiver on the top of the door frame at the entrance of a room. The UWB waveforms are propagated through the transmitter antenna, with a bandwidth of 3–6 GHz, into the detection zone of the radar. When people are passing the door, the UWB waveforms are scattered from human body and the surrounding objects, and are received at the receiver antenna via multiple paths. The UWB sensor is connected via USB to a PC to record the received signals.

For the task of human identification, eight subjects with different body shapes and heights participated in the experiments. The characteristics of these subjects

Table 1. Participating subjects

Subject	Gender	Height (cm)
A	Male	170
B	Male	171
C	Male	168
D	Male	177
E	Male	169
F	Female	165
G	Male	160
H	Male	176

Fig. 3. Data collection set-up

are listed in Table 1. To simulate how people pass the door in real life, each subject was requested to pass the door using four walking modes: straight walking, diagonal left-to-right walking, diagonal right-to-left walking, and random walking. For each walking mode, each subject continuously passed the door back and forth until he was asked to stop. Thus, we collected continuous UWB data sequences that contained multiple regions of interest that indicate the presence of participating subjects within the vicinity of the UWB radar.

For each UWB data sequence, we employed Canny edge detection [1] combined with density thresholding for efficiently localizing subjects in the sequence. The first step was to remove sparse signals to emphasize possible regions of interest where subjects might be detected. Second, we accumulated the densities along the time and frequency dimensions within a sliding window and used a threshold to find a center of a bounding box. With this we segmented regions of interest with dimension of 200×121, but due to the complexity in human's trajectory the localization was not perfect. This is why treating each data sample as a set of local batches is needed in this work. In total, we obtained 768 samples of UWB data, with 96 samples per subject.

For evaluation, we randomly partitioned the data into three sets, each maintaining the same class distribution. We used the first two sets for training, and the third set was held out for testing. We compare K-means clustering with two other feature learning algorithms, sparse auto-encoders and sparse RBMs. For each feature learning algorithm, we learned features from either raw data or whitened data, and trained a linear classifier using the L2-SVM objective function. The classification results are evaluated on the test data using accuracy.

For all feature learning algorithms, the number of pooling regions s is set to 4, and the number of random patches is set to 100,000. For sparse auto-encoders and sparse RBMs, the regularization term of sparsity penalty is set to 0.001.

5.2   Effect of Whitening and Patch Size

We first performed experiments to test the performance of all algorithms with respect to different patch sizes both with whitening and without whitening. We expected that larger patch sizes would allow us to discover more complex features that cover a larger region of the original data. On the other hand, this increases the dimensionality of patches to be processed and may require to use more data for learning or to learn more features. In this experiment, we tested the patch size of 4, 6, 8, 10, 12, and 14 pixels, and set the number of features to 100. The classification results are shown in Table 2.

Table 2. Effect of whitening and patch size

Patch size	4	6	8	10	12	14
Sparse RBM	43.04	47.47	53.16	58.23	60.75	46.83
Sparse RBM (white)	47.47	60.13	67.72	77.22	72.15	71.52
Sparse auto-encoder	40.51	48.73	53.80	55.06	55.06	50.82
Sparse auto-encoder (white)	58.23	68.35	**77.22**	70.25	48.73	53.80
K-means	77.85	77.85	73.41	75.94	74.68	74.68
K-means (white)	**79.11**	**80.34**	76.58	**77.85**	**76.58**	**77.31**

Table 2 clearly shows the benefit of whitening: all algorithms generally achieve higher classification accuracy by performing whitening on the random patches. This confirms that whitening is a crucial pre-processing step for all feature learning algorithms. We can see that, K-means with whitening achieves the highest accuracy in most cases, except that with the 8 pixel patch size, sparse auto-encoder performs slightly better than K-mean clustering. This is particularly notable because K-means is easy to implement and requires much less tuning, unlike sparse auto-encoders and sparse RBMs that require us to carefully choose several hyper-parameters to guarantee reasonable results. Despite of its simplicity, the feature vectors produced by K-means constitute a powerful sparse and localized basis for distinguishing different human identity. Overall, the 6 pixel patch size works best for K-means, which achieves an accuracy of 80.34%.

5.3   Effect of Number of Feature Bases

Since the number of feature bases K is an important parameter, we carried out experiments by varying the value of K from 50, 80, 100, 200, to 400. The value of K is equivalent to the number of centroids for K-means clustering, and the number of hidden units for sparse auto-encoders/RBMs. These experiments used the 6 pixel patch size and whitening for all algorithms.

Figure 4 shows classification accuracy of all algorithms with respect to different values of K. We can see that, at the very beginning, all algorithms generally

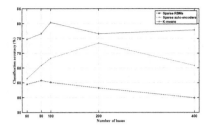

Fig. 4. Effect of number of feature bases

achieve higher performance by learning more features. This is expected because with more features learned, all algorithms have greater representative power, resulting in higher classification accuracy. However, at the later stage, learning more features decreases the accuracy, because the learned features are less distinguishable. Overall, K-means clustering achieves the best accuracy when the number of centroids is 100. Sparse RBMs and sparse auto-encoders performs best when the number of hidden units is 80 and 200, respectively. On the other hand, learning more features indeed increases the computational load and may require more data for training. Thus, carefully choosing an appropriate number of feature bases can help achieve a good trade-off between classification accuracy and computational efficiency.

5.4   Final Classification Results

We have shown through previous studies that whitening, 6 pixel patch size, and 100 feature bases work best in general across all feature learning algorithms. Using these parameters, we ran our full pipeline on the training set, trained an SVM classifier, and evaluated on the test set. We also compare against two other baselines: (1) Raw features + SVM, which trains an SVM classifier on the vectorized UWB data, (2) CNN, which is one state-of-the-art deep learning model trained on the original UWB data. Our final results are reported in Table 3.

Table 3. Classification accuracy

Algorithm	Test Accuracy (%)
Raw features + SVM	28.22
CNN	58.07
Sparse RBM	59.49
Sparse auto-encoder	68.35
K-means	**80.34**

From Table 3, we can see that the three feature learning algorithms, sparse-auto-encoders, sparse RBMs and K-means clustering outperform the other two baselines that directly work on the original UWB data. This confirms that local patterns learned from input patches have more discriminative power than original features. Although CNN has been demonstrated to render high performance on many image classification tasks, in the case that we only a limited number of training data, it produces unsatisfactory results for human identification. Again, K-means clustering achieves the highest performance with 80.34% accuracy.

Table 4. Confusion matrix with features learned by K-means clustering

Truth / predict	A	B	C	D	E	F	G	H	Accuracy (%)
A	17	0	2	0	1	0	0	0	85.00
B	0	20	0	0	2	0	0	2	83.33
C	2	0	16	0	0	2	0	0	80.00
D	0	0	0	13	0	1	0	0	92.86
E	1	1	0	2	12	0	1	0	70.59
F	0	0	2	5	2	13	1	1	56.52
G	0	0	0	0	1	1	17	0	89.47
H	0	0	0	0	2	1	0	17	85.00
Overall									80.34

Table 4 shows classification confusion matrix for identifying eight participants, with features learned by K-means clustering. As can be seen from the table, most of the errors occur when distinguishing participants with close heights, for example, for participants A (170 cm), B (171 cm), C (168 cm), and E (169 cm). Participants D (177 cm) and G (160 cm) are classified correctly with high accuracy of 92.86% and 89.47%, because they are the tallest and shortest among eight participants. Unexpectedly, the classifier sometimes has difficulty in distinguishing participant F (165 cm, female) from other male participants, yielding 56.52% accuracy only. This is probably because she walks by the UWB radar in a gait somehow similar to some other males, making it difficult to distinguish.

6   Conclusion and Future Work

In this paper, we proposed an automatic approach to human identification in multi-residential smart homes. We argued that, due to privacy or operational issues, previous approaches to human identification that rely on video surveillance or wearable devices are not suitable as a practical solution in real-life smart homes. Instead, we achieved the objective of human identification through properly processing and analyzing the received signals from the UWB radar. We investigated the use of unsupervised feature learning techniques to automatically learn local, discriminative features for human identification. We evaluated our proposed solution through extensive experiments using real data collected from eight participants. Our results showed that K-means based feature learning, coupled with whitening and pooling, yields the best performance. This suggested that, while more complex algorithms like sparse auto-encoders/RBMs may have greater representative power, they may not always be the best choice, given that in real-life smart homes there may exist only a limited number of training data.

In the future, we will explore other feature learning methodologies to improve the accuracy of human identification and generalize to online learning scenarios where unseen human identity can also be detected. We will also implement our solution in a real-time system and test its effectiveness and efficiency for recognizing the identity of a larger number of people in complex indoor environments.

Acknowledgment. The authors would like to thank Chad Hargrave and Ghassem Mokhtari for constructive discussions. The UWB data was collected with ethics approval from CSIRO Health and Medical Research Ethics Committee, LR 12/2017.

References

1. Canny, J.: A computational approach to edge detection. IEEE Trans. Pattern Anal. Mach. Intell. **8**(6), 679–698 (1986)
2. Chang, S.H., Mitsumoto, N., Burdick, J.W.: An algorithm for UWB radar-based human detection. In: Proceedings of the IEEE Radar Conference (2009)
3. Coates, A., Li, H., Ng, A.Y.: An analysis of single-layer networks in unsupervised feature learning. In: Proceedings of the AISTATS, pp. 215–223 (2011)
4. Goodfellow, I., Lee, H., Le, Q.V., Saxe, A., Ng, A.Y.: Measuring invariances in deep networks. In: NIPS, pp. 646–654 (2009)
5. He, X., Jiang, T.: Target identification in foliage environment using UWB radar with hybrid wavelet-ICA and SVM method. Phys. Commun. **13**, 197–204 (2014)
6. Hinton, G.E., Osindero, S., Teh, Y.-W.: A fast learning algorithm for deep belief nets. Neural Comput. **18**(7), 1527–1554 (2006)
7. Hornik, K., Feinerer, I., Kober, M., Buchta, C.: Spherical k-means clustering. J. Stat. Softw. **50**(10), 1–22 (2012)
8. Hyvarinen, A., Oja, E.: Independent component analysis: algorithms and applications. Neural Netw. **13**(4–5), 411–430 (2000)
9. Jarrett, K., Kavukcuoglu, K., Ranzato, M., LeCun, Y.: What is the best multi-stage architecture for object recognition? In: Proceedings of ICCV, pp. 2146–2153 (2009)
10. Klasnja, P., Consolvo, S., Choudhury, T., Beckwith, R., Hightower, J.: Exploring privacy concerns about personal sensing. In: Proceedings of PerCom, pp. 176–183 (2009)
11. Krizhevsky, A.: Learning multiple layers of features from tiny images. Master's thesis, Department of Computer Science, University of Toronto (2009)
12. Krumm, J., Harris, S., Meyers, B., Brumitt, B., Hale, M., Shafer, S.: Multi-camera multi-person tracking for easyliving. In: Proceedings of the 3rd IEEE International Workshop on Visual Surveillance, pp. 3–10 (2000)
13. Lazebnik, S., Schmid, C., Ponce, J.: Beyond bags of features: spatial pyramid matching for recognizing natural scene categories. In: Proceedings of CVPR, pp. 2169–2178 (2006)
14. LeCun, Y., Bottou, L., Bengio, Y., Haffner, P.: Gradient-based learning applied to document recognition. Proc. IEEE **86**(11), 2278–2324 (1998)

15. Lee, H., Ekanadham, C., Ng, A.Y.: Sparse deep belief net model for visual area V2. In: NIPS, pp. 873–880 (2008)
16. Lee, H., Grosse, R., Ranganath, R., Ng, A.Y.: Convolutional deep belief networks for scalable unsupervised learning of hierarchical representations. In: Proceedings of ICML, pp. 609–616 (2009)
17. Mokhtari, G., Zhang, Q., Hargrave, C., Ralston, J.: Non-wearable UWB sensor for human identification in smart home. IEEE Sensors, 99 (2017)
18. Olshausen, B.A., Field, D.J.: Sparse coding with an overcomplete basis set: a strategy employed by V1? Vis. Res. **37**(23), 3311–3325 (1997)
19. Zhai, S., Jiang, T.: Target detection and classification by measuring and processing bistatic UWB radar signal. Measurement **47**, 547–557 (2014)

Prescriptive Analytics Through Constrained Bayesian Optimization

Haripriya Harikumar[1,3]($\boxtimes$), Santu Rana[1], Sunil Gupta[1], Thin Nguyen[1],
Ramachandra Kaimal[2], and Svetha Venkatesh[1]

[1] Centre for Pattern Recognition and Data Analytics,
Deakin University, Geelong, Australia
{hharikum,santu.rana,sunil.gupta,thin.nguyen,svetha.venkatesh}@deakin.edu.au
[2] Computer Science and Engineering Department,
Amrita University, Kollam, India
mrkaimal@am.amrita.edu
[3] Amrita CREATE, Amrita University, Kollam, India

Abstract. Prescriptive analytics leverages predictive data mining algorithms to prescribe appropriate changes to alter a predicted outcome of undesired class to a desired one. As an example, based on the conversation of a reformed addict on a message board, prescriptive analytics may predict the intervention required. We develop a novel prescriptive analytics solution by formulating a constrained Bayesian optimization problem to find the smallest change that we need to make on an actionable set of features so that with sufficient confidence an instance can be changed from an undesirable class to the desirable class. We use two public health dataset, multi-year CDC dataset on disease prevalence across the 50 states of USA and alcohol related data from Reddit to demonstrate the usefulness of our results.

Keywords: Prescriptive analytics · Bayesian optimization
Linear and nonlinear classifiers · Constrained optimization

1 Introduction

Prescriptive analytics [3] is a machine learning approach to utilize prediction towards better outcomes. It stems from the often asked question of what we can change given our prediction of adverse events. If there are features which are actionable (can be changed) then prescriptive analytics can be used as an important decision making tool, for a variety of social media and public health applications.

Consider detecting and providing feedback for a group of people who want to quit binge drinking. Using social media participation data, we can build a classifier that can be used to identify participant risk using features that include different aspects of lifestyle, social life and family life. Then we may ask: what is the minimum lifestyle change that is required to avert the undesirable outcome? Such questions can be answered by our proposed prescriptive analytic framework.

© Springer International Publishing AG, part of Springer Nature 2018
D. Phung et al. (Eds.): PAKDD 2018, LNAI 10937, pp. 335–347, 2018.
https://doi.org/10.1007/978-3-319-93034-3_27

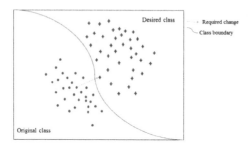

Fig. 1. Intuition of our proposed prescriptive analytics framework. Our goal is to find minimum required change to transform the data from original (red) to desired class (blue). (Color figure online)

Previous work in relation to prescriptive analytics often use inverse classification methods are [1,7]. Whilst most have been developed to explain classification decisions [2], some can be used for prescriptive analytics. The most closely related works are [7,8]. [8] assumes linear boundaries between classes and is unsuitable for general non-linear classifications. [7] formulates an optimization problem to maximize the probability of desired class given a budget on the amount of change, but does not directly minimize the change. To avoid such limitations, we propose a method for prescriptive analytics through constrained Bayesian optimization. We formulate an optimization problem to minimize the change in actionable feature sets such that the probability of belonging to the desired class reaches a desired confidence level (see Fig. 1). The probability requirement is included as a constraint and coupled with bound constraints on the allowable change in a global optimization framework. The functional form of the probability constraint is unknown and needs to be estimated during optimization. Bayesian optimization [4], a global optimizer can efficiently optimize a black box function. It can be applied irrespective of the nature of the classifier, sample-efficient (optimizes with as less sample as possible) and has good guarantee on the convergence [4,13]. We use a variant of the algorithm of Bayesian optimization with unknown constraints [5].

We apply our algorithm to two public health datasets, Centers for Disease Control and Prevention (CDC) and Reddit dataset. CDC dataset has the demographic and disease incidence rates of all the states of USA for 6 years. The diseases include diabetes, heart attack and stroke. We chose a set of actionable features in CDC, i.e. education, marital status and employment profile. Additionally, we put extra constraints in our formulation if some features can only be increased or decreased i.e. we cannot decrease the education status or the number of people who had never been married. We demonstrate the results based on both linear (logistic regression) and non-linear classifiers (kNN and Random Forest). We show that high education and less unemployment leads to healthier outcomes irrespective of the disease. We use Reddit dataset to demonstrate the utility of our algorithm for prescribing changes to people who are predicted to have relapsed on their drinking habit. Our method has potential to bring positive changes in public health applications.

2 Framework

2.1 Prescriptive Analytics

Consider we have a dataset with n instances $\{x^1, x^2, \ldots\ldots, x^n\}$. Each instance in the dataset has m features, i.e. $x^i \in \mathbb{R}^m$. We assume that each of these n instances are from binary class, $\{\mathcal{C}^-, \mathcal{C}^+\}$, where $\mathcal{C}^-$ is undesirable and $\mathcal{C}^+$ is desirable class. Our goal is to find the minimum changes that can transform an instance, which is originally in the undesirable class to the desirable class with sufficient confidence. Though we can prescribe changes to m features, it is practically impossible to change all if x^i is high dimensional and in situations when a feature cannot be directly changed (Table 1).

Table 1. Notations and definitions.

Notations	Definitions
x^i	i-th instance with m features
π_c	Probability constraint
$\mathcal{C}^+$	Desirable class
$\mathcal{C}^-$	Undesirable class
$\mathcal{A}$	Set of actionable features
$|\mathcal{A}|$	Number of actionable features
$\triangle x^i_{\mathcal{A}}$	Change in set of actionable features of x^i
$\triangle x^i$	Zero vector of size m with feature set $\mathcal{A}$ replaced by $\triangle x^i_{\mathcal{A}}$
$x^i + \triangle x^i$	Sum of original features and new values of actionable features
$\mathcal{D}_{train}$	Training dataset

More specifically, given an instance x^i, we want to find minimum changes in some *actionable features*, i.e $\triangle x^i_{\mathcal{A}}$, where $\mathcal{A}$ is the set of actionable features, in order to transform the data from $\mathcal{C}^-$ to $\mathcal{C}^+$. These feature set $\mathcal{A}$ could be identified through feature importance or could be user specified in some situations. The set of features essentially should satisfy the criteria of being actionable, hence in this paper we would assume it being supplied by a domain expert.

Since we want to prescribe changes such that the modified data vector, $x^i + \triangle x^i$ moves into the class $\mathcal{C}^+$ with probability greater than or equal to π_c, we formulate the following optimization problem:

$$\begin{aligned} \min \quad & f(\triangle x^i_{\mathcal{A}}) \triangleq \frac{||\triangle x^i_{\mathcal{A}}||_p}{|\mathcal{A}|} \\ \text{s.t.} \quad & \mathbb{P}(\mathcal{C}^+/x^i + \triangle x^i) \geq \pi_c, \\ & \triangle x^i_{\mathcal{A}} \in \left[\triangle x^i_{min}, \triangle x^i_{max}\right] \end{aligned} \qquad (1)$$

where (a). The first constraint (probability constraint) ensures that after applying the prescribed changes, the data instance belongs to $\mathcal{C}^+$ with sufficient confidence and (b). The second constraint is required to restrict the prescription vector search space, where $\triangle x^i_{min} = 0 - x^i_{\mathcal{A}}$ and $\triangle x^i_{max} = 1 - x^i_{\mathcal{A}}$. In the above objective function, any convex norm can be used. Use of $p = 1$ would provide parsimonious change vector, whereas using $p = 2$ would provide overall smallest change. For normalization, we divide the p-norm of prescription vector $\triangle x^i_{\mathcal{A}}$ by the size of the actionable set, $|\mathcal{A}|$.

2.2   Computing π_c

The computation of probability constraint, π_c is crucial in our framework. We can chose a probability constraint of 0.5 to give us the minimum change for class transformation near the boundary of the model. This value is sufficient enough for prescribing changes for balanced datasets. But, for unbalanced class problems, $\pi_c = 0.5$ may not correspond to the classification boundary. Since a custom computation of π_c is required.

We compute the probability constraint π_c from the data using the prediction model. We require π_c as the most common probabilities of data which is near the boundary of the classifier. So we use the median of the minimum probabilities assigned to the bottom 10% data from our desired class to set the value of π_c. Depending upon the classifier and the data topology, this value of π_c will vary.

2.3   Optimization

The goal of the optimization problem (Eq. 1) is to find the global minimum of the prescription vector, $\triangle x^i_{\mathcal{A}}$, sufficient enough to transform an instance to a desired class. There are several global optimization algorithms, but Bayesian optimization [4] is an efficient global optimization method with best possible guarantee on convergence for a global optimization problem.

Bayesian optimization works by setting a prior over the unknown objective function and combines it with the evidences to get a posterior function. It has mainly two components, Gaussian process as the prior over the function and an acquisition function. For $p = 1$ the objective function is not smooth, however, from our experience Gaussian process can approximate it well. We discuss the modeling of objective function with Gaussian process and construction of acquisition function below.

Bayesian Modeling of Functions. We use Gaussian process (GP) as prior to both the objective and the constraint function, since both of them are unknown. For convenience, let us describe GP with a generic vector u, where $u = \triangle x^i_{\mathcal{A}}$ for the objective and the probability constraint function. Gaussian process is a distribution over smooth function, and is fully defined by a mean function, $m(u)$ and variance function, $k(u_i, u_j)$. It can be represented as, $f(u) \sim \mathcal{GP}(m(u), k(u, u'))$. Since Gaussian process maintains a distribution over the function, instead of returning a scalar value as output, for an arbitrary $\boldsymbol{u}$ it returns the mean and variance of a normal distribution over the possible values of f at $\boldsymbol{u}$. We assume prior mean is the zero function, $m(\boldsymbol{u}) = 0$. The co-variance function, $k(.,.)$ we use in our method is squared exponential function, where the hyper parameter l controls the width of the kernel, $k(\boldsymbol{u}^i, \boldsymbol{u}^j) = \exp(-\frac{1}{2l^2} \parallel \boldsymbol{u}^i - \boldsymbol{u}^j \parallel^2)$. Consider the pairs, $\{\boldsymbol{u}^{1:t}, \mathbf{f}^{1:t}\}$, where $\mathbf{f}^{1:t} = f(\boldsymbol{u}^{1:t})$ and $\{\boldsymbol{u}^{1:t}\}$ are the sampled values at the indices from $\{1 : t\}$. Based on the Gaussian process, $\{\boldsymbol{u}^{1:t}\}$ follows a multivariate normal distribution $\mathcal{N}(0, \mathbf{K})$, where the kernel matrix is, $\mathbf{K}(i, i') = k(\boldsymbol{u}^i, \boldsymbol{u}^{i'})$. Given a new point $\boldsymbol{u}^{t+1}$, the predictive mean and the variance of the Gaussian process is, $\mu_t(\boldsymbol{u}^{t+1}) = \mathbf{k}^{\mathbf{T}}\mathbf{K}^{-1}\mathbf{f}^{1:t}$ and $\sigma_t^2(\boldsymbol{u}^{t+1}) = k(\boldsymbol{u}^{t+1}, \boldsymbol{u}^{t+1}) - \mathbf{k}^{\mathbf{T}}\mathbf{K}^{-1}\mathbf{k}$, where $\boldsymbol{k} = [k(\boldsymbol{u}^{t+1}, \boldsymbol{u}^1), ..., k(\boldsymbol{u}^{t+1}, \boldsymbol{u}^t)]$. Updating the Gaussian process with a new observation involves simply updating the observation set and updating the kernel matrix. We use two Gaussian processes: one for the objective function, denoted by $\mathcal{GP}_o$, and the other for the constraint function, denoted by $\mathcal{GP}_c$.

Construction of the Acquisition Function. The search for the global optimum given the GP models is guided by using an acquisition function. One of the most commonly used acquisition functions for Bayesian optimization is called Expected Improvement (EI) [9]. The idea is to seek the global optimum by seeking the location with highest expected improvement over the current best value in the observation. In absence of any constraints, the formulation for EI can be derived as:

$$\mathrm{EI}(\triangle \boldsymbol{x}_{\mathcal{A}}^i) = \begin{cases} \sigma\left(\triangle \boldsymbol{x}_{\mathcal{A}}^i\right) \tau\left(z\left(\triangle \boldsymbol{x}_{\mathcal{A}}^i\right)\right) & \text{if } \sigma\left(\triangle \boldsymbol{x}_{\mathcal{A}}^i\right) > 0 \\ 0 & \text{otherwise} \end{cases} \tag{2}$$

where $\tau(z) = z\Phi(z) + \phi(z)$, $z\left(\triangle \boldsymbol{x}_{\mathcal{A}}^i\right) = \frac{\mu(\triangle \boldsymbol{x}_{\mathcal{A}}^i) - f(\triangle \boldsymbol{x}_{\mathcal{A}}^{i^+})}{\sigma(\triangle \boldsymbol{x}_{\mathcal{A}}^i)}$ and $f(\triangle \boldsymbol{x}_{\mathcal{A}}^{i^+})$ is current best maximum over the previous observations. As defined earlier, $\mu\left(\triangle \boldsymbol{x}_{\mathcal{A}}^i\right)$ is the predictive mean at $\triangle \boldsymbol{x}_{\mathcal{A}}^i$ and $\sigma^2\left(\triangle \boldsymbol{x}_{\mathcal{A}}^i\right)$ is the predictive variance as computed based on $\mathcal{GP}_o$, $\Phi(.)$ and $\phi(.)$ are the standard normal CDF and PDF, respectively. We negate our objective function in order to convert the minimization to a maximization problem.

As our problem is constrained optimization, we use constraint weighted expected improvement function [5] for maximizing the acquisition function. The constraint weighted expected improvement function is obtained by modifying the expected improvement function through adding a constraint weight. Though there are some acquisition function which has constraint weight incorporated in it, we chose constraint weighted EI in our work. This will guarantee the maximum value of the acquisition function without violating the constraint. The weighted expected improvement acquisition function is,

$$\mathrm{EIC}(\triangle \boldsymbol{x}_{\mathcal{A}}^i) = \begin{cases} \mathrm{EI}(\triangle \boldsymbol{x}_{\mathcal{A}}^i) c\left(\triangle \boldsymbol{x}_{\mathcal{A}}^i\right) & \text{if } \sigma\left(\triangle \boldsymbol{x}_{\mathcal{A}}^i\right) > 0 \text{ and } \sigma_c\left(\triangle \boldsymbol{x}_{\mathcal{A}}^i\right) > 0 \\ 0 & \text{otherwise} \end{cases} \tag{3}$$

where $c\left(\triangle x^i_{\mathcal{A}}\right) = \Phi\left(z_c\left(\triangle x^i_{\mathcal{A}}\right)\right)$, $z_c\left(\triangle x^i_{\mathcal{A}}\right) = \frac{\mu_c\left(\triangle x^i_{\mathcal{A}}\right)}{\sigma_c\left(\triangle x^i_{\mathcal{A}}\right)}$, $\mu_c\left(\triangle x^i_{\mathcal{A}}\right)$ and $\sigma^2_c\left(\triangle x^i_{\mathcal{A}}\right)$ are the predictive mean and variance, respectively, as computed using $\mathcal{GP}_c$. The constrained Bayesian optimization (CBO) procedure which we use in our method is shown in Algorithm 1. The input to this algorithm is the instance x^i which belongs to $\mathcal{C}^-$, the actionable feature set $\mathcal{A}$, the bounds for $\triangle x^i_{\mathcal{A}}$ and the desired class $\mathcal{C}^+$. The CBO is a variant of Bayesian optimization in which we have to model the objective and constraint functions as separate Gaussian processes. The predictive mean and variance of both Gaussian processes, $\mathcal{GP}_o$ and $\mathcal{GP}_c$ are used in Eq. (3) for maximizing the acquisition function. The optimal changes can be computed by maximizing acquisition function using the Gaussian processes. In **Step** 3, $\triangle x^{i,t}$ is a zero vector of size m with feature set $\mathcal{A}$ replaced by $\triangle x^{i,t}_{\mathcal{A}}$, which is obtained from **Step** 2. The probability computation varies for different classifiers, e.g., the logistic function can be used for logistic regression classifier whereas mean of the predicted class probabilities of trees for random forest classifier. The output $\triangle \mathcal{D}_{1:t}$, of CBO is the set of prescription vectors and the probability of new vector, $x^i + \triangle x^{i,t}$ being in class $\mathcal{C}^+$. **Step** 5, is to augment the data for modeling objective and constraint Gaussian processes, $\mathcal{GP}_o$ and $\mathcal{GP}_c$, where $\triangle \mathcal{D}^f_{1:t}$ and $\triangle \mathcal{D}^c_{1:t}$ is the observations for objective and constraint functions. The objective ($f(\triangle x^{i,t}_{\mathcal{A}})$) and the constraint functions ($C(x^i + \triangle x^{i,t})$) are explained in Eq. 1. The $Maxiter$ is a user-specified value which decides the number of iterations required for the algorithm. As a thumb rule, Bayesian optimization converges very close to optima within $10 \times d$ iteration, where d is the dimension of the problem. The steps for our proposed prescriptive analytics method is illustrated in Algorithm 2. The input to this algorithm is a training dataset $\mathcal{D}_{train}$ and the actionable feature set $\mathcal{A}$. A predictor $\mathcal{M}$ can be generated with D_{train} using any linear or non-linear classifiers. This generated predictor, $\mathcal{M}$ is used for the computation of probability value of an instance x^i in $\mathcal{C}^-$, denoted as $\mathbb{P}(\mathcal{C}^-/x^i)$. If the class predicted by $\mathcal{M}$, for instance x^i is $\mathcal{C}^-$ we find the prescriptions vectors for it, by constrained Bayesian Optimization algorithm in Algorithm 1. The bound constraints, for limiting the values of $\triangle x^i_{\mathcal{A}}$, denote as minimum bound, $\triangle x^i_{min}$ and maximum bound, $\triangle x^i_{max}$ is computed in **Step** 3 and 4. In addition, we may have some extra known constraints for acquisition function optimization if some features are one way (explained more in Experiments section). We use NLopt [6] implementation of the derivative free constrained local optimizer, COBYLA [11] for acquisition function optimization with 100 random starts. From the output, in $\triangle \mathcal{D}_{1:t}$, we select the minimum change (prescription vector), $\triangle x^i_{\mathcal{A}}$ that satisfies the probability p which is greater than or equal to the constraint, π_c.

3   Experiments

We illustrate and validate our method with experiments conducted on various datasets. We use synthetic datasets to demonstrate the utility of Bayesian optimization compared to genetic algorithm based method. We use *Iris* dataset to visually illustrate the prescriptive analytic setting and the prescription vectors.

Algorithm 1. Constrained Bayesian Optimization

Inputs : x^i, $\mathcal{A}$, $\left[\triangle x^i_{min}, \triangle x^i_{max}\right]$, $\mathcal{C}^+$
Output : $\triangle \mathcal{D}_{1:t}$
1. **while** $(t \leq Maxiter)$ **do**
2.   Find $\triangle x^{i,t}_{\mathcal{A}}$ by optimizing the acquisition function in Eq. (3) using Gaussian Process $\mathcal{GP}_o$
       and $\mathcal{GP}_c$.
3.   Compute probability, p of $x^i + \triangle x^{i,t}$ in class $\mathcal{C}^+$.
4.   Augment prescription vector, $\triangle \mathcal{D}_{1:t} = \left\{\triangle \mathcal{D}_{1:t-1}, (\triangle x^{i,t}_{\mathcal{A}}, p)\right\}$
5.   Augment the data, $\triangle \mathcal{D}^f_{1:t} = \left\{\triangle \mathcal{D}^f_{1:t-1}, (\triangle x^{i,t}_{\mathcal{A}}, f(\triangle x^{i,t}_{\mathcal{A}}))\right\}$ and
       $\triangle \mathcal{D}^c_{1:t} = \left\{\triangle \mathcal{D}^c_{1:t-1}, (\triangle x^{i,t}_{\mathcal{A}}, C(x^i + \triangle x^{i,t}))\right\}$ and update $\mathcal{GP}_o$ and $\mathcal{GP}_c$.
6.   set $t = t + 1$.
7. **end while**

Algorithm 2. Prescriptive Analytics

Inputs : $\mathcal{D}_{train}$, x^i, $\mathcal{C}^-$, $\mathcal{C}^+$, $\mathcal{A}$
Output : $\triangle x^i_{\mathcal{A}}$
1. Generate model $\mathcal{M}$ with $\mathcal{D}_{train}$
2. **if** $(\mathbb{P}(\mathcal{C}^-/x^i)) \geq 0.5$ **then**
3. $\triangle x^i_{min} = 0 - x^i_{\mathcal{A}}$.
4. $\triangle x^i_{max} = 1 - x^i_{\mathcal{A}}$.
5.   Generate $\triangle \mathcal{D}_{1:t}$, by calling Algorithm 1 with bounds as $\left[\triangle x^i_{min}, \triangle x^i_{max}\right]$, x^i, $\mathcal{A}$ and $\mathcal{C}^+$
       as parameters.
6.   Find min $\triangle x^i_{\mathcal{A}}$ from $\triangle \mathcal{D}_{1:t}$ with $p \geqslant \pi_c$.
7. **end if**

Next, we use two real world datasets, one from the *Centers for Disease Control and Prevention (CDC)* disease incidence rates over 50 states of USA and another constructed from the *Reddit \ stopdrinking* message board. We use a linear, logistic regression with L_2 norm regularization and non-linear classifiers, kNN with 5 neighbors and random forest with 50 trees, to demonstrate the results.

3.1 Constrained Bayesian Optimization vs Genetic Algorithm

We use Improved Stochastic Ranking Evolution Strategy (ISRES) [12] as our baseline since it admits arbitrary nonlinear constraints. In order to compare these two methods, we use binary classification dataset in two dimensions. We generated data within ($\mathcal{C}^-$) and outside ($\mathcal{C}^+$) a circle. Our goal is to find minimal changes in the feature values of data in $\mathcal{C}^-$ to $\mathcal{C}^+$. The true minimum change required can be computed by the difference from the radius of the inner circle.

In real cases we have to generally deal with high dimensional data and complex models resulting in more than a few milliseconds of test time. We simulate that scenario by introducing a delay in function evaluation using sleep().

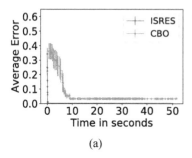
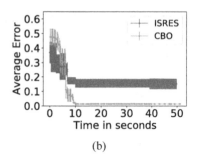

(a)   (b)

Fig. 2. Error vs optimization time for constrained Bayesian optimization and ISRES. (a) no function evaluation time, (b) function evaluation time of 100 ms.

We chose two different settings, no function evaluation time and a function evaluation time of 100 ms respectively. The performance comparison between CBO and ISRES is shown in Fig. 2. We observe that ISRES converges faster when there is no function evaluation, $\mathbb{P}(\mathcal{C}^+/x^i + \triangle x^i)$. In the case of expensive function evaluation of 100 ms ISRES fails to converge within the stipulated time, whereas Bayesian optimization is faster and converges. Hence, it validates our choice of using constrained Bayesian optimization to solve our prescriptive analytics optimization problem.

3.2 Iris Dataset

Iris dataset is available under UCI repository. It has data from three Iris flower species, Iris Setosa, Iris Versicolour and Iris Virginica. In order to show the prescriptive analytics for class inversion we selected data from two classes: Iris Setosa and Iris Versicolor. It has 100 instances with 4 features. We selected 2 features *petal width* and *sepal width* as actionable features, to demonstrate the intuition of our prescriptive analytic framework (in reality neither can be changed). Figure 3 shows the prescription vectors to change the class from Iris Setosa to Iris Versicolor. We used logistic regression and random forest as predictors. The original data (green points) in class Iris Setosa (light blue) and the transformed data (green points) to Iris Versicolor class (brown) with the red lines shows the L_1 norm change prescribed on the selected data points. This illustrate the differences in the prescription vectors suggested when we use a linear and a non-linear classifier. We set a probability constraint (π_c) of 0.5 for the logistic regression classifier to show the efficiency of constrained Bayesian optimizer to capture the minimal changes near boundary. From Fig. 3, it is clear that our method prescribes changes, which is near the boundary of the regression model target class (brown), that is able to find the minimum change to achieve the required confidence of $p(\mathcal{C}^+) = 0.5$.

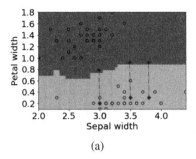

(a)

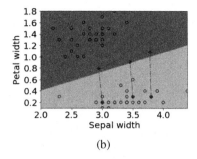

(b)

Fig. 3. Iris-setosa (light blue) and Iris-versicolor (brown). Chosen data points in Iris-setosa (green) for transforming to Iris-versicolor. (a) L_1 norm change (red line) for random forest and (b) L_1 norm change (red line) for logistic regression. (Color figure online)

3.3   Application to Policy Design for Better Community Health

CDC Dataset. The CDC data set contains state level prevalence of *diabetes, heart attack* and *stroke* diagnosis over 50 United State states from the year 2007–2012. We apply our prescriptive analytics solution to these three diseases. Each dataset has 24 features and 300 instances. Out of 24 features we have taken 7 features as actionable: *Bachelor degree or higher, High school graduate, Some college or associate degree, Unemployed, Separated, Divorced* and *Never married.* Some of these features can only be increased or decreased i.e. we cannot increase the *Never married* percentage. Whilst some of them are inter related e.g. when *High School Graduate* percentage decreases then we should only expect the reduction amount of population to move upwards in the education level i.e. totally absorbed in the increasing percentage of either *College* or *Bachelors degree.* We treat all such restrictions as 'known' constraints during optimization.

We first create binary classification datasets from the original CDC prevalence data for diabetes, heart attack and stroke. To create the label we find the median prevalence rate for each and then mark the states which have prevalence rate below that median rate as the positive class and the states with higher prevalence rate as the negative class. For example, diabetes median prevalence rate is 8.7 based on the training data from 2007–2011. Hence, the states below

Table 2. Accuracy (in %) and probability constraint (π_c) for *diabetes, heart attack* and *stroke* dataset.

Dataset	kNN		Logistic regression		Random forest	
	Accuracy	π_c	Accuracy	π_c	Accuracy	π_c
Diabetes	68%	0.6	70%	0.51	84%	0.68
Heart_attack	84%	0.6	76%	0.53	78%	0.78
Stroke	88%	0.6	84%	0.55	84%	0.8

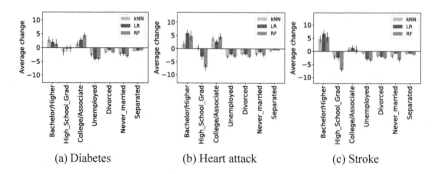

Fig. 4. Average L₁ norm change vs actionable features. (a) diabetes, (b) heart attack and (c) stroke for different classifiers.

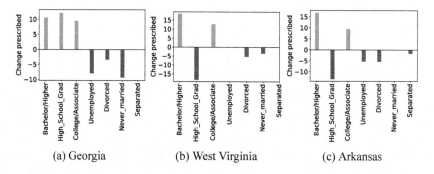

Fig. 5. Maximum L₁ norm change vs actionable features. (a) diabetes, (b) heart attack and (c) stroke.

this incidence rates falls in class $\mathcal{C}^+$ and the rest in class $\mathcal{C}^-$. We use data from 2012 as the test data.

The accuracy obtained after running k-Nearest Neighbor, logistic regression and random forest classifiers on the three datasest is shown in Table 2. The performance of classifiers on the stroke dataset is comparatively higher than other datasets.

The computed probability constraint (π_c) for each classifier is shown in Table 2. The most common probability of the bottom 10 percentage data when using k-Nearest Neighbor as classifier is 0.6 for all datasets, for logistic regression it is in the range of 0.5 and 0.55 and for random forest classifier it is between 0.68 to 0.8. Classifiers with higher capacity would generally have higher probability constraints, since they can explain the training data. This is evident by consistently higher probability constraint values for random forest.

From Fig. 4, it can be seen that the most influential factors across the board is *education* and *unemployment*. To reduce the prevalence of such diseases, *unemployment* is to be decreased and people should be educated more. We can see that the factors detrimental to family life such as *divorce* or *separated* are required to

Table 3. Accuracy (in %) and probability constraint (π_c) for *Reddit* dataset.

Measures	kNN	Logistic regression	Random forest
Accuracy (%)	66.57	68.0	71.42
Probability constraint (π_c)	0.6	0.51	0.72

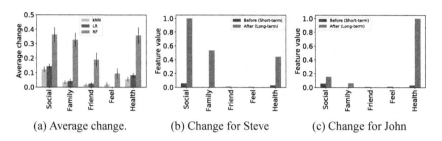

(a) Average change.   (b) Change for Steve   (c) Change for John

Fig. 6. L_1 norm change vs actionable features. (a) Average L_1 norm change vs actionable features. Actionable features before (short-term) and after (long-term) class change: (b) Steve *(name changed)* and (c) John *(name changed)* in *Reddit* dataset.

be lower. We also plot the maximum changes that has been suggested out of all the states based on the average of all three classifiers in Fig. 5. Similar factors are at play: adopting a policy to increase education and increase family life reduces the prevalence of these diseases. A more educated and working population is also a healthy one.

Reddit Dataset. Reddit (https://www.reddit.com/) is an online community having many 'subreddits', which are primarily based on specific topics. The '/r/stopdrinking' subreddit inspires alcoholics to control their drinking habit. We collected data from this subreddit using PRAW, a Python Reddit API Wrapper for collecting posts and its associated meta data such as the badge information, score, post title, post content, username from Reddit community. A redditor can request for a badge, which can be used to count the number of abstinent days. Although self-reported, this can be used as mostly correct ground truth of abstinence.

We used the badge information, for grouping the posts into two categories. A badge value of less than 30 days at the time of posting belongs to short-term group and greater than 365 days at the time of posting belongs to long-term group. The number of days, 30 versus 365 days time frame seems to be sufficient to identify the behavioral differences between the two groups. We collected 42,337 posts from years 2011–2016 and created short-term group which contains 6,910 posts and the long-term group which contains 3,641 posts from them. We use LIWC 2015 [10] package to extract LIWC feature set for representing the posts. LIWC feature represents percentage of words that reflect different emotions, thinking styles, social concerns, and even parts of speech using a 78 dimensional vector. We chose *social, family, friend, feel* and *health* as the actionable features. Any changes on these features should signify increasing or decreasing the involvement to those aspects of the social life and personal health.

The accuracy of the classifiers along with the probability constraint (π_c) is reported in Table 3. Figure 6a shows the average L_1 norm change along actionable features for Reddit dataset. Original feature values of two persons, *Steve* (name changed) and *John* (name changed) in the short-term abstainer group (C^-) and the new feature values they have got in the long-term abstainer group (C^+) by prescriptive analytics method is shown in Fig. 6b and c respectively. Our method suggests to Steve that, he has to be more social, family oriented and health conscious to become a long-term abstainer, same can be seen for John as well. Though it is prescribing positive changes along the three features, Steve needs to more social than John to become a long-term abstainer. On the other way, John has to be more health conscious than the other features to achieve his goal.

4   Conclusion

We have developed a novel prescriptive analytic solution by formulating a constrained optimization problem to find the smallest change that we need to make on actionable features so that with sufficient confidence an instance can be changed from an undesirable class to the desirable class. We used constrained Bayesian optimization to solve our problem where both the objective function and the constraint function are modeled using Gaussian process priors. Experiments demonstrate both suitability of our method over genetic algorithm based global optimizer and applicability in wide scenarios of public health data to find minimum prescriptive vectors on the actionable features to change the outcomes.

Acknowledgment. This research was partially funded by the Australian Government through the Australian Research Council (ARC) and the Telstra-Deakin Centre of Excellence in Big Data and Machine Learning. Professor Venkatesh is the recipient of an ARC Australian Laureate Fellowship (FL170100006).

References

1. Aggarwal, C.C., Chen, C., Han, J.: The inverse classification problem. J. Comput. Sci. Technol. **25**(3), 458–468 (2010)
2. Barbella, D., Benzaid, S., Christensen, J.M., Jackson, B., Qin, X.V., Musicant, D.R.: Understanding support vector machine classifications via a recommender system-like approach. In: Proceedings of the ICDM, pp. 305–311 (2009)
3. Basu, A.: Five pillars of prescriptive analytics success. Anal. Mag. 8–12 (2013)
4. Brochu, E., Cora, V.M., De Freitas, N.: A tutorial on Bayesian optimization of expensive cost functions, with application to active user modeling and hierarchical reinforcement learning. arXiv preprint arXiv:1012.2599 (2010)
5. Gelbart, M.A., Snoek, J., Adams, R.P.: Bayesian optimization with unknown constraints. arXiv preprint arXiv:1403.5607 (2014)

6. Johnson, S.G.: The NLopt nonlinear-optimization package (2014)
7. Lash, M.T., Lin, Q., Street, W.N., Robinson, J.G.: A budget-constrained inverse classification framework for smooth classifiers. arXiv preprint arXiv:1605.09068 (2016)
8. Mannino, M.V., Koushik, M.V.: The cost-minimizing inverse classification problem: a genetic algorithm approach. Decis. Support Syst. **29**(3), 283–300 (2000)
9. Mockus, J.: On Bayesian methods for seeking the extremum and their application. In: Proceedings of the Optimization Techniques IFIP Technical Conference, pp. 400–404 (1975)
10. Pennebaker, J.W., Booth, R.J., Boyd, R.L., Francis, M.E.: Linguistic Inquiry and Word Count: LIWC 2015 [Computer software]. Pennebaker Conglomerates, Inc. (2015)
11. Powell, M.J.: A view of algorithms for optimization without derivatives. Math. Today Bull. Inst. Math. Appl. **43**(5), 170–174 (2007)
12. Runarsson, T.P., Yao, X.: Stochastic ranking for constrained evolutionary optimization. IEEE Trans. Evol. Comput. **4**(3), 284–294 (2000)
13. Srinivas, N., Krause, A., Kakade, S., Seeger, M.: Gaussian process optimization in the bandit setting: no regret and experimental design. In: Proceedings of the ICML, pp. 1015–1022 (2010)

Neighborhood Constraint Matrix Completion for Drug-Target Interaction Prediction

Xin Fan, Yuxiang Hong, Xiaohu Liu, Yaogong Zhang, and Maoqiang Xie[✉]

College of Software, Nankai University, Tianjin, China
{nkufanxin,hongyuxiang,liuxiaohu,ygzhang}@mail.nankai.edu.cn,
xiemq@nankai.edu.cn

Abstract. Identifying drug-target interactions is an important step in drug discovery, but only a small part of the interactions have been validated, and the experimental determination process is both expensive and time-consuming. Therefore, there is a strong demand to develop the computational methods, which can predict potential drug-target interactions to guide the experimental verification. In this paper, we propose a novel algorithm for drug-target interaction prediction, named Neighborhood Constraint Matrix Completion (NCMC). Different from previous methods, for existing drug-target interaction network, we exploit the low rank property of its adjacency matrix to predict new interactions. Moreover, with the rarity of known entries, we introduce the similarity information of drugs/targets, and propose the neighborhood constraint to regularize the unknown cases. Furthermore, we formulate the whole task into a convex optimization problem and solve it by a fast proximal gradient descent framework, which can quickly converge to a global optimal solution. Finally, we extensively evaluated our method on four real datasets, and NCMC demonstrated its effectiveness compared with the other five state-of-the-art approaches.

Keywords: Low rank matrix completion · Neighborhood constraint
Drug-target interaction

1 Introduction

The prediction of drug-target interactions (DTIs) is an important part of drug discovery. But capturing the association between drugs and targets is an expensive and time-consuming procedure [16]. Many biochemical experiments have been made to discover new drugs in the past few years. However, the number of known DTIs remains quite low, there are only less than 7,000 compounds with confirmed target protein information in PubChem [17], one of the largest compound databases with currently around 35 million entries. Meanwhile, recent effort on the high-throughput experiments led to the creation of large open access databases of chemicals and associated bioactivity data [10], which provides us the

© Springer International Publishing AG, part of Springer Nature 2018
D. Phung et al. (Eds.): PAKDD 2018, LNAI 10937, pp. 348–360, 2018.
https://doi.org/10.1007/978-3-319-93034-3_28

chemical space of compounds and helps us understand the genomic spaces populated by the pharmaceutically useful protein targets. Therefore, there is a strong incentive to develop the computational methods for DTI prediction. These techniques can help discover the new unknown interactions for further experimental confirmation, which significantly improve the efficiency of drug discovery.

Docking simulation and machine learning are two major types of computational methods in DTI prediction. Docking simulation is well-accepted in biology because of its high prediction accuracy. However, simulation is computationally expensive and always needs 3D structural information of targets which are not unavailable in most cases [3]. Due to these difficulties, machine learning methods have been widely used in DTI prediction recently, which is much more efficient. An intuitive idea is to formulate the DTI prediction as a binary classification problem, where the observed drug-target pairs are treated as instances, and the chemical descriptors of drugs and the amino acid subsequences of targets are treated as the features. In [14], Support Vector Machine (SVM) was used to build the model. Bleakley et al. [1] used the information of drug/target similarity as kernel matrix and predicted DTIs by the SVM-based Bipartite Local Models (BLMs). Xia et al. [19] proposed a semi-supervised approach by Laplacian Regularized Least (LapRLS), and improved it by incorporating a new kernel established from the known DTI network (NetLapRLS). These methods have good performance on DTI prediction but face a common shortcoming, which the prediction process is decomposed into two steps (drug/target side) and the two sides are predicted independently. This indicates that the potential information of drug-target interaction network is not captured well in these methods.

The problem of DTI prediction can also be regarded as the recommendation task. As an effective recommendation method, matrix factorization has been used for DTI prediction in recent studies. For example, Cobanoglu et al. [4] introduced the Probabilistic Matrix Factorization (PMF) to predict unknown DTIs. Moreover, Zheng et al. [21] proposed a Collaborative Matrix Factorization (CMF) model, which exploited the drug/target similarities to constraint the low-dimensional represents of drug/target. Gönen [7] proposed a Kernelized Bayesian Matrix Factorization (KBMF2K) method, which extended the kernelized matrix factorization with a full-Bayesian treatment for DTI prediction. Ezzat et al. [6] proposed a factorization model with a dual graph regularization. However, the useful information can be lost with the low-dimensional represent of drug/target and various constraints might be not accurate any more. In addition, the optimization objective of these method are not-convex, which means it will convergence slowly and could only find the local optimal solution.

In this paper, we propose a new approach for the task of DTI prediction, namely Neighborhood Constraint Matrix Completion (NCMC). In order to analyse the DTI information, we construct the adjacency matrix R to represent the DTI network. Based on our observation that the matrix R is low rank, meaning the different drugs can be divided into several clusters with high cohesion while the different targets can also be divide into several groups through their corresponding drugs. In our method, the interaction matrix R is recovered by matrix completion with low rank constraint. Moreover, there is a challenge of this task lies in the rarity of observed entries, especially in prediction of new drugs which without any known targets. Unlike previous methods, which only focus on known pairs [13,19] or treat the unknown cases as negative instances [1,4]. We take advantage of the drug/target similarity information and define the concept of neighbors, then utilize the affine combination of neighbors to dynamically estimate and constraint the unknown cases. Experimental results on real-world datasets demonstrated the effectiveness of both neighborhood and low rank constraint. Besides, NCMC also performs better than other five state-of-the-art methods. In this context, our contributions are summarized as following:

- We transform the DTI prediction into a low rank matrix completion problem and improve the standard matrix completion to fit the binary matrix.
- We combine the original completion model with neighborhood constraint to adapt to the rarity of known entries and improve the recovery accuracy.
- We formulate the whole task into a convex optimization problem and solve it by a fast proximal gradient descent framework, which is guaranteed to obtain a global optimal solution with the convergence rate of $O(\frac{1}{k^2})$.

The remainder of this paper is organized as follows: In Sect. 2, we describe the materials and propose our method. In Sect. 3, we introduce the competing methods and experiment settings firstly, then present the experimental results and provide relevant discussion. Finally, we end with a conclusion in Sect. 4.

2   Materials and Method

2.1   Materials

We use four real benchmark datasets which correspond to four different target protein types, namely Nuclear Receptor, G-Protein Coupled Receptor (GPCR), Ion Channel, and Enzyme. These datasets are provided by [20][1] and frequently used in DTI prediction [6,7,12,19,21]. Some statistics of these datasets are listed in Table 1. Each dataset contains three parts: the observed DTIs, the chemical structure similarities of drug and the amino acid sequences similarities of targets. The observed DTIs were retrieved from four general databases: KEGG BRITE [10], DrugBank [18], BRENDA and SuperTarget, the drug similarities were measured by the SIMCOMP algorithm and the target similarities were computed by a normalized Smith-Waterman score [20].

2.2   Notation and Problem Description

In this paper, we use $D = \{d_i\}_{i=1}^m$ to denote the set of drugs, use $T = \{t_j\}_{j=1}^n$ to denote the set of targets, where m and n are the number of drugs and targets, respectively. The interactions between drugs and targets are represented by

---

[1] http://web.kuicr.kyoto-u.ac.jp/supp/yoshi/drugtarget.

binary matrix $\boldsymbol{R} \in \{0,1\}^{m \times n}$, where $\boldsymbol{R}_{ij}$ is 1 if drug d_i and target t_j are known to interact with each other and 0 otherwise. In addition, we use $\boldsymbol{S}^d \in \mathbb{R}^{m \times m}$ to represent the drug similarities and use $\boldsymbol{S}^t \in \mathbb{R}^{n \times n}$ to represent the target similarities. Let $\boldsymbol{X} \in \mathbb{R}^{m \times n}$ be a score matrix and assume $\boldsymbol{X}_{ij}$ represents the likelihood of interaction between drug d_i and target t_j. The object of this study is to estimate $\boldsymbol{X}$ by $\{\boldsymbol{S}^d, \boldsymbol{S}^t, \boldsymbol{R}\}$, then we can find out potential drug-target interactions through selecting the candidate drug-target pairs which have high scores in $\boldsymbol{X}$.

Table 1. Statistics of the used dataset

Dataset	#Drugs	#Targets	#Interactions
Enzyme	445	664	2926
Ion channel	210	204	1476
GPCR	223	95	635
Nuclear receptor	54	26	90

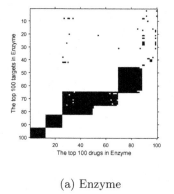

(a) Enzyme

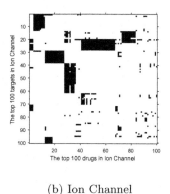

(b) Ion Channel

Fig. 1. Visualization of the low-rank pattern of drug-target interactions

2.3 Neighborhood Constraint Matrix Completion

To predict unknown interactions, we first analyze the benchmark datasets. Figure 1 shows the interaction matrix of datasets. Following [13], we only present the drugs/targets with the most number of interactions. Besides, in order to better explore the property of interaction network, we arrange the similar drugs/targets together. From Fig. 1, we can observe several blocks in Enzyme and Ion Channel datasets, and there is a similar situation on the GPCR and Nuclear Receptor datasets. It means according to therapeutic targets, the different drugs can be divided into several clusters and each drug in a cluster has the similar targets. Similarly, the different targets can be divided into several groups through their corresponding drugs and there is a strong association between the targets within each group. Due to these correlations between different drugs/targets, we assume the whole interaction matrix R is inherently low rank, which motivates us to use the standard matrix completion model to recover it. We use $\|X\|_* = \sum_i \sigma_i$ to denote the nuclear norm of X, where σ_i is the i-th singular value of X, low rank matrix R can be recovered by solving the following optimization problem:

$$\min_{X} \quad \tau\|X\|_* + \|X\|_F^2$$
$$s.t. \quad \Omega \circ X = \Omega \circ R, \tag{1}$$

where $\circ$ is the Hadamard product operator. $\Omega \in \{0,1\}^{m \times n}$ is the indicator matrix of R, that is $\Omega_{ij} = 1$ if R_{ij} is a known drug-target pair and $\Omega_{ij} = 0$ otherwise. $\|X\|_F$ is the Frobernius norm of matrix X, which is used to adapt the standard completion model to the binary matrix, and $\tau \geq 0$ is a tradeoff parameter which will be used in soft-thresholding operation.

Constraint by Neighborhood. With the help of matrix completion, we can get the new DTIs through the recovery results of R. However, because we only focus on known drug-target pairs in the objective function, the above matrix completion model can not predict for new drugs which do not have known interactions with any targets, or new targets which without any known drugs. In order to solve this problem and improve the accuracy of our model, we introduce the similarity information between drugs/targets and propose the neighborhood constraint to assist the prediction.

Previous studies [8,11] show that the interaction probability between d_i and t_j should be close to the interaction probabilities between d_i's neighbors and t_j's neighbors. Hence, many of the unknown cases in R can be constraint by their neighbors. Firstly, for drug d_i, we choose the K most similar drugs as its neighbors and use $\mathcal{N}(d_i)$ to denote the set of them. In this experiments, we empirically set the threshold value K to 5. We use adjacency matrix $\tilde{S}^d$ to represent the drug neighborhood information, where the (i, μ) element $\tilde{S}_{i\mu}^d$ is defined as following:

$$\tilde{S}_{i\mu}^d = \begin{cases} \frac{S_{i\mu}^d}{\sum_{p:d_p \in \mathcal{N}(d_i)} S_{ip}^d} & \text{if } d_\mu \in \mathcal{N}(d_i) \\ 0 & \text{otherwise.} \end{cases} \tag{2}$$

Similarly, we use $\mathcal{N}(t_j)$ to denote the set of t_j's neighbors, and calculate the adjacency matrix $\tilde{S}^t$ in the same way. Then, we notice that when R is completely recovered, the each element of final completion results X need to satisfy the foregoing assumption that $X_{ij} \approx X_{ij}^N$, where X^N denote the drug-target interaction likelihoods that calculated based on their nearest neighbors:

$$X^N = \omega\tilde{S}^d X + (1 - \omega)X(\tilde{S}^t)^\top, \tag{3}$$

and the tradeoff parameter $\omega(0 \leq \omega \leq 1)$ balances the impacts of drugs side and targets side. Stated differently, we want the loss $\sum_{i,j} \ell(X_{ij}, X_{ij}^N)$ to be small, so we add this loss as a constraint term into the original matrix completion problem:

$$\min_{\boldsymbol{X}} \quad \lambda \sum_{i,j} \ell(\boldsymbol{X}_{ij}, \boldsymbol{X}_{ij}^N) + \tau \|\boldsymbol{X}\|_* + \|\boldsymbol{X}\|_F^2 \tag{4}$$

$$s.t. \quad \boldsymbol{\Omega} \circ \boldsymbol{X} = \boldsymbol{\Omega} \circ \boldsymbol{R}.$$

To produce a concrete problem, we focus on the case where ℓ is squared loss in this work. We use $\mathcal{L}_N(\boldsymbol{X})$ to represent the neighborhood constraint of $\boldsymbol{X}$ and it can be written as:

$$\mathcal{L}_N(\boldsymbol{X}) = \sum_{i,j} \ell(\boldsymbol{X}_{ij}, \boldsymbol{X}_{ij}^N) = \|\omega(\boldsymbol{I}_m - \tilde{\boldsymbol{S}}^d)\boldsymbol{X} + (1 - \omega)\boldsymbol{X}(\boldsymbol{I}_n - \tilde{\boldsymbol{S}}^t)^\top\|_F^2, \tag{5}$$

where $\boldsymbol{I}_n$ denote the identity matrix of size n.

Moreover, following a method of solving equality constrained problems [2], we relax the equality constraint in (4), change it into an inequality constraint, to handle noise. Subsequently we can transform (4) to an unconstrained optimization problem:

$$\min_{\boldsymbol{X}} \quad \lambda \mathcal{L}_N(\boldsymbol{X}) + \rho \|\boldsymbol{\Omega} \circ (\boldsymbol{X} - \boldsymbol{R})\|_F^2 + \tau \|\boldsymbol{X}\|_* + \|\boldsymbol{X}\|_F^2. \tag{6}$$

Optimization Algorithm. The difficult point in optimization problem (6) is the $\|\boldsymbol{X}\|_*$. Since the non-smooth property of trace norm, we can't derive its gradient directly. In this paper, we utilize the accelerated proximal gradient descent algorithm [9] to solve it. Firstly, We define $\alpha = \omega\sqrt{\lambda}, \beta = (1 - \omega)\sqrt{\lambda}$ and let:

$$\boldsymbol{M} = \alpha(\boldsymbol{I}_m - \tilde{\boldsymbol{S}}^d), \boldsymbol{N} = \beta(\boldsymbol{I}_n - \tilde{\boldsymbol{S}}^t), \tag{7}$$

then let:

$$g(\boldsymbol{X}) = \rho \|\boldsymbol{\Omega} \circ (\boldsymbol{X} - \boldsymbol{R})\|_F^2 + \|\boldsymbol{M}\boldsymbol{X} + \boldsymbol{X}\boldsymbol{N}^\top\|_F^2 + \|\boldsymbol{X}\|_F^2. \tag{8}$$

The objective in (6) is also given by:

$$\min_{\boldsymbol{X}} F(\boldsymbol{X}) = g(\boldsymbol{X}) + \tau \|\boldsymbol{X}\|_*. \tag{9}$$

Then, we introduce a auxiliary matrix $\boldsymbol{Y} \in \mathbb{R}^{m \times n}$ and construct an approximation of $F(\boldsymbol{X})$ at a given point $\boldsymbol{Y}$ as:

$$Q_\eta(\boldsymbol{X}, \boldsymbol{Y}) = g(\boldsymbol{Y}) + tr((\boldsymbol{X} - \boldsymbol{Y})^\top \nabla g(\boldsymbol{Y})) + \frac{\eta}{2}\|\boldsymbol{X} - \boldsymbol{Y}\|_F^2 + \tau \|\boldsymbol{X}\|_*, \tag{10}$$

where η is a constant satisfies the inequality condition:

$$F(\boldsymbol{X}) \leq Q_\eta(\boldsymbol{X}, \boldsymbol{Y}). \tag{11}$$

Finally, given the initial $\boldsymbol{X}_0$, $\boldsymbol{Y}_0$ and η_0, the optimization problem (9) is solved by following alternative updating processes.

Computing $\boldsymbol{X}_k$: In the kth iteration, we fix $\boldsymbol{Y}_{k-1}$, η_{k-1} and update $\boldsymbol{X}_k$ as the unique minimizer of $Q_{\eta_{k-1}}(\boldsymbol{X}, \boldsymbol{Y}_{k-1})$:

$$
\begin{aligned}
\boldsymbol{X}_k &= \arg\min_{\boldsymbol{X}} Q_{\eta_{k-1}}(\boldsymbol{X}, \boldsymbol{Y}_{k-1}) \\
&= \arg\min_{\boldsymbol{X}} \left\{ \frac{\eta_{k-1}}{2} \|\boldsymbol{X} - (\boldsymbol{Y}_{k-1} - \frac{1}{\eta_{k-1}} \nabla g(\boldsymbol{Y}_{k-1}))\|_F^2 + \tau\|\boldsymbol{X}\|_* \right\} \quad (12) \\
&= \mathcal{D}_{\frac{\tau}{\eta_{k-1}}}(\boldsymbol{Y}_{k-1} - \frac{1}{\eta_{k-1}} \nabla g(\boldsymbol{Y}_{k-1})).
\end{aligned}
$$

where $\mathcal{D}_\lambda(\boldsymbol{C}) = \boldsymbol{U}\mathrm{diag}((\boldsymbol{\sigma} - \lambda)_+)\boldsymbol{V}^\top$; the $\boldsymbol{U}$, $\boldsymbol{V}$ and $\boldsymbol{\sigma}$ are the singular value decomposition (SVD) results of $\boldsymbol{C}$ such as $\boldsymbol{C} = \boldsymbol{U}\mathrm{diag}(\boldsymbol{\sigma})\boldsymbol{V}^\top$; and $(\cdot)_+ = \max(\cdot, 0)$.

Computing $\boldsymbol{Y}_k$: Following the fast convergence update scheme from [9], the auxiliary matrix $\boldsymbol{Y}_k$ is updated as:

$$
\begin{aligned}
s_k &= \frac{1 + \sqrt{1 + 4s_{k-1}^2}}{2} \\
\boldsymbol{Y}_k &= \boldsymbol{X}_{k-1} + \frac{s_{k-1} - 1}{s_k}(\boldsymbol{X}_k - \boldsymbol{X}_{k-1}).
\end{aligned} \quad (13)
$$

Computing η_k: We compute the value of $\bar{\eta}$ with multiplier γ to satisfy the inequality condition $F(\boldsymbol{X}_k) \leq Q_{\bar{\eta}}(\boldsymbol{X}_k, \boldsymbol{Y}_k)$:

$$
\bar{\eta} \leftarrow \eta_{k-1}, \text{ and } \bar{\eta} = \gamma\bar{\eta}, \quad (14)
$$

then assign $\eta_k \leftarrow \bar{\eta}$. According to [9], it can be confirmed that if we find an appropriate value for η at each iteration, then for any $k \geq 1$, we have:

$$
F(\boldsymbol{X}_k) - F(\boldsymbol{X}^*) \leq \frac{2\gamma\eta\|\boldsymbol{X}_0 - \boldsymbol{X}^*\|_F^2}{(k+1)^2}, \quad (15)
$$

where $\boldsymbol{X}^* = \arg\min_{\boldsymbol{X}} F(\boldsymbol{X})$. This conclusion shows that the sequence $\{\boldsymbol{X}_k\}$ will converge to the optimal solution of problem (9), and the convergence rate of our method can be $O(\frac{1}{k^2})$.

3   Results and Discussion

3.1   Comparison Methods

We compare our proposed method with other four popular DTI prediction algorithms as follow:

- NetLapRLS [19]: A manifold regularization semi-supervised learning method, which predicts the potential DTIs from the drugs/targets side respectively, and gets the final result by the average of two sides.

- BLM [1]: It uses drug similarity matrix as the kernel to train the SVM model as first. Then the same operation is done on targets side. The final prediction is combined by the results of two SVM models.
- CMF [21]: It's a prediction method based on matrix factorization. The object of CMF is find low dimension drug matrix and target matrix to reconstruct R. Besides, it requires the inner product of two drug/target latent vectors should be close to the corresponding score in drug/target similarity matrix.
- KBMF2K [7]: A Bayesian matrix factorization method which is similar to CMF. The innovation point of KBMF2K is the usage of fully Bayesian framework, and the variational methods is used to estimate two latent variables.

Besides the above baselines, we also consider a effective link prediction method which treats the side information as feature and integrates the linear model into low rank matrix completion, called Inductive Matrix Completion (IMC) [15], and compare its performance with our method on DTI prediction.

3.2 Experimental Settings

In this experiment, we conducted 5 repetitions of 10-fold cross validation (CV) to evaluate the performance of each method. Moreover, in order to test the different aspect of the prediction method, we considered two types of CV as following:

- CVD: CV on drugs where entire row in R were wiped for testing;
- CVT: CV on targets where entire column in R were wiped for testing.

We used both Area Under the ROC Curve (AUC) and Area Under the Precision-Recall curve (AUPR) [5] as the evaluation metrics. AUC is one of the effective metrics to evaluate the performance of interaction prediction method and it is widely used in recent studies. As a complement to AUC, AUPR is a more sensitive metric to assess the prediction result of sparse data, which is more applicable in this experiment.

NCMC has three trade-off parameters, α, β, τ, where α, β balance the impacts of the neighborhood constraint from drugs and targets, τ controls the impact of low rank constraint. We did parameter selection by using 10-fold cross-validation on the training set, and considered the following combinations: $\{0, 2^{-3}, 2^{-2}, 2^{-1}, 2^0, 2^1, 2^2, 2^3\}$ for α and β, $\{0, 2^{-2}, 2^{-1}, 2^0, 2^1, 2^2\}$ for τ. Besides, in this experiment, we set the soft approximation parameter ρ to 5 empirically.

3.3 Performance Results

Table 2 shows the result of AUC and AUPR under the setting CVD. As shown in Table 2, NCMC obtains the best results on all datasets except the Nuclear Receptor dataset, where NCMC performs slightly worse than CMF algorithm on AUC result. This is because the number of drug-target interactions is small, especially in the Nuclear Receptor dataset, most drugs only correspond to one target. In this case, the true positive rate is easy to reach a high value thus AUC is insufficient to distinguish the performance of methods. In contrast, AUPR

Table 2. AUC and AUPR results for DTI prediction under CVD

AUC						
Dataset	KBMF2K	CMF	BLM	NetLapRLS	IMC	NCMC
Enzyme	0.702 ± 0.006	0.832 ± 0.007	0.768 ± 0.009	0.831 ± 0.002	0.882 ± 0.002	**0.894 ± 0.001**
Ion channel	0.727 ± 0.009	0.808 ± 0.005	0.736 ± 0.006	0.848 ± 0.002	0.851 ± 0.004	**0.870 ± 0.006**
GPCR	0.756 ± 0.016	0.812 ± 0.007	0.730 ± 0.012	0.847 ± 0.004	0.849 ± 0.006	**0.861 ± 0.005**
Nuclear receptor	0.826 ± 0.011	**0.876 ± 0.010**	0.800 ± 0.020	0.813 ± 0.010	0.805 ± 0.006	0.858 ± 0.002
AUPR						
Enzyme	0.208 ± 0.004	0.437 ± 0.005	0.372 ± 0.008	0.482 ± 0.005	0.451 ± 0.005	**0.495 ± 0.008**
Ion channel	0.295 ± 0.006	0.430 ± 0.005	0.353 ± 0.008	0.434 ± 0.003	0.392 ± 0.007	**0.489 ± 0.010**
GPCR	0.373 ± 0.005	0.420 ± 0.006	0.353 ± 0.011	0.418 ± 0.003	0.367 ± 0.010	**0.463 ± 0.009**
Nuclear receptor	0.505 ± 0.017	0.481 ± 0.015	0.467 ± 0.032	0.460 ± 0.005	0.363 ± 0.007	**0.507 ± 0.015**

Table 3. The AUC and AUPR for DTI prediction under CVT

AUC						
Dataset	KBMF2K	CMF	BLM	NetLapRLS	IMC	NCMC
Enzyme	0.824 ± 0.008	0.877 ± 0.006	0.806 ± 0.007	0.895 ± 0.008	0.909 ± 0.003	**0.928 ± 0.003**
Ion Channel	0.896 ± 0.006	0.924 ± 0.001	0.913 ± 0.002	0.936 ± 0.001	0.934 ± 0.004	**0.963 ± 0.001**
GPCR	0.791 ± 0.010	0.844 ± 0.013	0.848 ± 0.016	0.794 ± 0.005	0.826 ± 0.005	**0.890 ± 0.005**
Nuclear Receptor	0.686 ± 0.036	0.762 ± 0.019	0.687 ± 0.034	0.603 ± 0.032	0.730 ± 0.020	**0.826 ± 0.022**
AUPR						
Enzyme	0.553 ± 0.010	0.604 ± 0.007	0.521 ± 0.012	0.625 ± 0.010	0.564 ± 0.009	**0.725 ± 0.007**
Ion Channel	0.675 ± 0.005	0.646 ± 0.007	0.688 ± 0.006	0.746 ± 0.005	0.691 ± 0.002	**0.777 ± 0.004**
GPCR	0.374 ± 0.006	0.488 ± 0.012	0.473 ± 0.007	0.431 ± 0.006	0.455 ± 0.004	**0.529 ± 0.009**
Nuclear Receptor	0.379 ± 0.021	0.421 ± 0.015	0.369 ± 0.012	0.412 ± 0.011	0.366 ± 0.011	**0.476 ± 0.013**

heavily punishes the highly ranked false positives. On the Nuclear Receptor dataset, the AUPR value of NCMC is 5.41% higher than that of CMF. There is a similar situation on the Enzyme and Ion Channel datasets, where the AUC value of NCMC is close to that of IMC, but NCMC achieves a better AUPR value, which is 9.76% and 24.74% higher than IMC on Enzyme and Ion Channel respectively. These results demonstrate NCMC has a higher accuracy on top ranked drug-target pairs, which is more meaningful in drug discovery process.

The results obtained under setting CVT are shown in Table 3. For target prediction, NCMC outperformed all five competing methods. Another phenomenon that can be observed is the both AUC and AUPR scores under CVT are higher than those under CVD on three dataset (Enzyme, Ion Channel and GPCR). That is, the different methods can achieve a better prediction result with the absence of target interaction profiles. It indicates that compared with the similarity information of drug structure, the target sequence similarity data is more informative, which is also a conclusion reached in previous studies [6,12].

Parameter Analysis. In order to confirm the impact of neighborhood and low rank constraint, we made the sensitivity analysis for three parameters α, β, τ. Firstly, we compared NCMC"s results against variants of NCMC where we set

$\alpha = 0$ for the first variant which means using only target similarity information, $\beta = 0$ for the second variant which means using only drug similarity information, the average AUPR scores for those variants under CVD and CVT are reported in Table 4. We can see that the NCMC has an advantage over all variants in most case, these results verify that the introduce of neighborhood constraint is useful for DTI prediction. In addition, note that the use of drug similarity information has little influence on target prediction, especially in Nuclear Receptor and Enzyme datasets, which also implies that the drug chemical structure similarity information might not be necessarily useful in these cases.

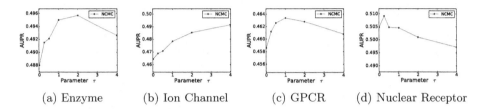

| (a) Enzyme | (b) Ion Channel | (c) GPCR | (d) Nuclear Receptor |

Fig. 2. AUPR with different settings of τ under CVD

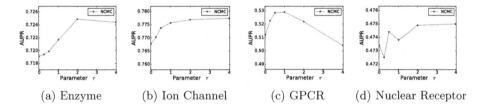

| (a) Enzyme | (b) Ion Channel | (c) GPCR | (d) Nuclear Receptor |

Fig. 3. AUPR with different settings of τ under CVT

Then, we tested NCMC under different low rank constraint parameters τ. As shown in Figs. 2 and 3, the AUPR results are improved after adding the low rank constraint. However, on the Nuclear Receptor dataset, the promotion is not significant. This is because the small size of the Nuclear Receptor dataset, and the low rank property of Nuclear Receptor is less evident than those of the other three datasets, which is causing the results to be unstable. Furthermore, the promotion is more pronounced on drug prediction, which means the low rank constraint is more effective when less reliable information is available.

Novel Interactions Prediction. In order to illustrate the ability of NCMC in real case, we selected the high probabilities interactions that do not occur in benchmark datasets, then tested them with the latest data in KEGG and DrugBank. Our method performed best on the GPCR dataset, and the top 10

Table 4. AUPR results for NCMC variants

Dataset	Drug prediction			Target prediction		
	NCMC $(\alpha = 0)$	NCMC $(\beta = 0)$	NCMC	NCMC $(\alpha = 0)$	NCMC $(\beta = 0)$	NCMC
Nuclear receptor	0.087	0.497	**0.507**	**0.476**	0.121	0.475
GPCR	0.147	0.454	**0.463**	0.519	0.112	**0.529**
Ion channel	0.123	0.476	**0.489**	0.773	0.263	**0.777**
Enzyme	0.024	0.487	**0.495**	0.724	0.077	**0.725**

Table 5. Novel interactions predicted by NCMC on GPCR

Rank	DrugID	TargeID	Probability	Evidence	Rank	DrugID	TargeID	Probability	Evidence
1	**D00604**	**hsa147**	0.805	DrugBank	6	**D00283**	**hsa1814**	0.709	DrugBank
2	**D02358**	**hsa154**	0.776	DrugBank	7	**D04375**	**hsa151**	0.704	KEGG
3	D02614	hsa154	0.757	\	8	**D00715**	**hsa1129**	0.674	KEGG
4	**D04625**	**hsa154**	0.753	KEGG	9	D00110	hsa1128	0.653	\
5	**D02147**	**hsa153**	0.727	DrugBank	10	**D01712**	**hsa136**	0.648	DrugBank

novel interactions predicted by NCMC are shown in Table 5. The interactions which exist in the reference databases are in bold while the fourth column shows the predicted probability of these pairs. The last column shows the databases evidence for each novel pair, and we found that 80% of predictions are confirmed in latest databases. In addition, there are 70%, 50%, 50% of top 10 novel interactions predicted by NCMC on the Enzyme, Ion Channel and Nuclear Receptor datasets have been confirmed respectively. These encouraging results indicate that NCMC is effective for finding new DTIs, which means it can provide reliable guidance for drug discovery and may reduce the cost of biological experiments.

4   Conclusion

In this paper, we proposed a novel matrix completion based method for drug-target interaction prediction, called Neighborhood Constraint Matrix Completion (NCMC). The novelty of NCMC comes from integrating low rank matrix completion with neighborhood constraint to predict the interaction probability of a unknown drug-target pair. Specifically, our method can capture the strong correlation between drugs/targets through the minimization of trace norm. Moreover, we proposed a neighborhood constraint, which provides a new idea for using side information to recover a matrix with fewer known entries. Furthermore, we formulate the whole task into a convex optimization problem which makes the algorithm more efficient. In our experiments, we find that NCMC has an outstanding performance over the five current state-of-the-art methods and achieves a high accuracy on novel interaction prediction, which makes NCMC a better tool for new drug discovery.

References

1. Bleakley, K., Yamanishi, Y.: Supervised prediction of drug-target interactions using bipartite local models. Bioinformatics **25**(18), 2397–2403 (2009)
2. Candes, E.J., Plan, Y.: Matrix completion with noise. Proc. IEEE **98**(6), 925–936 (2009)
3. Cheng, A.C., Coleman, R.G., Smyth, K.T., et al.: Structure-based maximal affinity model predicts small-molecule druggability. Nat. Biotechnol. **25**(1), 71–75 (2007)
4. Cobanoglu, M.C., Liu, C., Hu, F., et al.: Predicting drug-target interactions using probabilistic matrix factorization. J. Chem. Inf. Model. **53**(12), 3399–3409 (2013)
5. Davis, J., Goadrich, M.: The relationship between Precision-Recall and ROC curves. In: ICML 2006, pp. 233–240. ACM Press (2006)
6. Ezzat, A., Zhao, P., Wu, M., et al.: Drug-target interaction prediction with graph regularized matrix factorization. IEEE/ACM Trans. Comput. Biol. Bioinf. **14**(3), 646–656 (2017)
7. Gonen, M.: Predicting drug-target interactions from chemical and genomic kernels using Bayesian matrix factorization. Bioinformatics **28**(18), 2304–2310 (2012)
8. Jacob, L., Vert, J.P.: Protein-ligand interaction prediction: an improved chemogenomics approach. Bioinformatics **24**(19), 2149–2156 (2008)
9. Ji, S., Ye, J.: An accelerated gradient method for trace norm minimization. In: ICML 2009, pp. 1–8. ACM Press (2009)
10. Kanehisa, M., Goto, S., Hattori, M., et al.: From genomics to chemical genomics: new developments in KEGG. Nucleic Acids Res. **34**(Suppl. 1), D354–D357 (2006)
11. Keiser, M.J., Roth, B.L., Armbruster, B.N., et al.: Relating protein pharmacology by ligand chemistry. Nat. Biotechnol. **25**(2), 197–206 (2007)
12. Laarhoven, T.V., Nabuurs, S.B., Marchiori, E.: Gaussian interaction profile kernels for predicting drug-target interaction. Bioinformatics **27**(21), 3036–3043 (2011)
13. Liao, Q., Guan, N., Wu, C., Zhang, Q.: Predicting unknown interactions between known drugs and targets via matrix completion. In: Bailey, J., Khan, L., Washio, T., Dobbie, G., Huang, J.Z., Wang, R. (eds.) PAKDD 2016. LNCS (LNAI), vol. 9651, pp. 591–604. Springer, Cham (2016). https://doi.org/10.1007/978-3-319-31753-3_47
14. Nagamine, N., Sakakibara, Y.: Statistical prediction of protein-chemical interactions based on chemical structure and mass spectrometry data. Bioinformatics **23**(15), 2004–2012 (2007)
15. Natarajan, N., Dhillon, I.S.: Inductive matrix completion for predicting gene-disease associations. Bioinformatics **30**(12), i60–i68 (2014)
16. Novac, N.: Challenges and opportunities of drug repositioning. Trends Pharmacol. Sci. **34**(5), 267–272 (2013)
17. Wheeler, D.L., Church, D.M., Lash, A.E., et al.: Database resources of the National Center for Biotechnology Information. Nucleic Acids Res. **40**(D1), D13–D25 (2012)
18. Wishart, D.S., Knox, C., Guo, A.C., et al.: DrugBank: a knowledgebase for drugs, drug actions and drug targets. Nucleic Acids Res. **36**(Suppl. 1), D901–D906 (2008)

19. Xia, Z., Wu, L.Y., Zhou, X., et al.: Semi-supervised drug-protein interaction prediction from heterogeneous biological spaces. BMC Syst. Biol. **4**(Suppl. 2), S6 (2010)
20. Yamanishi, Y., Araki, M., Gutteridge, A., et al.: Prediction of drug-target interaction networks from the integration of chemical and genomic spaces. Bioinformatics **24**(13), i232–i240 (2008)
21. Zheng, X., Ding, H., Mamitsuka, H., et al.: Collaborative matrix factorization with multiple similarities for predicting drug-target interactions. In: SIGKDD 2013, pp. 1025–1033. ACM Press (2013)

Detecting Hypopnea and Obstructive Apnea Events Using Convolutional Neural Networks on Wavelet Spectrograms of Nasal Airflow

Stephen McCloskey$^{(\boxtimes)}$, Rim Haidar$^{(\boxtimes)}$, Irena Koprinska$^{(\boxtimes)}$, and Bryn Jeffries$^{(\boxtimes)}$

School of Information Technologies, University of Sydney, Sydney, Australia
{smcc7913,rhai6781,irena.koprinska,bryn.jeffries}@sydney.edu.au

Abstract. We present a novel approach for detecting hypopnea and obstructive apnea events during sleep, using a single channel nasal airflow from polysomnography recordings, applying a Convolutional Neural Network (CNN) to a 2-D image wavelet spectrogram of the nasal signal. We compare this approach to directly training a 1-D CNN on the raw nasal airflow signal. The evaluation was conducted on a large dataset consisting of 69,264 examples from 1,507 subjects. Our results showed that both approaches achieved good accuracy, with the 2-D CNN outperforming the 1-D CNN. The higher accuracy and the less complex architecture of the 2-D CNN show that converting biological signals into spectrograms and using them in conjunction with CNNs is a promising method for sleep apnea recognition.

Keywords: Sleep disorder · Obstructive Sleep Apnea
Convolutional Neural Networks · Wavelets · Supervised learning

1 Introduction

Sleep apnea is a disorder characterized by either a period of reduced breathing (hypopnea) or no breathing (apnea) during sleep. It affects 2–4% of the adult population [1]. Factors that contribute to sleep apnea include obesity, age, gender, smoking and drinking [2]. Sleep apnea has been found to be a major risk factor for motor vehicle accidents, increasing the risk 2–7 times due to daytime sleepiness, cognitive impairment and other associated factors [3]. Untreated sleep apnea has also been found to increase the risk of non-fatal and fatal heart events [4], stroke and also the risk of death, independent from other cerebrovascular risk factors [5]. There are three main types of apnea events: obstructive apnea, where breathing is obstructive but there is a continued respiratory effort; central apnea, where there is no respiratory effort, and mixed apnea which has a combination of obstructive and central apnea symptoms [6].

© Springer International Publishing AG, part of Springer Nature 2018
D. Phung et al. (Eds.): PAKDD 2018, LNAI 10937, pp. 361–372, 2018.
https://doi.org/10.1007/978-3-319-93034-3_29

Diagnosing sleep disorders such as sleep apnea traditionally uses polysomnography (PSG), for objective measurement of the abnormal activity. PSG involves overnight monitoring of the patient and measuring a number of signals with sensors attached to the body, e.g. respiratory, blood oxygen saturation, electroencephalography (EEG) brain activity and body movement signals.

After the PSG data is recorded, physicians and sleep experts inspect it, using statistical tools, to detect and classify sleep events such as sleep stages, hypopnea and obstructive apnea. This manual analysis of long, multi-channel PSG data is very labor intensive and time consuming. Numerous methods have been developed to identify sleep apnea and hypopnea events automatically using machine learning techniques, with a focus on PSG channels that are less intrusive such as the nasal airflow.

These techniques typically rely on manual feature selection and classifiers such as Support Vector Machine (SVM). However, manual feature selection is difficult to perform on the noisy and very long PSG recordings and requires high-level domain knowledge. Deep neural networks are an attractive alternative as they can learn the informative features effectively, without prior knowledge. In particular, CNNs are a popular class of deep neural networks that has had wide success analysing visual imagery [7], speech recognition [8] and text recognition [9]. They can identify motifs invariant of position in a signal or image and can merge semantically similar motifs together. In this paper, we evaluate the ability of CNNs to detect hypopnea, obstructive apnea and normal events, using two different types of signals: raw nasal and nasal wavelet spectrogram.

Wavelet analysis is a time-frequency decomposition technique suitable for the non-stationary and noisy PSG signals, and allowing better time localisation than Fourier analysis. It can be used to create spectrograms which are visualisations of the frequency spectrum of a signal, as it varies with time. Spectrograms are very useful in showing changes of events that occur at different frequencies. They have been used extensively in speech processing, music, radar and seismology research as visual representation of signals. Recently, spectrograms of EEG data have been used in [10] for sleep stages recognition. In this paper, we explore the use of spectrograms for sleep apnea detection.

The contributions of this paper are as follows: (1) we investigate the applicability of CNNs to detect hypopnea, obstructive apnea and normal events without manual feature engineering; (2) we investigate a novel CNN approach using Continuous Wavelet Transformation (CWT) spectrograms of a single respiratory channel (the nasal airflow), which can be recorded easily and non-intrusively; (3) we evaluate the performance of these approaches on a large dataset of 69,264 examples from 1,507 subjects.

2   Related Work

The methods used to detect both apnea and hypopnea events are typically based on SVM [11–13] and classical non-deep neural networks [14,15]. Other novel approaches include voice activity detection [16], EEG frequency variation [17], and ensemble methods [18]. The majority of research into sleep apnea investigates multiple respiratory signals such as abdominal, chest, nasal or thoracic.

The most similar work to our study is [19], where the performance of a 1-D CNN without feature selection was compared to a traditional SVM method with manual feature selection, for the recognition of two classes: normal and abnormal (obstructive apnea and hypopnea together). It was found that the CNN method outperformed SVM and was able to automatically learn the informative features from the raw signal, in a dataset of 100 subjects. We extend [19] with our investigation of CWT spectrograms. Furthermore we consider obstructive apnea and hypopnea events as distinct classes, and evaluate performance using a much larger dataset (1,507 subjects).

3 Data

We use data from the Multi-Ethnic Study of Atherosclerosis (MESA) collected by the National Sleep Research Resource (NSRR) [20]. This study investigated factors associated with the development and progression of cardiovascular disease. As part of it, 2,056 participants were enrolled in a sleep examination which included full night unattended PSG. The PSG recordings had durations of at least 8 h, and included airflow (by nasal-oral thermocouple and pressure recording from a nasal cannula), cortical EEG, bilateral electrooculograms, chin EMG, thoracic and abdominal respiratory inductance plethysmography, ECG, leg moments and finger pulse oximetry. Trained human scorers manually examined the PSG signals and marked the location and duration of the events (including apnea and hypopnea) and also marked every 30 s epoch of the signal as a wake or one of the four stages of sleep.

3.1 Data Preprocessing and Preparation

We used the PSG nasal airflow signal, recorded with a sampling rate of 32 Hz, along with the time and duration of each respiratory event identified by sleep experts. We classified every 30-s epoch as either obstructive apnea, hypopnea or normal from these respiratory events. An epoch was labeled as obstructive apnea if it had an obstructive apnea event lasting for more than 10 s. Similar labeling was applied for hypopnea events. An epoch was labeled as normal if it didn't have any apnea or hypopnea events (or any other types of abnormal events), or if it included apnea and hypopnea events with combined duration less than 10 s. Epochs that had both apnea and hypopnea events for combined duration more than 10 s were excluded.

After cleaning the data we had 534,840 normal events, 39,799 obstructive apnea, and 177,437 hypopnea events. To avoid bias we extracted a balanced set of normal, obstructive apnea and hypopnea epochs from each subject. Subjects that did not have both obstructive apnea and hypopnea events were therefore excluded, limiting our dataset to 1,507 subjects of the 2,056 subjects in MESA. This resulted in a total of 69,264 epochs (23,088 per event class).

4   Method

We built two competing CNN-based classifiers, trained and tested with a common dataset, and compared their performance. The two models are summarised in Fig. 1. Our first model used the 1-D nasal signal directly, while our novel 2-D approach used a CNN on wavelet spectrograms on nasal airflow for the classification of hypopnea, obstructive apnea and normal events.

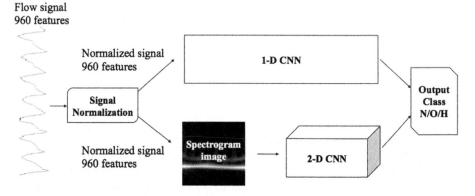

Fig. 1. Summary of the proposed CNN methods to classify 30-s epochs of nasal airflow as Normal (N), Obstructive apnea (O) and Hypopnea (H) events

To develop the CNN models and evaluate their performance, we followed the procedure described in [21]. The dataset was divided into stratified training (75%) and test (25%) sets. Both CNNs were developed using this training set with 10-fold cross validation for parameter selection. Each model was then built with the full training set using the selected parameters, and its performance was evaluated and compared on the hold-out test set.

Below we discuss the normalisation of the nasal signal, and the architecture of the two CNN models.

4.1   Signal Normalization

The airflow signals used in this study consist of all epochs between the sleep onset and the last sleep event. The nasal airflow signals were normalized by the mean and standard deviation of the normal epochs of the airflow signal for each subject. The normalized signal was calculated as:

$$\hat{S}_i = \frac{S_i - \bar{S}_{n,i}}{\sigma_{S_{n,i}}} \tag{1}$$

where S_i is the signal for subject i, $S_{n,i}$ is the signal of the normal epochs of subject i, $\bar{S}_{n,i}$ is the mean of the normal epochs signal for subject i and $\sigma_{S_{n,i}}$ is the standard deviation of the normal epochs signal.

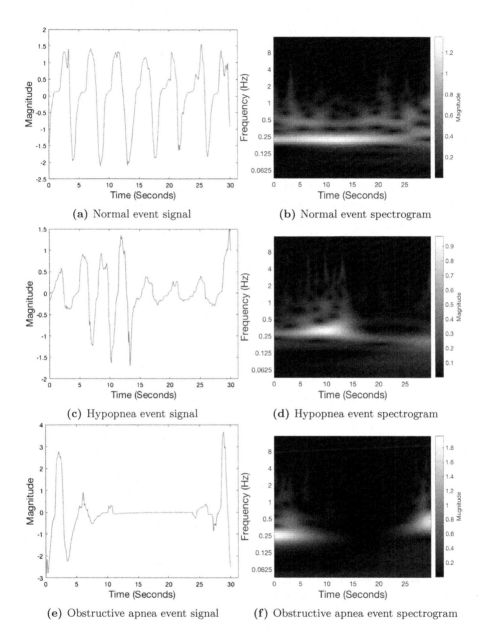

(a) Normal event signal

(b) Normal event spectrogram

(c) Hypopnea event signal

(d) Hypopnea event spectrogram

(e) Obstructive apnea event signal

(f) Obstructive apnea event spectrogram

Fig. 2. Examples of normal, hypopnea and obstructive apnea event epochs using the two representations: nasal signal and visual spectrogram of the nasal signal

4.2   CNN Design

A CNN consists of convolutional, pooling and fully connected layers. The convolutional layer includes a number of filters, which analyse certain portions of the data to identify features or motifs (i.e. presence of edges at particular orientations or locations). Typically each convolution layer is followed by a pooling layer, which seeks to systemically merge similar motifs together. This allows the identification of motifs with small variations in position or shape. After a number of convolutional and pooling layers, fully connected layers are used [22]. We used a single fully connected layer with softmax activation function, which produces the probability for each class. The input example is assigned to the class with highest probability.

Finding optimal configurations for neural networks like CNNs requires expert knowledge, estimation or exhaustive search [23]. There are four main approaches for selecting CNN hyper-parameters: manual search, grid search, random search [24] and Bayesian methods [23].

In this work the hyper-parameters were determined by performing a nonexhaustive manual search of parameter combinations, and using 10-fold cross-validation within the training set to select the best combination. The following hyper-parameters were investigated: number of layers, filter size and number of filters. The number of epochs was set to 50 and the training was stopped early if there was poor performance after the first fold. The CNN parameters that resulted in the highest average performance over the 10 folds were selected, and then the 1-D and 2-D CNNs with the selected parameters were re-trained on the full training set and their performance was evaluated on the hold-out test set.

4.3   1-D CNN: Nasal Signal

The input to the 1-D CNN was the normalized signal of each epoch. Every 30-s epoch included 960 features; no feature extraction or selection were applied. The 1-D CNN architecture is shown in Fig. 3 and consisted of 6 convolutional layers and 3 max pooling layers, followed by 1 fully connected softmax layer. Each convolutional layer had 32 filters with Rectified Linear Unit (ReLu) activation function, and each max pooling layer had a pool size of $[2 \times 1]$ nodes. Each convolutional layer of kernel size $[3 \times 1]$ with 3 strides is followed by another convolutional layer of kernel size $[2 \times 1]$ with 2 strides, and then a max pooling layer. The fully connected softmax layer consisted of 3 nodes with each node representing a probability output for each of the 3 classes.

The 1-D CNN model was trained using the backpropagation algorithm, optimising the categorical cross entropy function with the Adam optimizer [25] and a 50% dropout [26] was employed to prevent overfitting. The batch size was set to 100 samples.

4.4   2-D CNN: Nasal Signal Spectrogram

The input to the CNN was a spectrogram of the nasal airflow signal. The spectrograms were calculated by using CWT with the analytical Morlet wavelet. The

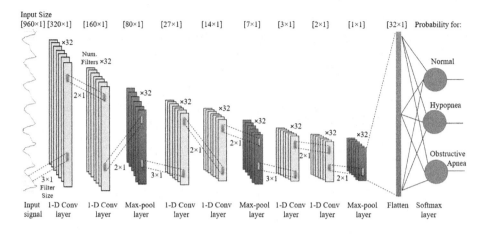

Fig. 3. Summary of 1-D CNN: nasal signal

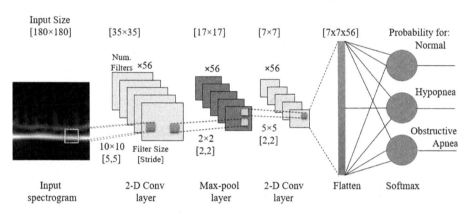

Fig. 4. Summary of 2-D CNN: nasal signal spectrogram

Morlet wavelet has good time localization properties which makes it well suited for detecting the transient properties of a signal. No feature selection was used to reduce the number of features. The frequency axes of the spectrogram images were scaled by $\log_2$ to show high frequency features with a similar size to the low frequency features. Examples of the spectrograms of normal, hypopnea and apnea events are shown in Fig. 2.

The 2-D CNN architecture is shown in Fig. 4. This CNN consisted of two convolutional layers with ReLu activation layers afterwards and 1 max 2-D pooling layers followed by a fully connected layer and softmax layer. The first convolutional layer used 56 filters of kernel size $[10 \times 10]$ with stride $[5, 5]$ and the second used 56 filters of kernel size $[5 \times 5]$ with stride $[2, 2]$. The 2-D max pooling layer had a size of $[2 \times 2]$ with stride $[2, 2]$. Similarly to the 1-D architecture, the fully connected layer and softmax layer consisted of 3 nodes, outputting the probability of each class.

The 2-D CNN was trained with the backpropagation algorithm, optimising the categorical cross entropy using the Stochastic Gradient Descent with Momentum (SGDM). The batch size was set to 128 samples per training epoch and 50% dropout was used at the end of each epoch.

Table 1. Performance (%) for obstructive apnea, hypopnea and normal events on the test set for the 1-D and 2-D CNN methods

Method	Event	Recall	Precision	F1 score
1-D CNN	N	76.1	73.7	74.9
	H	65.6	71.0	68.2
	O	91.2	87.4	89.3
	Average	77.6	77.4	77.5
2-D CNN	N	76.2	80.4	78.3
	H	73.6	68.8	71.1
	O	89.2	90.2	89.7
	Average	79.7	79.8	79.7

5   Results and Discussion

Table 1 summarizes the performance of the 1-D CNN and 2-D CNN models against the 25% test set, after training on the full 75% training set.

Both methods show excellent performance detecting obstructive apnea (O) with F1 scores close to 90%. This is likely due to such events being distinctly different to the others: as can be seen in Figs. 2f and 2e, obstructive apnea tends to have a total reduction in nasal airflow as breathing is completely stopped for more than 10 s. Although the performance of the CNNs in detecting the hypopnea and normal events was somewhat lower, it was significantly higher than the dataset baseline (33% for each class).

Comparison to other work is complicated by differences in measurement. The SVM work in [13] reports accuracies in detecting hypopnea and apnea events of 90.3% and 93.8%, respectively, but using a definition of accuracy that is the sum of recall and false negative rate. By the same criteria the 1-D CNN achieves respective scores of 83% and 103% (the accuracy formula used in [13] exceeds 100% when the number of detected events exceeds the true number), while the 2-D CNN achieves 98% for both hypopnea and apnea events. Importantly the SVM was tested on 26 recordings without balancing of event types, while our test set draws a balanced set from 1,507 recordings. The classifier in [11] reported a sensitivity (recall) of 87% for detecting apnea events of 12 subjects. Both 1-D and 2-D CNN models outperform this, at 91.2% and 89.2%, respectively. Our wavelet spectrogram CNN approach also shows significant improvement over the CNN method from [19], as we achieved 79.8% accuracy classifying normal, hypopnea and obstructive apnea events on 1,507 subjects compared to 75%

accuracy classifying normal and abnormal (combined hypopnea and obstructive apnea) events on 100 subjects.

Table 2. Confusion matrices with the predicted and actual classes for obstructive apnea, hypopnea and normal events on the hold-out test set for the 1-D and 2-D CNN methods

(a) 1-D CNN

Actual		Prediction			
		N	H	O	Sum
	N	4391	1188	193	5772
	H	1421	3785	566	5772
	O	148	358	5266	5772
	Sum	5960	5331	6025	

(b) 2-D CNN

Actual		Prediction			
		N	H	O	Sum
	N	4640	982	150	5772
	H	1325	3968	479	5772
	O	123	443	5206	5772
	Sum	6088	5393	5835	

The confusion matrices for the 1-D and 2-D CNNs are shown in Table 2. In both models there was significant misclassification of normal events as hypopnea (20.6% in the 1-D CNN, 17.0% in the 2-D CNN) and hypopnea as normal events (24.6%, 23.0%). This misclassification between normal and hypopnea events can be understood from examining the events in Fig. 2, in which it can be seen that a hypopnea event has a similar waveform as a normal event, but with a reduction of amplitude as breathing is reduced. A hypopnea event is also only a portion (>10 s) of the 30 s epoch which would be otherwise normal breathing.

Some misclassification may also be due to large artefacts in the signal during the night. There are also limits of the accuracy of the CNN based on the quality of the manual scoring, as there are examples of misclassification of normal and hypopnea events that were classified based on nasal pressure but actually contain little or no nasal pressure and should be classified as apnea events. Some of these events could be verified by examining other respiratory channels. In our experiments, only the nasal airflow signal was examined and adding more channels might improve the accuracy as artefacts in one channel could be identified.

Sleep apnea severity is commonly rated by the Apnea-Hypopnea Index (AHI), calculated from the average number of apnea and hypopnea events per hour during the night. It is important then to be mindful of the biases in the classification of each event. Both models show a bias towards classifying events (particularly hypopnea) as normal, which could lead to an artificially low AHI. The bias from falsely detected normal events is mostly offset by abnormal events (mostly hypopnea) misclassified as normal, leading to a 3% bias in the 1-D CNN and 5% in the 2-D CNN. This could possibly be compensated for by suitable scaling of the counts. In practice, as can be seen in Sect. 3, a sleep does not consist of a balanced set of events but is strongly skewed towards normal events so care would be needed in applying such an adjustment.

In terms of training time, the 2-D CNN was slower than the 1-D CNN, although it has a simpler architecture. In particular, the 2-D CNN spectrogram approach required 40 min to train with only two convolutional layers while the 1-D CNN required 5 min using six convolutional layers. However, as training is done off-line, both training times are acceptable for practical applications.

We also note that for the 2-D CNN approach (which examines 2-D spectrogram images), the optimal architecture found included only two convolutional layers, while the 1-D CNN approach included six convolutional layers. Since more convolutional layers are expected to correspond to higher complexity of the data, or more non-linearities, this demonstrates that the 2-D spectrograms are able to show the important features of the signal in a significantly less complex representation.

It is interesting to consider whether there would be a benefit to combining both approaches, using the raw 1-D in combination with the CWT spectrogram information. Within our test set the number of events correctly identified by both models was 3991, 3225 and 5010 for normal, hypopnea and apnea, respectively. This is significantly lower that the numbers along the diagonals of the confusion matrices in Table 2, suggesting that using information from both approaches could correctly identify a significantly larger number of events.

6   Conclusion

We proposed a novel CNN method using wavelet spectrograms to detect hypopnea, obstructive apnea and normal events, and demonstrated its applicability against a CNN using raw 1-D signal using a large dataset of 69,264 instances from 1,507 subjects. Both methods use a single respiratory signal from PSG - the nasal airflow, which is easy and non-intrusive to obtain. The first method uses the raw 1-D nasal airflow, while the second method uses 2-D spectrogram images of the nasal airflow. Both methods learn the informative features automatically, without prior feature extraction and selection.

Our results showed that both methods achieved good accuracy, with the 2-D CNN providing higher accuracy (79.8%) than the 1-D CNN (77.6%), and both considerably outperforming the baseline of 33.3%. An accuracy of 79.8% is a very promising result, especially given that only the nasal signal was used. Consistent with previous studies of sleep apnea, both methods performed better at identifying the obstructive apnea class than the other two classes, achieving F1 scores greater than 89%.

The less complex architecture of the 2-D CNN suggests that it is easier to identify the most significant features from spectrograms than from the 1-D signal. This demonstrates that converting biological signals into spectrograms can provide valuable information which is difficult to see in the original signal.

Further work will include investigating the use of other respiratory signals, e.g., thoracic, abdominal and chest, in addition to the nasal airflow. In this study, other apnea events including central apnea and mixed apnea were not investigated, and we plan to test the applicability of this method on detecting other types of apnea events in the future. Additionally, it would be interesting to compare the performance of other methods, including time series classifiers such as motif-based time series classifiers, to identify sleep apnea events [27]. We also plan to test the accuracy of our methods on other sleep apnea datasets. A high quality re-classification of the MESA dataset by several experts is also desirable. Our research also highlights the need to create a common, large benchmark dataset for comparing different apnea and hypopnea detection methods.

Acknowledgements. This research was supported by Sydney Informatics Hub's High Performance Computing Services, funded by the University of Sydney.

The Multi-Ethnic Study of Atherosclerosis (MESA) is supported by contracts N01-HC-95159 through N01-HC-95169 from the National Heart, Lung, and Blood Institute (NHLBI) at the National Institutes of Health. MESA Sleep was supported by NHLBI R01 L098433.

References

1. Epstein, L.J., Kristo, D., Strollo, P.J., et al.: Clinical guideline for the evaluation, management and long-term care of obstructive sleep apnea in adults. J. Clin. Sleep Med. **5**(03), 263–276 (2009)
2. Punjabi, N.M.: The epidemiology of adult obstructive sleep apnea. Proc. Am. Thorac. Soc. **5**(2), 136–143 (2008)
3. Sanna, A.: Obstructive sleep apnoea, motor vehicle accidents, and work performance. Chronic Respir. Dis. **10**(1), 29–33 (2013)
4. Marin, J.M., Carrizo, S.J., Vicente, E., Agusti, A.G.: Long-term cardiovascular outcomes in men with obstructive sleep apnoea-hypopnoea with or without treatment with continuous positive airway pressure: an observational study. Lancet **365**(9464), 1046–1053 (2005)
5. Yaggi, H.K., Concato, J., Kernan, W.N., Lichtman, J.H., Brass, L.M., Mohsenin, V.: Obstructive sleep apnea as a risk factor for stroke and death. N. Engl. J. Med. **353**(19), 2034–2041 (2005). PMID: 16282178
6. Yumino, D., Bradley, T.D.: Central sleep apnea and Cheyne-Stokes respiration. Proc. Am. Thorac. Soc. **5**(2), 226–236 (2008)
7. Krizhevsky, A., Sutskever, I., Hinton, G.E.: Imagenet classification with deep convolutional neural networks. In: Advances in Neural Information Processing Systems, pp. 1097–1105 (2012)
8. Abdel-Hamid, O., Mohamed, A.R., Jiang, H., Penn, G.: Applying convolutional neural networks concepts to hybrid NN-HMM model for speech recognition. In: 2012 IEEE International Conference on Acoustics, Speech and Signal Processing (ICASSP), pp. 4277–4280, March 2012
9. Wang, T., Wu, D.J., Coates, A., Ng, A.Y.: End-to-end text recognition with convolutional neural networks. In: Proceedings of the 21st International Conference on Pattern Recognition (ICPR2012), pp. 3304–3308, November 2012
10. Biswal, S., Kulas, J., Sun, H., et al.: SLEEPNET: automated sleep staging system via deep learning. arXiv preprint arXiv:1707.08262 (2017)
11. Maali, Y., Al-Jumaily, A.: Automated detecting sleep apnea syndrome: a novel system based on genetic SVM. In: 2011 11th International Conference on Hybrid Intelligent Systems (HIS), pp. 590–594, December 2011

12. Maali, Y., Al-Jumaily, A.: Hierarchical parallel PSO-SVM based subject-independent sleep apnea classification. In: Huang, T., Zeng, Z., Li, C., Leung, C.S. (eds.) ICONIP 2012. LNCS, vol. 7666, pp. 500–507. Springer, Heidelberg (2012). https://doi.org/10.1007/978-3-642-34478-7_61

13. Koley, B.L., Dey, D.: Automatic detection of sleep apnea and hypopnea events from single channel measurement of respiration signal employing ensemble binary SVM classifiers. Measurement 46(7), 2082–2092 (2013)

14. Waxman, J.A., Graupe, D., Carley, D.W.: Automated prediction of apnea and hypopnea, using a LAMSTAR artificial neural network. Am. J. Respir. Crit. Care Med. 181(7), 727–733 (2010)

15. Tagluk, M.E., Akin, M., Sezgin, N.: Classification of sleep apnea by using wavelet transform and artificial neural networks. Expert Syst. Appl. 37(2), 1600–1607 (2010)

16. Almazaydeh, L., Elleithy, K., Faezipour, M., Abushakra, A.: Apnea detection based on respiratory signal classification. Procedia Comput. Sci. 21, 310–316 (2013)

17. Hsu, C.C., Shih, P.T.: A novel sleep apnea detection system in electroencephalogram using frequency variation. Expert Syst. Appl. 38(5), 6014–6024 (2011)

18. Avcı, C., Akbaş, A.: Sleep apnea classification based on respiration signals by using ensemble methods. Bio-Med. Mater. Eng. 26(s1), S1703–S1710 (2015)

19. Haidar, R., Koprinska, I., Jeffries, B.: Sleep apnea event detection from nasal airflow using convolutional neural networks. In: Liu, D., Xie, S., Li, Y., Zhao, D., El-Alfy, E.-S.M. (eds.) ICONIP 2017. LNCS, vol. 10638, pp. 819–827. Springer, Cham (2017). https://doi.org/10.1007/978-3-319-70139-4_83

20. Dean, D.A., Goldberger, A.L., Mueller, R., et al.: Scaling up scientific discovery in sleep medicine: the National Sleep Research Resource. Sleep 39(5), 1151–1164 (2016)

21. Tibshirani, R., James, G., Witten, D., Hastie, T.: An introduction to statistical learning: with applications in R (2013)

22. LeCun, Y., Bengio, Y., Hinton, G.: Deep learning. Nature 521(7553), 436–444 (2015)

23. Snoek, J., Larochelle, H., Adams, R.P.: Practical Bayesian optimization of machine learning algorithms. In: Advances in Neural Information Processing Systems, pp. 2951–2959 (2012)

24. Bergstra, J., Bengio, Y.: Random search for hyper-parameter optimization. J. Mach. Learn. Res. 13, 281–305 (2012)

25. Kinga, D., Adam, J.B.: A method for stochastic optimization. In: International Conference on Learning Representations (ICLR) (2015)

26. Srivastava, N., Hinton, G.E., Krizhevsky, A., et al.: Dropout: a simple way to prevent neural networks from overfitting. J. Mach. Learn. Res. 15(1), 1929–1958 (2014)

27. Buza, K., Schmidt-Thieme, L.: Motif-based classification of time series with Bayesian networks and SVMs. In: Fink, A., Lausen, B., Seidel, W., Ultsch, A. (eds.) Advances in Data Analysis, Data Handling and Business Intelligence. Studies in Classification, Data Analysis, and Knowledge Organization. Springer, Heidelberg (2009). https://doi.org/10.1007/978-3-642-01044-6_9

Deep Ensemble Classifiers and Peer Effects Analysis for Churn Forecasting in Retail Banking

Yuzhou Chen[1], Yulia R. Gel[2(✉)], Vyacheslav Lyubchich[3], and Todd Winship[4]

[1] Southern Methodist University, Dallas, TX, USA
[2] University of Texas at Dallas, Richardson, TX, USA
ygl@utdallas.edu
[3] University of Maryland Center for Environmental Science, Cambridge, MD, USA
[4] Temenos, Surrey, BC, Canada

Abstract. Modern customer analytics offers retailers a variety of unprecedented opportunities to enhance customer intelligence solutions by tracking individual clients and their peers and studying clientele behavioral patterns. While telecommunication providers have been actively utilizing peer network data to improve their customer analytics for a number of years, there yet exists a very limited knowledge on the peer effects in retail banking. We introduce modern deep learning concepts to quantify the impact of social network variables on bank customer attrition. Furthermore, we propose a novel deep ensemble classifier that systematically integrates predictive capabilities of individual classifiers in a meta-level model, by efficiently stacking multiple predictions using convolutional neural networks. We evaluate our methodology in application to customer retention in a retail financial institution in Canada.

1 Introduction

Customer retention is crucial for company profitability and growth. Satisfied customers provide ongoing cross-sell and up-sell opportunities, and tend to refer a pool of new clients. Acquiring a new customer can be 5–25 times (depending on the industry) more expensive than retaining a current one [11]. In the saturated markets of retail banking, the intense competition pushes these costs toward the upper boundary. At the same time, there is a strong association between customer retention and profitability: long-term customers buy more and are less costly to serve, while new ones are likely to continue their churning behavior [16].

Loss of clients, also known as *churn* or customer attrition, is widely recognized as one of the most critical business challenges for a variety of companies, from telecommunication providers to financial institutions. While companies pursue new customers through acquisition marketing efforts, customer churn undermines the business growth. Voluntary turnover rates for banking and finance are the third largest (after hospitality and healthcare) among all industries [7]. Hence, analysis of customer characteristics, such as socio-demographics and activity patterns, is crucial for predictive identification of customers who are likely to churn, as well as for more efficient targeted application of marketing strategies for customer retention.

Through the theoretical and economic framework of customer retention strategy, [18] show that such reasons as purchase intention, proportion of category purchases and purchase regularity are strongly associated with loyalty decisions. Moreover, decisions of many customers tend to be strongly affected by customer's social neighbors. In the banking industry, 71% of customers turn to friends, family, and colleagues for information on bank products [10]. Still, most marketing tools primarily employ direct approaches, neglecting network effects and treating customers independently of their social network environment. As a result, banks lose invaluable information on the driving forces of customer's purchasing behavior and churn.

Despite the well-documented impacts of peer networks on customer behavior, still very few studies incorporate the network information of bank clientele in the retention models, and one of the reasons is the lack of the explicit network ties in the customer databases. For example, [3] use kinship information deliberately collected from bank customers for the study – this is a costly approach with a number of data quality and data privacy issues. As an alternative, [19] use information on bank transfers and joint loans to build customer networks. This approach, however, is applicable only for large banks, because a single bank with a moderate market share has a high portion of transfers being inter-bank transactions, where detailed information on the second customer is not available, thus, resulting in highly sparse data. In an attempt to enhance customer service and, possibly, the network database, HSBC has recently launched a social network for its business customers [14], which can be considered at this point as an experiment, rather than a standard practice. In this study, we adopt a different method of building the customer network, by taking advantage of information that is readily available at any bank – family name and address of each customer.

Furthermore, our customer dataset is highly unbalanced. That is, the number of non-churners is much larger than that of churners. In turn, most statistical and machine learning classifiers suffer from the inability to detect weak signals in such unbalanced datasets. In binary classification problems, such as customer retention, this phenomenon implies a low specificity or low sensitivity of a classifier. To address this issue, we propose a new deep ensemble classifier which harnesses powers of individual classifiers in a meta-level classifier, using convolutional neural networks. The rationale behind this novel framework is that convolutional neural networks can extract useful features by efficient stacking of multiple predictions. We demonstrate the performance of the new technique in binary classification tasks.

The main contributions of our study are as follows:

– We develop a novel predictive tool for customer churn in retail banking that accounts for the invaluable information on clientele social network effects.
– We introduce deep learning concepts to customer retention analysis in retail banking and propose a cost-effective way of building customer networks.

– We develop a novel deep ensemble classifier, which integrates predictive capabilities of single models in a meta-level classifier, using convolutional neural networks. Our studies indicate that the new deep ensemble classifier delivers a competitive performance, especially in largely unbalanced datasets, and hence has a potential for high utility in a wide variety of classification problems, well beyond customer retention.

The paper is organized as follows. Section 2 provides a background on the related work in social network analysis and customer retention modeling. Section 3 describes the data, and Sect. 4 presents the proposed methodology. Section 5 discusses the main results of the study. Section 6 summarizes the results and outlines directions for future research.

2   Related Work

Nowadays, there exists a plethora of machine learning approaches to customer data mining and retention modeling, ranging from classical regression to neural networks to random forests (e.g., see [12, 17, 20, 23] for a general topic overview). The experiments in [21] showed that neural networks typically outperform logistic regression and decision trees in churn prediction. Nevertheless, the performance of neural networks noticeably deteriorates under a lower monthly churn rate (unbalanced data), that is, the problem that we address in this paper using a new deep ensemble classifier.

Peer networks are known to influence a variety of customer decisions. Applications of social network analysis to customer retention, however, are often limited due to poor availability of data on customer peer networks. Most progress in this direction has been achieved in telecommunication industry, where social networks are naturally observed from the call and message records (e.g., see [1, 2, 13, 24, 28]). Constructing networks of bank customers requires additional steps, such as targeted surveys [3], mining the databases of customers and their transactions [19], and, potentially, employing big data approaches for harnessing customer information from disparate sources, including online social media [22]. Overall, the analysis of the impact of peer networks on customer behavior in retail banking remains largely at its infancy, comparing with other industries. We address this challenge by introducing a cost-effective way to collect peer information in retail banking and integrate these data with high predictive utility into customer analytics solutions.

Deep learning (DL) methods continue to attract increasing interest in customer churn prediction, while being a relatively new tool in customer analytics. DL architectures, like the multi-layer feedforward architecture, can effectively capture features of the underlying customer data and learn hierarchical clientele data structures [4, 25]. To our knowledge, this paper is the first one to introduce DL concepts into customer retention models in retail banking.

3   Data

The data used in this study comprise a database of all transactions, accounts, and (monthly) snapshots of a customer database of a retail financial institution in North America over a period of 3.5 years (2011.1–2014.6). The customer database contains information collected from the customers themselves (name, address, age, gender, etc.; some of these records are missing or outdated) and from the bank's records about each customer (tenure, number of accounts of each type, total amount owned, etc.; complete and up-to-date records). The customer data were redacted – first names completely removed, family names and addresses replaced by encrypted numeric IDs – so that customers' privacy was protected, but some information about their closeness (derived by matching family name and address IDs) was preserved.

From approximately 30 thousand customers in the sample dataset, we select customers who can make their own financial decisions (above 18 years old) and are likely alive (below 100 years old), then split the data into consecutive baseline and prediction periods. Information from a baseline period is used to predict whether a customer will churn in the nearest future, where *churn* is defined as inactivity (number of transactions is zero) during the prediction period. Hence, churn can be represented as a binary variable taking on the value 0 for customers who stay active in the prediction period and value 1 for those who churn. We use one-year baseline periods (2011, 2012, and 2013) with respective prediction periods of one year (2012 and 2013) and six months (2014.1–6).

3.1   Feature Engineering

Building a set of features for customers is an important step for capturing and quantifying nuances of customer behavior and achieving a superb predictive performance of the customer retention models. We use domain knowledge to create *individual features* (variables) that are potentially associated with customer's retention or churn: age, average time between transactions, time since last transaction, number of loans, number of past transactions, tenure, total savings and total credit balances. For example, middle-aged customers or those having large credit balances often are mortgage owners and will likely stay with the bank for some time. Conversely, older customers may become activity churners when they retire and direct their pension payouts to another bank.

Fig. 1. An example network of six bank customers (nodes), where edges connect people from the same family. Each node has individual and family network features identified

The matching address and family name IDs allow us to create *family networks* (Fig. 1) that join customers who have the same address and family name (the IDs do not reveal any other details, such as neighbor or co-worker relationships). These family networks are cliques, because we consider each family member as connected to everyone else in that family. To capture the dependence of customers on their family members, we apply the egocentric network approach and define family network features for each customer. The *network features* include the individual features aggregated within a family (average age, tenure, total savings, etc.), family size, and two variables indicating whether a family has had churners in the baseline period ("Presence of churners") and how many ("Number of churners in the family").

4   Methods

4.1   Deep Convolutional Neural Networks

Convolutional neural networks (CNN) is a class of artificial neural networks that are based on translational invariance and are weight-shared. These two characteristics increase learning efficiency and make CNN less prone to overfitting than simple artificial neural networks. Hence, for multi-label (binary) datasets, CNN can be trained as a feature extractor and perform better than other classification techniques. An attractive property of CNN is that CNN trained on large datasets have demonstrated an ability to capture high-quality features describing data.

CNN are widely used for image and natural language processing because they can handle static content, like an image or a sentence, well. We make the first attempt to apply CNN for churn prediction in retail banking. In contrast to 2D inputs in image classification, churn input data are 1D. We create a supervised feedforward neural network for binary predictive classification of customer retention. Using the features listed in Sect. 3.1, we found that CNN are able to efficiently mine interesting classification rules.

Architecture.   *Convolutional Filter Layer.* Let $\mathbf{X} = (X_1, X_2, \ldots, X_n)$ be a high-dimensional input matrix and Y be the output vector. Deep learning can be treated as learning a function, F, mapping input to output:

$$Y = F(\mathbf{X}), \text{ where } X_i \in \mathbb{R}^p, \ Y_i \in \{0, 1\}. \tag{1}$$

In the convolution process, set a filter $\mathbf{w}$ of size k. Then, to obtain a feature m_i of the feature map $\mathbf{m} = [m_1, m_2, \ldots, m_{p-k+1}]$, where $m_i = f(\mathbf{w} \cdot X_{i:i+k-1} + b)$ and b is its corresponding bias offset, apply activation function f to $X_{i:i+k-1}$.

Activation Layer. The role of activation function is to transform the input space of each layer in neural network in such a way that output units become linearly separable. Commonly used activation functions are ReLU $f(x) = max(0, x)$, logistic $\sigma(x) = 1/(1 + \exp(-x))$, and hyperbolic tangent $tanh(x) = 2\sigma(2x) - 1$.

Pooling Layer. Pooling (downsampling) layer decreases the computational complexity and prevents overfitting by reducing number or dimensions in a previous layer. It is done by applying sum-pooling, average-pooling, or max-pooling [24,27]. In our study, we apply max-pooling: $\hat{m} = \max\{\mathbf{m}\}$.

Loss Function and Regularization. The goal of our study is to train the learner (1) using a loss function $\mathcal{L}(\mathbf{y}, \mathbf{o})$, where $\mathbf{o}$ is the output from learner and $\mathbf{y}$ is the true output label. Typically $\mathcal{L}_1 = \|\mathbf{y} - \mathbf{o}\|$ and $\mathcal{L}_2 = \|\mathbf{y} - \mathbf{o}\|^2$ are applied to the regression problem and training process of neural networks. In our case, we use cross-entropy $\mathcal{L} = -\sum_{i=1}^{k} \mathbf{y}_i \log(\mathbf{o}_i)$, because it delivers stable good performance with softmax layer, which is the last layer of our CNN:

$$f(x)_i = e^{x_i} \Big/ \sum_{j=1}^{J} e^{x_j}, \quad \text{for } i = 1, \ldots, J.$$

When fitting a model on a relatively small training dataset, overfitting is always a problem for out-of-sample prediction. Neural networks have particularly many parameters that contribute together to building an excessively complex model, which may overfit the data. Dropout [26] is a regularization technique that helps to get an efficient final neural network architecture and to avoid overfitting. The dropout deletes some of the features in $\mathbf{X}$ and, in the training phase, sets the output of each hidden neuron to 0 with probability p. The feed-forward operation for layers $l = 1, \ldots, L - 1$ [26]:

$$d_k^{(l)} \sim Bernoulli(p); \quad \widetilde{y}^{(l)} = d^{(l)} \circ y^{(l)}; \quad y^{(l+1)} = f(W^{(l+1)}\widetilde{y}^{(l)} + b^{(l+1)}), \quad (2)$$

where d is a vector of Bernoulli random variables, $W^{(l)}$ and $y^{(l)}$ are the vectors of biases and outputs from layer l, and $\circ$ denotes the element-wise product.

4.2   Deep Ensemble Classifier

A single model cannot guarantee a uniformly optimal, or at least stable, performance in all cases for which we need to make predictions [29]. Some models are better than others in responding to specific patterns in the data, e.g., those mentioned in Sect. 3.1. A possible solution to this problem is training an ensemble of models and combining their results in some way to obtain more stable and accurate predictions. The stability of out-of-sample (generalization) errors is achieved in ensembles by aggregating information from many models that can potentially overfit the training data, but each model in its own way. Higher accuracy is often achieved even by simple averaging of the single model predictions, but more informed methods, which take into consideration specific strengths and weaknesses of each model, may lead to even better results.

The widely used ensemble methods include bagging, boosting, Bayesian model averaging (BMA), and stacking. Compared with stacking, BMA uses different posterior probabilities to weight each base-level model. The empirical

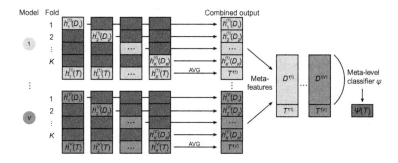

Fig. 2. Stacking with K-fold cross-validation

results in [6] showed that stacking consistently delivers more competitive performance than BMA. BMA works better only when the correct data generating model belongs to the set of model candidates and the noise is low, i.e., under the conditions that are very difficult to satisfy in applications. In turn, stacking outperforms other ensemble methods due to its ability to learn and flexibly account for the behaviors of other classifiers in a combining model [9].

The standard stacking technique is based on applying a logistic regression on the outputs of base-level models, which limits us to the case of monotonic relationships (also, with the same speed of approaching both asymptotes) between predictions from each base-level model and the response. We relax this condition and develop a new deep ensemble classifier for building the second layer of classifiers, based on more flexible machine learning methods. In particular, we propose and evaluate the performance of the following stacking approaches: stacking with CNN (StCNN); stacking with RF (StRF); stacking with XGB (StXGB); stacking with Extra-Trees (StET); stacking with NN (StNN), and stacking with KNN (StKNN). We also use K-fold cross-validation, which provides a good trade-off between variance and bias (see Algorithm 1 and Fig. 2).

5   Results

To compare the performance of single and stacked models and see the effect of adding network features, we design the following four scenarios: (i) single base-level model with individual features alone; (ii) single base-level model with both individual and network features; (iii) stacked models with individual features alone, and (iv) stacked models with both individual and network features.

We split the data into training ($\mathcal{D}$, 70%) and testing ($\mathcal{T}$, 30%) subsets and report results for predicting for the testing subset. On the dataset $\mathcal{D}$, we train five different single base-level models with the following methods: random forest (RF), extreme gradient boosting (XGB [5]), K-nearest neighbor algorithm (KNN), neural networks (NN), and CNN. Each of the considered five methods can provide several well-performing models with different tuning parameters, and we can use all of them when creating an ensemble (thus, each ensemble we

---

Algorithm 1. Stacking with K-fold cross-validation

INPUT: Training set $\mathcal{D} = \{\mathbf{x}_i, y_i\}_{i=1}^m$ and testing set $\mathcal{T} = \{\mathbf{x}_i, y_i\}_{i=1}^n$ ($\mathbf{x}_i \in \mathbb{R}^p, y_i \in \{0, 1\}$), number of folds K, and V different base-level classifiers

OUTPUT: A meta-level classifier Ψ

1: Randomly split $\mathcal{D}$ into K equal-size subsets: $\mathcal{D} \leftarrow \{\mathcal{D}_1, \mathcal{D}_2, \ldots, \mathcal{D}_K\}$
2: **for** v in 1 to V **do**
3:     **for** k in 1 to K **do**
4:         Train a classifier $h_k^{(v)}$ on $\mathcal{D} \setminus \mathcal{D}_k$
5:         Let $h_k^{(v)}(\mathcal{D}_k)$ be the out-of-fold predictions for the set $\mathcal{D}_k$
6:         Let $h_k^{(v)}(\mathcal{T})$ be predictions for the testing set
7:     **end for**
8:     Construct a new variable from out-of-fold predictions in the training set:

$$\mathcal{D}'^{(v)} \leftarrow \{h_1^{(v)}(\mathcal{D}_1), h_2^{(v)}(\mathcal{D}_2), \ldots, h_K^{(v)}(\mathcal{D}_K)\}$$

9:     The new variable in the testing set is an average of K predictions:

$$\mathcal{T}'^{(v)} \leftarrow \mathrm{AVG}\{h_1^{(v)}(\mathcal{T}), h_2^{(v)}(\mathcal{T}), \cdots, h_K^{(v)}(\mathcal{T})\}$$

10: **end for**
11: Train a meta-level classifier ψ on $\left\{ \left(\mathcal{D}'^{(1)}, \mathcal{D}'^{(2)}, \ldots, \mathcal{D}'^{(V)} \right), y_i \right\}_{i=1}^m$
12: Return $\Psi(\mathcal{T}) \leftarrow \psi \left(\mathcal{T}'^{(1)}, \mathcal{T}'^{(2)}, \cdots, \mathcal{T}'^{(V)} \right)$

---

created had more than five members). In the stacking Algorithm 1, we use 4-fold cross-validation for the first two time periods and 7-fold cross-validation for the third period. The optimal parameters for each base-level model were chosen through a grid search.

The CNN architectures used in this study are shown in Table 1. In the CNN training, we use tanh as an activation function, and ReLU in the second layer. The advantages of ReLU include faster model training and smaller chance of the gradient to vanish. We apply dropout with the probability $p = 0.6$ at the second layer before pooling and insert a batch normalization layer [15] (eps = 0.00001, momentum = 0.99) before applying the activation function in the second layer. Stochastic gradient descent was chosen as the CNN optimizer, with the learning rate of 0.001 and momentum value of 0.9 as optimal parameters.

For each of the scenarios (i)–(iv), Table 2 layouts a confusion matrix $\begin{bmatrix} TN & FP \\ FN & TP \end{bmatrix}$ for the subset $\mathcal{T}$, where TN is the number of non-churners classified as non-churners, FP is the number of non-churners classified as churners, FN is the number of churners classified as non-churners, and TP is the number of churners classified as churners. Table 2 shows that improved churn predictions can be achieved by leveraging the CNN architecture, novel stacking approach (StCNN), and customer network features.

Figure 3 reports the misclassification rates $R = (FN + FP)/N$ (where $N = |\mathcal{T}|$ is the size of the testing set) delivered by various base-level models on the

Table 1. Architectures of CNN

Layer	Layer type	Size of base level	Size of meta level
1	Convolution + tanh	1×2 20 filters	1×2 30 filters
1	Max pooling	1×3, stride 1	1×2, stride 1
2	Convolution + ReLU	1×3 50 filters	1×2 50 filters
2	Max pooling	1×5, stride 1	1×2, stride 1
3	Fully connected + tanh	500 hidden units	500 hidden units
4	Fully connected + tanh	2 hidden units	2 hidden units
5	Softmax	2 ways	2 ways

test datasets. The base-level models XGB, RF, and CNN outperform NN and KNN in each period.

Remarkably, CNN that include both individual and network features perform noticeably better than other baseline models for the 1st and 2nd periods. We also observe that RF performs better in the 3rd period when using both individual and network features. The results in Fig. 3 prove that most of our single models

Table 2. Confusion matrices[a] of predictive classifying of bank customers

Baseline period	Prediction period	Model	Individual features		Individual & network features	
			Single model	Stacked models	Single model	Stacked models
2011.1–12	2012.1–12	CNN	6634 13 108 240	6628 19 101 247	6629 18 101 247	**6630 17** **99 249**
		XGB	6618 29 101 247	6621 26 102 246	6617 30 104 244	6616 31 97 251
		RF	6627 20 103 245	6626 21 105 243	6623 24 102 246	6630 17 105 243
2012.1–12	2013.1–12	CNN	8305 20 29 280	8304 21 26 283	**8305 20** **26 283**	**8305 20** **26 283**
		XGB	8304 21 41 268	8303 22 34 275	8304 21 37 272	8305 20 30 279
		RF	8304 21 27 282	8304 21 31 278	8303 22 29 280	8304 21 30 279
2013.1–12	2014.1–6	CNN	8227 14 78 749	**8219 22** **54 773**	8219 22 66 761	8221 20 61 766
		XGB	8207 34 57 770	8212 29 54 773	8214 27 61 766	8214 27 59 768
		RF	8220 21 62 765	8216 25 59 768	8217 24 58 769	8214 27 56 771

[a]Each cell is a 2×2 confusion matrix. For each period, matrices with minimal sum $FP + FN$ are highlighted.

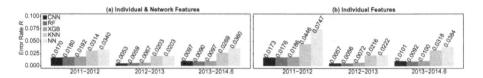

Fig. 3. Performance of base-level algorithms with different sets of features

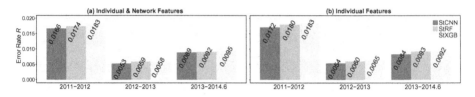

Fig. 4. Performance of meta-level algorithms with different sets of features

(especially CNN and RF) trained on both individual and network features are more accurate (lower R) than models trained exclusively on individual features.

Aggregation of results by stacking further improves the predictive performance. Among the six considered stacking algorithms, the best three are based on CNN, RF, and XGB – the algorithms that also show the best performance in the base-level scenarios (Fig. 3). Accuracy of these three methods is noticeably higher than of the other three (StET, StNN, and StKNN), while running time of XGB is considerably shorter. Figure 4 shows that StCNN always outperforms the other stacking schemes. Compared with Fig. 3, accuracy of the best-performing combinations changed as follows:

- improved from 98.30% (CNN with both individual and network features) to 98.34% (StCNN with both individual and network features), i.e., by 0.04 percentage points, for the period 2011–2012;
- stayed at about 99.47% (CNN and StCNN, each with both individual and network features) for 2012–2013;
- improved from 99.10% (RF with both individual and network features) to 99.16% (StCNN with individual features), i.e., by 0.06 points, for the period 2013–2014.6.

The results imply that the architecture of CNN can improve the performance of churn predictive classification with automatically capturing and extracting relevant features, especially after adding network features into the model. Furthermore, StCNN can simultaneously reduce false negative rates and yield the optimal true negative rate. Nevertheless, in the absence, to the best of our knowledge, of a formal statistical test applicable to a stacking scheme, more extensive experiments based on a cross-validation argument are needed to prove the statistical significance of the improvement of using StCNN.

In the StCNN, we use Adagrad [8] as the optimizer, and set learning rate, epsilon, and L_2 regularization coefficient (wd) to 10^{-2}, 10^{-10}, and 10^{-3} by tuning with a grid search. The accuracy of the above results is high (i.e., errors R are low) in part due to a very low proportion of churned customers (below 6%). The dataset is unbalanced, as well as the costs of losing a customer. Various studies suggest that such costs can be 5–25 times higher than the costs of retaining an existing one [11], but the results above assume equal costs of FP and FN.

		2011 – 2012			2012 – 2013				2013 – 2014.6			
Features:		Individual		Individual & network		Individual		Individual & network		Individual		Individual & network
Model:		Single	Stacked	Single	Stacked	Single	Stacked	Single	Stacked	Single	Stacked	Single	Stacked
Method	XGB	0.2165	0.2192	0.2212	0.2091	0.0857	0.0724	0.0782	0.0647	0.1087	0.1042	0.1161	0.1128
	RF	0.2229	0.2259	0.2198	0.2273	0.0588	0.0666	0.0627	0.0647	0.1185	0.1130	0.1114	0.1078
	CNN	0.2338	0.2198	0.2202	0.2171	0.0627	0.0568	0.0568	0.0568	0.1455	0.1048	0.1250	0.1169

Fig. 5. Misclassification rates R' (dark color shade means smaller) when churners weigh 20 times more than non-churners ($r = 20$)

In Fig. 5, we use cost ratio ($r = 20$) of FN to FP to upweight the errors in misclassifying churners ($FN + TP$):

$$R' = \frac{r \cdot FN + FP}{N + (r - 1)(FN + TP)}. \tag{3}$$

Figure 5 shows that the egocentric network approach and model stacking, in particular, the StCNN, improve the churn predictions.

6   Conclusion

We have proposed a novel predictive tool for customer retention in retail banking by introducing deep learning concepts into churn analysis. Our approach allows to systematically and consistently integrate invaluable information on customer peer effects into the customer analytics process. We have developed a new deep ensemble classifier that fuses predictive powers of individual classifiers in a meta-level model, by efficiently stacking multiple predictions using convolutional neural networks. The proposed deep ensemble classifier delivers competitive performance in largely unbalanced customer data and, hence, has a potential for a wide applicability in classification problems well beyond customer analytics.

Acknowledgements. This research was partially supported by NSF IIS 1633331 & 1633355, NSF DMS 1736368, and Simons Foundation. The work of V. Lyubchich was supported by Mitacs Accelerate Internship Awards with contributions from Temenos Canada.

References

1. Backiel, A., Baesens, B., Claeskens, G.: Mining telecommunication networks to enhance customer lifetime predictions. In: Rutkowski, L., Korytkowski, M., Scherer, R., Tadeusiewicz, R., Zadeh, L.A., Zurada, J.M. (eds.) ICAISC 2014. LNCS (LNAI), vol. 8468, pp. 15–26. Springer, Cham (2014). https://doi.org/10.1007/978-3-319-07176-3_2
2. Backiel, A., Baesens, B., Claeskens, G.: Predicting time-to-churn of prepaid mobile telephone customers using social network analysis. J. Oper. Res. Soc. **67**(9), 1135–1145 (2016)
3. Benoit, D.F., Van den Poel, D.: Improving customer retention in financial services using kinship network information. Expert Syst. Appl. **39**(13), 11435–11442 (2012)
4. Castanedo, F., Valverde, G., Zaratiegui, J., Vazquez, A.: Using deep learning to predict customer churn in a mobile telecommunication network (2014)
5. Chen, T., Guestrin, C.: XGBoost: a scalable tree boosting system. In: Proceedings of the 22nd SIGKDD International Conference on Knowledge Discovery and Data Mining, pp. 785–794. ACM (2016)
6. Clarke, B.: Comparing Bayes model averaging and stacking when model approximation error cannot be ignored. J. Mach. Learn. Res. **4**, 683–712 (2003)
7. Compensation Force: 2016 turnover rates by industry (2017)
8. Duchi, J., Hazan, E., Singer, Y.: Adaptive subgradient methods for online learning and stochastic optimization. J. Mach. Learn. Res. **12**, 2121–2159 (2011)
9. Džeroski, S., Ženko, B.: Is combining classifiers with stacking better than selecting the best one? Mach. Learn. **54**(3), 255–273 (2004)
10. Ernst & Young: The customer takes control. Consumer Banking Survey (2012)
11. Gallo, A.: The value of keeping the right customers. Harv. Bus. Rev. **5**, 2–6 (2014)
12. Han, S.H., Lu, S.X., Leung, S.C.: Segmentation of telecom customers based on customer value by decision tree model. Expert Syst. Appl. **39**(4), 3964–3973 (2012)
13. Hill, S., Provost, F., Volinsky, C.: Network-based marketing: identifying likely adopters via consumer networks. Stat. Sci. **21**(2), 256–276 (2006)
14. HSBC: HSBC launches global 'social network' for business customers (2017)
15. Ioffe, S., Szegedy, C.: Batch normalization: accelerating deep network training by reducing internal covariate shift. In: International Conference on Machine Learning, pp. 448–456 (2015)
16. Larivière, B., Van den Poel, D.: Predicting customer retention and profitability by using random forests and regression forests techniques. Expert Syst. Appl. **29**(2), 472–484 (2005)
17. Li, D.C., Dai, W.L., Tseng, W.T.: A two-stage clustering method to analyze customer characteristics to build discriminative customer management: a case of textile manufacturing business. Expert Syst. Appl. **38**(6), 7186–7191 (2011)
18. Macintosh, G., Lockshin, L.S.: Retail relationships and store loyalty: a multi-level perspective. Int. J. Res. Mark. **14**(5), 487–497 (1997)
19. Mao, H., Jin, X., Zhu, L.: Methods of measuring influence of bank customer using social network model. Am. J. Ind. Bus. Manag. **5**(4), 155 (2015)
20. Miguéis, V.L., Van den Poel, D., Camanho, A.S., e Cunha, J.F.: Modeling partial customer churn: on the value of first product-category purchase sequences. Expert Syst. Appl. **39**(12), 11250–11256 (2012)
21. Mozer, M.C., Wolniewicz, R., Grimes, D.B., Johnson, E., Kaushansky, H.: Predicting subscriber dissatisfaction and improving retention in the wireless telecommunications industry. IEEE Trans. Neural Netw. **11**(3), 690–696 (2000)

22. NG Data: Predicting and preventing customer churn by unlocking big data (2013)
23. Ngai, E.W., Xiu, L., Chau, D.C.: Application of data mining techniques in customer relationship management: a literature review and classification. Expert Syst. Appl. **36**(2), 2592–2602 (2009)
24. Scherer, D., Müller, A., Behnke, S.: Evaluation of pooling operations in convolutional architectures for object recognition. In: Diamantaras, K., Duch, W., Iliadis, L.S. (eds.) ICANN 2010. LNCS, vol. 6354, pp. 92–101. Springer, Heidelberg (2010). https://doi.org/10.1007/978-3-642-15825-4_10
25. Spanoudes, P., Nguyen, T.: Deep learning in customer churn prediction: unsupervised feature learning on abstract company independent feature vectors. arXiv:1703.03869 (2017)
26. Srivastava, N., Hinton, G.E., Krizhevsky, A., Sutskever, I., Salakhutdinov, R.: Dropout: a simple way to prevent neural networks from overfitting. J. Mach. Learn. Res. **15**(1), 1929–1958 (2014)
27. Szegedy, C., Liu, W., Jia, Y., Sermanet, P., Reed, S., Anguelov, D., Erhan, D., Vanhoucke, V., Rabinovich, A.: Going deeper with convolutions. In: Proceedings of CVPR, pp. 1–9. IEEE (2015)
28. Verbeke, W., Martens, D., Baesens, B.: Social network analysis for customer churn prediction. Appl. Soft Comput. **14**, 431–446 (2014)
29. Witten, I.H., Frank, E., Hall, M.A., Pal, C.J.: Data Mining: Practical Machine Learning Tools and Techniques. Morgan Kaufmann, San Francisco (2016)

GBTM: Graph Based Troubleshooting Method for Handling Customer Cases Using Storage System Log

Subhendu Khatuya[1](✉), Ajay Bakhshi[2], Jayanta Basak[2], Niloy Ganguly[1], and Bivas Mitra[1]

[1] Department of CSE, Indian Institute of Technology Kharagpur, Kharagpur, India
subhendu.cse@iitkgp.ac.in, {niloy,bivas}@cse.iitkgp.ernet.in
[2] NetApp Inc., Bangalore, India
{Ajay.Bakhshi,Jayanta.Basak}@netapp.com

Abstract. Present day computing environments consist of different bits of hardware and software that are associated with each other in a complex way. Hence, in case of failures of such system, it is very difficult to detect the exact module which has caused the problem. In such a situation, an automated technique which can pin down to (at least) a set of modules that may be responsible for the failure would be very useful for support engineers. This paper makes an important step towards that direction. We propose a graph based troubleshooting methodology exploring storage system logs (EMS) of around 4500 customer cases to troubleshoot customer problems. We provide a ranked list of modules to the support engineers which can significantly narrow down the troubleshooting process for around 95% cases with only 10% false positive rate whereas the competing baseline MonitorRank covers only 74% cases with 23% false positive rate.

1 Introduction

Efficient enterprise system should provide reliable and fast support service in face of anomalous events and failures; customers consistently rely on support engineers for troubleshooting. In effect, support engineers try to resolve issues within minimum case resolution time[1]. They typically rely on thumb rules (e.g. severity-based filtering for logs) or prior experience to identify the glitch & the responsible subsystems, and thereby the root cause. However, the study shows that about 50% of the cases take anytime between 3 to 20 days for resolution. Evidently, this approach is not scalable, accurate or fast enough, especially when it comes to managing multiple node clusters in large data centers.

Recently, attempts have been made to automate the timely detection of impending failures. However, the challenge of troubleshooting gets manifold if support

---

[1] Case resolution period is the duration between 'Case opened date' and 'Case closed date' from customer support database.

© Springer International Publishing AG, part of Springer Nature 2018
D. Phung et al. (Eds.): PAKDD 2018, LNAI 10937, pp. 386–398, 2018.
https://doi.org/10.1007/978-3-319-93034-3_31

engineers need to identify the exact system modules responsible for the glitch. This may be hard to exactly pinpoint the problem creating module in a large scale complex system; however an automated troubleshooter can potentially narrow down the search space, and even localize faulty components within a very short time. The challenge in developing such a troubleshooter comes from multiple sources. (a) Most of the real systems are complex as various constituent system components exhibit functional dependencies [7]. Although the support & monitoring team of a subsystem are equipped with deep domain knowledge, however this is extremely hard for them to have a holistic view of the complete system. This may be an artifact of the presence of a massive number of system components, poor coordination between the teams etc. (b) Individual components such as storage systems are themselves composed of many interconnected modules, each of which has its own failure modes. For example, a storage system failure can be caused by disks, physical interconnects, shelves, RAID controllers etc. (c) Typically in such a large evolving complex system the prior knowledge of dependency tree between modules is not available.

Complex enterprise systems generate massive amounts of telemetry data (counters, system log, etc.) for efficient troubleshooting. In [9,12], techniques have been developed leveraging on those system counters and storage logs respectively, for the accurate early warning capability for anomaly detection. Moving ahead, a bigger challenge is to provide the necessary tools and techniques for the operators to focus their attention to specific problematic subsystems (modules), thus reducing the complexity of the diagnostic process. Attempts have been made in bits and pieces in the gamut of troubleshooting and failure diagnosis using system log. [1] focused in the area of intrusion detection, where it learns popular problem signatures from past history of their corresponding system logs. It subsequently assigns a severity score per message extracting distributions of error messages. The proposed methods in [8,16,17] diagnosed the problem by connecting clues from system runtime logs to the static code, whereas in [4] problem symptoms are used to locate diagnostic logs. However, the performance of aforesaid supervised approaches heavily depends on the training signatures; they quickly tend to produce false positives if (a) the problem description is ill-defined, (b) it does not contain any correlation with the actual technical issue. For instance, [14] troubleshoots only a particular subset of cases (system misconfiguration), but in practice, the breadth of variety of cases are pretty broad. Moreover, these approaches are more relevant for finding software bugs as they offer a fine-grained view of the software code paths.

Importantly, the overall health of a system gets reflected by the frequency and type of message interactions between the constituent modules (e.g. wafl, raid, etc.). Graph appears to be an elegant way of representing the dependency information, where nodes and links denote the modules and corresponding message exchanges respectively. Substructures, embedded within the graph, represent the group of closely associated modules indicating component dependencies. Study [10] shows that, in the normal phase, due to the routine and regular interactions between the modules, a major fraction of substructures regularly appear in the graph. However, as the system enters into the anomalous phase, irregular and sporadic message exchanges result in formation of *anomalous substructures* in the graph, which can be quickly identified due to its occasional presence. Moreover, communities [3], present in the graph, represent the cohesive modules in terms of message interaction and dependencies. In a nutshell, anomalous substructure, coupled with the detected communities, may provide unique signature for identifying the modules responsible for the problem. Leveraging on the aforesaid observation, this paper takes an important step in developing an unsupervised troubleshooting infrastructure.

In this paper, we develop Graph Based Troubleshooting Methodology $GBTM$ - an unsupervised troubleshooting methodology for identifying modules responsible for system failure, termed as problem creating set (PCS). The methodology works in following two steps; first, we construct a sequence of time evolving dynamic graph from the collected system log and customer filed cases. We extract the initial set of problem creating modules identifying the anomalous substructure in the graph. In the second step, we extend this set of anomalous modules by detecting communities in the graph (Sect. 3). Apart from providing troubleshooting modules, we rank the modules in PCS based on confidence score, which might be helpful for support engineers. We demonstrate the performance of $GBTM$ through extensive evaluation (direct and indirect) and comparative study between baselines across metrics (Sect. 5). Obtaining the information of the real problem creating modules is a major challenge. Our support engineers extracted the trouble creating modules from domain knowledge for only 20.50% of cases, where evaluation becomes straightforward. However, for rest 80% cases, we leverage on an indirect method of evaluation, leveraging on the similarity between customer filed cases (Sect. 4). In this proposed method, actual problem causing module comes within top five rank in 86% cases covering 95% cases within top ten rank with only 9.71% as false positive rate.

We start this paper describing the details of the collected dataset and introducing the related terminologies (Sect. 2).

2   Background and Dataset

System generated data like counters, logs and command history etc. are critical for timely troubleshooting and improving customer service. Auto Support (ASUP) infrastructure provides an option of sending the daily logs from the individual machine to the support center's (here NetApp) server. While customers can opt out of this facility, most of them choose to go for it since it provides proactive system monitoring capabilities and hence faster customer support. When a customer faces certain problem, she files a case - the case history is filed in a **customer support database**. A support engineer attends the case by filing an instance in a bug report database. Note that bug reports are arranged according to the category of bugs (for example misconfiguration, performance, hardware failure etc.). The service engineer makes an entry in the database which she thinks to be relevant at first instance. This may not be exactly same as what she finally diagnoses which is stored in a **problem category database** - this she fills up after finishing the job. The customer is also advised to speak out the problem as technically as possible - the data of which is stored in another database called **sym-text**. Following is a sample example of sym-text. **OIN-ODE: Out of index node error; System Notification from eg-nascsi: node name in the cluster-h11.**

Corresponding to each filed case, the event message system (EMS) logs are available in the support center server (from ASUP) - we thus collect 18 weeks of EMS log data for our analysis. The last week is the week *after* the case has been filed. For our analysis we have assumed for the first 16 weeks, the system was in healthy state. The detail field of EMS log is stated next.

Table 1. EMS message structure

Field	Log entry example	Description
Event time	Sat Aug 17 09:11:12 PDT	Day, date, timestamp and timezone
System name	cc-nas1	Name of the node in cluster that generated the event
Event ID	filesystem1.scan.start	EMS event ID. Contains module name and event type
Severity	info	Severity of the event
Message string	Starting block reallocation on aggregate aggr0	Message string with argument values

2.1   Event Message System (EMS) Logs

Support infrastructure gives access to daily EMS logs[2]. Interpretation of fields of following typical EMS log is summarized in Table 1. An example of which is *Sat Aug 17 09:11:12 PDT [cc-nas1:filesystem1.scan.start:info]: Starting block reallocation on aggregate aggr0.*

Each log entry contains time of the event with fields like module name, day, date, timestamp and timezone, severity etc. Severity field can take a value from 'Debug', 'Info', 'Notice', 'Warning', 'Error', 'Critical', 'Alert', 'Emergency'.

2.2   Data Filtering

For the evaluation of the methodology, we identify most severe and impactful bugs (from bug report) filed by customers.

(a). We select bug reports (B_j) having a sufficient number of cases associated with them. This ensures adequate data to validate the model and also provides robustness and generality to the mechanism as these cases are spread across

---

[2] https://library.netapp.com/ecmdocs/ECMP1196817/html/event/log/show.html.

multiple systems, customers and configurations. (b). We filter out instantaneous failures - bugs that are race conditions[3], coding errors etc. (c). We filter cases based on the priority label. (d). We filter out cases with missing data. This produces 32 bug reports each having at least 70 customer support cases. There are 4827 cases from 4305 unique systems that span over 5.5 (January 2011 to June 2016) years.

2.3 Graph Construction from EMS Log

The snapshot of weekly log is abstracted as graph as discussed below.

(a) **Vertex:** From raw EMS log, the module name (m_u) can be extracted from *event message string*. Each module (m_u) is treated as a vertex in the graph. The vertex set of the graph $G_k^i : V_k^i \subset M$, where M is the set of all modules.

(b) **Edge:** If the timestamp difference between two event messages originated from two modules m_u and m_v lies within a threshold (set as 300 s), we incorporate a directed edge m_{uv} between them. The edge set of the graph G_k^i: $E_k^i \subset E$, where E represents all possible edges between modules.

(c) **Edge Weight:** Lower timestamp difference (i.e. messages generated from m_u and m_v within a close interval) denotes high dependency between modules, so higher weight will be assigned to that edge (m_{uv}). The edge weight of edge m_{uv} is formulated as $w_{uv} = r * \sum_{s=1}^{r} \frac{1}{t_s}$ where t_s denotes the time difference and r denotes the number of occurrence of edges between vertex m_u and m_v.

Corresponding to each case (C_k), as mentioned we collect 18 weeks of data - we construct a graph corresponding to each week $(T^1$ to T^{tmax} where $tmax = 18)$ - consequently, we get 18 graphs $(G_k^i, G_k^{i+1}, ... G_k^{tmax})$ from a single case. The last two graphs we assume is arising out of the anomalous state of the system.

3 Troubleshooting Methodology

Given a case filed by the customer, our aim is to shortlist a set of modules responsible for the specified issue. From a particular case, we construct 18 graphs as per the algorithm defined in the previous section. Taking a graph as an input, Graph Based System Troubleshooter (GBST) outputs an abnormal substructure. We consider the modules belonging to the resultant substructure as problem creating candidate set (PCCS). Further, the community structure of the graph is discovered (using a standard community detection algorithm) and the community which matches closest with the anomalous substructure is chosen. The nodes (modules) in that chosen community are added to PCCS and we get expanded problem creating set (EPCS). Consequently, we get 18 EPCS - one for each time window. We merge the first 16 (normal period - obtain NEPCS) and the last

---

[3] https://en.wikipedia.org/wiki/Race_condition.

two (abnormal period - obtain AEPCS) separately. Finally, we construct final problem creating set (PCS) and rank the modules based on causality score, i.e. if a module appears very frequent in AEPCS while it is rare in NEPCS we assign higher causality score to that module. We posit that the top ranking modules are the modules where certain failure has happened.

Algorithm 1. Graph Based Troubleshooting Methodology (GBTM)

1: Call GBST algo // Finding initial candidate set of the problem creating modules
2: Call Clustering // Discovering communities by Louvain algo
3: Call Set Expansion // Get expanded problem creating set (EPCS)
4: Create NEPCS, AEPCS // Merge all the problem creating sets in two sets
5: Call Ranking Modules and PCS construction // Construct final problem creating set and rank the modules based on causality score

---

3.1 GBST Algorithm

Inspired by the graph based anomaly detection technique SUBDUE [10], which uses MDL (minimum description length) principle as an evaluation method, we propose GBST algorithm. MDL states that the best description of a data set is the one that minimizes the description length of the entire data set - correspondingly in a graph the substructure which describes the entire graph in minimum length can be considered the best substructure. Whereas an anomalous substructure which won't be repeated much in the graph would require higher number of bits to represent the graph. GBST at first outputs a set of candidate substructures - we discard the ones with lower DLs - we use a graph encoding technique [6] to measure the description length. For each of the member of the reduced set, we calculate the number of instances the substructure matches in the entire graph - we use an approximate matching algorithm [5] to do so. The members with lower score are filtered in the candidate set. For each entity of the filtered list, we compute anomaly score by the transformation cost (using insertion and deletion of vertex and edges) to match [5] the entity with the best substructure. We finally shortlist only those abnormal substructure where anomaly score exceeds a certain threshold (0.95). Hence the problem creating candidate set (PCCS) is the union of the modules present in the shortlisted anomalous structure.

3.2 Clustering

The community structure of a graph states information of nodes which are 'close' (here functionally dependent) to each other. Taking $G_k^i = (V_k^i \subset M, E_k^i \subset E)$ as input, a standard community detection algorithm (we choose Louvain [2]) returns some set of clusters represented as $c^i = \{c_k^1, c_k^2, ..c_k^n\}^4$, where $|c^i| = n$. Further,

---

[4] Note that case index is denoted by C_k whereas cluster index by c_k.

to check consistency, we also ran various state of the art community detection algorithms on a subset of the graphs and compared the obtained community structure with that obtained through Louvain. We observe high NMI score [15] between the clusters which ensures the consistency of clusters.

3.3 Set Expansion

From the graph, GBST method produces the PCCS - let these modules be the initial members of PCS. In the second step, the vertex set of the graph is partitioned into some clusters c_k^i. We calculate normalized overlapping index denoted by O_I between PCCS and each community $c_k^p \in c_k^i$, $1 \leq p \leq n$. Mathematically, for two arbitrary sets S_1 and S_2, $O_I(S_1, S_2) = |S_1 \cap S_2| / |S_1 \cup S_2|$. If O_I exceeds some threshold (0.75) for a particular cluster (c_k^p), we expand PCS by incorporating the modules of that specific cluster, at the end of which we get the expanded problem creating set (EPCS).

3.4 Creation of NEPCS and AEPCS

NEPCS set is constructed taking the union of all EPCS of first 16 normal period graph. Merging the EPCS of abnormal period graph G_k^{17} and G_k^{18} we get AEPCS. Final set of AEPCS for case k denoted by $V_{eps_k} = V_{epcs_k}^{17} \cup V_{epcs_k}^{18}$. NEPCS for case k denoted by $V'_{eps_k} = \bigcup_{j=1}^{16} V_{epcs_k}^j$ where $V_{epcs_k}^j$ denotes EPCS of kth case at j^{th} timestamp.

3.5 Ranking Modules and PCS Construction

We measured how far a module (m_k) of AEPCS (V_{s_k}) is associated with abnormal period vs. normal operating condition. For a case, suppose we discover that module m_k appears n_1 times in abnormal set AEPCS out of total n_{abn} samples and it also appears in NEPCS n_2 times out of total n_{norm} normal samples. Then causality score (CS) is as follows

$$CS = \frac{X_1}{(X_1 + X_2)}; \ X_1 = \frac{n_1}{n_{abn}}, \ X_2 = \frac{n_2}{n_{norm}} \tag{1}$$

First, we cluster the list of causality scores of modules using x-means clustering [11]. Then we choose the cluster having high mean and low standard deviation. The modules belonging to that cluster forms final problem creating set (PCS). The final PCS of case k denoted by V_{s_k}. We rank the modules of PCS based on causality score - we posit top ranking modules are modules which are responsible for the failure.

4   Experimental Setup

Obtaining the ground truth information of the real problem creating modules is a major challenge for evaluation of the proposed model. In our context, support engineers extracted the trouble creating modules as ground truth from domain knowledge for only 20.50% of cases, where evaluation becomes straightforward. However, for rest 80% cases, we leverage on an indirect method of evaluation.

4.1 Evaluation Procedure

Our indirect method of evaluation relies on the following hypothesis; *'Similar cases will have an approximately similar problem creating modules set'*. Consider that set S_{sim} denotes the collection of similar cases ($S_{sim} = \{C_j : 1 \leq j \leq |S_{sim}|\}$) where V_{s_j} be the final problem creating module set (refer Sect. 3.5) of case C_j. According to our hypothesis, ideally, $V_{s_k} \setminus V_{s_{k'}} = \emptyset, \forall\, 1 \leq k, k' \leq |S_{sim}|$ where $k \neq k'$, indicating V_{s_k} and $V_{s_{k'}}$ both contains same set of modules.

Grouping Similar Cases: In the following, with the help of domain knowledge and rigorous analysis, we mine the similar cases from the corpus and construct the set S_{sim}^j. Since support engineers link each case with the corresponding bug report B_j, apparently, the set of all cases C_j under the bug report B_j should appear as similar. However, in order to further narrow down the group of similar cases under bug report B_j, we first populate two sets of similar cases S_{sim}^{sym} and S_{sim}^{esm} from **Sym-Text** and **EMS-Log** respectively and finally obtain S_{sim}^j considering only the overlapping cases for bug report B_j.

(a) **Sym-Text Based Similarity:** We transform the collected sym-text (see Sect. 2) of case C_k into a tf-idf [13] vector s_k^{vec}. We compute the pairwise cosine-similarity $cos_{k,k'}^{sym}(s_k^{vec}, s_{k'}^{vec}), \forall\, C_k, C_{k'} \in B_j$. If cosine-similarity $cos_{k,k'}^{sym}$ exceeds a threshold δ_{sim} (we set δ_{sim} as 0.80), we populate the set S_{sim}^{sym} with similar cases $C_k, C_{k'}$.

(b) **EMS-Log Based Similarity:** For each case $C_k \in B_j$, we collect all the event message string from the corresponding EMS-LOG (refer Sect. 2.1) and transform into a tf-idf vector e_k^{vec}. Next, like Sym-Text based similarity, we compute the pairwise cosine-similarity $cos_{k,k'}^{ems}, \forall\, C_k, C_{k'} \in B_j$ and populate the set S_{sim}^{ems} with similar cases $C_k, C_{k'}$ (where $cos_{k,k'}^{ems}(e_k^{vec}, e_{k'}^{vec})$ exceeds a threshold 0.65).

Finally, taking the overlapping cases appearing in S_{sim}^{sym} and S_{sim}^{ems}, we construct the final similar cases set (S_{sim}) for bug B_j as $S_{sim}^j = S_{sim}^{sym} \cap S_{sim}^{ems}$.

4.2 Evaluation Metrics

We validate the model with following metrics.

(a) **Overlapping score of two PCS:**

We compute $O_I(V_{s_k}, V_{s_{k'}}) = |V_{s_k} \cap V_{s_{k'}}| / |V_{s_k} \cup V_{s_{k'}}|$ as the overlap between two PCS, where V_{s_j} be the final problem creating module set (PCS) of case C_j. According to our hypothesis, $\forall\, C_k^j, C_{k'}^j \in S_{sim}^j$, for bug report B_j, we expect $O_I(V_{s_k}, V_{s_{k'}})$ to attain a high index. For all cases under bug report B_j, we compute

$$O_{I_{avg}}^{cases}(B_j) = \frac{\sum_{n=1}^{n=|S_{sim}^j|} O_I(V_{s_k}, V_{s_{k'}})}{|S_{sim}^j|}, \; C_k^j, C_{k'}^j \in S_{sim}^j.$$

(b) **Confidence score:** We assign a confidence score with each module m_k identified
in PCS set. For each bug report we observe how frequently a module m_k appears
in the corresponding PCS of associated cases. Suppose we discover that module
m_k appears in PCS for a set of cases under the bug B_j for n_a times out of total
n_b samples. Then we assign $\frac{n_a}{n_b}$ as confidence score of m_k.

(c) **False Positive Rate:** Intuitively, the problem creating modules should appear
only in the abnormal state. If a module m_k appears in both NEPCS and AEPCS
set we treat that module as a false positive. We measure false positive rate as
the fraction of problem creating modules of PCS appeared in both normal and
abnormal period.

4.3   Baseline Models

(a) MonitorRank: Taking system metrics as input, the batch-mode engine
of MonitorRank [7] generates call graph and random walk algorithm is applied
over the evolving graphs. MonitorRank assigns a root cause score for each sensor which reduces the time and human effort to find problem cause in service-
oriented architectures. We consider our constructed graph as call graph and
apply random walk (personalized page rank) and assume returned root cause
as problem creating set for the sake of comparison.

(b) GBST: We implement a variation of $GBTM$ model as baseline, where
we consider the outcome of GBST as the problem creating module PCS. Here
we don't explore the community structure.

5   Evaluation

We evaluate the performance of $GBTM$ in two steps, (a) direct validation, where
ground truth is available and (b) indirect validation, where ground truth can be
obtained from similar cases.

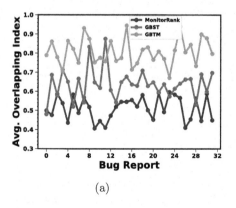

(a)

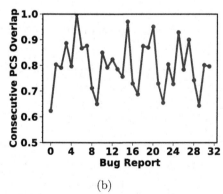

(b)

Fig. 1. (a) Overlapping of PCS between similar cases (b) Overlapping between consecutive PCS set; shows stability of $GBTM$

5.1 Direct Validation

We start the evaluation procedure with the small fraction of cases (20.50%), where the support engineers have reported the exact problem creating modules (one ground truth trouble creating module per case). We rank the modules in final problem creating set (PCS) recommended by our proposed methodology $GBTM$ based on confidence score and observe the position of the ground truth problem creating module. In Fig. 2(a) we exhibit the (cumulative) fraction of cases where the ground truth problem creating module appears at the top-k rank in the recommended PCS. Notably for only 9.66 % cases, the actual problem creating module appears at the top of the $GBTM$ recommended PCS, nevertheless, we observe $\approx 86\%$ cases the actual problem creating module appears within top-5 rank and $\approx 95\%$ cases within top-10 rank. As an example $GBTM$ discovers 'wafl' (matched with ground truth) as main problem creating module of a customer filed case mentioned as "There is some load from RAID domain workload in CPU, during back up the write latency becomes extremely high and slow speeds saving files". This result points to the fact that $GBTM$ would significantly help the support engineers to narrow down their troubleshooting space. On the other hand, in case of MonitorRank, only 2.55 % cases exactly matches with ground truth and only 66% cases are covered within top-5 rank and 74% cases within top-10 rank whereas GBST covers 78% cases within top-10 rank.

5.2 Indirect Validation

Next, we evaluate the performance of the $GBTM$ algorithm indirectly, considering only the similar cases identified in Sect. 4.1. We aim to validate the hypothesis introduced in Sect. 4. In Fig. 1(a), we show the overlapping score $O_{I_{avg}}^{cases}(B_j)$ for all the *similar* cases under each bug report B_j. We observe average overlap as high as 0.807 (SD 0.06) across all the bugs. This indicates the PCS set of *similar* cases are $\approx 81\%$ overlapping. However, in $GBST$, we observe average overlap as 0.638 (SD 0.07) and 0.553 (SD 1.03) in case of $MonitorRank$.

Table 2. Comparison with baseline model

Model	FPR (%)	APCS overlap (%)	Ground truth match (%)
MonitorRank	23	55.3	74.12
GBST	**9.15**	63.8	78.23
GBTM	9.71	**80.7**	**95.42**

5.3 Comparative Study Across Metrics

We evaluate the overall performance of $GBTM$ across the different metrics (Table 2). $MonitorRank$ gives 23% average FPR over all the bug reports whereas only 9.71% FPR in $GBTM$ (Fig. 2(b)). $GBST$ slightly gives better result in

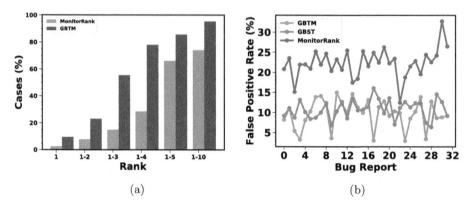

Fig. 2. (a) Cumulative distribution of case coverage % w.r.t rank (b) Comparison of baseline models and *GBTM* regarding FPR (%) over all bug reports

terms of false positive rate (9.15 %) as it does not consider set expansion. FPR of *MonitorRank* is high as it gives more importance to cluster the sensors (module in our case) rather than reducing FPR. But compared to *GBTM* (80.7% PCS overlap; ground truth match 95%) its variation (*GBST*) leads very low overlap (63.8%) between similar cases and covers less no of cases (78.23 %) under top ten ranking. The overlap (55.3%) and case coverage (74%) both are low in case of *MonitorRank* than *GBTM* and *GBST*. Finally, in Table 3, we demonstrated the performance of *GBTM* for three sample bug reports. For each bug report, we show three modules and their corresponding confidence score. Higher confidence score establishes the reliability of our model.

Table 3. Top three ranked modules based on confidence score

	Bug report 1			Bug report 2			Bug report 3		
	R1	R2	R3	R1	R2	R3	R1	R2	R3
Module	wafl	ems	cifs	nbt	api	disk	disk	raid	shelf
Confidence	0.90	0.88	0.85	0.93	0.92	0.90	0.97	0.96	0.92

5.4 Stability of *GBTM*

Finally, we evaluate the stability of the recommended problematic modules in *GBTM* methodology. We compute the overlap of *PCS* set between two consecutive anomalous time window. Essentially we observe whether the *PCS* set persist (from timestamp T_k^{17} to T_k^{18}) after the case filed date. We compute O_I between two problem creating set of an individual case at two consecutive timestamp T_k^{17} and T_k^{18} respectively.

$\forall C_k \in B_j$, we measure $O_I(V_{s_k}^{17}, V_{s_k}^{18})$ and report the mean of consecutive PCS overlap score in Fig. 1(b). The average value of 0.79 across all the bug reports prove the stability of $GBTM$.

6 Conclusion

Logs are challenging to analyze manually because they are noisy and the key events are often buried under hundreds of uninteresting messages. This paper presents $GBTM$, a troubleshooting tool which solely works on system logs. We abstracted the raw event message system log by a graph structure which efficiently incorporates dependency information and infers a probable set of malfunctioning modules with the help of community structure. Our proposed work ranks the modules based on the confidence score which further significantly reduces the number of candidate system modules. As we know customer problem troubleshooting is a time consuming and challenging task, we believe the tool developed here will immensely help support engineers to speed up the case resolution times.

Acknowledgement. This work has been partially supported by NetApp Inc, Bangalore funded research project "Real Time Fault Prediction System (RFS)". The authors also extend their sincere thanks to Siddhartha Nandi from Advanced Technology Group, NetApp for facilitating this research.

References

1. Abad, C., Taylor, J., Sengul, C., Yurcik, W., Zhou, Y., Rowe, K.: Log correlation for intrusion detection: a proof of concept. In: ACSAC, pp. 255–264 (2003)
2. Blondel, V.D., Guillaume, J.L., Lambiotte, R., Lefebvre, É.: The louvain method for community detection in large networks. JSTAT **10**, P10008 (2011)
3. Boccaletti, S., Latora, V., Moreno, Y., Chavez, M., Hwang, D.U.: Complex networks: structure and dynamics. Phys. Rep. **424**(4), 175–308 (2006)
4. Brodie, M., Ma, S., Lohman, G., Mignet, L., Wilding, M., Champlin, J., Sohn, P.: Quickly finding known software problems via automated symptom matching. In: Autonomic Computing, ICAC 2005, pp. 101–110. IEEE (2005)
5. Bunke, H., Allermann, G.: Inexact graph matching for structural pattern recognition. Pattern Recogn. Lett. **1**(4), 245–253 (1983)
6. Cook, D.J., Holder, L.B.: Substructure discovery using minimum description length and background knowledge. JAIR **1**, 231–255 (1994)
7. Kim, M., Sumbaly, R., Shah, S.: Root cause detection in a service-oriented architecture. ACM SIGMETRICS Perform. Eval. Rev. **41**, 93–104 (2013)
8. Lao, N., Wen, J.R., Ma, W.Y., Wang, Y.M.: Combining high level symptom descriptions and low level state information for configuration fault diagnosis. LISA **4**, 151–158 (2004)
9. Nair, V., Raul, A., Khanduja, S., Bahirwani, V., Shao, Q., Sellamanickam, S., Keerthi, S., Herbert, S., Dhulipalla, S.: Learning a hierarchical monitoring system for detecting and diagnosing service issues. In: ACM SIGKDD, pp. 2029–2038 (2015)

10. Noble, C.C., Cook, D.J.: Graph-based anomaly detection. In: KDD, pp. 631–636 (2003)
11. Pelleg, D., Moore, A.W., et al.: X-means: extending k-means with efficient estimation of the number of clusters. In: ICML, vol. 1, pp. 727–734 (2000)
12. Khatuya, S., Ganguly, N., Basak, J., Bharde, M., Mitra, B.: Anomaly detection from event log empiricism. In: IEEE Conference on Computer Communications, INFOCOM 2018. IEEE (2018)
13. Salton, G.: Automatic Text Processing: The Transformation, Analysis, and Retrieval of Information by Computer. AWL (Publishing), Boston (1989)
14. Talwadker, R.: Dexter: faster troubleshooting of misconfiguration cases using system logs. In: SYSTOR, p. 7 (2017)
15. Vinh, N.X., Epps, J., Bailey, J.: Information theoretic measures for clusterings comparison: variants, properties, normalization and correction for chance. JMLR **11**, 2837–2854 (2010)
16. Xu, W., Huang, L., Fox, A., Patterson, D., Jordan, M.I.: Detecting large-scale system problems by mining console logs. In: SIGOPS, pp. 117–132 (2009)
17. Yuan, D., Mai, H., Xiong, W., Tan, L., Zhou, Y., Pasupathy, S.: Sherlog: error diagnosis by connecting clues from run-time logs. ACM SIGARCH Comput. Architect. News **38**, 143–154 (2010)

Fusion of Modern and Tradition: A Multi-stage-Based Deep Network Approach for Head Detection

Fu-Chun Hsu[1] and Chih-Chieh Hung[2(✉)]

[1] Department of Electric Engineering, University of Melbourne, Melbourne, Australia
fcchsu@student.unimelb.edu.au
[2] Department of Computer Science and Information Engineering,
Tamkang University, Taipei, Taiwan
smalloshin@gms.tku.edu.tw

Abstract. Detecting humans in video is becoming essential for monitoring crowd behavior. Head detection is proven as a promising way to realize detecting and tracking crowd. In this paper, a novel learning strategy, called Deep Motion Information Network (abbr. as DMIN) is proposed for head detection. The concept of DMIN is to borrow the traditional well-developed head detection approaches which are composed of multiple stages, and then replace each stages in the pipeline into a cascade of sub-deep-networks to simulate the function of each stage. This learning strategy can lead to many benefits such as preventing many trial and error in designing deep networks, achieving global optimization for each stage, and reducing the amount of training dataset needed. The proposed approach is validated using the PETS2009 dataset. The results show the proposed approach can achieve impressive speedup of the process in addition to significant improvement in recall rates. A very high F-score of 85% is achieved using the proposed network that is by far higher than other methods proposed in literature.

Keywords: Head detection · Motion representation
Learning strategy · Deep architecture

1 Introduction

Monitoring large crowds using video cameras is an important task since it could improve the security and safety in large areas where people gather, such as train stations, stadium and other public places. However, occlusion and low resolution in the region of interest hinders accurate crowd segmentation. In such scenario, accurate detection and tracking of people in crowds is the promising option. However, recent study shows that it is difficult to perform accurate pedestrian detection for handling small instances because of insufficient resolution [8,27]. Based on the detailed observations [8] from the popular Caltech Pedestrian Dataset, it could be found that the head is the only visible part in 97% of

© Springer International Publishing AG, part of Springer Nature 2018
D. Phung et al. (Eds.): PAKDD 2018, LNAI 10937, pp. 399–410, 2018.
https://doi.org/10.1007/978-3-319-93034-3_32

occlusions. This strengthens the motivation of detecting pedestrians in crowds by finding their heads.

Pedestrian detections, the general case of head detection in video, have attracted many research efforts [2,6,7,27,28]. These approaches could be roughly categorized into three classes: multi-stages-based approaches [5,7,16], where they are majorly benefited by powerful boosting methods and simple low-level features (HOG, LUV); and deep-network-based approaches [2,17,21,27,28], where representative features are learned through non-linear deep transformation. Multi-stages-based approaches refer to the conventional approaches which tend to detect objects through multiple phases, as shown in the top of Fig. 1. With the rise of deep networks, deep-network-based approaches tends to build a end-to-end convolution deep network to extract features and an extra classifier for detecting heads as shown in the middle of Fig. 1. Another approach which has fewer research but has significant performance over single feature detector is joint-features detection [23]. Earlier research by Hsu et al. [14] belongs to this kind of approaches, which has demonstrated the advantage of applying motion information in the detection of human heads over color when the environment is cluttered and heavily occluded.

These two approaches are with their own advantages and disadvantages. The multi-stage-approaches are easy to understand due to the hand-crafted features and stages. Due to the independency of each stage, the training data can be prepared separately for each stage. However, this kind of approaches compute features and make decisions at each stage. For example, producing TV-L1 optical flow is independent from creating motion boundary histogram (MBH). Therefore, the poor performance in the previous stage will lead to that in the current stage. On the other hand, the deep-network approaches could adjust the weights of the whole networks when minimizing the loss functions. The disadvantages of this kind of approaches are in two-folds. First, the design of deep networks are usually dependent on the experience of the designer and need to make several trial until the better performance is made. Second, more and more training data are needed to train end-to-end deep networks when the size of deep networks become bigger and bigger, which may not be always available in practice.

This paper proposes a multi-stage-based deep network approach for head detection: Deep Motion Information Network (abbr. as DMIN). In the nutshell, the philosophy of DMIN is to use the modern technique (deep-network-based solution) with the traditional design (multi-stage-based solutions). The concept could be best understood in the bottom of Fig. 1. It could be seen that the deep network is composed of three sub-networks with a softmax layer, which simulate the functions of stages in the top of Fig. 1, respectively. Specifically, DMIN exploits two fully convolution networks FCN1 and FCN2 to learn the motion gradient images (corresponding to the stage of pre-processing features), a fully convolution network FCN3 to motion proposal (corresponding to the stage of advanced hand-crafted features), motion convolution network (corresponding to the stage of sophisticated feature engineering) to detect the head by merging the bottom layers and attaching a multi-layers-perceptron and a softmax layer.

The design philosophy of DMIN is to borrow the experience of well-developed multi-stages-based approaches.

1. The time for designing deep networks could be saved.
   We believe that the pipeline of stages which are refined by experienced researchers for years could be very sophisticated. Based on such a sophisticated design, the structure of the end-to-end deep network could be easily decided which could prevent time consuming trial-and-error process when building deep networks for head detection.
2. The global optimization among stages could be achieved.
   Since the approach implements the function of each stage and connect them into a deep network, each sub-network could be fine-tuned during the optimization of the final network cascade.
3. The size of training data needed could be reduced.
   With the specific function of each stage, pre-training of each stage could be easy achieved. First of all, many existing training dataset for each stage could be re-used for pre-training. Moreover, good pretraining could be beneficial to the convergence of learning when optimizing the whole end-to-end deep networks. Therefore, the size of training data can be further reduced.

The proposed approach DMIN is validated using the PETS2009 dataset. The results show that by adding the intermediate motion objectness proposal network ahead of the deep head classification network, impressive speedup of the process is achieved in addition to significant improvement in recall rates.

This paper is organized as follows. Section 2 describes the technical details of the proposed network DMIN. Section 3 presents the experimental results. Section 4 concludes this paper.

2   Deep Motion Information Network

Teney *et al.* [22] have demonstrated end-to-end training of CNNs on videos for high-level tasks like action recognition. They show limited capability for identifying intermediate representations of motion. The proposed motion information network can be seen as a cascade of motion feature extractors. In this network, first we extract dense optical flow using the first FCN network as in [9]. The 2-channel flow image is then input to the Motion Image Network to learn the motion boundary that includes MBX, MBY, and magnitude of Histogram of Optical Flow [14]. With the 3-channels learned motion images we then develop the motion proposal network to generate the candidate windows. The final deep network is then used to predicted heads/non-heads.

2.1   Motion Image and Its Representation

Extracting optical flow is considered the most general way to capture motion. Regardless of the image content, extraction of optical flow is not trivial and has been addressed by successful dense optical flow algorithms. In [14] the raw

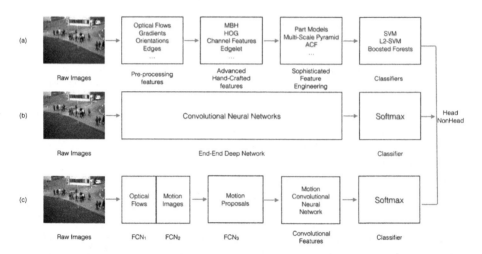

Fig. 1. Top: Traditional single task CNN. The weights and the abstraction of the knowledge in this network are strongly influenced and decided by the single task. Bottom: Architecture of the proposed Multi-Task Deep Motion Information Network. Output of intermediate tasks are also shown.

Table 1. The table shows the parameters of Motion Proposal FCN. Paddings are added in the bottom layers to avoid loss of important boundary information during early stages.

Layer	1	2	3	4	5	6	7	8	9
# Channels	64	64	-	96	96	-	128	64	2
Kernel size	3×3	3×3	2×2	3×3	3×3	2×2	3×3	4×4	1×1
Stride	1	1	2	1	1	2	1	1	1
Paddings	1	1	0	1	1	0	0	0	0
# Parameters	2368	36928	-	55392	83040	-	110720	131136	130
Map size	24×24	24×24	24×24	12×12	12×12	12×12	6×6	4×4	1×1

motion features used as input to detect heads are extracted using the $TV - L1$ algorithm [3]. The $TV - L1$ algorithm has the advantage of regulating varying illumination and the traditional optical flow constraints. However, getting such highly optimized dense optical flow requires massive amount of computation resource and time. Using a feed-forward convolutional network not only learns the various patterns of movement but also speedup the process in magnitudes. In this work, as the first intermediate task, optical flow is extracted using a full connected (FCN) architecture developed by Dosovitskiy *et al.* [9]. This results in motion images, which is a collection of gradient magnitude maps from the U and V channel and the magnitude of the optical flow. These maps are often used as input to the module that calculates histogram of optical flow (HOOF) and motion boundary histogram (MBH - MBX and MBY, which are motion boundary in X and Y directions respectively). They contains useful information such as determining the contour of a moving object, or the strength of movement within a group. For applications, Laptev et al. [15] presented a combination of Histogram of Oriented Gradients and Histogram of Flow (HOF) to capture shape and local motion. HOF is a popular feature and has been applied in many action recognition tasks [4,15,24]. In this work, a FCN of motion boundary capable of learning the structures of images of MBX, MBY, and HOOF is presented. These images are approximated using a small but fully convolutional network (parameters in Table 2).

Table 2. The table shows the parameters of Motion Image FCN. The total parameters is less 3,000.

Layer	1	2	3
# Channels	32	64	3
Conv. kernel size	3	1	1
Conv. stride	1	1	1
Zero paddings	1	0	0
Activations	ReLU	ReLU	ReLU
# Parameters	608	2112	195
Spatial input size	$H \times W$	$H \times W$	$H \times W$

Table 3. Comparison of different features detectors: Results from the proposed method were compared with three state-of-the art features: Wojek's et al. 's [26] multi-features and Dollar's et al. [7] channel features, and Maji's et al. [16] multi-layer pyramid features. The last part of the table compares the proposed method to Hsu's et al. [14] detector result. The top two are highlighted in each category.

Detector	TP	FP	FN	Recall (%)	Precision (%)	F-Score	Region	Motion	Static
Wojek [26]	2708	11272	2254	54.57	19.37	28.59	Only head	Yes	Yes
Dollar [7]	3054	10004	1908	61.55	23.39	33.89	Only head	No	Yes
Maji [16]	2076	160	2886	41.84	**92.84**	57.68	Full body	No	Yes
Hsu [14]	3184	969	1778	**64.17**	76.67	**69.88**	Only head	Yes	No
This work	3908	387	1054	**78.76**	90.99	**84.43**	Only head	Yes	No

2.2   Deep Motion Proposals Network

Convolutional Neural Networks has become the state-of-the-art among different object detection tasks from pedestrian detection [21] to large-scale object recognitions [20]. R-CNN, its successor faster R-CNN, and more recently YOLO [12,19,20] have shown a huge improvement in accuracy and speed by employing proposal detection in early stage of object detection. Despite the success of using proposal in object detection, in certain scenarios such as in a mass group of small objects, in cluttered and occluded environment, the performance degrades rapidly [19]. The main reason can be referred to too less information contained in a small window. Secondly, when targets are occluded, more data are corrupted and loss. Recently, motion boundary are viewed as an key to build a deep CNN network to find moving objects. In [10], authors learned the moving object proposals using hand-crafted motion boundary along with color features in a dual-CNN network. In our work, after extracting input from the prediction of motion boundary images, we extend the motion network by adding a deep proposal network to find the motion objectness or so called motion proposal. There are two types of existing approaches for generating object proposals: segment grouping methods and window scoring methods. Generating from low-level features and aiming for fast computation are two mostly approached strategies [13].

Ghodrati et al. [11] suggest that features in high-level convolutional layers is more capable of capturing objects and have good performance in recall rate. In contrast, the lower-layer features are better in detecting the location of the objects. Therefore, the first few layers of the network share the low-level object characteristics with the latter head detection network, and can be fine-tuned by head information. Two key aspects are highlighted in the design of our network: First, we built a deep motion proposal network which is a fully convolutional network that uses motion feature to perform prediction of moving proposals. The fully convolutional network can take a motion boundary image of arbitrary size and output a dense moving objectness map showing the probability of containing a moving object for each candidate region. The non-maximal suppression (NMS) is performed to remove redundant and low-ranking proposals having low prediction score. Secondly, benefited from FCN, time is significantly reduced compared to the traditional hand-crafted data flow strategies. The parameters and architecture of the network is summarised in Table 1.

The final resulting prediction map of our motion proposal network is a downsampled version of input images, and it has to be mapped back to the original co-ordinates to extract windows. To map back to the input proposal boxes from the dense output map, the key is to decide the area of the box in the input image that correspond to the output pixel (receptive field size). The inverse mapping requires the information of strides, padding, and size of convolutional kernel in each layer and use these to compute backward from the output to the first CONV layer. In layer L_i, with stride s_i, padding p_i, and kernel size k_i, the mapping of output in position (i, j) to layer L_i is: $s_i * (i, j) + (\lfloor \frac{k_i}{2} \rfloor - p_i)$. This process is recursively performed until reaches input layer.

2.3   Coarse-Fine Multi-level CNN for Head Detection

The outputs produced by Deep Motion Proposals Network are windows or hot areas that have high probabilities of moving objects. To detect heads, we further design a CNN extending from the DMPN to transform the hot areas to head regions. The network shares layer 1 to 7 with DMIN which is referred in Table 1. On top of it, we add a softmax layer to predict the probability of head/non-heads. The design of network is benefited from a good initialization that is transferred and shared with Deep Motion Proposals Network. Thus increases the possibility to successfully train a head network with small amount of data without the need to re-design a new network from scratch.

3   Results and Discussion

The dataset used throughout this paper is the popular PETS2009 dataset [18]. The PETS2009 dataset contains a mass group of small pedestrians in a cluttered and occluded environment. First we presented the qualitative and quantitative result of our Motion Information Network. Then we discussed the result of DMIN based on the observation of performance versus the design of network architectures. Finally, we compare our head detection results with the previous state-of-the-art shallow color features and our previous work on shallow motion features. All the networks are trained on Tensorflow framework [1], and the optimizer is Adagrad. Different training strategies among these sub-networks are mentioned correspondingly.

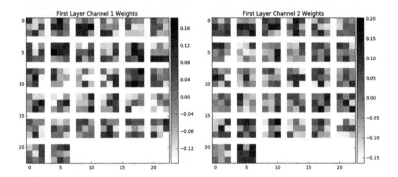

Fig. 2. Weights of the 1st layer in our Motion Image Network. The network try to approximate the motion images and the resulting filters represents the gradients of a single pixel to its neighborhood in all eight directions.

The Motion Images Network approximates the functions to produce these motion images from optical flows. We split the optical flow produced in last stage from the PETS2009 dataset into training and testing set in a 7:3 ratio. Mean Squared Error (MSE), Mean Absolute Error (MAE), and Binary Cross Entropy (BCE) were compared to find the loss function that could produce better image approximation under same FCN architecture. Three common metrics are used for comparison: Average PSNR, Structural SIMilarity index (SSIM) [25], and RMSE. In general, all three loss functions perform well in all three metrics. However, we found that BCE achieves highest score in two of the metrics and only slightly fall behind MSE in SSIM. It is intuitive that optimizes on cross-entropy in BCE could get higher PSNR, but in RMSE that cross-entropy loss also performs slightly better than MSE. Hence, BCE has been used in all our experiments. From the qualitative perspective: From Figure 2, the first 3×3 layer learn the gradient filters of a single pixel and its neighborhood in our Motion Image Network, which is exactly what motion histogram of gradients trying to do in their early stages. The final two 1×1 layers perform non-linear pixel-wise approximation of the square-root magnitude computation, which is the last stage of motion image histogram. We showed a sample prediction in Fig. 3, where our predictions can approximate the original motion images well. It is noted that the network is able to smooth the motion images, avoiding extreme values sometimes happen in rapid movements between targets.

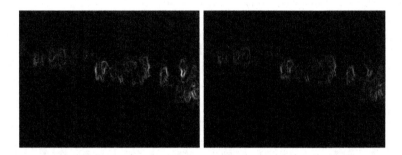

Fig. 3. Original Motion Boundary X-Image (MBX) - left and the predicted MBX - right. The predicted MBX becomes more smooth and extreme values (large movements) are cropped by the learned network

To determine the quality of proposals, people often apply IOU (Intersection-Over-Union) VS Recall rate to validate the quality of these generated proposals [13]. Three different setting based on two important factors of design the Convolutional Network were compared to assess DMIN: Pooling and padding, pooling but no padding, and no pooling and no padding. From Fig. 5, it was observed that by padding in early layers, the recall rate reaches almost 95%, much higher than others. We argue that when padding is applied, more spatial boundary information are kept in each layer. Whereas the IOU metric requires a better approximation of locations of these proposals, losing boundary information in early stages will sacrifice the precision of locations. In Fig. 3, we see the confidence and the density are high around the locations of pedestrian heads. Most body parts and backgrounds, are removed in this stage (Fig. 4).

As a final step, head detection was assessed and compared. The head data from PETS2009 dataset [18] were extracted and annotated. The average height of the ground truth bounding box is 18.49 pixels. The height/width of annotations span from (17, 27) due to different perspective distortions. Researches of proposal learning [13] often suggest to enlarge the dataset by creating more windows close to the Ground-Truth (GT) ones within the Intersection-Over-Union (IOU) larger than 0.5. In our case, we controlled the translation of the GT bounding box within a range of $[-3, 3]$ pixels. For each GT annotation, we generated 5-times more boxes in this range randomly. Background and body (apart from the

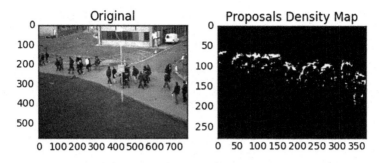

Fig. 4. The figure shows the result of the moving proposals. Most false-positives heads are from body parts similar to the heads which will be removed at the head classification stage

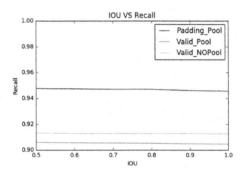

Fig. 5. Results of Recall *vs.* IOU between different convolutional kernel setting: Pooling with padding, Pooling without padding, and no Pooling.

heads) patches are annotated and generated three times and two times more respectively. We extracted 59,428 patches in total and divided into training, testing, and validation randomly in the ratio of 7:1:2. Early-stopping is performed on the loss of validation data.

We compared our result with three previously state-of-the-art shallow hand-crafted color features detectors, and a shallow hand-crafted motion feature detector [7,14,16,26]. We highlight the importance of learning the most discriminative features to detect heads in small region, and keep the comparison among the state-of-the-art features without other techniques such as boosted classifiers, bounding box regression(Faster R-CNN), and one shot detector(YOLO). The final results with comparison of different head detectors are summarised in Table 3. The F-Score of our system reaches 85%, and the precision reaches 91%. It should be noted that although the precision is slightly less than [16], our recall overwhelm their features significantly, mostly due to the presence of highly occluded crowds. Comparing with Hsu *et al.* [14], which they also learns features from motion, this work improves in all metrics by employing the proposal network first to removes background and non-heads moving objects in

Fig. 6. Result of the proposed multi-stages Deep Motion Head detector over a series of images from the PETS 2009 dataset (PETS 2009 S1-L1-View1-1357). Most of the heads were detected accurately, while some heads are missing due to the sensitive removal by the Deep Motion Proposal Network.

early stages, and keep as many positive windows as possible to perform the latter CNN network to refine the detection of heads. In Fig. 6, we show our results of correctly and accurately find the heads in small regions.

4   Conclusion

A new end-to-end convolutional architecture that uses motion information for head detection is proposed. The end goal of head detection is divided into a number of sub-networks that each of them can be used for multiple applications. The proposed network first calculates the optical flow (using FlowNet), followed by several motion features that is in turn used to develop motion proposals (similar to R-CNN). Finally, a convolutional network which extends the deep motion proposal network can produce accurate identification of head regions. A very high F-score of 85% is achieved using the proposed network that is by far higher than other methods proposed in literature.

Acknowledgment. This work was supported by the Ministry of Science and Technology, Taiwan, under Grant MOST 106-2218-E-032-004-MY2.

References

1. Abadi, M., Agarwal, A., Barham, P., Brevdo, E., Chen, Z., Citro, C., Corrado, G.S., Davis, A., Dean, J., Devin, M., et al.: Tensorflow: Large-scale machine learning on heterogeneous distributed systems. arXiv preprint arXiv:1603.04467 (2016)
2. Benenson, R., Omran, M., Hosang, J., Schiele, B.: Ten years of pedestrian detection, what have we learned? In: Agapito, L., Bronstein, M.M., Rother, C. (eds.) ECCV 2014. LNCS, vol. 8926, pp. 613–627. Springer, Cham (2015). https://doi.org/10.1007/978-3-319-16181-5_47
3. Chambolle, A., Pock, T.: A first-order primal-dual algorithm for convex problems with applications to imaging. J. Math. Imaging Vis. **40**(1), 120–145 (2011)
4. Chaudhry, R., Ravichandran, A., Hager, G., Vidal, R.: Histograms of oriented optical flow and binet-cauchy kernels on nonlinear dynamical systems for the recognition of human actions. In: 2009 IEEE Conference on Computer Vision and Pattern Recognition, CVPR 2009, pp. 1932–1939. IEEE (2009)
5. Dalal, N., Triggs, B.: Histograms of oriented gradients for human detection. In: 2005 IEEE Computer Society Conference on Computer Vision and Pattern Recognition, CVPR 2005, vol. 1, pp. 886–893. IEEE (2005)
6. Dollar, P., Belongie, S., Perona, P.: The fastest pedestrian detector in the west. In: BMVC, vol. 2, p. 7 (2010)
7. Dollár, P., Tu, Z., Perona, P., Belongie, S.: Integral channel features. In: British Machine Vision Conference, vol. 2, p. 5 (2009)
8. Dollar, P., Wojek, C., Schiele, B., Perona, P.: Pedestrian detection: an evaluation of the state of the art. IEEE Trans. Pattern Anal. Mach. Intell. **34**(4), 743–761 (2012)
9. Dosovitskiy, A., Fischery, P., Ilg, E., Hazirbas, C., Golkov, V., van der Smagt, P., Cremers, D., Brox, T., et al.: Flownet: Learning optical flow with convolutional networks. In: 2015 IEEE International Conference on Computer Vision (ICCV), pp. 2758–2766. IEEE (2015)
10. Fragkiadaki, K., Arbelaez, P., Felsen, P., Malik, J.: Learning to segment moving objects in videos. In: Proceedings of the IEEE Conference on Computer Vision and Pattern Recognition, pp. 4083–4090 (2015)
11. Ghodrati, A., Diba, A., Pedersoli, M., Tuytelaars, T., Van Gool, L.: Deepproposal: Hunting objects by cascading deep convolutional layers. In: Proceedings of the IEEE International Conference on Computer Vision, pp. 2578–2586 (2015)
12. Girshick, R., Donahue, J., Darrell, T., Malik, J.: Rich feature hierarchies for accurate object detection and semantic segmentation. In: Proceedings of the IEEE conference on computer vision and pattern recognition, pp. 580–587 (2014)
13. Hosang, J., Benenson, R., Dollár, P., Schiele, B.: What makes for effective detection proposals? IEEE Trans. Pattern Anal. Mach. Intell. **38**(4), 814–830 (2016)
14. Hsu, F.C., Gubbi, J., Palaniswami, M.: Head detection using motion features and multi level pyramid architecture. Comput. Vis. Image Underst. **137**, 38–49 (2015)
15. Laptev, I., Marszalek, M., Schmid, C., Rozenfeld, B.: Learning realistic human actions from movies. In: 2008 IEEE Conference on Computer Vision and Pattern Recognition, CVPR 2008, pp. 1–8. IEEE (2008)
16. Maji, S., Berg, A.C., Malik, J.: Classification using intersection kernel support vector machines is efficient. In: 2008 IEEE Conference on Computer Vision and Pattern Recognition, CVPR 2008, pp. 1–8 (2008)
17. Ouyang, W., Wang, X.: Joint deep learning for pedestrian detection. In: 2013 IEEE International Conference on Computer Vision (ICCV), pp. 2056–2063. IEEE (2013)

18. PETS2009: Performance Evaluation of Tracking and Surveillance Dataset (2013)
19. Redmon, J., Divvala, S., Girshick, R., Farhadi, A.: You only look once: unified, real-time object detection. In: Proceedings of the IEEE Conference on Computer Vision and Pattern Recognition, pp. 779–788 (2016)
20. Ren, S., He, K., Girshick, R., Sun, J.: Faster R-CNN: Towards real-time object detection with region proposal networks. In: Advances in neural information processing systems, pp. 91–99 (2015)
21. Sermanet, P., Kavukcuoglu, K., Chintala, S., Lecun, Y.: Pedestrian detection with unsupervised multi-stage feature learning. In: 2013 IEEE Conference on Computer Vision and Pattern Recognition (CVPR), pp. 3626–3633 (2013). https://doi.org/10.1109/CVPR.2013.465
22. Teney, D., Hebert, M.: Learning to extract motion from videos in convolutional neural networks. arXiv preprint arXiv:1601.07532 (2016)
23. Walk, S., Majer, N., Schindler, K., Schiele, B.: New features and insights for pedestrian detection. In: 2010 IEEE Conference on Computer Vision and Pattern Recognition (CVPR), pp. 1030–1037. IEEE (2010)
24. Wang, H., Klaser, A., Schmid, C., Liu, C.L.: Action recognition by dense trajectories. In: 2011 IEEE Conference on Computer Vision and Pattern Recognition (CVPR), pp. 3169–3176. IEEE (2011)
25. Wang, Z., Bovik, A.C., Sheikh, H.R., Simoncelli, E.P.: Image quality assessment: from error visibility to structural similarity. IEEE Trans. Image Process. **13**(4), 600–612 (2004)
26. Wojek, C., Walk, S., Schiele, B.: Multi-cue onboard pedestrian detection. In: 2009 IEEE Conference on Computer Vision and Pattern Recognition, CVPR 2009, pp. 794–801. IEEE (2009)
27. Zhang, L., Lin, L., Liang, X., He, K.: Is Faster R-CNN doing well for pedestrian detection? In: Leibe, B., Matas, J., Sebe, N., Welling, M. (eds.) ECCV 2016. LNCS, vol. 9906, pp. 443–457. Springer, Cham (2016). https://doi.org/10.1007/978-3-319-46475-6_28
28. Zhang, S., Benenson, R., Omran, M., Hosang, J., Schiele, B.: How far are we from solving pedestrian detection? In: Proceedings of the IEEE Conference on Computer Vision and Pattern Recognition, pp. 1259–1267 (2016)

Learning Treatment Regimens from Electronic Medical Records

Khanh Hung Hoang[(⊠)] and Tu Bao Ho

Japan Advanced Institute of Science and Technology,
1-1 Asahidai, Nomi, Ishikawa, Japan
{hung.hoang,bao}@jaist.ac.jp

Abstract. Appropriate treatment regimens play a vital role in improving patient health status. Although some achievements have been made, few of the recent studies of learning treatment regimens have exploited different kinds of patient information due to the difficulty in adopting heterogeneous data to many data mining methods. Moreover, current studies seem too rigid with fixed intervals of treatment periods corresponding to the varying lengths of hospital stay. To this end, this work proposes a generic data-driven framework which can derive group-treatment regimens from electronic medical records by utilizing a mixed-variate restricted Boltzmann machine and incorporating medical domain knowledge. We conducted experiments on coronary artery disease as a case study. The obtained results show that the framework is promising and capable of assisting physicians in making clinical decisions.

Keywords: Treatment regimen · Treatment learning
Treatment recommendation · Electronic medical records

1 Introduction

The two most important issues in healthcare are disease diagnosis and treatment. While many works have been conducted on the problem of diagnosis prediction, the problem of learning treatment regimens has not yet been extensively studied from the research community. This shortage becomes more serious when hospitals essentially need to make efforts to adopt treatment regimens that best fit their available resources. Additionally, it seems hard to have a fixed care plan for a particular disease due to its high dependency on various patient conditions. As a result, capturing treatment regimens in practice turns out to be meaningful for not only assisting physicians in making right clinical decisions but also helping hospitals manage their resources thoroughly.

In principle, treatment regimens could be learned based on the knowledge-driven approach which requires medical domain or expert knowledge. It can be a piece of information written in the literature or accumulated experience gained by physicians during their career. While this approach seems to be reliable, taking various domain knowledge into account is costly and not straightforward in

© Springer International Publishing AG, part of Springer Nature 2018
D. Phung et al. (Eds.): PAKDD 2018, LNAI 10937, pp. 411–422, 2018.
https://doi.org/10.1007/978-3-319-93034-3_33

reality. In contrast to the knowledge-driven approach, the data-driven approach derives treatment patterns from a large number of observations thanks to the availability of electronic medical records in recent years. Studies followed this approach could be found in [2, 5, 6, 8, 9].

Although many interesting results have been achieved, those studies have simply utilized a limited subset of features while many other kinds of data are usually omitted. Such data can be patients' demographics, laboratory test results or clinical notes consisting of signs and symptoms during patients' hospitalization. It is apparent that the more values from those data shared between two patients, the more possibility that the patients are treated with similar regimens. The lack of considering such valuable information simultaneously in current research could be attributed to the poor-feature data used in their experiments. Moreover, even when the above data is made available, it generally exists in form of numerical, binary, categorical, or text format. Such a heterogeneous data is not ready to use for many data mining methods. Another challenge stems from the fact that treatment regimen is typically defined over periods. Each period is distinct from others at milestones where major changes in a patient's health status happen that lead to a notable adjustment in subsequent prescriptions for the patient. Therefore, given a set of prescription records, identifying suitable treatment periods can considerably affect the learned treatment patterns.

This work aims to propose a treatment regimen learning framework which addresses both the above challenges. Our framework first divides patients into clusters from which treatment regimens over periods are discovered then. To overcome the challenges of learning from mixed-type data, we employ a mixed-variate restricted Boltzmann machine (MV.RBM) [11]. The advantage of this model is at its robustness in transforming heterogeneous objects to their homogeneous representations. The new latent representations are in the form of hidden binary vectors that could be further processed easily by clustering methods. To tackle the challenge of treatment period identification, we propose an algorithm which can relatively capture significant changes in prescription indications. Moreover, we also suggest another algorithm which derives treatment regimens from each cluster as a regimen tree. The tree can highlight frequently prescribed drugs and infrequently prescribed drugs inside each patient cluster which would be useful for recommending prescribed drugs to patients.

In short, the main contributions of our work are listed as follows. Firstly, we propose a generic framework which can exploit different kinds of relevant patient records. The framework is superior to others in terms of data utilization. Secondly, we employ both knowledge-driven approach and data-driven approach in our framework. The exploited medical domain knowledge is drug indications and their importance in the treatment for a particular disease. The combination approach used in our framework seems more feasible to deal with the longitudinal property inherent in prescription records. Lastly, we propose a new way to represent treatment regimens flexibly. Frequent drugs are learned from individual level to group level and organized as regimen trees which could be useful for recommending possible regimens to new patients.

2   Related Work

This section provides a brief review of studies about the treatment-related learning problem. Notable works can be found in [2,3,9]. In [3], the authors developed a process mining method to derive clinical pathway from medical behaviors. Their work, however, mainly focused on learning clinical procedures rather than a detailed treatment.

Inspired by the emergence of electronic medical records, recent studies have exploited prescription records which would provide more useful insights about patient treatment. In [2] the authors proposed a probabilistic model that linked patient features and treatment behaviors together to mine treatment patterns. Their model, however, employed many hyperparameters with almost no domain integration. This limitation undermines the model interpretability. Moreover, it was not explicitly described in that work how the chronological order among the learned treatment patterns related. In [9], the inspired work of our research, treatment regimens were derived solely from a set of prescription records. While many typical regimens could be described in an unsupervised mechanism, their prescription-based approach appears to lack of interpretability regarding patient profile and health conditions that lead to the derived regimens. Additionally, although the authors in [9] attempted to describe the chronological order between regimens with predefined treatment periods, their approach capture little medical domain knowledge as well as seems inflexible in dealing with the varying lengths of hospital stay. Regarding the treatment recommendation task, [9] also presented a way to recommend typical treatment regimen for a patient based on demographics and disease severity of patients. This approach, however, seems hard to be applied to new patients whose disease severity may not be recognized at the beginning dates of hospitalization.

3   Methods

In this section, we describe our framework of treatment learning problem. This generic framework is designed for a particular disease. Our approach is based on the assumption that a patient cohort may be divided further into groups of more homogeneous patients who share latent characteristics underlying in patient profile or health status. Patients in one group, therefore, are supposed to be treated by similar care plans that share many parts in common. Figure 1 illustrates the framework overview. It consists of two main tasks: clustering a cohort of patients and learning treatment regimens for each resulting cluster.

3.1   Data Collection and Preprocessing

Our framework takes medical records of cured patients as trained data. We are interested in the data that characterizes health conditions, for example, demographic information, discharge summary, and laboratory test results. It should be noted that for longitudinal data such as discharge and laboratory indicators,

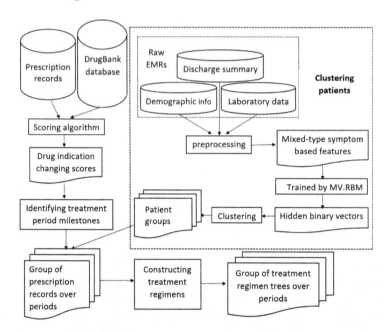

Fig. 1. An overview of the proposed framework for learning treatment regimens.

we only collected the records at the early stage of patients since such longitudinal data is usually not fully available for new patients at the time of admission. This solution is based on the intuition that patients who share initial signs, symptoms and laboratory indicators are likely to be treated in the same way.

After being filtered, patient medical records are encoded as one-hot vectors for categorical data or are normalized to zero-mean unit-variance for numerical data. For discharge summary, only text sections mentioning about the patient history of illness and description about their situation at admission are preferred. We note that segmenting these sections depends on how well-structured discharge summaries were written. In our experiment on MIMIC III database, some clue phrases enabled this solution to become implementable. For simplicity, signs and symptoms mentioned in the segmented text are extracted as new features of the trained patients. Our framework uses the collection processing engine (CPE) component with AggregatePlaintextFastUMLSProcessor provided in cTAKES [7], a well-known tool specifically designed for clinical text processing, to accomplish this task. It is worth noting that extracted signs and symptoms using this tool links to concepts in the Unified Medical Language System (UMLS) [1], the comprehensive ontology built for the biomedical domain.

3.2 Data Representation and Patient Clustering

The encoded data obtained from the previous step contains numerical, binary or categorical values. Such kind of mixed-type data is not ready to fit traditional clustering methods. Therefore, our framework employs MV.RBM, an extension of the restricted Boltzmann machine for data transformation and representation.

MV.RBM is a RBM where visible nodes are not restricted to binary units. Similar to the original RBM, each binary hidden unit in MV.RBM also tries to capture latent aspects in the imhomogeneous visible units. In other words, MV.RBM could be considered as a model to transform heterogeneous input to homogeneous space. Let $v = (v_1, v_2, .., v_N)$ denote the set of visible features and $h = (h_1, h_2, .., h_K)$ be the set of hidden units. The energy function of MV.RBM is defined more deliberately to handle the mixed variate input.

$$E(v, h) = -\left(\sum_i G_i(v_i) + \sum_k b_k h_k + \sum_{ik} H_{ik}(v_i) h_k\right)$$

where $b = (b_1, b_2, .., b_N)$ are biases vectors for hidden layer, $G_i(v_i)$ and $H_{ik}(v_i)$ are specified-type functions. By exploiting the conditional independence property within nodes in a layer of bipartite structure, we can get the following factorization equations: $P(v|h) = \prod_{i=1}^{N} P(v_i|h)$, $P(h|v) = \prod_{k=1}^{K} P(h_k|v)$.

The functions $G_i(v_i)$, $H_{ik}(v_i)$ and corresponding $P_i(v_i|h)$ for each kind of data are given as follows [11].

	$G_i(v_i)$	$H_{ik}(v_i)$	$P_i(v_i|h)$
Binary	$a_i v_i$	$w_{ik} v_i$	$\frac{\exp(a_i v_i + \sum_k w_{ik} h_k v_i)}{1 + \exp(a_i + \sum_k w_{ik} h_k)}$
Gaussian	$-v_i^2/2\sigma^2 + a_i v_i$	$w_{ik} v_i$	$\mathcal{N}(\sigma_i^2(a_i + \sum_k w_{ik} h_k), \sigma_i)$
Categorical	$\sum_m a_{im} \delta_m[v_i]$	$\sum_{m,k} a_{imk} \delta_m[v_i]$	$\frac{\exp(\sum_m a_{im} \delta_m[v_i]) + \sum_{m,k} w_{imk} \delta_m[v_i] h_k)}{\sum_l \exp(a_{il} + \sum_k w_{ilk} h_k)}$

where a_i, a_{im} are input bias parameters, w_{ik}, w_{imk} are input-hidden weighting parameters. Those with extra subscript m are dedicated for categorical features.

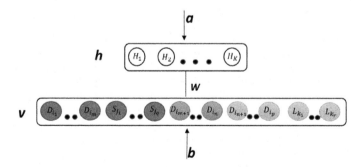

Fig. 2. A MV.RBM for patient records. The green, blue and orange circles represent for binary, categorical and continuous input units. The circles with labels D, S, L indicate demographic, signs/symptoms and laboratory data, respectively.

In our work, we assume features in the preprocessed data are mutually independent given their latent factors. Figure 2 illustrates our idea to utilize a MV.RBM. We suppose demographic data could receive numerical, binary or categorical values while extracted signs and symptoms are represented as one-hot vectors. Indicator values are assumed to take numerical values. Once the MV.RBM model has been learned, the computable hidden posteriors and hidden states are extracted as transforming features for input v. Those latent vectors could be used as input of well-known clustering algorithms. In this concrete work, we utilize the learned binary hidden vectors and select the hierarchical clustering algorithm to divide patients into groups. We use the Hamming distance as similarity measurement for binary vectors and the complete linkage which was reported to give low error rate for symmetric distance measurement [10].

3.3 Treatment Period Identification

For each resulting patient cluster obtained from the previous step, prescription records from its patients are collected to derive typical treatment regimens over periods. We represent every drug dr in prescription of patient p as a tuple $dr^p = (name, startdate, enddate, dosage)$ that describes drug name, starting date, ending date of usage and its dosage. Let $\Theta^p = \{dr^p\}$ be the set of drugs given to the patient, and $T^p = \{dr^p.startdate\}$ be the ordered set of dates the patient p was prescribed. As $|T^p|$ varies according to p, we propose an algorithm to split each T^p into the same number of treatment periods. The idea is for each timestamp in T^p, we compute an accumulated score that captures the changes in drug indications that have been delivered to the patient so far. We observe the plot of these scores for many patients in the clusters and decide an appropriate number of periods. The splitting dates for each period are the dates with significant changes in their associated scores.

It is worth noting that in our framework the scoring function takes into account newly prescribed drugs, re-prescribed drugs being stopped using for a while, recently stopped using drugs, or re-prescribed drug with changes in dosage. The aggregate score also gives different weights to those drugs based on their indication. Given a disease de and a set of its common symptoms $Symp^{de}$, we extract from DrugBank database [12] the drugs whose indication description directly mentions about de. We name those drugs as main drugs. Drugs with indication mentioned in $Symp$ are also extracted as symptom-healing drugs. Prescribed drugs for the patient p therefore are classified as main drugs, symptom-healing drugs, and unclassified drugs. The weight of each kind of drug is assigned decreasingly according to its importance for the treatment of de. We denote MDB, SDB as sets of main and symptom-healing drugs which are extracted from DrugBank; $w_{main}, w_{symp}, w_{unk}$ as the weight for main drugs, symptom-healing drugs, and unclassified drugs, respectively. The detailed algorithm for scoring changes in prescribed drug indications for a patient p is presented in Algorithm 1. For readability, we remove the superscript p and use Set notations in the pseudocode.

Algorithm 1. Scoring prescription records

Data: Θ, T, MDB, SDB
Result: return *scores* as a list of accumulated scores
Initialize U as an empty set ;   ▷ set of recently delivered drugs
Initialize *scores* as an empty list ;
$aScore := 0$;   ▷ the accumulated score
for *each* $d \in T$ **do**
    $D := \{dr \mid \forall dr \in \Theta \wedge dr.startdate == d\}$;   ▷ delivered drugs on date d
    $N := \{dr \mid \forall dr \in D \wedge dr.name \notin U.name\}$;   ▷ newly delivered drugs
    $DC := \{dr \mid \forall dr \in D, \exists dr' \in U$ such that $dr.name ==$
    $dr'.name \wedge dr.dosage <> dr'.dosage\}$;   ▷ dosage changed drugs
    $S := \{dr \mid \forall dr \in U \wedge dr.name \notin D.name \wedge dr.enddate < d\}$;   ▷ recently
    stopped using drugs
    for *each* $d' \in U$ **do**
        if $\exists d'' \in D$ such that $d'.name == d''.name$ **then**
            $d' := d''$;   ▷ update U with redelivered drugs

    $U := (U \setminus S) \cup N$;   ▷ update U with newly delivered drugs
    $CD := N \cup DC \cup S$;   ▷ considering drugs for calculating scores
    $CMD := CD.name \cap MDB$;   ▷ considering main drugs
    $CSD := CD.name \cap SDB$;   ▷ considering symptom-healing drugs
    $UD := CD.name \setminus (CMD \cup CSD)$;   ▷ unclassified drugs
    $aScore = aScore + |CMD| \times w_{main} + |CSD| \times w_{symp} + |UD| \times w_{unk}$;
    Add $aScore$ to *scores*

3.4 Learning Group Treatment Regimens

The previous section has demonstrated our domain integrated algorithm which allows prescription recorded to be divided into periods based on the associated scores which reflect the change in the indication of prescribed drugs. In this section, we describe how a treatment regimen over a period of a given patient cluster is derived. We relax the chronological order of delivered drugs in a period and restrict the element of constructed treatment regimens to drug names only. Other information such as dosage, route, is assumed to be decided by the physicians.

The learned regimens were organized in a tree form. Starting from the root, we assign the most frequently prescribed drug d to its left child node and extract prescribed drugs excluding d of the patients who were treated by d. The drug assignment for next right child nodes will follow the similar approach applied on prescribed drugs of those patients who were not treated by left-hand side nodes in the same level. We recursively perform this procedure on internal nodes. To avoid learning too complicated details of the derived tree, we only perform the procedure until a certain level of the tree or when the number of patients treated by the most frequent drug for the parent node is still greater than a threshold. Algorithm 2 presents our ideas to construct the treatment regimen tree for a particular group of patients in a period.

Algorithm 2. Procedure for the construction of a treatment regimen tree

---

REGIMEN-**Tree**(*depth, prescData, parent, traces*)

   if *prescData is empty or depth* $==$ *maxDepth* **then**
     └ **return**
   $d :=$ most frequent drugs from *prescData* ;
   $nPatients :=$ number of patients who were treated by d;
   $traces[parent, d] =$ " $\diagdown$ ";
   $cNodePresc :=$ prescribed records excluding d of patients treated by d ;
   $rNodePresc :=$ prescribed records of patients who were not treated by d ;
   if *nPatients* $<$ *threshold* **then**
     │ REGIMEN-**Tree**(*depth, rNodePresc, parent, traces*);
   else
     │ REGIMEN-**Tree**(*depth* $+ 1, cNodePresc, d, traces$);
     └ REGIMEN-**Tree**(*depth, rNodePresc, parent, traces*);

   return ;

---

4 Experimental Evaluation

This section presents our experimental evaluation of the proposed framework for deriving typical treatment regimens from electronic medical records. The obtained results of the clustering analysis, treatment period identification and learned treatment regimen trees are also given and analyzed. Lastly, we propose a method to evaluate the efficacy of the derived treatment regimen trees in recommending prescribed drugs for new patients.

4.1 Experimental Design

Our experimental evaluation was performed on MIMIC III, a freely accessible critical care database [4]. We considered the treatment regimen of patients who were diagnosed with coronary artery disease as a case study. Since a patient could be diagnosed with multiple ICD codes, to ensure the homogeneity of our patient cohort, we only selected those whose primary ICD is coronary artery disease and comorbidity scores are zero for other disease groups. In addition, patients who were prescribed fewer than three times were also excluded from the experimental evaluation. The number of extracted patients is 707 of which we randomly selected 687 patients for training and left 20 patients for testing the efficacy of the learned treatment regimens. We followed the approach described in the Data Representation section to preprocess raw data. A summary of preprocessed data with illustration features is given in Table 1.

We fit preprocessed data as input for MV.RBM with 200 hidden units since the trained error did not decrease significantly with a larger number of hidden units. The learned binary hidden states were then extracted as representation features for the subsequent clustering task. We employed hierarchical clustering with parameters are described in the previous section. For the task of treatment period identification, we extracted main drugs and symptom-healing drugs from

Table 1. A short summary of features in the dataset

Kind of data (no.features)	Sample features (data type)
Demographic info (11)	Age (numerical), gender (binary) admission type (categorical)
Laboratory data (175)	Arterial blood pressure (numerical) atrial pacemaker(numerical)
Signs and symptoms data (1466)	Abdominal discomfort (binary) ability to climb (binary) able to sleep (binary)

DrugBank database. The referred typical symptoms of coronary artery disease in the literature are "heart attack", "shortness of breath" and "chest pain". We assigned the weight of main drugs, symptom healing drugs, and unclassified drugs to 1, 0.5 and 0.1, respectively. The threshold of cutting node in treatment learning algorithm was set to 10 patients. In our experiment, we derived regimens until a certain level of the tree. The depth parameter was set to 4.

4.2 Results

Figure 3 describes a dendrogram of clustering results. It is noted that the trained patients themselves are homogeneous subjects in terms of diagnostic perspective. Thus, we preferred a relatively small number of clusters. Based on the visualization, we decided to group the trained patients into six clusters. The size of each cluster is 198, 69, 148, 43, 111 and 118 patients, respectively.

Figure 4 presents a few randomly taken line charts of accumulative scores for eight patients. Interestingly, most of the plots follow similar patterns. There is a slight increase in scores at the beginning and the end of every treatment compared to the significant change at the center interval. Therefore, we decided to divide prescription records of trained patients into three periods. Figure 5 illustrates an example of constructed regimen trees. Given a path in the tree, we note that the order of the nodes in this path should be understood as frequency order of drug use rather than chronological order of prescription time. It can be seen that the visualization can provide hint-drugs probably delivered together

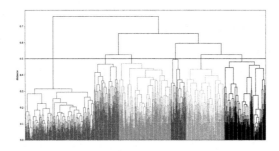

Fig. 3. Dendrogram of hierarchical cluster analysis

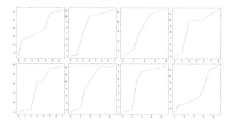

Fig. 4. Sample line charts of accumulative scores of randomly taken patients. For each line chart, the vertical axis represents the accumulative scores, the horizontal axis represents the timestamps when a patient was prescribed.

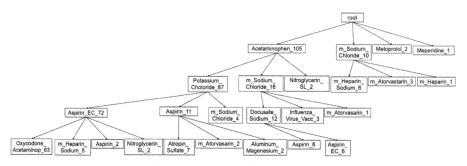

Fig. 5. Sample of learned regimen tree. The prefix "m" denotes for main drugs while the last integer denote the number of prescribed patients.

with a given drug. Therefore, physicians can use the learned trees as a checklist to decide which drugs are likely and unlikely to be prescribed.

4.3   Evaluation

We evaluate the efficacy of learned regimen trees in recommending prescribed drugs to new patients. It should be noted that patient records of the testing set are represented by the trained MV.RBM. We consider the patients in each resulting clusters as labeled data and assign the cluster index for test patients based on their nearest neighbors. Given a new patient p, let p' be his/her nearest neighbor which has been assigned to cluster c_i. The recommended drugs should be given to p in a particular period are drugs on the path of regimen tree of c_i in the same period such that p' was prescribed with each drug on that path.

Let T, nP, $\hat{D}_p^{t_j}$, and $D_p^{t_j}$ denote the test set, the number of periods, the recommended path, i.e., the set of recommended drugs for p over period t_j, and the set of prescribed drugs for p in that period, respectively. We propose two measures to evaluate the efficacy of learned regimens for the prescription recommendation task. These measures reflect how likely $\hat{D}_p^{t_j}$ is a subset of $D_p^{t_j}$. $\hat{D}_p^{t_j}$ is said "correct" if it is a non empty subset of $D_p^{t_j}$. In case $\hat{D}_p^{t_j}$ has non empty intersection with $D_p^{t_j}$ but not its subset, we say the set $\hat{D}_p^{t_j}$ is "approximately correct". We denote m_{cor} as the percentage of recommended paths which are "correct" and m_{app} as the percentage of recommended drugs actually prescribed in both "correct" and "approximately correct" paths. Let $I_A(B)$ define the indicator function which return 1 if $B \subset A$ or 0 otherwise. We have:

$$m_{cor} = \frac{1}{|T| \times nP} \sum_{p \in T} \sum_{j=1}^{nP} I_{D_p^{t_j}} (\hat{D}_p^{t_j}); \quad m_{app} = \frac{1}{|T| \times nP} \sum_{p \in T} \sum_{j=1}^{nP} \frac{|\hat{D}_p^{t_j} \cap D_p^{t_j}|}{|\hat{D}_p^{t_j}|}$$

We repeated our experiment 10 times for different training and testing sets. The obtained values of $\bar{m}_{cor}$ and $\bar{m}_{app}$ are 0.527 and 0.729, respectively. Although the obtained values of $\bar{m}_{cor}$ should be further improved, to some extent, these measures show the efficacy of the regimen trees derived from our proposed framework.

5 Discussion

Comparing to related works in the literature, our work obtained more interesting results in terms of domain exploitation and knowledge representation. Rather than defining a similarity metric by a frequency-based approach for complex objects [9], we tracked the change of drug indication in prescribed drugs as a hint to discover treatment periods. It can be seen that the idea fits our natural thinking on detecting patients' treatment periods given their prescription records. The common pattern found in Fig. 4 has reconfirmed the rationality of our proposed domain-based algorithm. Moreover, representing the learned regimens in form of trees not only fully reflects the usage-frequency of drugs but also allows doctors to quickly recognize groups of frequently and infrequently prescribed drugs in each patient sub-cohort. Therefore, in terms of knowledge representation, it could be said that our work is superior to [2,9] where the authors simply organized treatment patterns in flat form.

There are several reasons to explain the primitive results of our initial study on the task of treatment recommendation. Firstly, it is worth noting that we addressed the problem of treatment recommendation on MIMIC III, a practical and very challenging dataset. Even if it has been simplified to recommend in total up to 12 among many prescribed drugs for every patient, the problem is still not trivial as there are hundreds of different drugs given in the prescription records. Additionally, while our evaluation metrics directly assess whether the recommended drugs are prescribed to new patients, it is not clearly described in other studies how well the recommended treatments match the actual prescribed drugs. We leave the task of improving our prediction accuracy with a more deliberated framework for the future work.

6 Conclusion

In this paper, we have presented a generic framework to derive treatment regimens from electronic medical records. The proposed framework is novel in terms of data utilization, domain incorporation, and regimen representation. The experimental evaluation has shown the efficacy of learned treatments for the task of prescription recommendation. Although further improvement should be made such as data cleaning and normalizing for clinical features, this study is a pioneering work which encourages researchers to exploit medical domain knowledge and address the treatment learning problem more thoroughly.

Acknowledgement. This work is partially sponsored Asian Office of Aerospace R& D under agreement number FA2386-17-1-4094 and Vietnam National University at Ho Chi Minh City under the grant number B2015-42-02. We wish to thank Tu Dinh Nguyen for providing the implementation of the MV.RBM model.

References

1. Bodenreider, O.: The unified medical language system (UMLS): integrating biomedical terminology. Nucl. Acids Res. **32**(suppl_1), D267–D270 (2004)
2. Huang, Z., Dong, W., Bath, P., Ji, L., Duan, H.: On mining latent treatment patterns from electronic medical records. Data Min. Knowl. Discov. **29**(4), 914–949 (2015)
3. Huang, Z., Lu, X., Duan, H.: On mining clinical pathway patterns from medical behaviors. Artif. Intell. Med. **56**(1), 35–50 (2012)
4. Johnson, A.E., Pollard, T.J., Shen, L., Lehman, L.-W.H., Feng, M., Ghassemi, M., Moody, B., Szolovits, P., Celi, L.A., Mark, R.G.: MIMIC-III, a freely accessible critical care database. Sci. Data **3**, 160035 (2016)
5. Johnston, R.L., Carius, H.J., Skelly, A., Ferreira, A., Milnes, F., Mitchell, P.: A retrospective study of Ranibizumab treatment regimens for neovascular age-related macular degeneration (nAMD) in Australia and the United Kingdom. Adv. Ther. **34**(3), 703–712 (2017)
6. Merhej, E., Schockaert, S., McKelvey, T.G., De Cock, M.: Recommending treatments for comorbid patients using word-based and phrase-based alignment methods. In: 29th Benelux Conference on Artificial Intelligence (BNAIC 2017) (2017)
7. Savova, G.K., Masanz, J.J., Ogren, P.V., Zheng, J., Sohn, S., Kipper-Schuler, K.C., Chute, C.G.: Mayo clinical text analysis and knowledge extraction system (cTAKES): architecture, component evaluation and applications. J. Am. Med. Inform. Assoc. **17**(5), 507–513 (2010)
8. Song, R., Wang, W., Zeng, D., Kosorok, M.R.: Penalized Q-learning for dynamic treatment regimens. Stat. Sin. **25**(3), 901 (2015)
9. Sun, L., Liu, C., Guo, C., Xiong, H., Xie, Y.: Data-driven automatic treatment regimen development and recommendation. In: KDD, pp. 1865–1874 (2016)
10. Tamasauskas, D., Sakalauskas, V., Kriksciuniene, D.: Evaluation framework of hierarchical clustering methods for binary data. In: 2012 12th International Conference on Hybrid Intelligent Systems (HIS), pp. 421–426. IEEE (2012)
11. Tran, T., Phung, D., Venkatesh, S.: Mixed-variate restricted Boltzmann machines. arXiv preprint arXiv:1408.1160 (2014)
12. Wishart, D.S., Knox, C., Guo, A.C., Cheng, D., Shrivastava, S., Tzur, D., Gautam, B., Hassanali, M.: DrugBank: a knowledgebase for drugs, drug actions and drug targets. Nucl. Acids Res. **36**(suppl_1), D901–D906 (2007)

Human, Behaviour and Interactions
(Application)

Mining POI Alias from Microblog Conversations

Yihong Zhang[1(✉)] and Lina Yao[2]

[1] Graduate School of Informatics, Kyoto University, Kyoto, Japan
`yihong.zhang.8z@kyoto-u.ac.jp`
[2] School of Computer Science and Engineering, The University of New South Wales,
Sydney, Australia
`lina.yao@unsw.edu.au`

Abstract. In location-based analysis for microblogs, it is important to know if two toponyms refer to the same point-of-interest, i.e., alias. However, existing online knowledge bases are often incomplete or inaccurate for toponym alias data, especially for those used in informal conversations. In this paper, we propose a method for extracting compatible toponyms from microblog conversations. We first extract a number of coordinate-associated toponyms, then use compatibility measures to identify compatible toponyms. We propose three compatibility measures, namely, geographical closeness, surface name similarity, and association similarity. We show that by combining these measures and using particle swarm optimization for weight tuning, we can reach a high matching accuracy. The finding of this paper can be useful for improving location-based analysis as well as extending existing knowledge bases.

Keywords: Location analysis · Alias extraction
Compatibility measures · Microblog information processing

1 Introduction

Location analysis on microblog platforms such as Twitter has become an important research topic in recent years [3, 6, 11, 14, 17, 23]. Locations of microblog messages have been used in applications such as natural disaster monitoring [21], crime detection [16], and disease surveillance [4]. One crucial issue that hinders accurate location detection of messages and events is the lack of a comprehensive gazetteer that covers toponyms of various forms. Even though location databases such as Geonames[1] contain a large number of toponyms, they are still unable to cover various ways to name a location on microblog communications, which leads to limited detection rates for gazetteer-based location detection approaches [22]. For example, in an online conversation between two Singaporeans, *the statue* often means the same thing as the *merlion* sitting beside the Marina Bay. This kind of informal and generalized alias are usually not included in the location

---

[1] http://www.geonames.org/.

© Springer International Publishing AG, part of Springer Nature 2018
D. Phung et al. (Eds.): PAKDD 2018, LNAI 10937, pp. 425–436, 2018.
https://doi.org/10.1007/978-3-319-93034-3_34

database. In this paper, we deal with the problem of mining different toponyms that refer the same point-of-interest (POI), in other words, alias, from informal microblog conversations. The finding of this paper is beneficial for improving location detection rates on microblog location analysis, as well as extending existing location databases.

Given two toponyms, we are interested in discovering their *compatibility*. Two toponyms are considered alias to each other if they are compatible. The compatibility considers the geographical location as well as the meaning of the name. Two POIs can have the same name but are in different locations, for example, two outlets of a restaurant chain. Two POIs can also have the same location, for example, a building and a shop located within it. In this case the names of the building and the shop are not compatible because they refer to POIs of different meaning. But if a building is solely occupied by an institute, and the names of the building and the institute are used interchangeably in daily conversations, they are considered compatible. Two toponyms are considered compatible if they are similar in geographical and semantic measures. In this paper, we use *alias* and *compatible toponym* interchangeably.

There are many reasons one name is used instead of another to indicate a POI. The most common reason is to shorten the word for easier expression. In this case, abbreviations and shortened forms of a long name would be used. But there are also cases where two completely different names are used to indicate a same POI. The choice of alias in this case depends on the context and personal preferences. Also depending on the context, in some other cases, a codified name that includes digits can be used in place of the more common name. Table 1 shows examples of various types of alias. In this work, we do not treat different types of alias separately, and our method is designed to handle any type of alias.

Table 1. Examples of compatible toponyms

Type	Compatible toponyms	
Shortened	Republic polytechnic	Republicpoly
Shortened	Fatpapas burgers and shakes	Fatpapas burgers
Alternative	The statue	Merlion
Alternative	Theatres on the bay	Esplanade
Codified	Woodlands mrt station	ns9
Codified	Bus stop 76249	blk 370

The focus of our work is not in name entity recognition (NER). NER deals with identifying unlabeled keywords in text data as name entities, including persons, locations and events, as studied by a large number of existing researches [15,18,20]. We assume an effective way to extract location from tweet texts, given that microblog messages have geo-location information. In this paper, we use a method similar to the one propose by Abdelhaq et al. [1], which detects spatial keywords based on geo-tagged tweets. This method may not have high retrieval rate due to the limited availability of geo-location information, but the precision of detection is generally high. Given the location extraction method, we put our focus on providing a valid compatibility measurement for toponyms and an alias prediction method based on the measurements. To the best of our knowledge, this paper is the first to tackle to problem of mining POI alias used in microblogs based on compatibility measures. Our main contributions with this paper can be summarized as following:

- We quantify the compatibility of toponyms for efficient alias mining. Our measurement considers the geographical distance, surface name similarity, as well as the semantics of toponyms, taking advantage of the readily available geographical data and unlabeled text messages on platforms such as Twitter.
- We propose an algorithm for automatically tuning factor weights. In our experiments with real-world POIs and Twitter data, we found that our weight tuning method effectively improves the alias detection accuracy.

The remainder of this paper is organized as the following: in Sect. 2, we discuss related works. Section 3 introduces our method for selecting toponym candidates from raw tweets and associating them with geo-coordinates. Section 4 presents our toponym compatibility measurement and alias prediction methods. In Sect. 5, we present our experiments on real Twitter datasets and discuss the results. Finally, Sect. 6 concludes this paper.

2   Related Works

There is a number of works that focus on resolving an entity mention to a particular entity, effectively disambiguates when a name can mean multiple entities [7,8,12]. In this paper we do not deal such a problem of one-to-many linking and disambiguation. Instead our work aligns more closely to works that focus on many-to-one linking of compatible names.

Early works often detect alias based on the surface form of the name. For example, Hsiung et al. [9] first propose an alias detection method that considers association of names. It uses orthographic measures such as edit distance and normalized edit distance, as well as semantic measures that considers the co-occurrence of two names. While their method is relatively simple, it highlights the fundamental elements for alias detection in linked data, and some elements are used in this paper. Gelernter and Balaji [5] present a gazetteer-based method of geoparsing for tweets. A novelty in their work is that they include four different parsers, including machine learned parser and rule-based parser. They provide a module that matches misspelled names with the correct names in the gazetteer, with candidates ranking by frequency and edit distance. However, this kind of technique is not effective when two compatible names have completely different spellings, for example, "the statue" and "merlion".

Another group of works detects alias using language patterns that indicates alias usage. For example, Bollegala et al. [2] propose a web-search-based method for discovering alias. Their assumption is that when a name and its alias co-occur in a web document, some frequent patterns such as "also known as" would appear between them indicating the alias. However, in microblog environment, such cases are rare because with a limited length, a name and its alias are unlikely to appear in the same message. Huang et al. [10] propose a method to detect morphs used in censored microblogs. Their method considers orthographical, semantic, and social features. However their method cannot be directly applied to location alias detection, because unlike morphs, which are produced in censored environment, and therefore have an original name-morph relationship, location alias are compatible and interchangeable names used in the same environment. In our work, we use geographical distance and entity associations instead of language patterns as the basis for detecting alias.

3   Selecting Toponym Candidates

The first step to mine toponym alias is to select a number of candidate toponyms from the data. In existing work there are mainly two approaches to iden-tify toponyms from text data. First, there is gazetteer-based approach, which matches tokens in the text with an existing list of toponyms [4,22,24]. While this approach is generally accurate, it does not cover a large number of infor-mal spellings that are frequently used in social media, and not included in a formal gazetteer. Second, there are machine learning-based approaches, which tags token in the text with a trained language model [14]. The drawback of this approach is that it is accuracy may be poor depending on the training data, and does not provide information other than the category of the word. It is desirable to use an approach that detects informal spellings of toponyms and at the same time also extracts their geographical information.

Based on the fact that a portion of tweets are geo-tagged, we proposed an approach to detect spatial keywords based on its geographical correspondence. With this approach, we first extract noun phrases from geo-tagged tweets using existing part-of-speech (POS) tagging techniques. Then we associate each noun phrase with the coordinates of the tweets that mention it. As a result, each noun phrase w will be associated with a list of coordinates $L_w = \{c_1, ..., c_n\}$. Then we calculate the *median absolute deviation* (MAD) for each w:

$$MAD(w) = median(dist(c_i, median(L_w)), c_i \in L_w)$$

where $dist(.)$ measures geographical distance between two geographical points[2], and $median(L_w)$ is selecting as the point in L_w that has the minimum total distance to all other points.

A low MAD means the noun phrase has a high correspondence with a certain location. Another factor we consider is the *support*, which is the count of occur-rence of the phrase in the data. We select noun phrases with MAD and support that meet certain thresholds as the toponym candidates. In our experiments, we

---

[2] Calculation can be found at http://www.movable-type.co.uk/scripts/latlong.html.

set threshold for MAD and support as 0.5 and 3, respectively. At this point, each candidate is also associated with a corresponding coordinate as $median(L_w)$.

We need to note that not all candidates are valid toponyms. Some of these noun phrases may be the name of a local event, which is not a toponym but also have strong geographical correspondence. Some other noun phrases may be associated with a user. Because a user has a limited geographical area for their activities, some nouns associated with them can also appear to have a strong geographical correspondence but is not a valid toponym. Some of these non-toponym candidates, however, can be detected and ignored during the process of toponym alias detection.

4   Finding Compatible Toponyms

Given two toponym candidates, we calculate their compatibility in order to detect potential alias. The information we have for the toponyms include their string representation, corresponding geographical location, and a number of unlabeled microblog messages that mention them. Based on such information, we propose three measures for testing toponym compatibility, namely *Geographical Closeness*, *Surface Name Similarity*, and *Association Similarity*.

4.1   Compatibility Measures

The proposed compatibility measures are defined as the following.

Geographical Closeness. If two toponyms are referring to the same POI, their corresponding location must be close to each other. We calculate the geographical closeness of two toponyms $w1$ and $w2$ based on the geographical distance of their corresponding location:

$$f(w1, w2) = 1 - geo\_dist(w1, w2) = 1 - dist(median(L_{w1}), median(L_{w2}))$$

where $median(L_w)$ is the corresponding coordinate of the toponym described in the previous section. The $geo\_dist(.)$ is measured in kilometers, and we can generally ignore toponym pairs whose distance to each other is more than 1 km.

Surface Name Similarity. Surface name similarity is considered as an important factor for determining the compatibility of two concepts in existing works [23]. While we expect only a small portion of compatible toponyms to be similar in their surface form, we nevertheless include this in our compatibility calculation, because it is easy to obtain. The surface name similarity is calculated based on the edit distance of two names:

$$g(w1, w2) = surface\_sim(w1, w2) = 1 - \frac{edit\_distance(w1, w2)}{max(|w1|, |w2|)}$$

where the edit distance indicates the number of insertion, deletion and replacement required to change one string into the other[3].

---

[3] An algorithm for calculating edit distance can be found in https://nlp.stanford.edu/IR-book/html/htmledition/edit-distance-1.html.

Association Similarity. We now consider the semantics of toponyms. In existing works, semantic is often considered with respect to the linking of entities [19]. Following what Huang et al. [10] propose, we consider the semantic of a toponym with respect to its *co-occurrence* with other entities in the data. More specifically, for a toponym w, we generate a number of associated noun phrases Q_w, which are noun phrases co-occur with w in unlabeled tweets (geo-tagged and non-geo-tagged). After obtaining the associated noun phrases, we calculate the association similarity of two toponyms based on frequency and cosine similarity:

$$h(w1, w2) = assoc\_sim(w1, w2) = \frac{\sum_i tf(Q_{w1}, t_i) \times tf(Q_{w2}, t_i)}{\sqrt{\sum_i tf(Q_{w1}, t_i)^2} \times \sqrt{\sum_i tf(Q_{w2}, t_i)^2}}$$

where $t_i \in T$ is all the terms in $Q_{w1} \bigcup Q_{w2}$, and $tf(Q, t)$ is the term frequency of term t in set Q.

A simple way to reach an overall compatibility score for two toponyms is to combine all three factors using summation:

$$compatibility(w1, w2) = f(w1, w2) + g(w1, w2) + h(w1, w2) \tag{1}$$

However, a better way to measure the compatibility between toponyms than simple combinations is to assign a weight to each factor then find the suitable weights using some training examples. In the next section, we will propose one such method.

4.2   Tuning Factor Weights

A weighted version of formula (1) can be written as:

$$compatibility(w1, w2) = \alpha f(w1, w2) + \beta g(w1, w2) + \gamma h(w1, w2) \tag{2}$$

Suppose we have a training set where a list of toponyms $W = \{w_1, ..., w_l\}$ has a corresponding alias list $A = \{a_1, ..., w_l\}$, where w_i and a_i are also from a list of toponym candidate $T = \{t_1, ..., t_n\}$. The goal of training is to find the weights α, β, and γ so that $compatibility(w_i, a_i)$ will have the higher score than $compatibility(w_i, t_j)$, for any $t_j \neq a_i \in T$.

In this paper we propose a *particle swarm optimization* (PSO) [13] solution for tuning the parameters, due to its capability to tune continuous parameter with non-linear objective function. A PSO solution first randomly initializes a number of solutions. Then in each optimization iteration, it selects a optimal solution based on a certain objective function, and "moves" other solution towards the optimal solution. We define our objective function based on the ranking of a_i in the ranked candidate list T' with respect to their compatibility with w_i:

$$score(\alpha, \beta, \gamma) = \sum_{i=1}^{l} \frac{1}{rank(w_i, a_i, T, \alpha, \beta, \gamma)} \tag{3}$$

where $rank(w_i, a_i, T, \alpha, \beta, \gamma)$ gives the rank of a_i among T with compatibility calculated using α, β, γ and formula (2).

The steps for finding optimal parameter are shown in Algorithm 1. In addition to W, A, T, we need to supply three additional parameter, the number of particles $nPart$, the number of iterations $nIter$, and particle moving speed $speed$. In our experiments, we find that these parameters are not sensitive, and with high enough $nPart$, the particles always converge to the same optimal solution in a few iterations, regardless of random initial state. In the experiments, we set $nPart = 100$, $nIter = 10$, and $speed = 0.2$.

Algorithm 1. PSO for parameter tuning

INPUT: W, A, T, $nPart$, $nIter$, $speed$
OUTPUT: α, β, γ
1: randomly initialize $nPart$ particles $(\alpha_p, \beta_p, \gamma_p)$
2: **for** $nIter$ iterations **do**
3:     **for** each particle p **do**
4:         $score_p \leftarrow$ calculated score using formula (3)
5:     **end for**
6:     $best\_particle \leftarrow$ the particle with the highest $score_p$
7:     **for** each particle p **do**
8:         $p \leftarrow p + (best\_particle - p) \times speed$
9:     **end for**
10: **end for**
11: **return** $best\_particle$

---

5 Experimental Results

We conduct experiments to test the effectiveness of our alias mining method. Our experiments are conducted on real Twitter data we collected through its API. We analyze the effectiveness of individual factors as well as combinations of factors. In this section, we discuss our experimental data, evaluation metric, and accuracy results.

5.1 Dataset and Evaluation Metric

We focus our experiment on data and POIs from Singapore. We collect a dataset of tweets from Singapore as following. First we monitor Twitter's Sample API[4], which provides a small random sample of all public tweets. From the monitored tweets, we extract 5,000 users whose home location contains the word "Singapore". Then we use Twitter's timeline API[5] to collect up to 1,000[6] past tweets from each user. In total we collect a dataset of more than 3 million tweets, and it contains about 68k geo-tagged tweets. We then extract noun phrases using StanfordNLP POS tagger from the geo-tagged tweets, and generate a number of toponym candidates using the thresholds MAD <0.5 and support >3. The resulted number of toponym candidates and other statistics for the dataset are shown in Table 2.

---

[4] https://dev.twitter.com/streaming/reference/get/statuses/sample.
[5] https://dev.twitter.com/rest/reference/get/statuses/user_timeline.
[6] If a user has posted less than 1,000 tweets, we collect all past tweets.

Table 2. Statistics about experiment dataset

Number of tweets	3,826,259
Number of users	5,000
Number of geo-tagged tweets	68,618
Number of noun phrases from geo-tagged tweets	266,902
Number of toponym candidates	3,199

We manually label some pairs of toponyms as compatible toponyms. First we calculate pair-wise geographical distance and filter out pairs whose distance is more than 0.05 km, with the assumption that if two toponyms are compatible, they must be in the same location. Then we manually check each remaining pair and see if they refer to the same POI. The manual checking process involves applying common sense and experiences as a Singapore resident, examining Google search results and Google Map search results, looking up knowledge bases such as Wikipedia. Finally we determine 102 pairs of toponyms as compatible toponyms.

We use Accuracy@k to measure the effectiveness of an alias mining approach. For each toponym in the manually labeled pairs, we generate a ranked list of alias candidates by calculating the compatibility of the toponym to all other toponym candidates. Accuracy@k means the percentage of toponyms among labeled pairs that have their compatible toponym within first k in their respective ranked list of candidates. Although we have not found a compatible baseline method for comparison, it is clear that if chosen at random, the theoretic Accuracy@1 will be $1/102 \approx 0.01$.

5.2   Results and Discussion

Effectiveness of Single Factor. We first test the effectiveness of three individual factors, namely, geographical closeness (GC), association similarity (AS), and surface name similarity (SS). When tested an individual factor, the compatibility is calculated using only the respective measurement function, $f(.)$, $g(.)$, or $h(.)$. The accuracy results are shown in Table 3.

Table 3. Accuracy results for individual factors

	Acc@1	Acc@2	Acc@3	Acc@5	Acc@10
GC	0.58	0.76	0.88	0.97	1
AS	0.57	0.61	0.66	0.67	0.68
SS	0.17	0.25	0.33	0.37	0.46

Among individual factors, the geographical closeness is the most effective for identifying alias, which is consistent with our intuition that two compatible toponyms should have the same location. However, it is unable to achieve high Accuracy@1, mostly because some toponyms have the same location but are not compatible. For example, "national gallery singapore" is first matched by geographical closeness with "smoke & mirrors", a restaurant located within the gallery, instead of its alias. Therefore it is clear that individual factor is not sufficient to provide high mining accuracy.

Using association similarity achieves a comparable Accuracy@1. If two compatible toponyms are used in the same way, they can be detected at the top of the rank. However, it is unable to push the alias to higher rank when there is a substantial difference between the association of a name and its alias. This happens when two compatible toponyms are used in different context by habit, for example, the building name and the institute that occupies it. We are unable to achieve high detection accuracy using the surface similarity because it is obvious that most pairs of compatible toponyms in our labeled data do not appear to be similar on the surface.

Effectiveness of Combination of Factors. We then investigate the effectiveness of different factor combinations. The accuracy results are shown in Table 4. First we notice that, any combination of factors performs better than individual factors. Particularly, combining geographical closeness and association similarity significantly increases Accuracy@1 to over 70%, while individually they only achieve less than 60%. This is because the two measurements effectively mitigate the problem in each other. For example, when a restaurant is located within a building and causes a problem using geographical distance measures, we can distinguish two POIs using the association similarity. Adding surface name similarity measures further improves Accuracy@1 by two percentages. However, it decreases accuracy@10 by one percentage, because it introduces some noises when more candidates are considered.

Table 4. Accuracy results for factor combinations

	Acc@1	Acc@2	Acc@3	Acc@5	Acc@10
GC+AS	0.71	0.80	0.92	0.95	0.99
GC+SS	0.59	0.81	0.88	0.92	0.99
AS+SS	0.60	0.70	0.77	0.80	0.82
GC+AS+SS	0.73	0.87	0.92	0.94	0.98

Effectiveness of Weight Tuning. Finally, we test the effectiveness of automatic weight tuning using the PSO algorithm. We follow the leave-one-out approach, that for each toponym in the labeled pairs, we first find the weights using all other labeled pairs as the training data for the PSO, then using the tuned weights to detect the alias for the toponym. In other words, we have a training step and a prediction step for each labeled toponym. For the parameters, we set $nPart = 100$, $nIter = 10$, and $speed = 0.2$. Because the random initialization may affect the result, we run the experiment 10 times and record the mean and standard deviation. The results are shown in Table 5. We can see that the result variances for different runs are very low. In fact the results in different runs only differ by one or two percentage. Therefore we can consider the impact of random initialization negligible.

Table 5. Accuracy results using tuned parameters

	Acc@1	Acc@2	Acc@3	Acc@5	Acc@10
Mean	0.77	0.89	0.95	0.97	1
Stdev	0.007	0.004	0.005	0.003	0

Figure 1 shows the accuracy comparison between simple combination of all three factors and combination using tuned weights. We can see that by tuning the weights using the PSO algorithm, we effectively improve the alias detection accuracy in all ranking ranges. Particularly, the weight tuning algorithm improve Accuracy@1 by 4%.

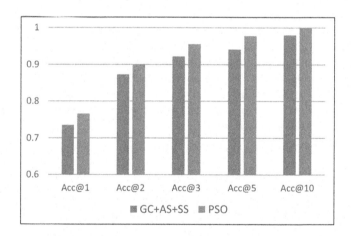

Fig. 1. Accuracy comparison of simple combination and tuned weight

Another interesting insight we gain in this experiment is that the optimal weights found by the algorithm center around a ratio of 8:2:3, which tells us that the most important factor for alias detection is the geographical closeness, followed by surface name similarity. Contradict to our initial expectation, the association similarity actually contributes less to the detection accuracy than the other two factors.

6   Conclusion

In this paper, we deal with the problem of mining POI alias from microblog conversations. One issue that hinders accurate location detection on microblogs is the lack of a comprehensive gazetteer that includes toponyms of various forms. Our work is beneficial for increasing location detection rates on microblogs as well as extending current POI knowledge bases. We propose three compatibility measures and a parameter tuning algorithm to further improve alias detection accuracies. To the best of our knowledge, this is the first work that aims to mine POI alias from microblogs using compatibility measures. Our work can serve as a baseline for future solutions.

In this paper, we mainly deal with two-to-one alias, in other words, two toponyms referring to the same POI. We have not deal directly the case of many-to-one alias, a group of toponyms referring to the same POI, although they can be dealt with similarly as the two-to-one alias. In the future we plan to extend our work to deal with many-to-one alias. We also plan to include more information that can be obtained from microblogs to further improve alias detection accuracy.

References

1. Abdelhaq, H., Sengstock, C., Gertz, M.: EvenTweet: online localized event detection from Twitter. Proc. VLDB Endow. **6**(12), 1326–1329 (2013)
2. Bollegala, D., Matsuo, Y., Ishizuka, M.: Automatic discovery of personal name aliases from the web. IEEE Trans. Knowl. Data Eng. **23**(6), 831–844 (2011)
3. Cheng, Z., Caverlee, J., Lee, K.: You are where you tweet: a content-based approach to geo-locating Twitter users. In: Proceedings of the 19th ACM International Conference on Information and Knowledge Management, pp. 759–768 (2010)
4. Dredze, M., Paul, M.J., Bergsma, S., Tran, H.: Carmen: a Twitter geolocation system with applications to public health. In: AAAI Workshop on Expanding the Boundaries of Health Informatics Using AI, pp. 20–24 (2013)
5. Gelernter, J., Balaji, S.: An algorithm for local geoparsing of microtext. GeoInformatica **17**(4), 635–667 (2013)
6. Graham, M., Hale, S.A., Gaffney, D.: Where in the world are you? Geolocation and language identification in Twitter. Prof. Geogr. **66**(4), 568–578 (2014)
7. Han, X., Sun, L., Zhao, J.: Collective entity linking in web text: a graph-based method. In: Proceedings of the 34th International ACM SIGIR Conference on Research and Development in Information Retrieval, pp. 765–774. ACM (2011)
8. Hoffart, J., Altun, Y., Weikum, G.: Discovering emerging entities with ambiguous names. In: Proceedings of the 23rd International Conference on World Wide Web, pp. 385–396. ACM (2014)
9. Hsiung, P., Moore, A., Neill, D., Schneider, J.: Alias detection in link data sets. In: Proceedings of the International Conference on Intelligence Analysis, vol. 4 (2005)
10. Huang, H., Wen, Z., Yu, D., Ji, H., Sun, Y., Han, J., Li, H.: Resolving entity morphs in censored data. In: Proceedings of the 51st Annual Meeting of the Association for Computational Linguistics, pp. 1083–1093 (2013)

11. Ikawa, Y., Enoki, M., Tatsubori, M.: Location inference using microblog messages. In: Proceedings of the 21st International World Wide Web Conference Companion, pp. 687–690 (2012)
12. Ji, Z., Sun, A., Cong, G., Han, J.: Joint recognition and linking of fine-grained locations from tweets. In: Proceedings of the 25th International Conference on World Wide Web, pp. 1271–1281 (2016)
13. Kennedy, J.: Particle swarm optimization. In: Encyclopedia of Machine Learning, pp. 760–766. Springer (2010)
14. Li, C., Sun, A.: Fine-grained location extraction from tweets with temporal awareness. In: Proceedings of the 37th International ACM SIGIR Conference on Research & Development in Information Retrieval, pp. 43–52 (2014)
15. Li, C., Sun, A., Weng, J., He, Q.: Tweet segmentation and its application to named entity recognition. IEEE Trans. Knowl. Data Eng. **27**(2), 558–570 (2015)
16. Li, R., Lei, K.H., Khadiwala, R., Chang, K.-C.: TEDAS: a Twitter-based event detection and analysis system. In: Proceedings of 28th International Conference on Data Engineering, pp. 1273–1276 (2012)
17. Lingad, J., Karimi, S., Yin, J.: Location extraction from disaster-related microblogs. In: Proceedings of the 22nd International World Wide Web Conference Companion, pp. 1017–1020 (2013)
18. Liu, X., Zhang, S., Wei, F., Zhou, M.: Recognizing named entities in tweets. In: Proceedings of the 49th Annual Meeting of the Association for Computational Linguistics: Human Language Technologies, vol. 1, pp. 359–367. Association for Computational Linguistics (2011)
19. Lucia, W., Ferrari, E.: Egocentric: ego networks for knowledge-based short text classification. In: Proceedings of the 23rd ACM International Conference on Information and Knowledge Management, pp. 1079–1088. ACM (2014)
20. Malmasi, S., Dras, M.: Location mention detection in tweets and microblogs. In: Hasida, K., Purwarianti, A. (eds.) Computational Linguistics. CCIS, vol. 593, pp. 123–134. Springer, Singapore (2016). https://doi.org/10.1007/978-981-10-0515-2_9
21. Sakaki, T., Okazaki, M., Matsuo, Y.: Earthquake shakes Twitter users: real-time event detection by social sensors. In: Proceedings of the 19th International World Wide Web Conference, pp. 851–860 (2010)
22. Schulz, A., Hadjakos, A., Paulheim, H., Nachtwey, J., Mühlhäuser, M.: A multi-indicator approach for geolocalization of tweets. In: Proceedings of the Seventh International Conference on Weblogs and Social Media, pp. 573–582 (2013)
23. Zhang, W., Gelernter, J.: Geocoding location expressions in Twitter messages: a preference learning method. J. Spat. Inf. Sci. **2014**(9), 37–70 (2014)
24. Zhang, Y., Szabo, C., Sheng, Q.Z.: Sense and focus: towards effective location inference and event detection on Twitter. In: Wang, J., Cellary, W., Wang, D., Wang, H., Chen, S.-C., Li, T., Zhang, Y. (eds.) WISE 2015. LNCS, vol. 9418, pp. 463–477. Springer, Cham (2015). https://doi.org/10.1007/978-3-319-26190-4_31

DyPerm: Maximizing Permanence
for Dynamic Community Detection

Prerna Agarwal[1], Richa Verma[2], Ayush Agarwal[2],
and Tanmoy Chakraborty[2($\boxtimes$)]

[1] IBM Research, New Delhi, India
prernaagarwal@in.ibm.com
[2] IIIT Delhi, New Delhi, India
{richa15054,ayush14029,tanmoy}@iiitd.ac.in

Abstract. In this paper, we propose DyPerm, the first dynamic community detection method which optimizes a novel community scoring metric, called *permanence*. DyPerm incrementally modifies the community structure by updating those communities where the editing of nodes and edges has been performed, keeping the rest of the network unchanged. We present strong theoretical guarantees to show how/why mere updates on the existing community structure lead to permanence maximization in dynamic networks, which in turn decreases the computational complexity drastically. Experiments on both synthetic and six real-world networks with given ground-truth community structure show that DyPerm achieves (on average) 35% gain in accuracy (based on NMI) compared to the best method among four baseline methods. DyPerm also turns out to be 15 times faster than its static counterpart.

Keywords: Incremental algorithm · Dynamic communities
Permanence

1   Introduction

Last one decade has witnessed tremendous advancement in the detection and analysis of community structure (densely connected groups containing homogeneous nodes) in different types of networks [7]. So far, major research has concentrated on detecting communities from static networks [9]. However, today's real-world networks, especially most of the social networks, are not always static – networks such as Facebook, Twitter are evolving heavily and expanding rapidly in terms of both size and complexity over time. This has recently led to turn the research focus from static networks to dynamic networks (where nodes and edges are added/deleted continuously) [5]. The evolving nature of network structure raises several new challenges to traditional community detection methods – on

---

Parts of the work were partially supported by the Infosys Center for AI, IIIT Delhi, India, and the Ramanujan Fellowship, SERB, DST, India.

one hand, the new community structure obtained due to certain changes in the network structure should not be drastically different from that in the previous time-stamp; on the other hand, the algorithm needs to guarantee that the communities has a dynamic adaptability to deal with the dynamic events.

Existing research on dynamic community detection either run static community detection method on different snapshot of the networks [11] and then correlate the community structures in two consecutive time-stamps, or adopt standard community goodness metrics such as modularity [7] and optimize them to obtain final communities [10,15]. In this paper, we propose DyPerm[1], *the first dynamic community detection method that adopts an effective community goodness metric, called* "permanence" [8] *and optimizes it to incrementally detect the community structure.* The benefits of adopting permanence as an optimization function are two-fold: (i) Permanence, being a local vertex-centric metric (as opposed to the global network-centric metrics such as modularity, conductance), allows us to reassign communities to only those nodes whose associated topological structure has changed, and guarantees that the remaining nodes do not affect the optimization. This leads to very low computing complexity in updating the community structure when the network changes dynamically. (ii) Incremental changes in the local portion of the community structure guarantee that the resultant communities are highly correlated with that in the previous time-stamp. We present theoretical justifications why/how mere changes in the community structure lead to maximize permanence.

We experiment with both synthetic and six real-world dynamic networks with known ground-truth community structure. A thorough comparative evaluation with four state-of-the-art baseline methods shows that DyPerm significantly outperforms all the baselines across different networks – DyPerm achieves up to 35% improvement in terms of Normalized Mutual Information (NMI) w.r.t. the best baseline method. Moreover, DyPerm turns out to be extremely fast, achieving up to 15 times speedup w.r.t. its static counterpart. In short, DyPerm is a fast and accurate dynamic community detection method.

2   Related Work

Community detection has been extensively studied in last one decade mostly for the static networks (see [7,9] for comprehensive reviews). However, due to the enormous growth of the network size and the evolving nature of the network structure, people turned their focus from static network to dynamic network. Major research on dynamic community detection can be divided into three categories [5]: (i) **traditional clustering** where a static community detection method is applied to different snapshots of the dynamic networks [11]; (ii) **evolutionary clustering** [6] where clustering at a particular time-stamp should be similar to the clustering of the previous time-stamp and should accurately reflect the data arriving during that time, and (iii) **incremental clustering** [10,15] where given the clustering result of the initial snapshot, it incrementally modifies

---

[1] For code and datasets, please visit https://tinyurl.com/dyperm-code.

clusters based on every occurrence of an event in the network. Modularity [7], a well-studied goodness metric for static communities, has recently been adopted for dynamic community detection [2,3,16]. QCA [12] is another such method which adopts modularity to identify and trace dynamic communities. Shang et al. [14] proposed GreMod which first uses Louvain algorithm [4] to detect the initial community structure, and then applies incremental updating strategies to track the dynamic communities. They further proposed LBTR [13] which uses machine learning classifiers to predict the vertices that need to be inspected for community assignment revision.

3 Methodology

DyPerm (**Dy**namic community Detection by maximizing **Perm**anence) is an incremental method which maximizes a vertex-centric community scoring metric, called *permanence* [8]. In this section, we start by providing a brief idea of permanence, followed by a detailed description of our proposed method.

Permanence: It measures the extent to which a vertex v remains consistent inside a community c based three factors [8]: (i) v's *internal connectedness*, measured by the ratio of its internal neighbors inside c, $I(v)$ to its degree $d(v)$; (ii) v's *cohesiveness*, indicating how connected its internal neighbors are and measured by $C_{in}(v) = \frac{E_{neig}(v)}{\binom{I(v)}{2}}$, the ratio of actual number of connections among its internal neighbors $E_{neig}(v)$ to the total number of possible connections among them; and (iii) v's *external pull*, measured by the maximum number of external connections of v to any of the external communities $E_{max}(v)$. These three factors are suitably combined to obtain permanence of v as follows:

$$Perm(v) = [\frac{I(v)}{E_{max}(v)} \times \frac{1}{d(v)}] - [1 - C_{in}(v)] \qquad (1)$$

Figure 1 shows an illustrative example to calculate permanence of a vertex. If $E_{max} = 0$, then $Perm(v) = \frac{I(v)}{d(v)}$.

Given a network $G(V, E)$ and its community structure C, permanence of the graph is obtained by $Perm(G) = \frac{1}{|V|}\sum_{v \in V} Perm(v)$. $Perm(G)$ always ranges between -1 (indicating weak community structure) to 1 (indicating strong community structure). We chose permanence as our objective function for two reasons: (i) it is a local vertex-centric metric, which enables us to inspect only the changes happened in a local portion of the network, instead of looking at all the changes as a whole, and (ii) it was already shown to outperform many well-studied local and global metrics such as modularity, conductance, SPart, significance etc. on different static networks (see a detailed survey in [7]).

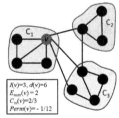

Fig. 1. A toy example showing permanence of vertex v. C_1, C_2 and C_3 are three communities.

Dynamic network: A dynamic network $\mathcal{G}(E, V)$ can be conceptualized by a time evolving process where the underlying network is continuously updated over time by either inserting or removing nodes/edges. Therefore, the atomic events can be of following types:

- $newNode(V \cup u)$: A node u is added to the network. It may or may not have one or more associated edges.
- $removeNode(V \setminus u)$: A node u is removed from the network along with its associated edges.
- $newEdge(E \cup e)$: A new edge e is added between two existing nodes in the network.
- $removeEdge(E \setminus e)$: An existing edge e is removed from the network.

Algorithm 1: NodeAddition
1 **Input:** Node u to be added; Set of edges $E(u,v)$ of node u; Current structure C_t.
2 **Output:** Updated community structure C_{t+1}
3 **if** $E{==}null$;
4     $C_{t+1} \leftarrow C_t \cup \{u\}$;
5 **else**
6     $C_{t+1} \leftarrow C_t \cup \{u\}$;
7   **for each** $e \in E(u,v)$ **do**
8         $C_{t+1} \leftarrow$ EdgeAddition(e, C_t); //Algorithm 3
9         $C_t \leftarrow C_{t+1}$;
10 **return** C_{t+1};

Algorithm 2: NodeDeletion
1 **Input:** Node u to be added; Set of edges $E(u,v)$ of node u; Current structure C_t.
2 **Output:** Updated community structure C_{t+1}
3 **if** $E{==}null$;
4     $C_{t+1} \leftarrow C_t - \cup\{u\}$;
5 **else**
6     $C_{t+1} \leftarrow C_t - \cup\{u\}$;
7   **for each** $e \in E$ **do**
8         $C_{t+1} \leftarrow$ EdgeDeletion(e, C_t); //Algorithm 6
9         $C_t \leftarrow C_{t+1}$;
10 **return** C_{t+1};

Therefore, the dynamic network $\mathcal{G}$ can be expressed as a collection of t static snapshots $\mathcal{G} = \{G_0, G_1, G_2, \cdots, G_t\}$, where $G_{i+1} = G_i \cup \Delta G_i$ indicates the static snapshot of $\mathcal{G}$ at $(i+1)^{th}$ time-stamp. G_{i+1} has evolved from G_i due to ΔG_i change in G_i, where ΔG_i is one of the four atomic events mentioned above.

3.1   The DyPerm Algorithm

DyPerm[2] requires the community structure C_0 (referred as *base community structure*) of the initial snapshot G_0, which can be obtained by running a static community detection method on G_0 or from an oracle who knows the ground-truth community structure of G_0. Depending upon the atomic event which causes the change in the network structure, DyPerm executes one of the following routines in order to maximize permanence:

(A) Addition of a new node: When a new node u is added into the network (i.e., case: $newNode(V \cup u)$), two scenarios may arise (see Algorithm 1):

- **Case A.1:** u does not have any associated edges. It then forms a new singleton community containing only itself.
- **Case A.2:** u has more than one associated edges. Adding u can be approached as inserting edge(s) associated with u, one by one, if the order of edge addition does not affect the final community structure (shown in Proposition 1).

Proposition 1. *The order in which the edges (both intra- and inter-community) associated with a node are inserted, is immaterial for permanence maximization.*

---

[2] See supplementary [1] for the proofs of all the propositions.

(B) Removal of an existing node: When an existing node u present in community C_u is removed (i.e., case: $removeNode(V \setminus u)$), its associated edges are also deleted (see Algorithm 2). Therefore, node removal can be handled by deleting the associated edges, one by one, if the order of edge deletion does not affect the final community structure (see Proposition 2).

Proposition 2. *The order in which the edges (both intra- and inter-community) associated with a node are deleted, is immaterial for permanence maximization.*

(C) Addition of a new edge: Let us consider adding an edge $e_{u,v}$ between two existing nodes u and v. There are two possible cases (See Algorithm 3):

- **Case C.1: Addition of an intra-community edge:** Both u and v belong to the same community C. Accordingly to Proposition 3, addition of $e_{u,v}$ will increase the permanence value of the entire network, and the community will not split into smaller communities (See Algorithm 4).

Proposition 3. *If C is a community in the current snapshot of $\mathcal{G}$, then adding any intra-community edge to C does not split it into smaller communities.*

- **Case C.2: Addition of an inter-community edge:** Let $e_{u,v}$ be the edge connecting communities C_u and C_v. Its presence could make either u or v leave its current community and join the new community (See Algorithm 5). Also, if u or v decides to change its membership, it can advertise its new community to all its neighbors and some of them might eventually want to change their memberships as a consequence. We first move u to its new community and consequently let its internal neighbors (both direct and indirect) determine their best modules to join in, using an algorithm similar to breadth first search. Similar steps are followed for v after moving it to its new community. Overall permanence for both the communities, C_u and C_v are calculated, once before changing the communities of u and v (lines 8–9, Algorithm 5), then after moving u and its neighbors, recursively to C_v. Finally, permanence of the two communities is computed again after moving v and its neighbors, recursively, to C_u (lines 13–23, Algorithm 5). The neighbors of u (and then v) are moved recursively to the other community till the move results in an increase in permanence of that node. The set of the moves that maximizes the overall permanence of the communities is finally accepted to determine the new community structure.

(D) Deletion of an existing edge: Let us consider the deletion of an edge $e_{u,v}$ connecting u and v which are a part of existing network. There are total 3 possible cases (Algorithm 6):

- **Case D.1: Single edge connecting only u and v:** In this case, u is only connected to v, and v is only connected to u. Let $Perm^C(u)$ and $Perm'^C(u)$ be the permanence of u before and after the edge removal respectively. If $E_{max}(u) = 0$ and $d(u) < 3$, permanence of u is calculated as $Perm^C(u) = \frac{I(u)}{d(u)}$

(as mentioned in the beginning of Sect. 3). Therefore, $Perm^C(u) = 1$ and $Perm'^C(u) = 0$, i.e., the permanence value decreases. Similarly, the permanence value of v will also decrease. And u and v form their own singleton communities. This case is handled in Algorithm 8.

- **Case D.2: Node v has unit degree, i.e., $d(v) = 1$:** In this case, v has only one neighbor in the entire network, and u can have more than one neighbors (See Algorithm 8). There are further two sub-cases:

Case D.2.1 u and v belong to two different communities C_u and C_v, respectively: There are further two sub cases as permanence of u is dependent upon $E_{max}(u)$, and it can be due to either C_v or some other community.

Algorithm 3: EdgeAddition

1 **Input:** Edge u, v to be added; Current structure C_t.
2 **Output:** Updated community structure C_{t+1}
3 $C_u \leftarrow Comm(G, u)$;
4 $C_v \leftarrow Comm(G, v)$;
5 **if** $C_u == C_v$ **then**
6    $C_{t+1} \leftarrow$ IntraEdgeAddition(e, C_t) //Algorithm 4
7 **else**
8    $C_{t+1} \leftarrow$ InterEdgeAddition(e, C_t) //Algorithm 5
9 **return** C_{t+1};

Algorithm 4: IntraEdgeAddition

1 **Input:** Graph G (V,E), Edge $e(u, v)$ to be deleted; Current structure C_t.
   Output: Updated community structure C_{t+1}
2 $C_u \leftarrow Comm(G, u)$;
3 $C_v \leftarrow Comm(G, v)$;
4 $G(V, E) \leftarrow G(V, E) + (u, v)$
5 $C_{t+1} \leftarrow C_t$
6 **return** C_{t+1};

Algorithm 5: InterEdgeAddition

1 **Input:** Edge u, v to be added; Current structure C_t.
2 **Output:** Updated community structure C_{t+1}
3 $visited \leftarrow null, queue \leftarrow null$;
4 $C_u \leftarrow Comm(G, u)$;
5 $C_v \leftarrow Comm(G, v)$;
6 $Cu_{new} \leftarrow C_u$;
7 $Cv_{new} \leftarrow C_v$;
8 $PermCu_{old} \leftarrow Perm(C_u)$;
9 $PermCv_{old} \leftarrow Perm(C_v)$;
10 $G(V, E) \leftarrow G(V, E) \cup (u, v)$
11 Append u to $queue$;
12 Add head($queue$) to $visited$;
13 **while** (length($queue$) > 0) **do**
14    $C \leftarrow$ current community of head($queue$);
15   **for** each unvisited node i in $queue$ **do**
16       $visited \leftarrow visited \cup$ head($queue$);
17       $P_1 = \text{Perm}(i)$ if i stays in Cu_{new};
18       $P_2 = \text{Perm}(i)$ if i moves to Cv_{new};
19   **If**($P_2 > P_1$) **do**
20   Move i to Cv_{new};
21   Remove i from Cu_{new};
22   **for** each node q in $queue$ **do**
23   Add unvisited internal neighbors of q to $queue$;
24 **if** $Perm(C_u) + Perm(C_v) > PermCu_{old} + PermCv_{old}$ **do**
25    $C_{t+1} \leftarrow C_t - C_u - C_v \cup Cu_{new} \cup Cv_{new}$;
26 **return** C_{t+1};

Case D.2.1.1: C_v is responsible for $E_{max}(u)$. Here, the new $E_{max}(u)$ i.e., $E'_{max}(u) < E_{max}(u)$ as the edge is deleted and one neighbor goes, while everything else remains constant. The new $Perm^{C_u}(u)$ i.e., $Perm'^{C_u}(u) > Perm^{C_u}(u)$. Permanence of v remains 0 before and after edge deletion as $I(u) = 0$.

Case D.2.1.2: C_v is not responsible for $E_{max}(u)$: In this case, the permanence values of both u and v will increase because the new degree of u i.e., $d'(u) = d(u) - 1$ and the new degree of v i.e., $d'(v) = d(v) - 1$ has decreased. Therefore, permanence increases as everything else remains constant.

Case D.2.2: Both u and v belong to the same community: Let us assume that both u and v belong to community C. $Perm^C(v) = 1$, and after deleting the edge, new permanence i.e., $Perm'^C(v) = 0$. $Perm^C(u) = \frac{I(u)}{E_{max}(u)} \frac{1}{d(u)} - (1 - C_{in}(u))$. The new degree of u becomes $d'(u) = d(u) - 1$ and the new $I(u)$ becomes $I'(u) = I(u) - 1$, therefore, $Perm'^C(u) =$

$\frac{I(u)-1}{E_{max}(u)}\frac{1}{d(u)-1} - (1 - C'_{in}(u))$. And, $C'_{in}(u) < C_{in}(u)$ because $I(u)$ has decreased; therefore $Perm'^{C}(u) < Perm^{C}(u)$. Algorithm 7 handles this case.

- **Case D.3:** u **and** v **belong to communities** C_u **and** C_v, **respectively and degrees of** u **and** v **are greater than 1:** (See Algorithm 8). There are further four sub-cases:

  Case D.3.1: C_v **is responsible for** $E_{max}(u)$, **but** C_u **is not responsible for** $E_{max}(v)$: The new $d(u)$ i.e., $d'(u) < d(u)$ and the new $E_{max}(u)$ i.e., $E'_{max}(u) < E_{max}(u)$ because one edge goes away. $I(u)$ remains the same. Therefore, the new permanence of u i.e., $Perm'^{C_u}(u) > Perm^{C_u}(u)$. Similarly, $Perm'^{C_v}(v) > Perm^{C_v}(v)$.

  Case D.3.2: C_u **is responsible for** $E_{max}(v)$ **but** C_v **is not responsible for** $E_{max}(u)$: The new $D(v)$ i.e., $D'(v) < D(v)$ and the new $E_{max}(v)$ i.e., $E'_{max}(v) < E_{max}(v)$ because one edge goes away. $I(v)$ remains the same. Therefore, the new permanence of v i.e., $Perm'^{C_v}(v) > Perm^{C_v}(v)$. Also, $Perm'^{C_u}(u) > Perm^{C_u}(u)$.

Algorithm 6: EdgeDeletion	**Algorithm 7:** IntraEdgeDeletion
1 **Input:** Graph G (V,E), Edge $e(u, v)$ to be deleted; Current structure C_t.	1 **Input:** Graph G (V,E), Edge u, v to be deleted; Current structure C_t.
Output: Updated community structure C_{t+1}	2 **Output:** Updated community structure C_{t+1}
2 $C_u \leftarrow Comm(G, u)$;	3 $visited \leftarrow null$, $queue \leftarrow null$;
3 $C_v \leftarrow Comm(G, v)$;	4 $C_{uv} \leftarrow Comm(G, u)$
4 **if** $C_u == C_v$ **then**	5 $G(V, E) \leftarrow G(V, E) - (u, v)$
5 $C_{t+1} \leftarrow$ IntraEdgeDeletion(G, e, C_t) //Algorithm 7	6 Append u to queue;
6 **else**	7 Add head$(queue)$ to $visited$;
7 $C_{t+1} \leftarrow$ InterEdgeDeletion(G, e, C_t) //Algorithm 8	8 **while** (length$(queue) > 0$) **do**
8 **return** C_{t+1};	9 **for each** unvisited node i in $queue$ **do**
	10 $C_u \leftarrow C_u \cup$ head$(queue)$;
	11 $visited \leftarrow visited \cup$ head$(queue)$;
Algorithm 8: InterEdgeDeletion	12 **for each** node q in $queue$ **do**
1 **Input:** Graph G (V,E), Edge $e(u, v)$ to be deleted; Current structure C_t.	13 Add unvisited internal neighbors of q to $queue$;
Output: Updated community structure C_{t+1}	14 Remove q;
2 $C_u \leftarrow Comm(G, u)$;	15 Follow step 5 to 13 to obtain C_v;
3 $C_v \leftarrow Comm(G, v)$;	16 **if** $Perm(C_{uv}) < Perm(C_u) + Perm(C_v)$ **do**
4 $G(V, E) \leftarrow G(V, E) - (u, v)$	17 $C_{t+1} = C_t - C_{uv} \cup C_u \cup C_v$;
5 $C_{t+1} \leftarrow C_t$	18 **else**
6 **return** C_{t+1};	19 $C_{t+1} = C_t$;
	20 **return** C_{t+1};

Case D.3.3: C_u **and** C_v **do not influence** $E_{max}(v)$ **and** $E_{max}(u)$, **respectively:** The new $d(u)$ i.e., $d'(u) < d(u)$ and $E_{max}(u)$, $I(u)$ remain the same. Therefore, the new permanence of u i.e., $Perm'^{C_u}(u) > Perm^{C_u}(u)$. Similarly, $Perm'^{C_v}(v) > Perm^{C_v}(v)$.

Case D.3.4: Both C_u and C_v influence $E_{max}(v)$ and $E_{max}(u)$, respectively: The new $E_{max}(u)$ i.e., $E'_{max}(u) < E_{max}(u)$ and the new degree decreases by 1 i.e., $d'(u) < d(u)$. Therefore, the new permanence $Perm'^{C_u}(u) > Perm^{C_u}(u)$. Similarly, the new $E_{max}(v)$ i.e., $E'_{max}(v) < E_{max}(v)$ and $d'(v) < d(v)$. Therefore, $Perm'^{C_v}(v) > Perm^{C_v}(v)$.

Case D.4: Both u **and** v **belong to the same community i.e., intra-community link:** Assume that both u and v belong to community C. After the edge between u and v is deleted, permanence of both the nodes decreases as shown in Proposition 4. Therefore, C may split. Algorithm 7 handles this case.

Proposition 4. *Deleting an intra-community edge between nodes u and v decreases the permanence value of the two nodes.*

The time complexity of DyPerm is $\mathcal{O}(E)$ (see Supplementary [1]).

4   Experimental Results

4.1   Datasets

Synthetic networks: We use the dynamic LFR benchmark model[3]. It allows users to specify different parameters – number of nodes (N), mixing coefficient (μ) which controls the ratio of external neighbors of a node to its degree, the average (k) and maximum degree (k_{max}), and the number of time-stamps s to generate the dynamic network. Here, we vary N from 500 to 3500, s from 10 to 30, μ from 0.10 to 0.80. The default values are considered for the rest of the parameters. However, unless otherwise mentioned, the default LFR network is generated with the following parameter setting: $N = 1000$, $s = 20$, $\mu = 0.2$.

Real-world networks: Four real-world dynamic networks are used whose ground-truth communities are known to us: (i) **Cumulative co-authorship network** (Coauth-C) [8], (ii) **Non-cumulative co-authorship network** (Coauth-N) [8], (iii) **2011 High school dynamic contact networks** (HS-11)[4], (iv) **2012 High school dynamic contact networks** (HS-12)(see footnote 4), (iii) **Primary school contact networks** (PS)(see footnote 4), and (iv) **Contact network in a workplace** (CW)(see footnote 4). Table 1 presents statistics of the datasets.

Table 1. Description of the real-world networks (notation: N (E): # of unique nodes (edges), $\bar{N}$ ($\bar{E}$): avg. # of nodes (edges) per time-stamp, $\bar{C}$: avg. # of communities per time-stamp, s: # of time-stamps).

Network	N ($\bar{N}$)	Node-type	E ($\bar{E}$)	Edge-type	$\bar{C}$	Community-type	s
Coauth-C	708497(41676)	Author	1166376(68610)	Coauthorship	24	Research area	17
Coauth-N	708497(41676)	Author	1166376(68610)	Coauthorship	24	Research area	17
HS'11	126(18)	Student	1710(244)	Contact	3	Class	7
HS'12	180(22.5)	Student	2220(225)	Contact	5	Class	8
PS	242(47)	Student	77602(323)	Contact	5	Class	6
CW	145(18)	Individual	1193(149)	Contact	5	Department	8

---

[3] http://mlg.ucd.ie/snam/.
[4] http://www.sociopatterns.org/.

4.2   Baseline Methods

We use the following state-of-the-art dynamic community detection methods to compare with DyPerm: (i) **Quick Community Adaptation (QCA)**: This framework uses a modularity-based approach for dynamic community detection [12]; (ii) **Fast Community Detection for Dynamic Complex Networks (FCDDCN)**: This is a community detection method for real-time dynamic networks. Modularity is optimized using heuristic search [3]; (iii) **GreMod**: It is an incremental algorithm that performs per-determined actions for every edge change to maximize modularity [14]; (iv) **Learning-based Targeted Revision (LBTR)**: It uses machine learning classifiers to predict the vertices that need to be inspected for community assignment revision [13].

4.3   Comparative Evaluation

We compare the obtained community structure with a given ground-truth structure based on the following metrics: Normalized Mutual Information (NMI) and Adjusted Rand Index (ARI). The value of NMI (*resp.* ARI) ranges from 0 (*resp.* −1) (no match) to 1 (perfect match).

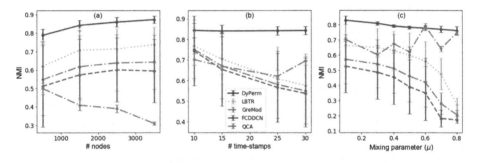

Fig. 2. Accuracy (average NMI and its standard deviation across different time-stamps for each network) of the competing methods with the change of LFR parameters for experimental setup I (similar pattern is observed for ARI, see Supplementary [1]).

Experimental setup I: Running best static community detection method to obtain base communities. In order to obtain the base community structure C_0 for DyPerm, we run MaxPerm (a permanence maximization algorithm for static networks) [8] on the initial snapshot of the network. Since all the baseline methods maximize modularity, we run Louvain algorithm (a modularity maximization algorithm for static network) [4] on the initial snapshot. In each time-stamp, we compare the output of each competing method with the ground-truth and report the average accuracy and the standard deviation.

Figure 2 shows the NMI value (and its standard deviation) of the competing methods with the change in different parameters of the LFR networks

Table 2. Accuracy (avg. NMI and ARI) of the competing methods on the default LFR and real-world networks for experimental setup I. Top results are in bold-face.

Dataset	QCA		LBTR		GreMod		FCDDCN		DyPerm	
	NMI	ARI	NMI	ARI	NMI	ARI	NMI	ARI	NMI	ARI
LFR (default)	0.55	0.41	0.65	0.41	0.57	0.34	0.53	0.36	**0.81**	**0.54**
Coauth-C	0.37	0.03	0.04	0.01	0.05	0.08	0.05	0.03	**0.49**	**0.11**
Coauth-N	0.39	0.04	0.04	0.01	0.04	0.05	0.03	0.02	**0.48**	**0.11**
HS'11	0.39	0.02	0.04	0.06	0.04	0.06	0.04	0.05	**0.59**	**0.13**
HS'12	0.43	0.19	0.02	0.05	0.02	0.05	0.02	0.04	**0.56**	**0.24**
PS	0.39	0.14	0.04	0.02	0.04	0.02	0.04	0.01	**0.53**	**0.25**
CW	0.41	0.01	0.02	0.07	0.03	0.01	0.03	0.03	**0.52**	**0.09**

(see Supplementary [1] for the same plot w.r.t ARI). We observe that the NMI value of DyPerm is consistently higher than those of the baseline methods irrespective of any LFR parameters. DyPerm outperforms the best baseline method (LBTR) by 20.6%, 26.74%, 35.75% on average with the increase of the number of nodes, time-stamps and μ respectively, which is significant according to the t-test with 95% confidence interval. The standard deviation of DyPerm is also less compared to that of LBTR, indicating that DyPerm is consistent in producing accurate community structure across different time-stamps of a network. Table 2 shows the results of the competing methods on the default LFR and real-world networks. Once again, we observe a significant gain in the performance of DyPerm compared to the other baselines, specially on the real-world networks. QCA turns out to be the bast baseline method for real-world networks. DyPerm outperforms QCA by 35.20% and 275.4% in terms of NMI and ARI respectively, averaged across all the networks.

Experimental setup II: Using ground-truth community structure as base communities. We would like to reiterate that all the dynamic community detection methods are highly dependent on the base community structure. The noise in the detection of the base community structure may propagate to the next stage of the algorithm and affect the overall performance. Therefore, one may argue that the baseline methods seem to be incompetent (as observed in Table 2) due to the inefficiency of the static community detection method applied on the initial snapshot, not due to the problem in dynamic community detection method itself. To verify this argument further, we use the ground-truth community structure of the initial network as the base community structure. This ensures that the base community structure is completely accurate. Following this, we run each competing dynamic method on the remaining snapshots and measure the accuracy.

Figure 3 shows the NMI value (and its standard deviation) of the competing methods with the change in different LFR parameters (see Supplementary [1] for the same plot w.r.t. ARI). Once again, we observe similar pattern – DyPerm

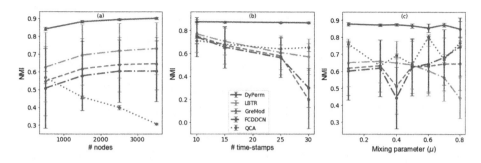

Fig. 3. Accuracy (average NMI and its standard deviation across different time-stamps for each network) of the competing methods with the change of LFR parameters for experimental setup II (similar pattern is observed for ARI, see Supplementary [1]).

significantly outperforms all other baseline methods. However, here both LBTR and QCA seem to be quite competitive. DyPerm beats LBTR by 27.11%, 30.88% and 45.03%, and QCA by 111.5%, 29.17% and 23.89% with the increase of the number of nodes, time-stamps and μ respectively, averaged over all the time-stamps. Table 3 shows the accuracy of the competing methods on the LFR and different real-world networks for experimental setup II. We again observe a significant improvement of the performance of DyPerm compared to the baselines. This implies that irrespective of the community detection method used on the initial snapshot of the network, our method always outperforms other baselines.

Interesting, while comparing Tables 2 and 3, we notice that the performance of the baselines does not improve much considering ground-truth as base community structure, specially for the real-world networks. However, DyPerm seems to achieve a significance performance gain in most cases – 4.41% and 1.23% in terms of NMI and ARI, averaged over all the datasets. This implies that with a better initialization, DyPerm can achieve even better performance.

Table 3. Accuracy (avg. NMI and ARI) of the competing methods on the default LFR and real-world networks for experimental setup II. Top results are in bold-face.

Dataset	QCA		LBTR		GreMod		FCDDCN		DyPerm	
	NMI	ARI	NMI	ARI	NMI	ARI	NMI	ARI	NMI	ARI
LFR (default)	0.57	0.47	0.76	0.55	0.62	0.28	0.55	0.41	**0.87**	**0.58**
Coauth-C	0.13	0.03	0.03	0.02	0.04	0.08	0.05	0.05	**0.53**	**0.12**
Coau-N	0.10	0.05	0.02	0.04	0.04	0.01	0.05	0.01	**0.53**	**0.12**
HS'11	0.08	0.05	0.04	0.06	0.04	0.06	0.03	0.05	**0.52**	**0.10**
HS'12	0.02	0.09	0.02	0.05	0.03	0.02	0.03	0. 03	**0.60**	**0.22**
PS	0.03	0.08	0.04	0.02	0.04	0.02	0.04	0.02	**0.53**	**0.23**
CW	0.03	0.02	0.02	0.03	0.04	0.06	0.05	0.04	**0.57**	**0.09**

Table 4. (a) Runtime (in minutes) of MaxPerm and DyPerm for different real-world networks. (b) Time complexity of the competing methods (N: # of nodes, E: # of edges, d: avg. degree of nodes).

(a)

Method	Runtime (in minutes) for different datasets					
	Coauth-C	Coauth-N	HS'11	HS'12	PS	CW
MaxPerm	3,420	3,020	180	192	17	40
DyPerm	300	204	45	48	1.5	5

(b)

QCA	LBTR	GreMod	FCDDCN	DyPerm
$\mathcal{O}(E^2)$	$\mathcal{O}(E)$	$\mathcal{O}(E)$	$\mathcal{O}(Ed\log N)$	$\mathcal{O}(E)$

4.4   Run-Time Analysis

Table 4(a) reports the runtime of MaxPerm and DyPerm, the static and dynamic community detection methods which maximize permanence, respectively. DyPerm seems to be 10 times faster than MaxPerm, averaged over all the real-world datasets. Maximum gain (15 times faster) is observed on Coauthor-N network. This result provides enough motivation to design an efficient dynamic community detection method. Note that we can not compare the runtime of other competing methods as the source codes were written in different languages. The theoretical time complexity of these methods is compared in Table 4(b).

5   Conclusion

In this paper, we proposed DyPerm, a novel dynamic community detection method that maximizes permanence (a local community scoring metric) in every snapshot of the network to detect the community structure. DyPerm significantly outperformed four state-of-the-art baselines on both synthetic and real-world networks – we observed a gain in NMI up to 35% compared to the best baseline method. Moreover, DyPerm turned out to be extremely faster than its static counterpart (MaxPerm), achieving up to 15 times speedup.

References

1. Anonymized supplementary (2018). https://tinyurl.com/dyperm-si
2. Aktunc, R., Toroslu, I.H., Ozer, M., Davulcu, H.: A dynamic modularity based community detection algorithm for large-scale networks: DSLM. In: ASONAM, pp. 1177–1183 (2015)
3. Bansal, S., Bhowmick, S., Paymal, P.: Fast community detection for dynamic complex networks. In: da F. Costa, L., Evsukoff, A., Mangioni, G., Menezes, R. (eds.) CompleNet 2010. CCIS, vol. 116, pp. 196–207. Springer, Heidelberg (2011). https://doi.org/10.1007/978-3-642-25501-4_20

4. Blondel, V.D., Guillaume, J.L., Lambiotte, R., Lefebvre, E.: Fast unfolding of communities in large networks. J. Stat. Mech. Theory Exp. **2008**(10), P10008 (2008)
5. Cazabet, R., Amblard, F.: Dynamic community detection. In: Alhajj, R., Rokne, J. (eds.) Encyclopedia of Social Network Analysis and Mining, pp. 404–414. Springer, New York (2014). https://doi.org/10.1007/978-1-4614-6170-8
6. Chakrabarti, D., Kumar, R., Tomkins, A.: Evolutionary clustering. In: SIGKDD, pp. 554–560 (2006)
7. Chakraborty, T., Dalmia, A., Mukherjee, A., Ganguly, N.: Metrics for community analysis: a survey. ACM Comput. Surv. **50**(4), 54:1–54:37 (2017)
8. Chakraborty, T., Srinivasan, S., Ganguly, N., Mukherjee, A., Bhowmick, S.: On the permanence of vertices in network communities. In: SIGKDD, pp. 1396–1405 (2014)
9. Fortunato, S.: Community detection in graphs. Phys. Rep. **486**(3–5), 75–174 (2010)
10. Li, X., Wu, B., Guo, Q., Zeng, X., Shi, C.: Dynamic community detection algorithm based on incremental identification. In: ICDMW, pp. 900–907 (2015)
11. Mitra, B., Tabourier, L., Roth, C.: Intrinsically dynamic network communities. Comput. Netw. **56**(3), 1041–1053 (2012)
12. Nguyen, N.P., Dinh, T.N., Xuan, Y., Thai, M.T.: Adaptive algorithms for detecting community structure in dynamic social networks. In: INFOCOM, pp. 2282–2290 (2011)
13. Shang, J., Liu, L., Li, X., Xie, F., Wu, C.: Targeted revision: a learning-based approach for incremental community detection in dynamic networks. Phys. A Stat. Mech. Appl. **443**, 70–85 (2016)
14. Shang, J., Liu, L., Xie, F., Chen, Z., Miao, J., Fang, X., Wu, C.: A real-time detecting algorithm for tracking community structure of dynamic networks. CoRR abs/1407.2683 (2014)
15. Xie, J., Chen, M., Szymanski, B.K.: Labelrankt: incremental community detection in dynamic networks via label propagation. CoRR abs/1305.2006 (2013)
16. Zhuang, D.: Modularity-based dynamic community detection. CoRR abs/1709.08350 (2017)

Mining User Behavioral Rules
from Smartphone Data Through
Association Analysis

Iqbal H. Sarker[1(✉)] and Flora D. Salim[2]

[1] Department of Computer Science and Software Engineering,
School of Software and Electrical Engineering,
Swinburne University of Technology, Melbourne, Australia
`msarker@swin.edu.au`
[2] School of Science (CS and IT), RMIT University, Melbourne, Australia
`flora.salim@rmit.edu.au`

Abstract. The increasing popularity of smart mobile phones and their powerful sensing capabilities have enabled the collection of rich contextual information and mobile phone usage records through the device logs. This paper formulates the problem of mining *behavioral association rules* of individual mobile phone users utilizing their smartphone data. Association rule learning is the most popular technique to discover rules utilizing large datasets. However, it is well-known that a large proportion of association rules generated are *redundant*. This redundant production makes not only the rule-set unnecessarily large but also makes the decision making process more complex and ineffective. In this paper, we propose an approach that effectively identifies the *redundancy* in associations and extracts a concise set of *behavioral association rules* that are non-redundant. The effectiveness of the proposed approach is examined by considering the real mobile phone datasets of individual users.

Keywords: Mobile data mining · Association rule mining
Non-redundancy · Contexts · User behavior modeling

1 Introduction

Now-a-days, mobile phones have become part of our life. The number of mobile cellular subscriptions is almost equal to the number of people on the planet [12]. The phones are, for most of the day, with their owners as they go through their daily routines. People use smart mobile phones for various activities such as voice communication, Internet browsing, apps using, e-mail, online social network, instant messaging etc. [12].

The sensing capabilities of smart mobile phones have enabled the collection of rich contextual information and mobile phone usage records through the device logs [20]. These are phone call logs [15], app usages logs [18], mobile notification logs [11], web logs [8], context logs [20] etc. The discovered behavioral association

© Springer International Publishing AG, part of Springer Nature 2018
D. Phung et al. (Eds.): PAKDD 2018, LNAI 10937, pp. 450–461, 2018.
https://doi.org/10.1007/978-3-319-93034-3_36

rules from such mobile phone data, can be used for building the adaptive, intelligent and context-aware personalized systems, such as smart interruption management system, intelligent mobile recommender system, context-aware smart searching, and various predictive services, in order to assist them intelligently in their daily activities in a context-aware pervasive computing environment.

In this paper, we mainly focus on *mining individual's phone call behavior* (*Accept|Reject|Missed|Outgoing*) utilizing their phone log data. In the real-world, mobile phone users' behaviors are not identical to all. Individual user may behave differently in different contexts. Let's consider a smart phone call handling service, a mobile phone user typically 'rejects' the incoming phone calls, if s/he is in a 'meeting'; however, 'accepts' if the call comes from his/her 'boss'. Hence, [reject, accept] are the user phone call behaviors, and [meeting, boss] are the associated contexts that have a strong influence on users to make decisions. Context is defined as *"any information that can be used to characterize the situation of a user"*, such as temporal (e.g., day, time), social activity or situation (e.g., meeting), location (e.g., office), social relationship (e.g., boss) etc. In this work, we aim to extract a concise set of *behavioral association rules* that are *non-redundant*, expressing an individual's phone call behavior in such multi-dimensional contexts for a particular confidence threshold preferred by individuals. The setting of this threshold for creating rules will vary according to an *individual's preference* as to how interventionist they want the call handling agent to be. Let's consider an example, one person may want the agent to reject calls where in the past he/she has rejected calls more than, say, 95% of the time - that is, at a threshold of 95%. Another individual, on the other hand, may only want the agent to intervene if he/she has rejected calls in, say, 80% of past instances. Such preferences may vary from user-to-user in the real world.

In the area of mobile data mining, association rule learning [2] is the most common techniques to discover rules of mobile phone users. In particular, a number of researchers [11,18,20] have used association rule learning to mine rules capturing mobile phone users' behavior for various purposes. However, the drawback is - Association rule learning technique discovers all associations of contexts in the dataset that satisfy the user specified minimum support and minimum confidence constraints. As a result, it produces a huge number of *redundant rules* (affects the quality and usefulness of the rules) because of considering all possible combinations of contexts without any intelligence. According to [5], association rule learning technique produces up to 83% redundant rules that makes the rule-set unnecessarily large. Therefore, it is very difficult for the decision making agents to determine the most interesting ones and consequently makes the decision making process ineffective and more complex.

In this paper, we address the above mentioned issues and propose an approach that effectively identifies the *redundancy* in associations and extracts a concise set of *behavioral association rules* that are *non-redundant*. In our approach, we first design an association generation tree, in which each branch denotes a test on a specific context value determining according to the precedence of contexts, and each corresponding node either interior or leaf represents the outcome, including the identified 'REDUNDANT' nodes, for the test. Once the tree has been generated, we extract rules by traversing the tree from root node to each rule producing node that satisfies the user preferred confidence threshold.

The contributions are summarized as follows:

- We effectively identify the *redundancy in associations* while producing rules rather than in post-processing.
- We propose an approach that extracts a concise set of *behavioral association rules* that are *non-redundant*.
- We have conducted experiments on real mobile phone datasets to show the effectiveness of our approach comparing with traditional association rule learning algorithm.

The rest of the paper is organized as follows. Section 2 reviews the background of association rule learning techniques. We discuss the redundancy in associations in Sect. 3. Section 4 presents our approach. We report the experimental results in Sect. 5. Finally, Sect. 6 concludes this paper highlighting the future work.

2   Association Rules: A Background

Association rule mining is one of the most important and well researched techniques in data mining. In this section, we introduce some basic and classic approaches for association rule mining. An association rule is an implication in the form of $A \Rightarrow C$, where, A is called antecedent while C is called consequent, the rule means A implies C.

The AIS algorithm, proposed by Agrawal et al. [1], is the first algorithm designed for association rule mining. The main drawback of the AIS algorithm is too many candidate itemsets that finally turned out to be small are generated, which requires more space and wastes much effort that turned out to be useless. At the same time this algorithm requires too many passes over the whole database. The SETM algorithm proposed by [10] exhibits good performance and stable behavior, with execution time almost insensitive to the chosen minimum support but has the same disadvantage of the AIS algorithm.

Apriori, Aprioiri-TID and Apriori-Hybrid algorithms are proposed by Agrawal in [2]. The performance is these algorithms are better than AIS and SETM. The Apriori algorithm takes advantage of the fact that any subset of a frequent itemset is also a frequent itemset. The algorithm can therefore, reduce the number of candidates being considered by only exploring the itemsets whose support count is greater than the minimum support count. All infrequent itemsets can be pruned if it has an infrequent subset. Apriori-TID and Apriori-Hybrid are designed based on Apriori algorithm. Another algorithm Predictive Apriori proposed by Scheffer [17] generates rules by predicting accuracy combining from support and confidence. So sometimes it produced the rules with large support but low confidence and got unexpected results.

Han et al. [9] have designed a tree based rule mining algorithm FP-Tree. However, FP-Tree is difficult to be used in an interactive mining system. During the interactive mining process, users may change the threshold of support according to the rules. The changing of support may lead to repetition of the whole mining process. Das et al. [3] have designed another tree based association rule mining method RARM that uses the tree structure to represent the original database and avoids candidate generation process. RARM is claimed to be much faster than FP-Tree algorithm but also faces the same problem of FP-tree [19]. Flach et. al [7] introduces an approach with learning first-order logic rules. This algorithm is able to deal with explicit negation. However, this algorithm can not learn rules in case of depth search.

Among the association rule mining algorithms, Apriori [2] is a great improvement in the history of association rule mining [19]. This is the most popular and common algorithm for mining association rules. The key strength of association rule mining is it's completeness. It finds all associations in the data that satisfy the user specified constraints. However, the main drawback is that - it produces a huge number of *redundant* associations, that makes the behavior modeling approach ineffective for mobile phone users.

Unlike these works, in this paper, we propose an approach that effectively identifies the *redundancy* in associations and extracts a concise set of *behavioral association rules* that are *non-redundant* for individual mobile phone users utilizing their mobile phone data.

3   Redundancy in Association Rules

Association rule learning algorithms produce many rules $(A \Rightarrow C)$ that have common consequent (C) 'behavior' but different antecedent (A) 'contexts'. Indeed many of those antecedent contexts are proper subset of others rules.

Let, two rules $R_1 : A_1 \Rightarrow C_1$ and $R_2 : A_2 \Rightarrow C_2$, we call the latter one redundant with the former one if $A_1 \subseteq A_2$ and $C_1 = C_2$. From this definition of redundancy, if we have a general rule $R_g : A_1 \Rightarrow C_1$ and there is no other more specific rule $A_1 B_1 \Rightarrow C_2$ in existence such that confidence of $A_1 B_1 \Rightarrow C_2$ is equal or larger than the confidence of $R_g : A_1 \Rightarrow C_1$ and $A_1 \subseteq A_1 B_1$, $C_1 = C_2$, then the rule $A_1 B_1 \Rightarrow C_2$ is said to be non-redundant with $R_g : A_1 \Rightarrow C_1$.

For example, typically a user rejects most of the incoming calls (83%), when she is in a meeting, i.g., the rule is $(meeting \Rightarrow reject)$ [say, user preferred confidence threshold 80%]. Another example is, the user rejects most of the incoming calls (90%) of her friends, when she is in a meeting, i.g., the rule is $(meeting, friend \Rightarrow reject)$. Both rules are valid in terms of confidence as the rules satisfy the user preferred confidence threshold. However, the later one is considered as redundant rule as the former one is able to take the same decision with minimal number of contexts. Additional context can play a significant role if it reflects different behavior. Table 1 shows an example of a set of association rules and their non-redundant production for a preferred minimum confidence 80%. According to Table 1 R_2, R_3, R_4, R_5 are redundant rules as only R_1 is able to take the same decision with minimal number of contexts. On the other hand, R_1 and R_6 are considered as non-redundant rule, in which we are interested in.

Table 1. Sample traditional association rules and corresponding non-redundant behavioral association rules of a sample user.

Association rules (Traditional)	Association rules (Non-redundant)
$R_1 : Meeting \Rightarrow Reject$	
$(conf = 83\%)$	
$R_2 : Meeting, Friend \Rightarrow Reject$	
$(conf = 90\%)$	
$R_3 : Meeting, Colleague \Rightarrow Reject$	$R_1 : Meeting \Rightarrow Reject$
$(conf = 88\%)$	$(conf = 83\%)$
$R_4 : Meeting, Friend, Monday[t1] \Rightarrow Reject$	$R_6 : Meeting, Boss \Rightarrow Accept$
$(conf = 100\%)$	$(conf = 100\%)$
$R_5 : Meeting, Colleague, Friday[t2] \Rightarrow Reject$	
$(conf = 98\%)$	
$R_6 : Meeting, Boss \Rightarrow Accept$	
$(conf = 100\%)$	

4 Our Approach

In this section, we present our approach for mining behavioral association rules of individual mobile user behavior utilizing their mobile phone data.

4.1 Association Generation Tree (AGT)

In this first step, we generate a tree based on multi-dimensional contexts and corresponding usage behavior of mobile phone users. As different contexts might have differing impacts in behavioral rules, we identify the precedence of contexts in a dataset while generating the tree.

Identifying the Precedence of Contexts: In order to identify the precedence of contexts in a dataset, we calculate information gain which is a statistical property that measures how well a given attribute separates training examples into targeted behavior classes. The one with the highest information is considered as the highest precedence context. In order to define information gain precisely, we need to define entropy first.

Entropy is a measure of disorder or impurity. The entropy characterizes the impurity of an arbitrary collection of examples. It reaches it's maximum when the uncertainty is at a maximum and vice-versa. Formally entropy is defined as [13]:

$$H(S) = - \sum_{x \in X} p(x) log_2 p(x)$$

Where, S is the current data set for which entropy is being calculated, X represents a set of classes in S, p(x) is the proportion of the number of elements in class x to the number of elements in set S.

Information gain (IG) measures how much "information" a feature gives us about the class. It is the expected reduction in entropy caused by partitioning the examples according to a given attribute. Features that perfectly partition should give maximal information. Unrelated features should give no information. To decide which attribute should be tested first, we find the one with the highest information gain. The formal definition of information gain is [13].

$$IG(A, S) = H(S) - \sum_{t \in T} p(t) H(t)$$

Where, H(S) is the entropy of set S, T represents the subsets created from splitting set S by attribute A such that $S = \cup_{t \in T} t$, p(t) is the proportion of the number of elements in t to the number of elements in set S, H(t) is the entropy of subset t.

Let's consider a sample dataset of three different contexts and corresponding call response behavior of a mobile phone user X. For example, the contexts might be ranked as follows:

$Rank1 : Social\ Activity/Situation(S) \in \{meeting, lecture, lunch\}$
$Rank2 : Social\ Relationship(R) \in \{boss, colleague, friend, unknown\}$
$Rank3 : Temporal(T) \in \{time\text{-}of\text{-}the\text{-}week\}$
Where,
$User\ phone\ call\ behavior(BH) \in \{Accept, Reject, Missed\}$

Tree Generating Procedure and Extracting Non-redundant Rules: A tree is a structure that includes a root node, branches, interior and/or leaf nodes [6]. Each branch denotes a test on a specific context value, and each node (interior or leaf) denotes the outcome containing the behavior class with confidence value of the test.

To build tree, we follow a top-down approach, starting from a root node. The tree is partitioned into classes distinguished by the values of the most relevant context according to the precedence. Once the root node of the tree has been determined, the child nodes and it's arcs are created and added to the tree with the associated contexts and corresponding dominant (highest occurrences) [16] behavior with confidence value. While creating a node, we check whether it is redundant ('REDUNDANT' node) or not.

"A child node in the tree is called 'REDUNDANT' node, if both the child node and it's parent node contain same behavior class and satisfy individual's preferred confidence threshold".

The algorithm recursively add new subtrees to each branching arc by adding child node one by one. If a node has 100% (maximum) confidence then there is no need to elaborate it's children, otherwise we continue this process according to the number of contexts in the datasets. The final result is a multi-level tree with various nodes including 'REDUNDANT' node according to their associated contexts. The overall process for constructing the tree is set out in Algorithm 1.

Algorithm 1. Association Generation Tree

Data: Dataset: $DS = X_1, X_2, ..., X_n$ // each instance X_i contains a number of nominal context-values and corresponding behavior class BH, confidence threshold $= t$

Result: An association generation tree

1 <u>Procedure AGT</u> $(DS, context\_list, BHs)$;
2 $N \leftarrow createNode()$ //create a root node for the tree
3 **if** *all instances in DS belong to the same behavior class BH* **then**
4 $\quad\mid$ return N as a leaf node labeled BH with 100% confidence.
5 **end**
6 **if** *context _list is empty* **then**
7 $\quad\mid$ return N as a leaf node labeled with the dominant behavior class and corresponding confidence value.
8 **end**
9 identify the highest precedence context C_{split} for splitting and assign C_{split} to the node N.
10 **foreach** *context value val* $\in C_{split}$ **do**
11 $\quad\mid$ create subset DS_{sub} of DS containing val.
12 $\quad\mid$ **if** $DS_{sub} \neq \phi$ **then**
13 $\quad\mid\quad\mid$ identify the dominant behavior and calculate the confidence value.
14 $\quad\mid\quad\mid$ create a child node with the identified dominant behavior.
15 $\quad\mid\quad\mid$ //check with it's parent node
16 $\quad\mid\quad\mid$ **if** *both nodes satisfy the confidence threshold* **then**
17 $\quad\mid\quad\mid\quad\mid$ **if** *both nodes represent same behavior class* **then**
18 $\quad\mid\quad\mid\quad\mid\quad\mid$ mark the child node as 'REDUNDANT' node.
19 $\quad\mid\quad\mid\quad\mid$ **end**
20 $\quad\mid\quad\mid$ **end**
21 $\quad\mid\quad\mid$ add a subtree with new node and associated context values.
22 $\quad\mid\quad\mid$ //recursively do this with remaining contexts
23 $\quad\mid\quad\mid$ $AGT(DS_{sub}, \{context\_list - C_{split}\}, BHs))$
24 $\quad\mid$ **end**
25 **end**
26 return N

Figure 1 shows an example of such an association generation tree containing 'REDUNDANT' nodes for the contexts (mentioned above) in phone call behaviors of a user, when the minimum confidence preference is 80%.

Once the tree has been generated, rules are extracted by traversing the tree. To do this, we first identify the valid rule generating nodes from the tree. *A node is taken into account as a valid rule generating node if it satisfies individual's preferred confidence threshold and not classified as 'REDUNDANT' node.* The followings are examples of produced behavioral association rules from the tree.

$R_1 : Lecture \Rightarrow Reject$ (conf $= 100\%$, using Node 2)
$R_2 : Meeting \Rightarrow Reject$ (conf $= 85\%$, using Node 3)
$R_3 : Lunch, Friend \Rightarrow Accept$ (conf $= 92\%$, using Node 4)
$R_4 : Lunch, Unknown \Rightarrow Missed$ (conf $= 95\%$, using Node 5)
$R_5 : Meeting, Boss \Rightarrow Accept$ (conf $= 100\%$, using Node 7)

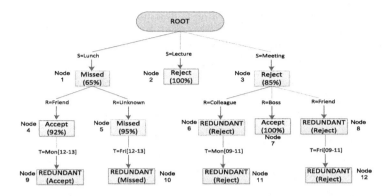

Fig. 1. An example of the tree (AGT) identifying 'REDUNDANT' nodes

Rule R_1 states that the user always rejects the incoming calls (100%) when she is in a lecture, which is produced from node 2 in the tree. Similarly, the other non-redundant rules R_2, R_3, R_4, R_5 are produced from node 3,4,5, and 7 respectively according to the tree shown in Fig. 1.

5   Experiments

In this section, we have conducted experiments on four individual mobile phone users' datasets that consist the phone call records in different contexts. We have implemented our approach in Java programming language and compare the output with the most popular association rule learning technique Apriori [2].

5.1   Dataset

We randomly select four individual mobile phone users' datasets from Massachusetts Institute of Technology (MIT) Reality Mining dataset [4]). These datasets contain three types of phone call behavior, e.g., incoming, missed and outgoing. As can be seen in the dataset, the user's behavior in accepting and rejecting calls are not directly distinguishable in incoming calls in the dataset. As such, we derive accept and reject calls by using the call duration. If the call duration is greater than 0 then the call has been accepted; if it is equal to 0 then the call has been rejected [14]. The contextual information includes temporal, locational, and social. We also pre-process the temporal data in mobile phone log as it represents continuous time-series with numeric timestamps values (YYYY:MM:DD hh:mm:ss). For this, we use BOTS technique [16] for producing behavior-oriented time segments, such as Friday[09:00–11:00], Monday[12:00–13:00] etc. Table 2 describes each dataset of the individual mobile phone user.

Table 2. Datasets descriptions

Datasets	Data collection period	Instances
Dataset-01	5 months	5119
Dataset-02	3 months	1229
Dataset-03	4 months	3255
Dataset-04	4 months	2096

5.2   Evaluation Results

Effect of Confidence: In this experiment, we show the effect of confidence on producing behavioral association rules using both approaches. For this, we first illustrate the detailed outcomes by varying the conference threshold from 100% (maximum) below to 60% (lowest) for different datasets. Since confidence is directly associated to the *accuracy of rules*, we are not interested to take into account below 60% as confidence threshold. To show the effect of confidence, Figs. 2 and 3 show the comparison of rule production for different confidence thresholds (accuracy level) for different datasets.

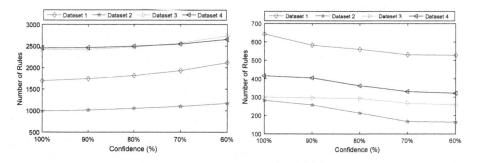

Fig. 2. Effect of confidence in "Apriori"   **Fig. 3.** Effect of confidence in our app-roach

If we observe Fig. 2, we see that the produced number of association rules using existing Apriori algorithm [2] increases with the decrease of confidence threshold. The reason is that it simply takes into account all combination of contexts while producing rules. Thus, for a lower confidence value, it satisfies more associations, and as a result, the output becomes larger. On the other hand, the produced number of behavioral association rules using our technique decreases with the decrease of confidence threshold, shown in Fig. 3. The main reason is that - for lower confidence threshold, more number of child nodes subsume in their parent node because of creating generalized nodes with the dominant behavior, and as a result, the number of produced rules decreases.

Effectiveness Analysis: To show the effectiveness of our approach, Figs. 4, 5, 6 and 7 show the relative comparison of produced number of rules for dataset-01, dataset-02, dataset-03 and dataset-04 respectively. For each approach, we use minimum support 1 (one instance) because no rules can be produced below this support [14]. Moreover, we have explored different confidence threshold, i.e., 100% (maximum) below to 60%.

From Figs. 4, 5, 6 and 7, we find that our approach significantly reduces the number of extracted rules comparing with traditional association rule learning algorithm for different confidence thresholds. The main reason is that existing approach Apriori [2] does not take into account redundancy analysis while producing rules and makes the rule-set unnecessarily larger. On the other-hand, we identify and eliminate the redundancy while producing rules and discovers only the non-redundant behavioral association rules. As a result, it significantly reduces the number of rules for a particular confidence threshold for each dataset.

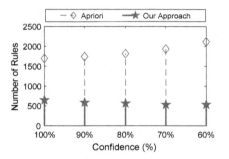

Fig. 4. Utilizing dataset 01

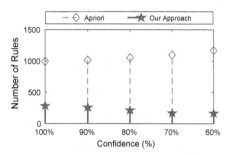

Fig. 5. Utilizing dataset 02

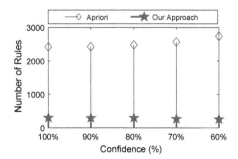

Fig. 6. Utilizing dataset 03

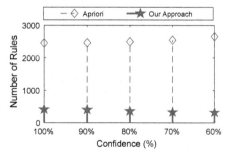

Fig. 7. Utilizing dataset 04

6   Conclusion and Future Work

In this paper, we have presented an approach to effectively identify the redundancy in association rules and to extract a concise set of behavioral association rules which are non-redundant, in order to model phone call behavior of individual mobile phone users. Although we choose phone call contexts as examples, our approach is also applicable to other application domains. We believe that our approach opens a promising path for future research on extracting behavioral association rules of mobile phone users.

In future work, we plan to conduct a range of experiments using additional mobile phone datasets and to use the discovered non-redundant rules in various predictive services. We have also a plan regarding efficiency analysis of our approach to use in real-time applications, in order to provide the personalized services for the end mobile phone users.

References

1. Agrawal, R., Imieliński, T., Swami, A.: Mining association rules between sets of items in large databases. In: ACM SIGMOD Record, vol. 22, pp. 207–216. ACM (1993)
2. Agrawal, R., Srikant, R.: Fast algorithms for mining association rules. In: Proceedings of the International Joint Conference on Very Large Data Bases, Santiago Chile, pp. 487–499, vol. 1215 (1994)
3. Das, A., Ng, W.K., Woon, Y.K.: Rapid association rule mining. In: Proceedings of the Tenth International Conference on Information and knowledge management, pp. 474–481. ACM (2001)
4. Eagle, N., Pentland, A., Lazer, D.: Infering social network structure using mobile phone data. In: Proceedings of National Academy of Sciences (2006)
5. Fournier-Viger, P., Tseng, V.S.: Mining Top-K non-redundant association rules. In: Chen, L., Felfernig, A., Liu, J., Raś, Z.W. (eds.) ISMIS 2012. LNCS (LNAI), vol. 7661, pp. 31–40. Springer, Heidelberg (2012). https://doi.org/10.1007/978-3-642-34624-8_4
6. Sarker, I.H., et al.: An approach to modeling call response behavior on mobile phones based on multi-dimensional contexts. In: Mobile Software Engineering and Systems, Argentina (2017)
7. Flach, P.A., Lachiche, N.: Confirmation-guided discovery of first-order rules with tertius. Mach. Learn. **42**(1–2), 61–95 (2001)
8. Halvey, M., Keane, M.T., Smyth, B.: Time based segmentation of log data for user navigation prediction in personalization. In: Proceedings of the International Conference on Web Intelligence, Compiegne, 19–22 September, pp. 636–640. IEEE Computer Society, Washington, DC (2005)
9. Han, J., Pei, J., Yin, Y.: Mining frequent patterns without candidate generation. In: ACM SIGMOD Record, vol. 29, pp. 1–12. ACM (2000)
10. Houtsma, M., Swami, A.: Set-oriented mining for association rules in relational databases. In: Proceedings of the Eleventh International Conference on Data Engineering, 1995, pp. 25–33. IEEE (1995)

11. Mehrotra, A., Hendley, R., Musolesi, M.: Prefminer: mining user's preferences for intelligent mobile notification management. In: Proceedings of the International Joint Conference on Pervasive and Ubiquitous Computing, Heidelberg, 12–16 September, pp. 1223–1234. ACM, New York (2016)

12. Pejovic, V., Musolesi, M.: Interruptme: designing intelligent prompting mechanisms for pervasive applications. In: Proceedings of the International Joint Conference on Pervasive and Ubiquitous Computing, Seattle, 13–17 September, pp. 897–908. ACM, New York (2014)

13. Quinlan, J.R.: C4. 5: Programs for machine learning. Morgan Kaufmann Publishers Inc., San Francisco (1993)

14. Sarker, I.H., Colman, A., Kabir, M.A., Han, J.: Behavior-oriented time segmentation for mining individualized rules of mobile phone users. In: Proceedings of the International Conference on Data Science and Advanced Analytics, Montreal, 17–19 October, pp. 488–497. IEEE Computer Society, Washington, DC (2016)

15. Sarker, I.H., Colman, A., Kabir, M.A., Han, J.: Phone call log as a context source to modeling individual user behavior. In: Proceedings of the 2016 ACM International Joint Conference on Pervasive and Ubiquitous Computing: Adjunct, pp. 630–634. ACM (2016)

16. Sarker, I.H., Colman, A., Kabir, M.A., Han, J.: Individualized time-series segmentation for mining mobile phone user behavior. Comput. J., 1–20 (2017)

17. Scheffer, T.: Finding association rules that trade support optimally against confidence. Intell. Data Anal. 9(4), 381–395 (2005)

18. Srinivasan, V., Moghaddam, S., Mukherji, A.: Mobileminer: mining your frequent patterns on your phone. In: Proceedings of the International Joint Conference on Pervasive and Ubiquitous Computing, Seattle, 13–17 September, pp. 389–400. ACM, New York (2014)

19. Zhao, Q., Bhowmick, S.S.: Association rule mining: a survey. Nanyang Technological University, Singapore (2003)

20. Zhu, H., Chen, E., Xiong, H., Kuifei, Y., Cao, H., Tian, J.: Mining mobile user preferences for personalized context-aware recommendation. ACM Trans. Intell. Syst. Technol. 5(4), 58:1–58:27 (2014)

A Context-Aware Evaluation Method of Driving Behavior

Yikai Zhai[1], Tianyu Wo[1(✉)], Xuelian Lin[1], Zhou Huang[1], and Junyu Chen[2]

[1] Beijing Advanced Innovation Center for Big Data and Brain Computing,
School of Computer Science and Technology, Beihang University, Beijing, China
{zhaiyk,woty,linxl,huangz}@act.buaa.edu.cn
[2] Department of Mechanical Engineering, Tsinghua University, Beijing, China
tsinghuachenjunyu@163.com

Abstract. As Uber-like chauffeured car services become more and more popular, many drivers have joined the market without special training. To ensure the safety and efficiency of transportation services, it is an important task to accurately evaluate the driving performance of individual driver. Most of the existing methods basically depend on the statistic of abnormal driving events extracted from individual vehicles. However, the occurrence of abnormal events can be affected by various factors, such as road conditions, time of day and weather. It can be bias to judge the driver's performance by merely counting the abnormal events without considering the driving context. In this paper, we analyze the influence of driving context over driving behaviors and propose a context-aware evaluation method. Instead of taking all the occurrence of driving events as the same, we adopt the TF-IDF to determine the risk weight of a driving event in a specific driving context. Based on the risk-weighted statistics, we evaluate the driving performance precisely and normalize it using the Z score model. An evaluation system is implemented. We evaluate the effectiveness of our method based on a real dataset with 3-year traces of 1000 drivers. The normalized score determined by our method have a greater correlation (0.611) with the accident records than that of the number of abnormal driving events (0.523).

Keywords: Driving behavior · Driving event · Driving context
Risk weight

1 Introduction

With the improvement of living standards, people's travel demand increases steadily and various driving service has become the indispensable part in people's daily life. At the same time, according to the statistics from the World Health Organization (WHO), traffic accidents become one of the top 10 leading causes of death in the world and the drivers' personal factors are the main reasons of traffic accidents. To improve driving safety and traffic efficiency, the demand of driving behavior evaluation is proposed in many fields. For instance, in the

© Springer International Publishing AG, part of Springer Nature 2018
D. Phung et al. (Eds.): PAKDD 2018, LNAI 10937, pp. 462–474, 2018.
https://doi.org/10.1007/978-3-319-93034-3_37

field of auto insurance [2], many insurance companies provide premiums based on personal driving behavior. For the tailored-taxi companies, such as Uber in America and Didi-taxi in China, driving behavior evaluation can be used as one of the criteria to assess driving skills and performance. In the field of environmental protection, since different driving behavior patterns may have different impact on vehicle fuel consumption and exhaust emissions, researchers try to identify more energy-efficient and environmental friendly driving mode [13]. All these driving behavior evaluation methods are helpful to correct poor driving behavior and improve driving safety and traffic efficiency.

The evaluation methods of driving behavior can be divided into two categories: classifying the driving style and counting the abnormal driving events. Methods in the first category [1,4,6,11,15] classify the drivers using the supervised learning algorithms based on the original vehicle data. Methods in the second category [3,8] give the evaluation of driving performance based on the number of abnormal driving events (e.g. sharp turnings, abrupt deceleration and acceleration), which is detected from the data of moving vehicles. However, it is not easy to get plenty of labeled training data in the first kind of methods. And for the second kind, it relies on simple counts of driving events which exhibit a limited view of drivers' intentions and behaviors.

The existing driving behavior evaluation methods do not take driving context into consideration and may present systematic bias in evaluation. Indeed, during driving, drivers control the vehicle by reacting to various external factors (e.g. traffic condition) and internal factors (e.g. aggressive driving habits). The same driving event may result from drivers' general tendency to take risks or could represent a protective measure in a hazardous situation. For example, a sudden brake can be deduced from either the tendency of aggressive driving style of the driver, or a necessary measure when the red traffic signal lights up suddenly. An effective evaluation method should be able to measure the actual driving skill (internal factors) and eliminate the influence of external factors.

To solve this problem, we present an evaluation method of driving behavior that not only relies on the number of events but also considers the driving context. In order to eliminate the influence of external factors, we analyze the external factors of driving behavior and demonstrate that these factors truly affect driving behavior. We proposed a driving behavior evaluation method, which not only based on the counts of abnormal diving events, but also considering the influence of driving context. The main contributions are summarized as follows:

- We provide an extensive data analysis of the factors which have impact on the driving behavior. Through the analysis of the periodicity, distribution and frequency of all driving events, we determine three significant factors: Time of day, Road segment and Weather (Sect. 4).
- We propose a novel method that applies TF-IDF to integrate driving context into the evaluation of driving behavior (Sects. 5 and 6). The IDF algorithm is an effective method to quantify the risk of different driving context and the TF-IDF algorithm provide a solution to the comprehensive evaluation problem.

– We evaluate the proposed algorithms extensively over three year real driving data (i.e., from1/1/2014–12/31/2016) in Beijing. The correlation coefficient between the driving behavior and the evaluation result increase from 0.523 to 0.611 after integrating the driving context into evaluation. (Sect. 7).

2   Related Work

In this section, we introduce some evaluation works for driving behavior, divided into two categories: classifying driving style and counting abnormal driving events.

Classifying Driving Style. These methods classify the drivers using the supervised learning approach based on the original driving data. The researchers choose some drivers and give different safety levels according to their driving behavior as their labels. Then the safety levels of other drivers are determined by the safety classification model created by SVM, HMM and other algorithms. [4,6,15] propose methods that make use of SVM, AdaBoost and HMM algorithms to create a driving behavior classification model, and finally could determine whether the driving behavior is safe or not. [11] models driving behaviors as car-following and pedal operation patterns with GMM. Several mathematical strategies are presented to analyze collected vehicle data for driver classification in [1].

Counting Abnormal Driving Events. SenseFleet system [3] uses fuzzy logic to detect driving behavior events like acceleration, deceleration and sharp turn based on the data collected by acceleration sensor, magnetic force sensor and GPS. It can provide different abnormal driving events threshold. This system can also score drivers subjectively by considering driving events, weather and timestamp. [8] is based on the vehicle movement data. They explore the relation between driving context and driving event using the real driving data and determine the measurement of road condition and traffic condition. It helps to provide a way for future research on integrating driving context into driving behavior evaluation.

However, it is not easy to get plenty of labeled training data in the classifying method and the counting method rely on simple counts of driving events which do not incorporate the effects of driving context fully.

3   Preliminary

In this section, we describe the evaluation problem and illustrate the dataset, important concepts and notations used in the work.

3.1   Problem Description

This study aims to propose an evaluation method of driving behavior which takes the influence of driving contexts in to consider.

Driving skill of a driver can be revealed from his abnormal driving events. In order to give a comprehensive evaluation, first, we should identify what kind of driving contexts are related to abnormal driving events. Then we should quantify the risk in different driving context and give a normalized score of driving behavior not only relies on the number of events but also the driving context.

The study is based on the fusing heterogeneous data, including all kinds of driving data and the information of different driving context. The output can be used to improve the driving behavior and ensure safety.

Table 1. Summary of notations.

	Terms	Description
t	time	24 h of a day
r	road segment	road segments in Beijing
w	weather	the weather types
E	event set	five negative driving events
e	event	e ∈ events_set
d	driver	a driver
rt	running time	the vehicle running time
pt	passing times	the passing times in a road segment
N	event number	number of events
F	event frequency	event frequency in a driving context
TF	driver's event frequency	event frequency of driver
IDF	inverse event frequency	inverse event frequency in a context
P	performance	driving performance of a driver

3.2   Driving Data and Driving Events

The dataset of driving data is collected in three years (i.e. 01/01/2014–12/31/2016) in Beijing and is offered by one of the largest car rental and car-booking service provider of China. The dataset contains 3.6 billion driving data produced by 1000 drivers at a frequency of 0.2 Hz. Each driving data contains a driver ID, a timestamp, a GPS point, a direction and a velocity to record the vehicle movement data at that moment.

From this dataset, we detect a set of abnormal driving events by a SVM-based multi-class classifier model [5]. These events are common in daily life and easy to cause safety problem. The five types of events we detected are *sudden acceleration, sudden deceleration, rapid swerving, rapid turning* and *fast U-turn*.

3.3   Driving Contexts

In this study, the external factors that affect driving behaviors are collectively called the driving contexts. Driving contexts mainly contain the time period of the day, driving road segment and the weather condition. In order to analyze the impacts of driving contexts toward driving events, we describe the driving contexts and analyze the effects of driving contexts on driving events using the time, location and weather information of driving events.

Time Period. In the factor of time, we divide one day into 24 time periods and the length of each is $1\,\mathrm{h}$. We define the time set of a day as $T = \{0, 1, 2, 3 \ldots 22, 23\}$.

Road Segments. The set of road segments is defined as $R = \{r_1, r_2, r_3, \ldots\}$, where r_i, $i \geq 1$, is a road segment. We use the road network within the fifth Ring Road of Beijing from Shapefile map, which contains 57852 road segments.

Weather Condition. The historical weather records are obtained by simply requesting the weather web [14]. We define the set of weather as $W = \{sunny, rainy, cloudy, haze\}$.

The notations used in this paper are summarized in Table 1.

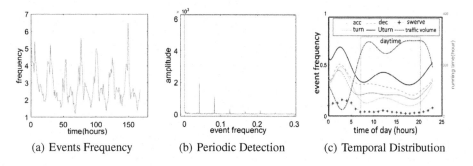

(a) Events Frequency   (b) Periodic Detection   (c) Temporal Distribution

Fig. 1. Impact of driving time.

4 Analysis of Driving Contexts

In this section, we analyze the effects of driving contexts on driving behaviors from three aspects: time period, road segment and weather.

Impact of Time Period. In order to conduct statistical analysis on abnormal driving events in different time periods, we calculate the events frequency per hour during three year (26,280 h). We conduct Fourier Transformation on events frequency to analyze the periodicity characteristics [7,9]. Figure 1(a) shows the frequency of all abnormal driving events in one week (168 h) and Fig. 1(b) shows the result of periodic detection in three years.

According to the analysis of Fourier Transformation, *events frequency has obvious periodicity of 24 h* (the sharp peaks around $0.0416h^{-1}$). So we can take one day as a cycle while analyzing the impacts of different time period on events frequency. The temporal distribution of all kinds of events frequency and traffic volume in a day is shown in Fig. 1(c). *The frequency of each driving event varies among different periods of a day and the frequency of different driving events in each hour of a day still has different distribution.*

Impact of Road Conditions. In order to analyze the influence of road conditions on driving behavior, we investigate the traffic volume and road topological structure with in the Fifth Ring in Beijing. Among the five driving events we have detected, the acceleration and deceleration are related to the traffic volume while the swerving, turning and U-turn are closely related to the road topological structure. Take sudden deceleration and rapid turning as examples and the corresponding thermodynamic diagram of event frequency is shown in Fig. 2.

The figure shows that each kind of driving events distribute unevenly in spatial domain. *Influenced by traffic conditions and topological structures, one road segment would have different effects on different driving events and different events have different high-risk areas.*

Impact of Weather. To verify the impacts of weather conditions to driving events, we calculate the frequency of driving events in all kinds of weather. The event frequency in rainy days is lower. From the perspective of safe driving, lower speed in rainy days than usual and cautious driving would lead to less anomalous events. The frequency in haze days is a little higher because of the low visibility.

Table 2. The correlation between frequency of driving events and weather condition

Weather	All events	Acc	Dec	Swerving	Turing	U-turn
Spearman correlation	.478**	.376**	.477**	.359**	.314**	.314**
Sig. (2-tailed)	.000	.000	.000	.000	.000	.000
N (day)	184	184	184	184	184	184

The spearman correlation coefficients are shown in Table 2. Significant Positive correlation coefficients indicate the positive relationship between weather condition and frequency of driving events. The sudden deceleration has the largest correlation with the weather condition. *The degree of correlation between weather conditions and driving events is different.*

5 Evaluation Method

In this section, we present a method to integrate driving context into driving behavior evaluation. We first introduce the method to measure the risk weights of event in driving context. Then, we give the evaluation method considering driving context.

5.1   Fundamental Idea

According to the analysis in Sect. 4, the number of driving events varies in each driving context among different drivers. Hence, if an event occurs in the driving context in which more events occur, the main reason of the event is probably the driving context rather than driving skill. This means if we simply compare the number or the frequency of events, it is unfair to drivers who have to work at the rush hours, black spot or bad weather.

In order to measure the effects of driving context on driving event, we adapt the TF-IDF, widely used in discipline of information retrieval. Where the term frequency (TF) and inverse document frequency (IDF) are two kinds of measure of the information quantity a word provides. The basic idea of the method is that a word is a keyword to distinguish a document from others in the corpus if it occurs many times in the document, which can be measured by TF, but occurs rarely in other documents in the corpus, which can be measured by IDF.

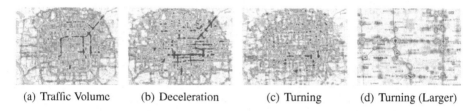

(a) Traffic Volume   (b) Deceleration   (c) Turning   (d) Turning (Larger)

Fig. 2. Spatial distribution of driving events in Beijing

In a similar way, if one type of driving event occurred frequently in a specific driving context, the risk weight of such event is low, because this type of event is more likely caused by driving context rather than the driver. Based on this consideration, we diminish the risk weight of driving event in driving context where the event occurs frequently, and increase the risk weight in driving context where the event occurs rarely. At last, the driving behavior can be evaluated by combining the driver's event frequency and the risk weights of them in driving context.

5.2   Weighting Event in Driving Context

We define the risk weights of different types of events in various driving context, by frequencies of these events in different hours of a day, road segments and weather. The calculating of risk weight of an event in driving context includes two steps:

Step 1. Getting frequencies of events in driving contexts. Using recorded driving data, we can detect the abnormal driving events. Then we can count the number of events in different driving context and collect the running time in different time of day, the passing times in different road segments and the running time in different weather conditions. The frequency of events in different driving context can be calculated.

Equation (1) are the formulas of frequency calculation of different types of events in the road segment.

$$F(r, e) = \frac{N(r, e)}{pt(r)} \tag{1}$$

where N represents the number of events in the road segment and pt represents the number of passing times in the road segment. For a road segment, considering that there may be no vehicle passing, the frequency of event in this road segment is depicted using the average frequency of the same events at all road segments. The F formulas of time and weather are similar.

Step 2. Determining the risk weight of events in driving contexts. The risk weights of different events in different driving context are given according to IDF formula. The formula which calculates the risk weight of a road segment is given in equation (2). The IDF (here we use the same symbol "IDF" and "TF" as they are in information retrieval for convenience) formulas represent the risk weight in specific driving context by quantify the frequencies of the events.

$$IDF_{r,e} = \lg\left(\frac{\sum_{r \in R} F(r, e)}{Min\{F\} + F(r, e)}\right) \tag{2}$$

where $F(r, e)$ represents the event frequency of an event type that occurs on a certain road segment and the numerator represents the total frequency of an event type occurring on all road segments; and $Min\{F\}$ is the minimum event frequency of this context to avoid the denominator becoming zero. When the frequency of events on a road segment is high and the frequency of events on other road segments is low, the risk weight is low and the IDF value of this specific driving context will be low, which means an event happening in this road segment is more likely caused by the context rather than the driver himself.

The IDF formulas of time and weather are similar.

5.3   Evaluating Driving Performance

Combining with drivers' events frequency and the event risk weight in different context, the evaluation of driving behavior can be given using the TF-IDF weighting technique. The formula of event frequency (TF) in different road segments context is given in Eq. (3) and the TF formula of time and weather are similar.

$$TF_{r,e}(d) = \frac{N(d, r, e)}{pt(r)} \tag{3}$$

The TF formula represents the event frequency of driver in specific driving context. Note that, in some cases, some drivers' historical records do not fully cover all of the contexts. Then the event frequency of the driver in these contexts can be depicted using all other drivers' average frequency of the same events in these contexts.

According to the driver's events frequency (TF) and corresponding risk weights (IDF) in specific contexts, the driver's driving behaviors in every driving contexts are evaluated. The driving performance of a driver is defined by Eq. (4).

$$P(d, r) = \sum_{e \in E} TF_{r,e}(d) \cdot IDF_{r,e} \tag{4}$$

In this definition, the higher frequency of a certain type of event in a particular context or the higher the risk weight of this context is, the worse the driving performance might be. Namely, the larger the driving performance value is, the worse the driver will be in this context.

Then, the comprehensive evaluation of all road segments is the sum of performance value of all road segments in Eq. (5), and the comprehensive evaluation of time and weather are similar.

$$P(R, d, e) = \sum_{r \in R} P(r, d, e) \tag{5}$$

All time of day, road segments and weather driving performance present drivers' driving performance of a certain dimension separately, and combining all three dimensions can demonstrate a more comprehensive evaluation of drivers' performance.

The level of a driver's driving skill in all drivers cannot be learnt directly if his or her driving performance value is given simply, the driver's performance can be normalized using the z-score normalization method.

6   System Design

We implement an evaluation system of driving behavior (Fig. 3), mainly including two components: Pre-Processing and Evaluation.

6.1   Pre-Processing

This part finishes the events detection and determines driving context measures based on the driving data, the road network and weather information.

Event Detection. This step takes the velocity and direction as input and detects abnormal driving events using a machine learning method (Support Vector Machine [5]). These events can reflect the driving behavior of drivers.

Map Matching. In this step, the system maps each GPS point onto the corresponding road segment. We consider the distance of the GPS point and road segment and the topological structures of the road network in the matching algorithm [10]. In order to improve the accuracy of turning and U-turn events, we adopt modifying-backward to solve the Y-junction problem [12].

Driving Context Measures. In this step, we calculate the running data of different driving context and different driver using the longitude, latitude and timestamp data. Running data includes the actual driving time of every hour, passing times of a road segment and driving time under certain weather. The running data is used for the subsequent calculation of event frequency.

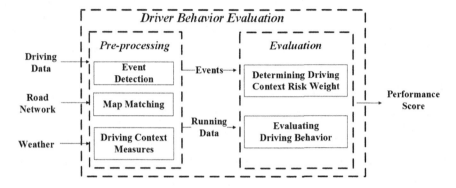

Fig. 3. The driving behavior evaluation system.

6.2 Evaluation

This part takes the events information and the running data of different driving context as input. The context-aware is implemented in this system.

Determining Driving Context Risk Weight. In this step, we use the spatio-temporal information of events and the running data of driving context to calculate the event frequency in each driving context. Then, we determine the risk weight of each driving context using the method depicts in Sect. 5.2.

Evaluating Driving Behavior. In this step, we calculate the event frequency of drivers and combined them with risk weight of each driving context to give a comprehensive evaluation (Sect. 5.3). Finally, the normalized score is given to show the professional level of the driver among others.

7 Experimental Study

In this section, we conduct extensive experiments to demonstrate the effectiveness of our evaluation method using the dataset described in Sect. 3. Besides, we choose 100 drivers randomly and get their accident records as an objective measure of their driving skills. We first illustrate the distribution of drivers' performance to prove the validity of the normalizing method of driving performance. Then, we investigate the relationship between the score of driving performance and the accident records to explore the effectiveness of our method. Finally, we demonstrate that our method is better than other evaluation methods.

7.1   Effectiveness of the Normalizing Method

In order to prove the validity of the use of normalized score, we check whether the driving performance of all drivers obeys the normal distribution. We get the driving performance of 1000 drivers using our evaluation method and calculate the Probability Density Function based on the performance value. And the result is shown in Fig. 4(a).

The classic bell curve in the figure verifies that the performance obeys a normal distribution. So we can give a normalized performance score for each driver.

7.2   Effectiveness of the Evaluation Method

Effectiveness. To demonstrate the effectiveness of our method, we pick 100 drivers randomly and get their accident records during the past three years, which can reflect their driving skills objectively. We calculate the driving performances of these 100 drivers and compare them with the number of accidents. Since we don't have the complete accident reports, we just use the number of abnormal driving events under a safe environment instead. The results are shown in Fig. 4.

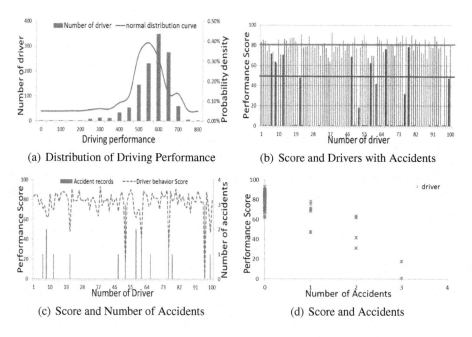

(a) Distribution of Driving Performance   (b) Score and Drivers with Accidents

(c) Score and Number of Accidents   (d) Score and Accidents

Fig. 4. Experiment results.

Figure 4(b) shows the relationship between the scores of driving performances and the number of accident records. 12 drivers have accident records, which are

Table 3. The correlation of performance score and accident records

Accident records	Events frequency	Performance score
Spearman Correlation	.523**	.611**
Sig. (2-tailed)	.000	.000
N (driver)	100	100

highlighted in red. We can find that all the drivers whose performance score is less than 50 have accident records. In contrast, all the drivers whose score of performance is higher than 80 do not have any accident records.

Figure 4(c) illustrates that the lower scores the drivers get, the more likely they are to have the accident records. The performance scores are especially low for those drivers who have lots of accident records. Figure 4(d) shows the scatter-plot of performance score and the number of accident records. In terms of the overall situation, there is a clear linear positive correlation between the driving performance scores and the number of accident records.

Comparative Experiment. In order to prove the evaluation method that considers the driving context is more effective than the events-frequency based methods, we compare the correlation between the number of accident records and driving performances generated from these two methods respectively. The spearman correlation coefficients are showed in Table 3.

The correlation coefficients are both positive and the P-values of the two-tailed significance test on irrelevance are all 0.00, which indicate extremely significant positive correlation between accident records and the results of two methods. While comparing the value of coefficients in the two methods, we find that our evaluation method obtains higher correlation. Accordingly, the context-aware method outperforms the event-frequency based method since the external factors of driving behaviors are weakened by the TF-IDF algorithm.

8   Conclusion

In this study, we propose an context-aware evaluation method of driving behavior and implement an evaluation system based on the fusion of vehicle data and the information of driving context. By using a large dataset collected from a chauffeured car service provider, we demonstrate the real influence of driving context. The correlations between the frequency of driving events and the environmental factors (road, time, weather) are prominent. By using TF-IDF, we can determine the risk weight of each specific driving context and highlight real abnormal behaviors. To demonstrate the effectiveness of our method, we calculate the correlation coefficient between the driving behavior under a safe environment (as the ground truth) and the evaluation score generated by our method. The results show that after integrating the driving context, more precise evaluation of driving behavior can be generated.

In the future, we plan to combine this method with the assignment of orders, the prediction of traffic condition and other scenarios to further improve the safety and efficiency of the transport system.

References

1. Alexander, M.J., Wagner, J.: Analysis of in-vehicle driver behaviour data for improved safety. IJVS **5**(3), 197–212 (2011)
2. Bordoff, J., Noel, P.: Pay-as-you-drive auto insurance: a simple way to reduce driving-related harms and increase equity. J. Risk Insur. **37**(2), 25–31 (2008)
3. Castignani, G., Derrmann, T., Frank, R., Engel, T.: Driver behavior profiling using smartphones: a low-cost platform for driver monitoring. IEEE Intell. Transp. Syst. Mag. **7**(1), 91–102 (2015)
4. Chen, S.H., Pan, J.S., Lu, K.: Driving behavior analysis based on vehicle OBD information and adaboost algorithms. In: Proceedings of the International Multi-Conference of Engineers and Computer Scientists, vol. 1, pp. 18–20 (2015)
5. Chen, Z., Yu, J., Zhu, Y., Chen, Y., Li, M.: D3: abnormal driving behaviors detection and identification using smartphone sensors. In: SECON, pp. 524–532 (2015)
6. Choi, S., Kim, J., Kwak, D., Angkititrakul, P., Hansen, J.H.: Analysis and classification of driver behavior using in-vehicle can-bus information. In: DSP for In-Vehicle and Mobile Systems, pp. 17–19 (2007)
7. Jindal, T., Giridhar, P., Tang, L.A., Li, J., Han, J.: Spatiotemporal periodical pattern mining in traffic data. In: Proceedings of the 2nd ACM SIGKDD International Workshop on Urban Computing, p. 11. ACM (2013)
8. Liang, Y., Mclaurin, E.J., Simmons, L.A., Verma, S.K., et al.: Considering traffic and roadway context in driver behavior assessments: a preliminary analysis. In: The 8th International Conference, pp. 87–92 (2016). https://www.researchgate.net/publication/312077623_Considering_Traffic_and_Roadway_Context_in_Driver_Behavior_Assessments_A_Preliminary_Analysis
9. Liu, W., Zheng, Y., Chawla, S., Yuan, J., Xing, X.: Discovering spatio-temporal causal interactions in traffic data streams. In: KDD, pp. 1010–1018 (2011)
10. Lou, Y., Zhang, C., Zheng, Y., Xie, X., Wang, W., Huang, Y.: Map-matching for low-sampling-rate GPS trajectories. In: ACM Sigspatial International Symposium on Advances in Geographic Information Systems, pp. 352–361 (2009)
11. Miyajima, C., Nishiwaki, Y., Ozawa, K., Wakita, T., Itou, K., Takeda, K., Itakura, F.: Driver modeling based on driving behavior and its evaluation in driver identification. Proc. IEEE **95**(2), 427–437 (2007)
12. Quddus, M.A., Ochieng, W.Y., Noland, R.B.: Current map-matching algorithms for transport applications: state-of-the art and future research directions. Transp. Res. Part C Emerg. Technol. **15**(5), 312–328 (2007)
13. Rui, A., Igreja, A., Castro, R.D., Rui, E.A.: Driving coach: a smartphone application to evaluate driving efficient patterns. In: Intelligent Vehicles Symposium, pp. 1005–1010 (2012)
14. Web: Weather (2017). http://www.tianqi.com/
15. Zhang, C., Patel, M., Buthpitiya, S., Lyons, K., Harrison, B., Abowd, G.D.: Driver classification based on driving behaviors. In: Proceedings of the 21st International Conference on Intelligent User Interfaces, pp. 80–84 (2016)

Measurement of Users' Experience on Online Platforms from Their Behavior Logs

Deepali Jain[1], Atanu R. Sinha[1(✉)], Deepali Gupta[2], Nikhil Sheoran[3], and Sopan Khosla[1]

[1] Adobe Research, Bangalore, India
{deepjain,atr,skhosla}@adobe.com
[2] Indian Institute of Technology Delhi, New Delhi, India
deepaligupta0737@gmail.com
[3] Indian Institute of Technology Roorkee, Roorkee, India
nikhilsheoran96@gmail.com

Abstract. Explicit measurement of experience, as mostly practiced, takes the form of satisfaction scores obtained by asking questions to users. Obtaining response from every user is not feasible, the responses are conditioned on the questions, and provide only a snapshot, while experience is a journey. Instead, we measure experience values from users' click actions (events), thereby measuring for every user and for every event. The experience values are obtained without-asking-questions, by combining a recurrent neural network (RNN) with value elicitation from event-sequence. The platform environment is modeled using an RNN, recognizing that a user's sequence of actions has a temporal dependence structure. We then elicit value of a user's experience as a latent construct in this environment. We offer two methods: one based on rules crafted from consumer behavior theories, and another data-driven approach using fixed point iteration, similar to that used in model-based reinforcement learning. Evaluation and comparison with baseline show that experience values by themselves provide a good basis for predicting conversion behavior, without feature engineering.

1 Introduction

Customer experience is the new "battleground" for firms [2], with experience-measurement as an imperative to benchmark actions and enhance user experience (UX) [2]. Also, personalization of UX pervades firms' aspirations, calling for individual level measurement of UX. Yet, there remains considerable reliance on explicit measurement through surveys, prone to large non-response rates (the

---

Electronic supplementary material The online version of this chapter (https:// doi.org/10.1007/978-3-319-93034-3_38) contains supplementary material, which is available to authorized users.

D. Phung et al. (Eds.): PAKDD 2018, LNAI 10937, pp. 475–487, 2018.
https://doi.org/10.1007/978-3-319-93034-3_38

gold standard ACSI reports upwards of 85% [1]), response biases, are retro-spective in nature and utilized at an aggregate level. While implicit measure-ment has been proposed in search satisfaction [19] to improve search, it has not been embraced in UX domains. Notably, experience is distinct from satisfaction. The latter is an outcome while experience is a journey [13]. In addressing these gaps we propose methods for computing UX values, that captures the process of experience as a journey.

The term UX covers experiences derived from any usage scenarios such as using knowledge software, or, using website/app for eCommerce, and includes *customer experience*. Experience is latent in users' mind and difficult to measure [13]. Explicit measurement by surveys supposedly avoid this difficulty. But, very few respond to surveys, responses are conditioned on questions asked and provide only a snapshot. Instead, using clickstream logs that reflect *actual* behavior we measure the *latent* user experience. We offer two methods, each with two steps. In step one, both simulate an off-line learned model of the environment. In step two, which computes UX values, they differ; whereas one method crafts rules using consumer psychology, the other uses value iteration by drawing from rein-forcement learning. We rely on an RNN with long short-term memory (LSTM) [7] units for modeling the environment. This allows us to use the multidimensional and continuous historical information encoded in the LSTM cell along with the current event to characterize states. Toward measurement of latent UX our con-tributions are:

1. Introducing formulations and rules based on consumer behavior theories toward computing UX values. Moreover, combining these rules with mod-eling dynamics of an on-line platform using an LSTM network.
2. Additionally, introducing a data-driven method without pre-defined rules, where we define experience in terms of the value of different events which elicit delayed rewards. This flexible framework allows generalizability across domains. Specifically, measuring the *value* of interaction sequences is new.
3. A novel application of state value iteration method, commonly used for solving Markov Decision Processes and in Reinforcement Learning, to the domain of click-stream data analysis.
4. Representation of partially observable states in the journey of user as the memory cell of an RNN pre-trained to predict next event.

Note that we exclude features available in clickstream data such as types of product, page content, etc. By relying only on click action sequences to measure UX values, our method has less dependence on feature engineering.

2   Related Work and Defining Experience Value

Drawing from the rich customer experience literature in Marketing and Con-sumer Psychology, [13] points out "what people really desire are not products but satisfying experiences". Customer experience is a process, or, a journey over time [13], which comprises three phases - pre-purchase, during, and post-purchase.

Measuring UX over the whole journey could be of interest [13]. UX comprises a "customer's behavioral, emotional, cognitive, and social responses to a firm's offerings ([13], pp. 71)." Thus, UX is a latent construct, in the mind of the user. Explicit measurement of UX by asking questions may capture *stated experience* on some aspects (e.g. emotion), but fails to capture actual actions. Our premise is that click action sequences observed in usage logs are crucial clues about *revealed experience*. Values of the latent construct UX can be computed for each phase based on phase-specific clickstream action sequence data. Considering the during-phase, footprints on a website or on a mobile app include click actions of filter, add to cart, for ecommerce; or, tasks performed during in-product usage of knowledge software. The actions are observed, but experiences are unobserved. By modeling these action sequences we assign values to latent experience, which drives actions observed in data.

We define experience as the value of being in a certain state of the environment in terms of proximity to a goal state. Consumer behavior literature highlights goal-orientation in online behavior and how goal-directed activities can achieve compelling experiences [17]. For the use case of e-Commerce, *Purchase* is an indicator of experience and consistent with goal attainment. Hence, we treat states in which *Purchase* event takes place as goal states. With a goal of making purchase, users go through several events on a site and incur transaction costs in search, time and psychological costs, which increase with efforts [15]. The events can be sequenced with respect to a goal; e.g., a sequence (*browsing, deliberate search, add to cart, purchase*). Moving forward from one stage to another in the sequence brings users closer to the goal and goal-gradient decreases [10], improving UX and encouraging behavior toward goal completion. Moreover, the process of purchase decision making itself contributes to experience [6]. From [16] we know the higher order event of *directed-buying sessions* has the highest conversion rate (12.94%), followed by the lower order stage of *search sessions* (8.02%).

The search literature in CS studies implicit measurement of satisfaction in order to improve search outcomes and finds that implicit measurement correlates with explicit, question based measures of satisfaction [8,19]. This provides support for our thesis. Deviating from metrics such as dwell time, search results click, [19] offers a latent structural learning model of search satisfaction, which recognizes action level dependencies and uses *rich structured features*. Other efforts examine *struggling* in search to obtain relevant information [18]. The problem we study is about decision making (e.g., whether to purchase) based on online platform interactions and sets our work apart from that of search which is about obtaining relevant information. Typically, a poor search has less consequences for a user than a poor purchase, making the goal orientation stronger in our context of browsing experience. In browsing there is a hierarchical structure imposed by the site, whereas in search a poor result leads to user formulating another query which may not have a hierarchical basis. Finally, our model does not rely on features unlike these papers in search. Clickstream mining for measuring UX has been used to provide visualizations of common paths for site visitors [14] and to infer personas of users [22], but none suggests a method to extricate UX metrics from user logs, which we do.

We draw upon the literature in use of RNN to understand consumer behavior from clickstream data. Usefulness of RNN to link individual click actions to predictions is shown in [12]. For improved purchase prediction [11] depict the benefit of using sequential input of tweets for RNN. A manifestation of RNN [23] is in predicting sequential clicks for sponsored search. None of these papers investigates experience, which is a continuous evolution from sequential behaviors [13]. Traditionally, HMMs are used to model latent states for obtaining insights into user behaviors [3, 4, 20]. Our RNN model of the environment is Markovian, but in histories of states [21] as described later. The multidimensional and continuous historical information encoded in the LSTM cell is a major departure from the finite, discrete values for HMM. Previous application of Markov Chain model to clickstream includes mapping of journals based on logs available in scholarly portal [5], but does not include decision-making which we do. In the class of sequential data modeling techniques we have not seen in the literature any existing method that specifically measures the *value* of an interaction sequence. In this regard, our data-driven approach of using value-iteration has been derived from classical literature in reinforcement learning and decision theory.

3   Framework

We model the browsing behavior of on-line users of an eCommerce Website as a first-order Markov process. Consider a state space, $\mathcal{S} = \{s_1, s_2, s_3, ...\}$ and a reward function $r : \mathcal{S} \to \mathbb{R}$. At time t, a user in state $S_t \in \mathcal{S}$ receives a reward $r(S_t)$. The transition probability function is $\mathcal{P}(s_i, s_j) = Pr(S_{t+1} = s_j | S_t = s_i)$. Let the sequence of events observed in a user's browsing journey till time t be $E_1, E_2, ..., E_t$ where $E_i \in \mathcal{E} = \{e_1, e_2, ..., e_{|\mathcal{E}|}\}$. Events can be actions or sets of actions. Let a vector $\boldsymbol{H}_{t-1}$ of d dimensions encode all the historical information from the sequence $E_1, E_2, ..., E_{t-1}$. Then, the state at t is represented as a tuple, $S_t = (\boldsymbol{H}_{t-1}, E_t)$. Consider the encoding function, $g : \mathcal{S} \to \mathbb{R}^d$ such that, $\boldsymbol{H}_0 = \boldsymbol{0}$ and $\boldsymbol{H}_t = g(\boldsymbol{H}_{t-1}, E_t)$. Also, let us define the operator $\oplus$ such that,

$$S_t \oplus E_{t+1} = S_{t+1}$$
$$(\boldsymbol{H}_{t-1}, E_t) \oplus E_{t+1} = (\boldsymbol{H}_t, E_{t+1}) \tag{1}$$
$$(\boldsymbol{H}_{t-1}, E_t) \oplus E_{t+1} = (g(\boldsymbol{H}_{t-1}, E_t), E_{t+1})$$

4   Learning Experience Values

We first build a model to simulate the dynamics of the environment and then apply two alternative methods for exploiting the learned model to extract latent experience values. The first method is based on predefined rules that experience values must satisfy. The second method is based on value-iteration, is data-driven and autonomous.

The environment is simulated using an RNN trained to predict the next event in the customer journey. The network encodes the information from the historical sequence of events in its d dimensional cell state. The gates of the LSTM unit of the RNN model the history encoding function g introduced above. The network estimates the transition probability function ($\hat{\mathcal{P}} \triangleq \mathcal{P}$) of the underlying Markov process of the environment. For every input sequence of events, a one step ahead sequence is predicted. The architecture of the model is as follows:

- **Input Layer:** The data are input in the form of sequences of events.
- **Embedding Layer:** The categorical variable, i.e. the event is then embedded into a latent space of dimension 150.
- **LSTM Layer:** The input is then fed into an LSTM layer with 200 hidden dimensions. The LSTM layer acts as the memory unit of the model. The hidden state of the LSTM is carried over as input to the future timestep, thus allowing the model to encode historical information.
- **Fully Connected Output Layer:** The output from the LSTM layer goes to a fully connected dense layer which produces the output of size $|\mathcal{E}|$ through softmax activation at each time-step of the sequence. The output at each time-step is a probability distribution vector over all possible next events.

The model is trained to minimize the categorical cross-entropy loss using Adam [9] optimization algorithm.

4.1 Rule-Based Method

For this first method we formalize the concepts of event base values (B) and event transition importance (TI). Then we outline intuitive rules that experience values ought to satisfy. While these rules are crafted from domain knowledge, some companies may prefer to impose own rules which conform to their specific situation. Later we show how the values B and TI along with the next event prediction model are used to compute final experience values (XV) at each state.

Drawing upon consumer behavior theories, a base value $B(e)$, is assigned to every event e, in the order of progression toward the goal task (*Purchase*, in this case). For example, a user who has added a product to cart is closer to completing the purchase-goal task than someone exploring products. Thus, we assign higher base value to the *Add to cart* event than the *Browsing* event.

An importance value, $TI(e_i, e_j)$ is assigned to a transition from any event e_i to another event e_j. This importance value captures how discriminative a transition is across purchase and non-purchase journeys. In other words, if a transition occurs equally frequently in both purchase as well as non-purchase journeys, then it is less important than a transition whose frequencies are unequal. Intuitively for example, transition from *Hedonic Browsing* to *Directed Search* is less important than that from *Directed Search* to *Add to Cart*, since the former likely occurs about as frequently in purchase and non-purchase journeys, while the latter occurs more frequently in purchase journey but less frequently in non-purchase journey. More formally, let

$$p = \sum_{k=1}^{K} \sum_{t=1}^{\tau} (E_t^k = e_i) \wedge (E_{t+1}^k = e_j) \wedge (E_\tau^k = Purchase)$$

$$np = \sum_{k=1}^{K} \sum_{t=1}^{\tau} (E_t^k = e_i) \wedge (E_{t+1}^k = e_j) \wedge (E_\tau^k \neq Purchase) \quad (2)$$

Then, $\quad TI(e_i, e_j) = \dfrac{|p - np|}{p + np}$

where, K is the number of event sequences and τ is the length of each sequence.

Let $S_t = (\boldsymbol{H}_{t-1}, e_i)$ and $S_{t+1} = (\boldsymbol{H}_t, e_j)$. The following rules characterize a desired property of experience value $XV(S_t)$:

$$\begin{aligned} \text{if} \quad & XV(S_t) \geq B(e_i) \quad \text{then} \quad \mathbb{E}(B(e_j)) \geq B(e_i) \quad \text{and} \\ \text{if} \quad & XV(S_t) < B(e_i) \quad \text{then} \quad \mathbb{E}(B(e_j)) < B(e_i) \end{aligned} \quad (3)$$

These rules imply that a user who is having a better experience than that indicated by the base value of the current event, is expected to transition to an event with higher base value and vice-versa. The objective is to find experience values that minimize the number of rules violated for a journey.

We propose alternative formulations for computing $XV(S_t)$. Later, we provide intuition for these formulations.

$$\Delta B_{i,j} = \omega_j \hat{P}(S_t, S_t \oplus e_j)(B(e_j) - B(e_i)) \quad (4)$$

Formulation 1: $\quad XV(S_t) = \omega_0 B(e_i) + \sum_{j=1}^{|\mathcal{E}|} \Delta B_{ij}$

Formulation 2: $\quad XV(S_t) = \omega_0 B(e_i) + \sum_{j=1}^{|\mathcal{E}|} TI(e_i, e_j) \Delta B_{ij}$

$\quad (5)$

Formulation 3: $\quad XV(S_t) = \omega_0 B(e_i) + T_z(S_t) \sum_{j=1}^{|\mathcal{E}|} TI(e_i, e_j) \Delta B_{ij}$

where, $\quad T_z(S_t) = \dfrac{T(S_t) - mean(T(e_i))}{std(T(e_i))}$

where, $W = \{\omega_0, \omega_1, ..., \omega_{|\mathcal{E}|}\}$ is a set of unknown parameters and $T(.)$ is the time spent in a state or event. To examine each of the proposed formulations in a simple manner, consider the special case when $\omega_i = 1 \quad \forall i$. In Formulation 1, XV is defined as weighted sum of the current base value and the expected change in base values from current to next time step (equivalently, the expected base value of the next event). In Formulation 2, the importance of the transition to next event is also taken into account. Formulation 3 builds upon Formulation 2 through the incremental inclusion of the effect of normalized time spent in the current state ($T_z(S_t)$). This recognizes that time spent may impact experience.

We estimate the optimal value for W by linear regression with the loss function as follows

$$\hat{y}_t = \sigma(XV(S_t) - B(e_i)) \quad \text{and} \quad y_t = \sigma(B(e_j) - B(e_i))$$

$$\mathcal{L}_W = \sum_{k=1}^{K} \sum_{t=1}^{\tau} (\hat{y}_t^k - y_t^k)^2 \tag{6}$$

where, K is the number of event sequences, τ is the length of each sequence and σ is a Sigmoid function with a high slope to simulate a unit step function. This is an implementation of number of rules violated in a differentiable form to facilitate gradient descent based parameter estimation.

4.2 Value Iteration Method

Our second method overcomes the deficiencies of hand-crafted rules which may not generalize to all domains. Herein, we need to use very little domain knowledge in the form of a reward function, r as follows

$$r(S_t) = \begin{cases} 1, & \text{if } E_t = Purchase \\ -\epsilon, & \text{otherwise} \end{cases} \tag{7}$$

where, $-\epsilon$ is a small penalty. Now, consider a user traversing the state space of the environment and assimilating rewards along the way according to the above reward function. She achieves high reward in *Purchase* event and a small penalty (ϵ) everywhere else. We can now define the experience value of any state, S_t as the total expected discounted reward after t.

$$XV(S_t) = \mathbb{E}(r(S_{t+1}) + \gamma r(S_{t+2}) + \gamma^2 r(S_{t+3}) + ...) \tag{8}$$

where, $\gamma \in (0, 1)$ is the discounting factor. The above expression can be written in the form of a Bellman Equation as follows

$$XV(S_t) = \mathbb{E}(r(S_{t+1}) + \gamma XV(S_{t+1}))$$

$$XV(S_t) = \sum_{i=1}^{|\mathcal{E}|} \hat{P}(S_t, S_t \oplus e_i)(r(S_{t+1}) + \gamma XV(S_{t+1})) \tag{9}$$

Since the state space is very large (all possible sequences of events), it is not feasible to get exact solution to this equation through methods such as dynamic programming or linear regression. To deal with this problem, we rely on a function approximation method. We define a simple linear estimation function f_θ with a set of parameters θ, to model the experience values.

$$f_\theta(S_t) = \hat{XV}(S_t) \doteq XV(S_t) \tag{10}$$

We use the fixed-point iteration method to find θ. Start with random initial values, θ_0. At iteration number n, experience values for all observed states in the training data are estimated using θ^{n-1}. Based on these estimates, expected values, XV^n are calculated using the Bellman Equation.

$$XV^n(S_t) = \sum_{i=1}^{|\mathcal{E}|} \hat{\mathcal{P}}(S_t, S_t \oplus e_i)(r(S_{t+1}) + \gamma X\hat{V}^{n-1}(S_{t+1})) \qquad (11)$$

The mean square error, $\mathcal{L}_\theta^n$ between expected $(XV^n(S_t))$ and estimated $(X\hat{V}^n(S_t) = f_\theta(S_t))$ values is used to update θ with gradient descent method until convergence. For a training dataset with K sequences with τ time-steps each,

$$\mathcal{L}_\theta^n = \sum_{k=1}^{K} \sum_{t=1}^{\tau} (f_\theta(S_t^k) - XV^n(S_t^k))^2 \qquad (12)$$

$$\theta^n = \theta^{n-1} + \alpha \frac{d\mathcal{L}_\theta^n}{d\theta}$$

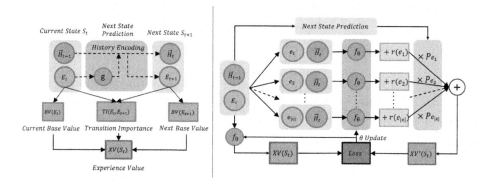

Fig. 1. Illustration of rule based (left) and value iteration (right) method

Table 1. Actions corresponding to each category

Category	Actions
Hedonic browsing (c_1)	Search, Search Filters, Product Details, Product Categories
Deliberate Search (c_2)	Reading Reviews, Product Comparison
Add to Cart (c_3)	Add to Cart, Add to List
Purchase (c_4)	Checkout, Payment, Place Order

Table 2. Category level transition frequency for sequences (Read from category in row to category in column)

No Pur./Pur	c_1	c_2	c_3	c_4
c_1	-	7153/2331	4075/8236	1694/4058
c_2	6626/2177	-	177/324	197/500
c_3	2302/3850	113/113	-	1903/5241
c_4	3375/4664	261/410	207/567	-

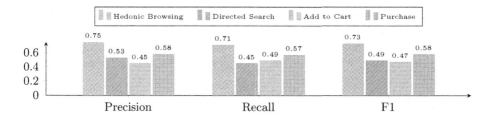

Fig. 2. Results for the next category prediction model

Table 3. Evaluation with purchase prediction

Method	Accuracy (%)	Precision	Recall	F1-Score	AUC
Category Sequences	66.52	0.63	**0.75**	**0.69**	**0.67**
Rule-based Form. 1 ($W = 1$)	66.64	0.69	0.65	0.67	0.66
Rule-based Form. 2 ($W = 1$)	66.48	0.71	0.64	0.67	0.66
Rule-based Form. 3 ($W = 1$)	66.64	0.73	0.64	0.68	**0.67**
Rule-based Form. 1 (optimal W)	**66.76**	0.58	0.69	0.63	**0.67**
Value iteration	63.16	**0.82**	0.59	**0.69**	0.65

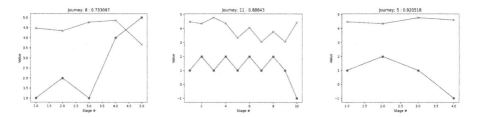

Fig. 3. Evolution of experience during journeys (Color figure online)

5   Experimentation

Click-stream data from an e-Commerce site, spanning a period of three months, are used. After cleaning the data only click actions corresponding to the *Appliances* category are retained. All click actions, for each user, are stitched together chronologically into a sequence of click actions. Altogether 31 relevant click actions such as *View product details, Apply search filter and Add to cart* are identified from the data. The set of unique actions is denoted $\mathcal{A} = \{a_1, a_2, ..., a_{31}\}$. As reasoned earlier, inspired by [16], click actions are categorized into a set of four categories i.e. *Hedonic Browsing, Directed Search, Add to Cart* and *Purchase*. Each category characterizes a different stage in a user's journey towards the goal state of purchase. The set of categories is denoted $\mathcal{C} = \{c_1, c_2, ..., c_4\}$. Categories and corresponding sample click actions are shown in Table 1. The algorithm for finding experience values is applied at category level, i.e. set of events $\mathcal{E}$ refer to the set of categories, $\mathcal{C}$. The final data are sets of sequences of events. In Table 2 we show the frequencies of transition among categories when journeys end in no purchase vs. end in purchase. Finally, the data are randomly split into two sets, training and testing, with a total of 12800 and 4600 sequences, respectively.

6   Results and Discussion

We have no access to survey based experience measurement scores of users whose usage logs we model. Firms do not share such scores. This obstacle of survey and the current use case of *goal fulfillment toward purchase* guide our evaluation. We compute a UX value for each user, for each event from usage log and then based solely on UX values predict the goal fulfillment (purchase), under the thesis that UX affects goal fulfillment. Purchase prediction is not the focus, but merely a way of evaluating the worth of derived UX values. We show that using UX values can give purchase prediction comparable to that of feature-engineered model. As exemplar of the latter, within the data we have, category (event) sequence based model is comprehensive since it captures the sequence along with frequency of sub-events and time spent on events and forms the baseline. True to our objectives, the methods are to be judged by how closely the accuracy of feature-based model can be reproduced by our methods.

A multi-layer RNN module performs purchase prediction by taking as input either sequence of events or experience values generated from one of the proposed methods. Prediction accuracies of models with fixed architecture and different inputs is then compared. The model architecture is similar to next event prediction model, with a difference in the final layer which produces a single output (purchase probability) per time-step through sigmoid activation. The training is done to minimize binary cross-entropy loss.

We evaluate both environment simulation and UX value generation models. Results for the former for next event prediction on test data are shown in Fig. 2. These measures are obtained by averaging across categories from which arrival into a category can occur. We find some variability in these measures across the categories. Results for UX value generation are shown (in Table 3) for four variations of UX value computation using the rule based method - Formulations 1-3 (AUC = 0.66, 0.66 and 0.67 respectively) and one with parameter tuning for Formulation 1 (AUC = 0.67). The results from the value iteration method (AUC = 0.65) are also compared. For each of these, purchase prediction is carried out by using the generated UX values as the only input. The baseline used is an event-sequence based prediction. We find that although the UX values are extracted based on rules, their performance in predicting purchase is very close to the baseline (AUC = 0.67), which uses features such as frequency of actions and time spent within each category. This suggests that computed experience values capture latent components of browsing experience, which explain purchase propensity as accurately as using information in raw data.

Figure 3 depicts UX values (red), as users move through states (green), for three users. Note the red UX values are leading indicators. E.g., from state 1 to state 2 if UX value decreases, it is expected the user moves from the stage in state 2 to a lower stage in state 3. For red lines, last segments are not interpretable. The left figure shows an upward drift consistent with higher stage attainment, and early sign of UX leading to higher stage. The leading indicator between states 2 and 3 suggests movement from stage 1 to stage 4 going from state 3 to state 4. The middle figure shows a user oscillates between stages 1 and 2 over states, without ever going to a higher stage. The overall downward drift is an early indicator of poor experience and no purchase. The right figure is less informative since the slightly upward tendency is not consistent with stage traversal.

7   Conclusion

We show that UX values can be uncovered from readily available user behavior logs. Drawing from theory we grouped actions into categories to build the model. An alternative thought could be to build a model directly from the raw actions. The action-level model using value iteration shows that for the task of purchase prediction we obtain accuracy (0.67), precision (0.87), recall (0.67), F1 (0.75) and AUC (0.57). Comparing with stage-level results from the last line of Table 3 we find that in AUC, the stage level model performs better.

Rules based method may fit customers who 'live' click to click or are myopic, while value iteration captures long-view customers' behaviors. Several challenges include how to do a direct evaluation based on experience metrics obtained in a direct way. Limitations also pertain to the generalizability of the approach to non-discretionary and low involvement products. The appliance category used here constitutes discretionary spending and a high ticket purchase engendering extensive browsing behaviors. Our use case is for the during phase of the whole customer journey. With data from the pre and post phases, future work can extend the approach to mine UX values for those phases. It is noted that our approach can ingest any goal, not just purchase. For example, information seeking. As well, other rewards and intermediate rewards can be provided. None of these applies in a purchase prediction model.

References

1. The American customer satisfaction index. http://www.theacsi.org/, http://www.theacsi.org/
2. Cx index - forrester. https://go.forrester.com/data/cx-index/
3. Anderson, C.R., Domingos, P., Weld, D.S.: Relational Markov models and their application to adaptive web navigation. In: Proceedings of the Eighth ACM SIGKDD International Conference on Knowledge Discovery and Data Mining, pp. 143–152. ACM (2002)
4. Balakrishnan, G., Coetzee, D.: Predicting student retention in massive open online courses using hidden markov models. Electrical Engineering and Computer Sciences University of California at Berkeley (2013)

5. Bollen, J., Van de Sompel, H., Hagberg, A., Bettencourt, L., Chute, R., Rodriguez, M.A., Balakireva, L.: Clickstream data yields high-resolution maps of science. PLoS ONE **4**(3), e4803 (2009)

6. Chylinski, M., Sinha, A., Lie, D.S., Neill, W.: Experience infusion: How to improve customer experience with incidental activities. Marketing Science Institute Working Paper Series (17–106) (2017)

7. Hochreiter, S., Schmidhuber, J.: Long short-term memory. Neural Comput. **9**(8), 1735–1780 (1997)

8. Kim, Y., Hassan, A., White, R.W., Zitouni, I.: Modeling dwell time to predict click-level satisfaction. In: Proceedings of the 7th ACM International Conference on Web Search and Data Mining, WSDM 2014. ACM (2014)

9. Kingma, D., Ba, J.: Adam: A method for stochastic optimization (2014). arXiv preprint arXiv:1412.6980

10. Kivetz, R., Urminsky, O., Zheng, Y.: The goal-gradient hypothesis resurrected: purchase acceleration, illusionary goal progress, and customer retention. J. Mark. Res. **43**(1), 39–58 (2006)

11. Korpusik, M., et al.: Recurrent neural networks for customer purchase prediction on twitter. In: ACM Conference on Recommender Systems (2016)

12. Lang, T., Rettenmeier, M.: Understanding consumer behavior with recurrent neural networks. In: International Workshop on Machine Learning Methods for Recommender Systems (2017)

13. Lemon, K.N., Verhoef, P.C.: Understanding customer experience throughout the customer journey. J. Mark. **80**(6), 69–96 (2016)

14. Liu, Z., Wang, Y., Dontcheva, M., Hoffman, M., Walker, S., Wilson, A.: Patterns and sequences: interactive exploration of clickstreams to understand common visitor paths. IEEE Trans. Vis. Comput. Graph. **23**(1), 321–330 (2017)

15. Masten, S.E., Williamson, O.E.: Transaction cost economics. Edward Elgar, Aldershot, Hants, England; Brookfield, Vt. (1995)

16. Moe, W.W.: Buying, searching, or browsing: differentiating between online shoppers using in-store navigational clickstream. J. Consum. Psychol. **13**(1–2), 29–39 (2003)

17. Novak, T.P., Hoffman, D.L., Duhachek, A.: The influence of goal-directed and experiential activities on online flow experiences. J. Consum. Psychol. **13**, 3–16 (2003)

18. Odijk, D., White, R.W., Hassan Awadallah, A., Dumais, S.T.: Struggling and success in web search. In: Proceedings of the 24th ACM International on Conference on Information and Knowledge Management, pp. 1551–1560. ACM (2015)

19. Wang, H., Song, Y., Chang, M.W., He, X., Hassan, A., White, R.W.: Modeling action-level satisfaction for search task satisfaction prediction. In: Proceedings of the 37th International ACM SIGIR Conference on Research & Development in Information Retrieval, pp. 123–132. ACM (2014)

20. Ypma, A., Heskes, T.: Automatic categorization of web pages and user clustering with mixtures of hidden Markov models. In: Zaïane, O.R., Srivastava, J., Spiliopoulou, M., Masand, B. (eds.) WebKDD 2002. LNCS (LNAI), vol. 2703, pp. 35–49. Springer, Heidelberg (2003). https://doi.org/10.1007/978-3-540-39663-5_3

21. Yu, H., Mahmood, A.R., Sutton, R.S.: On generalized bellman equations and temporal-difference learning. In: Mouhoub, M., Langlais, P. (eds.) AI 2017. LNCS (LNAI), vol. 10233, pp. 3–14. Springer, Cham (2017). https://doi.org/10.1007/978-3-319-57351-9_1

22. Zhang, X., Brown, H.F., Shankar, A.: Data-driven personas: constructing archetypal users with clickstreams and user telemetry. In: Proceedings of the 2016 CHI Conference on Human Factors in Computing Systems (2016)
23. Zhang, Y., Dai, H., Xu, C., Feng, J., Wang, T., Bian, J., Wang, B., Liu, T.Y.: Sequential click prediction for sponsored search with recurrent neural networks. In: AAAI, pp. 1369–1375 (2014)

Mining Human Periodic Behaviors Using Mobility Intention and Relative Entropy

Feng Yi[4], Libo Yin[2], Hui Wen[1(✉)], Hongsong Zhu[1], Limin Sun[1], and Gang Li[3(✉)]

[1] Beijing Key Laboratory of IOT Information Security,
Institute of Information Engineering, CAS, Beijing 100093, China
{wenhui,zhuhongsong,sunlimin}@iie.ac.cn

[2] China Industrial Control Systems Cyber Emergency Response Team,
Shijingshan District, Beijing 100040, China
yinlibo@etiri.org.cn

[3] School of Information Technology Deakin University, Geelong, Australia
gang.li@deakin.edu.au

[4] School of Computer Engineering, University of Electronic Science and
Technology of China, Zhongshan Institute, Zhongshan, China
yifeng@iie.ac.cn

Abstract. Human periodic behaviors is essential to many applications, and many research work show that human behaviors are periodic. However, existing human periodic works are reported with limited improvements in using periodicity of locations and unsatisfactory accuracy for oscillation of human periodic behaviors. To address these challenges, in this paper we propose a *Mobility Intention and Relative Entropy* (MIRE) model. We use mobility intentions extracting from dataset by tensor decomposition to characterize users' history records, and use subsequence of same mobility intention to mine human periodic behaviors. A new periodicity detection algorithm based on relative entropy is then proposed. The experimental results on real-world datasets demonstrate that the proposed MIRE model can properly mining human periodic behaviors. The comparison results also indicate that MIRE model significantly outperforms state-of-the-art periodicity detection algorithms.

1 Introduction

Periodic behaviour is one of the most common phenomena in human society. Nowadays, with the popularity of location-based services (LBSs) [4] and rapid development of Internet of Things (IOT) devices, a large amount of human footprints have been recorded as spatiotemporal datasets, and this provides an important resource for analyzing human periodic behaviors. Insights into human periodic behaviors can be useful in a variety of applications ranging from public security, privacy preserving [19] to mass movement prediction [17].

In general, the task of human periodic behavior mining refers to identify those human periodic activities that repeat with regular time intervals. In the past

© Springer International Publishing AG, part of Springer Nature 2018
D. Phung et al. (Eds.): PAKDD 2018, LNAI 10937, pp. 488–499, 2018.
https://doi.org/10.1007/978-3-319-93034-3_39

Fig. 1. A history footprint of a user

decade, a number of efforts have been devoted to mine human periodic behavior from spatiotemporal data, though in their context, the periodic behavior means that a user regularly visited certain locations. Unfortunately, because of the complexity of human behavior and the personality trait of neophilia [13], many periodic behaviors, like traveling or dining, may occur at different locations. Therefore, the appearances at one single location are not obvious periodic. Two significant barriers exist for more accurate human periodic behaviors mining.

- Locations related to one periodic behavior may be far away from each other.
- Multiple periodic behaviors could happen at the same location.

We use a running example in Fig. 1 to illustrate above two barriers. Suppose a fitness club and a store are located in the same shopping mall, and a supermarket, a beauty parlour and a swimming pool are scattered at different locations in downtown. A user regularly attends the recreational activities like gymming, hairdressing or swimming every Thursday night, and randomly goes to either the store or the supermarket every Friday after work. It's obvious that two periodic behaviors `recreation` and `shopping`, occur at the same location, but it is hard to differentiate them from locations only. Moreover, the user randomly goes to one of two markets which are far away from each other. Hence, it's impossible to detect the above two human periodic behaviors based on locations. From this example, we can see that existing periodic behavior detection methods remain unsatisfactory in handling this situation.

Generally speaking, *human mobility* is fundamentally driven by personal intentions [7], such as commuting. As intrinsic factor for human mobility, mobility intention shows more periodicity than location [16]. As shown in Fig. 1, it is obvious that recreation and shopping are two mobility intentions on which the user exhibits the periodic patterns, both with the period of one week. It implies that mobility intention is a promising factor in modelling human periodic behaviors.

From the perspective of human mobility intention, we propose a novel human periodic behavior detection model called the *mobility intention and relative entropy* (MIRE) model. We first discover the human mobility intentions from spatiotemporal dataset using the tensor decomposition. Through a comprehensive feature engineering, a multiclass classifier is trained to map every footprint into a mobility intention. Then a user's records are converted into a sequence of mobility intentions, from which the human periodic behaviors are to be detected.

For the task of periodicity detection, Li et al. [12] show that most observations concentrate in some time intervals when the timeline is segmented by the true period and the segments are overlayed. In other words, the observations are more disordered in incorrect period than in true period. Based on information theory [9], we use relative entropy which is a measure of discrepancy between two probability distributions as our periodic measurement.

The major contributions of this paper are as follows:

- To the best of our knowledge, this is the first work in explicitly using mobility intentions for the human periodic behavior detection.
- We propose a novel period detection method based on the relative entropy. We further provide the rigorous proof validity of conclusion for the method.
- Extensive experiments on both synthesis and real spatiotemporal datasets show that the proposed model is effective and more precise than the state-of-the-art methods.

The remaining of this paper is organized as follows. Section 2 presents the related works. Section 3 details the proposed model. Section 4 reports the extensive experiment results and analysis. Finally, we conclude the paper in Sect. 5.

2   Related Work

2.1   Human Periodic Behavior Mining

A number of studies have been proposed to analyze human periodic behaviors from spatiotemporal dataset. The key challenge lies on how to transform a two-dimensional movement sequence into one-dimensional symbolic time sequence. Mamoulis et al. replace the exact locations with the pre-defined regions to which they belong [14]. Many clustering algorithms are usually used for transforming, such as DBSCAN [2,14] and kernel density [11]. Location type [17] are also used for transforming and get good results. Yuan et al. propose a new method based on Dirichlet Process to automatically detect periodic regions [17]. Zhang et al. [18] utilize text of social network to obtain semantic information of locations and use them for transforming. After transformation, many traditional period detection methods in time series can then be adopted to discover the periodic patterns, such as cyclic association rules mining [15]and Frequent Pattern tree [3].

For the challenges mentioned above, there are limited periodic behaviors on locations and the results are often imprecise. The semantic information has been largely unexplored in human periodic behavior mining.

2.2   Periodicity Detection

The period detecting is a long-standing problem in data mining. Widely used period detecting methods include *Fast Fourier transform* (FFT) and Autocorrelation. However, they don't perform well on sparse, incomplete and noisy observations [11]. Researchers have also proposed many period detecting techniques, such as WARP [5], Lomb-Scargle periodogram [8], the combination of autocorrelation function and FFT [1,11]. Li et al. [12] proposed a probability measurement to evaluate the score of every potential period T and consider the T with the largest score as the true period. However, the measurement proposed by [12] favors shorter period, and performance badly when there is only one observation in a period. Yuan et al. [17] model the time gap between two consecutive records as a univariate Gaussian distribution and detect the periodicity using a probability generative model.

Existing periodicity detection algorithms are not suitable for detecting periodicity in sparse, incomplete and noisy spatiotemporal dataset. In this paper, we propose a novel periodicity detection method to address above challenges.

3 Mobility Intention Based Period Behaviors Mining

Given a spatiotemporal dataset D of N users, let O_i be the collection of records for user u_i, and each record $o_j^i \in O_i$ is 2-tuple $o_j^i = (loc_j^i, t_j^i)$ which indicates that u_i visited loc_j^i at time t_j^i. The location loc_j^i is a geographic coordinate. Let $o_k^i = (loc_k^i, t_k^i)$ and $o_{k-1}^i = (loc_{k-1}^i, t_{k-1}^i)$ be the k-th and the $(k-1)$-th record of u_i, respectively. Let $o_{k'}^i = (loc_k^i, t_{k'}^i)$ be the last record of o_k^i which means $o_{k'}^i$ and o_k^i are two continuous records at the same location loc_k^i. As aforementioned, we attempt to mine human periodic behaviors based on the mobility intentions rather than locations. Here, the *mobility intention* m refers to a common cause that explains why a user appeared in location loc at time t. Let $\mathcal{M} = \{m_i | 1 \leqslant i \leqslant M\}$ denote the set of M mobility intentions.

We use a binary sequence $X = \{I(t) | t = 0, \ldots, n-1\}$ to denote a time sequence, where $I(t) = 1$ if and only if user u_i has the mobility intention m_k^i at timestamp t, otherwise $I(t) = 0$.

A user u_i has a *periodic behavior* with period T_0 if u_i has a mobility intention every T_0 time span. However, in the real world, human periodic behaviors may not occur at exactly the same period in different cycles, and it may oscillate across different intervals. Hence, the human periodic behavior can be formally defined as follow.

Definition 1 (Human Periodic Behavior). *Suppose $T_0 > 1$ and $0 \leqslant t_0 < T_0$, for any $0 \leqslant t^\star < T_0$*

$$I(t^\star) = \begin{cases} 1, t^\star = t_0 \\ 0, \text{otherwise} \end{cases} \tag{1}$$

If there is one and only one timestamp $t' \in [t_0 - \delta + kT_0, t_0 + \delta + kT_0]$ of X which satisfies $I(t') = I(t_0)$ for $k = 0, 1, \ldots$, $\mod (n-1, T_0)$, the binary sequence X is a periodic behavior binary sequence with the period T_0.

Here t_0 is the average timestamp that m_k^i happened in a true period T_0. δ is a relative buffer that enables m_k^i to oscillate in the interval $[t_0 - \delta + kT_0, t_0 + \delta + kT_0]$ instead of being fixed at an exactly timestamp $t_0 + kT_0$.

Given the spatiotemporal dataset D, the aim of periodic behavior mining includes: (1) revealing a set of mobility intentions $\mathcal{M}$ and (2) detecting the true period T_0 associated with the mobility intention m_k^i.

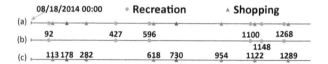

Fig. 2. The mix sequence corresponding to Fig. 1.

3.1 Mobility Intention Extraction

As aforementioned, mobility intention is latent but with a high degree of spatiotemporal regularity. For example, commuting, a basic mobility intention in many spatiotemporal datasets, can explain why a worker arrived at the work place around 9 a.m. every workday. Before utilizing the mobility intention, it is important to know what kinds of mobility intentions are embedded in a spatiotemporal dataset.

In this paper, we use the CANDECOMP/PARAFAC (CP) decomposition algorithm [10] to extract mobility intentions from a spatiotemporal dataset. CP decomposition is an effective tool for analyzing the relationship between spatial and temporal attributes in spatiotemporal datasets [6]. In order to utilize the CP algorithm, a three-dimensional tensor composed by location-hour-day is constructed. In the CP algorithm, the tensor $\mathcal{Y}$ is factorized into a sum of rank-1 component tensors $\mathbf{Y}_r$. After the decomposition, every rank-1 tensor $\mathbf{Y}_s$ is a mobility intention m_s. We use m_i to denote the i-th mobility intention and $\mathcal{M} = \{m_i | 1 \leqslant i \leqslant q\}$ to represent the set of q mobility intentions.

Every mobility intention can be considered as one class. A record f_i corresponds to one mobility intention, in other words, belongs to a class. Then mapping from f_i to m_j can be considered as a multi-class classification problem.

In order to acquire good performance of multi-class classification, we perform a comprehensive feature engineering and model training. After analyzing the vectors of rank-1 tensor and feature engineering, we propose three kinds of distinguishable features: spatial features, hour features and day features, such as location entropy and day type. Finally, with those three kinds of features, we train an Adaboost model to map the record into a mobility intention. Then, O_i can be characterized by the tuple set $\{(m_1^u, t_1), (m_2^u, t_2), \dots\}$.

For the running example, records in Fig. 1 can be characterized by a mixed sequence that contains two mobility intentions, recreation and shopping, as shown in Fig. 2(a). Then, as shown in Fig. 2(b) and (c), we obtain a time sequence X for each mobility intention. Though the Adaboost model, the tricky problem of separating mixed multiple periods can be solved.

3.2 Period Identification

Suppose a binary sequence of behavior X is periodic with the period T_0. According to Definition 1, the observations should fall in the compact interval $[t_0 - \delta, t_0 + \delta]$ when X is segmented by the true period T_0. However, they should

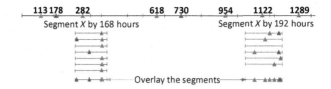

Fig. 3. Illustration example of folded time sequence X in Fig. 2(c).

dispersedly fall in the scattered intervals when X is segmented by an incorrect period T_f. For example, as shown in Fig. 3, the observations of the binary sequence X will fall around 114 when X is segmented by the true period 168 h (a week). The observations scatter in a wide time interval when X is segmented by 192 h. In general, suppose a binary sequence X with length n is segmented by a trial period T, we define

$$S_i(T) = \{t| \mod (t, T) = i \wedge \mathrm{I}(t) = 1\}, \ t = 0, 1, \ldots, n-1, \ i = 0, 1, \ldots, T-1 \tag{2}$$

The probability at each timestamp in T is then

$$p_i(T) = \frac{|S_i|}{\sum_{j=0}^{T-1} |S_j|}, \ i = 0, 1, \ldots, T-1 \tag{3}$$

For X, the distribution of observations is different from different candidate periods. It's more ordered when X is segmented by the true period T_0, and it's disordered when X is segmented by a incorrect period. An extreme case is that a mobility intention happens at every timestamp with nearly equal probability when X is segmented by a incorrect period T_f. There is no period in this extreme situation. Hence, we can use the discrepancy of entropy between $p_i(T)$ and uniform distribution on a potential period T as our periodic measurement.

$$\mathrm{KL}(T) = H(T) - \sum_{i=1}^{T-1} \frac{1}{T} \log \frac{1}{T} = \log T + H(T) \tag{4}$$

The periodic measurement $\mathrm{KL}(T)$ is indeed the relative entropy between $p_i(T)$ and uniform distribution on T. So, given a periodic behavior binary sequence X with unknown period, if the probability distribution $p_i(T)$ is more peaked and the discrepancy with uniform distribution on T is more obvious, T is more close to the true period T_0 .

We have the following lemma which states that the relative entropy will reach the maximum at the true period T_0.

Lemma 1. *If a binary sequence X is periodically generated according to a categorical distribution $\boldsymbol{\mu}_0'$ for some period T_0, then for any $T \geqslant 2, T \in \mathbb{N}$, we have*

$$\lim_{n \to \infty} \mathrm{KL}(T_0) \geqslant \lim_{n \to \infty} \mathrm{KL}(T) \tag{5}$$

Proof. Based on the Definition 1, we can suppose a periodic time sequnce X is generated according to a categorical distribution for some T_0, and the parameter of this categorical distribution is a vector $\boldsymbol{\mu}_0 = (\mu_0, \mu_1, \ldots, \mu_{T_0-1})$, where $\sum_{i=0}^{T_0-1} \mu_i = 1$. It's clear that $\mu_g \gg \mu_h, g \in I_v, h \in [1, T_0]/I_v$, where $I_v = [t_0 - \delta, t_0 + \delta] \subseteq [1, T_0]$.

We will use T_0 to denote the true period and use T to denote a candidate period. Suppose there is an interval $[0, T \cdot T_0 - 1]$, it's obvious that this interval contains T periods of T_0 and T_0 periods of T. Let $p_{i,j}$ be the i-th position of period T in j-th segment. Then, we have:

$$p_{i,j} = \mu_{(i+j \cdot T) \mod T_0}. \tag{6}$$

The i-th position's parameter of period T is:

$$\mu_i'(T) = \frac{1}{T} \sum_{j=0}^{T_0-1} p_{i,j}. \tag{7}$$

Then,

$$\mathrm{KL}(T_0) - \mathrm{KL}(T) \geqslant \ln T_0 + \sum_{k \in I_v} \mu_k \ln \mu_k - \frac{1}{T} \sum_{i=0}^{T-1} \sum_{j=0}^{T_0-1} p_{i,j} \left(\ln \sum_{j=0}^{T_0-1} p_{i,j} \right)$$

$$= \ln T_0 + \frac{1}{T} \sum_{i=0}^{T-1} \sum_{j=0}^{T_0-1} \left\{ \frac{1}{T_0} \sum_{k \in I_v} \mu_k \ln \mu_k - p_{i,j} \left(\ln \sum_{j=0}^{T_0-1} p_{i,j} \right) \right\}$$

$$\geqslant \ln T_0 + \frac{1}{T} \sum_{i=0}^{T-1} \sum_{j=0}^{T_0-1} p_{i,j} \ln \frac{1}{T_0} = \ln T_0 + \ln \frac{1}{T_0} = 0$$

Therefore, in order to find T_0, it is sufficient to compute $\mathrm{KL}(T)$ and select the one that maximizes $\mathrm{KL}(T)$ as the true period.

4   Experiment and Analysis

4.1   Periodicity Detection on Synthetic Time Series Data

In order to evaluate performance of the proposed periodicity detection method, we first generate synthetic datasets using the following four steps:

1. Given a fixed period T_0, a periodic segment X_F is a Boolean sequence of length T_0. The time t_i around t_0 when a mobility intention happens is modelled by Gaussian distribution

$$p(t_i) = \frac{1}{(2\pi\sigma^2)^{1/2}} \exp \left\{ -\frac{1}{2\sigma^2} (t_i - t_0)^2 \right\}, \tag{8}$$

where the σ^2 is the variance.

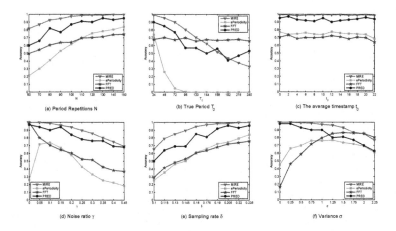

Fig. 4. The performance comparison with various parameter settings

2. X_F is repeated for N times to generate complete sequence X_C.
3. Sample X_C with sampling rate δ and get sequence X_S .
4. We randomly flip one value of 0 to 1 in each period segment with noise rate γ from X_S and obtain the synthetic dataset X_B.

In our experiments, the default values of above parameters are: $T_0 = 24$, $t_0 = 12$, $\sigma = 1$, $N = 120$, $\delta = 0.2$ and $\gamma = 0.1$. Three baseline models are chosen for performance comparison: FFT, ePeriodicity [12] and Periodic Region Detection(PRED) [17], and they are discussed in 2.2. The parameters of baseline methods are set to be their suggested values. For each experiment, one parameter is varied while others are set to the default values. For each parameter setting of synthetic time sequence, we repeat the experiment for 1000 times and report the percentage of correct period detections as *accuracy*.

The performance of compared methods on synthetic dataset is shown in Fig. 4, which indicates that the accuracy of our method is higher than that of baselines in most cases. Figure 4 shows that most methods have better performance when the data is with better quality, such as a larger number of period repetitions N, a greater sampling rate δ, lower noise rate γ and smaller variance σ^2.

Our method can detect more than 80% period when the observations only contains 60 periods which is very common in most spatiotemporal dataset, and baselines needs more periods to achieve same accuracy. The performance of all methods except FFT is worse when T_0 is greater, probably due to the increasing distraction of noise in longer periods. The center of observed time t has no effect on the performance of all methods. Figure 4(d) and (e) show that, with the decrease of noise rate γ and the increase of sampling rate δ, all methods achieve better results for periodicity detection. Figure 4(f) shows that it's much harder to detect the true period when the oscillation is large.

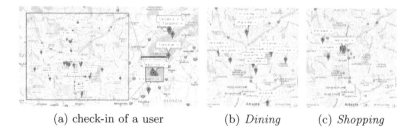

(a) check-in of a user   (b) *Dining*   (c) *Shopping*

Fig. 5. A example of human periodic behaviors mining using MIRE.

In conclusion, from those results, it's clear that the proposed periodicity detection method is more suitable for periodicity detection than other compared methods in real spatiotemporal dataset.

4.2   Performance Evaluation Using Real Dataset

We use a real spatiotemporal dataset `Gowalla` to illustrate the performance of our proposed human periodic behavior detecting model. `Gowalla` dataset is a public available social media check-in dataset [4] with $6,442,890$ check-ins of $196,591$ users from February 2009 to October 2010. After tensor decomposition, we extract 10 mobility intentions from this dataset.

We randomly select a user from Gowalla to discover the human periodic behavior. Figure 5a shows 75 check-ins of a randomly selected user #11838, and it's hard to identify the period behavior patterns directly from locations for history records are scattered in the map. However, five periodic behaviors including commuting, dinning, recreation, shopping and daily routine are mined by our proposed periodicity detecting method with the default sampling rate 1 h.

Figure 5b and c shows the locations where *Dining* and *Shopping* happened. The locations of *Dining* disperse across the north area of Atlanta City, as shown in Fig. 5b. Most *Shopping* intentions of #11838 appeared at a shopping centre. However, the user occasionally went to some new place for shopping. It's obvious that existing methods can not mine these two kinds of periodic behavior. On the contrary, the result of our proposed method based on mobility intention provides more satisfactory and intuitive results.

4.3   Location Prediction on Real Datasets

As aforementioned, human period behavior mining has a variety of applications. In this section, we use human period behavior to predict users' next locations.

Two real spatiotemporal datasets are used for location prediction. One is Gowalla check-ins dataset and the other is *Beijing Bus Smart Card* (BBSC) dataset. The BBSC dataset collects prepaid smart card records for public transportation in Beiing, China. The BBSC dataset contains 275,951,094 bus transaction records about 16,161,460 users in October of 2014. We divide users into 5

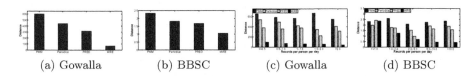

(a) Gowalla	(b) BBSC	(c) Gowalla	(d) BBSC

Fig. 6. Location prediction error distance.

groups based on their record density which refers to the record count per user per day. Each group has 500 users which are random selected from the two datasets. 90% records of each selected user in chronological order are for training and the left for testing.

There are three baseline for comparison in our experiments: PMM (Periodic Mobility Model) [4], Periodica [11] and PRED [17]. The PMM adopts a Gaussian mixture model which centers at home and work to learn user locations constrained by independent truncated Gaussian distribution temporal component. The Periodica model extract regions by kernel density estimation (KDE), and then estimate the period for each region using combination of FFT and autocorrelation. The PRED is discussed in Sect. 2.1. The parameters of above baseline methods are set to be their suggested values. The effectiveness is measured by error distance, which is the Euclidean distance between the true and predicted location of a testing record. The measurement result is the average value of all test records.

Figure 6a and b shows the performance on all testing records which are averaged by 5 groups. In general, for the two datasets, the error distance value of MIRE is much lower than that of baseline methods. PRED improve the cluster algorithm of Periodica by Chinese Restaurant Process (CRP) and use time model as one constrain for clustering, and its performance is prominent among all baseline models. Our proposed MIRE model outperforms the baseline models significantly, owing to that we use mobility intention to model spatial and temporal facts and detect periods by relative entropy.

We are also interested in the connection between records density with performance of human period behavior models. Figure 6c and d shows the error distance of different models in different data groups. With growth of record density, the performance of Periodica and PRED based on location cluster get better at the beginning and worse then. The reason is probably models obtain more information at the beginning and perform more well, and human period behaviors are simple. However, many complex human period behaviors are mixed in history records when record get more dense, and Periodica and PRED can not handle this situation. Our proposed model achieves better performance with increasing record density. The experiment results indicate that the proposed MIRE model outperform the baseline models.

5 Conclusions

In this paper, we address the important and challenging problem of human period behavior detection from spatiotemporal dataset. We first apply a method based on tensor decomposition to obtain mobility intentions from the spatiotemporal dataset and propose a novel human period behavior based on the mobility intention. Then, we design a novel measure based on relative entropy for periodicity and a practical algorithm to detect periods in real scenarios. We give a rigorous proof of its validity for our proposed method. The experiment results show that the proposed method is robust and significantly outperforms existing periodicity detection methods. The case study on real spatiotemporal dataset further demonstrates the effectiveness of our method and its capability in many applications, such as location prediction.

While our approach is designed for one observation in every period, one important extension is to handle the case that there are multi observations in every period. It's is about period pattern detection and is more general in reality. However, there is a problem that how to decide whether a binary sequence X is periodic or random. We consider this as interesting future work.

Acknowledgments. This work was supported by the National Natural Science Foundation of China (Grant No. 61472418, 61702508), the Major R&D Plan (Grant No. 2017YFB0802804), the International Cooperation Project of Institute of Information Engineering, Chinese Academy of Sciences (Grant No. Y7Z0511101).

References

1. Berberidis, C., Aref, W.G., Atallah, M.J., Vlahavas, I.P., Elmagarmid, A.K.: Multiple and partial periodicity mining in time series databases. In: Proceedings of the 15th Eureopean Conference on Artificial Intelligence, ECAI 2002, Lyon, France, July 2002, pp. 370–374 (2002)
2. Cao, H., Mamoulis, N., Cheung, D.W.: Discovery of periodic patterns in spatiotemporal sequences. IEEE Trans. Knowl. Data Eng. **19**(4), 453–467 (2007)
3. Chen, S.-S., Huang, T.C.-K., Lin, Z.-M.: New and efficient knowledge discovery of partial periodic patterns with multiple minimum supports. J. Syst. Softw. **84**(10), 1638–1651 (2011)
4. Cho, E., Myers, S.A., Leskovec, J.: Friendship and mobility: user movement in location-based social networks. In: Proceedings of the 17th ACM SIGKDD International Conference on Knowledge Discovery and Data Mining, San Diego, CA, USA, 21–24 August 2011, pp. 1082–1090 (2011)
5. Elfeky, M.G., Aref, W.G., Elmagarmid, A.K.: WARP: time warping for periodicity detection. In: Proceedings of the 5th IEEE International Conference on Data Mining (ICDM 2005), Houston, Texas, USA, 27–30 November 2005, pp. 138–145 (2005)
6. Fan, Z., Song, X., Shibasaki, R.: CitySpectrum: a non-negative tensor factorization approach. In: The 2014 ACM Conference on Ubiquitous Computing, UbiComp 2014, Seattle, WA, USA, 13–17 September 2014, pp. 213–223 (2014)

7. Giannotti, F., Pappalardo, L., Pedreschi, D., Wang, D.: A Complexity Science Perspective on Human Mobility, pp. 297–314. Cambridge University Press, Cambridge (2013)
8. Glynn, E.F., Chen, J., Mushegian, A.R.: Detecting periodic patterns in unevenly spaced gene expression time series using lomb-scargle periodograms. Bioinformatics **22**(3), 310–316 (2006)
9. Haroutunian, E.A.: Information theory and statistics. In: Lovric, M. (ed.) International Encyclopedia of Statistical Science. Springer, Heidelberg (2011). https://doi.org/10.1007/978-3-642-04898-2
10. Kolda, T.G., Bader, B.W.: Tensor decompositions and applications. SIAM Rev. **51**(3), 455–500 (2009)
11. Li, Z., Ding, B., Han, J., Kays, R., Nye, P.: Mining periodic behaviors for moving objects. In: Proceedings of the 16th ACM SIGKDD International Conference on Knowledge Discovery and Data Mining, KDD 2010, New York, NY, USA, pp. 1099–1108. ACM (2010)
12. Li, Z., Wang, J., Han, J.: Mining event periodicity from incomplete observations. In: The 18th ACM SIGKDD International Conference on Knowledge Discovery and Data Mining, KDD 2012, Beijing, China, 12–16 August 2012, pp. 444–452 (2012)
13. Lian, D., Xie, X., Zheng, V.W., Yuan, N.J., Zhang, F., Chen, E.: CEPR: a collaborative exploration and periodically returning model for location prediction. ACM Trans. Intell. Syst. Technol. **6**(1), 8:1–8:27 (2015)
14. Mamoulis, N., Cao, H., Kollios, G., Hadjieleftheriou, M., Tao, Y., Cheung, D.W.: Mining, indexing, and querying historical spatiotemporal data. In: Proceedings of the Tenth ACM SIGKDD International Conference on Knowledge Discovery and Data Mining, KDD 2004, New York, NY, USA, pp. 236–245. ACM (2004)
15. Ozden, B., Ramaswamy, S., Silberschatz, A.: Cyclic association rules. In: 1998 IEEE 14th International Conference on Data Engineering, pp. 412–421. IEEE (1998)
16. Ying, J.J.-C., Lee, W.-C., Tseng, V.S.: Mining geographic-temporal-semantic patterns in trajectories for location prediction. ACM TIST **5**(1), 2:1–2:33 (2013)
17. Yuan, Q., Zhang, W., Zhang, C., Geng, X., Cong, G., Han, J.: PRED: periodic region detection for mobility modeling of social media users. In: Proceedings of the Tenth ACM International Conference on Web Search and Data Mining, WSDM 2017, Cambridge, United Kingdom, 6–10 February 2017, pp. 263–272 (2017)
18. Zhang, C., Zhang, K., Yuan, Q., Zhang, L., Hanratty, T., Han, J.: Gmove: group-level mobility modeling using geo-tagged social media. In: Proceedings of the 22nd ACM SIGKDD International Conference on Knowledge Discovery and Data Mining, San Francisco, CA, USA, 13–17 August 2016, pp. 1305–1314 (2016)
19. Zhang, L., Cai, Z., Wang, X.: Fakemask: A novel privacy preserving approach for smartphones. IEEE Trans. Netw. Serv. Manage. **13**(2), 335–348 (2016)

Context-Uncertainty-Aware Chatbot Action Selection via Parameterized Auxiliary Reinforcement Learning

Chuandong Yin[1], Rui Zhang[1(✉)], Jianzhong Qi[1], Yu Sun[2], and Tenglun Tan[1]

[1] The University of Melbourne, Parkville, Australia
{chuandongy,tenglunt}@student.unimelb.edu.au,
{rui.zhang,jianzhong.qi}@unimelb.edu.au
[2] Twitter, Inc., San Francisco, CA, USA
ysun@twitter.com

Abstract. We propose a context-uncertainty-aware chatbot and a reinforcement learning (RL) model to train the chatbot. The proposed model is named *Parameterized Auxiliary Asynchronous Advantage Actor Critic* (PA4C). We utilize a user simulator to simulate the uncertainty of users' utterance based on real data. Our PA4C model interacts with simulated users to gradually adapt to different users' utterance confidence in a conversation context. Compared with naive rule-based approaches, our chatbot trained via the PA4C model avoids hand-crafted action selection and is more robust to user utterance variance. The PA4C model optimizes conventional RL models with action parameterization and auxiliary tasks for chatbot training, which address the problems of a large action space and zero-reward states. We evaluate the PA4C model over training a chatbot for calendar event creation tasks. Experimental results show that our model outperforms the state-of-the-art RL models in terms of success rate, dialogue length, and episode reward.

1 Introduction

Recently, personal assistants [10–12,14] become increasingly popular, such as Apple Siri, which can interact with human and provide with intelligent service. These personal assistants are also called task-oriented chatbots (*"chatbot"* for short in the rest of the paper) that help users complete tasks of certain domains, such as creating a calendar event. A task may consist of several *slots*, such as *time* and *location* for a calendar event. A chatbot needs to identify these slots correctly via a dialogue with a user. Due to the uncertainty of language fluency among different people, it is not a simple task to identify these slots accurately. This problem is particularly challenging if the user utterance confidence given by *automatic speech recognition* (ASR) and *natural language understanding* (NLU) systems is low (e.g., a user is new to the language spoken or has a heavy accent).

A straightforward approach to tackle this problem is to confirm with a user whenever his or her utterance confidence is lower than a predefined threshold.

© Springer International Publishing AG, part of Springer Nature 2018
D. Phung et al. (Eds.): PAKDD 2018, LNAI 10937, pp. 500–512, 2018.
https://doi.org/10.1007/978-3-319-93034-3_40

Table 1. Example dialogues of a rule-based chatbot and a context-uncertainty-aware chatbot.

Turn	Role	Rule-based chatbot	Context-uncertainty-aware chatbot
1	user	*inform* (title=[("dinner", 0.6)], invitee=[("Mike", 0.6)])	*inform* (title=[("dinner", 0.6)], invitee=[("Mike", 0.6)])
2	bot	*confirm* (title)	*confirm* (title)
3	user	*inform* (title=[("dinner", 0.7)])	*inform* (title=[("dinner", 0.7)])
4	bot	*confirm* (invitee)	*request* (time)
5	user	*inform* (invitee=[("Mike", 0.7)])	*inform* (time=[("6 p.m.", 0.6)])
6	bot	*request* (time)	*request* (location)
7	user	*inform* (time=[("6 p.m.", 0.6)])	*inform* (location=[("Korean BBQ", 0.6)])
8	bot	*confirm* (time)	*complete* ()
9	user	*inform* (time=[("6 p.m.", 0.7)])	
10	bot	*request* (location)	
11	user	*inform* (location=[("Korean BBQ", 0.6)])	
12	bot	*confirm* (location)	
12	user	*inform* (title=[("Korean BBQ", 0.7)])	
14	bot	*complete* ()	

However, in a practical scenario, this threshold may vary from person to person, and is also related to a specific dialogue context. Meanwhile, users would expect chatbots not only to complete the tasks required, but also complete in a limited *dialogue length*. A fixed threshold cannot adapt to various users' utterance confidences and may lead to lengthy dialogues, which might discourage the use of chatbots. To illustrate the problem, We use Table 1 to show two interaction sequences between a user and two different chatbots to create a calendar event for "dinner with Mike at 6 p.m. at Korean BBQ". The user input in these sequences is represented as an *"inform"* tuple, which contains slots including *title*, *invitee*, *time*, and *location* of the event. These slots are generated by ASR and NLU systems, which are beyond the scope of our study. Each slot is associated with a number representing the user utterance confidence proposed by the ASR and NLU systems. The "Rule-based chatbot" column showcases how a rule-based chatbot may interact with the user. In each turn, the rule-based chatbot takes one of three possible actions as a response to user input: (1) *confirm* (to request a confirmation of a slot previously captured from user input with low confidence), (2) *request* (to request a new slot from user), and (3) *complete* (to set up the calendar event and finish the conversation). The rule-based chatbot confirms with the user for each slot until its confidence reaches a fixed threshold 0.7. This has resulted in a lengthy dialogue with 14 turns.

We aim to overcome the problem of fixed confidence thresholds as illustrated above with a chatbot, which can adaptively choose a threshold according to the user and dialogue context. We call such a chatbot a *context-uncertainty-aware* chatbot. The "Context-uncertainty-aware chatbot" column of Table 1 illustrates how such a chatbot will interact with a user. This chatbot also has a starting confidence threshold of 0.7, and it needs to confirm with the user for the first slot (*title*) that has a confidence below this threshold (Turn 2). Once this is confirmed, the chatbot learns that only a threshold of 0.6 is sufficient to accept the input of this user. As a result, the *invitee* slot (and any slots afterwards) which also has a confidence of 0.6 does not need a confirmation anymore. This has shortened the dialogue to 8 turns and improved the user experience.

We take the above issues into account and propose an RL model named *PA4C* for chatbot training. This model addresses the following two problems of existing RL models. The first problem is that, in chatbot training, traditional RL models often have a large space for action selection, making it difficult to learn the best action to be selected with a large reward. The output of these models at a turn of the chatbot is a one-hot vector indicating which action should be selected. An action of the chatbot consists of two components: a function (the type of actions) and its parameter (slots). For example, the action *request(time)* has a function *request* and a parameter *time*. Traditional RL models simply list all possible combinations of functions and slots. Suppose that there are M action functions and N slots. Then the number of actions in these models will be $M \times N$. To reduce the action space and improve the reward, we introduce the *action parameterization* to separate the actions into two channels: one for functions and the other for parameters (cf. Fig. 1b). In this way, the action space can be reduced from quadratic (i.e., $M \times N$) to linear (i.e., $M + N$).

The second problem addressed is that only a few states in dialogues have positive rewards. This makes early discovery of states that may lead to large rewards difficult. RL models thus may encounter a bottleneck due to missing valuable states. Traditional methods only focus on the target task (e.g., predict the target action) without explicitly paying attention to estimate the reward of states. Inspired by [2], we propose to add additional tasks to the chatbot during training and guide it to discover large-reward states (detailed in Sect. 5.2). In particular, we design two auxiliary networks: (1) a reward prediction network for predicting the reward of dialogue states, and (2) a value function replay network for helping the RL model estimate the expected state value.

This paper makes the following contributions:

- To the best of our knowledge, we are the first to propose a context-uncertainty-aware chatbot that is self-adaptive to the uncertainty of users' utterance confidence in a dialogue context via reinforcement learning.
- To overcome the quadratic action space problem in chatbot training by reinforcement learning, we propose the action parameterization technique which learns the functions and slots in two separate channels.

– We further propose two auxiliary networks to guide our model to pay extra attention to valuable states, which is more robust to both short-term immediate rewards and long-term expected returns.

2   Related Work

Most commercial chatbots are based on hand-crafted rules design for dialogue state tracking and action selection, which naively choose the action with the largest NLU confidence [16]. It is non-trivial to create a large set of rules to cover diverse user utterances. To avoid manually developing rules, machine learning approaches have been used to build chatbots. Chatbot training has been modeled as a *Markov decision process* (MDP) [3] or a *partially observable Markov decision process* (POMDP) [17]. It is then formulated as a sequential labeling problem in *The Dialog State Tracking Challenge* (DSTC) [1,16].

Machine learning approaches, however, need a large amount of training data, and it is labor-intensive to prepare such data. *Reinforcement learning* (RL) then is used to reduce the amount of training data needed. A noticeable progress has been made on training chatbots with RL models [4,8,18], which demonstrates the feasibility of training chatbots via reinforcement learning. However, due to the "cold-start" problem in reinforcement learning, existing RL approaches have to use supervised learning as a bootstrap. These approaches may interfere with the action exploration of reinforcement learning and cause many ceilings on the success rate, length and episode reward of dialogues. To the best of our knowledge, no existing studies have tackled the problem of breaking the ceilings of RL models in chatbot training. Our study aims to address it.

3   Preliminaries

We start with basic concepts in *deep reinforcement learning* (DRL). A DRL model is essentially a *Markov decision process* (MDP). It can be defined as a tuple $\langle \mathcal{S}, \mathcal{A}, \mathcal{P}, \mathcal{R}, \gamma \rangle$, where $\mathcal{S}$ denotes the environment state; $\mathcal{A}$ denotes the action space; $\mathcal{P}$ denotes the transition probability $P(s_{t+1}|s_t, a_t)$; $\mathcal{R}$ denotes the expected immediate reward function $\mathcal{R}(s_t, a_t)$; γ denotes a discount factor, $\gamma \in (0, 1]$ [19]. The goal of a DRL model is to maximize the *return* (cumulative expected rewards from the state s_t) $G_t = \sum_{k=0}^{\infty} \gamma^k r_{t+k}$, where $r_t = \mathcal{R}(s_t, a_t)$. To maximize the return, there are two approaches in general: value-based DRL and policy-based DRL.

3.1   Value-Based DRL

Value-based DRL estimates the value of executing different actions in a state, and selects the action with the largest value. A typical value-based model is *Q-learning*, where "Q" represents the action value. Q-learning defines an action-value function $Q^\pi(s, a)$, where a is an action, s is a state, and π is a policy to be learned. It aims to find an optimal policy π with the maximum Q value according to $Q^*(s,a) = \max_\pi \mathbb{E}[G_t|s_t = s, a_t = a, \pi]$ and $\pi^*(s) = \text{argmax}_a Q^*(s,a)$ [6]. *Bellman equation* [15] is used to search for the optimal policy:

$$Q^*(s_t, a_t) = \mathbb{E}_{s_{t+1}} \left[r_t + \gamma \max_{a_{t+1}} Q^*(s_{t+1}, a_{t+1})|s_t, a_t \right] \tag{1}$$

Here, s_{t+1} is the next state and a_{t+1} is any possible actions for s_{t+1}.

3.2 Policy-Based DRL

Instead of directly optimizing the policy, value-based DRL models estimate a Q value for each action and chooses the action with the maximum Q value. This approach may cause some biases. For example, assume that there are two actions a_1 and a_2 with Q values 50 and 49, respectively. The action a_2 will not be selected, although it may have a larger long-term value than that of a_1.

Policy-based DRL models are proposed to tackle this problem by directly optimizing the policy [13]. A typical approach is called *policy gradient*, which can define a stochastic policy $a = \pi(a|s;u)$, where u is the weight. The total reward can be computed as:

$$L(u) = \mathbb{E} \left[r_1 + \gamma r_2 + \gamma^2 r_3 + ...|\pi(\cdot;u) \right] \tag{2}$$

The gradients can be updated via [13]:

$$\frac{\partial L(u)}{\partial u} = \mathbb{E} \left[\frac{\partial log\pi(a|s;u)}{\partial u} Q^\pi(s,a) \right] \tag{3}$$

4 User Simulator

We construct a user simulator for chatbot training. As an example application, we focus on training a chatbot for calendar event creation tasks, although the techniques proposed may be applied to train a chatbot for other tasks. All available user actions, system actions, and slots are defined in Table 2.

Table 2. Available actions and slots in the user simulator

User actions	*inform(slot), confirm_deny(slot), confirm_accept(slot), complete(), abort(), dont_care(slot)*
System actions	*greeting(), request(slot), confirm(slot), complete(), abort()*
Slots	title, time, invitee, location

Table 3. User intent database: the number in each tuple represents a confidence level, which is the product of ASR and NLU confidence

Id	Title	Time	Location	Invitee
1	("dinner", 0.9)	("7 p.m.", 0.95)	("Korean BBQ", 0.87)	("Michael", 0.54), ("Mike", 0.46)

Algorithm 1. User simulation

```
1: Initialize:
2:      T ← 0, complete ← False, terminal ← False
3:      Ω ← RandomSelect(IntentDB)              // user intent (a distribution)
4:      ω ← Sample(Ω)                           // user real goal
5:      ω' ← {}                                 // chatbot recorded goal
6:      z ← RandomSelect(NoiseDB)               // noise, a pair of (μ, σ)
7:      𝒜_user ← null                           // user action
8:      𝒜_bot ← Action.greeting                 // chatbot action
9: repeat
10:     𝒜_user ← UserRespond(𝒜_bot, Ω, z)       // randomly selected with 𝒜_bot, Ω, z
11:     𝒜_bot ← ChatbotRespond(𝒜_user)          // ouputted by RL model
12:     T ← T + 2
13:     ω' ← ω' + ParseEntity(𝒜_bot)
14:     if T > T_max or 𝒜_bot == Action.complete then
15:         terminal ← True
16:     reward ← R(terminal, complete, T)
17: until terminal
18: if ω == ω' then complete ← True
19: reward ← R(terminal, complete, T)
```

4.1 Data Preparation

We collect 300 calendar events from volunteers through a data collector website, including 300 titles, 300 time, 113 invitees and 173 locations. Since the uncertainty produced by ASR and NLU may cause a user's real intent to be distorted, such as "Mike" being misunderstood into "Michael", we construct a database *IntentDB* to store these intents with probabilities as shown in Table 3. In addition, we calculate the mean μ and the standard deviation σ of the ASR & NLU uncertainty for each dialogue in The Dialogue State Tracking Challenge 2 (DSTC2) dataset (in the domain of restaurant booking) [16]. We save the pairs of μ and σ into a *NoiseDB*. At each step of a dialogue simulation, a Gaussian noise with μ and σ will be introduced to augment the collected data (detailed in Sect. 4.2). In this way, we can simulate millions of dialogues by adding different noises to the 300 events collected from volunteers.

4.2 Dialogue Simulation

The simulation is based on the fact that a user's intent is known by himself (i.e., user simulator) but unknown by the chatbot. A step of a simulation includes two parts: (1) the chatbot asks the user a question, and (2) the user answers the question honestly. Each step will be assigned with a scalar reward according to the reward function $\mathcal{R}(terminal, complete, T)$ defined in the Eq. 4. The goal of a chatbot is to learn to ask a user questions which can maximize the total reward of a dialogue. The intuition of $\mathcal{R}$ is to give a higher reward to states that lead to a short dialogue that completes a user task successfully.

$$\mathcal{R}(terminal, complete, T) = \begin{cases} -0.1, & \text{if not } terminal \\ 2.0 \times \frac{T_{max}-T}{T_{max}}, & \text{if } terminal \text{ and } complete \\ -1.0, & \text{if } terminal \text{ and not } complete \end{cases}$$

(4)

Here, *terminal* indicates whether the simulation finishes, *complete* indicates whether the chatbot completes the task, T represents the length of the current dialogue, and T_{max} represents the predefined maximum length of a dialogue. A simulation terminates once T_{max} steps are executed regardless whether the user task has been completed. The reward function is Markovian because it only depends on the current dialogue state (i.e. *terminal*, *complete*, and T). At the end of each simulation, we compare the user's real goal ω (i.e., the *ground truth*) with the entities ω' captured by the chatbot. If ω is equal to ω', it means that the chatbot completes the user's task. Otherwise, the task fails. Algorithm 1 shows the detail of a simulation.

5   Proposed Model

As mentioned in Sect. 2, existing works usually use supervised learning as bootstrap before training with RL models. This approach may interfere with the action exploration and cause a ceiling on RL models because the data collected may produce some biases. In this paper, we propose a model named *parameterized auxiliary asynchronous advantage actor-critic* (PA4C) based on the A3C model [5]. Our PA4C model consists of two parts: (1) **Parameterized A3C** (PA3C), which solves the huge action space problem in traditional RL models for chatbot training; (2) **Auxiliary tasks**, which helps model discover the states with large rewards and enhances model robustness [2].

5.1   Parameterized A3C

PA3C is built on the vanilla A3C model, which is a hybrid value-based and policy-based DRL model. It can be expressed by two separate networks with shared weights (cf. Fig. 1a): (1) *Actor* network outputs the policy $\pi(a_t|s_t; u)$; (2) *Critic* network estimates a state value function $V(s_t; v)$ [5], where u and v denotes the weights of the actor and critic network, respectively.

$$V(s_t; v) = \mathbb{E}\left[r_{t+1} + \gamma r_{t+2} + \gamma^2 r_{t+3} + ...|s_t\right] \tag{5}$$

A function $A(s_t, a_t; v, u)$ is defined in Eq. 6 to estimate the advantage of executing a_t on s_t over the state value $V(s_t; v)$:

$$A(s_t, a_t; u, v) = \sum_{i=0}^{k-1} \gamma^i r_{t+i} + \gamma^k V(s_{t+k}; v) - V(s_t; v) \tag{6}$$

Therefore, the weights u of actor network can be updated as follow:

$$\frac{\partial L(u)}{\partial u} = \frac{\partial log\pi(a_t|s_t; u)}{\partial u} A(s_t, a_t; u, v) \tag{7}$$

The critic network can be optimized by minimizing the MSE loss [5]:

$$L(v) = (A(s_t, a_t; u, v))^2 \tag{8}$$

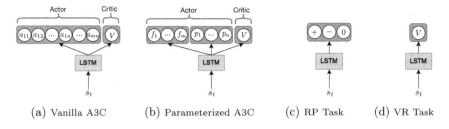

(a) Vanilla A3C   (b) Parameterized A3C   (c) RP Task   (d) VR Task

Fig. 1. Model structure: (a) Vanilla A3C model. (b) Parameterized A3C model. (c) Reward prediction sub-model. (d) Value function replay sub-model.

Like other RL models, vanilla A3C still cannot avoid the large action space problem. For example, assume that there are a set of M *functions* $\{f_1, f_2, ..., f_M\}$ and N *parameters* $\{p_1, p_2, ..., p_N\}$ available (functions and parameters are system actions and slots respectively in the user simulator). For vanilla A3C, the number of actions is $M \times N$, because its actor network simply outputs all combinations of functions and parameters (cf. Fig. 1a).

To address this problem, we propose PA3C by introducing action parameterization into the vanilla A3C model. Rather than listing all combinations, our model will split the policy π of vanilla A3C into two sub-policies π_f and π_p, which directly learns functions and parameters, respectively (cf. Fig. 1b). In this manner, the number of actions in our model can be reduced from $M \times N$ (quadratic) to $M + N$ (linear). Correspondingly, the loss of actor network in PA3C has a slight difference. We modify the loss defined in Eqs. 7–9, where u_f and u_p denote the weights of π_f and π_p respectively, and Λ is a mask vector for indicating whether the function f_m has a parameter.

$$
\begin{aligned}
\frac{\partial L(u)}{\partial u} &= \frac{\partial L(u_f)}{\partial u_f} + \Lambda \frac{\partial L(u_p)}{\partial u_p} \\
&= \left(\frac{\partial log \pi_f(f_t|s_t; u_f)}{\partial u_f} + \Lambda \frac{\partial log \pi_p(p_t|s_t; u_p)}{\partial u_p} \right) A(s_t, f_t, p_t; u_f, u_p, v) \quad (9)
\end{aligned}
$$

5.2 Auxiliary Tasks

A recent DRL model named *UNsupervised REinforcement and Auxiliary Learning* (UNREAL) [2] suggests that incorporating reasonable auxiliary tasks can improve the model robustness and performance. In chatbot training, the rewards are usually sparse. Only a few states can provide immediate rewards (e.g., large negative or positive rewards only occurs when the chatbot finishes the task). The rewards of most states during a conversation are zeros or very small numbers, making it difficult to learn the values of these states. This may cause many ceilings on RL models used in chatbot training, such as low success rate and lengthy dialogues. To address this problem, we design two auxiliary networks to assist the PA3C sub-model to take into account both short-term immediate rewards and long-term expected returns.

Reward Prediction (RP) is an auxiliary network to predict the immediate reward of the next unseen state given a historical context. It helps PA3C sub-model better evaluate the value of dialogue states, To train this task network, we sample a historical sequence $S_t = (s_{t-k}, s_{t-k+1}, ..., s_{t-1})$ from a replay buffer and predict the reward r_t of the state s_t. Here, we focus on whether the state is valuable rather than the specific reward. Instead of estimating the real value of r_t, RP only predicts the sign of r_t in three classes: *positive, negative* and *zero* (cf. Fig. 1c).

Value function Replay (VR) is an auxiliary network to enhance the state-value function $V(s_t; v)$ (Eq. 5) of the critic network in the PA3C sub-model. The function $V(s_t; v)$ is designed to estimate the long-term expected return of the current state s_t. Therefore, VR shares weights with the critic network in PA3C. The only difference is that the critic network is trained with on-policy, while VR can be trained with off-policy. It will sample a state sequence as input from a replay buffer like the RP network (cf. Fig. 1d). In this way, $V(s_t; v)$ combines the strength of both on-policy and off-policy training, which is more robust when estimating the expected return.

5.3 PA4C Model

Our full PA4C model integrates the PA3C sub-model and auxiliary networks. The full model is illustrated in Fig. 2. Firstly, the PA3C sub-model interacts with the user simulator, which generates a dialogue. The dialogue will be saved into a small replay buffer, where RP and VR can sample historical sequences to update their weights. The final loss of PA4C is the combination of PA3C, RP, and VR networks [2]:

$$L_{PA4C} = L_{PA3C} + \lambda_{RP} L_{RP} + \lambda_{VR} L_{VR} \tag{10}$$

where λ_{RP} and λ_{VR} are the weight factors of the RP and VR networks respectively; L_{PA3C} is the loss of the PA3C network defined in Eq. 9; L_{RP} is the cross entropy loss of the RP network; L_{VR} is the MSE loss (cf. Eq. 8) of the VR network.

Our PA4C model also extends the asynchronous training from vanilla A3C via multiple threads. First, PA3C will create a global network in the main thread

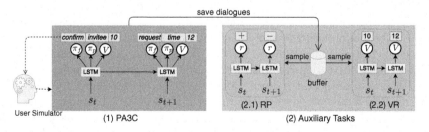

Fig. 2. Full PA4C model consists of (1) PA3C and (2) Auxiliary Tasks (RP and VR networks). All LSTM layers share the same weights.

and multiple local networks in several independent threads. The global network is responsible for dispatching gradients to each local network, which will run a user simulator to compute the local gradients. At the end of each simulation, the gradients of local networks will be aggregated back to the global network.

6   Experiments

In this section, we compare the performance of chatbots trained with our PA4C model and with baseline models in six metrics: *success rate* (SR), *dialogue length* (DL), *episode reward* (ER), and their standard deviations ("std" for short in the rest of the paper) over 10 runs, i.e., std SR, std DL, and std ER [7,9]. All the DRL models are trained on 180 dialogues with noises and evaluated on 120 dialogues. The result shows that our model outperforms the state-of-the-art models in these metrics for training a chatbot in the calendar event creation task.

6.1   Baseline Models and Hyperparameters

We implement 4 baselines, including a rule-based model and 3 existing DRL models:

- **Rule-based.** We implement a rule-based chatbot by *if-else* triggers. If a slot is informed by a user and its confidence is larger than a predefined threshold 0.724 (the average confidence of the collected events), the chatbot will request the next slot; Otherwise, the chatbots will confirm it with the user. When all slots are obtained, the chatbot will finish a simulation.
- **DQN.** We stack two consecutive states as the input s_t at step t, and use two fully-connected layers with 256 and 64 hidden units, respectively. The replay buffer size is 10^5 and batch size is 128. In the first 10^6 step, we use the *epsilon-greedy* policy to explore actions, with ε annealing from 1.0 to 0.1.
- **DRQN.** We only feed one single state into DRQN at step t. Unlike DQN, the replay buffer for DRQN stores the full episode of a dialogue. The *timestep* of LSTM is set to 10. The remaining hyperparameters are the same as those in DQN.
- **A3C.** We use LSTM in A3C, and update the gradients in local networks per episode rather than a fixed step in vanilla A3C [5]. We also set the regularization of policy entropy $\beta = 0.1$ to encourage action exploration and train with 16 threads.
- **A4C.** We only add auxiliary networks into A3C. Both λ_{RP} and λ_{VR} in Eq. 10 are set to 1.0. The replay buffer for auxiliary networks is 2000. When training the RP network, we clip the rewards whose absolute value is smaller than 0.1 to 0 to get their sign (i.e., positive, negative, zero). We assume these small rewards cannot guide to discover valuable states. Other settings are identical to those in A3C.
- **PA3C.** We only add action parameterization into A3C. The number of actions now becomes to 8. Other settings are identical to those in A3C.

– **PA4C.** In our full PA4C model, we integrate PA3C with A4C sub-models. The hyperparameters of these two sub-models are the same as those in A4C and PA3C.

The input feature is composed of a 52-dimensional vector provided by the user simulator, including the last system action, the current user action, and the current dialogue length. All LSTM layers have 256 hidden units, followed by an ReLU activation function. The discounted factor of reward γ is 0.99 for all models. We use the RMSProp optimizer for gradients computing with learning rate $\eta = 0.001$ and weight decay $\alpha = 0.99$. The maximum dialogue length T_{max} is set to 20. The action *greeting()* is removed from the set of actions because *greeting()* is always called firstly by the chatbot. The number of actions for non-parameterized models and parameterized models are $M_p \times N + M_{np} = 2 \times 4 + 2 = 10$ and $M + N = 4 + 4 = 8$, respectively, where M_p denotes the number of functions with parameters, i.e., {*request(slot)*, *confirm(slot)*}; M_{np} denotes the number of functions without parameters, i.e., {*complete()*, *abort()*}; N denotes the number of slots, i.e., {*title, time, location, invitee*}; $M = M_p + M_{np}$.

6.2   Results

In the section, we compare the performance of the PA4C model with baselines.

Comparison with the Rule-based Model: The result in Table 4 shows that PA4C outperforms the rule-based model in all metrics: 47% in success rate (SR), 52% in std SR, 42% in dialogue length (DL), 62% in std DL, 35% in episode reward (ER) and 54% in std ER. In particular, PA4C can achieve over 0.93 in SR, with less than 7.6 turns to complete the task. In contrast, the rule-based model just reaches 0.63 in SR while it takes more than 12 turns. This shows that PA4C can adapt to the utterance uncertainty according to the dialogue context. It can produce a high success rate and a short dialogue.

Table 4. Performance comparison. Larger *success rate (SR)* and *episode reward (ER)* indicate better performance. Smaller *dialogue length (DL)*, *std SR*, *std DL* and *std ER* indicate better performance. The number in the parenthesis shows the improvement over the rule-based model. When computing the improvement, we scale the value of *ER* to $[0, +\infty]$ because of negative rewards. The results are averaged over 10 runs

Model	SR	std SR	DL	std DL	ER	std ER
RULE	0.634	0.052	12.925	5.769	−0.286	1.018
DQN	0.794(+25%)	0.041(−21%)	**7.075**(−45%)	**1.889**(−67%)	0.443(+29%)	0.651(−36%)
DRQN	0.691(+09%)	0.046(−12%)	10.925(−15%)	2.429(−58%)	−0.194(+04%)	0.999 (−02%)
A3C	0.843(+33%)	0.037(−29%)	9.889(−23%)	3.075(−47%)	0.302(+23%)	0.560(−45%)
A4C	0.854(+35%)	0.034(−35%)	10.402(−20%)	3.675(−36%)	0.261(+22%)	0.669(−34%)
PA3C	0.900(+42%)	0.029(−44%)	8.382(−35%)	2.894(−50%)	0.519(+32%)	0.492(−52%)
PA4C	**0.932**(+47%)	**0.025**(−52%)	7.558(−42%)	2.216(−62%)	**0.585**(+35%)	**0.469**(−54%)

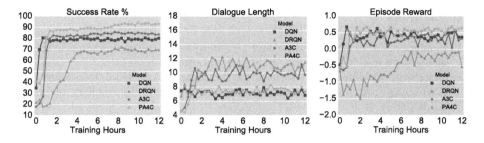

Fig. 3. Performance comparison with RL models: DQN, DRQN, A3C, and PA4C. Each figure is averaged over 10 runs.

Comparison with RL Models: Table 4 also shows that PA4C outperforms existing RL models, with 14%, 19%, and 12% improvement in SR, DL, and ER, respectively, compared with the most recent RL model A3C. We further illustrate the learning curves of SR, DL, and ER in Fig. 3. As the figure shows, although DQN has shorter dialogues, it sacrifices the success rate. It often fails when interacting with low-confidence users, while our PA4C model is more robust to these users.

Comparison within Sub-models: To verify the effect of action parameterization and auxiliary tasks, we compare A3C, A4C, PA3C, and PA4C. Table 4 shows that PA4C outperforms sub-models in all metrics. Using action parameterization achieves 9% improvement in SR, 12% in DL, 9% in ER over A3C. Although only adding auxiliary tasks to A3C (i.e., A4C) does not have such significant effect, the auxiliary tasks do help boost the model performance when integrated together with action parameterization to A3C (i.e., PA4C).

7 Conclusion

We presented a context-uncertainty-aware chatbot trained via reinforcement learning. A user simulator is designed to simulate the uncertainty of different users' utterance confidence. Our chatbot trained with this simulator can adapt to different users' utterance confidence based on the dialogue context. We proposed a reinforcement learning model named PA4C to optimize chatbot training, which can avoid a large action selection space via action parameterization and can discover valuable states via auxiliary tasks. We evaluate our model by training a chatbot for the calendar events creation task. Experimental results show that our PA4C model outperforms the state-of-the-art models in the metrics of success rate, dialogue length, and episode reward.

Acknowledgments. This work is supported by Australian Research Council (ARC) Future Fellowships Project FT120100832 and Discovery Project DP180102050.

References

1. Henderson, M., Thomson, B., Young, S.: Deep neural network approach for the dialog state tracking challenge. In: SIGDIAL (2013)
2. Jaderberg, M., Mnih, V., Czarnecki, W.M., Schaul, T., Leibo, J.Z., Silver, D., Kavukcuoglu, K.: Reinforcement learning with unsupervised auxiliary tasks. In: ICLR (2017)
3. Levin, E., Pieraccini, R., Eckert, W.: Learning dialogue strategies within the Markov decision process framework. In: Automatic Speech Recognition and Understanding, IEEE (1997)
4. Li, X., Chen, Y.N., Li, L., Gao, J.: End-to-end task-completion neural dialogue systems. arXiv preprint arXiv:1703.01008 (2017)
5. Mnih, V., Badia, A.P., Mirza, M., Graves, A., Lillicrap, T., Harley, T., Silver, D., Kavukcuoglu, K.: Asynchronous methods for deep reinforcement learning. In: ICML (2016)
6. Mnih, V., Kavukcuoglu, K., Silver, D., Rusu, A.A., Veness, J., Bellemare, M.G., Graves, A., Riedmiller, M., Fidjeland, A.K., Ostrovski, G., et al.: Human-level control through deep reinforcement learning. Nature **518**(7540), 529–533 (2015)
7. Pietquin, O., Hastie, H.: A survey on metrics for the evaluation of user simulations. Knowl. Eng. Rev. **28**(1), 59–73 (2013)
8. Rojas-Barahona, L.M., Gasic, M., Mrksic, N., Su, P., Ultes, S., Wen, T., Young, S.J., Vandyke, D.: A network-based end-to-end trainable task-oriented dialogue system. In: EACL (2017)
9. Schatzmann, J., Weilhammer, K., Stuttle, M., Young, S.: A survey of statistical user simulation techniques for reinforcement-learning of dialogue management strategies. Know. Eng. Rev. **21**, 97–126 (2006)
10. Sun, Y., Yuan, N.J., Wang, Y., Xie, X., McDonald, K., Zhang, R.: Contextual intent tracking for personal assistants. In: SIGKDD (2016)
11. Sun, Y., Yuan, N.J., Xie, X., McDonald, K., Zhang, R.: Collaborative nowcasting for contextual recommendation. In: WWW (2016)
12. Sun, Y., Yuan, N.J., Xie, X., McDonald, K., Zhang, R.: Collaborative intent prediction with real-time contextual data. ACM TOIS **35**(4), 30 (2017)
13. Sutton, R.S., McAllester, D.A., Singh, S.P., Mansour, Y.: Policy gradient methods for reinforcement learning with function approximation. In: NIPS (2000)
14. Wang, Y., Yuan, N.J., Sun, Y., Qin, C., Xie, X.: App download forecasting: an evolutionary hierarchical competition approach. In: IJCAI (2017)
15. Watkins, C.J., Dayan, P.: Q-learning. Mach. Learn. **8**(3–4), 279–292 (1992)
16. Williams, J., Raux, A., Ramachandran, D., Black, A.: The dialog state tracking challenge. In: SIGDIAL (2013)
17. Williams, J.D., Young, S.: Partially observable Markov decision processes for spoken dialog systems. Comput. Speech Lang. **21**(2), 393–422 (2007)
18. Williams, J.D., Zweig, G.: End-to-end LSTM-based dialog control optimized with supervised and reinforcement learning. arXiv preprint arXiv:1606.01269 (2016)
19. Zhao, T., Eskénazi, M.: Towards end-to-end learning for dialog state tracking and management using deep reinforcement learning. In: SIGDIAL (2016)

Learning Product Embedding
from Multi-relational User Behavior

Zhao Zhang[(✉)], Weizheng Chen, Xiaoxuan Ren, and Yan Zhang

Peking University, Beijing, China
zhangzhao199323@gmail.com, cwz.pku@gmail.com, renxiaoxuan@pku.edu.cn,
zhy@cis.pku.edu.cn

Abstract. Network embedding is a very important method to learn low-dimensional representations of vertexes in networks, which is quite useful in many tasks such as label classification and visualization. However, most existing network embedding methods can only learning embedding from single relational network, which only contains one type of edge relationship between two nodes. However, in real world, especially in product network, many information is presented in multi-relational network. Based on user behavior, edges in product network have many types: "co-purchasing", "co-viewing", "view after purchasing" and so on. Therefore, we propose a novel network embedding method aiming to embed multi-relational product network into a low-dimensional vector space. The results show that our method leads to better performance on label classification and visualization tasks in product network.

Keywords: Multi-relational network · Product network · Embedding

1 Introduction

With the rapid development of e-commerce, a large number of interactive relationships between products based on different types of user behaviors are generated. An effective way to deal with these huge and complex information is to construct a network [1,2]. Based on different user behaviors, we can construct a multi-relational product network, whose vertices refer to products and edges refer to the relationships between different products, such as "co-purchasing", "co-viewing", "view after purchasing" and so on. Figure 1 shows a simple multi-relational product network demo. Many researchers find that network embedding, which is used to represent each vertex of a network with a low-dimensional vector, is a very significant and useful approach to analyze complex information network, such as visualization [3], node classification [4], and recommendation [5].

In the past decades, many network embedding methods have been proposed, such as MDS [6], IsoMap [7], Laplacian eigenmap [8]. These methods suffer from huge computational cost as they are based on eigen-decomposition. In recent years, researchers, inspired by a natural language processing method: word2vec [9], establish analogies for networks, such as Deepwalk [10], node2vec [11] and

© Springer International Publishing AG, part of Springer Nature 2018
D. Phung et al. (Eds.): PAKDD 2018, LNAI 10937, pp. 513–524, 2018.
https://doi.org/10.1007/978-3-319-93034-3_41

LINE [12]. They perform well for large-scale network. However, all of the above methods can't handle multi-relational product network. Some researchers have focused on heterogeneous network embedding methods in some specific areas, such as text network [13], location information network [14]. The authors of [15] use metapath to learn embeddings from heterogeneous networks, but it is only effective for the networks with multiple types of nodes. All of these methods are not suitable for product network, which contain multi-relational links.

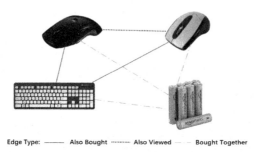

Edge Type: ——— Also Bought ·········· Also Viewed — — Bought Together

Fig. 1. A sample of multi-relational information network. There are four products and three types of edges based on different user behaviors: "Also Bought", "Also Viewed" and "Bought Together". In particular, "Bought Together" means two products are bought together in the same shopping basket. "Also Bought" means the two products are not limited to the same order to buy

There are three challenges to handle multi-relational product network: (1) the existence of directed/undirected and weighted/unweighted edges makes the product multi-relational network very complex, (2) in order to learn the latent similarity from the complex network, a model which is able to integrate multi-relational information is required, (3) the proposed model has to be scalable as the product multi-relational network is massive at most time.

In this paper, we present a novel multi-relational product network embedding method called "MRPNE (Multi-Relational Product Network Embedding)". We divide a multi-relational information network into multiple bipartite networks. We elaborately design objective function, which can compute different networks together. This function is an extension of the LINE algorithm [12]. For the sake of training MRPNE more efficiently, we use stochastic gradient descent and edge-sampling method using alias table for the optimization.

We evaluate MRPNE for label classification and visualization on real-world product data sets. Experiments show that MRPNE outperforms other methods. And the training step of MRPNE is trivially parallelizable, it can scale to large networks with millions of nodes in a few hours.

Overall our paper makes the following contributions:

1. We propose MRPNE, an efficient scalable algorithm for multi-relational product network embedding, which can efficiently optimize a novel neighborhood preserving objective using stochastic gradient descent.

2. MRPNE accelerates the objective function calculating by edge-sampling and updates weights by stochastic gradient, which can be run in parallel.
3. We evaluate MRPNE on real-world multi-relational product network. Experimental results prove the effectiveness and efficiency of our method.

The rest of this paper is organized as follows. In Sect. 2, we discuss related works in product network, network embedding methods and heterogeneous network. Section 3 formally defines the problem of multi-relational product network embedding and describes the details of our method. Section 4 shows our experiments, evaluation metrics, and results. Lastly, we conclude this study in Sect. 5.

2   Related Work

2.1   Product Information Network

Users' various behaviors in e-commerce sites, such as buying, viewing, etc., provide a lot of information. How to mine the commercial value from these information, for example finding the relationships between products and recommending products to consumers, attracts a lot of academic and industrial interests. In the area of business, researchers mainly focus on the characteristics of consumers [16,17], substitutional and complementary relationships between products [18] and so on. In the area of data mining, researchers focus on finding out hidden relationships and association rules in shopping records, such as Apriori [19] and FP-growth [20] algorithms in market basket analysis.

However, traditional techniques of market basket analysis fail to process huge amount of scattered data [2]. In recent years, more and more researchers have begun to apply network analysis in market and goods. By treating products as nodes in networks, this method no longer uses the specific contents of users and products. The key idea of this approach is that information is hidden in the relationship between them. Network-leveled analysis is expected to be more effectively and efficiently used in personalized services, such as cross selling, up selling, and personalized product display, utilizing the deep relation between products [1,21]. Thereby, analysing product information network is a very useful method for mining the relationships between products.

2.2   Network Embedding

It is not very easy to directly deal with network structure, therefore researchers want to find an embedding method to find feature vector representations for these nodes. In the past decades, many network embedding methods have been proposed, such as MDS [6], IsoMap [7], Laplacian eigenmap [8], which utilize spectral properties of various matrix representations of graphs, especially the Laplacian and the adjacency matrices. These methods perform well on small networks, but they are expensive for large real-world networks because these methods are all based on eigendecomposition.

With the development of representational learning for natural language processing, word2vec [9] is put forward to automatically learn the word representations, which uses the context of a word to predict the word (CBOW) or uses the word to predict its context (Skipgram). Inspired by word2vec, Deepwalk [10] is proposed, which uses many random walks on the node as sequences and these sequences are analogous to the contexts in the text. Then, Node2vec [11] develops Deepwalk method, which no longer perform pure random walks, but controls the depth and breadth of random walk with two parameters p and q. And LINE [12] method uses neighborhood nodes as supervisor to learn node feature vector representation. But all of above methods can't handle heterogeneous information network, which contains different kinds of nodes or multi-relational edges.

In some specific areas, researchers propose some heterogeneous network embedding methods. PTE [13], based on LINE [12], is a typical method, which constructs heterogeneous network through word-label, word-word and word-document relationships. However, PTE is a semi-supervised learning model that incorporates label information. Our method, only using the network structure information, is an unsupervised learning method. What's more, the heterogeneity in PTE comes from the text network, and our method is based on the user behavior information to the multi-relational product network.

Metapath2vec [15] is the latest study on heterogeneous networks embedding. It is also based on Deepwalk method [10] using random walk. The author propose to do random walks along metapath. For example, in the author-paper-conference network, the walks must follow "author-paper-conference-paper-author" path. Metapath2vec requires manual selection of a suitable metapath and can only apply to the network with heterogeneous types of nodes. In product networks, there are only one kind of nodes, so metapath does not exist in product networks. In multi-relational product network, Metapath2vec degenerates into the Deepwalk method.

3   Multi-relational Product Network Embedding

In this section, we firstly describe some definitions of our problem: Multi-Relational Product Network, Bipartite Network and Multi-relational Product Network Embedding. Then, we introduce our model and derive loss function. At last, we use two methods to accelerate our algorithm.

3.1   Problem Definition

We formally define the problem of product information network embedding using second-order proximities. We first define an information network as follows:

Definition 1 (*Multi-relational Product Network*). *A product network is defined as $G = (V, E_1, E_2, \cdots)$, where V is the set of vertices, each representing a product object and E_i is the set of edges between the vertices, each representing a relationship between two product objects based on user behavior. Each edge $e \in E_i$*

is an ordered pair $e = (u, v)$ and is associated with a weight $w_{uv} > 0$, which indicates the strength of the relation. If G is undirected, we have $(u, v) \equiv (v, u)$ and $w_{uv} \equiv w_{vu}$.

Definition 2 (Bipartite Network). A bipartite network is defined as $G = (V_A, V_B, E)$, where each edge $e_{ij} \in E$ is a pair $e = (v_i, v_j)$ $vi \in V_A$ and $vj \in V_B$. There is no edge inside V_A or V_B.

Definition 3 (Multi-relational Product Network Embedding). Given a multi-relational product network $G = (V, E_1, E_2, \cdots)$, the problem based on user behavior multi-relational product network embedding aims to represent each vertex $v \in V$ into a low-dimensional space $\mathbb{R}^d$, i.e., learning a function $f_G : V \longrightarrow \mathbb{R}^d$, where $d \ll |V|$. In the space $\mathbb{R}^d$, different types of interaction between the vertices are preserved.

Next, we introduce our model: MRPNE.

3.2 Learning Embedding

As dealing with the multi-relational product network directly is quite difficult, multi-relational networks can be regarded as the combination of several homogeneous networks and we can deal with each kind of edge separately. For example, we only focus on "Also Bought" edge in Fig. 1 and "Also Bought" network is a homogeneous network. Any kind of homogeneous network can be seen as a bipartite network, whose vertices can be divided into two disjoint sets V_A and V_B. And every edge connects a vertex in V_A to one in V_B. We can compute the objective function of each bipartite network. Finally, we combine the results of each homogeneous network together.

Bipartite Network. Because multi-relational product network has different types of edges, we can divide a multi-relational product network into several homogeneous sub-networks. And all these sub-networks can be treated as bipartite networks, which means we decompose the task of multi-relational product network embedding to several sub-tasks of bipartite network embedding.

There are several homogeneous network embedding methods, such as Deep-Walk [10], LINE [12] and node2vec [11]. However, all of them can only deal with one homogeneous network, which means we can't combine the information in several homogeneous sub-networks together. Inspired by LINE [12], we develop a new algorithm to learn the embeddings of multi-relational networks.

In [12], the authors propose a second-order proximity in the network, which assumes that vertices sharing many connections to other vertices are similar to each other. The neighbors of a vertex can be treated as "context" of the vertex and vertices with similar distributions over the "contexts" are assumed to be similar.

Given a bipartite network $G = (V_A \bigcup V_B, E)$, where V_A and V_B represent two separate sets of nodes and E is the set of edges between nodes, the conditional probability of observing vertex v_j in set V_B given the vertex v_i in set V_A is defined as:

$$p(v_j|v_i) = \frac{\exp(\boldsymbol{u}_j'^T \cdot \boldsymbol{u}_i)}{\sum_{k \in B} \exp(\boldsymbol{u}_k'^T \cdot \boldsymbol{u}_i)}, \tag{1}$$

where $\boldsymbol{u}_i$ is the embedding vector of node v_i in V_A, and $\boldsymbol{u}_j'$ is the embedding vector of node v_j in V_B. According to Eq. (1), the closer the embedding vectors of two vertices v_j and v_i are, the more likely observing v_j given v_i. And $p(\cdot|v_i)$ actually computes the conditional probability distribution of node v_i in V_A over all nodes in V_B. Thus the second-order proximity can be preserved if let $p(\cdot|v_i)$ close to its empirical distribution $\hat{p}(\cdot|v_i)$. Therefore, the objective function to be minimize is:

$$O = \sum_{i \in V} \lambda_i d(\hat{p}(\cdot|v_i), p(\cdot|v_i)), \tag{2}$$

where $d(\cdot, \cdot)$ is the KL-divergence between two probability distribution, λ_j is the importance of vertex v_j in the network and can be defined as the degree $degree_i = \sum_j w_{ij}$, and the empirical distribution can be defined as $\hat{p}(v_j|v_i) = \frac{w_{ij}}{degree_i}$. Simplify Eq. (2) and we can get the following formula:

$$O = - \sum_{(i,j) \in E} w_{ij} \log p(v_j|v_i). \tag{3}$$

Equation (3) is a bipartite network embedding loss function we will optimize.

Multi-relational Product Network Embedding. Because we can treat the multi-relational product network as some homogeneous sub-networks, we compute the loss functions of sub-networks separately.

$$O_{MRPNE} = - \sum_k O_k = - \sum_k \sum_{(i,j) \in E_k} w_{ij} \log p(v_j|v_i). \tag{4}$$

3.3   Model Optimization

Computing and optimizing loss function (3) is extremely expensive. Because it requires calculating the similarity between any two nodes in the network. A nice solution is using positive sampling for existent edge [12] and negative sampling for nonexistent edge [22]. The new loss function is as followed.

$$O = \sum_{(i,j) \in E} \left\{ \log \sigma(\boldsymbol{u}_j'^T \cdot \boldsymbol{u}_i) + \sum_{i=1}^{K} E_{v_n \sim P_n(v)}[\log \sigma(\boldsymbol{u}_n'^T \cdot \boldsymbol{u}_i)] \right\}, \tag{5}$$

where $\sigma(x) = \frac{1}{1+\exp(-x)}$ is the sigmoid function. The first term models positive edge samples, while the second term models the negative edge samples and K is the number of negative sampling. The same as in [22], we draw negative edges from the noise distribution $P_n(v) \propto d_v^{3/4}$ where d_v is the out degree of node v. Using negative sampling, we can greatly reduce the amount of calculation.

If we don't use negative edge sampling, we have no ability to choose a suitable learning rate for stochastic gradient descent because some edges have large weights and some edges have much small weights. To address the large variance of edge weights, we sample the edges according to their weights. The probability of edge e_{ij} is $\frac{w_{ij}}{\sum_{(k,l) \in E} w_{kl}}$. If we choose the Roulette wheel selection algorithm, it takes $O(log|E|)$ time to sample an edge, which is a little costly when $|E|$ is large. So we use the alias table method [23] to draw a sample according to the weights of the edges, which takes only $O(1)$ time when repeatedly drawing samples from the same discrete distribution.

Above is the method to calculate the objective function of one network. We want to calculate the total objective function Eq. (5), so we can sequentially sample from each network and then update the embedding vector.

3.4 Algorithm Complexity Analysis

Sampling an edge using alias method takes $O(1)$ time, and negative sampling takes $O(K)$ time, where K is the number of negative samples. Therefore, each step takes $O(K)$ time. So the overall time complexity of MRPNE is $O(KN)$, where N is the number of iterations. In practice, we find that N is usually proportional to the number of edges $|E|$. Therefore, the overall time complexity is $O(K|E|)$, which is linear to the number of edges E. Consequently, MRPNE is quiet efficient by using stochastic gradient descent with edge sampling.

4 Experiments

4.1 Data Sets and Experimental Setup

We use the data from [24]. The authors of the article crawled 9.4 million Amazon products information. These product data consists of 24 subcategories, such as Video Games, Kindle Store, Books, etc. As the data in these categories does not intersect, we conduct experiments with the data in each category. The data set contains four different user behaviors information: which goods users bought that product together with (Bought Together Relationship) in the same order, which goods users bought that product and also bought (Also Bought Relationship), which goods users viewed that product and also viewed (Also Viewed Relationship), which goods users bought that product after viewing (Bought after Viewing Relationship). We use these four user behaviors information to build multi-relational network.

Our experiments evaluate the embedding results on a visualizations task and a standard supervised learning task: label classification for nodes. We evaluate the performance of MRPNE against the following network embedding algorithm:

- DeepWalk [10]: This algorithm learns d-dimensional embedding vector by random walks. As mentioned in Sect. 2.2, Metapath2vec is the same as Deepwalk in product network.
- LINE [12]: This algorithm learns two representations separately, one preserving first-order proximity and the other preserving second-order proximity. Then, it directly concatenates the two vectors to form the final representation.
- Node2vec [11]: This algorithm can be seen as an extension of Deepwalk. Deepwalk only simulates uniform random walks, but Node2vec uses two parameters p and q to control the depth and breadth of random walk.

However, all the baseline algorithms ms can only deal with homogeneous networks. So we run our algorithm MRPNE on the multi-relational product network which contains four types of edges. Then we treat the multi-relational product network as four homogeneous subnetworks and run the three baseline algorithms on them independently to get four independent embedding vectors. At last, we use our embedding result to compare with the four embedding results respectively.

For the different tasks in the following experiments, we set the optimal parameters (p&q) manually. In Sect. 4.2, we choose the best performance parameters: $p = 4, q = 1$. In Sect. 4.3, for the "Also Bought" network, $p = 1$ and $q = 1$ are optimal, which means Node2vec is same as Deepwalk. For "Viewed Together" network, $p = 1$ and $q = 2$. For "Buy After Viewing" network, $p = 1$ and $q = 2$. For "Bought Together" network, $p = 4$ and $q = 1$.

4.2   Visualizations

Displaying a network in a 2-D space is a very important application of network embedding algorithm and also an intuitive method of evaluating the results of network embedding. We visualize "Bought Together" network from Video Game data set, which contains 20,654 nodes and 31,305 edges. We first get network embedding vectors through MRPNE and baseline algorithms. Then, we map the vectors of the vertices to a 2-D space with the t-SNE method [3]. As the video game data set contains more than 1,000 categories of the products and most of the categories are rare, it is not convenient to display all the products. So we choose the three most categories in Video Game "Bought Together" network: "PC games", "Nintendo DS Games" and "Wii Games" and randomly choose 500 products from each category. Figure 2 shows the visualization results with different embedding approaches. "Bought Together" network is relatively sparse and Deepwalk and LINE can not learn effective information, so their visualization are not very meaningful. Node2vec is able to avoid 2-hop redundancy in sampling and performs much better. However, the information that "Bought Together" network has is very limited. In the visualization of Node2vec, points with different category labels are not able to be separated from each other. MRPNE is a multi-relational network embedding method, and it can use much more information than homogeneous network embedding methods. Therefore, MRPNE performs quite well and generates meaningful results.

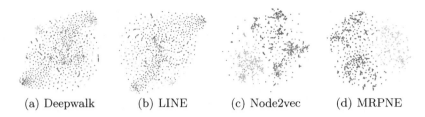

(a) Deepwalk	(b) LINE	(c) Node2vec	(d) MRPNE

Fig. 2. Visualization of the Video Game "Bought Together" network. Products are mapped into the 2-D space using the t-SNE method. Color of a node indicates the category of the product. Blue:"PC games", Red:"Nintendo DS Games", Yellow:"Wii Games" (Color figure online)

4.3 Label Classification

Label classification task is a very classic method for measuring the embedding performance [10–12]. Every node is assigned with one label from a finite set L. During the training phase, we observe a certain fraction of nodes and all their labels. This task is to predict the labels for the remaining nodes. In product network, the label of each product refers to its category. In particular, the category here refers to the most subdivided category. In other words, a product in Amazon has multi-level category, for example, a dance skirt has categories from big to small: "Clothes", "Sports Clothes", "Dance" and "Skirts". Only when all the subdivided categories of two products are the same can we say that they have the same label.

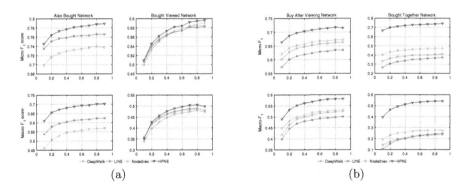

Fig. 3. Performance evaluation of different baselines on label classification task. The x axis denotes the fraction of labeled data, whereas the y axis in the top and bottom rows denote the Micro $-$ F_1 and Macro $-$ F_1 score.

Figure 3 shows the comparison of the results of our algorithm with the baseline algorithms in Video Games products network, which contains 50,953 nodes. The statistics of the four homogeneous sub-networks are as follows:

1. "Also Bought" network contains 28,965 nodes and 1,406,222 edges
2. "Viewed Together" network contains 14,839 nodes and 267,948 edges
3. "Buy After Viewing" network contains 39,905 nodes and 144,070 edges
4. "Bought Together" network contains 20,654 nodes and 31,305 edges

We can see that among the baseline algorithms Node2vec performs well in three networks, but choosing optimal parameters for Node2vec is an extremely time-consuming job. The result of MRPNE is the best, especially in "Bought Together" network. Because "Bought Together" network is a very sparse network, which makes it impossible for baseline algorithms to learn much useful information. However, MRPNE is able to learn the multi-relational information and not only uses "Bought Together" network but also uses "Also Bought","Viewed Together" and "Bought Together" network information. This also shows that considering a single homogeneous information network will result in the loss of a lot of information and learning directly from multi-relational network is much better.

Table 1. Results of label classification on two data sets: Kindle Data Set and Grocery and Gourmet Food Data Set

Metric	Alogrithm	Also Bought	Also Viewed	Buy After Viewing	Bought Together
		Kindle Data Set			
$Micro - F_1$	Deepwalk	67.01	72.61	57.67	34.07
	LINE	66.24	74.68	62.32	36.20
	Node2vec	67.48	74.02	63.23	48.52
	MRPNE	**69.68**	**75.67**	**66.57**	**66.40**
$Macro - F_1$	Deepwalk	44.95	55.26	46.91	33.90
	LINE	50.62	58.48	47.94	35.03
	Node2vec	52.70	58.89	48.08	41.72
	MRPNE	**57.87**	**60.03**	**56.76**	**57.41**
Metric	Alogrithm	Also Bought	Also Viewed	Buy After Viewing	Bought Together
		Grocery and Gourmet Food Data Set			
$Micro - F_1$	Deepwalk	69.00	58.67	63.91	47.07
	LINE	75.19	61.09	64.94	47.10
	Node2vec	73.89	64.66	66.08	48.56
	MRPNE	**75.97**	**54.36**	**66.76**	**55.67**
$Macro - F_1$	Deepwalk	54.11	47.53	45.89	43.28
	LINE	57.45	49.30	50.97	45.03
	Node2vec	56.61	53.12	51.36	44.72
	MRPNE	**59.47**	**67.12**	**51.89**	**51.41**

We also use other data sets to do experiments, but due to space constraints, the results can not be fully displayed. Table 1 shows the experiment results on a part of the data sets using a 50-50 split between labeled and unlabeled data. These results further show that our algorithm, through using the information of the whole multi-relational network, gets better results than only using a single homogeneous network information algorithm.

4.4   Parameter Sensitivity and Scalability

Unlike Node2vec method, MRPNE only uses little hyper-parameters. We investigate how the different choices of parameter dimension d affect the performance of MRPNE. Figure 4 shows the Macro $- F_1$ score w.r.t. the dimension d on the Video Games data set using a 50-50 split between labeled and unlabeled data.

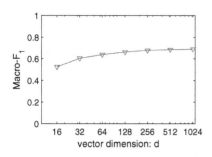

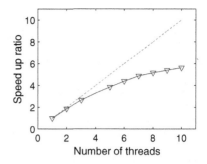

Fig. 4. Macro $-$ F$_1$ score w.r.t. vector dimension d

Fig. 5. Speed up w.r.t number of threads

We can see that with the increasement of the vector dimension, the label classification effect is also good. When the dimension grows from 16 to 128, Macro $-$ F$_1$ score increases very fast. When the dimension is greater than 128, Macro $-$ F$_1$ score is not sensitive to parameter d.

At last, we investigate the scalability of the MRPNE algorithm. Figure 5 reports the speed up ratio w.r.t. the number of threads on the Kindle data set. It shows good parallel performance. That is to say MRPNE algorithm is quite scalable.

5   Conclusions

This paper presents a novel multi-relational product network embedding model called "MRPNE". We carefully design objective functions which can handle complex product network containing different types of edges. Experimental results on multiple tasks prove the efficiency and effectiveness of MRPNE. In the future, we plan to investigate the general embedding method for heterogeneous information networks, which contains multiple types of not only edges but also vertices.

Acknowledgement. The authors would like to thank the anonymous reviewers for their valuable comments and helpful suggestions. This work is supported by NSFC under Grant No. 61532001, and MOE-China Mobile under Grant No. MCM20170503.

References

1. Kim, H.K., Kim, J.K., Chen, Q.Y.: A product network analysis for extending the market basket analysis. Expert Syst. Appl. **39**(8), 7403–7410 (2012)
2. Videla-Cavieres, I.F., Ríos, S.A.: Extending market basket analysis with graph mining techniques: a real case. Expert Syst. Appl. **41**(4), 1928–1936 (2014)
3. van der Maaten, L., Hinton, G.: Visualizing data using t-SNE. J. Mach. Learn. Res. **9**, 2579–2605 (2008)
4. Bhagat, S., Cormode, G., Muthukrishnan, S.: Node 123123123123 classification in social networks, pp. 115–148 (2011)

5. Yu, X., Ren, X., Sun, Y., Gu, Q., Sturt, B., Khandelwal, U., Norick, B., Han, J.: Personalized entity recommendation: a heterogeneous information network approach. In: WSDM, pp. 283–292. ACM (2014)
6. Cox, T.F., Cox, M.A.: Multidimensional Scaling. CRC Press, Boca Raton (2000)
7. Tenenbaum, J.B., De Silva, V., Langford, J.C.: A global geometric framework for nonlinear dimensionality reduction. Science 290(5500), 2319–2323 (2000)
8. Belkin, M., Niyogi, P.: Laplacian eigenmaps and spectral techniques for embedding and clustering. In: NIPS, pp. 585–591 (2002)
9. Mikolov, T., Chen, K., Corrado, G., Dean, J.: Efficient estimation of word representations in vector space. arXiv preprint arXiv:1301.3781 (2013)
10. Perozzi, B., Al-Rfou, R., Skiena, S.: DeepWalk: online learning of social representations. In: KDD, pp. 701–710. ACM (2014)
11. Grover, A., Leskovec, J.: node2vec: scalable feature learning for networks. In: KDD, pp. 855–864. ACM (2016)
12. Tang, J., Qu, M., Wang, M., Zhang, M., Yan, J., Mei, Q.: Line: large-scale information network embedding. In: International World Wide Web Conferences Steering Committee, WWW, pp. 1067–1077 (2015)
13. Tang, J., Qu, M., Mei, Q.: PTE: predictive text embedding through large-scale heterogeneous text networks. In: KDD, pp. 1165–1174. ACM (2015)
14. Xie, M., Yin, H., Wang, H., Xu, F., Chen, W., Wang, S.: Learning graph-based poi embedding for location-based recommendation. In: CIKM, pp. 15–24. ACM (2016)
15. Dong, Y., Chawla, N.V., Swami, A.: metapath2vec: scalable representation learning for heterogeneous networks (2017)
16. Fader, P.S., Lodish, L.M.: A cross-category analysis of category structure and promotional activity for grocery products. J. Mark. 54, 52–65 (1990)
17. Van den Poel, D., De Schamphelaere, J., Wets, G.: Direct and indirect effects of retail promotions on sales and profits in the do-it-yourself market. Expert Syst. Appl. 27(1), 53–62 (2004)
18. Hoch, S.J., Kim, B.D., Montgomery, A.L., Rossi, P.E.: Determinants of store-level price elasticity. J. Mark. Res. 32, 17–29 (1995)
19. Agrawal, R., Srikant, R., et al.: Fast algorithms for mining association rules. VLDB 1215, 487–499 (1994)
20. Han, J., Pei, J., Yin, Y.: Mining frequent patterns without candidate generation. In: ACM Sigmod Record, vol. 29, pp. 1–12. ACM (2000)
21. Yada, K., Motoda, H., Washio, T., Miyawaki, A.: Consumer behavior analysis by graph mining technique. New Math. Nat. Comput. 2(01), 59–68 (2006)
22. Mikolov, T., Sutskever, I., Chen, K., Corrado, G.S., Dean, J.: Distributed representations of words and phrases and their compositionality. In: NIPS, pp. 3111–3119 (2013)
23. Li, A.Q., Ahmed, A., Ravi, S., Smola, A.J.: Reducing the sampling complexity of topic models. In: KDD, pp. 891–900. ACM (2014)
24. McAuley, J., Pandey, R., Leskovec, J.: Inferring networks of substitutable and complementary products. In: KDD, pp. 785–794. ACM (2015)

Vulnerability Assessment of Metro Systems Based on Dynamic Network Structure

Jun Pu[1,2], Chuanren Liu[3], Jianghua Zhao[1,2], Ke Han[4],
and Yuanchun Zhou[1(✉)]

[1] Computer Network Information Center, Chinese Academy of Sciences,
Beijing, China
{pujun,zjh,zyc}@cnic.cn
[2] University of Chinese Academy of Sciences, Beijing, China
chuanren.liu@drexel.edu
[3] Decision Sciences and Management Information Systems, Drexel University,
Philadelphia, USA
[4] Department of Civil and Environmental Engineering, Imperial College London,
London, UK
k.han@imperial.ac.uk

Abstract. Invulnerable metro systems are essential for the safety and efficiency of urban transportation services. Therefore, it is of significant interest to systematically assess the vulnerability of metro systems. To this end, in this paper, we assess the vulnerability of metro systems with a data-driven framework in which dynamic travel patterns are considered. Specifically, we use effective attack strategies based on the topology structure of metro networks. The network structure depends on not only connectivity among metro stations but also dynamic passenger flow patterns. Thus, two data-driven metrics, satisfaction rate (SR) and satisfaction rate with path cost (SRPC), are proposed to quantify the vulnerability of metro networks after our attack strategies. Finally, we conduct experiments on Shanghai metro system. The results indicate that the metro system is vulnerable to malicious attacks while it shows strong robustness to random failures. Our results also highlight weak-points and bottlenecks in the system, which may bear practical managerial implications for policymakers to improve the reliability and robustness of the metro systems and the public transportation services.

Keywords: Metro systems · Network vulnerability · Node centrality
Travel patterns · Dynamic networks

1 Introduction

With the rapid development of the economy, public infrastructure such as metro system has played an essential role in public transportation services for large cities. Due to the severe and ever-increasing traffic pressure in the metropolis,

© Springer International Publishing AG, part of Springer Nature 2018
D. Phung et al. (Eds.): PAKDD 2018, LNAI 10937, pp. 525–537, 2018.
https://doi.org/10.1007/978-3-319-93034-3_42

more and more passengers consider the public transportation services especially the metro system as their primary choices for short-distance travels. For instance, the daily ridership of Shanghai metro system is about 10 million on average, and the annual ridership was 3.4 billion in 2016[1].

Developing at such a rapid pace, metro systems are facing new challenges and requirements on providing secure, reliable, and sustainable services. A number of attacks at city metro systems have occurred in the past few years around the world. Therefore, analyzing the vulnerability of metro systems will help policymakers and urban administrators to develop and operate metro systems in a well-informed way. In particular, the vulnerability analysis for a city metro network provides us with a better understanding of the robustness of the system in the case of unexpected events.

In this paper, we develop a data-driven framework to assess the vulnerability of metro systems based on dynamic topology structure of metro networks. To be specific, we construct the metro network and measure the vulnerability of the network under different ridership distributions. We develop the framework on the basis of various attack strategies and spectral analysis of network centrality. Particularly, we develop a weighted subgraph centrality method to incorporate dynamic passenger flow patterns. Further, we propose two data-driven metrics based on dynamic ridership data to quantify the vulnerability and the transportation efficiency of metro systems. With our framework, we can analyze the vulnerability of metro systems not only from the static topology perspective but also by incorporating dynamic travel patterns.

In the empirical study, we use Shanghai metro system and the daily ridership data to evaluate our framework. The Shanghai metro system, established in 1993, is now the world's largest urban metro traffic system measured by route length. Up to present, Shanghai urban metro system consists of 14 lines (not including magnetic lines) and covers 13 municipal districts with the total route length of 617 km. Our results indicate that the metro system is vulnerable to malicious attacks while it shows strong robustness to random failures. A detailed analysis based on our results provides policymakers and urban administrators with practical managerial implications to improve the reliability and robustness of the metro systems and the public transportation services.

The rest of our paper is organized as follows. Section 2 introduces related works about vulnerability assessment mainly from the view of topology structures of metro networks. Section 3 describes our framework with three parts: attack strategies, subgraph centrality, and vulnerability metrics. Finally, we present experimental results in Sect. 4 and conclude our work in Sect. 5.

2   Related Work

Network vulnerability has been the focus of research in various fields [1–4, 6–10, 16]. As one of the pioneering studies on network vulnerability, Albert and

---

[1] http://www.shmetro.com/.

Barabási [1] studied the robustness of two real networks under malicious attacks and random failures, and showed that the scale-free networks were more vulnerable to intentional attacks. Holme [7] et al. adopted four different attack strategies based on network betweenness to evaluate the vulnerability of the Internet and interpersonal relationship networks. As the transportations around us can be easily modeled with networks, many scientists introduced the vulnerability concept into the transportation field. Jenelius and Mattsson [9] analyzed the road network vulnerability and argued that an event which happened locally may spread on the network thus lead to serious disasters. Berche et al. [2] used two ways to construct the transportation networks, and then studied the resilience of public transportation networks by simulating different attack strategies. Derrible [3] looked at 33 metro systems and used theoretical network methods to analyze the robustness of metro systems. Jenelius and Cats [8] evaluated the network robustness with the consideration of passenger welfare under disruptions after adding new links to networks.

Those studies mentioned above mostly analyzed network vulnerability from the static view of network topology. However, transportation network structures often change with dynamic ridership distributions and passenger travel patterns. A network with different ridership distributions may bear different vulnerabilities and inefficiencies when faced with unexpected changes in the network topology. Therefore, in this paper, we study the problem of vulnerability assessment for transportation systems based on dynamic network structure.

3   Vulnerability Assessment Framework

One straightforward and effective approach to vulnerability assessment of a network system is to conduct a variety of attacks and measure the attack impacts on the network system. Obviously, implementing this approach in real-world is often too costly to be feasible. For example, disrupting the metro networks in cities could lead to social and economic losses of a large population. Therefore, in this paper, we develop a data-driven framework to simulate both random and malicious attacks on the network system, then estimate the attack impacts using vulnerability metrics.

3.1   Attack Strategies

The attack strategies basically fall into two types: random failure and malicious attack [12,18]. These two kinds of attacks are carried out by removing the nodes (e.g., metro stations) and all the edges linked to the nodes. The malicious attack aims at destroying the most important nodes so as to paralyze the functionality of the whole network (e.g., a metro system) to the greatest extent [11,15,17]. The random failure is specified as the dysfunction which results from the failures on some nodes with a random probability [15]. In this work, we adopt these two strategies to randomly and intentionally paralyze (remove) network nodes respectively. For malicious attacks, we target a set of important nodes, where the importance of each node is measured by subgraph centrality [5]. Particularly, we develop the weighted subgraph centrality by incorporating dynamic ridership distributions.

3.2 Subgrpah Centrality

Prior to malicious attacks, a crucial step is to measure the centrality of each station and identify the hubs. Several measurements, such as degree and betweenness, are usually adopted in many studies [2,15]. However, Yan et al. [14] argued that the measurement based on spectral method is a better indicator for network analysis, as it contains abundant characteristics of networks [13]. Therefore, in our study, we adopt a spectral method – subgraph centrality [5] to calculate the importance of each station.

The subgraph centrality is an improvement of degree centrality, and it better explains the centrality from the view of the topology of a network. It determines the importance of each node on the basis of its participation to the local subgraphs. Specifically, it is defined as the sum of closed walks that start and end at a certain node. Each closed walk represents a connected subgraph of the network, and a weight is assigned to it as smaller subgraph gets a higher weight. Thus, the centrality of a node can be measured by counting the sum of the different subgraphs that the node participates in, and smaller subgraph adds more importance to the node. The subgraph centrality is defined as follows:

$$SC(i) = \sum_{k=0}^{\infty} \frac{\mu_k(i)}{k!} \tag{1}$$

$$\mu_k(i) = (A^k)_{ii} \tag{2}$$

where $\mu_k(i)$ denotes the number of closed walks that start and end at node i through k steps. To avoid the case that $\sum_{k=0}^{\infty} \mu_k(i)$ diverges, it is divided by the factorial of length k. $(A^k)_{ii}$ is the i-th diagonal element of the k-th power of adjacency matrix A of a network.

Lemma 1. *Given an undirected network $G(V, E)$, where V is the set of N nodes and E denotes L edges. The upper bound of the subgraph centrality for each node i in V is:*

$$\lceil SC(i) \rceil = e^{\lambda_{main}}$$

where λ_{main} is the main eigenvalue of adjacency matrix A of network G.

Proof. The network G is an undirected graph and thus adjacency matrix A is a symmetric matrix. Therefore, the eigen decomposition for A can be written as: $A = Q\Lambda Q^T$, where Q is an orthogonal matrix (the orthonormal basis) with real and normalized eigenvectors $q_1, q_2, ..., q_N$ of A. The corresponding eigenvalues are $\lambda_1, \lambda_2, ..., \lambda_N$ and the diagonal matrix is written as $\Lambda = diag(\lambda_1, \lambda_2, ..., \lambda_N)$.

Thus we can obtain $A^k = Q\Lambda^k Q$ and $\Lambda^k = diag(\lambda_1{}^k, \lambda_2{}^k, ..., \lambda_N{}^k)$, so $(A^k)_{ii} = \|q_i\|_2^2 \lambda_i^k$. As q_i is an orthogonal and normalized real eigenvector, we can obtain $\|q_i\|_2^2 = 1$ and $(A^k)_{ii} = \lambda_i^k$ thus $SC(i) = \sum_{k=0}^{\infty} \frac{(A^k)_{ii}}{k!} = \sum_{k=0}^{\infty} \frac{\lambda_i^k}{k!}$. Based on Maclaurin series, it turns to: $\sum_{k=0}^{\infty} \frac{\lambda_i^k}{k!} = e^{\lambda_i}$. Thus we arrive at $SC(i) = e^{\lambda_i} \leq e^{\lambda_{main}}$.

Mathematically, the subgraph centrality can be calculated from the spectra of the adjacency matrix of the network, as shown in Theorem 1.

Theorem 1. *Let R^N be the space generated by the eigenvectors of the adjacency matrix A, and $(q_1, q_2, ..., q_N)$ be the orthonormal basis of space R^N. q_j denotes an eigenvector corresponding to the eigenvalue λ_j and $q_j(i)$ is the i-th element in q_j. Thus, the subgraph centrality $SC(i)$ can be obtained by:*

$$SC(i) = \sum_{j=1}^{N} [q_j(i)]^2 e^{\lambda_j} \tag{3}$$

Furthermore, in this paper, we not only measure the centrality based on static network topology, but also propose a weighted measurement by incorporating the ridership distribution. That is, the dynamic ridership matrix W is considered in the measurement instead of the adjacency matrix A. Thus, we modify SC to a weighted one – WSC, which is described as follows:

$$WSC(i) = \sum_{j=1}^{N} [q_{wj}(i)]^2 e^{\lambda_{wj}} \tag{4}$$

where $(q_{w1}, q_{w2}, ..., q_{wN})$ are the eigenvectors corresponding to the eigenvalues $(\lambda_{w1}, \lambda_{w2}, ..., \lambda_{wN})$ of the normalized weight matrix $\widetilde{W}$ which is obtained by normalizing the original ridership matrix W using min-max scaling:

$$\widetilde{W}_{i,j} = \frac{W_{i,j} - \min(W)}{\max(W) - \min(W)} \tag{5}$$

where element $W_{i,j}$ means the number of passengers who have to go through the edge between stations i and j to achieve their destinations.

3.3 Vulnerability Metrics

The attacks (i.e., the removal of stations) will usually have a significant impact on the metro system operation. Thus, we define vulnerability metrics by quantifying such an impact on the vulnerability of the metro network. Specifically, we use the real ridership data to define the evaluation metrics. To this end, we assume that all passengers choose the shortest paths as their travel paths from the start stations to the destination stations. After the attacks, the passengers will cancel their travels by metro if: (i) the length of the new travel path is n stations more than the path before; or (ii) there is no way to connect the source and destination stations.

On the basis of the assumptions, we can estimate the number of travels (ridership) which are not influenced by the attacks, as defined by satisfaction rate (SR):

$$SR = \frac{M - C}{M} = \frac{\sum\limits_{i,j \in G', i \neq j} I(i,j) \times r_{i,j}}{\sum\limits_{i,j \in G, i \neq j} r_{i,j}}, \qquad (6)$$

where

$$I(i,j) = \begin{cases} 0 & \text{if } l'(i,j) - l(i,j) > n \\ 1 & \text{otherwise} \end{cases}. \qquad (7)$$

Note that, M is the total ridership and C is the sum of cancelled travels. G and G' are connected graphs of the original metro network and the network after attacks, respectively. $r_{i,j}$ is the number of ridership that gets on at i and gets off at j. $l(i,j)$ measures the length of shortest path between i and j in G. According, $l'(i,j)$ is the shortest path length calculated in the network G' after attacks. In other words, $I(i,j)$ is the indicator function that indicates whether the new path length between station i and j is n or above longer than that before attacks.

The satisfaction rate (SR) effectively quantifies the changes in ridership due to the attacks on the metro network. However, different ridership often bears different travel cost, as measured by the length or the number of stations travelled by the passengers. To incorporate the travel cost in evaluating the network vulnerability, we define the satisfaction rate with path cost (SRPC), which combines the length of the shortest path and the ridership on the same path:

$$SRPC = \frac{\sum\limits_{i,j \in G', i \neq j} I(i,j) \times r_{i,j} \times l'(i,j)}{\sum\limits_{i,j \in G, i \neq j} r_{i,j} \times l(i,j)} \qquad (8)$$

Both SR and SRPC can be regarded as the efficiency of a metro system, as they estimate to which level the metro system can satisfy the demand of all passengers with efficient transportation services.

4 Experimental Analysis and Discussion

4.1 Data Preprocessing

In 2016, Shanghai Government Data Service Website[2] released several types of public transportation data, including the transaction data of public transportation cards generated in April 2015. The sensitive personal information has been kicked out in the transaction, and it mainly records the elements as: {*card ID, date, time, station, amount and type of transportation*}. According to this data,

---

[2] http://www.datashanghai.gov.cn/.

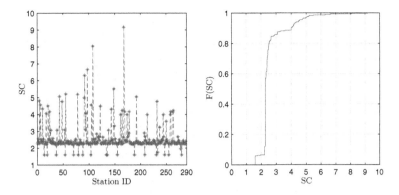

Fig. 1. The left panel shows the subgraph centrality (SC) for each numbered station, and the right one shows the cumulative distribution function (CDF) of SC values.

we first extract the transactions made on the metro system. As a cardholder only have to swipe the card when getting in and getting out of the metro system, we tease out the travel routes based on the records of stations over time for every cardholder. So each pair of stations ordered by time is regarded as one travel and the shortest path between them is the travel route. Thus the ridership on one travel route can be easily obtained by summing up the same pairs of source and destination stations from different cardholders. The map of Shanghai metro network operated in 2015 can be referenced from the website of Shanghai Metro[3].

4.2 Topological Properties of Metro Network

To give an overall understanding of Shanghai metro, some basic topological characteristics of the network are concluded. Firstly, the metro network is constructed as an undirected graph, in which the nodes represent the stations and the links mean that the two connected stations are neighbors. Then we calculate the characteristics, and there are 289 stations and 322 edges within the network. The diameter (the longest shortest path) of the network is 42. The average shortest distance $\langle l \rangle$ is 15.294 and the average degree $\langle k \rangle$ is 2.228, which means most of the stations are normal stations.

Prior to malicious attack simulations, we use the subgraph centrality (SC) to evaluate the importance of each station based on the network topology. The SC results and the cumulative distribution function (CDF) are shown in Fig. 1.

The left figure in Fig. 1 shows the SC values for the stations, and the values of most stations are concentrated in 2.0–2.5. The CDF indicates that nearly 90% stations are with an SC value lower than 4. Afterwards, we pick up the top-10 important stations as the targets of malicious attack. These stations are *Century Ave.* (SC = 9.166), *Xujiahui* (SC = 8.052), *People's Square* (SC =

---

[3] http://service.shmetro.com/en/.

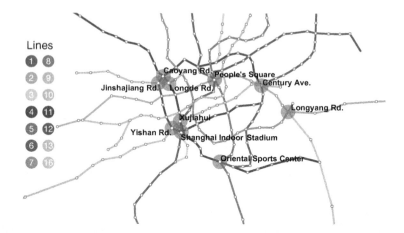

Fig. 2. The top-10 influential stations of Shanghai metro system.

6.642), *Shanghai Indoor Stadium* (SC = 6.291), *Yishan Rd.* (SC = 5.499), *Longde Rd.* (SC = 5.218), *Jinshajiang Rd.* (SC = 5.181), *Caoyang Rd.* (SC = 5.069), *Oriental Sports Center* (SC = 5.038) and *Longyang Rd* (SC = 4.978). Figure 2 depicts the locations of these important stations with red circles. As can be seen from Fig. 2, these stations are located in every direction of the city and they are indeed the important transfer stations in reality. Besides, these stations are all on the edge of downtown areas and serving as the connecting stations between suburban and downtown.

4.3   Vulnerability Analysis and Travel Patterns

Since most lines of Shanghai metro open at 5:30 and close at 23:30, we aggregate the ridership within every three hours from 3:00 to 24:00 for each day. Thus there are 7 aggregated periods on each day and totally 210 periods in April.

The ridership (upper panel in Fig. 3) presents a periodic pattern over time, with rush commuting hours from 6:00 to 9:00 and from 18:00 to 21:00 on every day. However, during the weekends and holidays (April 6th is the Tomb-sweeping Festival of China), the ridership has fallen by half and the rush hours turn to the period from 12:00 to 15:00. To analyze the vulnerability of each time period, we simulate the two attack strategies as described before on the metro system. Firstly, we suppose that the travel will be cancelled if the new shortest path is $n = 5$ (or above) stations longer than before. SR and SRPC are used to indicate the vulnerability of the metro system after attacks. In malicious attacks, we mainly attack and remove the top ten stations which are measured by subgraph centrality (SC). While in random failures, we randomly remove ten stations to calculate the vulnerability metric values and repeat this for ten times. Thus the results for random failures are averaged from ten simulations. The results are shown in the middle and bottom figures of Fig. 3.

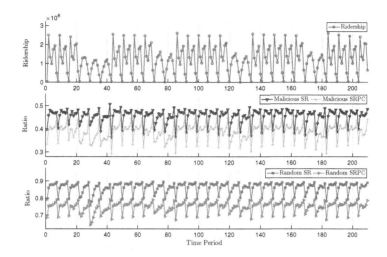

Fig. 3. The top panel shows the ridership aggregated for every three hours from 3:00 to 24:00 of every day in April. The lower two ones show the vulnerability ratios of the metro network after malicious attacks and random failures, respectively.

As can be seen from the middle figure of Fig. 3, SR metric indicates that the metro network appears to be extremely vulnerable to the malicious attacks. Only nearly 50% travels can be still carried on the metro, and the delivery efficiency indicated by SRPC is about 40%. However, from the bottom figure in Fig. 3, the metro network shows strong robustness to random failures. SR metric illustrates that about 90% travels can be satisfied after random failures, and SRPC shows that the delivery efficiency of the metro system reaches 80%. Two metrics both indicate that the most robust period is from 21:00 to 24:00 for each day after malicious attacks and random failures. Whereas, the weakest period is the morning rush-hour from 6:00 to 9:00, as this period is always with the lowest vulnerability value.

To better understand the correlation between the absolute amount of ridership and vulnerability, we pick out the ridership data during the period from 6:00 to 9:00 for each day and the corresponding SR and SRPC values. As shown in Fig. 4(a), SR and SRPC are positively correlated with the ridership after both random failures and malicious attacks.

Besides, the vulnerability results from two metrics exhibit a strong correlation after both random failures and malicious attacks. The reason for this phenomenon can be explained by the definitions of these two metrics. SR mainly considers how many travels are not affected by the attacks, while SRPC focuses on the delivery efficiency of the metro system rather than simply considers the number of the travels. In addition, as can be seen from Figs. 3 and 4(a), SRPC is always lower than SR after all kinds of attacks. This means the travels with length shorter than the average make up the majority of the unaffected travels, and the long-distance travels are more likely to be affected and cancelled by

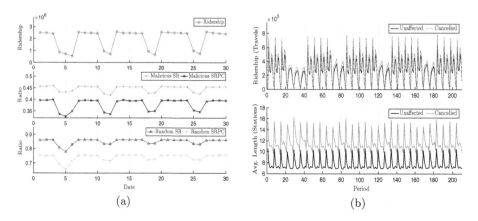

Fig. 4. (a) The correlation between ridership and vulnerability. (b) The upper one shows the number of unaffected and cancelled travels after malicious attacks. The lower one calculates that average distance for these two kinds of travels.

attacks. Specifically, to prove this, we further separately aggregate the unaffected travels and cancelled travels after malicious attacks and then calculate the average distance (the number of stations) of these two kinds of travels. Figure 4(b) indicates that the average distance of the cancelled travels is longer than that of the unaffected ones in all periods. Plus, another interesting travel pattern is that the people who travel from 3:00 to 6:00 (actually 5:30 to 6:00) always have relatively longer travel routes.

Table 1. The time periods for weekday and rest-day.

Weekday		Rest-day	
Period 1	5:00 a.m–11:00 a.m	Period 5	5:00 a.m–1:00 p.m
Period 2	11:00 a.m–5:00 p.m	Period 6	1:00 p.m–8:00 p.m
Period 3	5:00 p.m–8:00 p.m	Period 7	8:00 p.m–11:00 p.m
Period 4	8:00 p.m–11:00 p.m		

Due to the dynamic passenger flow, the metro network can be seen as a weighted one and its structure changes over time with the passenger flow. Therefore, the ridership should be considered when measuring the centrality of each station. As can be seen from Fig. 3, the peak of the ridership appears in different periods on rest-days (weekends and holidays) and weekdays. The peaks are always the rush-hours in weekdays, while in rest-days the peaks appear at period 12:00 to 15:00. To incorporate these facts, we re-aggregate the travels data to several periods for weekdays and rest-days separately, as shown in Table 1. This division of the periods is in accordance with our understanding of the daily traffic flow. The Period 1 and Period 3 are the morning and evening rush hours on

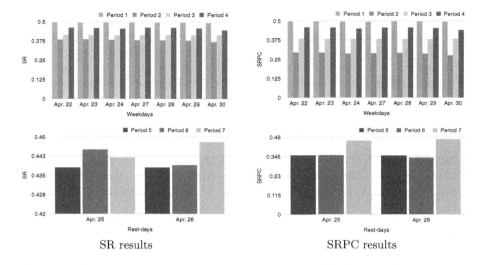

SR results   SRPC results

Fig. 5. The SR and SRPC results of dynamic malicious attacks on the last seven weekdays and two rest days in April, 2015.

weekdays. Thus the important stations in these periods may be different from Period 2 and Period 4 when considering the dynamic distribution of passengers.

We firstly calculate the average weight matrices for each of the seven periods based on the travel data of the first 21 days. The fourteen weekdays are used to calculate the average ridership matrices W_1, W_2, W_3 and W_4 for the four periods on weekdays. The seven rest-days (one holiday and three weekends) are used to calculate the matrices W_5, W_6 and W_7 for the three periods in rest-days. Afterwards, we use the proposed weighted subgraph centrality WSC to measure the top-10 stations for each of these seven periods. Last, the malicious attack strategy as described before is simulated on the seven periods of the remaining nine days (including seven weekdays and two rest-days). The results for dynamic malicious attacks are shown in Fig. 5.

As can be seen from it, on weekdays, more than half of the passengers have to cancel their travels on the metro after the dynamic attacks in every period. In each weekday, Period 1 seems the most robust period of the whole day. However, according to the SR results on two rest-days, every period presents almost the same vulnerability. But SRPC results show that Period 7 (8:00 p.m– 11:00 p.m.) has relatively stronger robustness on delivery efficiency when compared with the other two periods in rest-days.

5   Conclusion

We assess the vulnerability of metro system based on the dynamic network structure and travel patterns. Specifically, we develop a framework combining a series of attack strategies and two vulnerability metrics based on dynamic ridership data. By incorporating the ridership, we further develop a weighted subgraph centrality and simulate a dynamic malicious attack strategy. We conduct the empirical study on Shanghai metro system, and the results indicate that the metro system is vulnerable to malicious attacks while it shows strong robustness to random failures. Our results also high-light weak-points and bottlenecks in the system, which may bear practical managerial implications for policymakers to improve the reliability and robustness of the metro system. In the future, we will deploy our vulnerability analysis on more metro systems.

Acknowledgments. This work is supported by the National Key Research Program of China (No. 2016YFB0501900 and No. 2016YFB1000600). Jun Pu and Chuanren Liu contribute equally to this work. Yuanchun Zhou is the corresponding author.

References

1. Albert, R., Jeong, H., Barabási, A.L.: Error and attack tolerance of complex networks. Nature **406**(6794), 378–382 (2000)
2. Berche, B., Von Ferber, C., Holovatch, T., Holovatch, Y.: Resilience of public transport networks against attacks. Eur. Phys. J. B-Condensed Matter Complex Syst. **71**(1), 125–137 (2009)
3. Derrible, S., Kennedy, C.: The complexity and robustness of metro networks. Physica A: Stat. Mech. Appl. **389**(17), 3678–3691 (2010)
4. Ding, L., Zhang, L., Wu, X., Skibniewski, M.J., Qunzhou, Y.: Safety management in tunnel construction: case study of wuhan metro construction in china. Saf. Sci. **62**, 8–15 (2014)
5. Estrada, E., Rodríguez-Velázquez, J.A.: Subgraph centrality in complex networks. Phys. Rev. E **71**(5), 056103 (2005)
6. Flammini, F., Gaglione, A., Mazzocca, N., Pragliola, C.: Quantitative security risk assessment and management for railway transportation infrastructures. In: Setola, R., Geretshuber, S. (eds.) CRITIS 2008. LNCS, vol. 5508, pp. 180–189. Springer, Heidelberg (2009). https://doi.org/10.1007/978-3-642-03552-4_16
7. Holme, P., Kim, B.J., Yoon, C.N., Han, S.K.: Attack vulnerability of complex networks. Phys. Rev. E **65**(5), 056109 (2002)
8. Jenelius, E., Cats, O.: The value of new public transport links for network robustness and redundancy. Transportmetrica A: Transp. Sci. **11**(9), 819–835 (2015)
9. Jenelius, E., Mattsson, L.G.: Developing a methodology for road network vulnerability analysis. Nectar Cluster **1**, 1–9 (2006)
10. Lee, K., Jung, W.S., Park, J.S., Choi, M.: Statistical analysis of the metropolitan seoul subway system: network structure and passenger flows. Physica A: Stat. Mech. Appl. **387**(24), 6231–6234 (2008)
11. Sun, D.J., Guan, S.: Measuring vulnerability of urban metro network from line operation perspective. Transp. Res. Part A: Policy Pract. **94**, 348–359 (2016)
12. Wang, J.: Robustness of complex networks with the local protection strategy against cascading failures. Saf. Sci. **53**, 219–225 (2013)
13. Watanabe, T., Masuda, N.: Enhancing the spectral gap of networks by node removal. Phys. Rev. E **82**(4), 046102 (2010)
14. Yan, X., Li, C., Zhang, L., Hu, Y.: A new method optimizing the subgraph centrality of large networks. Physica A: Stat. Mech. Appl. **444**, 373–387 (2016)

15. Yang, Y., Liu, Y., Zhou, M., Li, F., Sun, C.: Robustness assessment of urban rail transit based on complex network theory: a case study of the beijing subway. Saf. Sci. **79**, 149–162 (2015)
16. Zhang, J., Xu, X., Hong, L., Wang, S., Fei, Q.: Attack vulnerability of self-organizing networks. Saf. Sci. **50**(3), 443–447 (2012)
17. Zhou, Y., Sheu, J., Wang, J.: Robustness assessment of urban road network with consideration of multiple hazard events. Risk Anal. **37**(8), 1477–1494 (2017)
18. Zhou, Z., Irizarry, J., Li, Q.: Using network theory to explore the complexity of subway construction accident network (scan) for promoting safety management. Saf. Sci. **64**, 127–136 (2014)

Visual Relation Extraction via Multi-modal Translation Embedding Based Model

Zhichao Li[(✉)], Yuping Han, Yajing Xu, and Sheng Gao

Beijing University of Posts and Telecommunications, Beijing, China
lzc6996@foxmail.com

Abstract. Visual relation, such as "person holds dog" is an effective semantic unit for image understanding, as well as a bridge to connect computer vision and natural language. Recent work has been proposed to extract the object features in the image with the aid of respective textual description. However, very little work has been done to combine the multi-modal information to model the subject-predicate-object relation triplets to obtain deeper scene understanding. In this paper, we propose a novel visual relation extraction model named Multi-modal Translation Embedding Based Model to integrate the visual information and respective textual knowledge base. For that, our proposed model places objects of the image as well as their semantic relationships in two different low-dimensional spaces where the relation can be modeled as a simple translation vector to connect the entity descriptions in the knowledge graph. Moreover, we also propose a visual phrase learning method to capture the interactions between objects of the image to enhance the performance of visual relation extraction. Experiments are conducted on two real world datasets, which show that our proposed model can benefit from incorporating the language information into the relation embeddings and provide significant improvement compared to the state-of-the-art methods.

Keywords: Visual relation extraction · Multi-modal network Translation embedding

1 Introduction

Knowledge graph contains abundant semantic relation information from multi-relational data, which is essential for reasoning and question answering. Multi-modal knowledge graph needs to extract more information from multi-modal data. Recently, many visual question answer and image captioning tasks attempt to make machine understand the sematic information in image and language. However ,most existing efforts only do a coarse scene-level understanding, but fall short in modeling and understanding the relationships among objects. As a result, visual relation detection task is receiving increasing attention [1–3]. The

© Springer International Publishing AG, part of Springer Nature 2018
D. Phung et al. (Eds.): PAKDD 2018, LNAI 10937, pp. 538–548, 2018.
https://doi.org/10.1007/978-3-319-93034-3_43

complex compositional structure of language makes problems at the intersection of vision and language challenging.

But language also provides a strong prior for image understanding. We propose a multi-modal framework to learn sematic information from language as well as extracting the relations between objects in images.

Recently, the Visual Translation Embedding Network is proposed for this task [1]. Using relational representation learning of knowledge bases [4], VTransE assumes embeddings of entities and relations being in the same space $\mathbb{R}^k$ and wants $h+r \approx t$ when triplet (h, r, t) holds. We discuss some mapping properties of relations which should be considered in embedding, such as *1-to-N*, *N-to-1*, and *N-to-N*. We note that VTransE does not do well in dealing with these properties. In image relation detection, there are more cases like this than Knowledge graphs such as *"person-ride-horse"*, *"person-ride-bike"*. Hence, we introduce TransR [5] to improve VtransE. TransR models entities and relations in distinct spaces. i. e. , entity spaces and multiple relation spaces (i.e., relation-specific entity spaces), and performs translation in the corresponding relation space. For example, there are $person+ride \approx bike$ and $person+ride \approx horse$ in "ride" space, but in "push" space person is not usually "push" a horse while $person+push \approx bike$ still holds.

Human have the remarkable ability of understanding image, because we can accumulate a lot of knowledge from free text. Inspired by this, we propose an extension to relation detection by adding an extra language module. We utilize pre-trained word vectors to help the model learn the similarity of relations from free text which allows our model to address zero-shot relations. For example, in word vectors space, *horse* and *elephant* are semantically similar, *"person-ride-elephant"* can be inferred by *"person-ride-horse"* , even though *"person-ride-elephant"* is not in the training set. Language prior also has been used to select visual attention. It suppresses non-relevant stimuli present in the visual field, helping model search for "goals".

In this work, we present a multi-modal framework for visual relation detection. For that, we first propose a novel visual relation extraction model to integrate the visual information and respective textual knowledge base, which places objects of the image as well as their semantic relationships in two different low-dimensional spaces where the relation can be modeled as a simple translation vector to connect the entity descriptions in the knowledge graph. Moreover, we also propose a visual phrase learning method to capture the interactions between objects of the image to enhance the performance of visual relation extraction. We then evaluate our model on two public benchmark visual relation datasets: Visual Relationship Dataset [2] and Visual Phrase Dataset [3]. We show that our model significantly outperforms several state-of-the-art visual relation models in visual dataset detection and zero-shot learning.

2   Related Work

Recently, the development of deep learning makes AI technology by leaps and bounds. Some tasks require multi-modal knowledge beyond a single sub-domain,

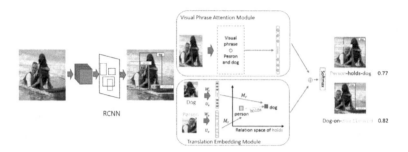

Fig. 1. An illustration of the architecture of our model. Given an image as input, Faster-RCNN outputs a set of detected objects. Then, every pair of objects and word embeddings are fed into relation prediction module.

| Input Image | RoI Feature | Phrase Feature |

Fig. 2. The difference between relation feature and phrase feature. The ROI feature treats two objects as two separate pictures.Visual phrase feature put the two objects in one picture.

such as VQA [6], image captioning [7], and complex query retrieval [8] targeting for "AI-complete" is still a difficult and open research problem. Our work is expected to produce a generalized semantic understanding of the image to underpin high-level vision-language tasks. Visual relation is not a new concept, Lu [2] build a dataset with millions of types of relationships. They present a "visual appearance and language module" modeling visual and language information respectively that can learn to detect visual relationships. Zhang [1] proposed an end-to-end relation detection network named as VTransE which models the visual relationships by TransE [4] and object detection. In particular, our method draws on recent Knowledge Graph Embedding and language module.

2.1   Knowledge Graph Embedding

Knowledge graph with multi-relational data is a very important tool for many high-level tasks, such as QA and retrieval. To complete a knowledge graph, we need to predict what kind of relationship between the two entities. Recently,a promising approach to solve this task is embedding entities and relationships in the knowledge graph in low-dimensional vector spaces.

TransE [4] represents relations and entities in low-dimensional spaces. For triplet (h,r,t), TransE wants $h + r \approx t$. This makes t the nearest neighbor of h + r. To learn such embeddings, TransE assumes the score function:

$$\mathcal{F}_r = ||h + r - t||_2^2 \tag{1}$$

is low if (h,r,t) holds, and high otherwise. TransE works well in 1-to-1 relations, but still has issues in 1-to-N, N-to-N and N-to-1 cases.

TransR [5] assumes that relations and entities are not in the same space. Each relation has its own space, when computing similarity between relation and entity, we should project embedding of entities to relations space firstly. This can address the issues in TransE.

2.2 Visual Relation Detection

Some papers explicitly collected relationships in images [9–13] and videos [11, 14, 15] and helped models map these relationships from images to language. There are some explicit relation models [3,9,16,17] which define a unique triplet (h,r,t) as a class. But in real world images, there is a long tail property of infrequent relationships. Therefore, separate model [1,2,18–20] which model objects and relations individually may perform better. Inspired by VTransE, which combines visual detection and large-scale knowledge representation [4,5], we propose a Multi-modal model to address some issues in VTransE.

Visual relation detection is based on object detection. R-CNN [21] combines selective search region proposals and convolutional network based post-classification brings a dramatic improvement in object detection. Fast-RCNN [22] and Fatser-RCNN [23] improve R-CNN in a variety of ways. In particular, we use Faster-RCNN to provide candidates of entity for relation detection.

3 Method

We propose a multi-modal framework to extract semantic relations between objects in an image. As illustrated in Fig. 1, our multi-modal framework consists of three parts:Objects Detection Module, Visual Phrase Attention Module,Translation Embedding Module. In Objects Detection Module, we detect a set of objects $\mathcal{D} = \{x_n\}_{n=1}^N$ and visual phrase feature $\mathcal{I} = \{I_n\}_{n=1}^M$ for the next two parts. Afterwards, we have projection matrices U and W which project objects and word embeddings to the same space. Visual Phrase Attention Module and Translation Embedding Module incorporate the visual information with the word embeddings. In Translation Embedding Module, a set of relation matrix $\mathcal{M} = \{M_{r1}, M_{r2}, ..., M_{rk}\}$ is defined to construct relation spaces.

Word embeddings are distributed representations of words that map each word in texts to a low-dimensional vector. Many works have shown its power in many NLP tasks. we also use pre-trained word vectors [24] to cast relationships into a vector space where similar relationships are optimized to be close to each other.

4   Objects Detection Module

Our framework is based on an object detection module composed of a region proposal network (RPN) and a classification layer.In practice, we use Faster-RCNN [23]) with the VGG-16 architecture to extract RoI from images. After that, we use a hierarchical structure to incorporate visual and semantic features to retain the objects' similarity in the word embedding space.

5   Visual Phrase Attention Module

We propose an attention structure incorporating the word embeddings into visual information which helps the model focus on objects interested. And we use visual phrase feature to extract more information to describe visual relation between two objects in an image. Visual phrase has shown benefit to improve object detection, i. e., detecting "a person riding a horse" improves the detection and localization of "person" and "horse" [3]. The difference between visual phrase and visual relation can be seen in Fig. 2. The RoI feature extracts the visual features of the detected individual object, and the visual phrase extracts the two objects that may be related as an interactive whole. We believe that a complete visual phrase contains more information about the interaction of two objects, while a single object feature can only provide graphic features of objects and ignoring the relationship. We argue that the complete visual phrase can describe interactions between objects while RoI pooling features of individual object only contains graphic information. As in Wang's work [25], we also use word embeddings to help model search for interactions between objects. Experiment shows that the framework can focus on the corresponding object regions when predicting relations.The score of visual phrase is defined as:

$$S_{p,s,o} = \delta(U_p(\text{word2vec}(o) \circ \text{word2vec}(s)) + U_I I) \tag{2}$$

where $\delta(.)$ is an activation function, $\circ$ means concatenate two vectors, U_p and U_I are the projection matrices for word2vec and visual phrase representation respectively, and I denotes visual phrase feature extracted by objects detection module.

6   Translation Embedding Module

We introduce TransR to address the issue of embedding space in VTransE [1] and we also improve on VTransE by leveraging language priors from pre-trained word embeddings.

VTransE transfers TransE in the visual domain. For triplet $subject - predicate - object$, VTransE wants to model the entities and relations within the same space $\mathbb{R}^k$ as vectors h, r and t which $h + r \approx t$. TransE will encounter a lot of problems when modeling 1-N, N-1 and N-N's relationships. For example, because there are "person-ride-bike" and "person-ride-horse" in the real world, there should be *person + ride ≈ bike* and *person + ride ≈ horse* which makes the vectors of horse and bike very similar. However, when the relation becomes "push", the translation becomes *person+push ≈ bike ≠ horse*. This will confuse our model and hinder the convergence of the model. We introduce TransR to solve the issue of embedding space in VTransE. In our model, we embed objects and relations respectively. When performing translation, we project objects from object space to relation space.

In the visual relation problem, we extract visual features $x_s, x_o \in \mathbb{R}^M$ from *subject-predicate-object* in an image which are deep learning features provided by Objects Detection Module. We should learn a relation translation vector $r \in \mathbb{R}^d$ and projection matrices $U_s, U_o \in \mathbb{R}^{r \times M}$. With word embeddings, we extend object embedding as:

$$h = \delta(U_s x_s \circ W_s \text{word2vec}(s)), \quad t = \delta(U_o x_o \circ W_o \text{word2vec}(o)) \qquad (3)$$

where $\circ$ denotes vector concatenation. we define triple as (h, r, t), h, $t \in \mathbb{R}^k$ are objects embeddings and relation embeddings is set as $r \in \mathbb{R}^d$. For each relation r, we set a projection matrix which projects objects embedding to relation space as $M_r \in \mathbb{R}^{k \times d}$. Thus, in r's relation space, triplet relation can be represented as:

$$h_{r,s} = hM_r, \quad t_{r,o} = tM_r \qquad (4)$$

The score function is correspondingly defined as:

$$\mathcal{S}_{r,s,o} = ||h_{r,s} + r - t_{r,o}||_2^2 \qquad (5)$$

Scoring and Testing. Following VTransE, instead of using large-margin metric learning, we use score only rewards the valid relations $r \in R$, R is the set of relations. We replace Eq. 6 with Eq. 5.:

$$\mathcal{S}_{r,s,o} = r(h_{r,s} - t_{r,o}) \qquad (6)$$

We define score of relations as:

$$\mathcal{S}_{r,p} = softmax(\mathcal{W}(\mathcal{S}_{r,s,o} + \mathcal{S}_{p,s,o})) \qquad (7)$$

$\mathcal{W}$ projection matrices. We use softmax to calculate the score of relationship prediction, inputting a set of training images containing three tuples (h, r, t) and the object bbox. The model will optimize all the parameters according to the loss function.When testing, for every input image, we use our learnt model to predict all visual relationships (h, r, t) in an image using:

$$S_{h,r,t} = (1 - \alpha)(S_h + S_t) + \alpha S_{r,p} \qquad (8)$$

S_h and S_t are the objection detection scores. α is hyper-parameter that were obtained though grid search to maximize performance on the validation set (Fig. 3 and Table 2).

Table 1. Comparison to related work in phrase, relation, predicate detection on VRD. Some of results are from [1,2]

Dataset	VRD					
Task	Phrase det.		Relation det.		Predicate det.	
Metric	R@50	R@100	R@50	R@100	R@50	R@100
Joint CNN	0.07	0.09	0.07	0.09	1.47	2.03
Visual Phrase [3]	0.04	0.07	-	-	0.97	1.91
Lu-VLK [2]	16.17	17.03	13.86	14.70	47.87	47.87
VTransE [1]	19.42	22.42	14.07	15.20	44.76	44.76
VTransR	21.55	24.22	15.10	16.11	49.32	49.32
VTransR with WEF	22.35	25.10	15.97	16.89	51.01	51.01
VTransR with VPF	22.19	24.96	15.74	16.71	50. 8	50.88
Full Model	**22.57**	**25.33**	**16.08**	**17.30**	**51.71**	**51.71**

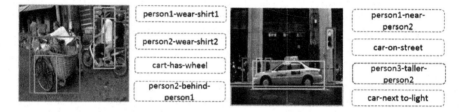

Fig. 3. Qualitative results of our model in VRD testing set. We show result of relation detection with top 4 detections. Red borders denote incorrect results. (Color figure online)

7   Experiments and Analysis

7.1   Datasets and Metrics

Datasets. We conduct experiments on two popular benchmark datasets:**Visual Relationships dataset(VRD)** [2], and **Visual Phrases dataset(VPD)** [3]. VPD contains 2,769 images with 5,067 bounding-box annotations and 17 kinds of visual phrases. VRD contains 5000 images with 37,993 triplets $subject - predicate - object$, 6,672 kinds of relations and 24.25 predicates per object category. There are 1,877 relationships for zero-shot task which only occur in test set.

Metrics. We also use the evaluation metrics in [1,2]. In relation detection task we use **recall@50** and **recall@100**. **R@K** represents the accuracy of the predicted relationships in the top K confident predictions in a image.

7.2   Comparison with State-of-the-Art

Because visual relationship detection includes two subtasks objects detection and relation prediction, we train three models in both **VRD** and **VPD** for predicate, phrase and relationship detection. Predicate detection allows us to study how difficult it is to predict relationships without the limitations of object detection.

Comparing Models. We compare our model with previously published methods and we get a new state-of-the-art in all the three tasks. And we further conduct a set of ablation studies to understand how each component affects our model in relation detection. (1) **Joint CNN**. We train a CNN model [26] to predict the three components of a relationship together. (2) **Visual Phrase**. A joint relation model [3] that treats a triplet as an individual class. (3) **Lu's-VLK**. A two-stage separate model [2] that combines Lu's Visual Appearance Module and language priors. (4) **VTransE**. An end-to-end relation detection network [1] with embedding translation and feature extraction layer. (5) **VTransR**. We only use the detection model with multi-space embedding translation described in Eq. 6 by optimizing L_{rel}. (6) **VTransR with word embeddings**. We use both VTransR model and word embeddings feature (Eq. 3). (7) **VTransR with visual phrase**. We use both VTransR model and visual phrase feature (Eq. 2). (8) **Full Model**. This is our full model. It contains the TransR (Eq. 6), word embeddings feature (Eq. 3) and visual phrase feature (Eq. 2).

Results. Visual Phrases [3] and Joint CNN [26] treat a triplet $object_1 - predicate - object_2$ as an individual relationship, the number of their classes is 6,672(there are 6,672 unique relationships in training set of VRD). Because of the long tail of infrequent relationships, there are insufficient samples to train good parameters. Therefore, both of them perform poorly in all the three tasks (Table 1). Lu's-VLK [2], VTransE [1] and VTransR are separate relation models, they work well in VRD. Using position feature, VTransE outperforms Lu's-VLK in relation and phrase detection. However, the language priors help Lu's-VLK work better in predicate classification. Word2vec brings sematic similarity from free text, *person* can *ride* a *horse*, so the *elephant* may be *ridden* too. VTransR improves the ability when modeling N-to-1, 1-to-N and N-to-N relations and gets some improved in R@100 and R@50. Word embeddings provide language priors like lu's-VLK and visual phrase module get global feature from RoI of relations. We infer that a *person-taller-person* with an image of two persons instead of two persons in separated image. So the Full Model get about 1% improved in R@100 on relation detection. In VPD, there is less phrase than VRD, Visual Phrases and Joint CNN are able to perform better. Full Model still shows its advantages in this dataset.

7.3   Zero-Shot Learning

In the real world images, if we have N object types and R relations, the total possible number of relationship predictions is N × R × N. It is difficult to build a dataset with every possible relationship. Hence, zero-shot prediction is an

Table 2. Comparison to related work in phrase, relation, predicate detection on VPD. We evaluate the models for visual relationship detection on 12 phrases in VPD that can be represented as a $object_1 - predicate - object_2$ relationship. Some of results are from [2]

Dataset	VPD	
Task	Phrase det.	
Metric	R@50	R@100
Joint CNN	49.3	52.7
Visual Phrase [3]	71.5	75.3
Lu-VLK [2]	78.1	82.7
Full Model	**89.79**	**94.10**

Table 3. Results for zero-shot visual relationship detection. Visual Phrases, Joint CNN are omitted from this experiment as they are unable to do zero-shot learning

Dataset	VRD			
Task	Phrase det.		Relation det.	
Metric	R@50	R@100	R@50	R@100
Lu-VLK [2]	3.36	3.75	3.13	3.52
VTransE [1]	2.65	3.51	1.71	2.14
VTransR	3.01	3.67	2.11	2.86
Full Model	**3.41**	**3.88**	**3.22**	**3.61**

important task in visual relation detection to address the long-tailed relation distribution.

In VRD, there are 1,877 relationships which only occur in test set(e. g. $elephant - standon - street$). We compare our model with previous models in zero-shot task in Table 3 for phrase and relationship detection.

We notice a huge drop in R@100 for all the task. For the purely visual models, VTransR performs a little better than VTransE. Lu's-VLK and our Full Model which are multi-modal model with language priors outperform visual models, because they can use sematic similarity from free text.

8   Conclusion

In this work we present a multi-modal framework to detect multiple visual relationships in a single image. We introduce the TransR network to address the problem in VTransE model, and show its effectiveness for visual relationships detection. We utilize a language prior and visual phrase feature to improve our prediction, and outperform previous state of art on Visual Relation Dataset. Future work on developing more sophisticated visual relationships models for understanding image and apply it to more complex tasks such as visual question answering will be conducted.

References

1. Zhang, H., Kyaw, Z., Chang, S.F., Chua, T.S.: Visual translation embedding network for visual relation detection. arXiv preprint arXiv:1702.08319 (2017)
2. Lu, C., Krishna, R., Bernstein, M., Fei-Fei, L.: Visual relationship detection with language priors. In: Leibe, B., Matas, J., Sebe, N., Welling, M. (eds.) ECCV 2016. LNCS, vol. 9905, pp. 852–869. Springer, Cham (2016). https://doi.org/10.1007/978-3-319-46448-0_51
3. Sadeghi, M.A., Farhadi, A.: Recognition using visual phrases. In: Computer Vision and Pattern Recognition, pp. 1745–1752 (2011)
4. Bordes, A., Usunier, N., Garcia-Duran, A., Weston, J., Yakhnenko, O.: Translating embeddings for modeling multi-relational data. In: Advances in Neural Information Processing Systems, pp. 2787–2795 (2013)
5. Lin, Y., Liu, Z., Sun, M., Liu, Y., Zhu, X.: Learning entity and relation embeddings for knowledge graph completion. In: AAAI, pp. 2181–2187 (2015)
6. Andreas, J., Rohrbach, M., Darrell, T., Dan, K.: Deep compositional question answering with neural module networks. Comput. Sci. **27**, 55–56 (2015)
7. Karpathy, A., Li, F.F.: Deep visual-semantic alignments for generating image descriptions. In: IEEE Conference on Computer Vision and Pattern Recognition, pp. 3128–3137 (2014)
8. Johnson, J., Krishna, R., Stark, M., Li, L.J., Shamma, D.A., Bernstein, M.S., Li, F.F.: Image retrieval using scene graphs. In: IEEE Conference on Computer Vision and Pattern Recognition, pp. 3668–3678 (2015)
9. Ramanathan, V., Li, C., Deng, J., Han, W., Li, Z., Gu, K., Song, Y., Bengio, S., Rosenberg, C., Fei-Fei, L.: Learning semantic relationships for better action retrieval in images. In: Proceedings of the IEEE Conference on Computer Vision and Pattern Recognition, pp. 1100–1109 (2015)
10. Guadarrama, S., Krishnamoorthy, N., Malkarnenkar, G., Venugopalan, S., Mooney, R., Darrell, T., Saenko, K.: YouTube2Text: recognizing and describing arbitrary activities using semantic hierarchies and zero-shot recognition. In: Proceedings of the IEEE International Conference on Computer Vision, pp. 2712–2719 (2013)
11. Regneri, M., Rohrbach, M., Wetzel, D., Thater, S., Schiele, B., Pinkal, M.: Grounding action descriptions in videos. Trans. Assoc. Comput. Linguist. **1**, 25–36 (2013)
12. Thomason, J., Venugopalan, S., Guadarrama, S., Saenko, K., Mooney, R.J.: Integrating language and vision to generate natural language descriptions of videos in the wild. In: Coling, vol. 2, p. 9 (2014)
13. Yao, J., Fidler, S., Urtasun, R.: Describing the scene as a whole: joint object detection, scene classification and semantic segmentation. In: 2012 IEEE Conference on Computer Vision and Pattern Recognition (CVPR), pp. 702–709. IEEE (2012)
14. Kulkarni, G., Premraj, V., Ordonez, V., Dhar, S., Li, S., Choi, Y., Berg, A.C., Berg, T.L.: BabyTalk: understanding and generating simple image descriptions. IEEE Trans. Pattern Anal. Mach. Intell. **35**(12), 2891–2903 (2013)
15. Zitnick, C.L., Parikh, D., Vanderwende, L.: Learning the visual interpretation of sentences. In: Proceedings of the IEEE International Conference on Computer Vision, pp. 1681–1688 (2013)
16. Atzmon, Y., Berant, J., Kezami, V., Globerson, A., Chechik, G.: Learning to generalize to new compositions in image understanding. arXiv preprint arXiv:1608.07639 (2016)

17. Farhadi, A., Hejrati, M., Sadeghi, M.A., Young, P., Rashtchian, C., Hockenmaier, J., Forsyth, D.: Every picture tells a story: generating sentences from images. In: Daniilidis, K., Maragos, P., Paragios, N. (eds.) ECCV 2010. LNCS, vol. 6314, pp. 15–29. Springer, Heidelberg (2010). https://doi.org/10.1007/978-3-642-15561-1_2

18. Desai, C., Ramanan, D., Fowlkes, C.C.: Discriminative models for multi-class object layout. Int. J. Comput. Vis. **95**(1), 1–12 (2011)

19. Gupta, A., Davis, L.S.: Beyond nouns: exploiting prepositions and comparative adjectives for learning visual classifiers. In: Forsyth, D., Torr, P., Zisserman, A. (eds.) ECCV 2008. LNCS, vol. 5302, pp. 16–29. Springer, Heidelberg (2008). https://doi.org/10.1007/978-3-540-88682-2_3

20. Sadeghi, F., Kumar Divvala, S.K., Farhadi, A.: VisKE: visual knowledge extraction and question answering by visual verification of relation phrases. In: Proceedings of the IEEE Conference on Computer Vision and Pattern Recognition, pp. 1456–1464 (2015)

21. Girshick, R., Donahue, J., Darrell, T., Malik, J.: Rich feature hierarchies for accurate object detection and semantic segmentation. In: Computer Vision and Pattern Recognition, pp. 580–587 (2013)

22. Girshick, R.: Fast R-CNN. In: Proceedings of the IEEE International Conference on Computer Vision, pp. 1440–1448 (2015)

23. Ren, S., He, K., Girshick, R., Sun, J.: Faster R-CNN: towards real-time object detection with region proposal networks. In: Advances in Neural Information Processing Systems, pp. 91–99 (2015)

24. Mikolov, T., Chen, K., Corrado, G., Dean, J.: Efficient estimation of word representations in vector space. arXiv preprint arXiv:1301.3781 (2013)

25. Wang, J., Yang, Y., Mao, J., Huang, Z., Huang, C., Xu, W.: CNN-RNN: a unified framework for multi-label image classification. In: Proceedings of the IEEE Conference on Computer Vision and Pattern Recognition, pp. 2285–2294 (2016)

26. Simonyan, K., Zisserman, A.: Very deep convolutional networks for large-scale image recognition. arXiv preprint arXiv:1409.1556 (2014)

Anomaly Detection and Analytics

Sub-trajectory- and Trajectory-Neighbor-based Outlier Detection over Trajectory Streams

Zhihua Zhu[1,2], Di Yao[1,2], Jianhui Huang[1], Hanqiang Li[3], and Jingping Bi[1(✉)]

[1] Institute of Computing Technology, Chinese Academy of Sciences, Beijing, China
{zhuzhihua,yaodi,huangjianhui,jpingbi}@ict.ac.cn
[2] University of Chinese Academy of Sciences, Huairou, China
[3] National Defence Key Laboratory of Blind Processing of Signals, Chengdu, China
cnpla@126.com

Abstract. Precisely and efficiently anomaly detection over trajectory streams is critical for many real-time applications. However, due to the uncertainty and complexity of behaviors of objects over trajectory streams, this problem has not been well solved. In this paper, we propose a novel detection algorithm, called STN-Outlier, for real time applications, where a set of fine-grained behavioral features are extracted from the sub-trajectory instead of point and a novel distance function is designed to measure the behavior similarity between two trajectories. Additionally, an optimized framework(TSX) is introduced to reduce the CPU resources cost of STN-Outlier. The performance experiments demonstrate that STN-Outlier successfully captures more fine-grained behaviors than the state-of-the-art methods; besides, the TSX framework outperforms the baseline solutions in terms of the CPU time in all cases.

Keywords: Outlier · Sub-trajectory · Trajectory streams

1 Introduction

Today, the location-acquisition devices such as GPS and smart phone, monitoring the behaviors of vehicles and people, are generating massive-scale high-speed trajectory streams. The applications like traffic management and security surveillance, need continuously detect the abnormal objects from high volumes of objects in such heavy data. Those abnormal objects such as drunk drive in traffic management or espionage in security surveillance, whose erratic behaviors are different from the majority in trajectory streams, must be detected efficiently based on the behaviors over a period of time and reported in time. Even a short time delay may lead to losses of huge funds.

This work has been supported by the National Natural Science Foundation of China (No. 61472403 and 61702470) and the Beijing Natural Science Foundation (No. 4182062).

© Springer International Publishing AG, part of Springer Nature 2018
D. Phung et al. (Eds.): PAKDD 2018, LNAI 10937, pp. 551–563, 2018.
https://doi.org/10.1007/978-3-319-93034-3_44

In existing studies, part of researchers use the techniques that normally are machine learning to get the discriminative model using the global characteristics of the dataset [3–8]. However, due to the concept drift in trajectory stream, which means the behaviors of the target object change over time in unforeseen ways, using one single pre-trained model to continuously detect outliers would lead to inaccurate results; besides, rebuilding the model periodically would result in expensive modeling costs for real-time applications. In addition, in a trajectory stream populated with massive scale moving objects, the moving patterns of high volumes of moving objects are more dynamic and complex than single object. That is, the moving pattern of one object over trajectory streams with a lot of objects is not suit to be modeled by the local continuity assumption [1,2,9]. Considering the above challenges, Yu et al. [6] proposed a novel method called TN-Outlier to detect the trajectory outliers over a real-time trajectory stream. However, the work only used the spatial distance between trajectory points to evaluate the relationships among trajectories. As a result, it is difficult to distinguish the difference of behaviors of moving objects especially when they are close to each other. Although the importance of continuously detecting such types of outliers, this problem has not been well solved.

Because distance-based outlier is robust against concept drift and amenable to swiftly evicting obsolete models of outlierness [16]; besides, many works [4,13,14] used sub-trajectories instead of points to measure the trajectory similarity and achieved better results, but they only considered the directional and spatial differences between two sub-trajectories which is hard to capture the detail differences of behaviors. In this paper, we propose a novel distance-based outlier definition, called Sub-trajectory- and Trajectory-Neighbor-based trajectory Outlier(STN-Outlier), to detect the complex abnormal behaviors in a trajectory stream. The definition not only considers the behavioral approximation of moving objects in a region, but also takes the duration of the behavioral similarity across time into account. Specially, a novel distance function that combines both the inter- and intra-trajectory features in an integrated manner is designed to measure the fine-grained difference between two trajectories. Then, a comprehensive framework, called the *temporal and spatial-aware examination* (TSX), is introduced to efficiently detect the outliers over high volume trajectory stream. The experimental studies on synthetic and real Taxi [6,10,11] datasets demonstrate that the STN-Outlier successfully captures the deviating behaviors effectively over other state-of-the-art methods; besides, the TSX framework outperforms the baseline solutions in terms of the CPU time in all cases.

2   Overview

2.1   Preliminary

We define the scenario of trajectory outlier detection. $O = \{o_1, o_2, \ldots, o_n\}$ denotes a dataset composed of n observed moving objects. Each moving object $o_m \in O$ is represented as an infinite sequence of trajectory points $Tr_m = \{p_m^1, p_m^2, \ldots p_m^i, \ldots\}$ at timebins $\{t_1, t_2, \ldots t_i, \ldots\}$, where the term "timebin" is referred to the smallest time granularity that ensures each trajectory has at least one point fall into each bin. Given n moving objects, a trajectory stream S is represented as n infinite sequences of trajectory points ordered by time-bins $S = \{p_1^1 p_2^1 \ldots p_n^1, p_1^2 p_2^2 \ldots p_n^2, \ldots, p_1^i p_2^i \ldots p_n^i, \ldots\}$, where $p_1^i p_2^i \ldots p_n^i$ are said to fall into the same timebin i in S. Then, a periodic sliding window W with a fixed window size w and slide length s is used to extract a finite sub-stream for processing.

Additionally, we use the tuple $(p_m^i, p_m^{i+1}, \boldsymbol{f}_{s_m^{i,i+1}})$ to represent a sub-trajectory $s_m^{i,i+1}$ of Tr_m, where $\boldsymbol{f}_{s_m^{i,i+1}}$ is a feature vector collected by other devices.

2.2   Problem Formulation

In Fig. 1, according to [6] Tr_6 is obviously recognized as an outlier since it changes its neighbors frequently. On the contrary, Tr_1 and Tr_2 would be detected as inliers, if they keep being neighbors, namely the Euclidean distances between their points are less than d (a distance threshold) at all 6 timebins. Likewise, Tr_{3-5} would be labeled as inliers w.r.t the same d. However, note that Tr_3 whose behavior is completely different from others should be detected as outlier not an inlier even though its points are close to the points of Tr_4 and Tr_5.

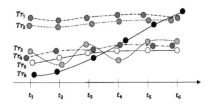

Fig. 1. Six trajectories in a window with size $w = 6$. Outlier: Tr_6, Tr_3; Inlier: Tr_1, Tr_2, Tr_4, Tr_5

To describe the trajectory outliers, e.g. Tr_3 in Fig. 1, classes of novel notions of distance-based outliers are defined referring to the semantics [6].

Definition 1 (*Sub-trajectory Neighbor*). *Given two sub-trajectories $s_m^{i,i+1}$ and $s_n^{i,i+1}$, they are said to be **sub-trajectory neighbors** if $dist(s_m^{i,i+1}, s_n^{i,i+1}) \leq d$ where $dist(s_m^{i,i+1}, s_n^{i,i+1})$ is a distance function and d is a distance threshold.*

The **sub-trajectory neighbor set** between Tr_m and Tr_n w.r.t a distance threshold d in a W is denoted as $\mathbb{N}_{mn}^d$, with $|\mathbb{N}_{mn}^d|$ denoting the size.

Definition 2 (*Trajectory Neighbor*). *In window W, given a distance threshold d and timebin count threshold thr_t, trajectory Tr_m is called a **trajectory neighbor** of Tr_n if $|\mathbb{N}_{mn}^d| \geq thr_t$.*

Definition 3 (*Trajectory Outlier*). *Given a distance threshold d, a neighbor count threshold k, and timebin count threshold thr_t, a trajectory Tr_m in the window W is a **trajectory-neighbor based trajectory outlier** if Tr_m has at most $k-1$ trajectory neighbors in W with trajectory neighbor as per Definition 2.*

Given the parameters d, k, thr_t and n trajectories in window W, *our goal is to detect and report all trajectory-neighbor based trajectory outliers in the window W with high accuracy and efficiency.*

3   Methodology

In this section, we introduce a novel distance function for sub-trajectories. The key of this function is the combination of two types of features that ensures a significant difference between outliers and inliers can be captured.

3.1   Feature Extraction

To describe the difference of behaviors of trajectories minutely, two types of features intra- and inter-trajectory features are extracted for each sub-trajectory.

Intra-trajectory Feature. Intra-trajectory features indicate the moving behavior of each trajectory that can be quantized by the differences of attributes between two consecutive trajectory points.

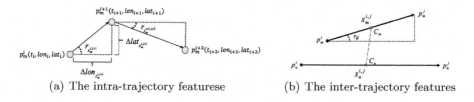

(a) The intra-trajectory featurese   (b) The inter-trajectory features

Fig. 2. The elements for intra- and inter-trajectory features

Let us consider a window with n 2-dimensions points belonging to Tr_m. Specifically, suppose the point is GPS point. Each point p_m^i can be denoted as $(t_i, lon_{p_m^i}, lat_{p_m^i})$ where t_i is the ith timebin, $lon_{p_m^i}$ and $lat_{p_m^i}$ denote the longitude and latitude respectively. For sub-trajectory $s_m^{i,i+1}$, four attributes that intuitively describe the behaviors of one trajectory are extracted as the intra-trajectory features. As shown in Fig. 2(a), the features are change of longitude $\Delta lon_{s_m^{i,i+1}} = lon_{p_m^{i+1}} - lon_{p_m^i}$, change of latitude $\Delta lat_{s_m^{i,i+1}} = lat_{p_m^{i+1}} - lat_{p_m^i}$, speed $v_{s_m^{i,i+1}} = \sqrt{(\Delta lon_{s_m^{i,i+1}})^2 + (\Delta lat_{s_m^{i,i+1}})^2} / (t_{i+1} - t_i)$ and change of rate of turn(ROT) $\Delta r_{s_m^{i,i+1}} = r_{s_m^{i,i+1}} - r_{s_m^{i+1,i+2}}$, where $r_{s_m^{i,i+1}} = arctan(\Delta lat_{s_m^{i,i+1}} / \Delta lon_{s_m^{i,i+1}})$. Specially, for the last sub-trajectory $s_m^{n-1,n}$ of Tr_m, we set $\Delta r_{s_m^{n-1,n}} = 0$.

In order to intuitively reflect the moving direction change of a trajectory, attribute $\hat{d}_{s_m^{i,i+1}}$ is defined to flag whether object changes its moving direction between two consecutive sub-trajectories or not. Given the moving direction of $s_m^{i+1,i+2}$ as the standard direction, $\hat{d}_{s_m^{i,i+1}} = 0$ if $|\Delta r_{s_m^{i,i+1}}| \leq \pi/4$ and $\hat{d}_{s_m^{i,i+1}} = 1$ otherwise. After computing these features for each sub-trajectory, we get a feature vector $\boldsymbol{f}_{s_m^{i,i+1}} = (\Delta lon_{s_m^{i,i+1}}, \Delta lat_{s_m^{i,i+1}}, v_{s_m^{i,i+1}}, \Delta r_{s_m^{i,i+1}}, \hat{d}_{s_m^{i,i+1}})$.

Inter-trajectory Feature. Inter-trajectory features reflect the spatial difference and directional difference of two trajectories. It ensures the outliers that always moves alone or always in the areas where other objects rarely visit can be detected. The inter-trajectory features consist of two components: the spatial distance d_c and angle distance d_θ. Suppose there are two sub-trajectories $s_m^{i,i+1}$ and $s_n^{i,i+1}$, and their elements are intuitively illustrated in Fig. 2(b).

The spatial distance between $s_m^{i,i+1}$ and $s_n^{i,i+1}$ is denoted as the Euclidean distance between center points of $s_m^{i,i+1}$ and $s_n^{i,i+1}$, namely $d_c(s_m^{i,i+1}, s_n^{i,i+1}) = \|C_m - C_n\|$. The angle distance between $s_m^{i,i+1}$ and $s_n^{i,i+1}$ is defined as Formula 1. It is the intersection angle between $s_m^{i,i+1}$ and $s_n^{i,i+1}$. Here, $\|s_m^{i,i+1}\|$ denotes the length of $s_m^{i,i+1}$.

$$d_\theta(s_m^{i,i+1}, s_n^{i,i+1}) = \begin{cases} arccos(\frac{s_n^{i,i+1} \cdot s_m^{i,i+1}}{\|s_m^{i,i+1}\| \|s_n^{i,i+1}\|}) & \text{if } \|s_m^{i,i+1}\| \neq 0 \ \& \ \|s_n^{i,i+1}\| \neq 0 \\ 0 & \text{otherwise} \end{cases} \tag{1}$$

3.2   Distance Function

Now based on the features denoted above, the distance function is defined to measure the similarity between $s_m^{i,i+1}$ and $s_n^{i,i+1}$. The function consists of two terms: similarity measure $sim(\boldsymbol{f}_{s_m^{i,i+1}}, \boldsymbol{f}_{s_n^{i,i+1}})$ and punishing item $\omega(d_c, d_\theta)$. Namely:

$$dist(s_m^{i,i+1}, s_n^{i,i+1}) = 1 - \omega(d_c, d_\theta) \times sim(\boldsymbol{f}_{s_m^{i,i+1}}, \boldsymbol{f}_{s_n^{i,i+1}}) \tag{2}$$

In particular, considering the differences of intra-trajectory features in semantic and value, we need a normalization function to normalize the features. However, due to the uncertainty caused by the concept drift over continuous trajectory streams, the minimum or maximum of each feature cannot be fixed. It is not suit to use the Min-Max scaling to normalize the features. Likewise, the Z-score standardization cannot be used.

Given two feature vectors $\boldsymbol{f}_{s_m^{i,i+1}}$, $\boldsymbol{f}_{s_n^{i,i+1}}$, a ratio $\rho^j = |f_{s_m^{i,i+1}}^j - f_{s_n^{i,i+1}}^j| / max\{f_{s_m^{i,i+1}}^j, f_{s_n^{i,i+1}}^j\}$ is computed to normalize the differences between features into $[0, 1]$, where f^j is the jth feature in feature vector. However, when $f^j \in \{\Delta lon, \Delta lat, \Delta r\}$, if one of $f_{s_m^{i,i+1}}^j$ and $f_{s_n^{i,i+1}}^j$ is equal to 0, the ρ^j would be amplified to 1 even if their difference is small; otherwise, we expect a small difference amplify to 1 when one of objects is static. Therefore, we use $\sigma = sin(d_\theta(s_m^{i,i+1}, s_n^{i,i+1}))$ as the constraint factor to limit the value of ρ^j. That is, the smaller d_θ is, the smaller ρ^j is. The new ratio $\hat{\rho}^j$ is defined as follows.

If $v_{s_m^{i,i+1}} = 0$ or $v_{s_n^{i,i+1}} = 0$, then $\hat{\rho}^j = \rho^j$. And if $v_{s_m^{i,i+1}} \neq 0$, $v_{s_n^{i,i+1}} \neq 0$ and $f^j \in \{\Delta lon, \Delta lat, \Delta r\}$, then $\hat{\rho}^j = \rho^j \times \sigma$.

Specially, for $max\{f^j_{s^{i,i+1}_m}, f^j_{s^{i,i+1}_n}\} = 0$, we set $\rho^j = 0$. Based on the formulas above, a new vector $\hat{\rho} = (\hat{\rho}_{\Delta lon}, \hat{\rho}_{\Delta lat}, \hat{\rho}_v, \hat{\rho}_{\Delta r}, \hat{\rho}_{\hat{d}})$ is computed for each sub-trajectory pair. Then, to map the similarity between two sub-trajectories into $[0, 1]$, the function $sim(\boldsymbol{f}_{s^{i,i+1}_m}, \boldsymbol{f}_{s^{i,i+1}_n})$ is defined as

$$sim(\boldsymbol{f}_{s^{i,i+1}_m}, \boldsymbol{f}_{s^{i,i+1}_n}) = 1 - \frac{\|\hat{\rho}\|_2}{\sqrt{|\hat{\rho}|}} \tag{3}$$

where $\|\cdot\|_2$ denotes the $L2$ norm and $|\hat{\rho}|$ is the size of $\hat{\rho}$. It means that the more similar $\boldsymbol{f}_{s^{i,i+1}_m}$ and $\boldsymbol{f}_{s^{i,i+1}_n}$ are, the bigger value of $sim(\boldsymbol{f}_{s^{i,i+1}_m}, \boldsymbol{f}_{s^{i,i+1}_n})$ is.

Next, considering the influence of spatial distance and moving direction on similarity comparison we define a punishing item $\omega(d_c, d_\theta)$ to control the value of $sim(s^{i,i+1}_m, s^{i,i+1}_n)$. Given a spatial distance limit ξ, the punishing item $\omega(d_c, d_\theta)$ is defined as:

$$\omega(d_c, d_\theta) = \begin{cases} e^{-\frac{|d_c - \xi|}{\xi}} \times cos(d_\theta) & \text{if } d_c > \xi \\ cos(d_\theta) & \text{otherwise} \end{cases} \tag{4}$$

Namely, given the spatial distance threshold ξ set by user, the farther apart two sub-trajectories the less similar they are. Likewise, the bigger difference of moving direction between two sub-trajectories the less similar they are.

4   Detection Framework

4.1   The Basic Framework

The basic framework of our trajectory outlier detection is shown as Fig. 3. In the basic framework, the STN-Outlier detects the outliers by first running a range query search for each trajectory at current window. The time complexity is $O(n)$ where n is the number of trajectories in current window. Then, it traverses all neighbor of Tr_m to determine the status of Tr_m, of which the worst case is to traverse all $n - 1$ trajectories. Thus, its worst complexity is $O(wn^2)$ where w is window size. And, it would fully reuses the neighbor relationships collected in the previous window, there is high computational costs when n is large.

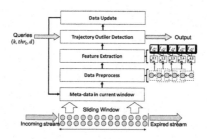

Fig. 3. The basic framework of STN-Outlier

Inspired by the minimal examination (MEX) framework [6] and the locality-sensitive hashing (LSH) algorithm [16], we design a temporal and spatial-aware examination (TSX) framework that efficiently solves this issue.

4.2 The Optimized Framework

A trajectory Tr_m will be labeled based on the neighbor evidence that has been acquired for this object. Note that those neighbors must be near to Tr_m in space. If a trajectory is far away from Tr_m, namely exceeding a radius R, then the trajectory must be non-neighbor of Tr_m. This fact leads to an important observation. That is, to identify whether a trajectory Tr_m is a neighbor-based inlier, it is unnecessary to compare with all other trajectories in window W. Instead a subset of the full trajectories that are near to Tr_m often can be sufficient to prove that it is an inlier. To acquired the small yet sufficient subset of trajectories for Tr_m a concept of *R-near Subset* with LSH is defined.

Specially, in TSX framework, the location of a trajectory is represented as an average of full points in the current window. Namely, the location l_m of a trajectory Tr_m in window W is denoted as $\frac{\sum_{j=1}^{w} p_m^j}{w}$, where $W = \{p_m^1, p_m^2, \ldots, p_m^w\}$.

Definition 4 (*R-near Subset*). *Given an LSH family $\mathcal{F}$ and a trajectory set DB_{Tr} in a window W, for a trajectory $Tr_m \in DB_{Tr}$, its R-near Subset TR_{Tr_m} is denoted as $\{Tr_n | h(l_n) = h(l_m), h \in \mathcal{F}, Tr_n \in DB_{Tr}\}$ with a collision probability at least $1 - \delta$, which denotes the probability that Tr_m, Tr_n collide for a hash function uniformly chosen from the family $\mathcal{F}$.*

In particular, we choose E^2LSH [17] that is defined for the case where the distances are measured according to the Euclidean norm to solve the R-near Subset problem. In E^2LSH, a new family $\mathcal{G}$ of hash functions g is defined. Each function g is obtained by concatenating K functions $h_1, \ldots, h_K$ from $\mathcal{F}$, i.e., $g(p) = [h_1(p), \ldots, h_K(p)]$. Finally, the algorithm constructs L hash tables, each corresponding to a different randomly chosen hash function g. Based on the definitions and theory analysis for E^2LSH, we get the following lemma.

Lemma 1. *Given a E^2LSH family $\mathcal{G}$ and L corresponding hash tables, a trajectory Tr_m could find a R-near subset $TR_{Tr_m} = \{Tr_n | g_j(l_m) = g_j(l_n), g_j \in \{g_1, \ldots, g_L\}\}$ with a collision probability $1 - \delta \geq 0.9$ iff $L \geq \frac{log10}{-log(1-P_1^K)}$ for a fixed K and P_1, where $P_1 = p(R) = Pr_{h \in \mathcal{F}}[h(p) = h(q)] = 1 - 2norm(-b/R) - \frac{2}{\sqrt{2\pi}b/R}(1 - e^{-b^2/2R^2})$ that b is the size of bucket.*

Proof. Consider a query trajectory Tr_m and an R-near trajectory Tr_n of Tr_m. According to the definitions of E^2LSH, we get $Pr_{g \in \mathcal{G}}[g(q) = g(p)] \geq P_1^K$. Thus, Tr_m and Tr_n fail to collide for all L functions g_j with probability at most $(1-P_1^K)^L$. Requiring that the trajectory Tr_m collides with Tr_n on some function g_j is equivalent to saying $1 - (1 - P_1^K)^L \geq 1 - \delta$. Therefore, given a fixed K and P_1, if setting $1 - \delta \geq 0.9$ namely $\delta \leq 0.1$, then we get: $1 - (1 - P_1^K)^L \geq 0.9 \Leftrightarrow (1 - P_1^k)^L \leq 0.1 \Leftrightarrow L \times log(1 - P_1^K) \leq log(1/10) = -log10 \Leftrightarrow L \geq \frac{log10}{-log(1-P_1^K)}$.

Lemma 1 shows the effectiveness of R-near Subset that guarantees the effectiveness and efficiency of STN-Outler. with the R-near subset, we are ready to propose one spatial-aware principle for optimizing the MEX framework.

Algorithm 1. STN-Outlier using TXS framework

Input: Trajectory Set DB_{Tr}, the current window W_c, hash tables $\mathcal{H}$, a E^2LSH family $\mathcal{G}$,
parameters: d, k, and thr_t
Output: Outliers
1 $DB_{lt} = []$
2 **for** each $Tr_i \in DB_{Tr}$ **do**
3 $\quad$ **if** $Tr_i.lifetime \leq W_c.start$ **then**
4 $\quad\quad$ $DB_{lt} \leftarrow Tr_i$

5 **for** each $Tr_i \in DB_{lt}$ **do**
6 $\quad$ **for** each $Tr_j \in Tr_i.NT$ **do**
7 $\quad\quad$ Time-aware Examination for Tr_i and Tr_j
8 $\quad\quad$ Minimal Support examination for Tr_i

9 $\quad$ **if** $|Tr_i.Neighbors| < k$ **then**
10 $\quad\quad$ $TR_{Tr_m} \leftarrow$ queryRnearSubset($\mathcal{H}$, $\mathcal{G}$, Tr_m)
11 $\quad\quad$ **for** each $Tr_j \in (TR_{Tr_m})$ **do**
12 $\quad\quad\quad$ Time-aware Examination for Tr_i and Tr_j
13 $\quad\quad\quad$ Minimal Support examination for Tr_i

14 $\quad\quad$ **if** $|Tr_i.getNeighbors| < k$ **then**
15 $\quad\quad\quad$ Tr_i is "outlier"

16 $\quad$ updating the lifetime of Tr_i

PRINCIPLE 1. *Spatial-aware examination*: *Given a query Q and the trajectory set DB_{Tr} in the current window W_c, for evaluating a trajectory Tr_m, the examination principle suggests that a R-near subset TR_{Tr_m} can replace the full trajectories DB_{Tr} to determination the status of Tr_m.*

This principle aims to prove the status of a given trajectory Tr_m by only discovering its R-near trajectories instead of searching through full trajectories. For space reasons, we omit the three principles(lines 2–4, 7 and 8 in Algorithm 1) of MEX framework and their proofs. See [6] for detail principles and proofs.

The new trajectory outlier detection algorithm is shown as Algorithm 1. The status of a trajectory will be re-examined when its lifetime expires(lines 2–4). And whenever a trajectory is being re-examined, the minimal support examination(lines 8 and 13) will cooperate with time-aware examination(lines 7 and 12) to re-establish the minimal support($=k$) in its R-near subset(line 10). It does so by only acquiring enough new evidence rather than building a new minimal support from scratch.

5   Experiments

All experiments are performed on a server with Intel Xeon CPU 2.10 GHz. In our implementation, we simulate the streaming manner by using a sliding window in the main memory.

Datasets. The experiments are performed on synthetic datasets and real taxi data [6]. The real taxi data contains 1k trajectories and a outlier set manually labeled by a user study. However, in the user study the behavior of a taxi driver is classified as abnormal by users only if he always operates in areas that other drivers rarely visit. Therefore, we apply two transformations [18], namely Add Noise Transformation and Random Shift Transformation, to the original trajectories and generate a series of new outliers to enrich the types of outliers.

Metrics and Measurements. For the effectiveness evaluation, we measure the quality of reported outliers by F1-Measure, at which $Precision = \frac{|R_0 \cap D_0|}{|D_0|}$ and $Recall = \frac{|R_0 \cap D_0|}{|R_0|}$ where R_0 denotes the set of outliers and D_0 is the outliers detected by the algorithm.

For the efficiency evaluation, we measure the CPU resources cost by recording the cost of the first window, at which each trajectory searches its neighbors over all trajectories and compares at least thr_t segments with other trajectories.

5.1   Effectiveness Evaluation

We generate a stream with a controlled number of outliers. Specifically, 200 normal trajectories are first randomly sampled from the real taxi dataset. A random offset of radius $R = 5\,\mathrm{m}$ is added to each sampled trajectory to emulate 3–10 trajectories. Then, 100 trajectories are randomly chosen from the sampled trajectories. Half of them are transformed by Add Noise Transformation, while the other half are transformed by Random Shift Transformation.

As shown in Table. 1, we evaluate the performance of STN-Outlier in comparison with TN-Outlier and a number of well-known sub-trajectory-based similarity measures, including CTraStream [14] and Hausdorff [13]. We execute

Table 1. Performance comparison by three evaluation metrics.

$w = 15$										
Method		$k = 1$			$k = 4$			$k = 8$		
		Precision	Recall	F1	Precision	Recall	F1	Precision	Recall	F1
STN-Outlier	$thr_t = 8$	0.9177	0.9602	**0.9385**	0.5488	0.9668	**0.7002**	0.1702	0.9867	**0.2904**
	$thr_t = 11$	0.7956	0.9801	**0.8783**	0.4639	0.9801	**0.6297**	0.1576	0.9867	**0.2718**
TN-Outlier	$thr_t = 8$	1.0	0.3907	0.5619	0.4957	0.3907	0.4370	0.1442	0.6688	0.2373
	$thr_t = 11$	1.0	0.4304	0.6018	0.5038	0.4304	0.4642	0.1478	0.6953	0.2439
STN-CTraStream	$thr_t = 8$	1.0	0.3576	0.5268	0.45	0.3576	0.3985	0.1443	0.6423	0.2357
	$thr_t = 11$	1.0	0.3841	0.5550	0.3295	0.3841	0.3547	0.1380	0.6754	0.2292
STN-Hausdorff	$thr_t = 8$	0.1604	0.4569	0.2375	0.0576	0.5430	0.1041	0.0747	0.7549	0.1359
	$thr_t = 11$	0.1579	0.5629	0.2467	0.0585	0.5695	0.1061	0.0742	0.7682	0.1353
$w = 30$										
Method		$k = 1$			$k = 4$			$k = 8$		
		Precision	Recall	F1	Precision	Recall	F1	Precision	Recall	F1
STN-Outlier	$thr_t = 15$	0.9423	0.9735	**0.9576**	0.5424	0.9735	**0.6966**	0.1710	0.9801	**0.2913**
	$thr_t = 22$	0.8333	0.9933	**0.9063**	0.5	0.9933	**0.6651**	0.1651	0.9933	**0.2832**
TN-Outlier	$thr_t = 15$	1.0	0.3509	0.5196	0.4690	0.3509	0.4015	0.1389	0.6357	0.2280
	$thr_t = 22$	1.0	0.4105	0.5821	0.4920	0.4105	0.4476	0.1456	0.6887	0.2404
STN-CTraStream	$thr_t = 15$	1.0	0.3377	0.5049	0.4690	0.3509	0.4015	0.1403	0.6357	0.2299
	$thr_t = 22$	1.0	0.3576	0.5268	0.4782	0.3642	0.4135	0.1434	0.6688	0.2362
STN-Hausdorff	$thr_t = 15$	0.184	0.4569	0.2623	0.0554	0.5298	0.1003	0.0723	0.7417	0.1319
	$thr_t = 22$	0.1611	0.5496	0.2492	0.0596	0.5827	0.1081	0.0744	0.7748	0.1358

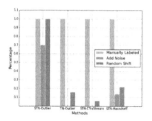 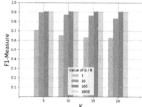 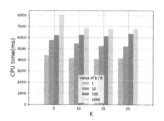

Fig. 4. The performance of methods on detecting three types of outlier

Fig. 5. The change of F1 with varying K and b/R

Fig. 6. The change of CPU time with varying K and b/R

multiple queries that vary the parameters k, thr_t and w to study how the metrics is impacted in parameter space. Referring the experimental parameters setting in [6], the distance threshold d is fixed as 0.1 and 300 m ($=\xi$) for STN-Outlier and other methods respectively. Besides, we set the weight set of Hausdorff to $(1, 1, 1)$ after tuning.

From Table. 1, STN-Outlier shows the better performance than other methods in F1-Measure, even if the inlier criteria is vary strict. In other words, STN-Outlier not only detects more outliers than others but also guarantees less false alarms. Furthermore, we study the performance of methods on detecting three types of outliers. When the inlier criteria is most relaxed where $k = 1, thr_t = 8$ and $w = 15$, the results are shown as Fig. 4. The most of outliers detected by TN-Outlier or CTraStream are manually labeled outliers, of which the detecting probability is near to 100%, while the outliers generated by the transformations are rarely or not detected. This is because the distance function in TN-Outlier or CTraStream is less focused on the differences of behaviors. In summary, compared with STN-Outlier, the TN-Outlier and other methods are worse in capturing and modeling more complex abnormal behaviors.

5.2 Efficiency Evaluation

Next we evaluate the efficiency of STN-Outlier with TSX. We vary the most important parameters, to (1) assess the impact of TSX framework versus the baseline, (2) evaluate sensitivity of parameter variations on STN-Outlier.

Varying Parameters of LSH. In this scenario, we vary the thresholds K and $b/R(P_1)$ to study how F1-Measure and CPU time are impacted. The other parameters are fixed as $k = 1, thr_t = 22, w = 30, R = \xi = 300$ and $d = 0.1$.

Figure 5 shows that the F1-Measure is directly proportional to b/R, because more trajectories are mapped into the same bucket with the size of bucket increasing. It ensures that STN-Outlier finds enough evidences to classify one trajectory as inlier. By contrast, the F1-Measure decreases when K enlarges. That is, with K increasing, the trajectories in the same bucket must have a more similar encoding $(g(l_i))$, causing that one trajectory lacks enough R-near

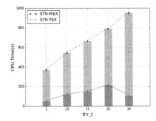 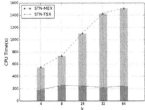 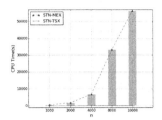

Fig. 7. The change of CPU time with varying thr_t

Fig. 8. The change of CPU time with varying k

Fig. 9. The change of CPU time with varying n

trajectories to determine its status. However, we notice that when the buckets (b/R) are enough big, the K has less influence on F1-Measure and the results are near to that of the solutions without the TSX framework in Table. 1.

Figure 6 shows that the CPU time of algorithm is also directly proportional to b/R due to one trajectory has to compare with more trajectories in one bucket to label its status. Then, the CPU time decreases with K increasing for a fixed b/R. That is, an increase on collision probability caused by K reduces the number of misclassification trajectories, and enables one trajectory not to compare with unnecessary trajectories.

In summary, the R-near subset could improve the efficiency of algorithm while ensure the effectiveness by selecting appropriate parameters.

Varying Other Parameters. Next we evaluate the efficiency of STN-Outlier using the real taxi data and synthetic datasets. We denote the MEX-based baseline solution for STN-Outlier as STN-MEX and the TSX-based solution as STN-TSX respectively. In particular, we fix the window size to 30, d to 0.2, ξ to 2,000 m and R to 4,000 m for the experiments on the real taxi data, while fix the window size to 30, d to 0.9, ξ to 300 m and R to 300 m for the synthetic datasets. In addition, the parameters K and b/R in LSH for all the cases are set to 10 and 100 respectively.

First, we evaluate the effect of varying the timebin count threshold thr_t from 1 to the full window size. As shown in Fig. 7, STN-TSX are superior to the corresponding basic solution w.r.t the CPU time in all cases. Especially when thr_t is set to the full window size, the STN-TSX outperforms the STN-MEX by a factor of 9x. We notice that the effect of STN-TXS decreases with thr_t increasing. There are two reasons for this trend: (1) the STN-TSX only needs a few non-neighbor sub-trajectories (Time-aware Examination in Algorithm 1) to label the relationship of two trajectories as non-neighbor with thr_t being big, and (2) the R-near subset of a trajectory is constant no matter how thr_t varies.

Then, we vary the neighbor count threshold k from 4 to 64. The results are shown as Fig. 8. The STN-TSX saves on average 76% of CPU time compared to the corresponding MEX solution. As the parameter k increases, the CPU time of the STN-MEX increases linearly due to more neighbors have to be acquired to determine the status of a trajectory. By contrary, for the STN-TSX instead we observe no sensitivity for varying k. This is because the STN-TSX finds the neighbors of Tr_m only by searching its R-near subset that always remains unchanged no matter how large k is.

Finally, we vary the number of trajectories n from $1k$ to $10k$. In this case, we generate five synthetic datasets containing 1k to 10k trajectories. To eliminate the effect of variations in the outlier rates, we stabilize the outlier rate in all cases to around 4% by slightly adjusting the number of outliers. As expected, Fig. 9 shows that the CPU time cost of STN-Outlier increases linearly as the number of trajectories increases, since a trajectory must compare with more trajectories in the current window until it finds k neighbors. Furthermore, it is obvious that STN-TSX exhibit much better performance than MEX-based solution. Especially, when n is up to 10k, the factor can be more than 300x.

6   Conclusion

In this work we focus on the outlier detection on trajectory streams. After analyzing the requirements of trajectory stream applications, we introduce a distance-based trajectory outlier definitions. Considering the complex behaviors of trajectories over streams, we select sub-trajectory as the analytic unit and design a novel distance function. We introduce an optimized TSX framework scalable to big data trajectory streams. The experiments on real taxi data and synthetic datasets show that STN-Outlier can effectively and efficiently detect the abnormal objects over high-volume trajectory stream.

References

1. Bu, Y., Chen, L., Fu, A. W.-C., Liu, D.: Efficient anomaly monitoring over moving object trajectory streams. In: Proceedings of SIGKDD 2009, pp. 159–168 (2009)
2. Aggarwal, C.C., et al.: A framework for clustering evolving data streams. In: International Conference on Very Large Data Bases 2003, pp. 81–92 (2003)
3. Knorr, E. M., Ng, R. T.: Algorithms for mining distance-based outliers in large datasets. In: Proceedings of VLDB 1998, pp. 392–403 (1998)
4. Lee, J.-G., Han, J., Li, X.: Trajectory outlier detection: a partition-and-detect framework. In: Proceedings of ICDE 2008, pp. 140–149 (2008)
5. Wang, H., Fan, W., Yu, P.S., Han, J.: Mining concept-drifting data streams using ensemble classifiers. In: Proceedings of SIGKDD 2003, pp. 226–235 (2003)
6. Yu, Y., Cao, L., Rundensteiner, E.A.: Detecting moving object outliers in massive-scale trajectory streams. In: Proceedings of SIGKDD 2014, pp. 422–431 (2014)
7. Li, X., Han, J., Kim, S.: Motion-alert: automatic anomaly detection in massive moving objects. In: Mehrotra, S., Zeng, D.D., Chen, H., Thuraisingham, B., Wang, F.-Y. (eds.) ISI 2006. LNCS, vol. 3975, pp. 166–177. Springer, Heidelberg (2006). https://doi.org/10.1007/11760146_15
8. Li, X., Han, J., et al.: Roam: rule- and motif-based anomaly detection in massive moving object data sets. In: SIAM International Conference on Data Mining (2007)
9. Aggarwal, C.C., Han, J., Wang, J., Yu, P.S.: A framework for clustering evolving data streams. In: Proceedings of VLDB 2003, pp. 81–92 (2003)

10. Yuan, J., et al.: T-drive: enhancing driving directions with taxi drivers intelligence. J. Trans. Knowl. Data Eng. **25**(1), 220–23 (2013)
11. Yuan, J., Zheng, Y., Zhang, C., Xie, W., Xie, X., Sun, G., Huang, Y.: T-drive: driving directions based on taxi trajectories. In: GIS, pp. 99–108 (2010)
12. Freedman, D., Pisani, R., et al.: Statistics. W. W. Norton and Company, New York (2007)
13. Lee, J., Han, J., Whang, K.: Trajectory clustering: a partition-and-group framework. In: Proceedings of the ACM SIGMOD 2007, pp. 593–604. ACM (2007)
14. Galić, Z.: Spatio-Temporal Data Streams. SCS. Springer, New York (2016). https://doi.org/10.1007/978-1-4939-6575-5
15. Knorr E.M, Ng R.T.: A unified notion of outliers: properties and computation. In: Heckerman, D., Mannila, H., Pregibon, D., Uthurusamy, R. (eds) Proceedings of KDD, Newport Beach, CA, pp. 219–222. AAAI Press, Menlo Park (1997)
16. Gionis, A., Indyk, P., Motwani, R.: Similarity search in high dimensions via hashing. In: Proceedings of VLDB (1999)
17. Datar, M., Immorlica, N.: Locality-sensitive hashing scheme based on p-stable distributions. In: Proceedings of the Symposium on Computational Geometry (2004)
18. Wang, H., Su, H., et al.: An effectiveness study on trajectory similarity measures. In: Twenty-Fourth Australasian Database Conference, vol. 137, pp. 13–22 (2013)

An Unsupervised Boosting Strategy for Outlier Detection Ensembles

Guilherme O. Campos[1,2(✉)], Arthur Zimek[2], and Wagner Meira Jr.[1]

[1] Department of Computer Science, Federal University of Minas Gerais,
Belo Horizonte, Brazil
{gocampos,meira}@dcc.ufmg.br
[2] Department of Mathematics and Computer Science,
University of Southern Denmark, Odense, Denmark
zimek@imada.sdu.dk

Abstract. Ensemble techniques have been applied to the unsupervised outlier detection problem in some scenarios. Challenges are the generation of diverse ensemble members and the combination of individual results into an ensemble. For the latter challenge, some methods tried to design smaller ensembles out of a wealth of possible ensemble members, to improve the diversity and accuracy of the ensemble (relating to the ensemble selection problem in classification). We propose a boosting strategy for combinations showing improvements on benchmark datasets.

Keywords: Outlier detection · Ensembles · Boosting
Ensemble selection

1 Introduction

The identification of outliers (i.e., data objects that do not fit well to the general data distribution) is very important in many practical applications. Application examples are the detection of credit card fraud in financial transactions data, the identification of measurement errors in scientific data, or the analysis of sports statistics data.

Recent research on the unsupervised problem of outlier detection advanced the area by applying ensemble techniques [40]. Ensemble methods, i.e., combining the findings or results of individual learners to an integrated, typically more reliable and better result, are well established in the supervised context of classification or regression [3,6,26,36]. In unsupervised learning, the theoretical underpinnings are less clear but can be drawn in analogy to the supervised context as it has been done for clustering ensembles [8,9,11,22,32,34,39].

Our focus in this paper is on ensemble selection, which has been well studied in supervised scenarios [5] (also called selective ensembles [38], or ensemble pruning [20,35,39]). Ensemble selection is also related to boosting [28], which is often used to change training conditions for additionally sought, yet to be trained ensemble members or to select the most suitable additional ensemble

© Springer International Publishing AG, part of Springer Nature 2018
D. Phung et al. (Eds.): PAKDD 2018, LNAI 10937, pp. 564–576, 2018.
https://doi.org/10.1007/978-3-319-93034-3_45

members from a pool of solutions. In this paper we transfer the supervised boosting technique to the unsupervised scenario of outlier detection ensembles. We thus propose a new outlier ensemble selection technique named BoostSelect.

In the remainder, we discuss related work in Sect. 2. We introduce our method in Sect. 3. We compare our method on a large collection of benchmark datasets against baselines and state of the art in Sect. 4. Finally, we conclude and summarize the paper in Sect. 5.

2   Related Work

The ensemble approach to learning has been studied in outlier detection several times. In analogy to supervised learning, an ensemble can be expected to improve over its components if these components deliver results with a reasonable accuracy while being diverse [40]. The two main challenges for creating good ensembles are, therefore, (i) the generation of diverse (potential) ensemble members, and (ii) the combination (or selection) of members to an ensemble.

Some strategies to achieve diversity among ensemble members have been explored, such as feature bagging (i.e., combining outlier scores learned on different subsets of attributes) [18], different parameter choices for some base method [7], the combination of actually different base methods [15, 21, 29], the introduction of a random component in a given learner [19], the use of different subsamples of the data objects (parallel [42] or sequential [25, 27]), adding some random noise component on the data ("perturbation") [41], or using approximate neighborhoods for density estimates [13]. Likewise, different combination procedures have been proposed based on outlier scores or on outlier ranks [7, 15, 18, 40].

Some methods have also been proposed to select the more diverse or the more accurate ensemble members [24, 29]. These unsupervised methods construct a target result vector from unfiltered results and then sequentially select individual results that somehow fit to the target vector while being different from already selected solutions.

The "Greedy ensemble" [29] target vector is produced considering the union of the top K ranked results of each potential ensemble member (i.e., a pool of various results) as true outliers, and the rest as inliers, placing 1 or 0 in the target vector for outliers or inliers, respectively. The similarity between result vectors is assessed through a weighted variant of Pearson correlation, where the weights are set to $\frac{1}{2K}$ for outliers and $\frac{1}{2(n-K)}$ for inliers (according to the estimated truth in the target vector), where K is the amount of outliers in the target vector (which is typically an estimate of the unknown amount of outliers) and n is the size of the dataset. Outlier score lists are then sorted in decreasing order w.r.t. weighted Pearson correlation to the target vector, and the first member is included into the ensemble. The remaining members of the list are then sorted in ascending order of weighted Pearson correlation similarity to the current ensemble, aiming at maximization of diversity. If the new member can improve the weighted Pearson correlation to the target vector, it is included into the ensemble and the order of the remaining list is updated, otherwise the member is discarded.

An obvious problem with this ensemble selection strategy is the target vector generation. Even though it uses all possible ensemble members to generate the target vector, selecting a union of top K objects as alleged outliers can be quite misleading. If one member has some inliers ranked among the top K positions (which typically is the case due to the imbalance between few outliers and many inliers), then these inliers will be labeled as outliers in the target vector. As a result, the construction of the ensemble selection procedure will focus on including those members that wrongly detect these inliers as outliers.

To overcome this downside of the "Greedy ensemble", "SelectV" [24] uses the combination of all outlier *scores* to produce the target vector, i.e., the target vector is not binary. Second, the weights for Pearson correlation are set as $1/r$, where r is the rank of each item after sorting the target vector in descending order of the outlier scores. However, "SelectV" does not maximize the diversity when selecting new members for the ensemble. As in the "Greedy ensemble", SelectV initializes the ensemble with the method exhibiting largest weighted Pearson correlation to the target vector. However, instead of ordering the list of remaining potential members to prefer those that are highly uncorrelated to the current ensemble (the strategy of maximizing diversity persued by the "Greedy ensemble"), "SelectV" uses the inverse approach, ordering the list in descending order w.r.t. weighted Pearson correlation. This way, "SelectV" prefers those remaining potential members that are highly correlated to those members that are already in the ensemble, i.e., the method values agreement (and as a consequence, accuracy, as given by the target vector) higher than diversity.

3   Boosting for Ensemble Selection

Starting from the ideas discussed for the "Greedy ensemble" [29] and for "SelectV" [24], we propose here an improved outlier ensemble selection method that is amenable to the application of boosting techniques. Boosting is well studied in supervised contexts [28]. We design and apply an equivalent technique in the unsupervised setting, to select good components for an ensemble of outlier detectors, resulting in our method BoostSelect.

3.1   Construction of the Target Vector

As a prerequisite for the combination of different outlier score lists (i.e., individual results, potential ensemble members), we normalize the scores following established procedures [15]. The target vector is constructed by combining the scores of all available results. Different combination methods could be used here, without further assumptions taking the average score is the most natural approach [40], i.e., the target vector lists the average scores of all individual results for each data object. From this target vector, we preliminarily assume the top $\lfloor n \cdot t \rfloor$ objects (ranked by their combined score) to be outliers, where n is the dataset size and $0 < t \ll 1$ is a parameter capturing the expected percentage of outliers in the dataset (i.e., there are $K = \lfloor n \cdot t \rfloor$ outliers assumed to be present).

Algorithm 1. BoostSelect

Input: $P :=$ set of normalized outlier score lists, $d :=$ drop rate (percentage), $t :=$ threshold (percentage), *combination* := combination technique

Output: $E :=$ ensemble members

1: $W := [n], E := \emptyset$
2: $target := combination(P)$ ▷ Generating the target vector
3: $target := convertBinary(target, t)$ ▷ Top $K = \lfloor n \cdot t \rfloor$ scores $\leftarrow 1$, others $\leftarrow 0$
4: $W := \left[out = \frac{1}{2K}, in = \frac{1}{2(n-K)} \right]$ ▷ $K =$ number of outliers, $n =$ size
5: Sort P by weighted Pearson Correlation (wPC) to $target$ ▷ Descending order
6: $f := getFirst(P)$ ▷ Remove f from P
7: $E := E \cup f$
8: **while** $P \neq \emptyset$ **do**
9:     $curr := combination(E)$ ▷ Current prediction
10:    sort P by wPC to $curr$ ▷ Ascending order
11:    $f := getFirst(P)$ ▷ Remove f from P
12:    **if** $wPC(combination(E \cup f), target) > wPC(curr, target)$ **then**
13:        $E := E \cup f$ ▷ Include into ensemble
14:        $Boosting(W, target, f, t, d)$ ▷ Adapt the weights
15:    **end if**
16: **end while**

The target vector thus becomes a binary vector, listing 1 for an (alleged) outlier and 0 for an (alleged) inlier and serves as pseudo ground truth for the boosting approach to ensemble selection.

3.2 Weights and Ensemble Diversity

Weighted Pearson correlation has been proposed as a similarity measure for outlier rankings [29]. We follow the procedure of Schubert et al. [29], setting the weights for Pearson correlation to $\frac{1}{2K}$ for outliers and $\frac{1}{2(n-K)}$ for inliers. Different from previous approaches, though, these values are only the initial weights. The weights will be updated by the boosting procedure.

The potential ensemble members are sorted according to their weighted Pearson correlation to the target vector. The candidate that is most similar to the target vector is chosen as the first ensemble member.

Remaining potential ensemble members are iteratively re-sorted in ascending order according to their similarity to the current prediction of the ensemble, resulting in a preference for the more different (i.e., most complementary) additional ensemble members. Potential members are included if their inclusion would increase the similarity of the ensemble prediction to the target vector, otherwise they are discarded. If the correlation improves, the ensemble is updated and the remaining lists are re-sorted by their weighted Pearson correlation to the updated prediction.

Algorithm 2. Boosting

Input: $W :=$ weight vector, $target :=$ target vector, $f :=$ new ensemble member, $t :=$ threshold (percentage), $d :=$ drop rate (percentage)

Output: $W :=$ Updated weights

1: $outliers := convertBinary(f, t)$
2: **for** $i \in 1 : size(target)$ **do**
3:     **if** $target(i) == 1$ & $outliers(i) == 1$ **then**
4:         $W(i) := W(i) * d$
5:     **end if**
6: **end for**

3.3 Boosting Procedure

The boosting is performed upon the inclusion of a new member into the ensemble. The idea is to reduce the weights for those outliers that have already been identified by any ensemble member. The weights are reduced by some specified parameter $0 < d < 1$ (drop rate).

The boosting effect is that the selection will prefer to include such additional ensemble members that detect those outliers that have not yet been detected by any ensemble member, while very easy outliers that have been detected by many ensemble members already will get assigned smaller and smaller weights.

Algorithm 1 lists the steps of the overall framework BoostSelect in pseudo code. The boosting procedure is detailed in Algorithm 2.

4 Experiments and Evaluation

4.1 Datasets

For evaluation, we use a benchmark data repository for outlier detection [4]. The repository is based on 23 basic datasets, processed in different ways mainly to provide variants with different percentage of outliers and with different handling of dataset characteristics such as duplicates, attribute normalization, and categorical values. As suggested for analysis [4], we focus on the normalized datasets without duplicates, which leaves us with 422 dataset variants.

4.2 Ensemble Members

As basic outlier detection results (i.e., potential ensemble members) we use the results provided along with the datasets [4], testing 12 neighborhood-based outlier detection algorithms changing the neighborhood size k from 1 to 100. The algorithms are: KNN [23], KNNW [1], LOF [2], SimplifiedLOF [31], LoOP [14], LDOF [37], ODIN [10], FastABOD [16], KDEOS [30], LDF [17], INFLO [12], and COF [33]. For LDOF and KDEOS, k must be larger than 1, for FastABOD, k must be larger than 2, resulting in 1196 results per dataset (less on some small datasets where k cannot reach 100). These results compose the set of potential ensemble members.

The outlier scores of these results are processed (following Kriegel et al. [15]) by applying an inverse logarithmic scaling on FastABOD results and an inverse linear scaling on ODIN results, since FastABOD and ODIN give inverse score results (i.e., the lower the scores, the higher is the chance of an observation to be an outlier). Then a simple linear scaling from 0 to 1 is applied to transform all scores into the same range.

4.3   Competitors and Settings

We compare BoostSelect against the Greedy [29] and SelectV [24] ensembles. We also generate a "Naïve" ensemble and Random ensembles as baselines. The "Naïve" ensemble is a combination of all individual outlier results (i.e., a full ensemble without selection procedure).

For each instance of an ensemble selection strategy (Greedy, SelectV, and BoostSelect, respectively, on each dataset), we generate 1000 "Random" ensembles consisting of the same number of members as the corresponding selective ensemble, where the ensemble members are randomly selected.

We used the Greedy ensemble rate parameter as 0.01, as suggested by the authors of the Greedy ensemble [29]. We test a range of parameters for BoostSelect: $d = [0.25, 0.5, 0.75]$ and $t = [0.05, 0.1, 0.15]$.

As combination technique for ensembles we use the average score.

4.4   Results

Figure 1 shows pairwise comparisons between all ensembles over all datasets, considering the ROC AUC evaluation measure (area under the curve of the receiver operating characteristic). We compare the ensemble selection techniques "Naïve", "Greedy", "SelectV", and "BS" (BoostSelect). We include random ensembles for each ensemble selection strategy and for each parametrization of BoostSelect: RG (Random Greedy), RS (Random SelectV), RBS (Random BoostSelect). The numbers represent on how many datasets the ensemble listed in the row has performed better than the ensemble listed in the column. Numbers representing the majority (more than 50%) of the datasets are white, smaller numbers black. The larger the number, the darker is its background. For the random ensemble, we take the average performance over the 1000 instances.

The best overall method is BoostSelect with $d = 0.75$ and $t = 0.05$, which has only more losses than wins when competing against BoostSelect with $d = 0.25$ and $t = 0.1$. The Greedy ensemble does not perform well in general, having more losses than wins against every other competitor. SelectV is better than all random variants and Greedy, but worse than Naïve and worse than all BoostSelect results. The Naïve ensemble behaves very consistent, as it beats by a large margin all random ensemble approaches, but still has more losses when compared to BoostSelect. Even though neither the threshold t nor the drop rate d has a strong impact on wins, setting a relatively large drop rate and a relatively small threshold overall seems to be a good choice of parameters for BoostSelect, although the optimal parameter choice differs from dataset to dataset.

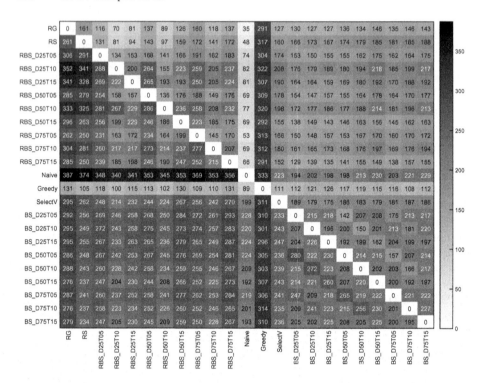

Fig. 1. Summarization of pairwise comparisons over all 422 datasets. The number counts the wins in term of ROC AUC (average ROC AUC in case of random ensembles) of the ensemble listed in the row against the ensemble listed in the column.

Looking at the top left quadrant of the heat map (Fig. 1), where the random ensembles compete against themselves, we also see a broad dominance by the random ensembles based on BoostSelect. This suggests that the number of ensemble members selected by BoostSelect is a better choice than those selected by the other strategies.

Figure 2 shows a comparison in terms of ROC AUC between BoostSelect (different parameter settings), the Naïve, the Greedy, and the SelectV ensemble averaged over all datasets (Fig. 2(a)) and on some dataset families as examples (i.e., seven sets of benchmark dataset variants generated from seven original datasets, Fig. 2(b)–(h)). While we see that BoostSelect behaves competitively over all datasets on average (Fig. 2(a)), the selected examples show different behavior regarding both parameters t and d. On Annthyroid (Fig. 2(b)) low threshold values achieves better ROC AUC overall, the reverse effect can be seen on Wilt (Fig. 2(e)), which is also an example where BoostSelect performs worse than the competitors.

These results highlight how close the SelectV ensemble and the Naïve ensemble are to each other. As SelectV generates the target vector using the Naïve result (i.e., combining all single results) and keeps methods sorted in decreasing

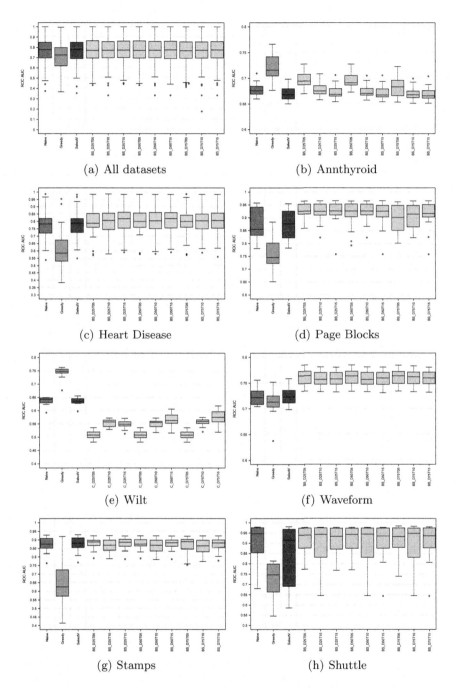

(a) All datasets

(b) Annthyroid

(c) Heart Disease

(d) Page Blocks

(e) Wilt

(f) Waveform

(g) Stamps

(h) Shuttle

Fig. 2. Comparison of the ensemble techniques in terms of ROC AUC distributions over some dataset families.

order w.r.t. weighted Pearson correlation to the current ensemble. While SelectV includes therefore those results that improve the correlation to the target vector, it does not promote diversification of ensemble members. Thus the SelectV results can be expected to be highly correlated to the Naïve ensemble results. Overall, the Naïve ensemble results (and also the SelectV results) are of reasonable quality, but not maximizing diversity between its ensemble members limits the full power of an ensemble.

On the same datasets, we also compare each ensemble selection strategy w.r.t. the distribution of results of the corresponding random ensembles (Naïve is compared over the distribution of the individual results). This comparison is depicted in Fig. 3. The z-score shows how many standard deviations each ensemble selection strategy is deviating from the mean, given the distribution of ROC AUC values of the corresponding random ensembles (or all results in case of Naïve).

Even though the random ensembles for BoostSelect perform better in general than the random ensembles corresponding to the other ensemble selection strategies (see Fig. 1), on average (Fig. 3(a)) and in most cases BoostSelect performs better than its corresponding random ensemble, in several cases by a large margin, e.g., 5-14 standard deviations on Page Blocks and 8-11 standard deviations on Waveform. However, there are also dataset families exhibiting the reverse effect. On Wilt, BoostSelect shows again poorer performance in general (Fig. 3(e)). On the Annthyroid dataset family we can observe again that choosing small t leads to considerably better results (Fig. 3(b)).

Naïve and SelectV does not improve too much over the individual results and the random ensembles, respectively. As discussed before, Naïve and SelectV can be expected to have very similar results. This can also be observed in Fig. 3. The Greedy ensemble performs worse than its random ensembles on most datasets (again, Annthyroid and Wilt are exceptions).

5   Conclusion

We proposed a new ensemble selection strategy for unsupervised outlier detection ensembles, using the unsupervised equivalent to a boosting strategy for ensemble selection. Experiments show the favorable behavior of the new ensemble selection strategy compared to existing methods (Greedy and SelectV) on a large set of benchmark datasets. Main differences between our method BoostSelect, the Greedy ensemble, and SelectV can be attributed to a different way of focusing on diversity and accuracy of ensemble members. Greedy goes all out for diversity and mostly disregards accuracy, while SelectV ignores diversity and maximizes accuracy of the ensemble members. Our new method BoostSelect considers both, diversity and accuracy, in a balanced manner and performs competitively on average over a large selection of benchmark datasets with strong improvements on many of the benchmark datasets.

The behavior of BoostSelect is robust to the parameters on many datasets but depends strongly on the choice of parameters on some datasets. We have

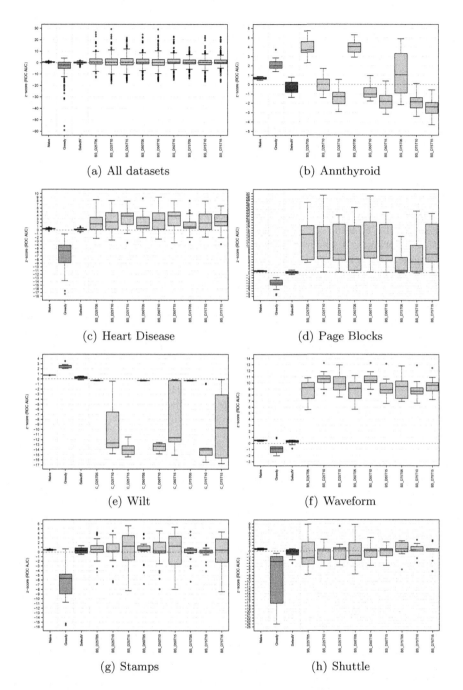

(a) All datasets

(b) Annthyroid

(c) Heart Disease

(d) Page Blocks

(e) Wilt

(f) Waveform

(g) Stamps

(h) Shuttle

Fig. 3. The z-scores of the ROC AUC achieved by the ensemble techniques w.r.t. the ROC AUC distribution of their corresponding random ensembles. The Naïve ensemble is compared against the ROC AUC distribution of all individual results.

to leave the exploration of this behavior and potential relation to properties of the datasets for future work. Also, BoostSelect as well as other outlier detection algorithms will be included in the Lemonade platform (http://www.lemonade.org.br).

Acknowledgments. This work was partially supported by CAPES - Brazil, Fapemig, CNPq, and by projects InWeb, MASWeb, EUBra-BIGSEA (H2020-EU.2.1.1 690116, Brazil/MCTI/RNP GA-000650/04), INCT-Cyber, and Atmosphere (H2020-EU 777154, Brazil/MCTI/RNP 51119).

References

1. Angiulli, F., Pizzuti, C.: Fast outlier detection in high dimensional spaces. In: Elomaa, T., Mannila, H., Toivonen, H. (eds.) PKDD 2002. LNCS, vol. 2431, pp. 15–27. Springer, Heidelberg (2002). https://doi.org/10.1007/3-540-45681-3_2
2. Breunig, M.M., Kriegel, H.-P., Ng, R., Sander, J.: LOF: identifying density-based local outliers. In: Proceedings SIGMOD, pp. 93–104 (2000)
3. Brown, G., Wyatt, J., Harris, R., Yao, X.: Diversity creation methods: a survey and categorisation. Inf. Fusion **6**, 5–20 (2005)
4. Campos, G.O., Zimek, A., Sander, J., Campello, R.J.G.B., Micenková, B., Schubert, E., Assent, I., Houle, M.E.: On the evaluation of unsupervised outlier detection: measures, datasets, and an empirical study. Data Min. Knowl. Disc. **30**, 891–927 (2016)
5. Caruana, R., Niculescu-Mizil, A., Crew, G., Ksikes, A.: Ensemble selection from libraries of models. In: Proceedings of ICML (2004)
6. Dietterich, T.G.: Ensemble methods in machine learning. In: Kittler, J., Roli, F. (eds.) MCS 2000. LNCS, vol. 1857, pp. 1–15. Springer, Heidelberg (2000). https://doi.org/10.1007/3-540-45014-9_1
7. Gao, J., Tan, P.-N.: Converting output scores from outlier detection algorithms into probability estimates. In: Proceedings of ICDM, pp. 212–221 (2006)
8. Ghosh, J., Acharya, A.: Cluster ensembles. WIREs DMKD **1**(4), 305–315 (2011)
9. Gionis, A., Mannila, H., Tsaparas, P.: Clustering aggregation. ACM TKDD **1**(1) (2007)
10. Hautamäki, V., Kärkkäinen, I., Fränti, P.: Outlier detection using k-nearest neighbor graph. In: Proceedings of ICPR, pp. 430–433 (2004)
11. Iam-On, N., Boongoen, T.: Comparative study of matrix refinement approaches for ensemble clustering. Mach. Learn. (2013)
12. Jin, W., Tung, A.K.H., Han, J., Wang, W.: Ranking outliers using symmetric neighborhood relationship. In: Ng, W.-K., Kitsuregawa, M., Li, J., Chang, K. (eds.) PAKDD 2006. LNCS (LNAI), vol. 3918, pp. 577–593. Springer, Heidelberg (2006). https://doi.org/10.1007/11731139_68
13. Kirner, E., Schubert, E., Zimek, A.: Good and bad neighborhood approximations for outlier detection ensembles. In: Beecks, C., Borutta, F., Kröger, P., Seidl, T. (eds.) SISAP 2017. LNCS, vol. 10609, pp. 173–187. Springer, Cham (2017). https://doi.org/10.1007/978-3-319-68474-1_12
14. Kriegel, H.-P., Kröger, P., Schubert, E., Zimek, A.: LoOP: local outlier probabilities. In: Proceedings of CIKM, pp. 1649–1652 (2009)
15. Kriegel, H.-P., Kröger, P., Schubert, E., Zimek, A.: Interpreting and unifying outlier scores. In: Proceedings of SDM, pp. 13–24 (2011)

16. Kriegel, H.-P., Schubert, M., Zimek, A.: Angle-based outlier detection in high-dimensional data. In: Proceedings of KDD, pp. 444–452 (2008)
17. Latecki, L.J., Lazarevic, A., Pokrajac, D.: Outlier detection with kernel density functions. In: Perner, P. (ed.) MLDM 2007. LNCS (LNAI), vol. 4571, pp. 61–75. Springer, Heidelberg (2007). https://doi.org/10.1007/978-3-540-73499-4_6
18. Lazarevic, A., Kumar, V.: Feature bagging for outlier detection. In: Proceedings of KDD, pp. 157–166 (2005)
19. Liu, F.T., Ting, K.M., Zhou, Z.-H.: Isolation-based anomaly detection. ACM TKDD 6(1), 3:1–3:39 (2012)
20. Margineantu, D.D., Dietterich, T.G.: Pruning adaptive boosting. In: Proceedings of ICML, pp. 211–218 (1997)
21. Nguyen, H.V., Ang, H.H., Gopalkrishnan, V.: Mining outliers with ensemble of heterogeneous detectors on random subspaces. In: Kitagawa, H., Ishikawa, Y., Li, Q., Watanabe, C. (eds.) DASFAA 2010, Part I. LNCS, vol. 5981, pp. 368–383. Springer, Heidelberg (2010). https://doi.org/10.1007/978-3-642-12026-8_29
22. Nguyen, N., Caruana, R.: Consensus clusterings. In: Proceedings of ICDM, pp. 607–612 (2007)
23. Ramaswamy, S., Rastogi, R., Shim, K.: Efficient algorithms for mining outliers from large data sets. In: Proceedings of SIGMOD, pp. 427–438 (2000)
24. Rayana, S., Akoglu, L.: Less is more: building selective anomaly ensembles. ACM TKDD 10(4), 42:1–42:33 (2016)
25. Rayana, S., Zhong, W., Akoglu, L.: Sequential ensemble learning for outlier detection: a bias-variance perspective. In: Proceedings of ICDM, pp. 1167–1172 (2016)
26. Rokach, L.: Ensemble-based classifiers. Artif. Intell. Rev. 33, 1–39 (2010)
27. Salehi, M., Zhang, X., Bezdek, J.C., Leckie, C.: Smart sampling: a novel unsupervised boosting approach for outlier detection. In: Kang, B.H., Bai, Q. (eds.) AI 2016. LNCS (LNAI), vol. 9992, pp. 469–481. Springer, Cham (2016). https://doi.org/10.1007/978-3-319-50127-7_40
28. Schapire, R.E., Freund, Y.: Boosting. Foundations and Algorithms. MIT Press, Cambridge (2012)
29. Schubert, E., Wojdanowski, R., Zimek, A., Kriegel, H.-P.: On evaluation of outlier rankings and outlier scores. In: Proceedings of SDM, pp. 1047–1058 (2012)
30. Schubert, E., Zimek, A., Kriegel, H.-P.: Generalized outlier detection with flexible kernel density estimates. In: Proceedings of SDM, pp. 542–550 (2014)
31. Schubert, E., Zimek, A., Kriegel, H.-P.: Local outlier detection reconsidered: a generalized view on locality with applications to spatial, video, and network outlier detection. Data Min. Knowl. Disc. 28(1), 190–237 (2014)
32. Strehl, A., Ghosh, J.: Cluster ensembles – a knowledge reuse framework for combining multiple partitions. J. Mach. Learn. Res. 3, 583–617 (2002)
33. Tang, J., Chen, Z., Fu, A.W., Cheung, D.W.: Enhancing effectiveness of outlier detections for low density patterns. In: Chen, M.-S., Yu, P.S., Liu, B. (eds.) PAKDD 2002. LNCS (LNAI), vol. 2336, pp. 535–548. Springer, Heidelberg (2002). https://doi.org/10.1007/3-540-47887-6_53
34. Topchy, A., Jain, A., Punch, W.: Clustering ensembles: models of concensus and weak partitions. IEEE TPAMI 27(12), 1866–1881 (2005)
35. Tsoumakas, G., Partalas, I., Vlahavas, I.: An ensemble pruning primer. In: Okun, O., Valentini, G. (eds.) Applications of Supervised and Unsupervised Ensemble Methods. SCI, vol. 245, pp. 1–13. Springer, Heidelberg (2009). https://doi.org/10.1007/978-3-642-03999-7_1

36. Valentini, G., Masulli, F.: Ensembles of learning machines. In: Marinaro, M., Tagliaferri, R. (eds.) WIRN 2002. LNCS, vol. 2486, pp. 3–20. Springer, Heidelberg (2002). https://doi.org/10.1007/3-540-45808-5_1

37. Zhang, K., Hutter, M., Jin, H.: A new local distance-based outlier detection approach for scattered real-world data. In: Theeramunkong, T., Kijsirikul, B., Cercone, N., Ho, T.-B. (eds.) PAKDD 2009. LNCS (LNAI), vol. 5476, pp. 813–822. Springer, Heidelberg (2009). https://doi.org/10.1007/978-3-642-01307-2_84

38. Zhou, Z., Wu, J., Tang, W.: Ensembling neural networks: many could be better than all. Artif. Intell. **137**(1–2), 239–263 (2002)

39. Zhou, Z.-H.: Ensemble Methods. Foundations and Algorithms. CRC Press, Boca Raton (2012)

40. Zimek, A., Campello, R.J.G.B., Sander, J.: Ensembles for unsupervised outlier detection: challenges and research questions. SIGKDD Explor. **15**(1), 11–22 (2013)

41. Zimek, A., Campello, R.J.G.B., Sander, J.: Data perturbation for outlier detection ensembles. In: Proceedings of SSDBM, pp. 13:1–13:12 (2014)

42. Zimek, A., Gaudet, M., Campello, R.J.G.B., Sander, J.: Subsampling for efficient and effective unsupervised outlier detection ensembles. In: Proceedings of KDD, pp. 428–436 (2013)

DeepAD: A Generic Framework Based on Deep Learning for Time Series Anomaly Detection

Teodora Sandra Buda$^{(\boxtimes)}$, Bora Caglayan, and Haytham Assem

Cognitive Computing Group, Innovation Exchange, IBM, Dublin, Ireland
{tbuda,haythama}@ie.ibm.com, bora.caglayan@ibm.com

Abstract. This paper presents a generic anomaly detection approach for time-series data. Existing anomaly detection approaches have several drawbacks such as a large number of false positives, parameters tuning difficulties, the need for a labeled dataset for training, use-case restrictions, or difficulty of use. We propose DeepAD, an anomaly detection framework that leverages a plethora of time-series forecasting models in order to detect anomalies more accurately, irrespective of the underlying complex patterns to be learnt. Our solution does not rely on the labels of the anomalous class for training the model, nor for optimizing the threshold based on highest detection given the labels in the training data. We compare our framework against EGADS framework on real and synthetic data with varying time-series characteristics. Results show significant improvements on average of 25% and up to $40-50\%$ in $F_1\text{-}score$, precision, and recall on the Yahoo Webscope Benchmark.

1 Introduction

A well-known characterization of an outlier is given by Hawkins as, "an observation which deviates so much from other observations as to arouse suspicion that it was generated by a different mechanism" [10]. An anomaly represents a non-conforming pattern that deviates from the expected behavior, and is often referred to as an outlier or exception [5]. Detecting and mitigating these anomalies is fundamental in various domains (e.g., health, performance, security), and translates to potentially saving lives by detecting critical conditions, revenue and reputation by avoiding downtime, or improvements in application performance.

A popular approach for anomaly detection is employing explicit generalization models [1], where a summarized model is created up front to capture the normal behavior of the monitored instance, and further using the deviation between the expected normal behavior and actual behavior as error metric for anomaly detection. Typically the deviation is then monitored and fitted to a particular distribution (e.g., Gaussian [13]) and then a threshold is identified based

This project has received funding from the European Union's Horizon 2020 research and innovation programme under the Grant Agreement No. 700381 (ASGARD) and No. 671625 (CogNet).

on optimizing the precision and recall in the training data through the use of past labelled anomalous instances. The use of the labels of the anomalous class, also referred to as *golden labels* is a requirement for most of the anomaly detection techniques, either for identifying a threshold or for building a classifier to detect anomalies based on anomalous patterns in the past. This however limits the applicability of these techniques to datasets where these labels have been collected and in addition, many times suffering from the class imbalance problem, since the normal instances typically overweight the abnormal ones. Moreover, besides the need for golden labels, existing anomaly detection approaches are typically suitable for a particular type of data or anomaly to capture, which makes their application more limited in practice [1,5].

This paper introduces a novel *Deep* learning-based *Anomaly Detection* framework, named DeepAD. The DeepAD framework discovers anomalies without the need of golden labels, while maintaining the highest levels of true anomaly detection, and reducing the number of false positives compared to the best available technique. DeepAD employs various explicit generalization models to learn the normal behaviour of the data and utilizes a dynamic sliding window for determining a dynamic threshold fitted for each time-series under analysis. The dynamic window is adjusted for each point to contain past rescaled squared errors to ensure the accuracy is highest. To the best of our knowledge, DeepAD represents the first framework of its kind that utilizes multiple advanced prediction models allowing multivariate inputs without the specific use of golden labels. The use of multiple models, combined with the dynamic threshold on rescaled errors increases F_1-*score*, precision and recall beyond the state of art. The key characteristics of DeepAD are identified below:

1. This framework leverages state-of-the-art *deep learning* models such as long short term memory (LSTM) neural networks, which are renown for their ability to remember relevant information in temporal sequence data even with large gaps in between using memory gates.
2. The model learns the normal behaviour of the monitored instance and deviations from this normal behaviour are signalled as anomalous data points. The framework does not use the *ground truth* of actual anomaly locations neither for training the model nor for determining the dynamic thresholds.
3. The framework does not set hard thresholds which makes it more adaptable to varying patterns in the dataset considering an *online* setting.
4. DeepAD supports *multivariate* analysis since it can receive as input more than one feature if needed, e.g., through LSTM, and hence can surpass the first limitation of approaches limited to univariate analysis.
5. The framework combines the predictions of multiple forecasting techniques, including autoregressive models and triple exponential smoothing, in order to offer a *generic* extensible approach for forecasting.

2   Related Work

Advanced anomaly detection techniques usually employ machine learning, which can be divided into three classes: supervised, semi-supervised and unsupervised.

Anomaly detection with supervised learning [9] requires a dataset where each instance is labelled and typically it involves training a classifier on training set. Semi-supervised algorithms such as [14] construct a model to represent the normal behaviour from an input training dataset; following the model is used to calculate the likelihood of the testing dataset to be generated by the model. Unsupervised models such as [3] do not require a labelled dataset and operate under the assumption that the majority of the data points are normal (e.g., employing clustering techniques [15]) and return the remaining ones as outliers.

LSTMs have captured the attention of researchers recently in anomaly detection. For instance, [13] utilize LSTM for predicting time series and use the prediction errors for anomaly detection. They assumed that the resulting prediction errors have a Gaussian distribution, which were used then to assess the likelihood of anomalous behavior. Then a threshold is learnt based on the validation dataset to maximize the F-score, which was calculated based on the golden labels within the validation dataset. The approach was validated on four time series. Moreover, [6] follows a similar approach applied to ECG time series, where the prediction errors are fit to a Gaussian distribution, and then the threshold is determined based on optimizing the F-score on the validation set, which similarly was calculated based on the given golden labels. Furthermore, [12] utilizes an LSTM-based encoder-decoder for multi-sensor anomaly detection. When enough anomalous sequences are available, a threshold is learnt by maximizing precision and recall. The use of recurrent neural networks is also common for intrusion detection, such as in [2], with the aim of detecting and classifying attacks. However, the approaches identified above utilize the golden labels for optimizing the threshold against the prediction errors or building classifiers.

Two major limitations exist in current techniques: (1) Most approaches, such as statistical and probabilistic models, are typically suitable only for univariate datasets where a single metric is monitored at a time. This can be extended to multiple metrics by building a model for each metric. However, this would not consider any correlations between metrics. Hence these approaches cannot easily be extended to multivariate analysis where correlations among metrics can be used to identify potential anomalous behaviour. This is avoided as DeepAD can receive as input multiple features, since it can use a single LSTM model that can capture anomalies across multiple features, which makes it multivariate. (2) Existing approaches typically rely on datasets that contain the ground truth labels, where the anomalies are specifically pin pointed to a data point. This can be difficult to gather in real-life scenarios as labelled data is expensive and requires expert knowledge which yet might be affected by human errors in labelling the data. Moreover, the amount of data to be monitored and labelled would be unrealistic. In addition, the initial model might not generalize to new types of anomalies unless retrained and hence requiring expert knowledge for the entire duration of the deployment of the anomaly detection model, making these approaches unrealistic to be deployed in dynamic environments. This is avoided with our dynamic threshold-based anomaly detection approach since no labels are required for training or detecting the thresholds.

3   DeepAD Framework

The DeepAD framework is illustrated in Fig. 1 and has three main phases, detailed in the following subsections:

1. **Time Series Forecasting (TSF):** The first phase employs various different explicit generalization models. We train the probabilistic and statistical models and the LSTM models utilizing different architectures for learning the normal behaviour of the monitored environment and then apply them on incoming streaming data for scoring. Through this approach, our framework supports plugging in different TSF models and can leverage *multivariate* models for forecasting.
2. **Merge Predictions (MP):** The second phase combines the predictions of the multiple models, since some techniques provide better results than others depending on the dataset characteristics. This phase is crucial as it enables DeepAD to be a *generic* framework in the sense that it does not depend on a specific time series forecasting model.
3. **Anomaly Detector (AD):** The third phase employs extreme value analysis for computing a dynamic threshold, as follows: it compares the actual values and the predicted values and when the distance is above a certain threshold the framework reports the current value as anomalous. The distance represents the squared error between the actual and predicted value, normalized between 0 and 1, and the threshold is computed at each time step on the past scaled squared error. Through this approach, our framework is *independent* of the golden labels and hence can be applied to any time series data irrespective of them containing anomalous labels in the past.

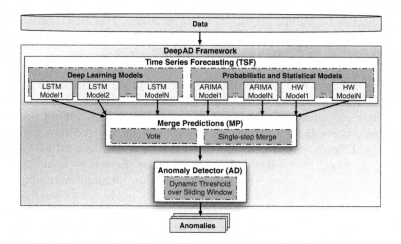

Fig. 1. DeepAD framework overview.

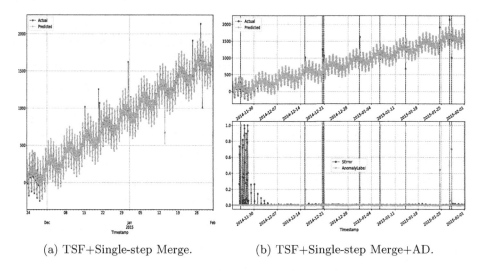

(a) TSF+Single-step Merge.   (b) TSF+Single-step Merge+AD.

Fig. 2. DeepAD$_{Merge}$: Time series forecasting with single-step merge and AD output on a sample time series (#90) from A3 benchmark. (color figure online)

3.1   Time Series Forecasting (TSF)

Given a dataset D, the TSF phase aims to learn the normal behavior of the system under analysis. The output of each TSF model is a one-step ahead prediction which will contain what the value is expected to be at the next timestamp. For this purpose, DeepAD supports plugging in different models to enable the prediction. Currently, DeepAD utilizes the following techniques: Long-short term memory (LSTM), autoregressive integrated moving average (ARIMA) and triple exponential smoothing, also commonly referred to in the literature as Holt-Winters (HW), as the models can complement each other depending on the dataset. For instance, deep neural networks such as LSTM may provide best results given large training data, whereas given small datasets, ARIMA and HW may provide better forecasts.

In the case of LSTM, the *look_back* parameter needs to be specified, which represents the number of previous time steps to use as input values to predict the next time step value. DeepAD utilizes the following LSTM architectures: (i) LSTM simple: 1 hidden layer with n neurons. The following three variations of this architecture were plugged into DeepAD: $n = \{4, 10, 16\}$, (ii) LSTM wide: 3 hidden layers with 64, 256, and 100 neurons, respectively, and (iii) LSTM deep: 7 hidden layers with 16, 48, 48, 96, 96, 48, and 16 neurons, respectively. The objective is to use simple, wide and deep architectures. For the each architecture we have trained two models, one with a *look_back* of 1 and another with a *look_back* of 3, respectively, and for all we have used $rmsprop^1$ as optimizer, since these resulted in the lowest RMSE. We also evaluated the following *look_back* variations: 1, 3, 12, 24, 60.

---

[1] http://ruder.io/optimizing-gradient-descent/index.html#rmsprop.

Furthermore, in the case of ARIMA and HW, DeepAD utilizes the past $24 \cdot 5$ values for forecasting, in case of hourly measurements, which leads to utilizing the past 5 days of data for the next prediction. In particular for ARIMA we utilize the following values for building the different models: $p = \{0,1\}$, $d = \{1\}$, and $q = \{1,2\}$, where p is the number of time lags of the autoregressive model, d is the degree of differencing, and q is the order of the moving-average model. Moreover, for HW we utilize: $\alpha = 1$, $\beta = 0$, $\gamma = 0.7$ and $\alpha = 0.716$, $\beta = 0.029$, $\gamma = 0.993$, since these resulted in the lowest RMSE. For both ARIMA and HW, more models can be plugged in with other parameters values combinations.

We illustrate the outputs of the TSF phase in Fig. 2a, where the *Actual* values are highlighted with orange, and the *Predicted* with blue. In this phase, we can observe that the predicted values typically follow the actual values, except for most of the sudden spikes in the data.

3.2   Merging Predictions (MP)

Similarly to an ensemble, the second phase combines the predictions of the multiple models following two distinct approaches:

1. *Single-step merge ($DeepaAD_{Merge}$)*: This strategy aims to combine the outputs of multiple models in order to get a more accurate forecast for a single dataset. For this purpose, this strategy compares the predicted values produced by each individual model with the actual value and selects the prediction with the lowest RMSE to forward to the AD phase at each timestamp.
2. *Vote ($DeepaAD_{Vote}$)*: This strategy aims to select the use of a single model for a given dataset. For this purpose, this strategy follows a voting approach, keeping only the model that provided the most accurate predictions in terms of RMSE for the training dataset to be utilized further for forecasting.

3.3   Anomaly Detector (AD)

Once the predictions are merged, a dynamic threshold is determined based on the squared error as follows: for each predicted value, a queue representing the sliding window of the previous squared errors is maintained. A scaler is applied to fit and transform the past squared errors from the sliding window between 0 and 1. In order to ensure DeepAD is not bound to the underlying distribution of the errors, we leverage Chebyshev's inequality [7]. In contrast to the 68-95-99 rule, also referred to the empirical rule [8], which applies to normal distributions only, the Chebyshev's inequality guarantees that, for a wide class of probability distributions, no more than a certain fraction of values can be more than a certain distance from the mean. In order to allow our framework to work with a variety of distributions, we utilize this inequality to determine the threshold. We identify that 99%(i.e., $1 - \frac{1}{10^2}$) of the values must lie within 10 times the standard deviation, and hence to identify the $<1\%$ that might lie outside, we use 10 times the standard deviation of the errors as dynamic threshold. This

---

Algorithm 1. isAnomaly(*actualValue, pastValues, squaredErrors, predictedValue, look_back, slidingWindow*)

---

1 $scaler \leftarrow$ MinMaxScaler($feature\_range = (0,1)$));
2 //Rescale errors from sliding window for dynamic threshold fitting;
3 $scaledSErrors \leftarrow scaler.$fitTransform($squaredErrors[-slidingWindow :]$);
4 //Compute dynamic threshold as 10 times standard deviation;
5 $dynamic\_thresh \leftarrow 10 \cdot$ numpy.std($scaledErrors$);
6 //Calculate squared error and apply transformation on current error
   $crtSError \leftarrow (actualValue - predictedValue) \wedge 2$;
7 $crtScaledSError \leftarrow scaler.$transform($crtSError$);
8 //If current error bigger than dynamic threshold signal return *True*;
9 **if** $crtScaledSError \geq dynamic\_thresh$ **then return** *True* // Otherwise add non scaled squared error to the queue;
10 $squaredErrors \leftarrow squaredErrors.$put($crtSError$);
11 **return** *False*

---

confirmed optimum results for detecting anomalies across the 367 time series analysed in Sect. 4.

Following, if the squared error of the predicted value is higher than 10 times the standard deviation of the previous squared scaled errors then the module signals the instance as anomalous. Hence the squared errors and threshold are dynamic and generally change at every prediction to adapt for the new values and increase accuracy. The module is set to wait for a period of 50 timestamps before calculating the standard deviation in order to make sure the standard deviation calculated has sufficient values to derive it and also that there are not too many false positives reported at the beginning runtime of AD. This wait period is a tuneable parameter, however we observed that waiting for 50 timestamps was sufficient for the considered datasets. The step is described in Algorithm 1. Moreover, we illustrate the output of the AD phase in Fig. 2b, where the upper part of the diagram illustrates the TSF outputs (i.e., the actual and predicted values), and the lower part of the diagram illustrates AD outputs, i.e., the squared error (SError) and the anomaly label (AnomalyLabel), which is 1 for detected anomalous data points and 0 for normal points. The dashed vertical lines represent the actual anomalous instances from the ground truth. We observe that the AnomalyLabel produced by DeepAD$_{Merge}$ follows the dashed lines either at the time of the anomaly or slightly after.

4 Evaluation

This section presents the evaluation of our proposed framework DeepAD. We compare our framework to a recently published generic and scalable anomaly detection framework called EGADS [11], since it follows similar steps to DeepAD for detecting anomalies. The framework compares against the Anomaly

Detection R library[2] released by Twitter, change point methods, and outlier detectors with static threshold, on the Yahoo Webscope Benchmark, claiming to provide highest accuracy levels, irrespective of the dataset.

In addition, we compare DeepAD$_{Merge}$ and DeepAD$_{Vote}$ against the results of three of the individual TSF models coupled with the AD based on dynamic threshold. In this way, we illustrate the benefits of the MP phase of our framework compared to each individual TSF model. Since ARIMA+AD and HW+AD showed similar results across all evaluation metrics, we only illustrate the results of ARIMA+AD, further denoted by DeepAD$_{ARIMA}$. In addition, we illustrate the results of the simple and deep LSTM architectures, denoted further by DeepAD$_{LSTM-S}$ and DeepAD$_{LSTM-D}$, as each was more suitable for a particular dataset, based on the evaluation metric.

Finally, we ranked the performances of the six compared approaches based on the evaluation metrics. We chose *modified competition ranking* as ranking methodology (also known as "1334" ranking). In this ranking methodology, a model's rank is equal to the lowest rank of the model(s) it has a tie with. The modified competition ranking approach guarantees that: (a) The results of the ranking would be deterministic, (b) The best model would be ranked 1st and the worst model would be ranked 6th for all of the datasets, thus making it possible to aggregate the results.

4.1 Dataset

We utilized the Yahoo Webscope Benchmark[3] for our evaluation since this benchmark has been widely referenced in the community and consists of a wide set of time-series with tagged anomaly points. The benchmark is suitable for testing the detection accuracy of various anomaly-types including outliers and change-points. The benchmark consists of a total of 367 time series, split into four main benchmarks. The A1 Benchmark is based on the real production traffic to some of the Yahoo properties. The other three benchmarks are based on synthetic time-series. A2 and A3 Benchmarks include outliers, while the A4 Benchmark includes change-point anomalies. The synthetic time-series generated have varying length, magnitude, number of anomalies, anomaly type, anomaly magnitude, noise level, trend and seasonality. The real dataset is comprised of Yahoo Membership Login (YML) data and it tracks the aggregate status of user logins to the Yahoo network. Both the synthetic and real time-series contain 3000 data-points each, which for the YML data represents 3 months worth of data-points.

4.2 Evaluation Metrics

We evaluate the techniques based on the standard measures of precision, recall and F_1-*score*. Furthermore, we evaluate the early detection of a technique with

---

[2] https://github.com/twitter/AnomalyDetection.
[3] Yahoo! Webscope dataset ydata-labeled-time-series-anomalies-v1_0. http://webscope.sandbox.yahoo.com.

the *Ed-score* defined in [4]. The *Ed-score* evaluates how early an anomaly was detected relative to the anomaly window. The *Ed-score* is between 0 and 1, where 1 represents that the anomaly was discovered at the beginning of the interval and 0 at the end of the interval. In this way, the techniques are compared against even if they discover the anomaly after it had occurred (i.e., *Ed-score* less than 0.5). The *Ed-score* is relative to the time interval, i.e., a 10% increase in *Ed-score* means that a technique detected an anomaly 10% of the time interval earlier on average.

4.3 Results

Figure 3a, b, and c present the DeepAD results compared to EGADS for F_1-*score*, precision and recall, respectively. First, we observe that DeepAD achieves an improvement on average across all datasets as follows by metric: *(i)* F_1-*score*: 26%, with a median improvement from 2% in A1 to 40% and 44% in A3 and A4, respectively, *(ii)* precision: 25%, with a median improvement from −13% in A1 to 50% in A4, and *(iii)* recall: 24%, with a median improvement from 0 in A2 to 53% in A4. Note that only for A1 in the case of precision, EGADS

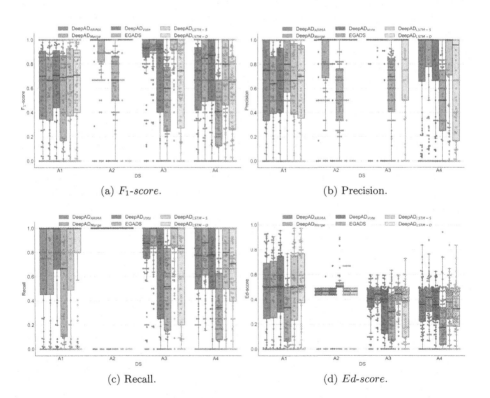

(a) F_1-*score*.   (b) Precision.

(c) Recall.   (d) *Ed-score*.

Fig. 3. Evaluation results in terms of F_1-*score*, precision, recall and early detection score.

achieves a higher median by 13% compared to DeepAD. This suggests that the framework may be biased towards some datasets than others. However, it can be observed from Fig. 3c that the higher median in precision resulted in a less stable and lower median for recall for EGADS in A1. Second, we observe that the performance of some individual TSF models is unstable across different datasets for various evaluation metrics: e.g., for the A1 benchmark consisting of real time series, $DeepAD_{LSTM-D}$ provides better results than $DeepAD_{LSTM-S}$ in terms of recall in Fig. 3c, however it provides worse results for the other benchmarks. $DeepAD_{Merge}$ and $DeepAD_{Vote}$ aim to address this commonly found challenge of instability through their ensemble strategy by employing multiple prediction models and results show a more stable performance across datasets and evaluation metrics. Third, depending on the requirements, different MP strategy can be followed: *(i)* $DeepAD_{Merge}$ typically maintains a higher level of recall than $DeepAD_{Vote}$ for all datasets due to picking the closest prediction to the actual value at each timestamp, since for the true anomalies typically the TSF predictions are far from the actual value which is expected, and *(ii)* $DeepAD_{Vote}$ typically maintains a higher level of precision than $DeepAD_{Merge}$ for all datasets, since it avoids the case of low RMSE TSF models that don't quite learn the underlying patterns but report close to actual values at each time stamp (e.g., a model that learns that the next timestamp has a close value to the current one).

Furthermore, Fig. 3d illustrates the early detection score for all techniques. We observe that for the A1 benchmark, the models powered by AD have reached a median of 0.51, compared to 0.34 for EGADS, as the A1 corresponds to the real dataset contain more dynamic realistic patterns. In A2, the performance of the models was very close, with EGADS reaching an *Ed-score* higher with 0.04 than the rest of the models. However, for A3 and A4 none of the models managed to reach a higher value than 0.5, with a median up to 0.44 in A3 and 0.42 in A4 for DeepAD and 0.3 in A3 and 0.17 in A4 for EGADS, leading to the observation that most anomalies have been detected slightly after their occurrence. We observe that in general DeepAD outperforms EGADS in terms of early detection score across all benchmarks reaching the highest difference of 0.24 in A4.

Figure 4 shows the distribution of ranks for the four performance measures and for all datasets. The figure illustrates the number of datasets for which a model scored a rank between 1 and 6, where rank 1 represents the best model and rank 6 represents the worst model for a given dataset. It should be noted that each model has one or more wins (i.e., rank 1) and one or more lowest rank (i.e., rank 6) for all of the performance measures. This result shows that there is no model that categorically perform best or worst. However, the distribution illustrates the probability of lower and higher rankings. EGADS had the lowest number of wins for and highest number of lowest ranks among the six models based on F_1-*score*, precision and recall. Surprisingly, for *Ed-score*, EGADS has both the highest number of wins and highest number of lowest rank cases. This suggests once again that EGADS may be biased towards certain datasets. For all the performance measures, EGADS has the lowest median and mean rank

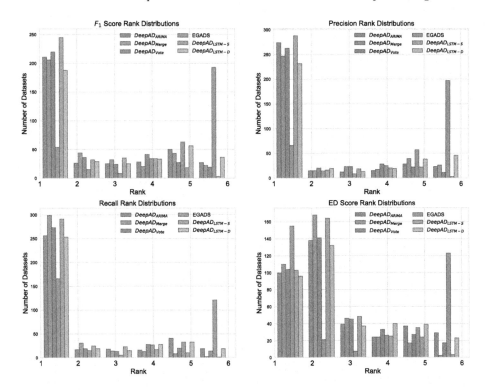

Fig. 4. Modified competition ranking of the models for all datasets

overall. EGADS had a mean rank of 4.67 for F_1-$score$, 3.30 for recall, 4.59 for precision and 3.36 for Ed-$score$. EGADS had a median rank of 6 for F_1-$score$ and precision, 3 for recall and 4 for Ed-$score$. Lastly, we found that the rank distribution of EGADS is significantly lower than all the other models based on DeepAD using Wilcoxon test ($P < 0.001$). This result shows that on the considered benchmark datasets, picking EGADS would not be the optimal choice. Moreover, the median rankings for all the DeepAD models are 1 for precision, recall and F_1-$score$ and 2 for Ed-$score$. The mean ranking difference between the best and worst DeepAD model is less than 1, which shows similar ranking across all DeepAD models.

5 Conclusion

This paper presented a generic anomaly detection framework based on deep-learning (DeepAD) that does not utilize the prior knowledge of the anomalous class neither for training the model nor for determining the threshold. We compared our framework against a state-of-the-art anomaly detection framework EGADS [11] on the Yahoo Webscope Benchmark. We observed that DeepAD generally outperformed and outranked the EGADS framework in terms of early detection score, precision, recall and F_1-*score*. As future work, we plan to plug in other TSF models into the framework, such as convolutional neural networks which can be leveraged in spatiotemporal datasets.

References

1. Aggarwal, C.C.: An introduction to outlier analysis. In: Outlier analysis, pp. 1–40. Springer, New York (2013)
2. Al-Jarrah, O., Arafat, A.: Network intrusion detection system using neural network classification of attack behavior. J. Adv. Inf. Technol. 6(1) (2015)
3. Amer, M., Goldstein, M., Abdennadher, S.: Enhancing one-class support vector machines for unsupervised anomaly detection. In: ACM SIGKDD, pp. 8–15 (2013)
4. Buda, T.S., Assem, H., Xu, L.: Ade: an ensemble approach for early anomaly detection. In: IFIP/IEEE IM, pp. 442–448 (2017)
5. Chandola, V., Banerjee, A., Kumar, V.: Anomaly detection: a survey. ACM Comput. Surv. (CSUR) 41(3), 15 (2009)
6. Chauhan, S., Vig, L.: Anomaly detection in ECG time signals via deep long short-term memory networks. In: IEEE DSAA, pp. 1–7 (2015)
7. Dixon, W.J., Massey Frank, J.: Introduction to Statistical Analsis. McGraw-Hill Book Company Inc., New York (1950)
8. Dunlop, N.: Statistical Calculations, pp. 203–224. Apress, Berkeley (2015)
9. Görnitz, N., Kloft, M.M., Rieck, K., Brefeld, U.: Toward supervised anomaly detection. J. Artif. Intell. Res. 46, 235–262 (2013)
10. Hawkins, D.M.: Identification of outliers, vol. 11. Springer, Netherlands (1980)
11. Laptev, N., Amizadeh, S., Flint, I.: Generic and scalable framework for automated time-series anomaly detection. In: ACM SIGKDD, pp. 1939–1947 (2015)
12. Malhotra, P., Ramakrishnan, A., Anand, G., Vig, L., Agarwal, P., Shroff, G.: LSTM-based encoder-decoder for multi-sensor anomaly detection (2016). arXiv preprint arXiv:1607.00148
13. Malhotra, P., Vig, L., Shroff, G., Agarwal, P.: Long short term memory networks for anomaly detection in time series. In: ESANN (2015)
14. Noto, K., Brodley, C., Slonim, D.: Frac: a feature-modeling approach for semi-supervised and unsupervised anomaly detection. Data Min. Knowl. Discov. 25(1), 109–133 (2012)
15. Rajasegarar, S., Leckie, C., Palaniswami, M.: Hyperspherical cluster based distributed anomaly detection in wireless sensor networks. J. Parallel Distrib. Comput. 74(1), 1833–1847 (2014)

Anomaly Detection Technique Robust to Units and Scales of Measurement

Sunil Aryal[(✉)]

School of Engineering and Information Technology, Federation University,
Mount Helen, VIC, Australia
sunil.aryal@federation.edu.au

Abstract. Existing anomaly detection methods are sensitive to units and scales of measurement. Their performances vary significantly if feature values are measured in different units or scales. In many data mining applications, units and scales of feature values may not be known. This paper introduces a new anomaly detection technique using unsupervised stochastic forest, called 'usfAD', which is robust to units and scales of measurement. Empirical results show that it produces more consistent results than five state-of-the-art anomaly detection techniques across a wide range of synthetic and benchmark datasets.

Keywords: Anomaly detection · Scales of measurement
Local Outlier Factor · Isolation Forest · Unsupervised stochastic forest

1 Introduction

The data mining task of anomaly detection is to detect unusual data instances which do not conform to normal or expected data automatically. The unusual data are called anomalies or outliers. Anomaly detection has many applications such as detecting fraudulent transactions in banking and intrusion detection in computer networks. The task of automatic detection of anomalies has been solved using supervised, unsupervised or semi-supervised learning [1].

In supervised techniques, a classification model is learned to classify test data as either anomaly or normal. They require labelled training data from both normal and anomaly classes. Obtaining labelled training data from anomaly class is challenging in many applications [1]. Unsupervised techniques do not require labelled training data and rank test data based on their anomaly scores directly. They assume that most of the test data are normal and anomalies are few. They may perform poorly when the assumption does not hold [1]. Semi-supervised techniques learn a model representing normal data from labelled normal training data only and rank test data based on their compliance to the model. Majority of data in anomaly detection problems are normal, and thus labelled normal training data can be obtained easily in many applications [1]. This paper focuses on the semi-supervised anomaly detection task.

© Springer International Publishing AG, part of Springer Nature 2018
D. Phung et al. (Eds.): PAKDD 2018, LNAI 10937, pp. 589–601, 2018.
https://doi.org/10.1007/978-3-319-93034-3_47

Most existing unsupervised and semi-supervised anomaly detection techniques assume that anomalies are few and different, i.e., anomalies have feature values that are very different from those of normal instances and lie in low density regions [1–6]. This assumption may not be always true in data mining applications where the units and scales of measurement of feature values are often not known. An anomalous instance may appear to be a normal instance when feature values are measured in different scales. For example, the instance represented by red dot in Fig. 1(a) is clearly an anomaly but it looks like a normal point if the data are measured as $x' = 1/x$ (represented by red dot in Fig. 1(b)). Many existing anomaly detection methods fail to detect the anomaly in Fig. 1(b). In other words, their performances vary significantly if feature values are measured in different units or scales, i.e., they are sensitive to units and scales of measurement.

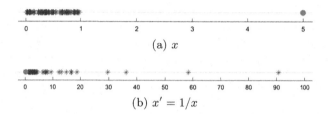

Fig. 1. An example of data represented in two scales. The data point represented by red dot in case (a) is clearly appeared to be an anomaly whereas the corresponding point in case (b) is more like a normal data.

In real-world applications, feature values can be measured in different units and/or scales. For example, fuel efficiency of vehicles can be measured in km/litre or litre/km and annual income of individuals can be measured in an integer scale like $x = 100000$ or using a logarithmic scale of base 10 like $x' = 5$. Unfortunately, units and scales of feature values are often not provided when data are given for anomaly detection where only magnitudes of feature values are available. Many existing anomaly detection methods may perform poorly if feature values are not measured in appropriate units or scales for the task.

Recently, the impact of units and scales of feature values in the context of pairwise similarity measurement of data has been studied [7,8]. Fernando and Webb (2017) introduced a scale invariant similarity measure using a variant of unsupervised random forest called 'Unsupervised Stochastic Forest' (USF) [7]. Each tree in USF partitions the space into regions using a small subsamples of data and the partition is robust to units and scales of feature values.

In this paper, we introduce an anomaly detection technique robust to units and scales of measurement using USF, called 'usfAD'. In each tree, the space is partitioned using a small subsamples of labelled normal training data. Then in each node of trees, normal and anomaly regions are defined based on the labelled normal training data falling in the node. In the testing phase, anomaly score of a test instance is computed in each tree based on the depth of the first node where the test instance lies in the anomaly region. The overall anomaly score is computed by aggregating anomaly scores over a collection of trees. Our empirical results over a wide range of synthetic and benchmark datasets show that it is robust to units and scales of feature values and it produces more consistent results in comparison to five state-of-the-art anomaly detection techniques.

The rest of the paper is organised as follows. Preliminaries and previous work related to this paper are discussed in Sect. 2. The proposed semi-supervised anomaly detection technique of 'usfAD' is discussed in Sect. 3 followed by the empirical evaluation in Sect. 4 and concluding remarks in the last Section.

2   Preliminaries and Related Work

We assume that data are represented by vectors in an M-dimensional real domain $(\mathbb{R}^M)$ where each dimension represents a feature of data. Each data instance $\mathbf{x}$ is an M-dimensional vector $\langle x_1, x_2, \cdots, x_M \rangle$ where each component $x_i \in \mathbb{R}$ represents its value of the ith feature. Let D be a collection of N training data of normal instances and Q be a collection of n test instances which is a mixture of normal and anomalous data. The task in semi-supervised anomaly detection is to learn an anomaly detection model from D and rank instances in Q based on their anomaly scores.

Popular nearest neighbour (NN) based methods [2,9,10] rank a test instance $\mathbf{x} \in Q$ based on its kNNs in D. Being very different from normal data, anomalies are expected to have larger distances to their kNNs than normal instances. **Local Outlier Factor (LOF)** [2] is the most widely used kNN-based anomaly detection method. It does not require any training. Test instances are ranked based on the ratio of their local reachability distance (lrd) to the average lrd of their kNNs in D. The lrd of an instance is estimated using the distance to its kth NN. Euclidean distance is a common choice of distance measure.

Another distance or similarity based anomaly detection technique is **One-Class Support Vector Machine (SVM)** [3]. It learns a model of normal data based on pairwise similarities of training instances using kernel tricks [11]. It requires a kernel function to compute pairwise similarities of instances. Gaussian kernel is a common choice of kernel function that uses Euclidean distance. Test instances are ranked based on their deviation from the model of normal data.

Both NN-based and SVM-based methods can be computationally expensive when training data size $N = |D|$ is large. Though the time complexity of NN search can be reduced to $O(\log N)$ from $O(NM)$ by using indexing schemes such as k:d-trees [12], their effectiveness degrades as the number of dimensions increases and become useless in high dimensional spaces. Recently, Sugiyama and Borgwardt (2013) introduced a simpler and efficient NN-based anomaly detector called **Sp** [5] where test instances are ranked based on their distances to their nearest neighbours (1NN) in a small random subsamples of training data, $\mathcal{D} \subset D$, $|\mathcal{D}| = \psi \ll N$. They have shown that Sp with ψ as small as 25 performs better than or competitive to LOF but runs several orders of magnitude faster.

Liu et al. (2008) introduced an efficient anomaly detector using unsupervised random forest called **Isolation Forest (iforest)** [4] which does not use distance measure. It constructs an ensemble of random trees where each tree is constructed from a small subsamples of training data ($\mathcal{D} \subset D$). It attempts to isolate instances in $\mathcal{D}$ through recursive axis-parallel random split of feature space in each tree. Because anomalies are few and different, they are expected to have shorter average path lengths than those of normal instances over a collection of trees.

Another efficient anomaly detection method which does not require distance measure is based on histograms [6,13]. It discretises feature values in each dimension into a fixed number of equal-width bins and frequency of training data in each bin is recorded. Being few and different, anomalies are expected to fall in bins with small frequencies in many dimensions. Aryal et al. (2016) introduced a simple histogram-based anomaly detection method called **Simple Probabilistic Anomaly Detector (SPAD)** [6] which is more robust to skewed training data because bin width in each dimension depends on the data variance in that dimension.

All these existing methods discussed above rely on the assumption that anomalies have feature values significantly different from normal instances. As discussed in Sect. 1 (Fig. 1), this may not be always true because the distribution of feature values depends on the units and scales of measurement. Existing methods may not perform well if feature values are not measured in appropriate scales so that this assumption holds. Therefore, existing methods are sensitive to units and scales of measurement.

Very recently, the impact of units and scales of measurement of feature values in distance-based pairwise similarity measurement of data has been studied [7,8]. When feature values are measured in different units or scales, the ordering of feature values is either preserved or reversed. Exploiting this characteristic, Fernando and Webb (2017) introduced a non-distance based similarity measure which is robust to units and scales of measurement. The similarity of two instances is defined as the number of shared leaves in a collection of t trees called **Unsupervised Stochastic Forest (USF)** [7]. Each tree is constructed from a small subsamples of data, $\mathcal{D} \subset D$ where $|\mathcal{D}| = 2^h$ and h is a user-defined parameter that determines the height of trees. At each internal node in a tree, subsamples are partitioned into two equal subsets by splitting at the median of values in a randomly chosen attribute. Because of the median split, the similarity measure is robust to units and scales of measurement.

In the next section, we combine the ideas of USF and iforest to introduce a new effective and efficient anomaly detection method which is robust to units and scales of measurement.

3   New Method Robust to Units and Scales of Measurement

iforest [4] attempts to isolate instances in data subsamples using random splits resulting in unbalanced binary trees. Anomalies are expected to fall in leaves

---

Algorithm 1. node($\mathcal{D}$)

Input: $\mathcal{D}$ - Subsamples of training data

```
1 if |D| = 1 then                      /* check if leaf node is reached */
2 |  return ;                                              /* return */
3 self · a ← select(1, 2, · · · , M) ;        /* randomly select an attribute */
4 S ← sort(D_self·a) ;              /* sort values of the selected attribute */
5 self · s ← (S[|D|/2] + S[1 + |D|/2])/2 ;            /* median split point */
6 D_L ← F(D_self·a ≤ self · s); D_R ← F(D_self·a > self · s) ;   /* filter data */
7 self · lNode ← node(D_L); self · rNode ← node(D_R) ;  /* build sub-trees */
8 return ;                                                 /* return */
```

---

with shorter pathlengths in many trees. However, the implementation of iforest is sensitive to units and scales of feature values. At each internal node of a tree, the space is partitioned by selecting a random split between the range of sample values in a randomly selected dimension. The probability of having a split between two consecutive points is proportional to their distance which is sensitive to units and scales of measurement.

USF [7] isolates instances in data subsamples using median splits resulting in balanced binary trees. The median split makes it robust to units and scales of measurement. However, the concept of pathlength can not be used to detect anomalies because all leaves are at the same height.

We propose the following extensions to USF so that pathlengths in trees can be used as a measure to rank test instances to detect anomalies. Once a balanced binary tree is constructed from $\mathcal{D} \subset D$, the entire training data D are passed through the tree to define normal and anomaly regions in each node. In each internal node, the normal range is defined by the minimum and maximum of feature values of the normal training data falling in the node in the dimension j selected to partition the space. In each leaf node, the normal range is defined by the bounding hyper-rectangle covered by the training data falling in the leaf node i.e., minimum and maximum values of training data in all M dimensions. Regions outside of the normal range is considered as anomaly regions in each node. The number of training data falling in each leaf is also recorded.

While a test instance $\mathbf{x}$ is traversing i^{th} tree during testing, first we check whether it lies within the defined normal range at each node. We traverse further down the tree only if it is within the range, otherwise we terminate and return the pathlength of the node where it lies outside of the normal range as the anomaly score of $\mathbf{x}$ in i^{th} tree (let's say $p_i(\mathbf{x})$). If $\mathbf{x}$ traverses to a leaf and lies in the normal region, the anomaly score is defined as the pathlength augmented by the training data mass in the leaf (let's say m) as $p_i(\mathbf{x}) = h + \log_2 m$. The second term is the height of a binary search tree constructed from m data instances and $p_i(\mathbf{x})$ will the be the total height if the leaf node was allowed to grow further until all instances are isolated. This augmentation is important to differentiate leaf nodes with high data mass from those with low data mass because their anomaly scores should be different. Similar adjustment was done in iforest [4].

Algorithm 2. update(D)

Input: D - Training data

1 if $leafNode$ then   /* if it is leaf node */
2      $self \cdot m \leftarrow |D|$;   /* training data mass */
3      $self \cdot range \leftarrow rangeAll(D)$;   /* min & max values in all M dims. */
4   return ;   /* return */
5 $self \cdot range \leftarrow range(D_{self.a})$;   /* min & max values in dim. $self.a$ */
6 $D_L \leftarrow F(D_{self.a} \leq self \cdot s)$; $D_R \leftarrow F(D_{self.a} > self \cdot s)$;   /* filter data */
7 $self \cdot lNode \cdot update(D_L)$; $self \cdot rNode \cdot update(D_R)$;   /* do on sub-trees */
8 return ;   /* return */

Algorithm 3. score($\mathbf{x}, p$)

Input: $\mathbf{x}$ - A test data, p - pathlength so far ($p = 0$ for the root)

1 if $leafNode$ then   /* if leaf, check range in all dimensions */
2   if $inRange(\mathbf{x})$ then   /* if within the range in all dimensions */
3   return $p + \log_2(self \cdot m)$;   /* return augmented pathlength */
4   return p ;   /* out of range, return pathlength */
5 if $inRange(x_{self.a})$ then   /* if non-leaf, check range in dim. $self.a$ */
6      $p \leftarrow p + 1$;   /* increase pathlength */
7   if $x_{self.a} \leq self \cdot s$ then   /* go to respective child */
8   return $self \cdot lNoded \cdot score(\mathbf{x}, p)$
9   else
10   return $self \cdot rNode \cdot score(\mathbf{x}, p)$
11 return p ;   /* out of range, return pathlength */

Algorithms to construct a tree from $\mathcal{D}$ (a random subsamples of D of size 2^h), updating ranges and data mass using D and computing score of a test instance $\mathbf{x}$ are provided in Algorithms 1, 2 and 3, respectively.

The overall anomaly score of $\mathbf{x}$ is estimated by aggregating pathlengths over t trees, $score(\mathbf{x}) = \frac{1}{t} \sum_{i=1}^{t} p_i(\mathbf{x})$. Anomalies will have smaller score than normal instances. We call the proposed unsupervised stochastic forest based anomaly detection method 'usfAD'. It is based on the same idea of isolating anomaly regions from normal regions as used in iforest [4] but using different mechanism of isolation.

As distance is not involved and trees are construct using median splits, usfAD is robust to units and scales of measurement. Even though the size of ranges can be changed with the change in units or scales of measurement, the ordering of values is either preserved (e.g., logarithmic scale) or reversed (e.g., inverse). If a point u lies in the range $[x, y]$ in one scale, the corresponding point u' is expected to lie in the corresponding range $[x', y']$ in another scale. Because of the split at the mid point of two values in the middle (median in the case of even data), there will be small variations in the definition of regions in different scales resulting in small differences in the anomaly detection accuracy.

Figure 2 shows the contour plots of anomaly scores of every point in a two-dimensional space using iforest ($t = 100, \psi = 256$) and usfAD ($t = 100, h = 5$) in a dataset in two scales: x and $x' = 1/x$. It shows that though iforest can detect the anomaly in the original space (Fig. 2(b)), it fails to detect the same anomaly after inverse transformation (Fig. 2(e)). But usfAD has no problem detecting the anomaly in both scales (see Figs. 2(c) and (f)).

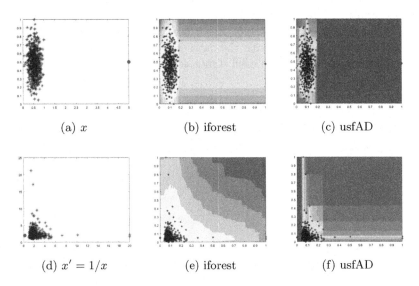

Fig. 2. Anomaly contours of iforest and usfAD in a two-dimensional dataset in two different scales. Note that data are normalised to be in the unit range of [0, 1] in each dimension in all contour plots. The darker the colour, the higher the chances of being anomaly. Note that the anomaly point represented by the red dot is not considered as a part of training data D or D' which includes only the normal instances represented by blue asterisks. (Color figure online)

In the training phase, usfAD requires to create t trees and update normal data range in each tree using the entire training data. It's training runtime complexity is $O(Nth + \psi M)$. Note that $\psi = 2^h$. It needs $O(t\psi M)$ space to store t trees and normal range for all M dimensions in each leaf node. In the testing phase, the runtime complexity of ranking n test instances is $O(n(th + M))$. Because testing time is independent of training data size N, it runs faster than LOF and SVM in datasets with large N. It runs slower than iforest due to the overhead to check range in each node from the root to a leaf in each tree.

4   Empirical Evaluation

In this section, we present the results of experiments conducted to evaluate the performance of usfAD against five state-of-the-art anomaly detectors: LOF, one-class SVM, iforest, Sp and SPAD. We used synthetic and benchmark datasets in our experiments. All experiments were conducted in semi-supervised setting where half of the normal instances in a dataset were used as labelled training data and the remaining other half of normal data and anomalies are considered as test data as done in [14]. Anomaly detection model was learned from the training data and tested on the test data. Area under the ROC curve (AUC) was used as the performance evaluation measure. For random methods: iforest, Sp and usfAD, each experiment was repeated 10 times and reported the average AUC. A significance test was conducted using the confidence interval based on the two standard errors over 10 runs. The same training and test sets of a dataset were used for all experiments with the dataset. Feature values are normalised to be in the unit range of [0, 1] in each dimension.

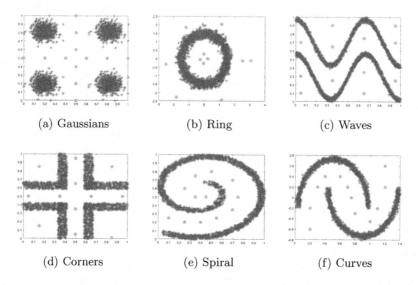

(a) Gaussians   (b) Ring   (c) Waves

(d) Corners   (e) Spiral   (f) Curves

Fig. 3. Two-dimensional synthetic datasets. Each dataset contains 2000 normal instances represented by blue asterisks and 12 anomalies represented by red dots. (Color figure online)

We used the implementation of LOF and SVM included in the Scikit-learn machine learning library [15]. Other methods and experimental setups were also implemented in Python using the Scikit-learn library. All the experiments were conducted in a Linux machine with 2.27 GHz processor and 8 GB memory. Parameters in algorithms were set to suggested values by respective authors: $k = \lfloor \sqrt{N} \rfloor$ in LOF; subsample size $\psi = 25$ in Sp; number of bins $b = \lfloor \log_2 N \rfloor + 1$ in SPAD; and $t = 100$ and $\psi = 256$ in iforest. We used the default settings of SVM. For usfAD, default values of $h = 5$ and $t = 100$ were used.

4.1   Synthetic Datasets

We used six two-dimensional datasets as shown in Fig. 3 to evaluate the robustness of anomaly detection algorithms with different scales of measurement. We used four order preserving and order reversing transformations of data using square, square root, logarithm and inverse, where each feature value x was transformed as x^2 and $\sqrt{x}$, $\log x$ and $\frac{1}{x}$, respectively. Because $\frac{1}{x}$ and $\log x$ are not defined for $x = 0$, all transformations were applied on $\hat{x} = c(x + \delta)$ where $\delta = 0.0001$ and $c = 100$. Note that the original feature values in both dimensions were normalised to the unit range of $[0, 1]$ before applying the transformations to ensure the same effect of δ and c in both dimensions. Once the feature values were transformed, they were renormalised to be in the unit range again. We used exactly the same procedure of transformation as employed by [7].

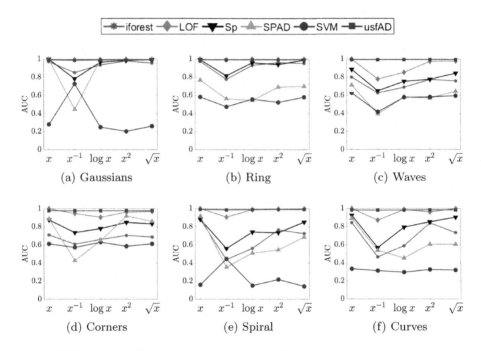

Fig. 4. AUC of contending methods in the six synthetic datasets with order preserving and order reversing transformations of data.

AUC of all contending measures in six synthetic datasets with and without transformations are presented in Fig. 4. It shows that usfAD produced best or equivalent to the best results in all cases. It produces similar results in all datasets with the original feature values and all four transformations. This results show that it is robust to units and scales of measurement.

All five existing measures were sensitive to transformations of data. Among them, LOF is the least sensitive. It could be because of the use of relative k^{th}NN distance of **x** to its kNNs' k^{th}NN distances which captures the contrast in the locality well even though overall variance of data is changed due to transformations. Its performance also dropped with inverse and logarithmic transformations in Waves and Corners. Other four contenders failed to detect all anomalies correctly even in the original scale in four datasets. It is interesting to note that some existing methods produced better results with a transformation than in the original space, e.g., SVM produced best results with the inverse transformation in Gaussians and Spiral.

Table 1. Benchmark datasets

ID.	Name	#dim (M)	Training size (N)	Test data	
				Total (n)	#Anomalies
1.	Arrhythmia	274	193	259	66
2.	Covertype	10	141650	144398	2747
3.	Ionosphere	32	112	239	126
4.	Ism	6	5461	5722	260
5.	Kddcup99	31	30296	34463	4166
6.	Mammography	6	5461	5722	260
7.	Miniboone	49	46554	53446	6892
8.	Mnist	96	9884	10560	676
9.	Mulcross	4	117965	144179	26214
10.	Musk2	166	2790	3082	291
11.	Pima	8	250	518	268
12.	Satellite	36	2199	4236	2036
13.	Shuttle	9	22793	26304	3511
14.	Smtp	3	47563	47593	30
15.	U2r	33	30296	30525	228

4.2   Benchmark Datasets

We used 15 benchmark datasets from UCI machine learning data repository [16], many of which were used in the iforest and SPAD papers. The properties of datasets are provided in Table 1. Data in each dimension were normalised to be in the unit range of $[0, 1]$. To demonstrate the robustness of usfAD to scales of measurement, we also evaluated the performance of contending measures in benchmark datasets with the inverse transformation ($x' = 1/x$) which was done as discussed in Sect. 4.1.

The AUC of all contenders in the 15 benchmark datasets is provided in Table 2. In the original scale, usfAD produced best or equivalent to the best result in seven datasets followed by LOF in five, iforest in four, SPAD and SVM in three each and Sp in one dataset only. usfAD produced significantly better

AUC than the closest contender in Musk2 (ID 10) - AUC of 0.908 by usfAD vs that of 0.700 by LOF. The average results in the last row show that usfAD produced more consistent results than existing methods across different datasets.

Table 2. Anomaly detection performance (AUC) in benchmark datasets in the given original scale (x) and inverse transformation ($x' = 1/x$). First column is the dataset ID from Table 1. The average AUC over the 15 datasets is provided in the last row. Best or equivalent to the best results based on the two standard errors confidence interval of random methods over 10 runs in each scale are underlined.

ID.	Given original scale (x)						Inverse transformation ($x' = 1/x$)					
	LOF	SVM	SPAD	Sp	iforest	usfAD	LOF	SVM	SPAD	Sp	iforest	usfAD
1.	0.800	0.810	0.823	0.807	<u>0.826</u>	0.812	0.714	0.751	<u>0.841</u>	0.746	0.816	0.811
2.	<u>0.992</u>	0.921	0.828	0.870	0.848	<u>0.990</u>	0.949	0.622	0.902	0.818	0.968	<u>0.996</u>
3.	0.958	0.816	0.721	0.949	0.896	<u>0.969</u>	0.922	0.834	0.898	0.901	0.929	<u>0.969</u>
4.	<u>0.893</u>	0.843	0.790	0.816	0.883	<u>0.896</u>	0.833	0.457	0.485	0.776	0.645	<u>0.897</u>
5.	0.895	<u>0.997</u>	0.978	0.996	0.995	0.994	0.870	0.986	0.988	0.968	0.993	<u>0.996</u>
6.	<u>0.889</u>	0.850	0.786	0.826	0.880	<u>0.900</u>	0.830	0.463	0.555	0.777	0.652	<u>0.905</u>
7.	0.700	0.685	0.702	0.599	<u>0.750</u>	0.717	0.727	0.548	<u>0.741</u>	<u>0.746</u>	<u>0.745</u>	0.708
8.	<u>0.877</u>	0.824	0.799	0.810	0.835	0.853	0.579	0.721	0.766	0.716	0.771	<u>0.849</u>
9.	<u>1.000</u>	<u>1.000</u>	0.998	<u>1.000</u>	0.999	<u>1.000</u>	<u>1.000</u>	0.973	0.928	0.983	0.973	<u>1.000</u>
10.	0.700	0.195	0.604	0.552	0.427	<u>0.908</u>	0.700	0.579	0.597	0.631	0.585	<u>0.911</u>
11.	0.704	0.706	0.743	0.741	<u>0.754</u>	0.667	0.409	0.579	0.436	0.405	0.454	<u>0.653</u>
12.	0.837	0.651	<u>0.867</u>	0.837	0.793	0.821	<u>0.834</u>	0.655	0.806	0.816	0.799	0.821
13.	0.991	0.987	<u>0.999</u>	0.989	0.997	<u>1.000</u>	0.991	0.979	0.998	0.987	0.996	<u>1.000</u>
14.	0.868	0.728	<u>0.932</u>	0.841	0.883	0.873	0.840	0.785	<u>0.960</u>	0.863	0.929	0.875
15.	0.886	<u>0.987</u>	0.977	0.982	<u>0.986</u>	0.926	0.860	0.976	<u>0.988</u>	0.976	<u>0.987</u>	0.923
Av.	0.866	0.800	0.836	0.841	0.850	0.889	0.804	0.727	0.793	0.807	0.816	0.888

With the inverse transformation, the performance of all existing methods dropped in many cases. usfAD produced best or equivalent to the best result in 10 datasets followed by SPAD in four, iforest and LOF in two each, and Sp in one dataset only. SVM did not produce best or equivalent to the best result in any dataset. It is interesting to note that some existing methods produced better results with the inverse transformation than in the original space, e.g., LOF, SPAD and Sp in Miniboone (ID 7); iforest, Sp and SVM in Musk2 (ID 10) etc.

In terms of runtime, usfAD was one order of magnitude faster than LOF and SVM in large and/or high dimensional datasets. For example, to complete one run of experiment in Miniboone, usfAD took 440 s whereas LOF and SVM took 2308 s and 1187 s, respectively. However, it was up to one order of magnitude slower than Sp, SPAD and iforest. In Miniboone, Sp took 22 s, SPAD took 43 s and iforest took 83 s.

5   Concluding Remarks

Existing anomaly detection methods largely rely on spatial distances of data to identify anomalous instances. They may fail to detect anomalies which are masked due to the use of inappropriate units or scales of measurement. In many data mining applications, units and scales of feature values are often not provided where only magnitudes of feature values are given. Thus, an anomaly detection method which is robust to units and scales of measurement is preferred. In this paper, we introduce one such technique using unsupervised stochastic forest. Our empirical results in synthetic and benchmark datasets suggest that the proposed method is robust to units and scales of measurement and it's performance is either better or competitive to existing methods. It produces more consistent and stable results across a wide rage of data with different order preserving and order reversing transformations.

References

1. Chandola, V., Banerjee, A., Kumar, V.: Anomaly detection: a survey. ACM Comput. Surv. **41**(3), 15:1–15:58 (2009)
2. Breunig, M.M., Kriegel, H.P., Ng, R.T., Sander, J.: LOF: identifying density-based local outliers. In: Proceedings of ACM SIGMOD Conference on Management of Data, pp. 93–104 (2000)
3. Schölkopf, B., Platt, J.C., Shawe-Taylor, J., Smola, A.J., Williamson, R.C.: Estimating the support of a high-dimensional distribution. Neural Comput. **13**(7), 1443–1471 (2001)
4. Liu, F., Ting, K.M., Zhou, Z.H.: Isolation forest. In: Proceedings of the Eighth IEEE International Conference on Data Mining, pp. 413–422 (2008)
5. Sugiyama, M., Borgwardt, K.M.: Rapid distance-based outlier detection via sampling. In: Proceedings of the 27th Annual Conference on Neural Information Processing Systems, pp. 467–475 (2013)
6. Aryal, S., Ting, K.M., Haffari, G.: Revisiting attribute independence assumption in probabilistic unsupervised anomaly detection. In: Proceedings of the 11th Pacific Asia Workshop on Intelligence and Security Informatics, pp. 73–86 (2016)
7. Fernando, T.L., Webb, G.I.: SimUSF: an efficient and effective similarity measure that is invariant to violations of the interval scale assumption. Data Min. Knowl. Disc. **31**(1), 264–286 (2017)
8. Aryal, S., Ting, K.M., Washio, T., Haffari, G.: Data-dependent dissimilarity measure: an effective alternative to geometric distance measures. Knowl. Inf. Syst. **35**(2), 479–506 (2017)
9. Ramaswamy, S., Rastogi, R., Shim, K.: Efficient algorithms for mining outliers from large data sets. In: Proceedings of the 2000 ACM SIGMOD Conference on Management of Data, pp. 427–438 (2000)
10. Bay, S.D., Schwabacher, M.: Mining distance-based outliers in near linear time with randomization and a simple pruning rule. In: Proceedings of the Ninth ACM SIGKDD Conference on Knowledge Discovery and Data Mining, pp. 29–38 (2003)

11. Vapnik, V.: Statistical Learning Theory. Wiley, New York (1998)
12. Bentley, J.L., Friedman, J.H.: Data structures for range searching. ACM Comput. Surv. **11**(4), 397–409 (1979)
13. Goldstein, M., Dengel, A.: Histogram-based outlier score (HBOS): a fast unsupervised anomaly detection algorithm. In: Proceedings of the 35th German Conference on Artificial Intelligence, pp. 59–63 (2012)
14. Boriah, S., Chandola, V., Kumar, V.: Similarity measures for categorical data: a comparative evaluation. In: Proceedings of the Eighth SIAM International Conference on Data Mining, pp. 243–254 (2008)
15. Pedregosa, F., Varoquaux, G., Gramfort, A., Michel, V., Thirion, B., Grisel, O., Blondel, M., Prettenhofer, P., Weiss, R., Dubourg, V., Vanderplas, J., Passos, A., Cournapeau, D., Brucher, M., Perrot, M., Duchesnay, E.: Scikit-learn: machine learning in Python. J. Mach. Learn. Res. **12**, 2825–2830 (2011)
16. Bache, K., Lichman, M.: UCI machine learning repository. University of California, Irvine, School of Information and Computer Sciences (2013). http://archive.ics.uci.edu/ml

Automated Explanations of User-Expected Trends for Aggregate Queries

Ibrahim A. Ibrahim$^{(\boxtimes)}$, Xue Li, Xin Zhao, Sanad Al Maskari,
Abdullah M. Albarrak, and Yanjun Zhang

The University of Queensland, Brisbane, Australia
{i.ibrahim,xueli,x.zhao,s.almaskari,a.albarrak,yanjun.zhang}@uq.edu.au

Abstract. Recently, a deeper level of data exploration has emerged enabling users to infer anomalies in their queries. This exploration level strives to explain why a particular anomaly exists within a query result by providing a set of explanations. These explanations are precisely a set of alterations, such that when applied on the original query cause anomalies to disappear. Trends are pattern changes in business applications generated based on SQL aggregated queries. Additionally, a user expected trend is a particular pattern change in data was supposedly happen based on businesses studies.

In this paper, we generalize this process to automatically produce explanations for users expected trends. We propose User Trend Explanations (UTE) framework which provides insightful explanations by taking a set of user-specified points (called *prospective trend*), and finds a top explanation that produce this trend. We develop a notion of *uniformity* of a predicate on a given output, and implement a set of algorithms to search the data space efficiently and effectively. The key idea is harnessing the linear search space rather than the exponential space to enable accurate explanations that are possible with tuples. Our experiments on real datasets show significant improvements UTE provides when compared with state-of-the-art related algorithms.

1 Introduction

The explosion of big data drives users to use a diverse set of visualization tools to efficiently discover trends and patterns while exploring data. Examples of these tools are Tableau, ShowMe, Fusion Tables [6,9]. Although various visualization tools effectively discover outliers but, most of them are unable to explain why a given set of outputs are outliers or identifies reasons behind such outliers. Thus, a deeper level of exploration that explains reasons behind a particular trend or pattern found is needed.

To illustrate, a mobile applications company is interested in studying the behavior of three new released apps: (Chat, Music, and Video). Figure 1b, shows a bar chart visualizing the average downloads of these apps (blue-color bars).

© Springer International Publishing AG, part of Springer Nature 2018
D. Phung et al. (Eds.): PAKDD 2018, LNAI 10937, pp. 602–614, 2018.
https://doi.org/10.1007/978-3-319-93034-3_48

However, the analyst team is expecting different averages as shown in the red-color line. It is quite essential to explain why a specific pattern exists, or alternatively describe the cause(s) which are responsible of the deviation in the query outcome and user specified (expected) trend.

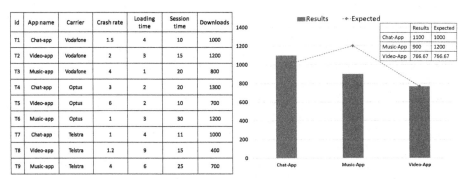

a) table R of a mobile applications data   b) Blue bars shows Q_1 results and the red line plots user expected trend.

Fig. 1. Mobile Apps visualization example (Color figure online)

Example 1. Using the relation R in Fig. 1a, an analyst is interested in exploring the performance of different mobile applications. Each tuple in R holds details about an application such as: name, carrier, crash rate, loading time, session time, and number of downloads. Assume she is interested in the average downloads per application, hence she formulates the following SQL query:

$Q1$: *SELECT avg(downloads), App_name FROM R GROUP BY App_name;* □

Then, she visualises the result of the query using a bar chart, such as the one in Fig. 1b. The analyst is expecting the average downloads for *Chat_App, Music_App, and Video_App* to be 1000, 1200, and 766.67 respectively. However, the result of $Q1$ is far from her expectations. Hence, she wants to understand why the result in the chart does not match her expectations, or alternatively, what are the reasons of this deviation. Specifically, what are the predicates that can be added to Q_1 in order to achieve her expected trend (i.e., the red-colored line)?

There are many reasons behind the results divergence from her expected trend: it might be one of the carriers caused this divergence, due to a certain crash rate of apps, or others. The predicate which explains the divergence Example 1 is: *(Crash_rate NOT BETWEEN 3 and 4) AND (session_time NOT BETWEEN 20 and 25).*

Applying this predicate to the original query Q_1 will return the user's expected result.

This problem is non-trivial and challenging due to the following factors:

- **Exponential Search Space:** Each single aggregate value needs to consider various combinations of input tuples affecting the original results, which depends on properties of the aggregate function itself. In the worst case, we cannot predict how combinations of input tuples interact with each other, and resolve to evaluating all possible predicates (exponential search space).
- **CPU and I/O Intensive:** An efficient and effective measure is to be designed to determine which subset of input tuples cause the value to be deviated. Such measures are challenging to design because it involves iterating over all possible subsets of input tuples and applying the aggregate operator from scratch. In addition, Executing this number of modified aggregate queries (predicates) is expensive and costly.
- **Specification of User Expected Values:** Unlike previous work [11,12,16], as users only specify how an outlier result looks wrong (i.e. the outcome is too high/low compared to their expectations), however in reality, instead of the binary choice of 'too high' or 'too low', users want to go further to specify how much is an outlier far from their expectations which makes the problem more complex.

The paramount challenge in this problem and the limitation of current approaches is the exponential search space of the predicates. To illustrate, assume in a relational table R with d categorical dimensions which are not involved with the aggregate function nor the group by that are used to construct the explanations. Then, there are $\prod_{i=1}^{d}(2^{n_i} - 1)$ possible predicates, where n_i is the number of distinct values of a dimension d_i. To manage such expensive search space, we propose to utilize a linear search space $|R|$, rather than an exponential one in previous works [12,16]. According to mobile data in Fig. 1a, the size of the search space based on predicate-level is $(2^3 - 1) * (2^7 - 1) * (2^6 - 1) * (2^6 - 1) = 3,528,441$ possible explanations (predicates), while our proposed approach evaluates at most only $|R| = 9$ possible explanations.

In this paper, we introduce *UTE* framework which provides insightful explanations based on the intervention of tuples *not predicates*. *UTE* doesn't require to evaluate all tuples and evaluates only tuples with distinct values in the aggregated attribute. *UTE* shares the common philosophy in [12,16] of measuring explanations using intervention or influence of tuples. However, it searches for explanations based on the *highest granularity* using tuples, rather than predicates. Finally, we summarize our contributions as follows:

- We formalize a notion of *uniformity* over predicates and develop a system that searches for uniform predicates in a single relation and utilizes the linear tuple-level search space rather than the exponential predicate-level.
- We provide a set of splitting algorithms: *XTrend-Basic, XTrend-Advanced and XTrend-Fast* which split attribute domain linearly.
- We proposed merging algorithms which exploit geometric properties in the distribution of predicates uniformity to group predicates in an efficient manner.

- We present XTrend Transformation process that generates meaningful explanations (predicates) to the user.
- We run experiments on real datasets showing the efficiency and effectiveness of our algorithms compared with state-of-the-art algorithms.

The paper is structured as follows: Sect. 2 illustrates related work and Sect. 3 describes the formal problem formulation, Sect. 4 presents UTE architecture and several algorithms and Experiments are described in Sect. 5.

2   Related Work

Several projects from databases and data mining research communities were proposed to provide explanations and to support interactive data exploration, e.g., user ratings on sites like Yelp and IMDB [4], access log and security permissions [2]. Kanagal et al. [7] studied the problem of computing top-k influential variables and top-k explanations to answer "why a tuple is in the output", or "why the probability of an output tuple is greater than another one" in probabilistic databases.

Explanations in Databases: Wu and Madden [16] proposed the Scorpion system to provide explanations for user aggregated queries over a single relation. Scorpion uses a set of user-specified outlier and hold-out points in a given aggregate query result to output top-K predicates which make outliers disappear. Wu and Madden [16] developed a notion of predicate influence based on sensitivity analysis [13] in the context of data provenance [10] to define how the probability of a result tuple is influenced by an input tuple. Although Scorpion provides explanations in terms of predicates but, it fails to provide a concrete method of how much high or low outliers should be bounded.

The optimization techniques used in Scorpion [1,3] are based on the incremental properties of aggregate functions such as *sum, avg, stddev*. Though, multiple performance limitations arise for user-specified aggregate functions or when an aggregate function is considered as a black-box. Roy and Suciu [12] proposed a formal framework for defining explanations to complex SQL queries over database schemas involving multiple relations and foreign key constraints. However, their framework limits predicates to be conjunctive predicates of atomic equality predicates on tables.

OLAP Data Cubes: Sarawagi et al. [14] proposed *Diff* operator to generate summary tuples that explain why two subcubes values differ assist users to find reasons for drops or increases observed at an aggregated level. It is based on an information theoretic formulation for expressing the reasons of why subcubes' values are different. RELAX [15] assists users to propagate from a specific problem case in multidimensional hierarchal data cubes (e.g. drop in sales of a store in a specific region) and returns a wider context in which the problem occurs. MRI [5] automatically provides a meaningful interpretation of ratings based on the idea of data cube. This work uses randomized hill exploration algorithms to discover interesting cuboid in OLAP rating cubes.

PerfXplain [8] explains relative performances questions (i.e., why some jobs have faster runtimes than others) of pairs of MapReduce jobs. PerfXplain provides a query language named PXQL for expressing performance queries and produces explanations from past MapReduce executions logs.

3   Problem Formulation

UTE attempts to find a predicate over a dataset that achieves most uniformity to a user specified trend. We start by defining some notations that are used throughout the paper.

Let D be a flat relational table with a set of attributes $A = \{a_1, a_2, ..., a_m\}$. A visualization query Q on D is an SPJ aggregate query which returns a result set $\mathcal{R} = Q(D)$ grouped by attributes $a_{gby} \in A$ and aggregated by a single aggregation function F over $a_{agg} \in A$, where $a_{gby} \neq a_{agg}$. The result set $\mathcal{R}$ is a collection of t pairs: $\mathcal{R} = \{(r_1, v_1), (r_2, v_2), .., (r_t, v_t)\}$ where $r_i \in a_{gby}$ is a distinct value in a_{gby}, and v_i is the aggregated value associated with r_i. Let $A_{Rest} = A - a_{gby} - a_{agg}$ be the set of the remaining attributes which are not involved in Q and used to construct explanations.

In Example 1, query Q_1 contains a single group-by attribute, $a_{gby} = \{$App_name$\}$, and an aggregate attribute, $A_{agg} = \{$downloads$\}$. The user is interested in combinations of $A_{rest} = \{$Carrier, Crash_rate, Session_time, Loading_time$\}$ values that are responsible for the anomalous average downloads.

3.1   Prospective Trend Problem

Informally, the prospective trend problem is a user question: *why my query is returning different values than the ones I'm expecting?* Formally:

Definition 1. *Prospective Trend Problem (PTP):* *Given a user prospective trend T containing a set of user expected values $\{(r_1, e_1), (r_2, e_2), .., (r_t, e_t)\}$ where $|\mathcal{R}| = |T|$.*

A user question $\mathcal{U}$ is defined as $\{Q, T\}$, where Q is an SPJ aggregate query with a group-by clause. The objective of PTP is to find a top explanation that minimize the difference between T and $\mathcal{R}$.

Definition 2. *Explanation (e):* *An explanation is a set of conjunctive predicates $P = \{p_1, p_2, ..., p_h\}$, where $p_i \in P$ is either a range of a continuous attribute or a set of containment clauses over a discrete attribute.*

Our objective in PTP is to find the optimal explanation e^* such that when e^* is applied to Q, the query will return exactly T. Though, e^* in practice is quite difficult to find hence, we relax this objective to look for the explanation that minimizes the distance to T (referred as *Predicate Uniformity*). We use the Euclidean distance to measure that distance. Let $\mathcal{R}'$ be the result of Q after adding e, then our objective is to minimize the following:

$$Uniformity(\mathcal{R}', T) = \sqrt{\sum_{i=1}^{t}(v_i - e_i)^2} \tag{1}$$

Equation 1 ranges from zero to ∞, such that smaller values indicate better fit to $\mathcal{T}$. *predicate uniformity* is obtained by calculating the distance between two vectors which is commonly used to find the deviation in various data exploration approaches. UTE is independent of using any distance measures such as Earth Mover distance, cosine distance, ... etc.

4   UTE Architecture

This section illustrates UTE system architecture we have developed to solve the problem of finding prospective predicates defined in Sect. 4.2. It describes naive implementations of the main system components then, discusses reasons of why this implementations are not efficient. These implementations do not assume anything about the aggregates so can be used on various user defined aggregates to find the most *uniform* predicate. *UTE* system is implemented in Java as part of an end-to-end data exploration tool[1] Users can select databases and execute aggregate queries whose results are visualised as charts in Fig. 1b. Users can select arbitrary results and specify their expected values or plot expected trend. Both expected trend values and the aggregate query are sent to the *UTE* backend to construct explanations. *UTE* consists of two main processes: *Splitting* process which is responsible for generating predicates which uses the Scorer module in parallel to compute the objective function i.e. L_2norm and returns a ranked list of scored predicates. Highly scored predicates are greedily combined during the *Merging* process as long as the uniformity of merged predicates improves. Finally, the top predicate is returned to the user.

4.1   Naive Splitter

NAIVE algorithm defines all distinct single-attribute clauses, then enumerates all conjunctions of up to one clause from each attribute. Clauses over a discrete attribute, A_i are of the form, $A_i in(Ax_1, Ax_2, ..., Ax_s)$ where $Ax_1, Ax_2, ..., Ax_s$ is replaced with all possible combinations of the attribute's distinct values. In Example 1, permuting all predicate for *Carrier* attribute such as *Carrier* in $('Vodafone')$, *Carrier* in $('Telstra')$, *Carrier* in $('Vodafone', 'Optus')$, ... etc. Clauses over continuous attributes are constructed by splitting the attribute's domain into a fixed number of equisized ranges, and enumerating all combinations of consecutive ranges. Then it computes the uniformity of each predicate by sending it to the *Scorer.*

Naive Splitting Complexity. To discuss the complexity analysis of predicates generation problem by *naive*. As denoted in Sect. 3, A_{Rest} is a set of remaining attributes in the dataset D which used to construct the explanations We denote A_C as a set of all categorical attributes in A_{Rest} and similarly, A_R is a set of all numeric attributes in A_{Rest}, where $A_C \cap A_R = \phi$ and $A_C \bigcup A_R = A_{Rest}$. Firstly,

---

[1] https://github.com/ibrahimDKE/UTE_Xtrends.

the number of generated predicates $P_{count}(a_i)$ for each categorical attribute $a_i \in A_C$ is equivalent to number permutations of all distinct values dN_i in a_i.

$$P_{count}(a_i) = 2^{dN_i} - 1$$

The number of generated predicates in a set A_C of categorical attributes is increasing both *linearly* with number of categorical attributes $|A_C|$ and **exponentially** with the number of distinct values in attributes. The time complexity is $O(|A_C|) = |A_C|^{dN_{Max}}$ where, dN_{Max} is the maximum number of distinct values found in all attributes in $|A_C|$. Secondly, To compute the number of generated predicates over numeric attribute $a_j \in A_R$, it requires to specify a splitting ratio β over the attribute domain which used to identify the partition size. For example, $\beta = 0.5$, this means that each partition size is 50% of domain range. For any numeric attribute $a_j \in A_R$ with a domain range $[0-1]$ and splitting ratio β, the number of generated predicates $P_{count(a_j)}$ is:

$$P_{count(a_j)} = \sum_{i=0}^{|\frac{1}{\beta}|} \left(|\frac{1}{\beta}| - i \right)$$

Naive algorithm is inefficient since the number of single-attribute clauses increases exponentially for a discrete attribute as its cardinality increases. Additionally, the space of possible conjunctions is exponential with the number of attributes [16], such issues make the problem indefensible and unacceptable for even small datasets.

4.2   Basic Merger (BM)

BM takes as input a ranked list of predicates which is produced during splitting process and merges subsets of the predicates. Two predicates are merged by finding the minimum bounding box of the continuous attributes and the union of all values for each discrete attribute. *BM* repeatedly extends the existing predicates in ascending order of their scores. Each predicate is expanded by greedily merging it with adjacent predicates until the resulting uniformity does not improve.

For a list of splitted predicates $\mathcal{L}$ and a current number of merged predicate so far $\mathcal{X}$. Then, the complexity of *BM* is $O(|\mathcal{L} - \mathcal{X}|)$. In the worst case, *BM* keeps merging the top predicate *(most uniform one)* with all remaining predicates in the list and its time complexity will be $O(|\mathcal{L}|)$. The merging behavior of *BM* implies a trade off between efficiency and effectiveness, since some of predicates combinations will be missed and not evaluated. However, if all combinations are evaluated by *BM*, then it ends up with an exponential complexity of merging.

4.3   *UTE* Splitting Approaches

In this section, we describe our proposed algorithms to solve predicates generation problem linearly and present optimizations for both splitting and merging processes.

X Trend Basic Splitter. The first proposed splitting approach *XTrend Basic* starts with reading meta-data information for all attributes in A_{Rest} and returns attribute $A_{MaxSplit}$ that contains the maximum number of distinct values. Then, it uses $A_{MaxSplit}$ to generate a set of splitted predicates $SP(r_i)$ for each result id r_i by retrieving all tuples in result id r_i. For each predicate P_j in result id r_i, it calculate internal score that equals the difference between expected value e_i and the result $P_j(v)$ of predicate P_j.

XTrend Basic utilizes the aggregate feature of query Q to limit evaluating all tuples in the dataset. It evaluates only tuples which contain distinct values in the aggregated attribute A_{Agg}. Tuples that have similar A_{Agg} value in each result id are not evaluated and have same score. In Example 1, XTrend Basic uses attribute *id* as $A_{MaxSplit}$ (since it contains 9 distinct values) to generate predicates set $SP(Chat\_App)$ for result id *(Chat_App)*. It contains three predicates $id = T1$, $id = T4$, and $id = T7$, only two predicates are being evaluated since *T1* and *T7* have same downloads value 1000. Removing tuple *T1* or *T7* changes the average downloads to $\frac{(1300+1000)}{2} = 1150$, and similarly for *T4*. Predicates $id = T1$ and $id=T7$ have highest internal score to the target value of result id *Chat_App=1000*, $|1000 - 1150| = 150$. While removing $id = T4$ changes the result to the exact target value $|1000 - 1000| = 0$.

In parallel, *XTrend Basic* generates a predicates set for each result id then, each sorted set $SP(r_i)$ is passed to a basic Merger *BM* which merges top predicates according to their internal score as discussed above in Sect. 4.2. *XTrend Basic* is basically designed to evaluate predicates in tuple-level as it sets $A_{MaxSplit}$ to *ROW ID* as the default splitting attribute. However, when the number of distinct values in $A_{MaxSplit} < |D|$ and ignoring *ROW ID*. It modifies predicate clause by adding one more condition $A_{gby} = r_i$. The total number of splitted predicates $sPred$ is calculated as follows: Let dn_i is number of distinct values of aggregated attribute in result id r_i. Then, total generated predicates is $sPred = \sum_{i=1}^{|r|} dn_i$.

The time complexity of splitting process is linear $O(|sPred|)$, while $sPred \ll |D|$. In the worst case, when the aggregated attribute contains only unique values, the complexity will be $O(|D|)$. *XTrend Advanced* is similar to *XTrend Basic* except that the former applies transformation process during the Merging process.

Xtrend Fast Splitter. This splitting approach works in tuple-level and computes the result of splitted predicates in memory. It completely avoids both executing queries of predicates on the database and calling Scorer when the aggregate function F of query Q is incrementally removable *(e.g., SUM, AVG, and STDDEV)*. An incrementally removable aggregate can directly evaluates predicate P from its tuple. If the updated result of removing a tuple t from the inputs D, can be computed by only reading t. *SUM* is incrementally removable because $SUM(D - t) = SUM(D) - SUM(t)$,

Fast Splitter cashes all input tuples of $A_{MaxSplit}$, A_{gby}, and A_{agg} attributes, and directly computes the internal score of applying each predicate. In Example 1, computing the result of predicate $Tuple\_id \neq' T7'$ affects only on the its result id *Chat_App*, it can be easily calculated $\frac{1300+1000-900}{3-1} = 700$ and directly calculates its internal score $|1300 - 700| = 600$.

4.4 XTrend Transformation

The final output of Merging process is a set of ranked predicates contains only all clauses generated from a single attribute. This because only $A_{MaxSplit}$ attribute is used during both splitting and merging processes. Thus, such predicates are usually meaningless and difficult to understand by the user. XTrend transformation process involves changing a merged predicate resulted from merging one or more predicates to a meaningful and understandable predicate. The process finds alternative predicates by exploring other attributes that are not involved during splitting process. To illustrate, Let P_{Mrg} is a merged predicate with a set of merged clause attributes P_{Atts} and A_{Rest} is a set of remaining attributes in database D where $P_{Atts} \subseteq A_{Rest}$. The number of available transformed predicate equals $|A_{Rest} - P_{Atts}|$. Transforming predicate Pa_c for categorical attributes is done by adding all distinct values of attribute a_c of all tuples satisfy predicate P_{Mrg}. For a numeric attribute a_r, transformation is done by finding the bounding box of all tuples satisfy predicate P_{Mrg} in attribute a_r.

In Example 1, suppose predicate $P_{Mrg} = Tuple\_id\ not\ in\ (T1, T5, T7)$ is high scored merged predicate, to transform P_{Mrg} we look for remaining attributes {*Carrier, Crash_rate, Loading_time, Session_time*} as they haven't used before. Transformation of predicate P_{Mrg} is done by finding the bounding box for *Tuple_id in (T1, T5, T7)* over attribute *Session_time* and outputs new predicate (Session_time) *Not between 10 and 11*. Each new predicate is evaluated by the *Scorer* and add only predicates that improve the score or at least has similar score with the original predicate.

Although, transformation process has extra merging cost but it drops rapidly while increasing merged clause attributes $|P_{Atts}|$. The time complexity for any merged predicate P_{Mrg} is linear and equals $O(|A_{Rest}| - |P_{Atts}|)$. However, In the worst case when only a single attribute in the predicate clause, the complexity will be $O(|A_{Rest}|)$.

4.5 Xtrend Merging

In this section we present *Geo Merger* optimizations that employs geometric features of the predicates list generated by *UTE* splitters. As illustrated in Sect. 4.3, *UTE* splitters produce a ranked list of predicates based on tuple-level and *Geo Merger* takes the list as input then generates a histogram for all predicates' scores to measure the skewness of predicates. Computing skewness of predicates' scores helps *Geo Merger* to specify a threshold in order to limit merging predicates. For instance, negative skew means the mass of predicates distribution are concentrated on the right of the histogram and this implies that most predicates' scores are high and far from the target. Thus, *Geo Merger* doesn't need to merge inefficient predicates and it puts the mean as a threshold.

Finally, *Geo Merger* keeps predicates merging as long as the uniformity of predicates doesn't improve and reaches the skewness threshold. This distinct improvement of *Geo Merger* is specifying a threshold to limit merging to the half of predicates list in the worst case, while *Basic Merger* keeps merging to the end of the list. In contrast, Scorpion's *Merger* [16], limits the number of predicates that needed to be merged by only expanding the predicates whose scores within the top quartile of the list. Scorpion *Merger* approximates predicate's influence by estimating the number of cached-tuples in each predicate. However, it assumes that tuples are distributed uniformly within the partitions and such assumption rarely exits in real world datasets.

5   Experiments

This section presents our experiments that compare *UTE* splitting algorithms *XTrend Basic*, *XTrend Advanced* and *Fast Splitter* against Scorpion Decision Tree (*DT*) algorithm. The experiments provide a clear illustration to show how these algorithms compare in terms of performance and quality of explanations.

5.1   Datasets

GoCard: A transportation dataset contains 4.4 million tuples with 12 dimensions describing Brisbane city transportation from January to March 2013. Each tuple represents a trip such as (route, journey length, no. passengers, boarding and lighting stop, ..etc). GoCard contains both discrete and numeric attributes varying cardinality from 2 to 170214. We study the average passengers per boarding stop for two buses namely *411 and 412*. It shows a single stop *Coldridge Street* recorded 20 times higher than average passengers in all boarding stops. It's found that 32630 passengers used these buses in one day. We defined the ground truth all tuples where the operation date *=06-Mar-13* and alighting stop *in('Adelaide St', 'George St')*.

Expense: dataset contains all campaign expenses between July 2014 and August 2016 from the 2016 US Presidential Election[2]. The Expense dataset contains 219579 rows and 14 attributes mostly are categorical attributes (e.g., recipient name, amount, state, and city), 12 attributes are used to create explanations. The attributes are varying between 2 to 221490 (recipient names) distinct values. The SQL query sums the total expenses per state of Hilary Clinton campaign. It's found that more than \$4.2M spent for Media purchases in Washington DC only. We defined the ground truth as tuples where *desc ='MEDIA BUY'* and the expense was greater than \$1.2M. *UTE* generated the explanation state ='DC' & recipient nm ='GMMB INC.'

Experiments evaluate both efficiency and effectiveness along different metrics of precision, recall and F-score. The experiments run on a single threaded PC

---

[2] http://www.fec.gov/disclosurep/PDownload.do.

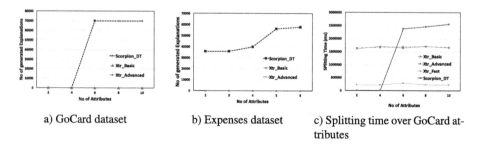

a) GoCard dataset   b) Expenses dataset   c) Splitting time over GoCard attributes

Fig. 2. Cost of splitting process as dataset dimensionality increases

(Windows 7, 16 GB RAM). Scorpion splitters were configured to split each continuous attributes domain into 15 equi-sized ranges. We evaluate the efficiency of UTE splitting algorithms along runtime and number of generated explanations across different number of attributes.

5.2   Comparing Splitting Algorithms

The following experiments compare the runtime and number of generated explanations using *UTE* splitters and Scorpion DT algorithms. Figure 2 shows the number of generated explanations by *XTR_Basic, XTR_Advanced, XTR_Fast, and Scorpion*. As shown in Figs. 2a and b, *XTR_Basic, XTR_Advanced, and XTR_Fast* generate a fixed number of explanations as the dimensionality increases of *Gocard* and *expenses* datasets. However, number of explanations generated by *Scorpion* increase rapidly while data dimensionality increases gradually. In Fig. 2a, *Scorpion* produced almost same number of explanations (142) on the first 4 attributes as *UTE* splitters but, extending dimensionality to 6 attribute by adding a single discrete attribute makes this number jumping around 500 times and *Scorpion* outputs 69766 explanations. This increase is obviously affects on execution time as shown in Fig. 2c. Additionally, *XTR_Fast* shows the lowest execution time among all splitting algorithms since it evaluates explanations in memory as discussed in Subsect. 4.3.

5.3   Comparing Merging Algorithms

The following experiments compare the runtime and number of merged explanations using *Geo Merger* splitters and Scorpion *Basic Merger*. As shown in Fig. 3, the number of merged predicates generated by *XTR_Basic* and *Scorpion* is stable along attributes for both *GoCard* and *Expenses* dataset in Figs. 3a and b. The reason is *Geo Merger* merges predicates produced by *XTR_Basic* using only a single attribute $A_{MaxSplit}$ and don't apply any transformation. Since *Basic Merger* in *Scorpion* merges predicates whose scores within the top quartile of the list, this makes the probability of merging predicates low. In Fig. 3c, *XTR_Advanced and XTR_Fast* show higher merging execution time because both of them apply transformation on merged predicates and it clearly shown in

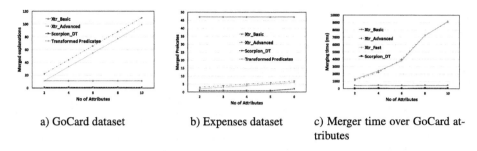

| a) GoCard dataset | b) Expenses dataset | c) Merger time over GoCard attributes |

Fig. 3. Cost of merging process as dataset dimensionality increases

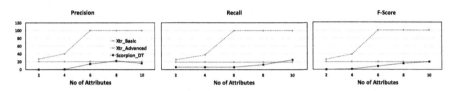

Fig. 4. Accuracy measures *precision, recall, and F-score* as dimensionality increases

Figs. 3a and b the linear increase of transformed predicates while the dimensionality increases.

Effectiveness Evaluation. We discuss the quality of results produced by proposed algorithms according to different metrics: precision, recall, and F-score as the dimensionality varies from 2 to 10 in Fig. 4. As the dimensionality increases, the quality of explanations produced through *XTR_Basic* remains fixed about 20% for all metrics since it uses only a single attribute for splitting and merging processes. *Scorpion* show a slight improvement in precision and F-score from 16% to 20% between attributes 6 to 10. *XTR_Advanced* remains competitive with *XTR_Basic and Scorpion* since its accuracy is increasing rapidly in all metrics. *XTR_Advanced* scored 100% precision, recall, and F-score between attributes 6 to 10. Finally, *UTE* algorithms have linear execution times and achieved better quality compared with *Scorpion DT* along different accuracy metrics and not sensitive to the dimensionality of datasets same as *Scorpion DT*.

6   Conclusions

An efficient UTE framework has been presented to help users in understanding origins of outliers and discovering reasons behind the deviation of their exceptions in aggregate results. UTE generates human readable predicates to help explain outliers aggregate groups based on attributes of tuples that contribute to the value of those groups. We introduced a notion of uniformity for computing the impact of a tuple on an output value and a transformation process to output meaningful explanations. Furthermore, proposed splitting approaches are based on the *(Tuple Level)* has been described and show a linear complexity instead of exponential complexity. Experimental results have proven high efficiency and effectiveness of the presented approaches according to different metrics.

References

1. Agrawal, R., Gehrke, J., Gunopulos, D., Raghavan, P.: Automatic subspace clustering of high dimensional data for data mining applications. In: Proceedings of the 1998 ACM SIGMOD International Conference on Management of Data, SIGMOD 1998, pp. 94–105 (1998)
2. Bender, G., Kot, L., Gehrke, J.: Explainable security for relational databases. In: International Conference on Management of Data, SIGMOD 2014, Snowbird, UT, USA, 22–27 June 2014, pp. 1411–1422 (2014)
3. Breiman, L.: Classification and regression trees. Wadsworth International Group (1984)
4. Das, M., Amer-Yahia, S., Das, G., Yu, C.: MRI: meaningful interpretations of collaborative ratings. PVLDB 4(11), 1063–1074 (2011)
5. Das, M., Amer-Yahia, S., Das, G., Yu, C.: MRI: Meaningful interpretations of collaborative ratings. Am. Soc. Mech. Eng. (Paper) 4(11), 1063–1074 (2011)
6. Ibrahim, I.A., Albarrak, A.M., Li, X.: Constrained recommendations for query visualizations. Knowl. Inf. Syst., 1–31 (2016). https://doi.org/10.1007/s10115-016-1001-5
7. Kanagal, B., Li, J., Deshpande, A.: Sensitivity analysis and explanations for robust query evaluation in probabilistic databases. In: Proceedings of the ACM SIGMOD International Conference on Management of Data, SIGMOD 2011, Athens, Greece, 12–16 June 2011
8. Khoussainova, N., Balazinska, M., Suciu, D.: Perfxplain: debugging mapreduce job performance. Proc. VLDB Endow. 5(7), 598–609 (2012)
9. Mackinlay, J.D., Hanrahan, P., Stolte, C.: Show me: automatic presentation for visual analysis. IEEE Trans. Vis. Comput. Graph. 13(6), 1137–1144 (2007)
10. Meliou, A., Gatterbauer, W., Halpern, J.Y., Koch, C., Moore, K.F., Suciu, D.: Causality in databases. IEEE Data Eng. Bull. 33(3), 59–67 (2010)
11. Roy, S., Orr, L., Suciu, D.: Explaining query answers with explanation-ready databases. Proc. VLDB Endow. 9(4), 348–359 (2015)
12. Roy, S., Suciu, D.: A formal approach to finding explanations for database queries. In: International Conference on Management of Data, SIGMOD 2014, Snowbird, UT, USA, 22–27 June 2014, pp. 1579–1590 (2014)
13. Saltelli, A.: The critique of modelling and sensitivity analysis in the scientific discourse. TAUC (2006)
14. Sarawagi, S.: Explaining differences in multidimensional aggregates. In: Proceedings of the 25th International Conference on Very Large Data Bases, VLDB 1999, pp. 42–53. Morgan Kaufmann Publishers Inc., San Francisco (1999)
15. Sathe, G., Sarawagi, S.: Intelligent rollups in multidimensional olap data. In: Proceedings of the 27th International Conference on Very Large Data Bases, VLDB 2001, pp. 531–540. Morgan Kaufmann Publishers Inc., San Francisco (2001)
16. Wu, E., Madden, S.: Scorpion: explaining away outliers in aggregate queries. Proc. VLDB Endow. 6(8), 553–564 (2013). https://doi.org/10.14778/2536354.2536356

Social Spammer Detection: A Multi-Relational Embedding Approach

Jun Yin[1,2], Zili Zhou[2,3], Shaowu Liu[2], Zhiang Wu[1(✉)], and Guandong Xu[2]

[1] School of Information Engineering,
Nanjing University of Finance and Economics, Nanjing, China
zawuster@gmail.com
[2] Advanced Analytics Institute, University of Technology, Sydney, Australia
{Jun.Yin-2,Zili.Zhou}@student.uts.edu.au,
{Shaowu.Liu,Guandong.Xu}uts.edu.au
[3] School of Computer Engineering and Science, Shanghai University, Shanghai, China

Abstract. Since the relation is the main data shape of social networks, social spammer detection desperately needs a relation-dependent but content-independent framework. Some recent detection method transforms the social relations into a set of topological features, such as degree, k-core, etc. However, the multiple heterogeneous relations and the direction within each relation have not been fully explored for identifying social spammers. In this paper, we make an attempt to adopt the *Multi-Relational Embedding (MRE)* approach for learning latent features of the social network. The *MRE* model is able to fuse multiple kinds of different relations and also learn two latent vectors for each relation indicating both sending role and receiving role of every user, respectively. Experimental results on a real-world multi-relational social network demonstrate the latent features extracted by our MRE model can improve the detection performance remarkably.

Keywords: Social spammer · Social networks
Heterogeneous relations · Graph embedding · Classification

1 Introduction

Social networks have played a huge role in information dissemination and communication. While the social media is favoring both organizations and individuals with great facilities, it has become an emerging and effective platform on which malicious users overwhelm other users with unwanted content [8]. It has been shown that around 83% of users have received more than one unwanted friend requests or messages in social networking platforms and one in 200 social messages contain spam [5,26]. These spammers and the misleading contents released by them are seriously threatening the sustainable development of online social networks.

In the literature, an extensive body of research has been devoted to identify various kinds of spam, such as email spam [22], Web spam [4,31], review/reviewer

© Springer International Publishing AG, part of Springer Nature 2018
D. Phung et al. (Eds.): PAKDD 2018, LNAI 10937, pp. 615–627, 2018.
https://doi.org/10.1007/978-3-319-93034-3_49

spam on e-commerce sites [6, 29], and consequently, social spam [5, 14]. The main research stream within spammer detection adopts the two-phase approach: constructing multifold features to indicate the abnormal behavior, and developing supervised classifiers or unsupervised ranking algorithms. Finding right features largely determines the detection performance, and it is both *data-specific* and *task-specific*. That is, a right feature should be computable on the available data and it should also be qualified for the specific detection task. Along this line, feature construction towards identifying spam from online reviews in e-commerce has been widely studied. Researchers have designed a variety of features for reviews, users, or even user groups, by fully exploiting the metadata of the review such as rating, timestamps and review text [15, 16, 23, 29]. Nevertheless, spammer detection in social networks is much different from that in e-commerce. The metadata in social networks, especially the contents, is relatively scarce, because the whisper contents should not be exposed due to user privacy. By contrast, the topological relation becomes the inherent attribute of social networks, but it exhibits weakly in e-commerce platforms. Therefore, social spammer detection calls for the *relation-dependent* but *content-independent* framework.

There is limited research on spammer detection framework solely on social relations. Fakhraei *et al.* [5] make a useful attempt in this area: for each relation, a topological graph is generated to describe the interactions among single relation in a topological way, with the underlying assumption that spammers are more important in the graph. Moreover, for each user, they use the sequence of relations based on the time it happened to partly disclose the relevance among relations. Then a framework is combined with these two aspects. However, both graph and sequence are extracted from single relation and individual user, the inter-activities between two users cross different relations have been neglected.

Graph structured embedding method has been widely used in the area like knowledge graph [18]. It excavates the latent information with the utilizing of both edges and vertices, which can exactly make up for the shortcoming of previous researches. Hence, we shall propose our *Multi-Relational Embedding* (*MRE*) model to trade on the preponderance of graph and remedy the limitation of it with graph-embedding method. The main contributions of this work are summarized as follows:

- To the best of our knowledge, this is the first attempt to model different types of relations among all users in a single model for multi-relation spammer detection.
- The *MRE* model is made scalable with the option to set the embedding space size, thus, both small and large number of relation types can be accommodated.
- We conducted empirical experiments on a large real-world social network dataset and provided interesting findings and discussions.

The following sections will be organized as follows. In Sect. 2, we formulate the problem and outline the previous methodologies along with its limitations. We technically address details of our *MRE* model in Sect. 3. In Sect. 4, we exhibit experiment results, and present related work in Sect. 5. Finally, we conclude our work and give future plan in Sect. 6.

2   Preliminaries

In this section, we define the problem of identifying spammers from the multi-relational social network, and briefly summarize existing approaches as well as their limitations that motivated our research.

2.1   Formulating Multi-Relational Spammer Detection

Let $\mathcal{U} = \{u_1, \cdots, u_n\}$ be the set of n users who are connected by m kinds of relations denoted as $\mathcal{R} = \{r_1, \cdots, r_m\}$. In this multi-relational network, assume the relation of type $r_k \in \mathcal{R}$ exists between two users u_i and u_j is encoded as π_{ijk}, where the first two subscripts indicate users and the third subscript tells the type of relation. Examples of relations include "add friend" and "block user". Note that the relation has a direction, therefore π_{ijk} and π_{jik} are different, where the first one is relation r_k from user u_i to user u_j and the second one is the same relation but from user u_j to user u_i. The collection of all relations π_{ijk} is denoted as Π.

The goal is to learn from provided relations Π to predict the probability of being spammer for each user $u \in \mathcal{U}$. In practice, the probability is often unnormalized, thus the goal becomes ranking the users correctly instead of estimating the exact probability, i.e., spammers are ranked higher than normal users but the ordering among spammers does not matter. Let further divide the set of users $\mathcal{U}$ into a set $\mathcal{S}$ of spammers and a set $\mathcal{L}$ of legitimate users, i.e., $\mathcal{U} = \mathcal{S} \cup \mathcal{L}$. The ultimate goal of the spammer detection is to learn an order function for all users, denoted as $O(\mathcal{U})$. Then, we can define an indicator variable $I_{ij} = 1$ to represent that $u_i \in \mathcal{S}, u_j \in \mathcal{L}, O(u_i) > O(u_j)$, otherwise for $I_{ij} = 0$. Hence, one possible formulation of multi-relational spammer detection is:

$$\operatorname*{argmax}_{O(\mathcal{U})} \sum_{u_i \in \mathcal{S}} \sum_{u_j \in \mathcal{L}} I_{ij}. \tag{1}$$

2.2   Feature Design from Multi-Relational Data

While many quality classifiers are available, the main challenge is how to design effective features. Unlike traditional spammer detection models that make use of textual data, the multi-relational social network focuses on topological information. The main features design approaches to multi-relational data are graph-based and sequence-based approaches.

Graph-Based Features. Graph-based features are extracted by converting relations into a directed graph $\mathcal{G}$, where the vertices $\mathcal{V}$ represent the users and the edges $\mathcal{E}$ represent interactions among users. When there exist multiple types of relations, a graph is usually generated for each of them: $\{\mathcal{G}_1, \ldots, \mathcal{G}_m\}$ for m types of relations. Each graph is then feed into a feature extraction function $\mathbf{X}_m^{\text{graph}} = \psi(\mathcal{G}_m)$ to convert a directed graph into either numerical or categorical feature matrix $\mathbf{X}_m^{\text{graph}}$ for each type of relation. Existing literature has defined many feature extraction functions $\psi(\cdot)$, and we list a few popular choices:

- *Triangle Count* [24] computes how many times each vertex involves in subgraphs with three vertices, i.e., a triangle structure.
- *k-core* [1] measures the centrality of each vertex by gradually removing the least connected vertices. The earlier a vertex was removed the lower the centrality.
- *Graph Coloring* [9] assigns a set of colors to vertices with no adjacent vertices having the same colors, and the assigned colors are used as a categorical feature.
- *Page Rank* [19] similar to measuring the importance of Web page by counting the number of incoming links, the incoming edges are counted for each vertex.
- *Weakly Connected Components* [20] counts the number of subgraphs each vertex involves without considering the direction of edges.

Despite of their effectiveness, existing graph feature extraction techniques often assume a separated graph for each type of relation, or aggregation is performed by simple addition. The interactions among relations have been largely overlooked.

Sequence-Based Features. Sequence-based features are extracted by converting relations into a user-wise sequence $\mathcal{T}_i = \{t_1, \cdots, t_q\}$ for each user u_i, where $t_j \in [1, m], 1 \leq j \leq q$, is the relation type and the length q of the sequence depends on the user. The sequence of each user is then fed into a feature extraction function $\mathbf{x}_i^{\text{seq}} = \psi(\mathcal{T}_i)$ to convert the sequence of user u_i into a feature vector $\mathbf{x}_i^{\text{seq}}$. Sequence-based feature extraction has also been used in spammer detection:

- *Sequential k-gram Features* [5] considers the activity order of users by counting the frequency of each length k sub-sequences for each user.
- *Mixture of Markov Models* [21] can be used to overcome the limitation of small k in k-gram models by identifying a small set of important and long sequence chains.

Unlike graph-based features, sequence-based feature can capture interactions among different types of relations to some extent. Nevertheless, user interactions are not captured properly as the sequence features are extracted independently for each user.

In this work, we take the graph-based approach, however, all types of relations are modeled simultaneously in a single graph instead of separated graphs for each type of relation. By embedding the users and relations at the same time, the proposed model overcomes the limitations of traditional graph-based and sequence-based feature extraction methods.

3   Methodology

In this section, we propose the *MRE* model to capture the interactions among different types of relations. The rest of this section defines the multi-relation learning problem, followed by a detailed description of the *MRE* model. In what follows, we shall use u and r as identity of user and type of relation, and use the bold-faced notation **u** and **r** to represent the latent vectors for user and relation respectively.

3.1   Multi-Relational Embedding

The prediction problem itself has been well-studied in literature, and mature classifiers are available in open source libraries. However, the main issue is that off-the-shelf classification algorithms expect numerical variables as input and do not accept input format such as relations defined in Sect. 2.1. Therefore, the main challenge is to learn a vector representation **u** for each user $u \in \mathcal{U}$ from relations such that the new representation is in numerical format while discriminative information is preserved. Attempts [5] were made in literature to learn such representations, but all of them learn the representations for each type of relation independently. While informative interactions may exist among relations, we propose to learn from all types of relations simultaneously.

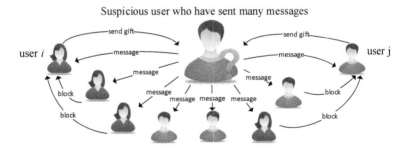

Fig. 1. The suspicious user in the middle who have sent messages to too many users looks like a spammer. However, he has received gifts from users i and j, which is a strong indicator of good user. But we realized that the users who sent gifts are actually low-credit users who have been blocked by others, thus the fact might be the spammer is trying to fool the detection system.

Learning from all types of relations at the same time provides more insights into user behaviors than looking at each individual relation type. For example, the simplest method of encoding relations into numerical representation is counting, i.e., how many times a user has sent/received each type of relation. Despite of its simplicity, this approach does encode important information such as *"users who have sent more messages are more likely to be spammers"*. However, interactions among relations are ignored. A toy example of interactions among relations is shown in Fig. 1 for three types of relations: *"send message"*, *"block user"*, *"send gift"*. In multi-relational embedding, the interactions among relations can be learned as latent factors.

To be specific, we model all users and all types of relations in a shared embedding space. Given the set of all relations Π, we can construct a graph $\mathcal{G}$ where users are the vertices and relations are the edges. Then each user $u_i \in \mathcal{U}$ is represented as a numerical vector $\mathbf{u}_i \in \mathbf{R}^z$ and each type of relation $\mathbf{r}_k \in \mathbf{R}^z$ is represented as a numerical vector $\mathbf{r}_k \in \mathbf{R}^z$. The shared embedding space has a user-defined dimension z. Unlike traditional matrix factorization, the multi-relational embedding has a graph structure, and representation of type of edges (relations) must be learned. Formally, we aim to learn all $\mathbf{u} \in \mathbf{R}^z$ and $\mathbf{r} \in \mathbf{R}^z$ such that

$$\mathbf{u}_i \cdot \mathbf{r}_k + \mathbf{u}_j \cdot \mathbf{r}_k \approx \pi_{ijk}. \tag{2}$$

The above model has not considered the direction of relations yet. For the same type of relation, the sending node (*src*) and the receiving node (*dest*) often delivery different semantic meanings. For instance, the spammer tends to propagate the unwanted content to a large number of users, where the user who usually acts as the sending node should be embedded as the spam user. Therefore, it is a good idea to model them separately. To do so, we define two vectors $\mathbf{r}_k^{\mathrm{src}}$ and $\mathbf{r}_k^{\mathrm{dest}}$ for each type of relation $r_k \in \mathcal{R}$. Similarly, we define $\mathbf{u}_i^{\mathrm{src}}$ and $\mathbf{u}_i^{\mathrm{dest}}$ for each user $u_i \in \mathcal{U}$. Then jointly, we aim to learn $\mathbf{r}^{\mathrm{src}}$, $\mathbf{r}^{\mathrm{dest}}$, $\mathbf{u}^{\mathrm{src}}$, and $\mathbf{u}^{\mathrm{dest}}$ for all types of relations and all users such that

$$\mathbf{u}_i^{\mathrm{src}} \cdot \mathbf{r}_k^{\mathrm{src}} + \mathbf{u}_j^{\mathrm{dest}} \cdot \mathbf{r}_k^{\mathrm{dest}} \approx \pi_{ijk}. \tag{3}$$

Figure 2 illustrates the proposed *MRE* model considering both sending node and receiving node with two relations. As can be seen, we have a source user vector $\mathbf{u}_i^{src}$ and a destination user vector $\mathbf{u}_j^{dest}$, which are mapped to the shared

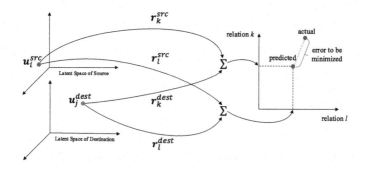

Fig. 2. Illustration of the MRE model on two relations.

Algorithm 1. *Multi-Relational Embedding* Algorithm

Input: List of triples (source user, destination user, relation type).
Preparing:
Step 1: Draw pairwise user pair set $\mathcal{S}$, each pair contains a relation sender user and a relation receiver user.
Step 2: Collect all relation types as $\mathcal{R}$.
Embedding Model Training:
Step 3: Repeat
for each user pair $(u_i, u_j) \in \mathcal{U}$ **do**
    Draw relation frequency vector $\mathbf{r}$ for (u_i, u_j),
    Each value of $\mathbf{r}$ counts the frequency of a type relation sent from u_i to u_j.
    for each $r_k \in \mathcal{R}$ **do**
        Measure the real frequency value with the value predicted by user embedding and transfer matrices with relation k.
        error $= \|\pi_{ijk} - (\mathbf{u}_i^{\mathrm{src}} \cdot \mathbf{r}_k^{\mathrm{src}} + \mathbf{u}_j^{\mathrm{dest}} \cdot \mathbf{r}_k^{\mathrm{dest}})\|_2^2$
        minimize the error between real value and predict value *error* by updating the parameters $\mathbf{u}_i^{\mathrm{src}}, \mathbf{r}_k^{\mathrm{src}}, \mathbf{u}_j^{\mathrm{dest}}, \mathbf{r}_k^{\mathrm{dest}}$

Until stopping criteria met

---

embedding space of two types of relations r_k and r_l. The learning task is to estimate the latent vectors of users and relations such that the prediction error is minimized.

3.2 Parameter Estimation

In general, Multi-Relational Embedding models cannot be determined by convex optimization, instead, approximation techniques are often used in practice. In this work, we adopt the Adam [12] optimizer, the parameters are learned by minimizing the loss function as follows:

$$\underset{(\mathbf{u}^{\mathrm{src}}, \mathbf{u}^{\mathrm{dest}}, \mathbf{r}^{\mathrm{src}}, \mathbf{r}^{\mathrm{dest}})}{\mathrm{argmin}} \sum_{(u_i, u_j) \in \mathcal{U}} \sum_{r_k \in \mathcal{R}} \|\pi_{ijk} - (\mathbf{u}_i^{\mathrm{src}} \cdot \mathbf{r}_k^{\mathrm{src}} + \mathbf{u}_j^{\mathrm{dest}} \cdot \mathbf{r}_k^{\mathrm{dest}})\|_2^2, \quad (4)$$

where $\| \cdot \|_2^2$ is the L2 norm. The overall learning algorithm is summarized in Algorithm 1. We follow common practice by setting the stopping criteria as $error \leq 10^{-4}$.

Table 1. Statistics of dataset

Dataset	#User	#Spammer	#Legitimate	#Relations
Tagged.com	4, 111, 179	182, 939	3, 928, 240	85, 470, 637

4   Experimental Results

To evaluate the effectiveness of the proposed Multi-Relational Embedding model, experiments were conducted on a large real-world dataset from *Tagged.com*. Comparisons were made against several graph-based and sequence-based methods. Our algorithm was implemented on TensorFlow and experiment was conducted on a computer with 28 CPU cores and 256 GB of memory.

4.1   Experimental Setup

Dataset. The dataset used in this experiment was from *Tagged.com*, which is a website for people to meet and socialize with new friends. The dataset contains 7 types of directed relations, including *Message, Pet Game, Meet-Me Game Add Friend, Give a Gift, Report Abuse*, and *View Profile* However, the semantic meaning of each relation is not utilized as multi-relational spammer detection models should learn the importance of each relation from training data. The ground truth label is provided by domain experts to mark each user as legitimate or spam. The data is stored as quad-tuples: $\langle timestamp, u_i^{\text{src}}, u_j^{\text{dest}}, r_k \rangle$, where user u_i^{src} performs action (relation) r_k on user u_j^{dest}. We extracted all relations of a day, resulted in a dataset containing 85M interactions among 4M users. Out of these users, 182 K of them are labeled as spammers, i.e., 4.45%. Statistics of the dataset is shown in Table 1.

Test Data. Among the 7 types of relations, there exists a reporting relation that is provided by the *Report Abuse* mechanism. In this reporting relation, the user u_i^{src} reports user u_j^{dest} for violating the terms of conditions. However, a user who has been reported may or may not be a spammer. The collective detection framework [5] combines the classification results with the report relation using the probabilistic soft logic (PSL) rule, in order to improve the security team's efficiency. Two important PSL rules proposed are:

$$\text{Legitimate}(u_i^{\text{src}}) \wedge \text{Report}(u_i^{\text{src}}, u_j^{\text{dest}}) \rightarrow \text{Spammer}(u_j^{\text{src}}),$$
$$\text{Spammer}(u_j^{\text{dest}}) \wedge \text{Report}(u_i^{\text{src}}, u_j^{\text{dest}}) \rightarrow \text{Legitimate}(u_i^{\text{src}}). \tag{5}$$

The PSL rules limit the evaluation to users who appear in the reporting relation. To be consistent with related research, we adopted the same testing scheme by extracting users appeared in the reporting relation as our test data.

Evaluation Metrics. Since the ground-truth label of each user is provided by the dataset, we adopt standard metrics (P-R-F), including precision (P), recall (R) and F-measure (F) to do evaluate the effectiveness the models. Furthermore, all metrics are computed on the class of spammers:

$$R = \frac{TP}{TP + FN}, \quad P = \frac{TP}{TP + FP}, \quad F = \frac{2PR}{P + R}, \tag{6}$$

where TP is the number of spammers that have been identified correctly, on contrast, FP is the number of spammers that have been identified mistakenly, and FN is the number of spammers that have been missed by the model. Depending on the application scenario, a trade-off can be made on these metrics. High precision represents for catching more spammers. Meanwhile, it will do harm to legitimates, as it takes more users as spammers. While, high recall represents for higher confidence on detected spammers, but may lead to more missing of some spammers. F-measure balances between precision and recall, and is suitable for general scenarios. As the main focus is to evaluate the quality of features extracted from multi-relational data instead of new classification algorithm, two classic but simple supervised models are selected: Logistic Regression (LR) and Gaussian Naive Bayes (GNB).

4.2   Performance Comparison

Several state-of-the-art graph-based and sequence-based features are chosen as the baselines, including k-core [1], Graph Coloring [9], Page Rank [19], Weakly Connected Components [20], Degree [5], and Sequential k-gram Features [5].

Graph-based features are computed using Graphlab Create[1] on each type of relation and resulted in a total of 56 graph-based features, i.e., 8 for each type of relation. For sequence-based features, we compute them using bigram sequence. With 7 relations in the dataset, we ended up with 49 bigram sequence-based features. In our multi-relational embedding features, we generated 30 features to do the overall comparison. Other scale of multi-relational embedding features will discuss later in this section.

After getting the baseline features, we split train and test dataset with 10 different random seeds for evaluation on LR and GNB classifiers. First, we compare our multi-relational embedding features with them separately. Then, we combine the baseline methods together to show the effectiveness of our proposed method.

Table 2. Comparison of two classifiers with different kinds of features

Features	Logistic regression			Gaussian Naive Bayes		
	Precision	Recall	F-measure	Precision	Recall	F-measure
Graph	0.4537	0.6390	0.5308	0.5978	0.3840	0.4675
Sequential	0.4907	0.8620	0.6253	0.4168	0.9320	0.5759
Graph+Sequential	0.5316	0.8600	0.6570	0.4571	0.9260	0.6120
MRE ($z = 30$)	**0.6138**	0.7730	**0.6844**	**0.6165**	0.7020	**0.6566**

Table 2 shows the comparison performance of different kinds of features. As can be seen, our multi-relational embedding features have shown a significant

___
[1] https://turi.com/.

performance advantage over other features on F-measure both with *LR* and *GNB*, which means we can catch the spammer more accurately with the least harm to legitimates. Encouragingly, the precisions of embedding features consistently are the highest ones, giving the proof that the proposed features can reveal the most of spammers with a little loss in recalls. In terms of recall, although sequential features enjoy the highest position, they show the worst performance on precision as the price, which means they treat more users as spammers and greatly affect the legitimates.

To throughly examine the performance of the *Multi-Relational Embedding* model, we analyze its performance by varying the size of embedding space from 10 to 40. Figure 3 shows the performance of each embedding features on precision, recall and F-measure separately. Obviously, the recall rate increases with the raise of dimension, giving the sign that more spammers will be disclosed when increasing the dimension of our *MRE* model. While, the precision and the F-measure reach their peaks at the dimension of 30, followed by a decline. That is to say, if the dimension keeps growing after reaching 30, the *MRE* model will lose its preciseness by listing more users as spammers. In general, it shows that the most effective performance has been achieved on 30 embedding features. Nevertheless, the number of embedding features depends on the dataset. One recommendation is that the number of embedding features should be increased alongside the number of types of relations, because more type of relations implies more complex interactions.

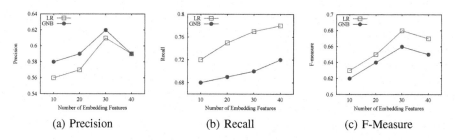

(a) Precision   (b) Recall   (c) F-Measure

Fig. 3. Impact of the number of dimensions (z) in our *MRE* model.

5   Related Work

In the literature, an extensive researches have been developed to extract abnormal behavior as features in social media, including e-commerce sites [6,29] and social networks sites [15,16]. The indicating features of spammers are depending on the available metadata, e.g., timestamps, text content, ratings, etc. Generally, it can be categorized into three parts: *content-based*, *behavior-based* and *topological* features. In early studies of email spams and e-commence spams, reviews/emails containing similar content have a high probability to be spams [10,11]. Various of content-based features are designed to detect such spams in e-commerce and emails, e.g. average length in number of words [17], ratio of objective words [13]. In addition, behavior-based features are mainly generated considering the timestamps, sequence of time, ranks, distributions, etc. For instance, Fei *et al.* [7] suggest that the ratio of Amazon verified purchases will somehow track spammers. Arjun *et al.* [15,16] proposed other behavior-based features focusing on timestamps and ranks. Fakhraei *et al.* [5] raised a k-gram sequential feature with the help of Mixture Markov Model. Beside of individual spammers, groups of spammers also attract researchers' attention [16,27,28,30], with the assumption that spammers within a group are more likely to attack legitimates together, which indicates that the relationships in social media might be useful to detect spammers. Along such mentality, topological features have been proposed in recent literature [5,6], which usually consists of *degree*, *score of Page Rank*, *k-core* etc. However, the existing topological feature extraction methods often assume the data to be homogeneous, i.e., different types of user relations need to be modeled separately. This assumption limits the potential of topological methods as the interactions among different types of relations are not captured.

Graph structured embedding can help with the utilization of interactions among different relations, as it leverages relational learning methods [18] to extract the latent information of graph elements including both vertices and edges. Depending on the assumptions, each relational learning method proposes a different model to represent graph triple: two vertices and one edge. The models can be categorized into three categories: direct vector space translating, vector space translating with relation subspace, and tensor factorization. Considering of graph is a multi-relational heterogeneous network, Bordes et al. [3] proposed a bi-directed relation subspace mapping based model, which maps head vertex and tail vertex by two different matrices of one relation. Bordes et al. [2] proposed another model using direct vector space translating model, which ignore multi-relation problem but make the model much more efficient in training speed. Nickel et al. [18] and Socher et al. [25] proposed a new type of relational learning methods based on tensor factorization, which is efficient in both speed and accuracy. In present work, we extended the graph structured embedding method to our special case of a small number of relation types.

6   Conclusions

In this work we tacked the multi-relational spammer detection problem from graph perspective of view by proposing the *Multi-Relational Embedding* model. The *MRE* model takes advantages of both the representational power of graph and the ease of modeling higher order interactions of embedding. Experiment results on public dataset have demonstrated the effectiveness of the *MRE* model by achieving improved spammer detection performance. For future work, the computational efficiency of *MRE* can be further improved by parallelization. This is feasible due to the fact that the full graph consists of many isolated subgraphs, i.e., the graph is not fully connected.

Acknowledgment. This work was supported in part by National Key Research and Development Program of China under Grant 2016YFB1000901, the National Natural Science Foundation of China (NSFC) under Grant 71571093, Grant 91646204, Grant 71372188, Grant 71701089, and the National Center for International Joint Research on E-Business Information Processing under Grant 2013B01035.

References

1. Alvarez-Hamelin, J.I., Dall'Asta, L., Barrat, A., Vespignani, A.: Large scale networks fingerprinting and visualization using the k-core decomposition. In: Advances in Neural Information Processing Systems, pp. 41–50 (2006)
2. Bordes, A., Usunier, N., Garcia-Duran, A., Weston, J., Yakhnenko, O.: Translating embeddings for modeling multi-relational data. In: International Conference on Neural Information Processing Systems, pp. 2787–2795 (2013)
3. Bordes, A., Weston, J., Collobert, R., Bengio, Y., et al.: Learning structured embeddings of knowledge bases. In: AAAI, vol. 6, p. 6 (2011)
4. Cheng, Z., Gao, B., Sun, C., Jiang, Y., Liu, T.Y.: Let web spammers expose themselves. In: ACM International Conference on Web Search and Data Mining, pp. 525–534 (2011)
5. Fakhraei, S., Foulds, J., Shashanka, M., Getoor, L.: Collective spammer detection in evolving multi-relational social networks. In: ACM SIGKDD International Conference on Knowledge Discovery and Data Mining, pp. 1769–1778. ACM (2015)
6. Fayazi, A., Lee, K., Caverlee, J., Squicciarini, A.: Uncovering crowdsourced manipulation of online reviews. In: International ACM SIGIR Conference on Research and Development in Information Retrieval, pp. 233–242 (2015)
7. Fei, G., Mukherjee, A., Liu, B., Hsu, M., Castellanos, M., Ghosh, R.: Exploiting burstiness in reviews for review spammer detection. In: ICWSM, pp. 175–184 (2013)
8. Hu, X., Tang, J., Liu, H.: Online social spammer detection. In: AAAI, pp. 59–65 (2014)
9. Jensen, T.R., Toft, B.: Graph Coloring Problems. Wiley, Hoboken (1995)
10. Jindal, N., Liu, B.: Analyzing and detecting review spam. In: International Conference on Data Mining (ICDM), pp. 547–552. IEEE (2007)
11. Jindal, N., Liu, B.: Review spam detection. In: International Conference on World Wide Web, pp. 1189–1190. ACM (2007)
12. Kingma, D., Ba, J.: Adam: a method for stochastic optimization. Comput. Sci (2014)
13. Li, F., Huang, M., Yang, Y., Zhu, X.: Learning to identify review spam. In: International Joint Conference on Artificial Intelligence (IJCAI), vol. 22, pp. 2488–2493 (2011)
14. Liu, H., Zhang, Y., Lin, H., Wu, J., Wu, Z., Zhang, X.: How many zombies around you?. In: International Conference on Data Mining (ICDM), pp. 1133–1138. IEEE (2013)
15. Mukherjee, A., Kumar, A., Liu, B., Wang, J., Hsu, M., Castellanos, M., Ghosh, R.: Spotting opinion spammers using behavioral footprints. In: ACM SIGKDD International Conference on Knowledge Discovery and Data Mining, pp. 632–640 (2013)
16. Mukherjee, A., Liu, B., Glance, N.: Spotting fake reviewer groups in consumer reviews. In: International Conference on World Wide Web, pp. 191–200. ACM (2012)

17. Mukherjee, A., Venkataraman, V., Liu, B., Glance, N.S.: What yelp fake review filter might be doing?. In: ICWSM (2013)
18. Nickel, M.: Tensor Factorization for Relational Learning. Ludwig-Maximilians-Universitt Mnchen, Munich (2013)
19. Page, L., Brin, S., Motwani, R., Winograd, T.: The pagerank citation ranking: Bringing order to the web. Stanford Digital Libraries Working Paper, pp. 1–14 (1998)
20. Pemmaraju, S.V., Skiena, S.S.: Computational Discrete Mathematics: Combinatorics and Graph Theory with Mathematica. Cambridge University Press, New York (2009)
21. Peng, F., Schuurmans, D., Wang, S.: Augmenting Naive Bayes classifiers with statistical language models. Inf. Retrieval **7**(3–4), 317–345 (2004)
22. Pitsillidis, A., Levchenko, K., Kreibich, C., Kanich, C., Voelker, G.M., Paxson, V., Weaver, N., Savage, S.: Botnet judo: fighting spam with itself. In: Network and Distributed System Security Symposium (2010)
23. Rayana, S., Akoglu, L.: Collective opinion spam detection: bridging review networks and metadata. In: Proceedings of the 21th ACM SIGKDD International Conference on Knowledge Discovery and Data Mining, pp. 985–994. ACM (2015)
24. Schank, T.: Algorithmic aspects of triangle-based network analysis. Ph.D. in Computer Science, University Karlsruhe (2007)
25. Socher, R., Chen, D., Manning, C.D., Ng, A.: Reasoning with neural tensor networks for knowledge base completion. In: International Conference on Neural Information Processing Systems, pp. 926–934 (2013)
26. Stringhini, G., Kruegel, C., Vigna, G.: Detecting spammers on social networks. In: Computer Security Applications Conference, pp. 1–9. ACM (2010)
27. Wang, Y., Wu, Z., Bu, Z., Cao, J., Yang, D.: Discovering shilling groups in a real e-commerce platform. Online Inf. Rev. 62–78 (2016)
28. Wu, L., Hu, X., Morstatter, F., Liu, H.: Adaptive spammer detection with sparse group modeling. In: The International AAAI Conference on Web and Social Media, pp. 319–326 (2017)
29. Wu, Z., Wang, Y., Wang, Y., Wu, J., Cao, J., Zhang, L.: Spammers detection from product reviews: a hybrid model. In: IEEE International Conference on Data Mining, pp. 1039–1044. IEEE (2016)
30. Yu, R., He, X., Liu, Y.: GLAD: group anomaly detection in social media analysis. In: International Conference on Knowledge Discovery and Data Mining, pp. 372–381. ACM (2014)
31. Zhou, B., Pei, J.: Link spam target detection using page farms. ACM Trans. Knowl. Discov. Data (TKDD) **3**, 1–38 (2009)

Opinion Mining and Sentiment Analysis

Learning to Rank Items of Minimal Reviews Using Weak Supervision

Yassien Shaalan$^{(\boxtimes)}$, Xiuzhen Zhang, and Jeffrey Chan

School of Science (Computer Science), RMIT University, Melbourne, Australia
{yassien.shaalan,xiuzhen.zhang,jeffrey.chan}@rmit.edu.au

Abstract. Customer reviews and star ratings are widely used on E-commerce and reviewing sites for the public to express their opinions. To help the online public make decisions, items (e.g., products, services, movies, books) are typically represented and ordered by an aggregated star rating from all reviews. Existing approaches simply average star ratings or use other statistical functions to aggregate star ratings. However, these approaches rely on the existence of large numbers of reviews to work effectively. On the other hand, many new items have few reviews. In this paper, we argue that at the core of review aggregation is ranking items, hence, we cast the problem of ranking a set of items as a learning to rank (L2R) problem to address the issue of reviews scarcity. We devise a rank-oriented loss function to directly optimize the ranking of groups of items. Standard L2R models require ranking labels for training, but item ranking ground-truth information is not always available. Therefore, we propose to aggregate star ratings for items with large numbers of reviews to automatically generate weak supervision ranking labels for training. We further propose to extract features from review contents, rating distributions and helpfulness information to train the ranking model. Extensive experiments on an Amazon dataset showed that our model is very effective compared to state-of-the-art heuristic aggregation approaches, regression and standard L2R approaches.

Keywords: Ranking · Learning to rank · Weak supervision

1 Introduction

Everyday millions of online users read reviews for different items to compare, evaluate and to help make informed buying decisions. However, some item categories of interest, e.g., television, could have hundreds of items and each item can have thousands of reviews, making it impractical for customers to read them all.

To help the online population make informed purchase decisions while not needing to read too many reviews, items are typically represented and ordered by an aggregated star rating that uses the ratings from all reviews. In fact, the average of star ratings is the de facto standard approach to aggregate reviews. Existing research focuses on devising more robust aggregate functions, including

© Springer International Publishing AG, part of Springer Nature 2018
D. Phung et al. (Eds.): PAKDD 2018, LNAI 10937, pp. 631–643, 2018.
https://doi.org/10.1007/978-3-319-93034-3_50

different weighting schemes [1,5,9,13], or extracting sentiment polarity from review contents [19,20]. Most importantly, they all require the ranked items to have many reviews in-order to be accurate and avoid biases.

However, in a typical e-commerce website, there are many items that have few reviews. This could be due to the item been new (cold-start problem), or they have few sales hence few reviews (long tail problem). This leads to the need to design new approaches for aggregating reviews of such items with few reviews.

In this paper, we argue that at the core of review aggregation is to rank items based on reviews. Rather than just generating an aggregate rating score for items, the ultimate goal for review aggregation is to rank items based on reviews. This is because the scores on its own are not useful unless they are compared to one another for preference reasons. We cast the problem of ranking a group of items as a learning to rank (L2R) problem to address the issue of lacking reviews. With the L2R approach, a model for ranking objects is constructed by supervised learning [10]. L2R has been successfully used to rank documents in information retrieval (IR) [3,8] and other ranking tasks [16]. In fact, structurally, ranking a group of items is yet a very similar task -with few differences- which suggests that L2R can be one possible solution.

To apply L2R for ranking items, we face several challenges: (1) Different from ranking documents in IR where position in the list and level of relevance is important for evaluation, the learning objective for our setting is to produce an "optimal" ranked group of items based on customer reviews that truthfully reflects the reputation for items with no relevancy to a specific query. (2) The standard L2R approach requires ranking labels for training, but item ranking ground-truth information is not always available. (3) Generic features -not specific to certain reviewing platforms- that can characterize various types of items are needed to effectively train a ranking model.

We propose L2RI (Learning to Rank Items), a model to rank a set of items based only on the reviews of the items. We make the following contributions:

- We devise a rank-oriented loss function to directly optimize the ranks for a group of items.
- We propose to use a heuristic aggregation method that accounts for time and quality of reviews for items with large numbers of reviews to automatically generate weak supervision ranking labels for training.
- We further propose to extract features from reviews contents, rating distributions and helpfulness information to train the ranking model.

Extensive experiments on an Amazon datasets showed that L2RI is consistently more effective in ranking products than state-of-the-art heuristic aggregation approaches, regression and standard L2R models.

2   Literature Review

Related work lies in three areas, aggregating reviews for items, learning to rank and weak supervision.

Several statistical models for quantifying a product's quality based on an aggregated star rating of its reviews were introduced in [13]. These models include average rating, median rating, lower bound on normal and binomial confidence. On the other hand, probability-based approaches are mostly focused on removing the impact of outliers or partial information of reviews [9]. Heuristic-based models obtain the aggregate score using different weighting heuristics, including review influence [5], reviewer influence and review posting time [15], and the distribution of ratings [1]. All of these approaches depend on the availability of many reviews per item to be effective.

There has also been research on mining opinions from reviews textual contents and aggregating their polarities to rank products. These approaches are based on the observation that the star rating may be biased and textual contents are more reliable and detailed source for opinions. As an example, [20] presented a feature-based product ranking technique that mine reviews' text, then a weighted directed graph model is used to evaluate items' relative quality.

Another line of related work is learning to rank (L2R), originated from the task of predicting document rankings for queries. It has also been applied to other tasks [16]. L2R approaches can be categorized into three main categories. The point-wise approach [4] which maps the ranking task to a regression task and uses the predicted scores for ranking. The pairwise approach [3,8] which is aimed at optimizing the relative ordering for document pairs so as to order a list of documents. For the listwise approach [14,18] the learning objective directly optimizes the ranking for lists of documents. The three approaches have their strengths and weaknesses, and where they are effective is task dependent [10].

The difficulty of getting annotated data for supervised learning is behind the emergence of weak supervision, also often called distant supervision, where learning benefits from the huge amounts of data available, but unlabeled. It is applied to various applications in NLP relation extraction [6], Twitter account classification [2] and knowledge base completion [7].

3 L2RI: Learning to Rank Items with Weak Supervision

In this section, we outline our L2RI approach, including the ranking loss function, weak supervision and the features used.

The problem of item ranking given a group of items and their reviews, is to rank them based on a learnt model that takes into account the review-based features and training rankings. We aim to learn a ranking model to effectively rank items even if some items have scarce reviews.

3.1 Learning to Rank Items Model

To rank a group of items c from reviews, the L2R approach learns a scoring function S(.) with parameters θ that produces a ranking π_c that minimizes the difference between it and the ground-truth ranking y_c. Formally, given a loss function $L(X_c, \pi_c, y_c, \theta)$ that measures the ranking differences in the ordered sequences of y_c and π_c and at the same time measures the predictability of π_c using the feature vectors for items X_c. The L2RI model aims at learning a scoring function with parameters θ, $S(\theta)$, so that:

$$\hat{\pi}_c = \operatorname*{argmin}_{\pi_c} L(X_c, \pi_c, y_c, \theta)$$

In the learning process, parameter θ is learned from the training data with rankings scores so as to minimize the loss function, that is the difference between the predicted rank π_c from the ranking scores by the function $S(.)$ and the labeled rank y_c.

Although the form of this loss function is the same as existing L2R approaches, we note that the main difference – the loss function for existing L2R models (such as NDCG and MAP) is defined in terms of both the rank and relevance level of documents. In our setting, we do not have nor need relevance levels, hence traditional loss functions are not appropriate for our problem. We next present our loss function that only consider the ordering and not relevancy.

3.2   A Rank-Oriented Loss Function

The difference between the labeled ranking y_c and candidate ranking π_c can be measured by a ranking loss function $L(y_c, \pi_c)$. Then, the L2RI model essentially learns a scoring function $S(.)$ that produces candidate rankings which minimizes the ordering difference between itself and the labeled orderings $(L(.))$. Therefore, the choice of the measure for calculating the difference is an important factor for an accurate scoring function.

Ranking correlation coefficients like Kendall's τ or Spearman ρ measures the pairwise rank consistencies between two ranked groups and hence are good candidates for measuring the ordering difference. Generally, the Kendall's τ correlation is preferred to Spearman ρ correlation because of its small gross error sensitivity (more robust) and its small asymptotic variance (more efficient). However, in terms of computation, Kendall correlation has $O(n^2)$ computation complexity comparing with $O(nlogn)$ of Spearman correlation, where n is the sample size.

Kendall's τ is calculated based on the predicted rank π_c and the labeled rank y_c as follows:

$$\tau(\pi_c, y_c) = \frac{nc(\pi_c, y_c) - nd(\pi_c, y_c)}{n * (n-1)/2}$$

where $nc(\pi_c, y_c)$ is the number of concordant pairs, $nd(\pi_c, y_c)$ is the number of dis-concordant pairs, n is the number of items in the group. Given two ranked groups of the items, a pair of items (x, y) is said to be concordant if the rank for both items agree in the two ranked groups, otherwise (greater or less than) it is said to be dis-concordant.

Since the correlation functions measure similarities, we convert them to a loss function via $L(\pi_c, y_c) = 1.0 - \tau(\pi_c, y_c)$, which is then transformed to:

$$L\tau(\pi_c, y_c) = \frac{2 * nd(\pi_c, y_c)}{n * (n-1)/2}$$

Also, we can use Spearman ρ as a loss function for our ranking problem. It measures the ranking correlation strength with no relevance score required and computationally fast especially for long groups of items. It is calculated as follows:

$$\rho(\pi_c, y_c) = 1 - \frac{6 * \sum_{i=1}^{n} d}{n * (n^2 - 1)}$$

where $d_i = \pi_c(i) - y_c(i)$ is the difference between the two ranks. Similarly, Spearman ρ can be converted to a loss function as follows:

$$L\rho(\pi_c, y_c) = \frac{6 * \sum_{1}^{n} d}{n * (n^2 - 1)}$$

Considering the typical user behavior where they tend to only focus on the top few items in a ranked list [10,13], we can equally define a $\tau@k(.)$ and $\rho@k(.)$, where we only consider the top k items in the input ordering. Generally k can be set by any size (for example, top 10) up to the length of the whole group. In Sect. 4, we will examine all the subtle differences between these loss functions.

We next explain how to automatically generate ranking scores for weak supervision by aggregating similar items with large numbers of reviews.

3.3   Generating Ranking Scores for Weak Supervision

Our L2RI framework requires labels for items. But ranking labels are not readily available. Using human annotation is not only costly but also not practical, as research shows that human annotators have significant variance when ranking items by reviews [20]. We propose to apply an unsupervised aggregate function to aggregate the star ratings of reviews and then use the output to automatically generate ranking labels for items within a given group and later use as weak supervision signals to train our L2RI ranking model.

The average ratings for products can be used as weak supervision labels for L2RI. The average star rating model is a widely used aggregate function of star ratings. However, the average aggregated ratings tend to lead to all items having very similar aggregated ratings known as "the all good reputation" phenomenon [20]. Thus, the average ratings may be very noisy as weak supervision labels. Our experiments indeed show that the model L2RI(AVG), L2RI with average ratings as weak supervision, can not effectively rank items.

We propose to use TQRank [15] as an aggregate function of star ratings, which was proven to rank products better than probabilistic and other heuristic approaches [9,19]. TQRank takes into account star rating, posting time and review quality to generate ranking scores for items. The aggregated ranking score for item p, is defined as:

$$S_p = \sum_{r=1}^{n} (W_r * S_r)/n \tag{1}$$

where $S_r \in [1..5]$ for review r, $r = 1..n$. W_r is defined in terms of posting time t, item category g and reviewer u:

$$W_r = W_r(t, u, g) = e^{\beta(T_t + Q(u,g))} \qquad (2)$$

where β is a decay factor $\in [10..50]$ based on reviews scarcity. T_t is the elapsed time from item release. $Q(u, g)$ is an accumulated measure for opinion quality

$$Q(u, g) = \sum_{i=1}^{m} H(i, u, g) / \sum_{i=1}^{m} A(i, u, g) \qquad (3)$$

$H(i, u, g)$ and $A(i, u, g)$ are respectively the number of helpfulness votes and the total number of votes, user u gets for his review i for an item under g category. S_p values are transformed into a rank labels and later used for training L2RI. This model that uses TQRank scores as weak supervision labels is denoted as L2RI(TQRank).

3.4   The Features

Given a set of items partitioned into groups (categories), an item belongs to a group and is characterized by a feature vector. The feature vectors for items comprise item-level features as well as group-level to characterize items. We only extract these features from reviews to be generally applicable to all categories.

Previous research [15,19] shows that the posting time of reviews is an important factor when evaluating the quality of product and recent reviews carry more weight. Therefore, we define the *time factor* for reviews when extracting item-level features from reviews. For item p with reviews $\{r_1, ..., r_n\}$ where the posting time of r_i is before that of r_{i+1} for $i = 1..n$. Let review r has a posting time T_r, and let T_1 and T_n respectively denote the posting time of the first and last review, then we can define the time factor for review r as $W_r = (T_r - T_1)/(T_n - T_1)$.

Item Level Features. We construct four types of item-level features: Aggregated rating, Helpfulness level score, Review text polarity score, and Activeness level score.

- *Aggregated rating score (R):* To capture the rating level of reviews and their posting time, the aggregated rating score R is defined as the average of the exponential function for the time weighted star rating for all reviews: $R \equiv \sum_{r=1}^{n} (e^{W_r * S_r})/n$, where W_r and S_r are the time factor and star rating level for review r. Moreover, the number of reviews n is also included as a separate feature to complement the aggregated rating.
- *Helpfulness level score (H):* It is well acknowledged that opinions expressed in reviews are of varying qualities [11]. Moreover, research has shown that helpfulness votes of reviews in online communities are strong indicators of quality of reviews. The helpfulness level of reviews for an item is defined as the average of the exponential function weighted by time and helpfulness votes for each review: $H \equiv \sum_{r=1}^{n} (e^{W_r * H_r * S_r})/n$ where H_r denotes the portion of helpfulness votes for review r.

- *Review text polarity score (P):* In addition to star ratings, customers often read reviews textual contents to get detailed opinion about items. It is shown in [19,20] that the star ratings should not be considered as an overall aggregate score, but aggregated sentiments expressed in the text. Thus, we define the review polarity score as follows: $P \equiv \sum_{r=1}^{n} (e^{W_r*(P-N)/(P+N)})/n$, where P and N are respectively the positive and negative polarity scores of all sentences in review r. We use SentiStrength [17] to compute polarity scores, which estimates the strength of positive and negative sentiments by analyzing the linguistic patterns.
- *Activeness level score (A):* Based on the observation that the lifespan for items varies significantly, the activeness level is defined to capture this information. The activeness level for an item is defined as the total number of reviews normalized by the lifespan for an item $A \equiv n/(T_n - T_1)$.

Group Level Features. Group-level features are important for capturing differences among item groups. By empirical analysis, we select six statistics that show significant variance across different groups.

- The minimum and maximum number of reviews for items in the group.
- The minimum and maximum lifespans for items in the group.
- The minimum and maximum activeness level for items in the group.

As will be shown in Sect. 4, both the item-level and group-level features are important for characterizing items and training an effective ranking model.

4   Experiments

In this section, we describe our dataset, evaluation measures, baselines and report experiment results.

4.1   Experiment Setup

Experiments were performed on an Amazon dataset [12] of 143.7 million reviews from May 1996 to July 2014. We performed preprocessing to remove corrupted reviews, crawl missing ones, group reviews by items[1] and group items by categories and sub-categories (groups). The final dataset used comprises 74k items from 9 categories (Amazon subcategories) as shown in Table 1. Groups that have items of few reviews are used for testing. The division of groups was based on the median number of reviews as it is a good indicator of central tendency for the number of reviews per item in this dataset. Our task is to rank groups of products to demonstrate the effectiveness of our L2RI model.

Since there is no readily available ground-truth ranks for products based only on reviews, we use Amazon Sales Rank[2] as the proxy. In fact, Amazon

---

[1] Technically for Amazon these are products, but we wish to maintain consistent terminology and hence use items.

[2] https://www.amazon.com/gp/help/customer/display.html?nodeId=525376.

Table 1. Amazon dataset statistics of customer reviews

Categories	#SubCategories	# Products	# Reviews	Med#R	#Votes	#Helpful votes
Testing set						
Arts & Crafts	99	3,107	179,117	35	521,073	449,087
Industrial	56	2,235	131,961	36	347,684	288,505
Jewelry	94	4,689	253,740	35	3,12234	2,60477
Training set						
Toys	229	20,260	1,277219	38	2,887232	2,310573
Computers	15	941	85,428	45	233,488	176,889
Video Games	77	10,659	1,029574	49	4,555411	2,701881
Electronics	79	6,666	678,429	45	2,493038	1,943635
Software	52	2,271	237,709	56	1,224112	968,020
Cell Phones	69	23,355	2,557402	46	3,109081	2,385656

Sales Rank has been used for product ranking evaluation for similar Amazon datasets in previous studies [15, 19]. Sales Rank is a good indicator for products popularity under a category, but can also reflect a product's quality and how customers perceive and rank it.

We compare our L2RI model against the baselines depicted in Table 2. L2RI $\lambda_\tau@k$ is based on LamdaMart [18] and is our proposed approach. It uses gradient boosted decision trees to form model that is a weighted combination of an ensemble of weak learners. We use it to optimize Kendall's τ for top-k items using TQRank as weak supervision signal. Coordinate Ascent (CA) [14] is another similar model. It is a linear feature based model that directly optimizes evaluation measures using the coordinate ascent optimization method.

4.2   Results

Ranking Performance: Table 3 shows the ranking correlation for ranking models against the ground-truth ranking in terms of Kendall's τ and Spearman's ρ, for ranking a group and for ranking the top 10 items in a group. Large values indicate higher correlation coefficients and better performance for the relevant ranking model. When an item has a few reviews, the reviews are noisy and can be easily biased, heuristic aggregation of ratings can be ineffective for ranking items. This is evident by looking at the AVGRank ranking performance in the table. This model performs the worst, as it only considers the aggregate score of the star ratings. However, TQRank performs much better than AVGRank, as its aggregated score takes into account time, helpfulness votes and star ratings. The results show that it still suffers from the lack of reviews because it requires a reasonable number of reviews to work accurately.

Point-wise models, SVMReg and Mart produce better ranking than AVGRank, but still less accurate than TQRank. The main reason is that they ignore the group structure of ranking problem and deal with it as a point prediction. On the other hand, pair-wise models, show some progress over TQRank,

Table 2. Baselines models to evaluate our model L2RI $^\lambda\tau$@k

Model	Description & parameters
Heuristic based Models	
AvgRank	Ranking products by their average ratings
TQRank [15]	A time and quality heuristic aggregation model
L2R models	
Point-wise L2R models	
SVMReg [8]	SVM regression model, @ Params: trade_off error = 0.01
Mart [4]	Multiple additive regression trees model, @ Params: #trees = 1000, #leaves = 10, learning_rate = 0.1, max_iter = 100
Pair-wise L2R models	
SVMRank [8]	SVM ranking model, @ Params: t_error = 0.01
RankBoost [3]	Boosting machine learning ensemble model, @ Params: rounds = 300, threshold = 10 candidates to search
List-wise L2R models	
Coordinate Ascent (CA) [14]	Linear feature-based model, @ Params: random-restarts = 5, search-iterations = 25, tolerance = 0.001, Optimization = NDCG and τ

Table 3. Ranking performance of L2RI vs other approaches, ▼ means statistically significant worse than L2RI, using a two-sided t-test with significance level $p < 0.05$

Model	Arts & Crafts				Jewelry				Industrial			
	τ@all	τ@10	ρ@all	ρ@10	τ@all	τ@10	ρ@all	ρ@10	τ@all	τ@10	ρ@all	ρ@10
AVGRank	0.050 ▼	0.024 ▼	0.067 ▼	0.039 ▼	0.026 ▼	0.047 ▼	0.034 ▼	0.054 ▼	0.091 ▼	0.096 ▼	0.130 ▼	0.124 ▼
TQRank	0.333	0.275 ▼	0.451 ▼	0.347 ▼	0.425 ▼	0.312 ▼	0.574 ▼	0.380 ▼	0.363 ▼	0.367 ▼	0.473 ▼	0.459 ▼
SVMReg	0.234 ▼	0.217 ▼	0.321 ▼	0.277 ▼	0.407 ▼	0.347 ▼	0.547 ▼	0.428 ▼	0.282 ▼	0.317 ▼	0.369 ▼	0.406 ▼
Mart	0.215 ▼	0.106 ▼	0.303 ▼	0.155 ▼	0.378 ▼	0.187 ▼	0.507 ▼	0.254 ▼	0.339 ▼	0.331 ▼	0.446 ▼	0.422 ▼
RankBoost	0.327 ▼	0.299 ▼	0.445 ▼	0.391 ▼	0.438 ▼	0.342 ▼	0.596 ▼	0.461 ▼	0.363 ▼	0.399 ▼	0.468 ▼	0.509 ▼
SVMRank	0.367	0.333	0.492	0.410	0.456 ▼	0.403	0.606 ▼	0.475 ▼	0.348	0.378 ▼	0.455 ▼	0.493 ▼
CA($NDCG$@n)	0.365	0.332	0.489	0.408	0.459 ▼	0.379 ▼	0.61 ▼	0.475 ▼	0.34	0.379 ▼	0.455 ▼	0.495 ▼
CA(τ@k)	0.369	0.338	0.491	0.410	0.470	0.391	0.630	0.490	0.383	0.398 ▼	0.495 ▼	0.515 ▼
L2RI$^\lambda\tau$@k	**0.371**	**0.342**	**0.492**	**0.414**	**0.473**	**0.403**	**0.633**	**0.495**	**0.399**	**0.431**	**0.519**	**0.543**
Win\Draw\Loss	4\4\0	5\3\0	5\3\0	5\3\0	7\1\0	5\3\0	7\1\0	7\1\0	5\3\0	7\0\0	8\0\0	8\0\0

specifically SVMRank. Employing a classifier for the ranking orders of pairs is proven more effective than point-wise models.

Our model L2RI$^\lambda\tau$@k optimizing only the top-k products, is shown to produce significant ranking correlation results compared to all other baselines. Also, CA τ@k is highly performing against other learning models and comparable to our model. The main reason is that we replaced its loss function with our proposed Kendall's τ correlation (in contrast to the default using NDCG). Moreover, L2RI shows high significant correlation in terms of both τ and ρ in ranking of the top 10 products, which is the most practical case from user perspective to best rank the top 10 products. Also, we see that ρ values are always higher value than τ which is a good sign that there is no extreme ranking errors.

Weak Supervision: In this section, we investigate the effect of the weak supervision signal and how it can help improving the performance of L2RI. Table 4 shows the ranking performance of using two different weak supervision signals. L2RI(AVG) model uses the average star rating as weak supervision signal. We can see the high significant degradation in ranking performance when using such model. Also, we can see in categories like Jewelry and Industrial the correlation values are even with negative sign which indicates inverse correlation. This demonstrates that a poor supervision signal can definitely lead to very low ranking performance. Thus, the weak supervision signal should be selected carefully.

Table 4. Ranking performance of L2RI using different weak supervision labels

Model	Arts & Crafts				Jewelry				Industrial			
	τ@all	τ@10	ρ@all	ρ@10	τ@all	τ@10	ρ@all	ρ@10	τ@all	τ@10	ρ@all	ρ@10
L2RI(AVG)	0.016	0.011	0.04	0.032	−0.149	−0.132	−0.209	−0.195	−0.185	−0.162	−0.254	−0.219
L2RI(TQRank)	**0.371**	**0.342**	**0.492**	**0.414**	**0.473**	**0.403**	**0.633**	**0.495**	**0.399**	**0.431**	**0.519**	**0.543**

Table 5. Performance of L2RI using different optimization functions & depths

Model	Arts & Crafts				Jewelry				Industrial			
	τ@all	τ@10	ρ@all	ρ@10	τ@all	τ@10	ρ@all	ρ@10	τ@all	τ@10	ρ@all	ρ@10
L2RIτ@all	0.368	0.332	0.492	0.409	0.472	0.399	0.634	0.493	0.398	0.424	0.52	0.542
L2RIτ@k	0.371	0.342	0.492	0.414	0.473	0.403	0.633	0.495	0.399	0.431	0.519	0.543
L2RIρ@all	0.370	0.338	0.492	0.414	0.471	0.397	0.632	0.494	0.399	0.430	0.519	0.544
L2RIρ@k	0.371	0.341	0.492	0.414	0.470	0.393	0.631	0.494	0.398	0.431	0.517	0.542

Ranking Loss Function: We show the effect of using different ranking loss functions namely Kendall's τ and Spearman ρ. We further show the effect of limiting the optimization to a certain depth of the group. Table 5 shows the ranking performance of four models, optimizing Kendall's τ for the whole group size, optimizing for the top-k items, and the other two for Spearman ρ. The average number of products per group in the 3 testing categories is 70, for training is 125 products and for the whole dataset on average is 100 products per group. First, we can see no significant difference between optimizing Kendall's τ versus Spearman ρ, which suggests that they both work equally well for product ranking. Also, we can see optimizing Kendall's τ for top k products is slightly better than other models specially for τ@10, but not significantly better. The reason is that the average number of products in our dataset is not very high. Thus, optimizing for only the top 10 from the whole group in this case is no different from the whole group. However, with longer groups there might be evident performance increments, which we leave to future work. In terms of computation time, Kendall correlation has a $O(n^2)$ computation complexity comparing with $O(nlogn)$ of Spearman correlation (where n is the sample size). This suggests that the average number of products per group is an important factor in selecting the ranking optimization function.

Feature Importance: In this section, we show the effect of each feature category on the learning performance as summarized in Table 6. We show the individual contribution of using each feature category of the product level features. We can see that the most important individual features are the star ratings followed by activeness of reviews of a given product. However, the aggregate sentiment polarity of reviews of an item has shown poor results. This may be due to the inability of the sentiment extraction module to identify more than one sentiment describing the whole sentence, which might not be correct. Also, with a few reviews per product and 3 sentences on average per review, there is not enough data. The Helpfulness feature performance is comparable to the sentiment feature in the Arts & Craft category and worse in the Industrial category which is due to the low number of helpfulness votes for this category.

Table 6. Ranking performance of L2RI using different feature categories

Model	Arts & Crafts				Jewelry				Industrial			
Item level features	$\tau@all$	$\tau@10$	$\rho@all$	$\rho@10$	$\tau@all$	$\tau@10$	$\rho@all$	$\rho@10$	$\tau@all$	$\tau@10$	$\rho@all$	$\rho@10$
Aggregated Ratings (R)	0.333	0.270	0.450	0.343	0.425	0.314	0.574	0.383	0.360	0.365	0.472	0.460
Helpfulness (H)	0.088	0.021	0.117	0.025	0.092	0.025	0.125	0.015	0.054	0.046	0.072	0.051
Sentiment Polarity (P)	0.082	0.088	0.118	0.131	0.124	0.135	0.169	0.165	0.115	0.094	0.156	0.133
Activeness (A)	0.287	0.271	0.389	0.357	0.371	0.316	0.502	0.396	0.294	0.328	0.384	0.423
H+R	0.331	0.269	0.449	0.34	0.423	0.308	0.572	0.377	0.363	0.369	0.471	0.457
P+H+R	0.364	0.326	0.484	0.402	0.466	0.394	0.623	0.487	0.377	0.402	0.493	0.515
A+P+H+R	0.365	0.328	0.487	0.401	0.472	0.399	0.632	0.495	0.375	0.407	0.487	0.52
Group+Item features	**0.371**	**0.342**	**0.492**	**0.414**	**0.473**	**0.403**	**0.633**	**0.495**	**0.399**	**0.431**	**0.519**	**0.543**

Only after adding sentiments to the ratings and helpfulness (P+H+R) in Table 6), we started to see a noticeable performance gain, especially for the top-10 ranking. The same observation applies to the case of adding the activeness features (A+P+H+R).

The group-level features (Group+Item features) added around 5% to Kendall's τ and 7% to Spearman's ρ of ranking performance gain of the top-10 products. The Group-level features are very important, because they capture the differences between groups of items, especially in terms of items activeness dynamics. Note that we did not evaluate using only group level features, as our L2R models require item level features to differentiate items within groups.

5   Conclusion and Future Work

The substantial importance of item (e.g. product, service, movie, books, etc.) reviews for customers making purchase decisions motivated our research. Most studies on ranking items using reviews have focused on statistical and heuristic approaches to aggregate star ratings, but they often fail on items with few reviews. In this paper, we formulated the problem as a L2R problem – to rank a group of items. A ranking model is trained on a repository of reviews grouped by item groups. We overcame several technical challenges and proposed our L2RI model. To bootstrap the L2R process from unlabeled reviews, we proposed to aggregate star ratings and use as weak supervision labels for learning, taking into account the posting time, textual contents sentiments and helpfulness votes of reviews. We further proposed the use of a suitable loss function based on the number of pairwise ranking errors to effectively rank a group of items. Experiments on a real-world Amazon dataset showed that our model L2RI is more effective than the state-of-the-art heuristic aggregation approaches, regression and standard L2R models.

Regarding future work, we will further examine the textual contents of reviews to extract fine-grained aspect level features to enhance the L2RI model. Moreover, we will explore applying our L2RI framework to wider applications without explicit or implicit ranking labels.

References

1. Abdel-Hafez, A., Xu, Y., Josang, A.: A normal-distribution based rating aggregation method for generating product reputations. Web Intell. **13**(1), 43–51 (2015)
2. Cue, L., Zhang, X., Qin, A., Wu, L.: CDS: collaborative distant supervision for Twitter account classification. Expert Syst. Appl. **83**(15), 94–103 (2017)
3. Freund, Y., Iyer, R., Schapire, R., Singer, Y.: An efficient boosting algorithm for combining preferences. JMLR **4**, 933–969 (2003)
4. Friedman, J.: Greedy function approximation: a gradient boosting machine. Ann. Stat. **29**(5), 1189–1232 (2001)
5. Garcin, F., Flaing, B., Jurca, R.: Aggregating reputation feedback. In: ICORE, pp. 62–74 (2009)
6. Han, X., Sun, L.: Global distant supervision for relation extraction. In: AAAI, pp. 2950–2956 (2016)
7. Homann, R., Zhang, C., Ling, X., Zelemoyer, L., Weld, D.S.: Knowledge-based weak supervision for information extraction of overlapping relations. In: ACL, pp. 541–550 (2011)
8. Joachims, T.: Optimizing search engines using clickthrough data. In: KDD, pp. 133–142 (2002)
9. Josang, A., Haller, J.: Dirichlet reputation systems. In: ARES, pp. 112–119 (2007)
10. Li, H.: A short introduction to learning to rank. IEICE TIOS **E94**, 1854–1862 (2011)

11. Liu, J., Cao, Y., Lin, C., Huang, Y., Zhou, M.: Low-quality product review detection in opinion summarization. In: EMNLP-CoNLL, pp. 334–342 (2007)
12. McAuley, J., Targett, C., Shi, J., Van den Hengel, A.: Image-based recommendations on styles and substitutes. In: SIGIR, vol. 14, pp. 43–52 (2015)
13. McGlohon, M., Glance, N., Reiter, Z.: Star quality: aggregating reviews to rank products and merchants. In: ICWSM, pp. 1844–1851 (2010)
14. Metzler, D., Croft, B.: Linear feature-based models for information retrieval. Inf. Retr. **10**(3), 257–274 (2007)
15. Shaalan, Y., Zhang, X.: A time and opinion quality-weighted model for aggregating online reviews. In: Cheema, M.A., Zhang, W., Chang, L. (eds.) ADC 2016. LNCS, vol. 9877, pp. 269–282. Springer, Cham (2016). https://doi.org/10.1007/978-3-319-46922-5_21
16. Shi, Y., Larson, M., Hanjalic, A.: List-wise learning to rank with matrix factorization for collaborative filtering. In: Proceedings of the Fourth ACM Conference on Recommender Systems, pp. 269–272. ACM (2010)
17. Tehelwall, M., Buckley, K., Paltoglou, G.: Sentiment strength detection for the social web. JASIST **63**(1), 63–173 (2012)
18. Wu, Q., Burges, C., Svore, K., Gao, J.: Adapting boosting for information retrieval measures. Inf. Retr. **13**(3), 254–270 (2010)
19. Zhang, K., Cheng, Y., Liao, W., Choudhary, A.: Mining millions of reviews: a technique to rank products based on importance of reviews. ICEC **12**, 1–8 (2011)
20. Zhang, X., Cui, L., Wang, Y.: CommTrust: computing multi-dimensional trust by mining E-commerce feedback comments. IEEE TKDE **26**(7), 1631–1643 (2014)

Multimodal Mixture Density Boosting Network for Personality Mining

Nhi N. Y. Vo[1], Shaowu Liu[1], Xuezhong He[2], and Guandong Xu[1(✉)]

[1] Advanced Analytics Institute, University of Technology Sydney, Ultimo, Australia
Nhi.Vo@student.uts.edu.au, {Shaowu.Liu,Guandong.Xu}@uts.edu.au
[2] Business School, University of Technology Sydney, Ultimo, Australia
Tony.He1@uts.edu.au

Abstract. Knowing people's personalities is useful in various real-world applications, such as personnel selection. Traditionally, we have to rely on qualitative methodologies, e.g. surveys or psychology tests to determine a person's traits. However, recent advances in machine learning have it possible to automate this process by inferring personalities from textual data. Despite of its success, text-based method ignores the facial expression and the way people speak, which can also carry important information about human characteristics. In this work, a personality mining framework is proposed to exploit all the information from videos, including visual, auditory, and textual perspectives. Using a state-of-art cascade network built on advanced gradient boosting algorithms, the result produced by our proposed methodology can achieve lower the prediction errors than most current machine learning algorithms. Our multimodal mixture density boosting network especially perform well with small sample size datasets, which is useful for learning problems in psychology fields where big data is often not available.

Keywords: Personality mining · Mixture density boosting network
Deep learning

1 Introduction

Personalities denote the individual variances in characteristics patterns of thinking, feeling and behaving. People with different personalities tend to conduct themselves in varied ways and have different cognitive processes. Knowing one's traits and understanding the differences in their preferences would help with communicating and connecting to the person on a more individual level. One of the most well-known measurements of personality traits is the Five-Factor Model of Personality (Big Five) [12]. As shown in Fig. 1, the Big Five model contains the five fundamental underlying personality dimensions: agreeableness, conscientiousness, extraversion, neuroticism, and openness to experience. These personality dimensions are stable across time, cross-culturally shared, and explain a substantial proportion of behavior [7]. Therefore, the Big Five model has been the

© Springer International Publishing AG, part of Springer Nature 2018
D. Phung et al. (Eds.): PAKDD 2018, LNAI 10937, pp. 644–655, 2018.
https://doi.org/10.1007/978-3-319-93034-3_51

standard measurement for personality mining in current literature. Personality Mining is the process of identifying a person's traits by mining the information in different types of individual data. The main techniques to identify the Big Five personalities of an individual have been the qualitative methods of surveys [13]. Recently, there have been some applications of machine learning in personality mining, mostly through text mining [16] using standard algorithms, e.g. support vector regression or decision tree [8].

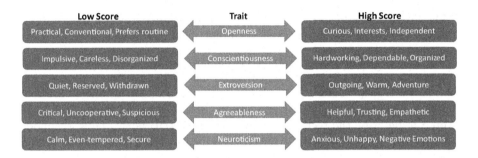

Fig. 1. The Five-Factor Model of Personality

Though current methodologies have shown the feasibility of personality mining, they have some critical limitations preventing for broader adoption. To be specific, the qualitative approaches are not practical, time-consuming, costly and might contain subjective errors. On the other hand, the standard machine learning algorithms can quickly mine the personalities of a large number of people at once without conducting surveys [10]. However, the prediction accuracy of these quantitative techniques suffers from small data size. Considering ethical reasons, using up-sampling techniques to increase the number of observations will not be acceptable.

In addition, most researchers have been approaching the personality mining problem using textual data only [11,15]. However, human characteristics are explicitly expressing not only in the spoken words but also in their facial expression and the way they speak as well. More research has incorporated these sensory information into their predictive models. Some research have showed that sensory data would significantly improve the prediction accuracy of one's traits [1]. These motivates us to look for a deep learning method which can learn from multimodal data.

Realizing these research gaps, we would like to propose here one of the very first multimodal approaches in personality mining using information from videos, audio and text data. Our Multimodal Mixture Density Boosting Network (MMDB) combines advanced deep learning techniques to build a multi-layer neural network. From small size of personality datasets. We will have an initial feature fusion layer to avoid over-weighting of one type of input data. Afterwards, we construct a combined neural network consisting of mixture density layers to avoid over-fitting and dynamic cascade gradient boosting layers to improve our prediction accuracy. In addition, our MMDB neural network has a general structure which can be applied flexibly to other similar multimodal deep learning problems.

There would be three main contributions of our paper in personality mining.

- To the best of our knowledge, this is the first research attempt to use deep neural network to analyze multimodal data for personality prediction model.
- Our MMDB model was built to adapt both small and large dataset, which is extremely useful in psychology research where data collecting is costly.
- The final contribution is the mixture density approach which makes it easy to transfer learning cross-dataset with different input features.

The rest of this paper is organized as follows. Section 1 introduces the basic concepts personality mining, current research limitations and our proposed solution. In Sect. 2, we review the literature on personality mining with focus on recent methodologies as the motivation for our research work. Section 3 is devoted to describe the technical details of our MMDB neural networks model. In Sect. 4, the proposed MMDB model is applied to public datasets to perform personality mining. Finally, conclusions are drawn in Sect. 5.

2   Preliminary

Personality mining has been mainly studied by psychologist for decades using primarily descriptive and qualitative methodologies [19]. With the growth of data analytics using machine learning algorithms, there are more quantitative efforts to estimate one's traits. However, due to the cost of collecting data, most personality datasets are relatively small in sample size and contain only text data [17]. Therefore, most research in this fields are based on textual data only, which yield a low accuracy on the results. Since the availability of multimedia data [2,21] on personality mining in recent years, we can now apply advanced neural network approach to build a better prediction model which utilizes multimodal features.

Current literatures on personality mining mainly focus on feature extraction and selection using different analysis [22,25]. For textual features, most of current papers use Linguistic Inquiry and Word Count (LIWC) [18], Bag of Words and other text sentiment analysis techniques. Regarding audio visual features, there are many different approaches using *Python* or *MATLAB* packages for prosody cue, speaking activity, scenery and face recognition [4,20]. There is also an application of a multimodal feature extraction technique called Doc2Vec [6]. The result shows improvement in prediction accuracy of some but not all Big Five traits. The variety of extracted feature sets tend to have significantly different correlation to the personality scores [24], which makes it difficult to compare the methodologies and empirical results even with the same dataset.

There have been some applications of machine learning methodologies to build prediction model [5]. Researchers have also looked predicting personality scores both separately or together as a multivariate problems using support vector machines and decision tree algorithms with different stacking models [8,9].

According to their results, the differences between univariate and multivariate model are not significant. Therefore, we will not approach personality mining as a multi-label prediction problem and will compare our model with two single stacking models from these papers. They also suggest that the cross-datasets transfer learning would not help with prediction accuracy. We will test this hypothesis again with our neural network using both two multimodal personality datasets that are publicly available [2,21].

Personality prediction has been commonly approached as regression problem. Even though we can convert the personality scores to binary labels for classification model using a certain threshold, many researchers have proven that it is not a good practice to determine human characteristics. Most classification models have also showed a pretty low prediction accuracy around 52% to 65% only [10]. Moreover, the sample size might not be equally distributed in each of binary classes. Therefore, we will only focus on building the regression model for personality mining within the scope of this paper.

As far as we concern, there have been no application of neural network in multimodal personality mining. Even though neural network still doesn't significantly outperform machine learning algorithms regarding regression problems, we believe neural network would have certain advantage in psychology fields such as human thinking and behavior. Our proposed MMDB neural network would be the first attempt to estimate one's traits using this advanced approach. It helps solve the challenge in personality mining with limitations in sample size and multimodal data. The research would contribute to the current literature with the shifting trend to use deep learning techniques.

3   Methodology

We propose here a multimodal neural network that can combine different type of input data at different sizes with our Discriminant Correlation Analysis (DCA) Feature Fusion layer. Then the fused features will be used as inputs and target for layers in our Mixture Density Network to adjust for the information loss due to feature fusion without over-fitting the model. Last but not least, the output of Mixture Density Network layers will be the input for layers in our Dynamic Cascade Boosting Network to regress the final prediction with high accuracy.

3.1   DCA Feature Fusion Layer

The standard DCA Feature Fusion algorithm [14] considers the class associations in feature sets. It eliminates the between-class correlations and restricts the correlations to be within classes. DCA maximizes the correlation of corresponding features across the two feature sets and in addition, decorrelates features that belong to different classes within each feature set. It also solves small sample size (SSS) problem, where the number of samples is less than the number of features which makes the covariance matrices singular and non-invertible. Within our multimodal neural network, the DCA Feature Fusion Layer will first fuse the video, audio and text features pairwise, then it will aggregate to compute the final fused feature of three modal inputs as

$$ff_{vst} = ff_{vs} + ff_{vt} + ff_{st} \tag{1}$$

where $ff_{vst}, ff_{vs}, ff_{vt}, ff_{st}$ are the DCA fused features of video-sound, video-text, sound-text and video-sound-text features accordingly. We will use ff_{vst} as target scores and $ff_{vs}, ff_{vt}, ff_{st}$ as inputs in our Mixture Density Neural Network.

3.2   Mixture Density Network

Mixture Density Networks (MDN) [3] predicts not a single output value but an entire probability distribution for the output. This help us get the inference between each fused feature and the aggregated ff_{vst}, reduce the loss of information from the DCA Feature Fusion Layer without over-fitting the model. The MDN will predict Mixture Gaussian distributions, where the output value is modeled as a sum of many Gaussian random values, each with different means and standard deviations. So for each input x, we will predict a probability weighted sum of smaller Gaussian probability distributions

$$P(Y = y | X = x) = \sum_{k=0}^{K-1} \Pi_k(x)\phi(y, \mu_k(x), \sigma_k(x)) \tag{2}$$

where $\phi(y, \mu_k(x), \sigma_k(x))$ is the probability distribution function (pdf) of Gaussian distribution k with predicted mean $\mu_k(x)$ and predicted deviation $\sigma_k(x)$. $\Pi_k(x)$ is the predicted weight of Gaussian distribution k, and $\sum_{k=0}^{K-1} \Pi_k(x) = 1$ to ensure that the pdf integrates to 1. Each of the parameters $\Pi_k(x), \mu_k(x), \sigma_k(x)$ will be determined by the neural network, as a function of the input x. We construct our MDN leveraging TensorFlow Slim, with three fully-connected hidden layers of 10 nodes each and Adam Optimizer for training. This feed-forward neural network will parameter $1,000$ Gaussian mixture components as outputs after $1,000$ iteration rounds.

3.3   Dynamic Cascade Boosting Network

Boosting algorithms have been one of the most effective machine learning methodologies for regression problems. Therefore, we believe the incorporation of boosting algorithms into our multimodal neural network would help increase the prediction accuracy for personality mining. In this model, we use gradient boosting regression from Scikit-Learn with 100 estimators as our base learner algorithm. For testing purpose, we only use two variants of hyperparameters in our model set up, where learning rates are 0.001 and 0.01 respectively. Other parameters are the same for all models, where max depth is 1 and loss function is least absolute deviations (LAD). To avoid over-fitting, we did not perform any specific form of parameter tuning either manually or automatically.

Our dynamic boosting network, inspired by gcForest [26], is built with a cascade structure, where each layer is embedded with multiple boosting algorithms. These algorithms will estimate the personality scores separately, then the average scores will be evaluated using mean accuracy (MA) before constructing the next cascade layer.

$$MA = 1 - 1/N \sum_{i=1}^{N} |y_{pred} - y_{true}| \tag{3}$$

After feeding the output of previous cascade layer to a new layer, the network will automatically assess the prediction accuracy of the model, and the training procedure will stop if there is no significant increase in performance. For our experiment, we set the tolerance rate to zero, which means new cascade layer will be constructed even with the smallest increase in MA values. The number of layers in this dynamic cascade boosting network will be implicitly constructed depending upon how fast the model learn (see Fig. 2). During our experiment, the number of cascade layers constructed is ranging from 2 layers to 8 layers. Since our dynamic boosting network can reactively chooses the number of cascade layer and decides on early stopping, it can efficiently handle different dataset sizes without wasting computing power.

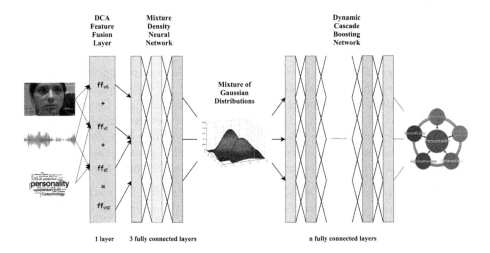

Fig. 2. The complete architecture of our MMDB Neural Network

4   Experiment

To study the performance of the MMDB model, experiments were conducted on two public datasets: First Impressions dataset and YouTube Personality dataset. Comparisons were made against several baselines.

4.1   Datasets

First Impressions dataset (FI) [21]. This dataset comprises $10,000$ clips with an average duration of 15 s. Each clip is a video of people facing and speaking English to the camera. The *gender, age, nationality,* and *ethnicity* information can be observed from clips. Beside sensory data, the dataset also contains the text transcription of the speakers' words. In total, $435,984$ words were transcribed ($183,861$ non-stopwords), which corresponds to 43 words per clip on average. Each clip is labeled with Big Five personality traits scores from $[0,1]$.

YouTube Personality dataset (YT)   [2]. This dataset consists of 404 YouTube clips when Video bloggers explicitly show themselves in front of the a webcam talking about a variety of topics. The text transcriptions are provided in raw text and contain 10 K unique words and 240 K word tokens. The personality impressions consist of Big Five scores that were collected using Amazon Mechanical Turk (AMT) and the Ten-Item Personality Inventory (TIPI) [13]. The scores are rescaled into range $[0,1]$.

4.2   Features Extraction

As the feature set of the two datasets are different, feature importances would be varied. Therefore, we will not perform any correlation analysis and feature selection. All extracted features will directly be inputs for our multimodal network and baseline models.

Video Features. Videos in FI dataset are showing one person speaking directly to the camera, therefore we are more interested in their facial movements and gestures than other general scenic data. To extract video features, we use OpenCV [4] to extract the landmarks data with face detection, alignment and tracker for every single frame of each clip. From these image-base data points, we then extract major facial features such as Yaw, Roll, Eyes and Lip movements, etc. Finally, we average these facial features across all frames of each clip.

Audio Features. Several speech features are extracted from audio, such as pitch and energy. These features are extracted at the interval of 5000 Hz each by using the Hidden Markov Models. Similar to video features, we compute the average of speech features for each audio file, which resulted 21 audio features in total.

Text Features. The LIWC 2015 [18] dictionary is used to extract text features from the transcription of each video clips. Even though there are many text analysis tools available such as Bag of Words, Word Sentiment, etc., these approaches might not suitable for our corpus with short text and variety of topics. Therefore, we will use only LIWC as the standard approach, which covers many topic-related features (e.g. work, family, friend, money), sentiment (e.g. possemo, negemo) and even speech related features (non-fluent). A total 93 text features were extracted.

4.3   Evaluation

Evaluation Metrics. Similar to related works, the Mean Absolute Error (MAE) and Root Mean Squared Error (RMSE) as our evaluation metrics: $\text{MAE} = 1/N \sum_{i=1}^{N} |y_{\text{pred}} - y_{\text{true}}|$, $\text{RMSE} = \sqrt{1/N \sum_{i=1}^{N} (y_{\text{pred}} - y_{\text{true}})^2}$ where N is the size of data, and y_{pred} and y_{true} are the predicted and true personality scores, respectively. We also perform rank evaluation by ranking data instances for each type of personality and comparing against the true ranking using Spearman Rank Correlation Coefficient (Spearman's rho) [23].

Baselines. We will also build prediction models using some models in current literature to compare the performances of our MMDB neural network. Specifically, we will build the baseline models using the Gaussian Process from [11] and two single stacking models with base learner support vector regression and decision tree from [8,9], which will be denoted as GP, SVR and DT respectively. We also build a simple neural network (NN) for comparison with our MMDB neural network, using TensorFlow framework to construct three fully-connected ReLU layers with 10 nodes each. The deep neural network regressors will use Adam Optimizer as solver and will train for 10, 000 iterations. The input features for these baseline models would include all text, auditory and visual features as in our MMDB model. We will also not perform any specific parameter tuning on these baseline models.

4.4   Results

For individual dataset evaluation, we perform 10-fold cross validation and compare the results with baseline models using the extracted features as input. For cross-datasets evaluation and component testing, we split each dataset into train set and test at ratio 8 : 2. For FI dataset, there are 8, 000 and 2, 000 instances in train and test set. For YT dataset, there are 323 and 81 instances in train and test set. The 10-fold cross validation results are shown in Table 1, with agreeableness as AGR, conscientiousness as CON, extraversion as EXT, neuroticism as NEU and openness as OPN. On the MAE, our MMDB neural network performs better or on the same level of accuracy to some current methods on both datasets. On rank evaluation, our model performs significantly better in most personality dimensions for both tested datasets. It has the highest Spearman's rho together with NN model for EXT on YT dataset. It performs slightly worse than SVR and NN for EXT and NEU on FI dataset. In general, our model has better prediction accuracy than other personality mining models. Especially in dealing with small data size case in psychology field like the YT dataset, our MMDB neural network helps improve prediction accuracy significantly compared to other mentioned models.

We also want to test whether we can perform transfer learning the personality mining models using cross datasets. Unlike most machine learning models where we need the same number of input features from different dataset to perform

Table 1. 10-fold cross-validation on YouTube and first impression datasets

	Model	MAE					rho				
		AGR	CON	EXT	NEU	OPN	AGR	CON	EXT	NEU	OPN
YouTube Dataset	MMDB	**.1070**	**.0980**	**.1285**	**.0973**	**.0957**	**.4352**	**.3030**	**.3233**	**.3398**	.0949
	NN	2.19	3.68	2.22	1.55	1.08	.1423	.0838	.2156	−.0627	.0253
	GP	.6138	.5829	.6042	.6277	.6108	.0338	−.1029	.0021	−.0223	.0488
	SVR	.1211	.1008	.1354	.1040	.0960	.1218	.1739	.0799	.0943	.0397
	DT	.1363	.1291	.1669	.1295	.1338	.2972	.2481	.1578	.1861	.0962
First Impression Dataset	MMDB	**.1042**	**.1228**	**.1184**	**.1186**	**.1136**	**.2616**	**.3025**	**.3356**	**.3643**	**.3118**
	NN	.1254	.1646	.1380	.1314	.1254	.1191	.1793	.1457	.2000	.0846
	GP	.5495	.5243	.4767	.5209	.5667	.0626	.0493	.0836	.0847	.0759
	SVR	.1070	.1261	.1226	.1235	.1171	.0590	.0833	.0650	.0739	.0478
	DT	.1441	.1612	.1552	.1585	.1522	.1117	.1587	.1864	.1762	.1562

cross-data learning, our model use DCA and mixture density to fuse features and compute the inference with Gaussian distribution to create an equal input nodes for cascade boosting network. This allows our model to transfer learning easily between datasets with multimodal features. The reported results in Table 2 are in line with current literature that transfer learning does not improve prediction accuracy. This is explainable as the trait dimensions' scores were denoted by different author using various techniques and scales. However, when comparing the MAE between Tables 1 and 2, the transfer learning from the FI to YT dataset still performs better than some baseline models trained on YT dataset for different personality dimensions. This positive result of transfer learning could help in specific case where one dataset is much smaller than the other dataset, then transfer learning would have better result than in the vice versa case.

Last but not least, we believe the integration of gradient boosting algorithms into neural network would significantly improve the prediction accuracy, especially in the case of small sample size. We test this hypothesis using the YouTube Personality dataset by running two separated prediction models. The first one is our proposed MMDB neural network with the full layers. The second one contains only the DCA fusion layer and the mixture density neural network (MMD). The results in Table 3 show that the performance of MMD is quite satisfactory, which can still outperform other baseline models. As observed in

Table 2. Transfer learning.

Direction	MAE					RMSE				
	AGR	CON	EXT	NEU	OPN	AGR	CON	EXT	NEU	OPN
YT→FI	.1348	.1291	.1457	.1640	.1250	.1706	.1599	.1805	.2042	.1568
FI→YT	.1258	.1219	.1782	.1344	.0972	.1486	.1539	.1893	.1579	.1155

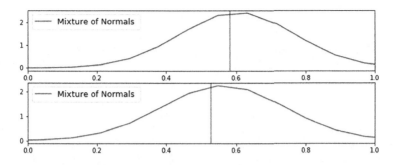

Fig. 3. Sample distribution predictions for Openness from MMD neural network (where the black vertical line is the ground truth personality scores)

Table 3. MMDB and MMD evaluation with YouTube Personality dataset

Model	MAE					RMSE				
	AGR	CON	EXT	NEU	OPN	AGR	CON	EXT	NEU	OPN
MMDB	.10940	.10286	.15039	.09190	.09449	.13455	.13386	.16981	.12121	.11699
MMD	.11244	.11589	.15276	.13414	.11643	.14369	.14530	.18018	.17003	.14822

Fig. 3), the predicted means of the Gaussian distributions are very close to the ground truth personality scores, which give us a lower MAE and RSME with MMD model only. However, the dynamic cascade boosting network reduces the MAE and RMSE further. This proves our hypothesis on the effectiveness of gradient boosting algorithms in these regression problems. Our proposed MMDB neural network and its component layers together can solve better the challenges in personality mining.

5 Conclusions

Within the scope of this paper, we have identified the current research gap in personality mining, which was dominantly using costly qualitative methods e.g. surveys. Most quantitative research only use text mining techniques to predict personality, while we believe sensory data can contain useful information about one's traits. The recent approaches using machine learning algorithms have limitations when it comes to small sample size, which have low prediction accuracy. Our proposed MMDB neural network has been proven to be an effective model in solving personality mining challenges.

The MMDB neural network is a quantitative methodology, which is the first research work to use deep learning approach in personality mining. It consists of three main components. The first one is a DCA feature fusion layer to fuse multimodal features from visual, auditory and textual data. The second component is a mixture density neural network to predict the full distribution of personality scores. This solves the common problem of over-fitting due to small sample size. Finally, a dynamic cascade boosting network will significantly improve the accuracy of finally prediction.

Our MMDB neural network has outperformed other baseline models from current literature. The experiments with cross-datasets have showed the transfer learning for personality mining is not effective in general, but can still help in case of predicting on small sample size using our model from bigger dataset. Last but not least we test the components of our MMDB neural network individually. The results confirm our hypothesis of integrating the dynamic cascade boosting will improve prediction accuracy of the mixture density network. In conclusion, our MMDB neural network has great performances, especially with small datasets in personality and psychology fields.

In future work, we would like to explore more options to improve the prediction model for personality mining using particularly deep learning techniques. We would work on a more intuitive neural networks that can perform personality prediction using the videos as raw input data without the intermediate step of features extraction and feature fusion. With the current progress in computer vision and multimodal neural networks, we believe it will bring further breakthroughs in personality mining research.

Acknowledgment. This work was financially supported by our industry partner at UTS and was technically supported by our faculty with the infrastructures and computing power for empirical works. We would like to send our appreciation to fellow researchers who provided the datasets for this research. We would also like to thanks our colleagues, family and friends who have been supportive during the experimenting and paper writing process.

References

1. Alam, F., Riccardi, G.: Predicting personality traits using multimodal information. In: Proceedings of the 2014 ACM Multi Media on Workshop on Computational Personality Recognition, pp. 15–18. ACM (2014)
2. Biel, J.I., Gatica-Perez, D.: The youtube lens: crowdsourced personality impressions and audiovisual analysis of vlogs. IEEE Trans. Multimed. **15**(1), 41–55 (2013)
3. Bishop, C.M.: Mixture density networks (1994)
4. Bradski, G.: The OpenCV library. Dr. Dobb's J. Softw. Tools Prof. Programmer **25**(11), 120–123 (2000)
5. Buettner, R.: Innovative personality-based digital services. In: PACIS, p. 278 (2016)
6. Chen, L., Feng, G., Leong, C.W., Lehman, B., Martin-Raugh, M., Kell, H., Lee, C.M., Yoon, S.Y.: Automated scoring of interview videos using Doc2Vec multimodal feature extraction paradigm. In: Proceedings of the 18th ACM International Conference on Multimodal Interaction, pp. 161–168. ACM (2016)
7. Costa, P.T., McCrae, R.R.: Four ways five factors are basic. Personality Individ. Differ. **13**(6), 653–665 (1992)
8. Farnadi, G., Sitaraman, G., Sushmita, S., Celli, F., Kosinski, M., Stillwell, D., Davalos, S., Moens, M.F., De Cock, M.: Computational personality recognition in social media. User Model. User-Adapted Interact. **26**(2–3), 109–142 (2016)

9. Farnadi, G., Sushmita, S., Sitaraman, G., Ton, N., De Cock, M., Davalos, S.: A multivariate regression approach to personality impression recognition of vloggers. In: Proceedings of the 2014 ACM Multi Media on Workshop on Computational Personality Recognition, pp. 1–6. ACM (2014)
10. Farnadi, G., Zoghbi, S., Moens, M.F., De Cock, M.: Recognising personality traits using facebook status updates. In: Proceedings of the Workshop on Computational Personality Recognition (WCPR 2013) at the 7th International AAAI Conference on Weblogs and Social Media (ICWSM 2013). AAAI (2013)
11. Golbeck, J., Robles, C., Turner, K.: Predicting personality with social media. In: CHI 2011 Extended Abstracts on Human Factors in Computing Systems, pp. 253–262. ACM (2011)
12. Goldberg, L.R.: An alternative description of personality: the big-five factor structure. J. Pers. Soc. Psychol. **59**(6), 1216 (1990)
13. Goldberg, L.R., Johnson, J.A., Eber, H.W., Hogan, R., Ashton, M.C., Cloninger, C.R., Gough, H.G.: The international personality item pool and the future of public-domain personality measures. J. Res. Pers. **40**(1), 84–96 (2006)
14. Haghighat, M., Abdel-Mottaleb, M., Alhalabi, W.: Discriminant correlation analysis: real-time feature level fusion for multimodal biometric recognition. IEEE Trans. Inf. Forensics Secur. **11**(9), 1984–1996 (2016)
15. High, R.: The Era of Cognitive Systems: An Inside Look at IBM Watson and How It Works. IBM Corporation, Redbooks, New York (2012)
16. Kosinski, M., Matz, S.C., Gosling, S.D., Popov, V., Stillwell, D.: Facebook as a research tool for the social sciences: opportunities, challenges, ethical considerations, and practical guidelines. Am. Psychol. **70**(6), 543 (2015)
17. Mairesse, F., Walker, M.: Personage: personality generation for dialogue. In: Annual Meeting-Association For Computational Linguistics, vol. 45, p. 496 (2007)
18. Pennebaker, J.W., Francis, M.E., Booth, R.J.: Linguistic Inquiry and Word Count: LIWC 2001, vol. 71. Lawrence Erlbaum Associates, Mahway (2001)
19. Pennebaker, J.W., King, L.A.: Linguistic styles: language use as an individual difference. J. Pers. Soc. Psychol. **77**(6), 1296 (1999)
20. Pentland, A.: Social dynamics: signals and behavior. In: International Conference on Developmental Learning, vol. 5 (2004)
21. Ponce-López, V., et al.: ChaLearn LAP 2016: first round challenge on first impressions - dataset and results. In: Hua, G., Jégou, H. (eds.) ECCV 2016. LNCS, vol. 9915, pp. 400–418. Springer, Cham (2016). https://doi.org/10.1007/978-3-319-49409-8_32
22. Sarkar, C., Bhatia, S., Agarwal, A., Li, J.: Feature analysis for computational personality recognition using youtube personality data set. In: Proceedings of the 2014 ACM Multi Media on Workshop on Computational Personality Recognition, pp. 11–14. ACM (2014)
23. Spearman, C.: The proof and measurement of association between two things. Am. J. Psychol. **100**(3/4), 441–471 (1987)
24. Verhoeven, B., Daelemans, W., et al.: Evaluating content-independent features for personality recognition. In: Proceedings of the 2014 ACM Multi Media on Workshop on Computational Personality Recognition, pp. 7–10. ACM (2014)
25. Vinciarelli, A., Mohammadi, G.: A survey of personality computing. IEEE Trans. Affect. Comput. **5**(3), 273–291 (2014)
26. Zhou, Z.H., Feng, J.: Deep forest: towards an alternative to deep neural networks. arXiv preprint arXiv:1702.08835 (2017)

Identifying Singleton Spammers via Spammer Group Detection

Dheeraj Kumar[1], Yassien Shaalan[2], Xiuzhen Zhang[2(✉)], and Jeffrey Chan[2]

[1] Lyles School of Civil Engineering, Purdue University, West Lafayette, USA
kumar299@purdue.edu
[2] School of Science (Computer Science), RMIT University, Melbourne, Australia
{yassien.shaalan,xiuzhen.zhang,jeffrey.chan}@rmit.edu.au

Abstract. Opinion spam is a well-recognized threat to the credibility of online reviews. Existing approaches to detecting spam reviews or spammers examine review content, reviewer behavior and reviewer-product network, and often operate on the assumption that spammers write at least several if not many fake reviews. On the other hand, spammers setup multiple sockpuppet IDs and write one-time, singleton spam reviews to avoid detection. It is reported that for most review sites, a large portion, sometimes over 90%, of reviewers are singletons (identified by the reviewer ID). Singleton spammers are difficult to catch due to the scarcity of behavioral clues. In this paper, we argue that the key to detect singleton spammers (and their fake reviews) is to detect group spam attacks by inferring the hidden collusiveness among them. To address the challenge of lack of explicit behavioral signals for singleton reviewers, we propose to infer the hidden reviewer-product associations by completing the review-product matrix by leveraging the product and review metadata and text. Experiments on three real-life Yelp datasets established that our approach can effectively detect singleton spammers via group detection, which are often missed by existing approaches.

Keywords: Opinion spam · Singleton spammers · Sockpuppet IDs
Inductive matrix completion

1 Introduction

Online shoppers are ever increasing and the product reviews influence their buying decisions to a great extent. According to the *Local Consumer Review Survey 2016* by BrightLocal[1], 84% of the online shoppers trust product reviews as much as personal recommendations. Positive reviews and higher star ratings results in substantial financial gains for the businesses, while negative reviews can cause reputation damage and financial losses. As a result of such financial incentives,

---

D. Kumar—performed this research while working as a Research Officer at RMIT University, Melbourne, Australia under the supervision of Prof. Xiuzhen Zhang.

[1] https://www.brightlocal.com/learn/local-consumer-review-survey/.

opinion spam (fake reviews which deliberately mislead readers) is prevalent. There is an estimate from 2012 that one-third of consumer reviews on the Internet were fake[2]. Similarly, the number of fake reviews on yelp.com rose from 5% in 2006 to 20% in 2013 [8].

Opinion spam detection has attracted significant research [1,5–7,11,13–15,17]. Existing approaches to detecting spam reviews and spammers focus on extracting spam signals from review texts [5,11,14], reviewer behaviour [7,15], or the reviewer-product networks [1,13]. Many approaches assume spammers write at least several if not many fake reviews for multiple products. It is reported, however, that the majority of reviewers in most of the opinion websites are singletons, i.e. writes only one review. According to [17], over 90% of the reviewers of resellerratings.com write only one review. Indeed, our analysis showed that a majority, in the range of 65% to 70%, of the reviewers of Yelp datasets [13] are singletons (See Table 1 for details). Detecting singleton spam reviews and singleton spammers is challenging due to the lack of obvious spamming signals. Existing approaches based on reviewer behavior and reviewer-product networks are not effective for detecting singleton spammer reviewers [14,17]. Sandulescu and Ester [14] argued that the key to catch singleton spammers can only be found in the review texts. A drawback of their text-based approach is that the collusiveness among reviewers is overlooked and therefore may not be effective.

To address the challenge of scarcity of spam signals for singleton reviewers, we argue that the key to effective detection of singleton spammers is via identifying spammer groups. A spammer group is *"a group of reviewers writing fake reviews together to promote or to demote some target products"* [9]. Here reviewers are defined by the reviewer ID. The actual person behind different IDs could be a single person, multiple persons, or a combination of both. There is a great incentive for opinion spammers to create multiple sockpuppet IDs to write singleton reviews since it helps them avoid detection and provide higher revenues by writing many fake reviews. However, existing approaches to detecting spammer groups [9,18–20] are not directly applicable for singleton spammers since they assume that participants of spammer groups frequently write reviews for multiple products together (hence, non-singleton).

In this paper, we propose SSGD (*singleton spammer group detection*), a novel approach to detecting singleton spammers via spammer group detection. The intuition behind SSGD is that given the purpose of group spamming is to promote or demote the reputation of products within a short time window, a burst of changes in signals such as rating and number of reviews can indicate the occurrence of spam attacks. From the likely spam attacks, the candidate singleton spammers and their targeted products are identified. We further examine the review textual content, rating and time to infer latent reviewer-product associations and uncover the collusiveness among singleton reviewers. We formulate the problem of inferring (hidden) reviewer-product association as a review-product matrix completion problem. The sparse review-product associa-

---

[2] http://www.nytimes.com/2012/08/26/business/book-reviewers-for-hire-meet-a-demand-for-online-raves.html.

tion matrix is *completed* using the additional information such as review text and metadata (star rating and date), the product description text and the product "also bought" and "also viewed" network. Lastly, the inferred reviewer-product associations are clustered to detect spammer groups consisting of (mostly) singleton spammers.

We conducted experiments on three real-world opinion spam datasets on yelp.com (YelpChi, YelpNYC, and YelpZIP) [13] to evaluate the effectiveness of SSGD for detecting singleton spammers by detecting them in groups. We benchmarked SSGD against five approaches in the literature for detecting spammer groups and individual spammers based on reviewer behavior, reviewer-product network and review text. SSGD outperformed all these approaches in terms of both recall and precision for singleton spammer detection. To the best of our knowledge, this paper is the first attempt at identifying singleton spammers via detecting hidden collusiveness among them for group spam attacks.

2   Related Work

Existing studies in the literature on detecting spam reviews and spam reviewers can be broadly classified as reviewer behaviour based [7,15], reviewer-product network based [1,13], and review text based [5,11,14]. A detailed survey of these techniques can be found in [4]. Reviewer behavior-based and reviewer-product network-based approaches are not effective for detecting singleton spammers, as they focus on reviewers with multiple reviews. Text-based approaches [11, 14] can address the challenge of lacking behavioral clues for one-time singleton reviewers by examining the psycholinguistic features in review contents [11] or by examining the pairwise content similarity between reviews [14]. However, the text-based approaches totally ignore the collusiveness among reviewer IDs and other review metadata, and our experiments show that they are not very effective in detecting singleton reviewers.

There are some recent studies on detecting singleton spam reviews [14,17]. Xie et al. [17] constructed a multidimensional time series consisting of the average rating, number of all and singleton reviews for time windows of fixed duration. It then detects spam attacks by finding abnormal sections in each time series. Though not specifically for singleton reviews, [21] monitors a list of carefully selected indicative signals of opinion spam over time and design efficient techniques to both detect and characterize abnormal events in real-time. The indicative features used for temporal spam detection in [21] are a superset of the time series used in [17]. One thing to note is that the approaches presented in [17,21] does not label individual singleton reviews as spam or genuine, but predict the time when a product is most likely to be a victim of a spam attack.

Our research is also related to studies concerning spammer group detection [9, 18–20]. Mukherjee et al. [9] first proposed an approach to detect spammer groups using *frequent itemset mining* (FIM) to find a set of candidate groups. It then uses several behavioral models derived from the collusion phenomenon among suspected fake reviewers to detect fake reviewer groups. The approach proposed in [20] uses only the network footprint information (user-product graph) to detect spammer groups. The approaches proposed in [18,19] uses pairwise features of reviewers to detect spammer groups which are defined for only those reviewer pairs who have adequate reviewing histories. Recently, Li et al. [6] proposed an HMM-based approach to detect spammers' co-bursting behavior to detect spammer groups. For all these approaches, the assumption about the spammer groups is that reviewers in a group write fake reviews for multiple products together and detection is based on the explicit reviewer-product associations, hence all of these approaches would miss the singleton spammers.

More generally, our research is related to spam detection on social media [16], where, most studies focus on finding clusters of linked nodes and the singleton spammers are likely ignored. In addition, our approach of inferring hidden associations is related to inferring network structures from data [2], which is an important data mining task in many domains.

3   The Proposed Approach: SSGD

We propose a novel approach to catch singleton spammers by discovering hidden collusiveness among them and detecting spammer groups. Detecting singleton spammer group is challenging as singleton reviewers write only one review and there is little information available about each individual singleton spammer. However, there are still (hidden) signals available for detection of singleton spammer groups. Such spammer groups aim to influence average ratings and impressions of target products by using following two tactics [14,15]:

- Inject enough fake reviews to affect the average ratings;
- Flood the most recent review pages with fake reviews as most buyers read only top several reviews before forming an opinion about a business.

Both these approaches require a (singleton) spammer group to generate a relatively abnormal number of positive or negative reviews over a short period of time, depending on what influences they wish to exert. Figure 1 shows the framework of our proposed approach SSGD. We first apply the spam attack detection approach [21] to identify the target products and the attack time in the multivariate time series for a set of indicative signals: average rating, number of positive/negative reviews, rating entropy, the ratio of singletons and first-timers, youth score, and temporal gap entropy to detect abnormal changes/bursts as potential spam attacks. The magnitude of abrupt change is used to assign an anomaly score for each detected attack. Further details about this step are well documented in [21] and are not reproduced here for brevity.

The spam attacks produce a review-product subgraph of target products and (potentially) spam reviews. The matrix representing the review-product network is very sparse without meaningful review-product associations. This data sparsity problem is similar to the one encountered in the recommendation systems, where matrix completion has shown great success in dealing with such sparsity. A detailed survey of the novel techniques to infer hidden network structure from

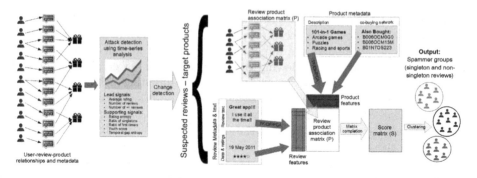

Fig. 1. Schematics of the proposed approach (SSGD)

the data for matrix completion is given in [2]. For the problem of inferring hidden collusiveness among reviewers, we use the *inductive matrix completion* (IMC) algorithm [10], which uses additional information such as review text and metadata (star rating and date) and product metadata such as its description text. Once the "completed", enriched review-product matrix is obtained from IMC, we cluster the reviews (using the inferred associations among reviews as the feature vector) to discover groups of similar reviews (reviewers) targeting the attacked products. We next describe these steps in details.

3.1   Inferring Hidden Reviewer-Product Associations

The spam attacks define a review-product bipartite graph of N_r suspicious reviews and N_p target products as nodes. An edge is present between review $i, 1 \leq i \leq N_r$ and product $j, 1 \leq j \leq N_p$ if review i belongs to the product j. This bipartite review-product graph is represented as a review-product associations matrix $P \in \mathbb{R}^{N_r \times N_p}$, where $P_{ij} = 1$ if review i belongs to product j, otherwise, $P_{ij} = 0$. The sparse review-product association matrix P is then "completed" using the IMC algorithm (described next) to learn hidden reviewer-product associations.

The IMC algorithm was used in [10] to predict *gene-disease associations by combining multiple types of evidence (features) for diseases and genes to learn latent factors that explain the observed gene-disease associations*. The spammer group detection is similar to this problem since we have a (sparse) review-product association matrix and based on the features of reviews and products, the aim is to discover groups of users who are most likely to write fake reviews for a group of products. IMC can be interpreted as a generalization of the transductive multi-label learning formulation: *low rank empirical risk minimization for multi-label learning* (LMEL) [22] and assumes that the associations matrix is generated by applying feature vectors associated with its rows as well as columns to a low-rank matrix Z (representing actual person behind various sockpuppet IDs). The goal is to recover Z using observations from P. Let $x_i \in \mathbb{R}^{f_r}$, and $y_j \in \mathbb{R}^{f_p}$ denotes the feature vector for review i, and product j respectively. Let $X \in \mathbb{R}^{N_r \times f_r}$

denote the training feature matrix of N_r reviews, where the ith row is the review feature vector x_i, and let $Y \in \mathbb{R}^{N_p \times f_p}$ denote the training feature matrix of N_p products, where the jth row is the product feature vector y_j. The inductive matrix completion problem is to recover a low-rank matrix $Z \in \mathbb{R}^{f_r \times f_p}$ using the observed entries from P. Denote the set of observed entries (i.e., training review-product associations) by Ω. The entry P_{ij} of the matrix is modeled as $P_{ij} = x_i^T Z y_j$ and the goal is to learn Z using the observed entries Ω. Z is of the form $Z = WH^T$, where $W \in \mathbb{R}^{f_r \times k}$ and $H \in \mathbb{R}^{f_p \times k}$, and k is small. The low-rank constraint on Z is NP-hard to solve. The standard relaxation of the rank constraint is the trace norm, i.e., sum of singular values. Minimizing the trace-norm of $Z = WH^T$ is equivalent to minimizing $\frac{1}{2}(||W||_F^2 + ||H||_F^2)$. The factors W and H are obtained as solutions to the following optimization problem:

$$\min_{\substack{W \in \mathbb{R}^{f_r \times k}, \\ H \in \mathbb{R}^{f_p \times k}}} \sum_{(i,j) \in \Omega} \ell(P_{ij}, x_i^T WH^T y_j) + \frac{1}{2}\lambda(||W||_F^2 + ||H||_F^2), \qquad (1)$$

The loss function ℓ penalizes the deviation of estimated entries from the observations. A common choice for loss function is the squared loss function ($\ell_{sq}(a, b) = (a - b)^2$). The regularization parameter λ trades off accrued losses on observed entries and the trace-norm constraint. IMC adapt the LEML solver [22] for solving (1). The solver uses alternating minimization (fix W and solve for H and vice versa) to optimize (1). The resulting optimization problem in one variable (W or H) is solved using the conjugate gradient iterative procedure. The features used to learn hidden reviewer-product association using IMC are described next.

We use the *term frequency - inverse document frequency* (TF-IDF) as the text feature for review/product description. TF-IDF formally measures how concentrated the occurrences of a given word is into relatively few documents [12]. The terms with the highest TF-IDF scores are often the terms that best characterize the topic of the document. Before extracting the TF-IDF feature for the review texts and product descriptions, we filter out the stop words as they are extremely common words which would appear to be of little value in deciding review text similarity. We project the review and product TF-IDF features to a lower dimensional space. In particular, we use *principal component analysis* (PCA) that performs a linear mapping of the data onto the lower dimensional space by maximizing the variance of the data in the new representation. We choose the leading 200 eigenvectors of the covariance matrix as the text features for reviews and product description. Another set of features for reviews consists of the date and the star rating associated with each review. For product feature, apart from the TF-IDF features obtained from the review content and product description, the product "also bought" and "also viewed" network is useful in identifying similar products which could be the common target of a group of spammers. In case this information is not available, identity matrix (I) is used as the product features representing each product as independent of others.

3.2 Finding and Ranking Spammer Groups

The output of the IMC algorithm is a completed review-product association matrix called as *score matrix* ($S \in \mathbb{R}^{N_r \times N_p}$). Higher values of S_{ij} indicate a greater likelihood of review i being written for product j. The learned product association for each review is used as the review feature vector to cluster them to detect spammer groups. We consider the N_r rows of the score matrix S as the N_p- dimensional feature vector for each review. To detect spammer groups, we need to find the set of reviews which are most likely being written by a group of a few spammers (sockpuppet IDs). These reviews would form a dense cluster in the N_p- dimensional feature representation given by score matrix S. The genuine reviews are expected to be at a large distance from other genuine reviews and spammer groups.

We choose the popular density-based clustering algorithm DBSCAN [3] to cluster spammer groups since it does not require the number of clusters to seek as an input, which is not known for our problem. Also, the reachability distance (ϵ) parameter of DBSCAN provides an easily tunable parameter for detecting a dense cluster of spammers, leaving out noise points (genuine reviews which may not belong to any spammer group). Most of the reviews in the detected groups are singleton reviews, hence review group is similar to reviewer group, however, in some cases, multiple reviews belonging to the same user can be a classified in different spammer groups (some spammers write a few genuine looking reviews to camouflage their campaign or may be part of multiple spam campaigns). The candidate spammer groups are then ranked based on the average intra-cluster distance between the reviews belonging to a group in the feature vector space representation. The groups whose members are close to each other are given high ranking as compared with the groups whose members are sparse.

4   Experiment Setup

We performed experiments on three publicly available online review datasets that are widely used in the opinion spam literature: YelpChi, YelpNYC, and YelpZIP [13]. These datasets contain reviews for restaurants in Chicago and NYC and in areas defined by a zip code in the NY state. All datasets contain review metadata such as star rating and date as well as review text. Their basic statistics is given in Table 1. It can be seen that all datasets contain a high percentage of singleton reviews, ranging from 65.35% to 70.55%.

The Yelp datasets have "near" ground truth labels for spam reviews based on the fake/suspicious filtering algorithm used at Yelp.com. The author of a spam review is labeled a spammer. Although the Yelp anti-fraud filter is not perfect, it was found to produce accurate results, and the spam reviews and spammers thus labeled were used as ground truth for evaluating opinion spam detection algorithms [13]. Table 2 also shows the portion of singleton spammers in the ground truth spammers in the three Yelp datasets. It can be seen that a significant portion of spammers on the Yelp datasets are singleton spammers.

Table 1. Basic statistics of the three datasets (In the parlance of customer reviews, restaurants are the products.)

	# Reviewers	# Products	#Reviews per reviewer	#Reviews per product	#Singleton reviewers (%)
YelpChi	38,063	201	1.77	335.30	70.55%
YelpNYC	160,225	923	2.25	389.00	66.15%
YelpZIP	260,277	5,044	2.34	120.66	65.35%

Table 2. Singleton spammers in the Yelp datasets

	# Spammers	Singleton spammers (% of # Spammers)
YelpChi	8,919	76.03%
YelpNYC	36,885	63.91%
YelpZiP	80,466	62.97%

We compare SSGD against five state-of-the-art baseline approaches for detecting spammer groups and individual spammers – two are spammer group detection approaches, one is a spam review detection algorithm utilizing network as well as metadata and two are text-based approaches, one being specialized for detecting singleton spammers, as described below:

1. Spammer group detection baselines:
   (a) FIM (Frequent Itemset Mining) [9]: This approach assumes that spammer groups are groups where reviewers (identified by reviewer IDs) frequently write reviews together. The candidate groups are then ranked based on the group spam features described in [9], which are found effective in distinguishing spammer and non-spammer groups.
   (b) NFS-GroupStrainer [20]: This approach detects spammer groups based on the footprint of reviewers on the reviewer-product network. It first finds targeted products using a graph-based measure *Network Footprint Score* (NFS) which quantifies the statistical distortion caused by spamming activities in the reviewer-product graph. A hierarchical clustering algorithm called GroupStrainer is then applied on the two-hop subnetwork of the targeted products to find spammer groups.
2. Spam review detection utilizing network as well as metadata baseline:
   (a) SpEagle: The SpEagle algorithm [13] utilizes clues from all review metadata (text, timestamp, rating) as well as the reviewer-review-product network to find suspicious users and reviews and the targeted products.
3. Review text-based baselines: The text-based approaches use only the textual review contents to detect spam reviews and accordingly singleton reviewers.
   (a) Ott: Ott et al. [11] built a supervised classification model based on a comprehensive set of psycholinguistic features extracted from review text.

(b) DSR [14]: This approach detects singleton spam reviews by computing the semantic similarity among pairs of reviews.

5   Results and Discussion

We next report the results of SSGD for detecting spammers and spammer groups on the three Yelp datasets.

5.1   Recall and Precision for Singleton Spammer Detection

We evaluated the average precision and recall of the singleton spammers (a majority among all the spammers in all of the three datasets) detected by each approach as a measure of its effectiveness. The results are given in Table 3 where the maximum value of achieved average precision/recall is shown in bold.

Table 3. Average precision and recall (%) of singleton spammers detected by each approach

		SSGD	FIM [9]	NFS - Group-Strainer [20]	SpEagle [13]	Ott [11]	DSR [14]
YelpChi	Precision	**23.50**	7.23	18.24	20.14	16.20	1.03
	Recall	**87.57**	22.21	63.60	50.00	49.80	10.20
YelpNYC	Precision	**21.15**	5.21	16.87	18.28	16.24	5.12
	Recall	**74.25**	21.17	60.42	55.89	72.81	28.15
YelpZip	Precision	**24.39**	6.25	18.69	17.57	22.63	5.08
	Recall	**88.79**	25.78	66.15	59.37	71.61	13.24

SSGD achieves the highest average precision and recall compared to other approaches across all datasets. The results in Table 3 show that the groups detected by SSGD contain more ground-truth singleton spammers. Spammer group detection approaches FIM and NFS-GroupStrainer obtain poor precision and recall as they cannot capture singleton spammers effectively. SpEagle, like SSGD, which is also based on the review-product network and review and product metadata, is not that effective in detecting singleton spammers as indicated by its low average precision/recall scores. A possible reason may be that SpEagle infers the spam probability for reviewers based on the possible targeted restaurants, rather than on the spam attacks as in SSGD, due to which many genuine reviews are also labeled as spammers resulting in higher false negatives and hence lower value of recall. The text-based approaches: Ott [11] and DSR [14] show very different performance for detecting spammers. The Ott [11] approach shows much better performance than DSR, which confirms that supervised learning based on psycholinguistic features from review contents is effective for detecting spam reviews. Still, the Ott approach does not perform as well as SSGD in terms of recall or precision for detecting singleton spammers. This indicates that the collusiveness among singleton reviewers during spam attacks embed strong signals complementary to the linguistic features for detecting singleton spammers.

Table 4. Singleton reviewers in spam attacks

	#products	#reviewers	#Singleton reviewers (%)
YelpChi	48	5,026	93.12%
YelpNYC	45	18,243	91.84%
YelpZIP	48	26,926	92.57%

We next investigate the fraction of singleton spammers in the spammer groups detected by SSGD. Table 4 lists number of restaurants (products), number of unique reviewers and percentage of singleton reviewers in the top 50 spam attacks detected by SSGD. Comparing the percentage of singleton reviewers in Tables 1 and 4, it clearly shows the sharp increase in the number of singleton reviewers during spam attacks. This result reaffirms that spammers tend to write singleton reviews from multiple IDs (sockpuppets) to avoid being caught. This also shows the effectiveness of the first step of SSGD for identifying spam attacks. Our approach to identifying candidate spam activities via examining the temporal dynamics of multiple signals at the review level as well as the review meta-data is effective for detecting spam activities from singleton reviewers.

5.2   Qualitative Analysis of Detected Spammer Groups

The spammer groups detected by SSGD (under default settings of DBSCAN of reachability distance parameter $\epsilon = 0.01$ and a minimum of 3 reviewers in a cluster) mostly consist of singleton reviewers who either gave all high (4–5) or all low (1–2) star ratings and wrote nearly identical reviews for a set of restaurants within a short time duration. Table 5 lists statistics for the top 5 groups detected by SSGD on the Yelp datasets. Among the top 5 spammer groups, the group consists of 20–90 reviewers (many are singletons) targeting 4–9 products. The timestamps and rating distribution of most groups are concentrated for maximum impact. The group spam targeted restaurants exhibit some common characteristics such as of the same cuisine, located in the same locality, etc.

6   Conclusions

Opinion spam is a prevalent problem hampering the credibility of online reviews. Existing methods often focus on reviewers who have written multiple reviews and spammer are detected by their abnormal behaviors. However, majority of

Table 5. Summary of the statistics of the top 5 SSGD detected groups in the Yelp datasets (#P: number of products, #U: number of users, Time & Rating distribution (s: scattered, c: concentrated))

	ID	#P	#U	Time	Rating	(near) Duplicate	Restaurant Description
YelpChi	1	4	24	s	c	10/30	#2 (same cuisine)
	2	5	25	c	c	12/29	In same area
	3	4	21	c	s	7/25	Hot dog
	4	6	55	s	c	32/64	Two attacks 2 month apart
	5	5	40	c	c	21/48	#3 (same cuisine)
YelpNYC	1	6	48	c	c	28/53	5 in same area
	2	5	39	s	c	22/43	#3 (same cuisine)
	3	6	37	s	c	28/47	breakfast
	4	7	52	c	c	45/59	-
	5	8	60	s	s	38/71	3 in same area
YelpZIP	1	8	63	c	s	42/74	4 in same area
	2	7	54	c	c	36/67	Fine dining
	3	9	87	s	s	47/109	Pizza
	4	7	41	c	c	13/52	#4 (same cuisine)
	5	6	42	c	c	28/49	-

reviewers are singleton reviewers, and are often overlooked by existing opinion spam detection approaches. In this paper we proposed a novel approach to detecting spammer groups consisting of singleton reviewers. Our approach comprises several strategies to address the challenge of scarcity of explicit signals for singleton reviewers. Especially we focus on identifying the collusiveness among singleton reviewers via detecting coordinated spam attacks. We experimented on three real-life Yelp datasets to evaluate our approach. Our results showed that the problem of singleton spam is widespread – many online review sites have mostly singleton spammers, and many group spam attacks involve singleton reviewers. Experiments show that our approach can more accurately capture singleton spammers than existing approaches and can detect spammer groups of singleton spammers overlooked by existing approaches. For future work, we will investigate approaches that make use of more hidden signals for accurate singleton spam detection.

Acknowledgments. This work was supported by the Australian Research Council (ARC) linkage project grant LP120200128.

References

1. Akoglu, L., Chandy, R., Faloutsos, C.: Opinion fraud detection in online reviews by network effects. In: ICWSM 2013 (2013)
2. Brugere, I., Gallagher, B., Berger-Wolf, T.Y.: Network structure inference, a survey: motivations, methods, and applications. arXiv preprint arXiv:1610.00782 (2016)

3. Ester, M., Kriegel, H.P., Sander, J., Xu, X.: A density-based algorithm for discovering clusters a density-based algorithm for discovering clusters in large spatial databases with noise. In: KDD 1996 (1996)
4. Heydari, A., Tavakoli, M., Salim, N., Heydari, Z.: Detection of review spam: a survey. Expert Syst. Appl. **42**(7), 3634–3642 (2015)
5. Jindal, N., Liu, B.: Opinion spam and analysis. In: WSDM 2008 (2008)
6. Li, H., Fei, G., Wang, S., Liu, B., Shao, W., Mukherjee, A., Shao, J.: Bimodal distribution and co-bursting in review spam detection. In: WWW 2017 (2017)
7. Lim, E.P., Nguyen, V.A., Jindal, N., Liu, B., Lauw, H.W.: Detecting product review spammers using rating behaviors. In: CIKM 2010 (2010)
8. Luca, M., Zervas, G.: Fake it till you make it: reputation, competition, and yelp review fraud. Manag. Sci. **62**(12), 3412–3427 (2016)
9. Mukherjee, A., Liu, B., Glance, N.: Spotting fake reviewer groups in consumer reviews. In: WWW 2012 (2012)
10. Natarajan, N., Dhillon, I.S.: Inductive matrix completion for predicting gene-disease associations. Bioinformatics **30**(12), i60–i68 (2014)
11. Ott, M., Choi, Y., Cardie, C., Hancock, J.T.: Finding deceptive opinion spam by any stretch of the imagination. In: ACL 2011 (2011)
12. Rajaraman, A., Ullman, J.D.: Mining of Massive Datasets. Cambridge University Press, Cambridge (2011)
13. Rayana, S., Akoglu, L.: Collective opinion spam detection: Bridging review networks and metadata. In: KDD 2015 (2015)
14. Sandulescu, V., Ester, M.: Detecting singleton review spammers using semantic similarity. In: WWW 2015 Companion (2015)
15. Savage, D., Zhang, X., Yu, X., Chou, P., Wang, Q.: Detection of opinion spam based on anomalous rating deviation. Expert Syst. Appl. **42**(22), 8650–8657 (2015)
16. Wu, L., Hu, X., Morstatter, F., Liu, H.: Adaptive spammer detection with sparse group modeling. In: ICWSM, pp. 319–326 (2017)
17. Xie, S., Wang, G., Lin, S., Yu, P.S.: Review spam detection via time series pattern discovery. In: WWW 2012 Companion (2012)
18. Xu, C., Zhang, J.: Combating product review spam campaigns via multiple heterogeneous pairwise features. In: SIAM International Conference on Data Mining, pp. 172–180 (2015)
19. Xu, C., Zhang, J., Chang, K., Long, C.: Uncovering collusive spammers in Chinese review websites. In: CIKM 2013 (2013)
20. Ye, J., Akoglu, L.: Discovering opinion spammer groups by network footprints. In: ECML PKDD 2015 (2015)
21. Ye, J., Kumar, S., Akoglu, L.: Temporal opinion spam detection by multivariate indicative signals. In: WSDM 2016 (2016)
22. Yu, H.F., Jain, P., Kar, P., Dhillon, I.S.: Large-scale multi-label learning with missing labels. In: ICML 2014 (2014)

Adaptive Attention Network for Review Sentiment Classification

Chuantao Zong[1], Wenfeng Feng[1], Vincent W. Zheng[2],
and Hankz Hankui Zhuo[1(✉)]

[1] School of Data and Computer Science, Sun Yat-Sen University, Guangzhou, China
{zongcht,fengwf}@mail2.sysu.edu.cn, zhuohank@mail.sysu.edu.cn
[2] Advanced Digital Sciences Center (ADSC), Singapore, Singapore
vincent.zheng@adsc.com.sg

Abstract. Document-level sentiment classification is an important NLP task. The state of the art shows that attention mechanism is particularly effective on document-level sentiment classification. Despite the success of previous attention mechanism, it neglects the correlations among inputs (*e.g.*, words in a sentence), which can be useful for improving the classification result. In this paper, we propose a novel Adaptive Attention Network (AAN) to explicitly model the correlations among inputs. Our AAN has a two-layer attention hierarchy. It first learns an attention score for each input. Given each input's embedding and attention score, it then computes a weighted sum over all the words' embeddings. This weighted sum is seen as a "context" embedding, aggregating all the inputs. Finally, to model the correlations among inputs, it computes another attention score for each input, based on the input embedding and the context embedding. These new attention scores are our final output of AAN. In document-level sentiment classification, we apply AAN to model words in a sentence and sentences in a review. We evaluate AAN on three public data sets, and show that it outperforms state-of-the-art baselines.

1 Introduction

Sentiment classification [12] is an important task in NLP. Document-level sentiment classification attracts a lot of research interests. In general, review sentiment classification task is modeled as either a binary (*i.e.*, "positive" or "negative"), or multi-class classification (*e.g.*, ratings from "one star" to "five stars") problem.

Earlier work on sentiment classification relies on engineering useful features from the data to build the classification models. Some may also consider user features and product features [2]. With the development of neural networks, recent study starts to explore using automatic feature learning for the review sentiment classification. For example, the state of the art uses various hierarchical neural networks to model the words and the sentences in a review [1,17]. In particular, Chen *et al.* propose to model the representation of a review through

© Springer International Publishing AG, part of Springer Nature 2018
D. Phung et al. (Eds.): PAKDD 2018, LNAI 10937, pp. 668–680, 2018.
https://doi.org/10.1007/978-3-319-93034-3_53

a word-sentence-review hierarchy. To differentiate the importance of each word and each sentence in generating the whole review's representation, they introduce the *attention* mechanism [21] to model a weight for each word. Then they sum up the word embeddings in a sentence with the resulting weights as the sentence's embedding. Similarly, they apply attention for a weighted sum of the sentences as the review's representation.

Although the attention mechanism has shown to significantly improve the classification results [1], we notice that it assumes each input (word or sentence) as independent. Thus it overlooks the input correlations, which can be useful for the classification. Take Fig. 1(a) as an example. The review consists of one sentence. It shows the *word attention scores* for some salient words in a review. As we can see, "high" has a relatively large word attention score, indicating its high importance in representing the whole review. However, as "high" in the shop reviews is often associated with price, it usually holds a negative polarity. Therefore, given a large attention score for "high", we tend to assign a negative polarity to the whole review. This results in a conflict between the predicted rating of "two stars", and the ground truth rating of "four stars". The fundamental reason of having such a conflict is that the attention mechanism treats each word's attention independently. This overlooks the *context* in this specific review, *i.e.*, what this review is mainly about and what the leading polarity is. From a human's perspective, we can easily tell that this review is mainly about an endorsement of this dessert shop's yogurt, and the leading polarity is highly positive (*i.e.*, "happy", "amazing") despite the "high" price.

NSC+LA (predicted rating: 2; gold rating: 4): always **happy** (0.074) here great yogurt and toppings brownies are **amazing** (0.076), the price is kinda **high** (0.193) or I would be here more often.	AAN (no U, no P) (predicted rating: 4; gold rating: 4): always **happy** (0.109) here great yogurt and toppings brownies are **amazing** (0.123), the price is kinda **high** (0.112) or I would be here more often.
(a) Results from NSC+LA (state of the art) [1]	(b) Results from AAN (ours)

Fig. 1. Salient words (in boldface) with highest *word attention scores* in a sample review.

Is it possible for us to take the context of a specific review into account, and *adaptively* assign the attention for each salient word (*e.g.*, discount the weight of "high")? Our answer is yes! In this paper, we propose a novel *Adaptive Attention Network* (AAN) to explicitly model the correlation between the inputs (*e.g.*, words and sentences in the review domain) in the attention definition. Our AAN has a deeper two-layer attention hierarchy. Take the word attention in a sentence as an example. AAN first computes an attention score for each word in the sentence, by employing the outstanding attention mechanism of Chen *et al.* [1]. Then it introduces a *context embedding*, which aggregates all the words' embedding vectors with the attention scores by a weighted sum. To model the correlation, it computes another attention score for each word, based on how much the word matches the context embedding. These new attention scores are the final outputs of AAN for the words in a sentence. Similarly, we also apply AAN to model the sentence attention in the review. By utilizing the input correlations among words and sentences, AAN is able to improve the review comprehension and thus the classification results. As shown in Fig. 1(b), under our AAN mechanism, the word attention scores of salient words "happy" and "amazing" all significantly increase, whereas that of "high" decreases, which eventually helps us generate a perfect rating prediction of "four stars". Note that in this sample review we only have one sentence, thus the sentence attention score from AAN is one.

We summarize our contributions as follows.

- We identify an important limitation of the existing attention mechanism, and develop an adaptive attention network to model the input correlations in attention modeling.
- We evaluate AAN with three public, real-world review sentiment datasets. We show that AAN outperforms state-of-the-art baselines.

2   Related Work

Sentiment classification is usually seen as a special case of text classification. As the performance of text classifiers heavily relies on the extracted features, early work on sentiment classification mostly focuses on designing useful features from text content [12], sentiment lexicons [4], social network [2] and so on. With the development of neural networks, some recent studies start to explore the application of deep learning in sentiment classification to avoid engineering the features. For example, to model the text's syntactic structure, Socher et al. explored a set of recursive neural networks models such as Recursive Auto-Encoder [13] and Recursive Neural Tensor Networks [14]. To leverage the dependency parsing information, Tai et al. proposed a tree-structured Long Short-Term Memory (LSTM) [16] in learning the representation of a document. To model the n-gram patterns, Lai et al. [8] and Kalchbrenner et al. [6] both explored using Convolutional Neural Networks (CNN) over the words in a sentence. To model the word-sentence-document hierarchy, Tang et al. proposed to first use CNN over the words to embed each sentence, then aggregate all the sentences by either simple pooling or Gated Recurrent Neural Network to embed the whole document [18]. Compared with our AAN model, these above neural network methods do not study the attention mechanism.

To incorporate the different importances of the words in each sentence, as well as the sentences in each document, attention mechanism [21] was introduced into text representation learning. For example, Yang et al. [22] proposed a hierarchical attention mechanism, which leverages the local semantic information in both word and sentence levels. In document-level sentiment classification, Chen et al. further extended the hierarchical attention mechanism to incorporate the user and product information with the attention design for words and sentences [1]. Tang et al. explored modeling attentions for different types of signals, including text content and text location [19]. Compared with AAN, these above attention methods often assume the inputs as independent. As a result, their attention definitions only have one single layer, which is from the inputs directly to the attention score output. Unlike these works, our AAN exploits the correlations among the inputs (to our knowledge this is the first work).

3 Adaptive Attention Network

We first formulate the problem of review sentiment classification. As inputs, we have a set of training documents $\mathcal{D} = \{(d_1, y_1), ..., (d_n, y_n)\}$, where each d_i is a document and $y_i \in \mathcal{Y}$ is the sentiment class (e.g., $\mathcal{Y} = \{1, ..., 5\}$ indicating the ratings from "one star" to "five stars"). As output, we want to build a model $\mathcal{M}$, which can take a test document d as inputs and predict a rating class in $\mathcal{Y}$. Inspired by the pioneer work [1,17], we choose to model each document as a sequence of sentences and each sentence as a sequence of words. Formally, we denote a document as $d_i = \{s_{i,1}, ..., s_{i,m_i}\}$, where each $s_{i,j}$ is a sentence and m_i is the number of sentences in d_i. We denote each sentence $s_{i,j} = \{w_{i,j,1}, ..., w_{i,j,m'_{ij}}\}$, where $w_{i,j,k} \in \mathcal{V}$ is a word from the vocabulary $\mathcal{V}$ and m'_{ij} is the number of words in $s_{i,j}$.

3.1 Two-Layer AAN Architecture

Next we develop the AAN model. We begin with reviewing existing attention mechanism in [1]. As shown in Fig. 2(a), the existing attention mechanism generally takes a set of vectors $\{\mathbf{h}_1, ..., \mathbf{h}_m\}$ as inputs, and tries to compute an attention score α_i for each vector $\mathbf{h}_i \in \mathbb{R}^{K_1}$ by

$$f_i = \mathbf{v}^\top \tanh(W\mathbf{h}_i + \mathbf{b}), \tag{1}$$

$$\alpha_i = \frac{\exp(f_i)}{\sum_{j=1}^m \exp(f_j)}, \tag{2}$$

where $\mathbf{v} \in \mathbb{R}^{K_1}$, $W \in \mathbb{R}^{K_1 \times K_1}$ and $\mathbf{b} \in \mathbb{R}^{K_1}$ are learnable parameters. Based on the attention scores, the output $\mathbf{z} \in \mathbb{R}^{K_1}$ is

$$\mathbf{z} = \sum_{i=1}^m \alpha_i \mathbf{h}_i. \tag{3}$$

As we can see, the above attention definition treats each input $\mathbf{h}_i$ independently. In practice, the correlation between the inputs can be useful. For example, in Fig. 1(a), we can see the necessity to consider the correlation between the salient word "high" with the other words, so as to ensure its attention score to be fully aware of the context in this specific review.

In order to take the input correlation into account, we develop a two-layer attention hierarchy as shown in Fig. 2(b). Let us use the example in Fig. 1(a) again to illustrate how we design such a hierarchy. For word attention, we denote each input $\mathbf{h}_i$ as an embedding for the i-th word in a sentence. To assign an

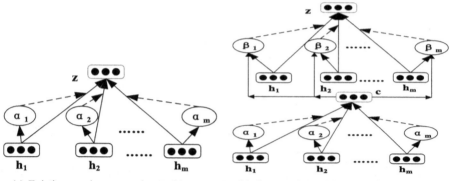

(a) Existing one-layer attention architecture   (b) Our two-layer AAN attention architecture

Fig. 2. Comparing our two-layer AAN with the existing one-layer attention architecture.

appropriate attention score to the salient word "high", we first need to understand its context in the sentence. We represent such a sentence context by a *context embedding*, and we compute it by Eq. 3 as

$$\mathbf{c} = \sum_{i=1}^{m} \alpha_i \mathbf{h}_i, \tag{4}$$

where α_i is estimated by Eq. 2 and $\mathbf{c} \in \mathbb{R}^{K_1}$. This context embedding estimation corresponds to our first layer of AAN in Fig. 2(b), where the α_i's are the first-layer attention scores. Ideally, we want to prompt those salient words that "match" the context. Therefore, we introduce a second layer of attention in Fig. 2(b), which measures how well $\mathbf{h}_i$ matches the context embedding $\mathbf{c}$ by a score b_i and then outputs a final attention score β_i:

$$b_i = \mathbf{h}_i^\top \mathbf{c}, \tag{5}$$

$$\beta_i = \frac{\exp(b_i)}{\sum_{j=1}^{m} \exp(b_j)}. \tag{6}$$

The β_i's are the second-layer, and the final, attention scores for AAN. Once having the final attention scores of AAN, we compute the representation of the whole sentence as

$$\mathbf{z} = \sum_{i=1}^{m} \beta_i \mathbf{h}_i. \tag{7}$$

Remark: to help understand why mathematically Eq. 7 models the correlation among the inputs, we can do some simple expansion:

$$\mathbf{z} \overset{1}{=} \sum_{i=1}^{m} \frac{\exp(b_i)}{\sum_{j=1}^{m} \exp(b_j)} \mathbf{h}_i,$$

$$\overset{2}{=} \sum_{i=1}^{m} \frac{\exp(\mathbf{h}_i^\top \mathbf{c})}{\sum_{j=1}^{m} \exp(\mathbf{h}_j^\top \mathbf{c})} \mathbf{h}_i,$$

$$\overset{3}{=} \sum_{i=1}^{m} \frac{\exp\left(\sum_{k=1}^{m} \alpha_k \mathbf{h}_i^\top \mathbf{h}_k\right)}{\sum_{j=1}^{m} \exp\left(\sum_{k=1}^{m} \alpha_k \mathbf{h}_j^\top \mathbf{h}_k\right)} \mathbf{h}_i, \tag{8}$$

where at step 1, we plug in Eq. 6; at step 2, we plug in Eq. 5; at step 3, we plug in Eq. 4. As we can see in Eq. 8, the attention score for each input $\mathbf{h}_i$ now becomes aware of the correlation between $\mathbf{h}_i$ and the other $\mathbf{h}_k$'s.

3.2   AAN for Review Modeling

We customize AAN for document-level sentiment classification. As suggested by [1,17], we model each review as a hierarchy from words to sentences and finally a document. Therefore, we can assign attention to both the words in the sentence level and the sentences in the document level. Next, we illustrate how to take a review's content, as well as its user (who publishes this review) and product (which this review is about), as inputs, and finally predict a sentiment class as output. We summarize our deep neural network architecture of using AAN for document-level sentiment classification in Fig. 3. The architecture consists of three parts, as we shall introduce one by one next.

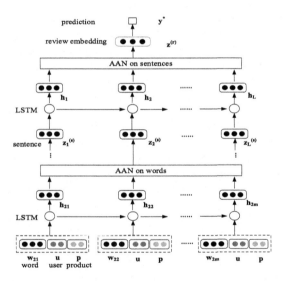

Fig. 3. Using AAN for both words and sentences in document-level sentiment classification.

• **Embedding from word to sentence.** For each sentence, we aim to generate a sentence embedding vector, from its words. First of all, for each word $w_{j,k}$ ($k = 1, ..., m_j$) in sentence s_j, we assign an embedding vector $\mathbf{w}_{j,k} \in \mathbb{R}^{K_0}$. We pre-train these word embedding vectors by word2vec [11]. Secondly, in order to incorporate the user and product information, we also introduce an embedding vector

$\mathbf{u} \in \mathbb{R}^{K_3}$ for each user u and an embedding vector[1] $\mathbf{p} \in \mathbb{R}^{K_3}$ for each product p. We choose to concatenate each word embedding $\mathbf{w}_{j,k}$ with the user embedding $\mathbf{u}$ and the product embedding $\mathbf{p}$ as the inputs for sentence embedding. It is worth noting that, such a concatenation design is significantly different from the previous designs. For example, in [17], each user (product) is represented with a matrix, which is multiplied with each $\mathbf{w}_{j,k}$ to get the input for sentence embedding. This matrix representation is likely to suffer from the data insufficiency for those users (products) with limited reviews. In [1], neither user nor product is used as input for sentence embedding, thus missing the opportunity of enriching the sentence semantics with user and product information. Thirdly, given the embedding concatenation for each word in the sentence, we employ an LSTM [15] to generate a hidden output $\mathbf{h}_{j,k} \in \mathbb{R}^{K_1}$ for each word.

Finally, we apply AAN to assign an attention score to each word. To incorporate the user and product information, we start with extending the first layer attention score definition as

$$f_{j,k}^{(w)} = \mathbf{v}^\top \tanh(W_1^{(w)} \mathbf{h}_{j,k} + \mathbf{b}_1) + g_{j,k}^{(w)},$$
$$g_{j,k}^{(w)} = \mathbf{h}_{j,k} W_1^{(u)} \mathbf{u} + \mathbf{h}_{j,k} W_1^{(p)} \mathbf{p},$$
$$\alpha_{j,k} = \frac{\exp(f_{j,k}^{(w)})}{\sum_{l=1}^{m} \exp(f_{j,l}^{(w)})}, \tag{9}$$

where $g_{j,k}^{(w)}$ is an extra term we introduce to indicate the interactions between word and user, as well as word and product. $W_1^{(w)} \in \mathbb{R}^{K_1 \times K_1}$, $W_1^{(u)} \in \mathbb{R}^{K_1 \times K_3}$, $W_1^{(p)} \in \mathbb{R}^{K_1 \times K_3}$ and $\mathbf{b}_1 \in \mathbb{R}^{K_1}$ are parameters. Then, we compute the second layer attention score $\beta_{j,k}$ for each word in the same way as Eq. 6. In the end, we aggregate all the words with their embedding vectors and the second-layer attentions to output a sentence embedding $\mathbf{z}_j^{(s)} \in \mathbb{R}^{K_1}$ by Eq. 7.

• **Embedding from sentence to review.** Given the embedding of each sentence, we take a similar procedure as embedding from word to sentence. Specifically, we take all the sentence embeddings as input, and employ an LSTM to generate a hidden output $\mathbf{h}_j \in \mathbb{R}^{K_2}$ for each sentence s_j (for $j = 1, ..., L$). Then, we apply AAN to assign an attention score to each sentence embedding vector. We use the similar extension in Eq. 9 to the first layer attention score definition for sentence. Denote $W_2^{(s)} \in \mathbb{R}^{K_2 \times K_2}$, $W_2^{(u)} \in \mathbb{R}^{K_2 \times K_3}$, $W_2^{(p)} \in \mathbb{R}^{K_2 \times K_3}$ and $\mathbf{b}_2 \in \mathbb{R}^{K_2}$ as parameters. Thus we compute the first-layer attention score for each sentence as

$$f_j^{(s)} = \mathbf{v}^\top \tanh(W_2^{(s)} \mathbf{h}_j + \mathbf{b}_2) + g_j^{(s)},$$
$$g_j^{(s)} = \mathbf{h}_j W_2^{(u)} \mathbf{u} + \mathbf{h}_j W_2^{(p)} \mathbf{p}.$$
$$\alpha_j = \frac{\exp(f_j^{(s)})}{\sum_{l=1}^{L} \exp(f_l^{(s)})}. \tag{10}$$

---

[1] Generally, user and product can have different dimensions, but we set them as the same to control the number of hyperparameters.

Then, we compute the second-layer attention score β_j for each sentence in the same way as Eq. 6. Finally, we aggregate all the sentences with their embedding vectors and the second-layer attentions in Eq. 7 to output a review embedding $\mathbf{z}_j^{(r)} \in \mathbb{R}^{K_2}$.

• **Sentiment class prediction**. Given the review embedding, we aim to generate a prediction of the review's sentiment class. In our design, we feed the review embedding $\mathbf{z}_j^{(r)}$ into a multi-layer perceptron (MLP) with one hidden layer, which outputs the probabilities of this review belonging to each class in $\mathcal{Y}$:

$$\mathbf{y}^* = \text{softmax}(W'\mathbf{z}_j^{(r)} + \mathbf{b}'), \tag{11}$$

where $W' \in \mathbb{R}^{|\mathcal{Y}| \times K_2}$ are $b_c \in \mathbb{R}^{\mathcal{Y}}$ are parameters. We can link $\mathbf{y}^*$ with the ground truth label of this review, so as to supervise the model training. Denote Θ as the set of parameters, including the AAN parameters on words and on sentences, the LSTM parameters on words and on sentences, the sentiment class prediction MLP parameters. For the training data set $\mathcal{D}$, we design the objective function as

$$\mathcal{L} = -\sum_{i=1}^n \log P(y_i|\mathcal{D}) + \lambda\Omega(\Theta), \tag{12}$$

where $P(y_i|\mathcal{D})$ is the probability of predicting review d_i as class y_i, and it can be computed by Eq. 11. $\Omega(\cdot)$ is a regularization function, *e.g.*, it sums up the ℓ_2-norm of each parameter in Θ. $\lambda > 0$ is a trade-off parameter.

4   Experiments

We evaluate AAN on three public benchmark data sets, including IMDB, Yelp 2013 and Yelp 2014 which are review texts including user/product information developed by [17]. Each record in the data sets is composed of a user ID, a product ID, a review and a rating. Table 1 list the statistics of the datasets.

Table 1. Statistics of three public data sets. doc/user denotes the average number of documents per user, sen/doc denotes the number of sentences (average per document).

Data	User	Product	Class	Doc	Doc/User	Doc/Product	Sen/Doc	Word/Sen	Voc
IMDB	1,310	1,635	10	84,919	64.82	51.94	16.08	24.54	105,373
Yelp 2013	1,631	1,633	5	78,966	48.42	48.36	10.89	17.38	48957
Yelp 2014	4,818	4,194	5	231,163	47.97	55.11	11.41	17.26	93197

We follow [17] to employ two evaluation metrics: (1) accuracy $Acc = \frac{T}{N}$, where T is the number of ratings predicted correctly and N is the size of the testing set; (2) root mean square error $RMSE = \sqrt{\frac{\sum_i (gd_i - pr_i)^2}{N}}$, where gd_i and pr_i are the gold rating and the predicted rating for document i, respectively.

We learn the word embedding by word2vec [10] and set the embedding dimension as $K_0 = 200$. We set the dimensions of hidden states and cell memory states in LSTM, sentence embedding and review embedding, user and product embeddings as $K_1 = 100$, $K_2 = 100$, $K_3 = 50$ respectively, which fit well to our GPU memory. We organize the reviews into batches for training. For varying length of sentences and reviews in each batch, we do zero padding in using LSTM. We set the batch size as 32. We set the regularization weight as $\lambda = 1E-5$. We use the data splits provided by [17], which separate each data set into training, development and testing sets with a 80/10/10 split. We use adadelta [23] for stochastic gradient descent.

4.1 Comparison with Baselines

We compare AAN with the state-of-the-art baselines, as listed below. **Majority:** it assigns the majority sentiment category in the training set to each test review. **Trigram:** it uses trigrams as features to train a Support Vector Machine (SVM) [12] for review classification. **TextFeature:** it extracts several text features, such as word and character n-grams, sentiment lexicons to train a SVM [7]. **UPF:** it was introduced by [17]. It extracts user and product features like [5], and concatenates them with the features in **Trigram** and **TextFeature** for SVM. **AvgWordvec:** it learns 200-dimensional word embeddings by word2vec [11] and uses the average word embeddings of each review for SVM training. **SSWE:** it learns sentiment-specific word embeddings and thus the review embedding for SVM classification [20]. **RNTN + Recurrent:** its learns an RNTN [14] for sentence embedding, and a recurrent neural network (RNN) for review embedding. **Paragraph Vector:** it uses the paragraph structure to learn the embedding for varying-length sentences and documents [9]. **JMARS:** it combines user and review aspects by collaborative filtering and topic modeling for review sentiment classification [3]. **UPNN:** it takes user-text and product-text consistency matrices as additional inputs, to assist the embedding for words, sentences and reviews [17]. **NSC & NSC + LA & NSC + UPA:** these three neural network models all explore the word-sentence-review hierarchy [1]. NSC uses a mean pooling in sentence and review embedding. NSC + LA improves NSC with a local semantic attention (LA) [22]. NSC + UPA improves NSC by considering user and product in the attention definitions.

Table 2 shows the performance. We test all the methods in two settings: (1) with user and product information, denoted by "with U and P"; (2) without them, denoted by "no U, no P". Note that, because we use exactly the same data set splits and experimental settings with [17], we can directly borrow some of their results in the "no U, no P" setting, including "Majority", "Trigram", "TextFeature", "AvgWordvec + SVM", "SSWE + SVM", "Paragraph Vector", "RNTN + Recurrent" and "UPNN (no U, no P)".

In the setting of "no U, no P", AAN outperforms AvgWordvec + SVM, SSWE + SVM, Paragraph Vector and RNTN + Recurrent, which shows the necessity to differentiate the words and sentences in the review representations for sentiment classification. Besides, AAN outperforms both NSC and NSC + LA, which justifies our motivation to capture correlations among words or

Table 2. Results of all the approaches on IMDB, Yelp2013 and Yelp2014 datasets. *Acc* (the higher, the better) and *RMSE* (the lower, the better) are two evaluation criteria. The best performances in each group are in boldface.

Settings	Models	IMDB		Yelp13		Yelp14	
		Acc	RMSE	Acc	RMSE	Acc	RMSE
no U, no P	Majority	0.196	2.495	0.411	1.060	0.392	1.097
	Trigram [12]	0.399	1.783	0.569	0.814	0.577	0.804
	TextFeature [12]	0.402	1.793	0.556	0.845	0.572	0.800
	AvgWordvec + SVM [11]	0.304	1.985	0.526	0.898	0.530	0.893
	SSWE + SVM [20]	0.312	1.973	0.549	0.849	0.557	0.851
	Paragraph Vector [9]	0.341	1.814	0.554	0.832	0.564	0.802
	RNTN + Recurrent [14]	0.400	1.764	0.574	0.804	0.582	0.821
	UPNN (no U, no P) [17]	0.405	1.629	0.577	0.812	0.585	0.808
	NSC [1]	0.438	1.495	0.628	0.703	0.635	0.687
	NSC + LA [1]	0.474	1.391	0.631	0.708	0.641	0.683
	AAN (no U, no P)	**0.483**	**1.385**	**0.636**	**0.694**	**0.643**	**0.681**
with U and P	Trigram + UPF [17]	0.404	1.764	0.570	0.803	0.576	0.789
	TextFeature + UPF [17]	0.402	1.774	0.561	1.822	0.579	0.791
	JMARS [3]	N/A	1.773	N/A	0.985	N/A	0.999
	UPNN (U + P) [17]	0.435	1.602	0.596	0.784	0.608	0.764
	NSC + UPA [1]	0.513	1.299	0.645	0.689	0.666	0.655
	AAN (U + P)	**0.538**	**1.243**	**0.662**	**0.663**	**0.670**	**0.646**

sentences, since NSC and NSC + LA both only model the words and sentences individually in the attention estimation.

In the setting of "with U and P", AAN still outperforms all the baselines in all of the three datasets. This confirms the superiority of our AAN model in exploring the attention mechanism, as well as the user and product information. It is worth noting that the improvement of our AAN (U + P) model over NSC + UPA is larger than those of AAN (no U, no P) over UPNN (U + P) and NSC + LA. This means, in addition to the benefit we obtained from our novel two-layer attention mechanism, our new design of incorporating the user and product information (*i.e.*, concatenating each word embedding with the user and product embeddings) can indeed help improve the classification results. Such an observation justifies our conjecture of how to model the user and product information in Sect. 3.2.

4.2 Impact of User and Product Embeddings

We further study the impacts of using user and product embeddings in our AAN model. We perform AAN in four different settings, depending on whether we use the user or product embedding or not. We summarize the results in Table 3.

As we can see, AAN (P + U) outperforms the other three models, which suggests both user and product information can help improve the sentiment classification results. This is consistent with our intuition that, the more information we exploit, the better result we will generally have. We also observe that, AAN (only U) outperforms AAN (only P), which implies that user preferences seems to be more important than the product properties in sentiment classification. We may understand it as that the reviews are subjective, thus the user preferences play a more important role in sentiment classification. Finally, we also see that both AAN (only U) and AAN (only P) outperform AAN (no U, no P), which suggests both user and product are useful for sentiment classification.

Table 3. Impacts of using user and product information in AAN.

Models	IMDB		Yelp13		Yelp14	
	Acc	RMSE	Acc	RMSE	Acc	RMSE
AAN (P + U)	**0.538**	**1.243**	**0.662**	**0.663**	**0.670**	**0.646**
AAN (only U)	0.527	1.264	0.653	0.674	0.666	0.652
AAN (only P)	0.485	1.373	0.632	0.695	0.641	0.670
AAN (no U, no P)	0.483	1.385	0.636	0.694	0.643	0.681

4.3   Impact of Adaptive Attention Mechanism

We also study the impacts of using AAN in both word and sentence levels. In Table 4, "AA" indicates AAN, and "MP" indicates simple mean pooling. From Table 4, we can see that employing our attention mechanism in both word and sentence levels (*i.e.*, "AA + AA" in the third row of Table 4) outperforms all the other settings. This means: (1) it is important to use adaptive attention in both the word and sentence level; (2) adaptive attention is more effective than the simple mean pooling, since it tries to differentiate the importances of different words and sentences. We also make another interesting observation, by comparing "AA + MP" and "MP + AA". The results show that, "AA + MP" seems to slightly outperform "MP + AA"; *i.e.*, using AAN in the word level seems to be better than in the sentence level. The possible reason is that the number of words is much larger than that of sentences, thus the noise is bigger in

Table 4. Impacts of using AAN in both word and sentence levels.

Attention mechanisms		IMDB		Yelp13		Yelp14	
Word-level	Sentence-level	Acc	RMSE	Acc	RMSE	Acc	RMSE
AA	AA	**0.538**	**1.243**	**0.662**	**0.663**	**0.670**	**0.646**
AA	MP	0.519	1.276	0.650	0.682	0.664	0.650
MP	AA	0.513	1.281	0.648	0.690	0.665	0.656
MP	MP	0.496	1.339	0.643	0.685	0.661	0.659

the word level. Besides, since there are more words than sentences, we may have more data to better learn the attentions in the word level than in the sentence level.

5   Conclusion

In this paper, we identify that the existing attention mechanisms often suffer from a significant limitation, which assumes the inputs as independent. Therefore, we propose a novel Adaptive Attention Network to model the correlation among the words and the sentences in their attention definitions. We also customize AAN for document-level sentiment classification, especially incorporating the user and product information. We evaluate AAN on three public benchmark data sets and show that it outperforms the state-of-the-art baselines. In the future, we plan to extend AAN with syntactic structure of the text, such as dependency trees, so as to further improve the classification.

Acknowledgments. We thank the National Key Research and Development Program of China (2016YFB020 1900), National Natural Science Foundation of China (U1611262), Guangdong Natural Science Funds for Distinguished Young Scholar (2017A030306028), Pearl River Science and Technology New Star of Guangzhou, and Guangdong Province Key Laboratory of Big Data Analysis and Processing for the support of this research. Zheng thanks the support of the National Research Foundation, Prime Ministers Office, Singapore under its Campus for Research Excellence and Technological Enterprise (CREATE) programme.

References

1. Chen, H., Sun, M., Tu, C., Lin, Y., Liu, Z.: Neural sentiment classification with user and product attention. In: EMNLP, pp. 1650–1659 (2016)
2. Cheng, K., Li, J., Tang, J., Liu, H.: Unsupervised sentiment analysis with signed social networks. In: AAAI, pp. 3429–3435 (2017)
3. Diao, Q., Qiu, M., Wu, C.Y., Smola, A.J., Jiang, J., Wang, C.: Jointly modeling aspects, ratings and sentiments for movie recommendation (JMARS). In: SIGKDD, pp. 193 202 (2014)
4. Ding, X., Liu, B., Yu, P.S.: A holistic lexicon-based approach to opinion mining. In: WSDM, pp. 231–240 (2008)
5. Gao, W., Yoshinaga, N., Kaji, N., Kitsuregawa, M.: Modeling user leniency and product popularity for sentiment classification. In: IJCNLP, pp. 1107–1111 (2013)
6. Kalchbrenner, N., Grefenstette, E., Blunsom, P.: A convolutional neural network for modelling sentences. In: ACL, pp. 655–665 (2014)
7. Kiritchenko, S., Zhu, X., Mohammad, S.M.: Sentiment analysis of short informal texts. J. Artif. Intell. Res. **50**, 723–762 (2014)
8. Lai, S., Xu, L., Liu, K., Zhao, J.: Recurrent convolutional neural networks for text classification. In: AAAI, pp. 2267–2273 (2015)
9. Le, Q.V., Mikolov, T.: Distributed representations of sentences and documents. CoRR. arXiv:1405.4053 (2014)

10. Mikolov, T., Chen, K., Corrado, G., Dean, J.: Efficient estimation of word representations in vector space. CoRR. arXiv:1301.3781 (2013)
11. Mikolov, T., Sutskever, I., Chen, K., Corrado, G.S., Dean, J.: Distributed representations of words and phrases and their compositionality. In: NIPS, pp. 3111–3119 (2013)
12. Pang, B., Lee, L., Vaithyanathan, S.: Thumbs up?: sentiment classification using machine learning techniques. In: EMNLP, pp. 79–86 (2002)
13. Socher, R., Pennington, J., Huang, E.H., Ng, A.Y., Manning, C.D.: Semi-supervised recursive autoencoders for predicting sentiment distributions. In: EMNLP, pp. 151–161 (2011)
14. Socher, R., Perelygin, A., Wu, J.Y., Chuang, J., Manning, C.D., Ng, A.Y., Potts, C.: Recursive deep models for semantic compositionality over a sentiment treebank. In: EMNLP, p. 1642 (2013)
15. Sutskever, I., Vinyals, O., Le, Q.V.: Sequence to sequence learning with neural networks. In: NIPS, pp. 3104–3112 (2014)
16. Tai, K.S., Socher, R., Manning, C.D.: Improved semantic representations from tree-structured long short-term memory networks. In: ACL, pp. 1556–1566 (2015)
17. Tang, D., Qin, B., Liu, T.: Learning semantic representations of users and products for document level sentiment classification. In: ACL, pp. 1014–1023 (2015)
18. Tang, D., Qin, B., Liu, T.: Document modeling with gated recurrent neural network for sentiment classification. In: EMNLP, pp. 1422–1432 (2015)
19. Tang, D., Qin, B., Liu, T.: Aspect level sentiment classification with deep memory network. In: EMNLP, pp. 214–224 (2016)
20. Tang, D., Wei, F., Yang, N., Zhou, M., Liu, T., Qin, B.: Learning sentiment-specific word embedding for Twitter sentiment classification. In: ACL, pp. 1555–1565 (2014)
21. Xu, K., Ba, J., Kiros, R., Cho, K., Courville, A.C., Salakhutdinov, R., Zemel, R.S., Bengio, Y.: Show, attend and tell: neural image caption generation with visual attention. In: ICML, pp. 2048–2057 (2015)
22. Yang, Z., Yang, D., Dyer, C., He, X., Smola, A.J., Hovy, E.H.: Hierarchical attention networks for document classification. In: NAACL, pp. 1480–1489 (2016)
23. Zeiler, M.D.: ADADELTA: an adaptive learning rate method. CoRR. arXiv:1212.5701 (2012)

Cross-Domain Sentiment Classification via a Bifurcated-LSTM

Jinlong Ji$^{(\boxtimes)}$, Changqing Luo, Xuhui Chen, Lixing Yu, and Pan Li

Case Western Reserve University, Cleveland, OH, USA
{jxj405,cxl881,xxc296,lxy257,pxl288}@case.edu

Abstract. Sentiment classification plays a vital role in current online commercial transactions because it is critical to understand users' opinions and feedbacks in businesses or products. Cross-domain sentiment classification can adopt a well-trained classifier from one source domain to other target domains, which reduces the time and efforts of training new classifiers in these domains. Existing cross-domain sentiment classification methods require data or other information in target domains in order to train their models. However, collecting and processing new corpora require very heavy workload. Besides, the data in target domains may be private and not always available for training. To address these issues, motivated by multi-task learning, we design a Bifurcated-LSTM which takes advantages of attention-based LSTM classifiers along with augmented dataset and orthogonal constraints. This Bifurcated-LSTM can extract domain-invariant sentiment features from the source domain to perform sentiment analysis in different target domains. We conduct extensive experiments on seven classic types of product reviews, and results show that our system leads to significant performance improvement.

1 Introduction

Sentiment classification plays a vital role in current online commercial transactions because it is essential to understand users' opinions and feedbacks in businesses or products. It identifies the overall sentiment polarity (e.g., positive or negative) of a text. In 2002, Bo et al. [22] were the first pioneers to utilize machine learning techniques to tackle the sentiment classification problem. Since then, many researchers have shown their interests in this field [9,21]. Noticeably, most of them try to obtain sentiment classifiers by assuming there are sufficient training data in a specified domain. In practice, consumers are usually interested in a number of different types of product, and sentiment is expressed differently in various domains. When we apply previous sentiment classification techniques, large amounts of labeled data are required each time when we need to conduct sentiment analysis for a new product. To alleviate this issue, cross-domain sentiment classification [4], which utilizes labeled data from related domains, has attracted people's attention. It is to adapt a well-designed sentiment classifier,

© Springer International Publishing AG, part of Springer Nature 2018
D. Phung et al. (Eds.): PAKDD 2018, LNAI 10937, pp. 681–693, 2018.
https://doi.org/10.1007/978-3-319-93034-3_54

which is trained on the data in one domain, to classify the sentiment of data in other domains.

Although in the literature, several cross-domain sentiment classification schemes have been proposed [11,13], all of them need target domain data, which is not always available. Specifically, when a new domain emerges, it costs a lot of efforts to collect and process its data, especially for supervised methods where the labels have to be added manually. Besides, there may be sensitive information in the new domain data, such as reviews for beta version products, which cannot be leaked or made public.

To address these problems, we design a novel Bifurcated-LSTM for cross-domain sentiment classification. Particularly, we notice that there are two crucial points a user's review tries to convey: topic and sentiment. Topic, which is different from one domain to another, describes the product or service that the customer comments on. Sentiment is the opinion of the customer about the topic, which is common in all the domains, such as "positive" or "negative". By eliminating the topic-related features, we can decrease the topic-conglutination influence from the source domain to the target domain. Motivated by the idea of multi-task learning, which can detach each task's private feature space from the shared space among several tasks [16], the proposed Bifurcated-LSTM divides the review representation feature space into topic subspace and sentiment subspace. After that, the extracted domain-invariant sentiment features from the source domain can be utilized to perform sentiment analysis in different target domains. To better capture domain-dependent topic features from the source domain training dataset, we apply the dataset augmentation method to improve the performance. Besides, to prevent the topic and sentiment feature spaces interfering with each other, we introduce orthogonal constraints strategies. The experiment results show that our approach can improve sentiment classification in each target domain.

The main contributions of this paper are four-folds:

- We design a novel Bifurcated-LSTM that divides a sentence feature space into domain-dependent topic feature space and domain-independent sentiment feature space.
- We use dataset augmentation to better extract domain-dependent topic features from the source domain, which can help separate these features from sentiment features.
- We employ orthogonal constraint technique to avoid interference between topic and sentiment features.
- Different from other cross-domain sentiment classification models, our system no longer needs any target domain data or other related information.

2 Related Works

2.1 Cross-Domain Sentiment Classification

Cross-domain sentiment classification, a subclass of domain adaptation, is to first learn a sentiment classifier for a source domain by training on this domain's data and then apply the learned classifier into other domains (i.e., target domains) for sentiment classification. To achieve high accuracy, one main challenge is how to analyze data from the source domain and identify its feature space that happens to be related to the feature space of a target domain.

Previous works have studied the problem of feature space mapping from a source domain to target domains [3,4]. However, those works require the data from target domains and need a lot of efforts to label data manually.

2.2   Multi-task Learning

Multi-task learning is to learn multiple related tasks in parallel so as to improve the learning performance. In particular, the representations of all tasks are effectively combined by neural-based models. The architecture of multi-task learning is shown in Fig. 1. Specifically, multiple tasks have several shared layers that are used to detach common feature space. Then, the output of the shared layers is split into multiple branches that are utilized to capture private features for each task [16].

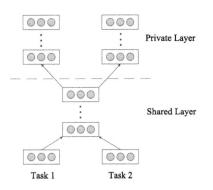

Fig. 1. The architecture of multi-task learning.

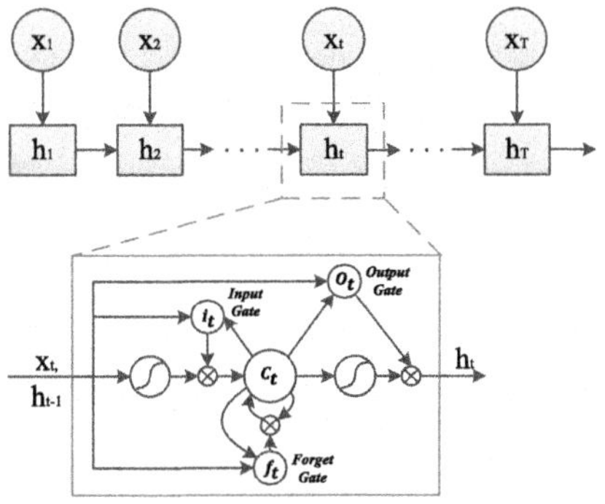

Fig. 2. The structure of RNN with LSTM units.

3   Recurrent Neural Network Models for Text Classification

So far, deep learning comes into play in many area, and achieves high performance [5,6,8,14]. Many researchers have developed many neural network based sentence models [18,21], which can be applied to conduct sentiment classification. In this paper, we adopt a recurrent neural network (RNN) with long short-term memory (LSTM) units [12] due to its great performance in handling multiple natural language processing (NLP) tasks [15].

3.1 Long Short-Term Memory

LSTM is very effective in learning long-term dependencies. It has been proposed to address the issue that standard RNN suffers from severe gradients vanishing or exploding when dealing with long sequential data. The mathematical description of the LSTM structure is as follows:

$$
\begin{bmatrix} \tilde{\mathbf{c}}_t \\ \mathbf{o}_t \\ \mathbf{i}_t \\ \mathbf{f}_t \end{bmatrix} = \begin{pmatrix} \tanh \\ \sigma \\ \sigma \\ \sigma \end{pmatrix} \left(\mathbf{W}_p \begin{bmatrix} \mathbf{x}_t \\ \mathbf{h}_{t-1} \end{bmatrix} + \mathbf{b}_p \right)
\tag{1}
$$

$$
\mathbf{c}_t = \tilde{\mathbf{c}}_t \odot \mathbf{i}_t + \mathbf{c}_{t-1} \odot \mathbf{f}_t
\tag{2}
$$

$$
\mathbf{h}_t = \mathbf{o}_t \odot \tanh\left(\mathbf{c}_t\right)
\tag{3}
$$

where $\mathbf{x}_t \in \mathbb{R}^e$ is the input at the current time step, d denotes the number of the LSTM units, $\mathbf{W}_p \in \mathbb{R}^{4d \times (d+e)}$ and $\mathbf{b}_p \in \mathbb{R}^{4d}$ are parameters of affine transformation, σ denotes the logistic sigmoid function and $\odot$ denotes elementwise multiplication.

The update of each LSTM unit can be briefly summarized as follows:

$$
\mathbf{h}_t = LSTM(\mathbf{h}_{t-1}, \mathbf{x}_t, \theta).
$$

Function $LSTM$ is a combination of Eqs. (1)–(3), and θ represents all the parameters in the LSTM network. The structure of RNN with LSTM units is shown in Fig. 2.

3.2 Text Classification with LSTM

Basically, for a given text sequence $x_t = \{x_1, x_2, \ldots, x_T\}$, the embedding layers [17,20] are used to find the representation vectors $\mathbf{x}_t$ for all words. Then, the representation vectors are input into the LSTM layers to output a representation vector $\mathbf{h}_T$. Finally, $\mathbf{h}_T$ is input into a fully connected layer to generate a probability distribution over all classes.

$$
\hat{\mathbf{y}} = softmax\left(\mathbf{W}\mathbf{h}_T + \mathbf{b}\right)
$$

where $\hat{\mathbf{y}} = \{\hat{\mathbf{y}}^1, \hat{\mathbf{y}}^2, \ldots, \hat{\mathbf{y}}^C\}$ represents the prediction probabilities for each class $j \in [1, C]$, $\mathbf{W}$ is the learned weights, and $\mathbf{b}$ is the bias.

For a given classic classification task, the loss function is defined as the cross-entropy between predicted and ground-truth distribution.

$$
L\left(\hat{\mathbf{y}}, \mathbf{y}\right) = -\sum_{i=1}^{N}\sum_{j=1}^{C} y_i^j \log\left(\hat{y}_i^j\right)
\tag{4}
$$

where y_i^j is the ground-truth label for sample i regrading class j, N is the number of samples in the dataset, and C is the number of classes.

4   Bifurcated-LSTM for Cross-Domain Sentiment Classification

Motivated by multi-task learning, we design a Bifurcated-LSTM for cross-domain sentiment classification, which can divide some domain's reviews' feature space into domain-independent sentiment space and domain-dependent topic space.

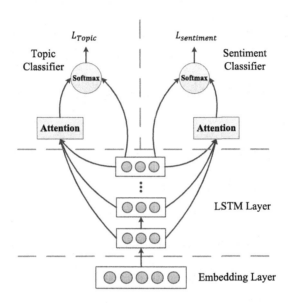

Fig. 3. The structure of Bifurcated-LSTM

The structure of a Bifurcated-LSTM is shown in Fig. 3. First, a sentence passes through the embedding layer and LSTM layers to obtain the representation vector $\mathbf{h}_T$, which is the entire feature space of the text. Then, we simultaneously input $\mathbf{h}_T$ into two LSTM classifier branches, which have the same structure but for different objectives. One is for topic classification and the other is for sentiment classification. Topic features are needed to help the system distinguish source domain reviews from texts in other domains. To better achieve this, we augment the original dataset to obtain a more complete dataset for the system. Moreover, to accurately capture the features, we integrate the attention mechanism to the standard LSTM-based classifier for improving the categorization performance. In addition, to further enhance the performance of our model, we use orthogonal constraints strategy to separate the sentiment and topic features thoroughly. In the following, we describe dataset augmentation, Bifurcated-LSTM, and orthogonality constraints, respectively.

4.1   Dataset Augmentation

Our model aims at extracting topic-related and sentiment-related features from sentence representations. It is obvious that topic feature space varies in different domains. As a result, a model needs to be capable of obtaining distinct topic features from multiple domains.

Therefore, we reconstruct our training dataset by applying the dataset augmentation technique [8]. Specifically, we add some "noisy" data into the training dataset during the step of data collecting and preprocessing. These "noisy" data are text sequences picked from other domains, which have different topics from the ones in original dataset. After dataset augmentation, each data instance has two labels, and is denoted by (x, y^{Se}, y^{To}), where x is the text sequence, and y^{Se} is the sentiment label. $y^{To} \in \{0, 1\}$ is a binary label, where 1 indicates that the instance is from the current domain, and 0 means that the instance is a "noisy" sample.

4.2   Bifurcated-LSTM

As shown in Fig. 3, the Bifurcated-LSTM is composed of the sentiment classifier, the topic classifier and the feature bifurcation. We describe them respectively in the following.

Attention-Based LSTM Sentiment Classifier. We integrate word embeddings and attention mechanism into the standard LSTM model to improve the performance of capturing the representative features from text sentences. Particularly, for a word x_t, we employ word embedding, like GloVe [20] and Word2Vec [17], to transform it into a representation vector $\mathbf{x_t}$. In addition, we adopt a word-level attention mechanism [1], which can identify the crucial part of a sentence, to improve the performance of our sentiment classifier.

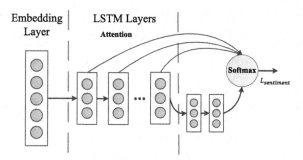

Fig. 4. Attention-based LSTM sentiment classifier.

As shown in Fig. 4, in upper branch of the sentiment classifier, we apply attention mechanism at the common LSTM layers that are shared with the topic classifier, so that the information from the original training data can be further extracted and still exploited at the sentiment classifier. Let $\mathbf{H}_a \in \mathbb{R}^{d \times T}$ denote a matrix consisting of hidden vectors $[\mathbf{h}_1, \ldots, \mathbf{h}_T]$ produced by the LSTM, where d is the number of hidden layers and T is the length of a given sentence. The attention mechanism produces an attention weight vector $\mathbf{a}$ and a hidden representation $\mathbf{s}$ which is a weighted representation of a sentence with the given word. Both of them can be calculated as follows:

$$\mathbf{a} = softmax\left(\mathbf{w}^T \tanh(\mathbf{W}_h \mathbf{H}_a)\right)$$
$$\mathbf{s} = \mathbf{H}\mathbf{a}^T$$

where we have $\mathbf{a} \in \mathbb{R}^T$, $\mathbf{s} \in \mathbb{R}^d$, $\mathbf{W}_h \in \mathbb{R}^{d \times d}$, and $\mathbf{w} \in \mathbb{R}^d$ are projection parameters.

In the lower branch of the sentiment classifier, particularly following common LSTM layers, we add several LSTM layers to extract sentiment features from the whole sentence feature space. The output of these LSTM layers is as follows:

$$\mathbf{h}_{output} = LSTM(\mathbf{h}_T, \mathbf{x}_t, \theta)$$

The final sentiment representation vector of the sentence, denotes by is given by:

$$\mathbf{h}^* = \tanh\left(\mathbf{W}_{Attention}\mathbf{s} + \mathbf{W}_{output}\mathbf{h}_{output}\right)$$

where $\mathbf{W}_{Attention}$ and $\mathbf{W}_{output}$ are projection parameters on the two branches of the sentiment classifier to be learned during the training process. Then, a $softmax$ layer is followed to transform $\mathbf{h}^*$ to the conditional probability distribution, i.e.,

$$\hat{\mathbf{y}} = softmax\left(\mathbf{W}_{softmax}\mathbf{h}^* + \mathbf{b}_{softmax}\right)$$

where $\mathbf{W}_{softmax}$ and $\mathbf{b}_{softmax}$ are the parameters for $softmax$ layer.

Based on Eq. (4), the loss of sentiment classification can be computed as follows:

$$L_{Se}\left(\hat{\mathbf{y}}^{Se}, \mathbf{y}^{Se}\right) = -\sum_{i=1}^{N}\sum_{j=1}^{C} y_i^{jSe} \log\left(\hat{y}_i^{jSe}\right) \tag{5}$$

where $\hat{\mathbf{y}}^{Se} = [\hat{\mathbf{y}}^{1Se}, \hat{\mathbf{y}}^{2Se}, \ldots, \hat{\mathbf{y}}^{jSe}]$ represents the predicted probabilities for each sentiment classification class $j \in [1, C]$, and $\mathbf{y}^{Se} = [\mathbf{y}^{1Se}, \mathbf{y}^{2Se}, \ldots, \mathbf{y}^{jSe}]$ represents the ground-truth labels, and N is the number of samples.

Attention-Based LSTM Topic Classifier. Note that both classifiers for topic and sentiment analysis share the same structure, and slightly difference lies in the objective function. Therefore, we simply show the loss function of the topic classifier as follows:

$$L_{To}\left(\hat{\mathbf{y}}^{To}, \mathbf{y}^{To}\right) = -\sum_{i=1}^{N'}\sum_{j=1}^{C'} y_i^{jTo} \log\left(\hat{y}_i^{jTo}\right) \tag{6}$$

where, similarly, $\hat{\mathbf{y}}^{To} = [\hat{\mathbf{y}}^{1To}, \hat{\mathbf{y}}^{2To}, \ldots, \hat{\mathbf{y}}^{jTo}]$ represents the predicted probabilities for each topic classification class $j \in [1, C']$, and $\mathbf{y}^{To} = [\mathbf{y}^{1To}, \mathbf{y}^{2To}, \ldots, \mathbf{y}^{jTo}]$ represents the ground-truth labels, and N' is the number of samples.

Feature Bifurcation. The feature representation bifurcation is constructed by merging the sentiment classifier and the topic classifier. The shared attention-based LSTM layers condense an input sentence into a representation vector, which includes all features of the sentence. Each classifier only extracts the features that it is interested in according to the considered loss function.

4.3   Orthogonality Constraints

We notice that it is possible that the domain-dependent topic features and domain-independent sentiment features may interfere with each other. Inspired by recent work on multi-task learning [16] and shared-private latent space analysis [5], we employ the orthogonality constraint technique in our proposed feature divider. Specifically, it enables the divider to penalize commonly shared features in sentiment feature space and topic feature space and encourage to extract the independent sentiment topic features as purely as possible. To achieve this goal, we define the optimal loss function as follows:

$$L_{orth} = \sum_{i=1}^{N} \left\| \mathbf{H}_i^{Se^{\mathrm{T}}} \mathbf{H}_k^{To} \right\|_F^2 \tag{7}$$

where $\|\cdot\|_F^2$ is the squared Frobenius norm, $\mathbf{H}^{Se}$ and $\mathbf{H}^{To}$ are two matrices whose rows are parameters from the private LSTM layers of sentiment classifier and topic classifier, respectively.

4.4   Training and Testing

Combining Eqs. (5)–(7), the final loss function of our features divider model can be summarized as follows:

$$L = L_{Se} + L_{To} + \gamma L_{orth} \tag{8}$$

where γ is a hyperparameter.

In the training process, we feed the augmented dataset to the whole neural-based model to train the classifier. After training, we can obtain a Bifurcated-LSTM. For cross-domain sentiment classification task, we only focus on the sentiment classifier branch of the Bifurcated-LSTM. Therefore, in the testing process, we mainly transfer the well-trained sentiment classifier to other domains.

Table 1. Statistical knowledge of the 7 datasets. The columns 2–4 denote the number of samples in training, development and testing sets. The columns 5 and 6 represent the average length and vocabulary size of corresponding dataset.

Dataset	Train	Dev.	Test	Avg.L	Vocab.
Books	1400	200	400	159	62K
Electronics	1400	200	400	103	30K
DVD	1400	200	400	172	69K
Kitchen	1400	200	400	88	28K
Baby	1300	200	400	105	26K
Magazine	1300	200	400	113	30K
Software	1400	200	400	130	26K

5 Experiments Setting

5.1 Dataset

We collect product reviews of 7 different domains from Amazon [2]. First, we extract the comment sentences and corresponding labels from raw data, and then use keras [7] to perform the tokenization. After text preprocessing, we randomly partition all the datasets into a training set, a development set, and a testing set with the proportion of 70%, 10%, 20%, respectively. Table 1 shows the statistical information of all considered datasets.

5.2 Dataset Augmentation

In experiments, we randomly choose reviews from domains other than the considered source and target domains as "noisy" datase to conduct dataset augmentation. Meanwhile, we control the size of "noisy" dataset to be half size of the original training dataset. After combining the original and "noisy" training datasets, we have the augmented dataset.

5.3 Hyperparameters

We apply 200d GloVe vectors [20] to initialize the input sentence sequences, and $\gamma = 0.03$ in Eq. (8). Other parameters in the neural networks are initialized by randomly generated from a uniform distribution in $[-0.1, 0.1]$. We employ Adam to optimize our loss function shown in Eq. (8) with mini-batch size 24.

6 Performance Evaluation of Bifurcated-LSTM

6.1 Performance Evaluation

Table 2 shows the average error rate achieved by the proposed model, and compares it with that achieved by the one without domain adaptations. The LSTM networks in Bifurcated-LSTM are vanilla LSTM networks.

Table 2. Error rates of Bifurcated-LSTM for cross-domain classification. In "Bifurcated-LSTM" columns, the numbers in brackets represent the improvements relative to same domain classification results without domain adaptation.

Source domain	Transferring to target domains without domain adaptation							Avg.
	Book	Elec.	DVD	Kitc.	Baby	Maga.	Soft.	
Book	20.8	21.3	22.7	23.2	23.0	23.3	21.7	22.29
Elec.	24.6	19.8	23.5	22.7	22.0	22.5	25.7	22.97
DVD	24.0	25.1	17.9	23.0	25.7	20.0	24.3	22.86
Kitc.	22.9	25.6	22.5	22.0	25.2	24.1	24.9	23.89
Baby	24.9	25.5	20.7	25.8	15.8	18.9	19.4	21.57
Maga.	24.4	23.0	24.6	21.3	21.6	11.2	18.1	20.60
Soft.	22.6	22.4	23.5	23.1	19.7	19.0	16.3	20.94
Source domain	Transferring to target domains with bifurcated-LSTM							Avg.
	Book	Elec.	DVD	Kitc.	Baby	Maga.	Soft.	
Book	17.6(−3.2)	18.3	19.1	19.4	19.2	18.1	19.0	18.67(−3.62)
Elec.	19.8	15.4(−4.4)	18.2	19.5	17.4	16.9	16.1	17.61(−5.36)
DVD	21.2	19.7	14.3(−3.6)	18.1	16.4	18.1	17.5	17.90(−4.96)
Kitc.	19.8	18.6	17.3	15.7(−6.3)	17.6	16.5	16.9	17.48(−6.41)
Baby	19.7	17.3	16.7	17.4	11.6(−4.2)	13.9	18.0	16.37(−5.20)
Maga.	20.1	16.2	16.9	18.1	14.3	6.9(−4.3)	17.1	15.65(−4.95)
Soft.	18.9	19.8	17.5	18.5	17.9	17.0	12.1(−4.2)	17.38(−3.56)

From this table, we can find that our proposed model can reduce the average error rate. Compared with the one without domain adaptation, our proposed model can reduce the error rate by 6.41%. Moreover, Table 2 also illustrates that our developed model can improve the performance of the classifier trained in its own domains, and the value can be up to 6.3%.

6.2   Performance Comparison

The baseline methods in the comparison include:

- **SCL**: Blitzer et al. proposed Structural Correspondence Learning (SCL) to learn a low-dimensional feature representation for source and target domains [2].
- **SFA**: Pan et al. proposed Spectral Feature Alignment (SFA) to build a bridge between source and target domains by aligning pivots with non-pivots [19].
- **DANN**: Ganin et al. applied the shallow version of Domain Adversarial Neural Networks (DANN) to the cross-domain sentiment classification [10].

We perform twelve domain adaptation tasks, and the results are in Table 3. We can find that our proposed model can achieve best performance on most tasks. For specific source domain, our proposed Bifurcated-LSTM always achieve the best average performance.

Table 3. Error rates of SCL, SFA, DANN, and Bifurcated-LSTM for cross-domain sentiment classification.

Source	Target	SCL	SFA	DANN	Bifurcated-LSTM
Kitc.	Book	33.9	25.2	29.1	**19.8**
Kitc.	Elec	16.3	**14.9**	15.7	18.6
Kitc.	DVD	24.6	23.0	26.0	**17.3**
Avg.		24.93	21.03	23.60	**18.57**
Book	Kitc.	21.3	21.2	22.1	**19.4**
Book	Elec.	22.5	27.5	26.7	**18.3**
Book	DVD	26.0	**18.6**	21.6	19.1
Avg.		23.27	22.43	23.47	**18.93**
Elec.	Kitc.	15.6	**13.3**	14.6	19.5
Elec.	Book	24.6	24.3	28.7	**19.8**
Elec.	DVD	25.7	22.8	26.2	**18.2**
Avg.		21.97	20.13	23.17	**19.17**
DVD	Kitc.	20.6	19.2	21.7	**18.1**
DVD	Book	23.2	22.5	27.7	**21.2**
DVD	Elec.	25.9	23.3	24.6	**19.7**
Avg.		23.23	21.67	24.67	**19.67**

7   Conclusion and Future Work

In this paper, we propose a Bifurcated-LSTM for cross-domain sentiment classification. In particular, this Bifurcated-LSTM can separate reviews' feature space into sentiment and topic feature subspaces. To enhance the performance of the Bifurcated-LSTM, we employ an attention mechanism to extract sentiment and topic features. Moreover, we also apply data augmentation and orthogonal constraints techniques to further improve the performance. We conduct extensive experiments to evaluate the performance of the proposed system.

Acknowledgments. This work was partially supported by the U.S. National Science Foundation under grants CNS-1602172 and CNS-1566479.

References

1. Bahdanau, D., Cho, K., Bengio, Y.: Neural machine translation by jointly learning to align and translate. arXiv preprint arXiv:1409.0473 (2014)
2. Blitzer, J., Dredze, M., Pereira, F.: Biographies, bollywood, boom-boxes and blenders: domain adaptation for sentiment classification. In: ACL, vol. 7, pp. 440–447 (2007)

3. Bollegala, D., Mu, T., Goulermas, J.Y.: Cross-domain sentiment classification using sentiment sensitive embeddings. IEEE Trans. Knowl. Data Eng. **28**(2), 398–410 (2016)
4. Bollegala, D., Weir, D., Carroll, J.: Cross-domain sentiment classification using a sentiment sensitive thesaurus. IEEE Trans. Knowl. Data Eng. **25**(8), 1719–1731 (2013)
5. Bousmalis, K., Trigeorgis, G., Silberman, N., Krishnan, D., Erhan, D.: Domain separation networks. In: Advances in Neural Information Processing Systems, pp. 343–351 (2016)
6. Chen, X., Ji, J., Loparo, K., Li, P.: Real-time personalized cardiac arrhythmia detection and diagnosis: a cloud computing architecture. In: 2017 IEEE EMBS International Conference on Biomedical and Health Informatics (BHI), pp. 201–204. IEEE (2017)
7. Chollet, F.: Keras (2015). https://github.com/fchollet/keras
8. DeVries, T., Taylor, G.W.: Dataset augmentation in feature space. arXiv preprint arXiv:1702.05538 (2017)
9. Dong, L., Wei, F., Tan, C., Tang, D., Zhou, M., Xu, K.: Adaptive recursive neural network for target-dependent twitter sentiment classification. In: ACL, vol. 2, pp. 49–54 (2014)
10. Ganin, Y., Ustinova, E., Ajakan, H., Germain, P., Larochelle, H., Laviolette, F., Marchand, M., Lempitsky, V.: Domain-adversarial training of neural networks. J. Mach. Learn. Res. **17**(59), 1–35 (2016)
11. He, Y., Lin, C., Alani, H.: Automatically extracting polarity-bearing topics for cross-domain sentiment classification. In: Proceedings of the 49th Annual Meeting of the Association for Computational Linguistics: Human Language Technologies, vol. 1, pp. 123–131. Association for Computational Linguistics (2011)
12. Hochreiter, S., Schmidhuber, J.: Long short-term memory. Neural Comput. **9**(8), 1735–1780 (1997)
13. Li, T., Sindhwani, V., Ding, C., Zhang, Y.: Knowledge transformation for cross-domain sentiment classification. In: Proceedings of the 32nd International ACM SIGIR Conference on Research and Development in Information Retrieval, pp. 716–717. ACM (2009)
14. Liao, W., Salinas, S., Li, M., Li, P., Loparo, K.A.: Cascading failure attacks in the power system: a stochastic game perspective. IEEE Internet Things J. **4**(6), 2247–2259 (2017)
15. Liu, P., Qiu, X., Chen, J., Huang, X.: Deep fusion LSTMs for text semantic matching. In: ACL, vol. 1 (2016)
16. Liu, P., Qiu, X., Huang, X.: Adversarial multi-task learning for text classification. arXiv preprint arXiv:1704.05742 (2017)
17. Mikolov, T., Chen, K., Corrado, G., Dean, J.: Efficient estimation of word representations in vector space. arXiv preprint arXiv:1301.3781 (2013)
18. Mikolov, T., Karafiát, M., Burget, L., Cernocký, J., Khudanpur, S.: Recurrent neural network based language model. In: Interspeech, vol. 2, p. 3 (2010)
19. Pan, S.J., Ni, X., Sun, J.T., Yang, Q., Chen, Z.: Cross-domain sentiment classification via spectral feature alignment. In: Proceedings of the 19th International Conference on World Wide Web, pp. 751–760. ACM (2010)
20. Pennington, J., Socher, R., Manning, C.: Glove: global vectors for word representation. In: Proceedings of the 2014 Conference on Empirical Methods in Natural Language Processing (EMNLP), pp. 1532–1543 (2014)

21. Socher, R., Pennington, J., Huang, E.H., Ng, A.Y., Manning, C.D.: Semi-supervised recursive autoencoders for predicting sentiment distributions. In: Proceedings of the Conference on Empirical Methods in Natural Language Processing, pp. 151–161. Association for Computational Linguistics (2011)
22. Turney, P.D.: Thumbs up or thumbs down: semantic orientation applied to unsupervised classification of reviews. In: Proceedings of the 40th Annual Meeting on Association for Computational Linguistics. pp. 417–424. Association for Computational Linguistics (2002)

Author Index

Printed in the United States
By Bookmasters